Organic Reactions in Water

Organic Reactions in Water

Principles, Strategies and Applications

Edited by

U. Marcus Lindström

Assistant Professor
Department of Chemistry
McGill University

Blackwell
Publishing

Blackwell Publishing editorial offices:
Blackwell Publishing Ltd, 9600 Garsington Road, Oxford OX4 2DQ, UK
Tel: +44 (0)1865 776868
Blackwell Publishing Professional, 2121 State Avenue, Ames, Iowa 50014-8300, USA
Tel: +1 515 292 0140
Blackwell Publishing Asia Pty Ltd, 550 Swanston Street, Carlton, Victoria 3053, Australia
Tel: +61 (0)3 8359 1011

First published 2007 by Blackwell Publishing Ltd

ISBN: 978-1-4051-3890-1

Library of Congress Cataloging-in-Publication Data

Organic reactions in water:principles, strategies and applications / edited by U. Marcus Lindström.
p. cm.
Includes bibliographical references and index.
ISBN-13: 978-1-4051-3890-1 (hardback:alk. paper)
ISBN-10: 1-4051-3890-4 (hardback:alk. paper)
1. Water chemistry. 2. Solvents–Environmental aspects. 3. Organic compounds–Synthesis–Environmental aspects. I. Lindström, U. Marcus

QD 169.W3O74 2007
547′.2–dc22

2006022512

A catalogue record for this title is available from the British Library

Set in 10/12 pt Minion
by TechBooks
Printed and bound in Singapore
by Fabulous Printers Pte Ltd

The publisher's policy is to use permanent paper from mills that operate a sustainable forestry policy, and which has been manufactured from pulp processed using acid-free and elementary chlorine-free practices. Furthermore, the publisher ensures that the text paper and cover board used have met acceptable environmental accreditation standards.

For further information on Blackwell Publishing, visit our website:
www.blackwellpublishing.com

Contents

Contributors

Ronald Breslow	Department of Chemistry, Columbia University, New York, NY 10027, USA
Boy Cornils	Kirschgartenstraße 6, D-65719 Hofheim, Germany
C.A. Eckert	School of Chemistry, Georgia Institute of Technology, 311 Ferst Drive NW, Atlanta, GA 30332-0400, USA
Jan B.F.N. Engberts	Physical Organic Chemistry Unit, Stratingh Insitute, University of Groningen, Nijenborgh 4, 9747 AG Groningen, The Netherlands
Valery V. Fokin	Department of Chemistry, The Scripps Research Institute, 10550 North Torrey Pines Road, La Jolla, CA 92037, USA
Francesco Fringuelli	Dipartimento di Chimica, Università di Perugia, Italy, Via Elce di Sotto, 8, 06123 Perugia, Italy
J.P. Hallett	School of Chemistry, Georgia Institute of Technology, 311 Ferst Drive NW, Atlanta, GA 30332-0400, USA
Shū Kobayashi	Graduate School of Pharmaceutical Sciences, The University of Tokyo, The HFRE Division, ERATO, Japan Science Technology Agency (JST), Hongo, Bunkyo-ku, Tokyo 113-0033, Japan
Chao-Jun Li	Department of Chemistry, McGill University, 801 Sherbrooke St. West, Montreal, Quebec, Canada H3A 2K6
Charles L. Liotta	School of Chemistry, Georgia Institute of Technology, 311 Ferst Drive NW, Atlanta, GA 30332-0400, USA
Tomoko Matsuda	Department of Bioengineering, Tokyo Institute of Technology, 4259 Nagatsuta, Midori-ku, Yokohama 226-8501, Japan
Kaoru Nakamura	Institute for Chemical Research, Kyoto University, Uji, Kyoto 611-0011, Japan
Sridhar Narayan	Department of Chemistry, The Scripps Research Institute, 10550 North Torrey Pines Road, La Jolla, CA 92037, USA
Chikako Ogawa	Graduate School of Pharmaceutical Sciences, The University of Tokyo, The HFRE Division, ERATO, Japan Science Technology Agency (JST), Hongo, Bunkyo-ku, Tokyo 113-0033, Japan
Oriana Piermatti	Dipartimento di Chimica, Università di Perugia, Italy, Via Elce di Sotto, 8, 06123 Perugia, Italy

Ferdinando Pizzo Dipartimento di Chimica, Università di Perugia, Italy, Via Elce di Sotto, 8, 06123 Perugia, Italy

Pamella Pollet School of Chemistry, Georgia Institute of Technology, 311 Ferst Drive NW, Atlanta, GA 30332-0400, USA

T.V. RajanBabu Department of Chemistry, The Ohio State University, 100 W. 18th Avenue, Columbus, OH 43210, USA

K. Barry Sharpless Department of Chemistry, The Scripps Research Institute, 10550 North Torrey Pines Road, La Jolla, CA 92037, USA

Roger A. Sheldon Department of Biotechnology, Julianalaan 136, 2628 BL Delft, The Netherlands

Seunghoon Shin Department of Chemistry, Hanyang University, 17 Haengdang-dong, Seongdong-gu, Seoul, Korea 133-791

Denis Sinou Laboratoire de Synthèse Asymétrique, associé au CNRS, ESCPE Lyon, Université Claude Bernard Lyon 1, 43, boulevard du 11 novembre 1918, 69622 Villeurbanne Cédex, France

Luigi Vaccaro Dipartimento di Chimica, Università di Perugia, Italy, Via Elce di Sotto, 8, 06123 Perugia, Italy

Ernst Wiebus Celanese Europe GmbH, D-46128 Oberhausen, Germany

Preface

Water is the most abundant molecule on Earth and the universal solvent in which the chemistry of the life processes mostly occur. Nevertheless, the modern synthetic chemist rarely uses, or even considers, water as a medium in which to perform organic reactions.

Students on all levels of organic chemistry are being trained in the art of performing reactions under strictly anhydrous conditions. The diligent synthetic chemist keeps glassware, syringes, reagents and solvents free from traces of water. From my own experience, I remember being told as a young student of organic chemistry: 'in organic synthesis water is a contaminant'. This book, through the words of its outstanding contributors, addresses the important academic question that perhaps should have been asked much sooner: Water, why not?

Water has, however, not always been considered incompatible with organic synthesis. In fact, water was once frequently used for synthetic reactions. Wohler's synthesis of urea in 1828, commonly referred to as the starting point for organic synthesis, was performed by heating an aqueous solution of ammonium isocyanate. Many of the name reactions developed in the 19th century, and which formed the foundation of organic synthesis as we know it today, were first developed in aqueous media. Prominent examples include the Baeyer-Villiger oxidation, the Curtius rearrangement, the Hofmann degradation, the Lieben haloform reaction, the Pictet-Spengler reaction, the Sandmeyer reaction, and the Wolff-Kischner reduction. The list can be made much longer. The rapid transition to organic solvents in the early years of the last century can to a significant extent be ascribed to the development of organometallic chemistry. The first organometallic reagents developed were extremely sensitive to hydrolysis and their use was restricted to dry aprotic solvents. Nevertheless, the impact of the Grignard reagent on synthetic chemistry was so significant that in the first decade after its discovery ca. 700 papers were published on studies of its use. In 1912, Victor Grignard stated in his Nobel Lecture that 'The compound prepared in this way presents all the characteristics of an organometallic compound: water decomposes it with violence; it fixes oxygen and carbon dioxide gas, and it reacts vigorously with almost all the functional groups of chemistry'. We know today that there are many organometallic reagents that water does not 'decompose with violence' but which are in fact remarkably stable under aqueous conditions, as well as compatible with various functional groups.

With the instant success of the Grignard reagent, as well as other hydrolytically labile organometallic reagents, water was supplanted by synthetic solvents that were amenable with the powerful new synthetic methods, as well as readily available through the rapidly expanding petroleum industry. Throughout the better part of the last century, synthetic methods were developed for use in organic solvents. It was not until the early 1980's that the use of water was re-evaluated when it was found that the rates and selectivities of Diels-Alder reactions may be greatly enhanced in water compared to the same reactions in organic solvents. The interest in water as solvent was further invigorated in the 1990's with the introduction of the concepts of Green Chemistry. Water, being cheap, safe, non-toxic, and environmentally benign, was soon recognized as perhaps the ultimate 'green' solvent. In view of the potential reward of replacing hazardous organic solvents with water, researchers took up the academic challenge of developing new synthetic methods that were compatible with the aqueous medium. Progress has been quite dramatic. A recent review listed nearly 1000 references on aqueous carbon-carbon bond forming reactions alone.

This book exposes the current status of aqueous synthesis through the most comprehensive one-volume coverage of the field to date. In this book we have made the important distinction between reactions that are performed in 'the presence of water' and reactions that are run in water alone or in water with only a small amount of co-solvent. Unless there is some conceptual relevance that warrants discussion, examples of reactions performed with less than 50% water content are not within the scope of this book and thus not included.

In the opening chapter, R. Breslow shares his long-standing interest in aqueous chemistry with us through a beautiful perspective on fifty years of research into the use of water as solvent for organic reactions. In the second chapter, J.B.F.N. Engberts gives a generous account of the physical and chemical properties of water, which are so recognizably different from other solvents, and how these properties relate to our understanding of organic reactivity in water. Discovering new catalysts for aqueous synthesis is obviously of considerable importance. In chapter three, C. Ogawa and S. Kobayashi explains important concepts of acid catalysed reactions in water and provides a broad survey of recent developments in the area. C.J. Li introduces chapter four by explaining the reactivity of organometallic compounds in water and how rational design can lead to water-compatible metal-based reagents. In addition, this chapter's comprehensive summary of current research progress should help overcome the conventional view that metal-mediated reactions require strict exclusion of moisture and air. In chapter five, F. Fringuelli, O. Piermatti, F. Pizzo, and L. Vaccaro describe the use of pericyclic reactions in water, a class of reactions that may be particularly suitable for synthesis in water as beneficial effects of water on both reactivity and selectivity are frequently observed under aqueous conditions. Chapter six is written by T.V. RajanBabu and S. Shin and deals with catalyzed reductions in aqueous media. Because of its industrial relevance catalytic hydrogenation, in particular, has been studied extensively in water. The promising concept of 'heterogenizing' homogeneous catalysts in the aqueous phase for facile recovery and reuse is widely discussed in the context of this chapter. Then, in chapter seven, R. Sheldon goes on to describe a variety of oxidation processes that proceed in water. Although oxygen has low aqueous solubility and many known transition metal-based oxidants are deactivated in water, there has been considerable progress in developing oxidation reactions in water. D. Sinou devotes chapter eight to describing recent progress in performing nucleophilic additions and substitutions in water. Obviously, such reactions are challenging because hydrolysis of the electrophile may compete with the desired nucleophilic attack. Under the right conditions, however, even reactive electrophiles can be used in water. In chapter nine, C.L. Liotta, J.P. Hallett, P. Pollet and C.A. Eckert discusses reactions that proceed in high-temperature, near-critical water, a medium with different physical properties than ambient water thus expanding the scope of water as process solvent. In the tenth chapter, K. Nakamura and T. Matsuda outlines the remarkable potential of using biocatalysts for the synthesis of fine chemicals in aqueous media. Many enzymes are now commercially available as synthetic reagents and serve as increasingly efficient complements to chemical catalysts. Chapter 11 takes a deeper look into the complex relationship between solubility and reactivity in water. S. Narayan, V. V. Fokin and K. B. Sharpless describe recent developments in the area of chemistry 'on water', where highly efficient reactions are performed simply by stirring insoluble substrates in an aqueous suspension. Finally, in chapter twelve, E. Wiebus and B. Cornils provides us with case studies of water as a large scale process solvent providing insight into the current status, as well as future potential, of water as a solvent for chemical manufacturing.

On a final note, it is hoped that the contents of this book will serve to rectify some of the misconceptions that persist about the inadequacy of water as reaction medium, and that water will soon become a not only viable but also attractive option to the synthetic chemist in the planning of new synthetic processes.

U. Marcus Lindström
Editor

Foreword

It really wasn't so long ago. Not long ago at all really when most of us were trained in synthetic organic chemistry at a time when water was rejected as a solvent. The logic seemed solid enough if you didn't think too hard about it. Since there was limited solubility for many organic compounds in aqueous systems and you wanted homogeneous solutions to bring about desirable synthetic conversions, it was clear that water would naturally be a disadvantageous solvent in which to conduct organic reactions. The logic was clear and wrong.

What this excellent book, *Organic Reactions in Water*, shows us is all of the possibilities that conducting syntheses in water presents to the synthetic chemist. It is appropriate that Prof. Ronald Breslow contributes to this volume because his early recognition, articulation, and advocacy of organic chemistry in aqueous systems can be seen as catalyst for a renaissance in investigations in this important field. The editor, Prof. Lindstrom, should be complimented in gathering the leaders of this field of study together for this significant reference work.

It is often recognized that the use of water as a solvent has tremendous benefits as a green chemistry solvent. Certainly, it is obvious that water is nontoxic, nonflammable, cheap, and available. When compared to the typically used organic solvents based on petroleum feedstocks, one would argue that water is the pinnacle of the green solvents. However, that would be only telling part of the story. While the advantages of water are clear from a green chemistry perspective, it is also necessary to consider potential limitations of organic synthesis in water. One of the greatest of these limitations is the problem of product separation and the generation of an aqueous-based waste stream. It is precisely because of this recognition of both opportunities and potential limitations that this book is so important.

This volume illustrates that the value of all new green chemistry approaches, like organic reactions in water, comes from the design and discovery of new and innovative chemistry. If this book consisted of merely a collection of well-demonstrated, known chemistry that has been conducted in water, it would not be as compelling and important as it is. It is because this collection of work represents new chemistries, new transformations, new mechanisms, new catalysts, and new synthetic pathways that were not known previously and bring about green chemistry benefits that make it the essential resource for those interested in synthetic organic chemistry. It demonstrates that at the heart of green chemistry is innovation and the discovery of new chemistries and not merely the incremental improvement of traditional synthetic methodologies.

If the word 'green' in green chemistry refers to 'fresh, young, and new,' *Organic Reactions in Water* is a presentation of excellent research that emphasizes the spirit of invention and design from some of the leading chemists of our time.

Paul T. Anastas
Yale University
December 2006

Chapter 1

A Fifty-Year Perspective on Chemistry in Water

Ronald Breslow

For 50 years I have been pursuing the use of water as a solvent in organic chemistry. The reasons are several. First of all, in biochemistry enzymatic reactions are normally performed in water, except for those enzymes that are bound to biological membranes. In water, such biochemical reactions normally use hydrophobic energies, in large part, to achieve substrate binding into enzyme pockets. Such hydrophobic binding is also a major force in the association of antibodies with antigens, and of medicines and hormones into biological receptors. It also drives the association of proteins into aggregates, as in transcription complexes and enzyme dimers and tetramers. Thus I wanted to use this ubiquitous force, special to water solution, in chemical processes.

In our earliest work, we studied enzyme models and mimics in water solution. Later we studied the reactions of small molecules in water, and achieved selectivities that reflected the geometries of their hydrophobic association. Finally, we saw that we could use the hydrophobic effect to achieve one of the greatest goals of modern mechanistic chemistry – the determination of the structures of transition states. In this opening chapter I will describe our work in these areas. We have written a number of reviews of this work previously.[1–32]

1.1 Enzyme mimics and models

1.1.1 Thiamine

Our earliest studies concerned the mechanism by which thiamine pyrophosphate (**1**) acts as a coenzyme (Fig. 1.1).[33–38] We discovered that the C-2 hydrogen on the thiazolium ring of thiamine was able to ionize to form a thiazolium ylide (**2**) that has the important resonance structure (**3**), which can be called a 'stabilized carbene'. We pointed out this carbene contribution to its structure, related to the well-known hybrid structure of carbon monoxide, and also saw that an imidazolium ion (**4**) and an oxazolium ion (**5**) could form such ylide/carbene resonance structures. Recently such structures, usually referred to as 'stabilized carbenes', have proven to be useful ligands for catalytic metal ions. The carbenes are of course 'stabilized' by electron donation from the heteroatoms, forming the ylide resonance forms.

The thiazolium ylide is the intermediate in the action of thiamine pyrophosphate as a coenzyme; it is intellectually related to cyanide ion, and just like cyanide it is able to catalyze the benzoin condensation. However, later there were assertions in the literature that the

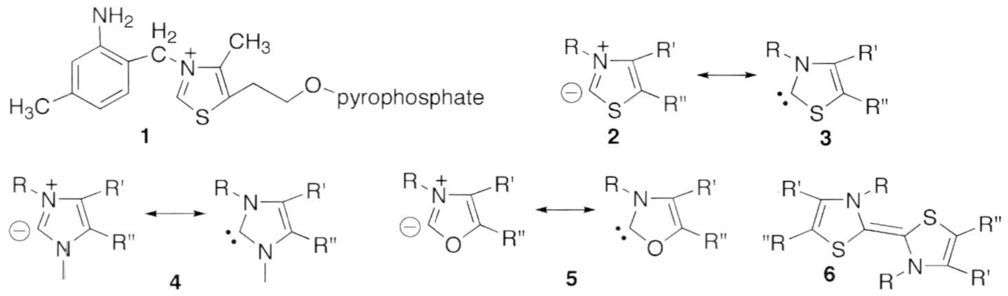

Figure 1.1 Thiamine pyrophosphate **1** and the thiazolium ylide **2** that is the key to its function as a coenzyme. The ylide **2** has a second resonance form **3**, a stabilized carbene. We also demonstrated such ylide/carbene resonance hybrids for imidazolium cations **4** and oxazolium cations **5**. Such 'stabilized carbenes' have proven very useful as metal ligands. Compound **6** is a dimer of the thiazo ium ylide.

true catalyst in the benzoin condensation was a *dimer* **6** of this thiazolium species, not the monomer.[39] It was even asserted that the thiamine pyrophosphate in the enzyme functioned as a dimer, although it is known that the enzyme has a single-bound thiamine pyrophosphate, well buried. We investigated the thiazolium-catalyzed benzoin condensation with careful kinetics, and were able to show that the catalyst was indeed the monomeric ylide/carbene, not its dimer.[40,41] The mechanism of the thiazolium-catalyzed benzoin condensation is shown in Fig. 1.2. It will play a role, discussed later, in models for the enzyme that use hydrophobic binding in water.

1.1.2 Cyclodextrins

Friedrich Cramer[42] and Myron Bender[43] had studied the binding in water of small hydrophobic species into the cavity of naturally occurring macrocyclic rings composed of glucose units, the α-, β-, and γ-cyclodextrins (Fig. 1.3). Water is special for such binding, although we did see that some solvophobic binding of hydrocarbon substrates into cyclodextrins could also occur with polar solvents such as dimethylsulfoxide.[44] Bender et al. also examined some reactions of the cyclodextrin with the bound substrates, such as acetylation of a cyclodextrin hydroxyl group by bound *m*-nitrophenyl acetate.[45] Stimulated by their work, we took up cyclodextrins as components of artificial enzymes.

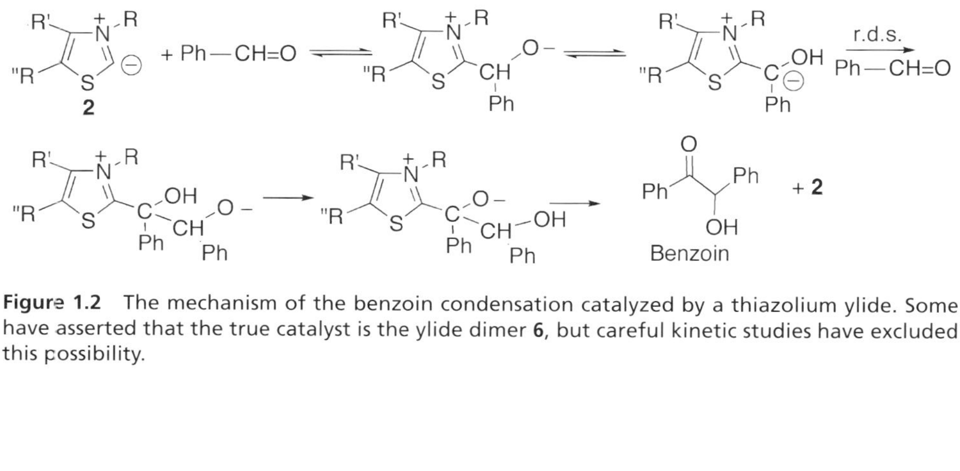

Figure 1.2 The mechanism of the benzoin condensation catalyzed by a thiazolium ylide. Some have asserted that the true catalyst is the ylide dimer **6**, but careful kinetic studies have excluded this possibility.

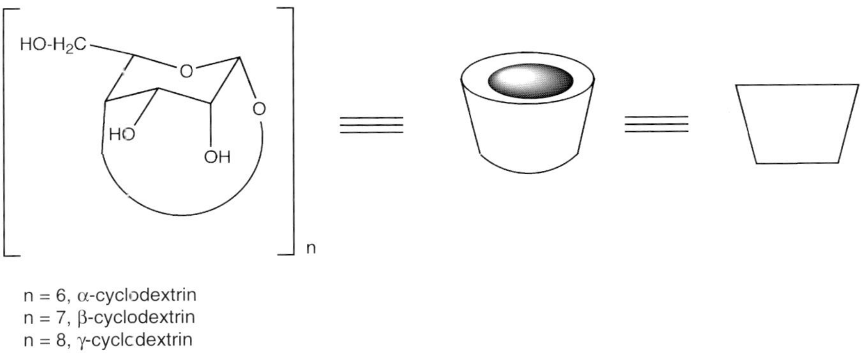

Figure 1.3 The principal cyclodextrins with 6, 7 and 8 glucose residues in a ring. Two other ways of representing them will be used in the remaining figures.

In our first study, we examined the chlorination of anisole in water by HOCl, with and without α-cyclodextrin.[46,47] Without the cyclodextrin, the product is 60% *para*- and 40% *ortho*-chloroanisole, but with 9 mM α-cyclodextrin (cyclohexaamylose) the ratio was 96% *para*- and only 4% *ortho*-chloroanisole. Furthermore, the anisole was only 72% bound in the cyclodextrin. From detailed kinetic studies we showed that the para position was 5.3 times as reactive in the complex as in free solution while the ortho position was completely blocked. The reaction was also first order in [HOCl], while in simple water solution it was second order. This showed that binding did not simply block the ortho positions, it also catalyzed the chlorination of the para position, by reversibly forming a cyclodextrin hypochlorite and delivering the chlorine to the accessible para position (Fig. 1.4). This was the first example in which simple cyclodextrin acted as a selective turnover catalyst for a reaction in water solution because of hydrophobic binding of the substrate into the cavity. We also examined other aromatic chlorinations with cyclodextrins.

In later work, we also extended the studies to methylated cyclodextrins, identifying the hydroxyl group that acted to deliver the chlorine atom to the bound substrate.[48] In that work, we also made a polymer of the cyclodextrin, and showed that it could work as a flow-through catalyst for the chlorinations, with high selectivity. All of this depended on the use of water as a solvent, promoting hydrophobic binding of the substrate into the cavity of the cyclodextrin.

In Bender's original study of the acetylation of a cyclodextrin by *m*-nitrophenyl acetate he saw a modest rate acceleration of only 250-fold, relative to the hydrolysis of the substrate in the

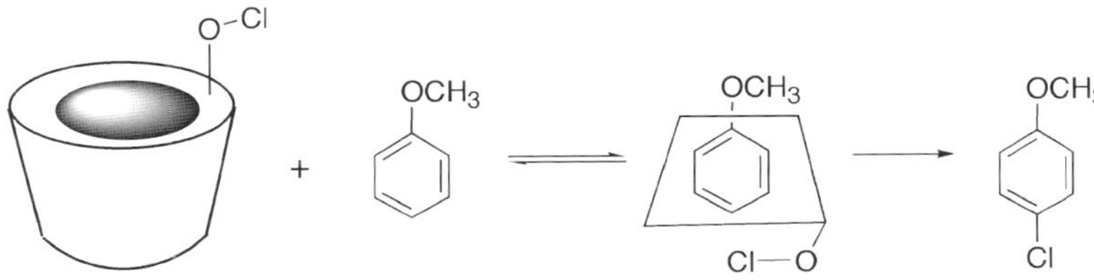

Figure 1.4 Cyclodextrin catalyzes the selective chlorination of bound anisole in water. The reagent HOCl reacts with a cyclodextrin hydroxyl group to produce a cyclodextrin hypochlorite, which transfers the chlorine to the bound anisole.

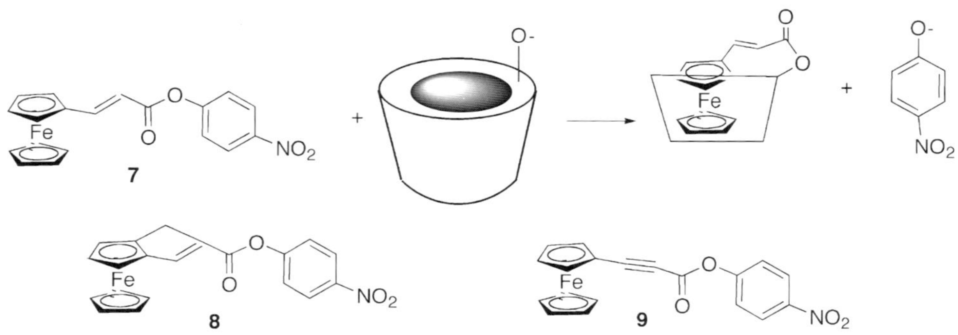

Figure 1.5 Ferrocene derivatives **7**, **8** and **9** bind to a cyclodextrin in water and acylate a hydroxyl on the secondary face of the cyclodextrin. With **8**, there was an acceleration of 3,200,000-fold and an enantioselectivity of 20:1.

buffered water solvent.[45] This is of course well below the accelerations achieved by enzymes. We saw with molecular models that the substrate could bind well into the cavity, but that most of the binding would be lost in the tetrahedral intermediate for the acetylation process. This is undesirable, since the transition state will resemble the high-energy intermediate. Thus we synthesized a series of new substrates that could remain fully bound throughout the acylation process. (The best would of course be that they bind even more strongly in the transition state. As we will describe later, this was achieved with cyclodextrin catalysts that used metal ions.)

In our first case, a derivative of ferrocene, *p*-nitrophenyl ferrocinnamate (**7**), was used (Fig. 1.5). It bound strongly to the β-cyclodextrin and acylated it with a 51,000-fold acceleration relative to the hydrolysis rate.[49] With molecular modeling, we investigated the geometry of the process.[50] Then we modified the cyclodextrin by giving it a hydrophobic floor, and the rate acceleration rose to 750,000-fold.[51] Then we immobilized the substrate further in **8**, freezing out undesirable degrees of freedom, and the acceleration was 3,200,000-fold, with an enantioselectivity of 20-fold for the racemic substrate.[52] In a further study, we saw even higher rates and enantioselectivities.[53] We also correlated our predicted geometry for the transition state with the effect of pressure on these acylations.[54]

Acylation reactions go through interesting geometric changes. The nucleophile must first attack the carbonyl group perpendicular to its plane, forming the tetrahedral intermediate. Then the departure of the leaving group brings the nucleophile into the plane of the carbonyl. We showed that this second step could be rate limiting with weaker leaving groups than *p*-nitrophenoxide ion, so it was necessary to allow for the geometric change in the position of the nucleophile. When we did this, by putting in one degree of flexibility in substrate **9**, we saw large acylation rate accelerations even with weak leaving groups.[55] We have also done a number of other studies of acylations by cyclodextrins, including reactions related to cocaine,[56,57] and a reinvestigation of a purported chymotrypsin model.[58]

1.1.3 Cyclodextrins with bound metal ions

We had created an enzyme mimic in which a metal ion was used to hold a substrate and to catalyze its hydrolysis in water.[59] We extended this to a case in which cyclodextrin binding

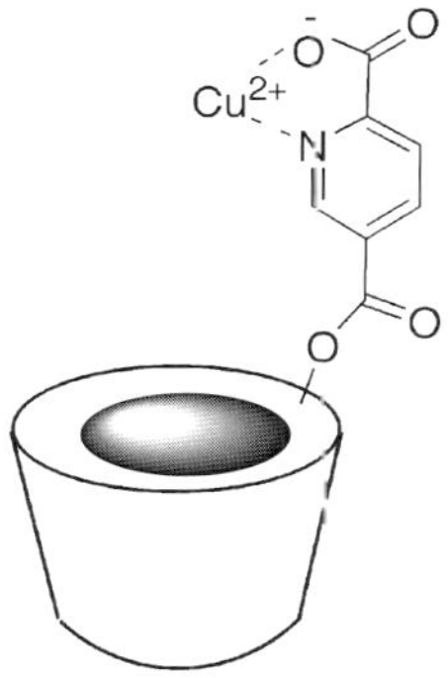

10

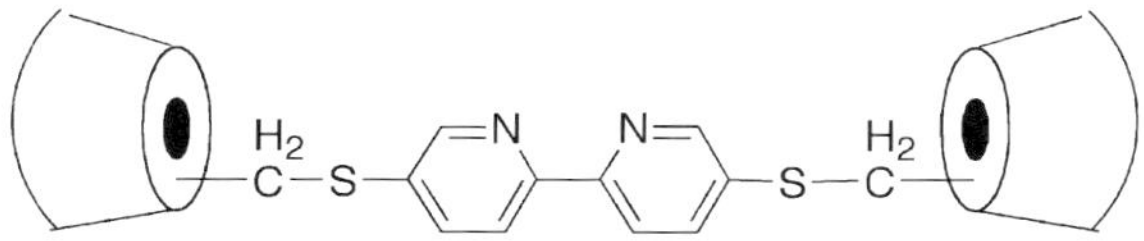

11

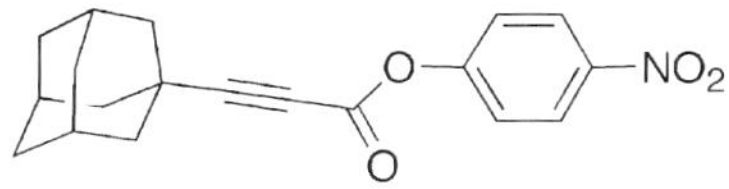

12

Figure 1.6 Reactions in which cyclodextrins combine with metal ions in catalysis in water. Compound **10** was the first species described as an 'artificial enzyme' in the literature. Compound **11** binds both ends of a substrate such as **12** and catalyzes its hydrolysis with a metal ion bound to the bipyridyl linker of **11**.

was used instead of metal coordination, so substrates could be used which did not bind to metal ions (Fig. 1.6).[60] This (**10**) was the first example of a catalyst that was called an 'artificial enzyme' in the published literature. The major catalysis resulted from the metal ion, but the substrate was hydrophobically bound as in many metalloenzymes.

We have used this plan – cyclodextrin binding in water solution, but catalysis by metal ions – in many subsequent studies. For example, we created a dimer **11** of cyclodextrin with a bipyridyl group in the linker, which would bind metal ions.[61] We then examined its use with Cu^{2+} as a catalyst for hydrolyzing esters such as **12** that could doubly bind into both cyclodextrin groups in water. We saw a 220,000-fold acceleration of the hydrolysis of such a doubly binding ester. As expected, the product fragments could not doubly bind, so they did not inhibit the catalytic hydrolysis process.

We have done many studies with metal ion catalysis in water solution, often imitating the role that the metal ions play in enzymes.[62–78] These will not be discussed in detail, for lack of space.

1.1.4 Cyclodextrin dimers

We had studied the binding of substrates to such dimers in water earlier. In the first work, we had seen that dimeric substrates such as **13** could bind to cyclodextrin dimers such as **14** with binding constants as large as 100,000,000 M^{-1}, compared with 10,000 M^{-1} for the monomeric binding of the same species to single uncoupled cyclodextrins (Fig. 1.7).[79] This corresponds to doubling the Gibbs energy of binding. Interestingly, we saw in some related cases that the increased strength of binding in the dimers was the result of enthalpy advantages, not entropies (which changed in the wrong direction).[80] We also saw that some cyclodextrin dimers with short linkers could strongly bind cholesterol in water, by including ring A into one cyclodextrin and ring D and its side chain into the other.[81] Then we saw that making two links between the two cyclodextrins in compound **15** – decreasing the freedom of the system – increased the binding energy for rigid substrates even more.[82,83] We also

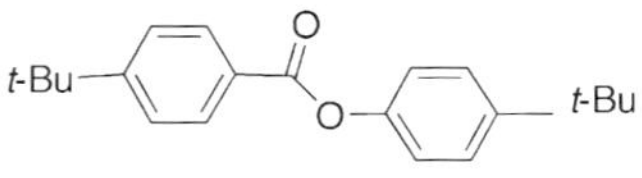

13

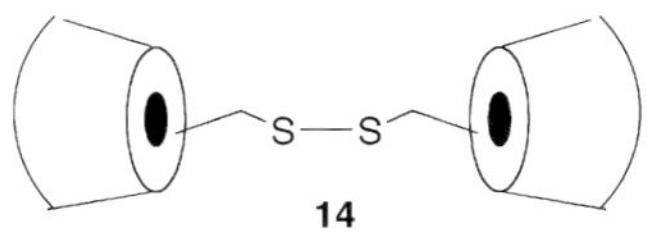

14

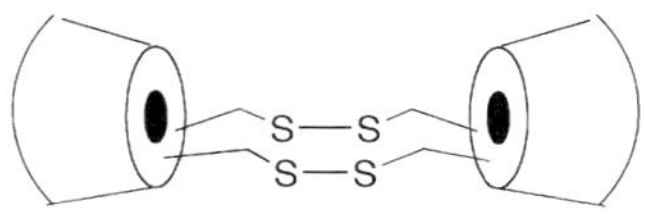

15

Figure 1.7 The rigid substrate **13** binds in water to cyclodextrin dimer **14** with extremely strong affinity, and the doubly linked cyclodextrin dimer **15** binds substrates even more strongly.

examined some trimers of cyclodextrins, but did not see strong cooperative effects.[84] (See below, however, the later cases of triple binding to cytochrome P-450 mimics.)

Sequence-selective binding of peptides and small proteins is of considerable interest. We saw that some cyclodextrin dimers could selectively doubly bind peptides in water with appropriately placed hydrophobic side chains.[85,86] This built on our earlier collaborative work on the selective binding of peptides by simple cyclodextrin.[87] We then showed that we could break up a protein dimer and a protein tetramer with appropriate cyclodextrin dimers in water, since such protein aggregation ordinarily involved hydrophobic side chains that our dimers could bind to.[88] In perhaps the most striking example, our cyclodextrin dimers and trimers were able to inhibit the protein aggregation involved in the formation of Alzheimer's plaques.[89]

Some synthetic macrocycles can bind hydrophobic groups in water similar to the binding into cyclodextrins. We examined the selective binding of some substrates by dimers of such synthetic macrocycles.[90] We have also examined where catalytic groups should be placed on cyclodextrins.[91,92]

1.1.5 Ribonuclease mimics

The enzyme ribonuclease A performs a two-step cleavage of RNA (Fig. 1.8). In the first step the $2'$ hydroxyl of one nucleotide piece cyclizes on the $3'$-$5''$ bridging phosphate ester, forming a $2'$-$3'$ cyclic phosphate and liberating the $5''$ hydroxyl group of the other nucleotide end. Then the enzyme hydrolyzes the cyclic phosphate to a simple $3'$ phosphate monoester, liberating the $2'$ hydroxyl group. The principal catalytic groups of the enzyme are the imidazoles of His-12 and His-119, although a lysine side chain ammonium group also plays a role. We have built and studied mimics of this process, learning much that is relevant to the enzyme itself.[91,92]

In our simplest study, we examined the cleavage/cyclization of uridyluridine **16**, abbreviated UpU, with a $3'$-$5''$ phosphate diester link (Fig. 1.9).[93,94] We used a concentrated buffer consisting of imidazole and imidazolium cation, mimicking the state of the two imidazoles in the enzyme. Indeed we saw that the buffer catalyzed the cyclization/cleavage reaction, forming uridine $2'$-$3'$ cyclic phosphate and liberating uridine. However, we also saw that there was some isomerization of the starting material, from the $3'$-$5''$ phosphate diester to the

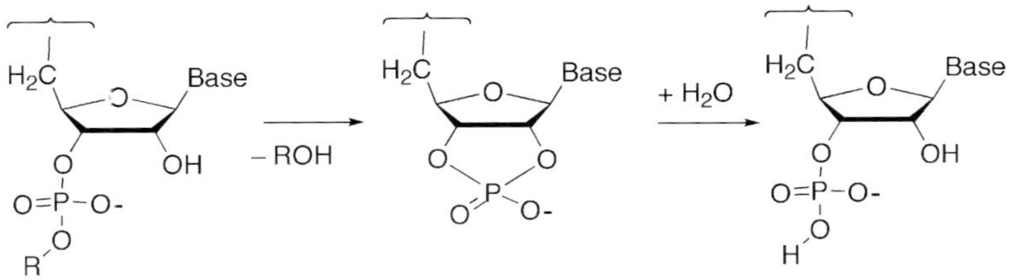

Figure 1.8 The two-stage mechanism by which the enzyme ribonuclease A hydrolyzes RNA. More details about each stage are discussed in the text.

2′-5″ phosphate diester. The reaction showed a bell-shaped pH vs rate profile, indicating that both the basic imidazole and the acidic imidazolium ion were catalytic, as with the enzyme. In the enzyme the two catalytic groups operate simultaneously, while in this imidazole buffer system they operated sequentially (the rate was first order in buffer concentration). There was a first step catalyzed by imidazolium ion, and then a second step catalyzed by imidazole.

In a two-step mechanism there must be an intermediate, a five-coordinate phosphorane **17**. We confirmed this from the observation that this intermediate could branch, either to the cleavage product **18** or to the isomer **19** of the starting material. Detailed kinetics showed how each step was catalyzed.[95–98] We did a related study for the cleavage by a combination of imidazole and Zn^{2+} in water.[99]

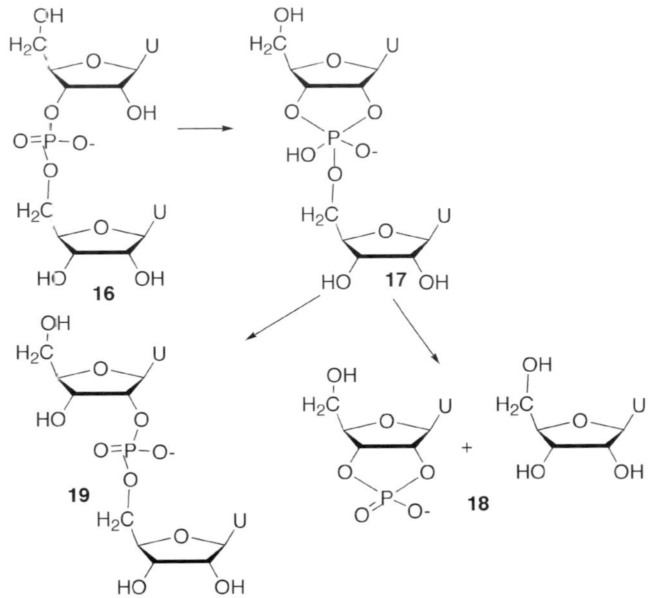

Figure 1.9 The detailed mechanism by which imidazole buffer in water catalyzes the hydrolysis and isomerization of a simple RNA piece, uridyluridine.

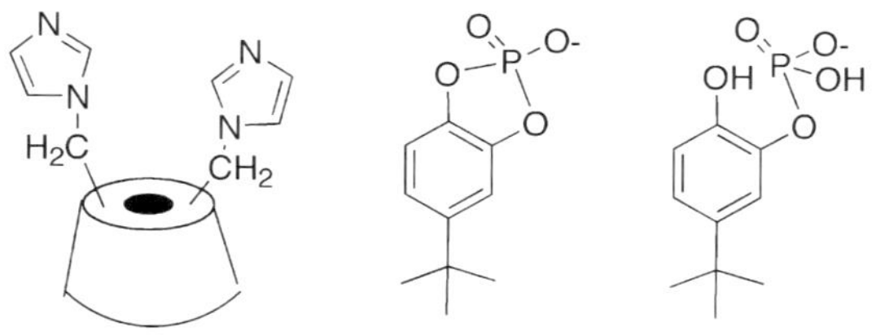

20 21 22

Figure 1.10 A β-cyclodextrin bisimidazole **20** with the imidazoles attached to the primary carbons of glucose residues A and D, as far apart as possible. It catalyzes the hydrolysis of substrate **21** in water to form product **22**, and with bifunctional catalysis by an imidazole and an imidazolium ion. Compound **21** is a model for the RNA cyclic phosphate in Fig. 1.8.

Even before this work was started, we had prepared a mimic **20** of ribonuclease consisting of β-cyclodextrin with two imidazole rings attached to the C-6 primary methylenes of two different glucose residues of the cyclodextrin.[100] The substrate was not an RNA, but a cyclic phosphate **21** whose cleavage mimicked the cleavage of the 2′-3′ cyclic phosphate of RNA. This substrate, 4-*t*-butylcatechol cyclic phosphate, bound well into the cyclodextrin in water, and the catalyst hydrolyzed the cyclic phosphate with a bell-shaped pH vs rate profile, showing that both the imidazole and the imidazolium ion were playing a catalytic role.

The first catalyst had imidazoles on the C-6 carbons of the farthest apart glucose residues A and D, contaminated to some extent by the A,C isomer. We saw that the hydrolysis was quite regiospecific (Fig. 1.10), cleaving the bond between the phosphorus and the oxygen on carbon 1 of the substrate to form product **22**. This was as expected from a mechanism in which a water molecule is delivered perpendicular to the cyclodextrin axis as models predicted. In a later catalyst the imidazoles were mounted further from the ring and significantly cleaved the P—O bond to carbon 2, again consistent with models.[101]

In both of these cases we made the A,D isomers since we assumed that the function of the imidazolium ion was to protonate the leaving group oxygen. This is what is usually written for the mechanism of the real enzyme. However, when we did a detailed study of the three isomers, we saw that the A,B isomer of bisimidazole cyclodextrin was the best catalyst of all.[102] This is not consistent with a mechanism in which the function of the imidazolium ion is to protonate the leaving group, which requires that the two catalytic groups be as far apart as possible. It is consistent with a mechanism in which the imidazolium ion protonates the phosphate anion of the substrate, promoting the formation of a phosphorane intermediate. Models show that this process is best done with the A,B catalyst isomer. This is the same role we later discovered for the imidazolium component of the imidazole buffer studies described above.

In an enzyme the protonation of a phosphate anion and the deprotonation of an attacking hydroxyl group can be simultaneous, since the two catalytic groups are fixed in space. In fact, the true mechanism for ribonuclease A, the enzyme, involves such a simultaneous process. A tool called 'proton inventory' has been created to detect such simultaneous proton transfers. The reaction is run in H_2O, in D_2O, and in mixtures of the two. If a single proton is 'in flight' during the rate-determining step, the kinetic isotope effect will be proportional to the mole fraction of deuterium in the solvent. However, if two protons are 'in flight' during

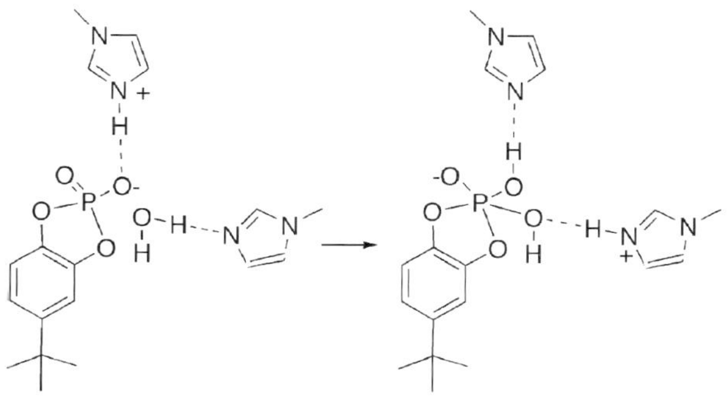

Figure 1.11 The simultaneous bifunctional mechanism, indicated by isotopic studies, by which the A,B β-cyclodextrin bisimidazole catalyzes the cleavage of compound **21** in water by first forming a phosphorane.

this step, the kinetic isotope effect will be proportional to the square of the mole fraction of deuterium in the solvent. With ribonuclease such a square dependence has been observed, showing that both removal of the proton in the nucleophile (the C-2′ hydroxyl in cyclization, the H_2O in ring opening of the cyclic phosphate) by imidazole and the protonation by the imidazolium ion are occurring simultaneously.[103]

We applied this test to our ribonuclease mimic, the 6A,6B isomer of cyclodextrin bisimidazole, cleaving the bound cyclic phosphate **21**. We found that there was indeed a square dependence of the kinetic isotope effect on the mole fraction of deuterium in the water solvent, and interestingly the values of the isotope effect for the two protons in flight were almost identical with those that had been seen with the enzyme itself and its normal substrate.[104,105] As described above, the protonation in the model system involves an imidazolium ion putting a proton on the substrate phosphate anion as the imidazole delivers a water molecule to the phosphorus.

The picture is more detailed than that (Fig. 1.11).[97,106] There is evidence that the imidazolium ion is hydrogen bonded to the phosphate anion, but that anion is not basic enough to force a full proton transfer. Also the imidazole is surely hydrogen bonded to a water molecule, but the imidazole is not basic enough to force a full proton transfer from the water. However, as the oxygen of the water starts to add to the phosphorus the water proton becomes more acidic, and approximately halfway through the O—P bond formation the proton will be equally shared between the incipient hydroxyl group and the imidazole, and later it will 'transfer' in the sense that the new hydrogen bond has a weak O—H bond and a strong imidazole—H bond. At the same time, the phosphate anion becomes more basic as it is being turned into a phosphorane anion, and again approximately halfway through the addition process the proton will be equally shared between the incipient phosphorane and the incipient imidazole group. Again after full 'transfer' the proton is still shared by the phosphorane oxygen and the new imidazole group, but with a strong O—H and a weak imidazole—H bond. That is, the hydrogen bonds are unsymmetrical, except at the transition state when the protons are equally bonded to their two partners each. This is a more detailed description of how general acid and general base catalysis operate in water solution.

We have proposed that a similar mechanism operates for the enzyme itself, in which the imidazolium ion acts first to protonate the phosphate oxygen of RNA, not the leaving group.[107] There is not yet full agreement on this proposal.

The cyclodextrin bisimidazoles we created have general use in determining the detailed geometries of reactions in water with bifunctional catalysts. As one example, we showed that the enolization of ketones can be catalyzed by such cyclodextrin bisimidazoles, but with a different geometric preference from that for phosphate ester hydrolysis.[108] We also showed that a number of aldol reactions could be catalyzed in water with selectivity.[109–112]

1.1.6 Transaminase mimics

The class of enzymes called transaminases use the coenzyme pyridoxamine phosphate **23** to convert a substrate α-keto acid to an α-amino acid, while the pyridoxamine itself is converted to pyridoxal phosphate **24** (Fig. 1.12). Later the pyridoxal phosphate coenzyme is converted back to the pyridoxamine form by transaminations in the reverse direction from a sacrificial amino acid. We set out to mimic this interesting process by attaching pyridoxamine to a cyclodextrin, so there would be a preference for transaminating keto acids that could bind to the cyclodextrin in water.

In our first example (Fig. 1.13), the first in which a coenzyme was linked to a cyclodextrin, we synthesized compound **25**, with a pyridoxamine covalently linked to a C-6 of β-cyclodextrin.[113] We saw transaminations of pyruvic acid, of phenylpyruvic acid, and of indolepyruvic acid to form alanine, phenylalanine, and tryptophan, respectively, and with high selectivity for the hydrophobic phenyl and indole derivatives relative to simple pyruvic acid. Relative to its reaction with simple pyridoxamine in solution, without an attached or unattached cyclodextrin, the indolepyruvate reacted 50 times faster with compound **25**.[113] Also, we saw a 5:1 preference for the formation of the L-phenylalanine relative to the D enantiomer in transamination by compound **25**.[113]

In compound **25** we had attached the pyridoxamine to the primary C-6 position of the cyclodextrin, but in an earlier study we had seen that there were sometimes, but not always, advantages to attaching catalytic groups to the secondary face of a cyclodextrin.[114] Thus we attached pyridoxamine to the secondary face of β-cyclodextrin, and saw again a preference for the transamination of indolepyruvic acid and of phenylpyruvic acid, but the preferences were only approximately half as large as those with compound **25**. Thus here the original attachment of the cofactor to the primary face of the cyclodextrin was actually the best.

Another comparison substituted the cyclodextrin-binding group by a synthetic macrocycle that also strongly binds hydrophobic substrates in water solution. We synthesized

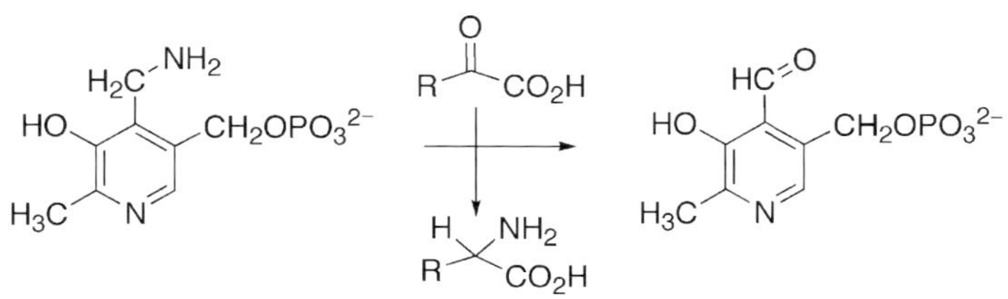

Figure 1.12 Pyridoxamine phosphate is the coenzyme for the enzymatic conversion of α-keto acids to amino acids while itself being converted to pyridoxal phosphate.

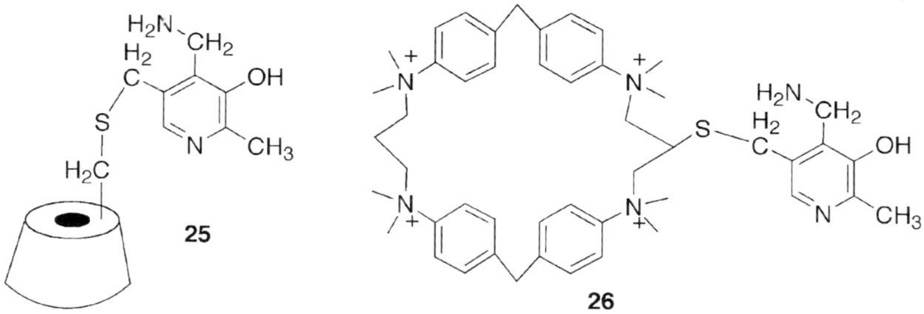

25 26

Figure 1.13 Pyridoxamine was attached to a cyclodextrin on compound **25** and to a synthetic macrocycle in **26** that also binds substrates and promotes their transaminations.

compound **26** and examined its ability to transaminate keto acids.[115] We found that it was comparable to the cyclodextrin-based compounds, although of course it did not induce any enantiomeric excesses in the products. We also prepared an analog of compound **25** in which all the primary hydroxyl groups were reduced, so the hydrophobic pocket of the cyclodextrin was deeper.[116] Again there proved to be no significant advantage to this modification over the original compound **25**. This work was summarized in a full paper.[117]

In the compounds described so far, the pyridoxamine units were attached to the cyclodextrin, or macrocycle, by a single covalent linkage. This of course introduces degrees of freedom, which could be a disadvantage in reactions that require cooperative interaction of a substrate with both the binding group and the pyridoxamine group. To test this, we prepared two compounds, **27** and **28**, in which a pyridoxamine was linked to two neighboring glucose residues in β-cyclodextrin, related to the 6A,6B bisimidazole cyclodextrin derivative described earlier (Fig. 1.14). Each was a pair of diastereomers, reflecting either an A to B orientation or a B to A sequence.[118]

We saw that the *endo* pair of isomers, with the pyridoxamine directed over the cyclodextrin cavity (e.g. **27**), was particularly effective in transaminating a substrate, *p*-*t*-butylphenylpyruvic acid, that is oriented along the cyclodextrin axis. By contrast, the exo compounds (e.g. **28**) were more effective with *m*-*t*-butylphenylpyruvic acid, which holds

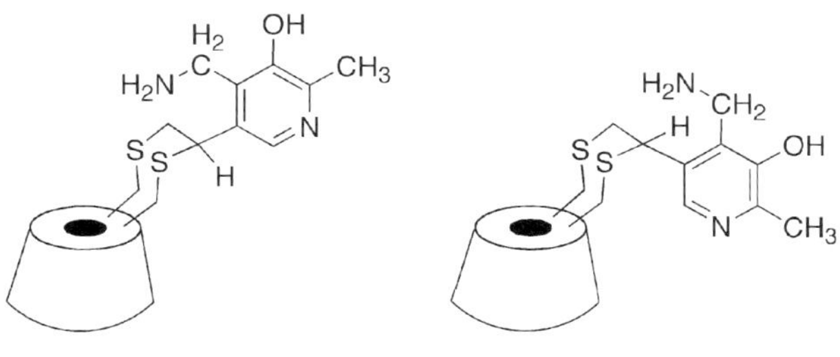

27 28

Figure 1.14 Two isomers in which pyridoxamine is doubly bound to a cyclodextrin. They show very large rate accelerations of transaminations in water and striking differences in substrate selectivity.

its keto acid group away from the cyclodextrin axis. Thus the fixed geometries of the new transaminase mimics are reflected in sensible geometric preferences.

As an additional bonus, the *p*-*t*-butylphenylpyruvic acid was a particularly good substrate even for our simple transaminase **25** with a single link to the cyclodextrin. It preferred the *p*-*t*-butylphenylpyruvic acid substrate over the substrate without the *t*-butyl group by a factor of over 150-fold, presumably because the *t*-butylphenyl group is particularly well bound in water. The preference of this substrate relative to pyruvic acid itself was over 15,000-fold.

Our original transaminase mimic **25** produced phenylalanine with a significant enantiomeric excess of the L form, with chirality induced by the chiral cyclodextrin. However, we were also interested in transaminase mimics in which chirality was induced by a properly placed basic group to deliver the proton to the alpha carbon of the product amino acid. Since some of these studies were done in organic solvents, without hydrophobic binding, I will simply list their references here – Refs. 119–122. However, in a few cases we examined transaminations in water by cyclodextrin derivatives carrying not only a pyridoxamine, but also a chirally mounted base. In a review of this general area we cited a number of our examples along with one claimed by Tabushi to have a particularly high enantiomeric excess.[123] However, in later work we found that the Tabushi claim could not be confirmed.

In all this work we used water solvent to promote the hydrophobic binding of a substrate into a cyclodextrin. In more recent work we addressed a different question: What is the effect of the fact that in enzymatic reactions the processes are formally carried out in water solution, but actually occur in the interior of the enzyme away from the water? Can this have large rate advantages, since in water there are water solvents bound to acids and to bases of the enzyme catalysts, and to substrate groups as well? In other words, has the large size of proteins produced a way in which reactions formally in water solution are actually performed in the equivalent of dimethylformamide? To address this, we initiated studies of catalysts using a special group of polymers.

We used polyaziridines, compounds with one nitrogen atom and two methylene groups in a polymer unit. The pioneer in this area was Irving Klotz, who had used these commercially available polyamines to catalyze various cleavage reactions.[124,125] We set out to use them in transaminase mimics.[32,126–129] We also made transaminase mimics with dendrimers.[129,130] The polymers themselves have the interesting feature that the nitrogens are so close to each other that they do not fully titrate until almost at pH 3. That is, at pH 8 or so they have an equal proportion of free amine and of protonated ammonium ions. Thus they contain the strongest possible bases and acids at that pH and exhibit outstanding general acid/general base catalysis.

In addition, in our earliest catalysts **29** we had a pyridoxamine unit covalently linked to the polymer, and we also added long-chain alkyl groups so as to form a hydrophobic core (Fig. 1.15). We saw that the pyridoxamine unit in this catalyst transaminated α-keto acids with high accelerations – with pyruvic acid forming alanine, the acceleration was 10,000-fold, while with indolepyruvic acid forming tryptophan, the acceleration was 240,000-fold. The catalyst/substrate combination showed Michaelis–Menten kinetics, and k_{cat} was increased for both substrates. This we ascribe both to the high general acid/general base catalytic contributions from the polymer and to the hiding of the reaction zone from solvent water, since k_{cat} was greatly increased by the addition of hydrophobic side chains. The hydrophobic side chains diminished the binding constant for pyruvate, which is not hydrophobic, but increased the strength of binding of indolepyruvate.[131]

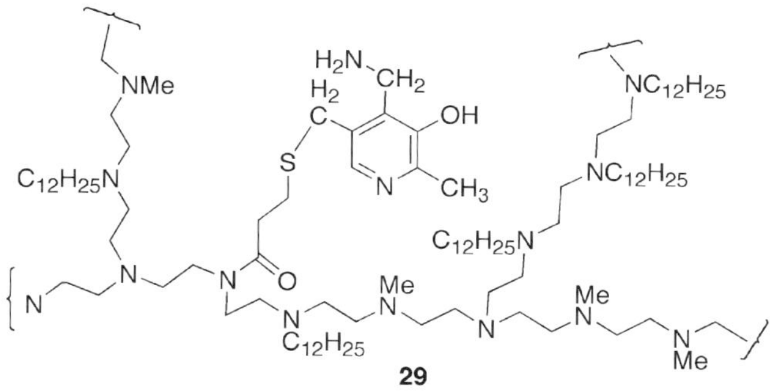

Figure 1.15 A pyridoxamine residue attached to a polymer that also carries hydrophobic chains so that substrates are strongly bound in water and transaminated with very large accelerations.

In a second approach, we used the polymer with its attached hydrophobic chains, but had the pyridoxamine unattached, but carrying hydrophobic side chains itself. Thus in water the coenzyme bound into the hydrophobic core of the polymer, and so did the indolepyruvic acid. The acceleration in the synthesis of tryptophan by this system was 725,000-fold, relative to the rate with free pyridoxamine alone.[128]

In enzymatic transaminations the product pyridoxal is reverted to the pyridoxamine form by reversing the transamination process with a second, sacrificial, amino acid. However, the process in this direction is generally endothermic and very slow in simple model systems, so we devised an alternative. Pyridoxal with α,α-disubstituted glycines performs a decarboxylation to afford pyridoxamine and a ketone, and the loss of CO_2 makes this process irreversible (Fig. 1.16). Thus we coupled this decarboxylation with our normal transaminations to produce multiple turnovers of the catalytic process. Again, all of this is in water solution. We have studied the structure–rate relationship in this decarboxylative transaminations.[132]

One of the interesting goals in this field is to produce amino acids with high enantiomeric excesses (ee) as well as high rate accelerations and catalytic turnovers. An opening effort in

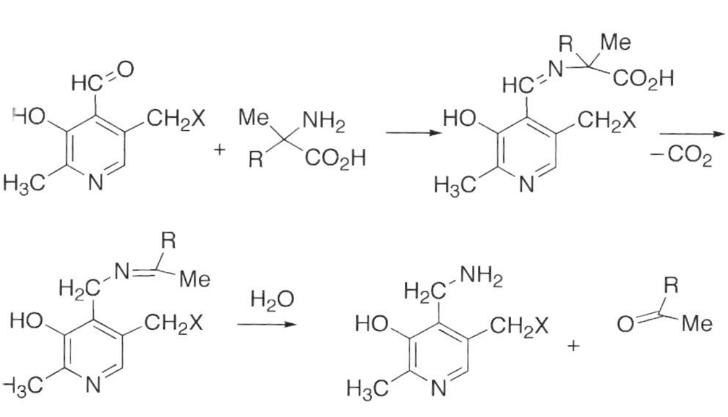

Figure 1.16 Pyridoxal species can be converted to pyridoxamines in water by transaminations involving irreversible decarboxylation of C,C-dialkylglycine derivatives.

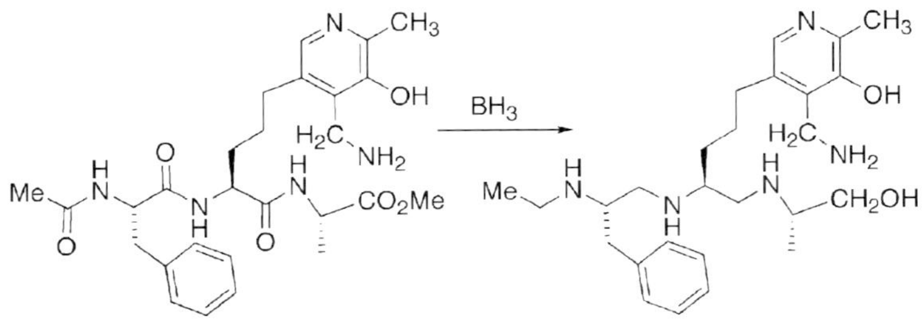

Figure 1.17 A tripeptide carrying a pyridoxamine unit can be reduced to the chirally pure triamine unit, which induces chirally selective transaminations in water.

this direction involved the use of pyridoxamines covalently attached to short oligoamines produced by the reduction of oligopeptides (Fig. 1.17).[133] With such a system we achieved a 64% ee in the conversion of pyruvic acid to L-alanine in water, but only an 18% ee in methanol. However, with some other keto acids the methanol system was better. Work on this approach is still underway.

The pyridoxal/pyridoxamine system is not only involved in transaminations. We examined catalysts for the synthesis of tryptophan by coupling reactions with indole[134] and for pyridoxal-catalyzed aldol condensations.[122,135] We also examined optical induction in transaminations[122,136] and changing transaminations into selective racemization.[137]

1.1.7 Cytochrome P-450 mimics

One of the most challenging goals in the artificial enzyme field is to learn how to perform selective oxidations on unactivated carbons, imitating that which is routinely done by the class of enzymes called cytochrome P-450 species. We had earlier done a number of studies to show that it was possible to perform such selective functionalizations of unactivated carbons directed by the geometry of a substrate with an attached reagent or template, but this work was performed in organic solvents, not in water.[3] In order to make the process more general, we moved it into water and used hydrophobic binding to coordinate a substrate to a catalyst in a geometrically defined manner.

We produced metalloporphyrins **30** carrying iron or manganese atoms and with four cyclodextrins linked to the porphyrins, and saw that they could catalytically direct oxidation of hydrophobically doubly bound substrates.[138,139] Then we showed that the system would hydroxylate a bound steroid **31** with complete selectivity for the C-6 equatorial hydrogen, producing the 6-α-hydroxy steroid **32** (Fig. 1.18).[140] By adding some fluorine atoms to this system in catalyst **33** to stabilize the catalyst against self-oxidation, we greatly improved the catalytic turnovers.[141,142]

A more interesting target was the C-9 hydrogen, whose hydroxylation would produce an attractive lead into useful corticosteroids. By using hydrophobic triple binding of a steroid substrate to our catalyst, we changed the geometric preference to completely selective C-9 hydroxylation, again in water.[143–146] We replaced the cyclodextrin-binding groups with synthetic cyclophanes that could also bind the steroid hydrophobically in water, and saw similar selectivities.[147] In some related work, we also showed that we could use cytochrome

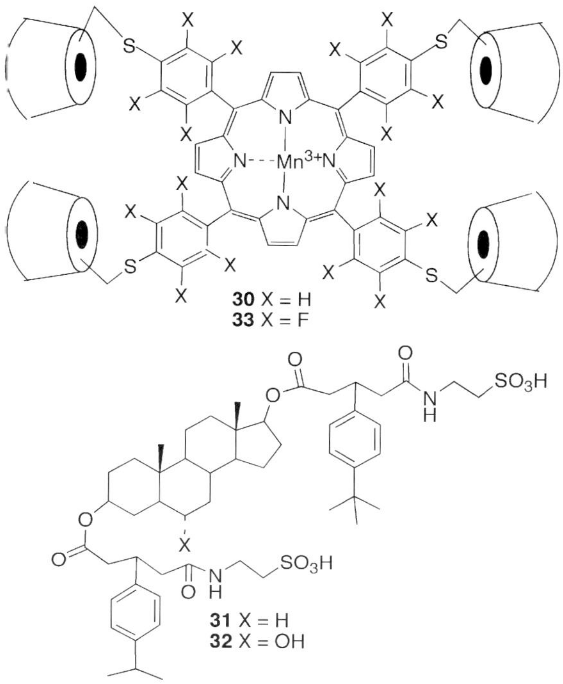

Figure 1.18 A porphyrin carrying four β-cyclodextrins **30** and its highly fluorinated analog **33** bind the steroid substrate **31** in water and catalyze its hydroxylation at the C-6 alpha position with very high selectivity.

P-450 mimics to transfer nitrogen atoms, not just oxygen.[148,149] Expansion of this approach to enzyme mimicry is still underway.

1.2 Reactions in water promoted by hydrophobic binding of small molecules

In the work described above we have used hydrophobic binding in water solution to hold a substrate next to a catalyst. However, we also found that reactions in water could show important selectivity and rate effects if two reacting small molecules were hydrophobic.[150] In one case we examined addition reactions, in another series we examined atom transfer processes.

1.2.1 Diels–Alder reactions

We were interested in the possibility that Diels–Alder additions could be promoted if the two components bound into the same cavity in a cyclodextrin. We examined three examples: 1,3-cyclopentadiene plus butenone, 1,3-cyclopentadiene plus acrylonitrile, and anthracene-9-carbinol plus *N*-ethylmaleimide (Fig. 1.19).[151] We saw several interesting effects. First of all, relative to the second-order rate constant in water, the addition of 10 mM β-cyclodextrin

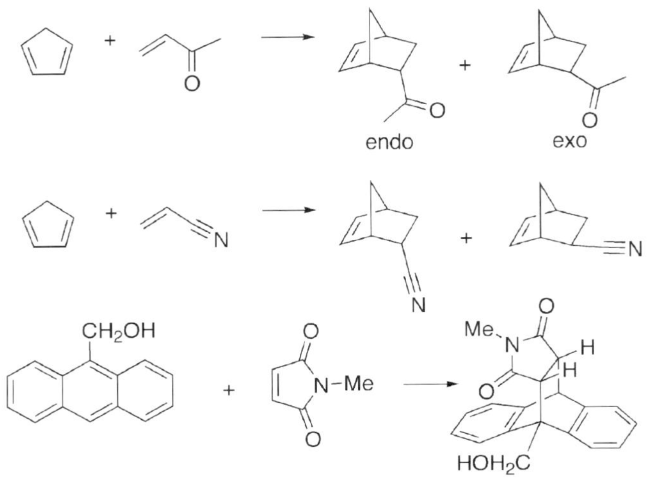

Figure 1.19 Three Diels–Alder reactions that were performed and studied in water solution.

accelerated the butenone reaction by a factor of 2.5, while 10 mM α-cyclodextrin slowed it by a factor of 1.7. This was consistent with the indication, from models, that both components could bind into the cavity of β-cyclodextrin, but only the cyclopentadiene could bind into the smaller α-cyclodextrin cavity. This binding of only one component slows the reaction.

Similarly, with the acrylonitrile reaction the 10 mM β-cyclodextrin increased the water rate by a factor of 9, while 5 mM α-cyclodextrin slowed it by a factor of 1.2. With the larger anthracene case, even β-cyclodextrin slowed the water reaction, by a factor of 1.6, since the anthracene filled the cavity and did not permit simultaneous binding of the dienophile. However, the more striking rate effects were seen with the water solvent alone.

In the butenone reaction, the rate constant in water was 740 times as large as in isooctane and 58 times as large as in methanol. In the acrylonitrile reaction the rate constant in water was 31 times as large as in isooctane and 15 times as large as in methanol. In the anthracene reaction the rate constant in water was over 200 times that in acetonitrile, but only 28 times that in isooctane. Actually, the rate constant in methanol was less than that in isooctane, by a factor of 2.3. Thus the huge water effect did not simply reflect the polarity of the solvent, it also reflected a hydrophobic effect. Supporting this, the water rate for the butenone reaction was increased by LiCl and decreased by guanidinium chloride. As we will discuss below, such solute effects are diagnostic of hydrophobic effects.[152,153]

In further studies of the remarkable water effects on Diels–Alder reactions, we examined the exo–endo selectivity of the processes.[154,155] We saw that butenone added with a 95.7% preference for endo addition in water, but only an 80% endo preference in cyclopentadiene as solvent. Thus the endo addition is favored not only by 'secondary orbital overlap', it is even more strikingly favored by the hydrophobic effect. In the transition state for the addition reaction, the endo geometry diminishes the amount of water/hydrocarbon interface more than does the exo geometry. The high energy of a hydrocarbon/water interface is the cause of hydrophobicity, the tendency of nonpolar materials and segments to cluster in water so as to diminish the interface with water.

Perhaps more strikingly, we saw related effects even when the reactants were at concentrations much above their solubility limits in water. In fact, the title of our paper referred to both 'solutions and suspensions' in water. We saw that vigorous stirring could increase the rate in such suspensions and that they gave endo preferences not quite as large as those of the true solution but still quite large, such as 95.5% or 94.9% with increasing amounts of insoluble cyclopentadiene. As we pointed out in this paper and in the review of this work,[153] these observations made the use of water in suspensions more practical than requiring true solutions of insoluble compounds.

We will return to the Diels–Alder reactions later, when we discuss the use of the hydrophobic effect to deduce the detailed geometry of the transition states for some chemical reactions.

1.2.2 The benzoin condensation

The reaction of two molecules of benzaldehyde to form benzoin is generally referred to as the benzoin condensation. It is normally catalyzed by cyanide ion, although thiazolium ions will also catalyze it, as we have discussed above and shown in Fig. 1.2. The normal solvent for the benzoin condensation is ethanol, to dissolve all the components of the reaction. However, it seemed to us likely that there would be overlap of the phenyl rings in the transition state for the benzoin condensation, and thus that reaction in water could lead to hydrophobic accelerations. This proved to be the case. We saw that the rate of the cyanide-catalyzed benzoin condensation was 200-fold faster in water than in ethanol.[156] Also, we saw that added LiCl increased the reaction rate, while added lithium perchlorate decreased it. Such salt effects are diagnostic of the presence of some acceleration by hydrophobic packing in the transition state for the reaction.

We did a study of the origin of the prohydrophobic and antihydrophobic salt effects.[157] It was well understood that the prohydrophobic effect – reflected in a decreased solubility of hydrocarbons in water when salts like lithium chloride were added – was caused by electrostriction of the water, raising the energy needed to produce a cavity into which the hydrocarbon could enter. However, the antihydrophobic effects of additives such as urea, guanidinium ion, and perchlorate ion were less well understood. Sometimes these substances were called 'water structure breakers' to reflect one idea about how they might work. Sometimes they were called 'denaturants', since they could break up the hydrophobic associations responsible for the natural folding of proteins, for instance. I choose to call them simply 'antihydrophobic additives' – denaturants is too narrow a word for their general properties, and they do not work by simply breaking up water structure.

When an additive to water increases the solubility of hydrocarbon species and segments, there are two choices as to how they function. One possibility is that they make cavitation easier, which is the idea behind 'water structure breakers'. The other possibility is that they help solvate the hydrocarbon surface, by bridging between that surface and the water. This is what we found. Simply, we excluded the possibility that easier cavitation was involved, by showing that all the antihydrophobic additives actually increased the surface tension of water.[157] Making a hole in water involves increasing the surface area of that hole, and an additive that increases surface tension would fight that increased surface area. We later used the quantitative magnitude of such antihydrophobic effects to determine the shapes of transition states, as we will discuss below.

As part of our exploration of these effects, we also looked at solvents that are 'waterlike' and the effects of additives in them.[158] We saw that these additives did not show antihydrophobic effects in ethylene glycol, in acetonitrile, or in dimethylsulfoxide. Apparently the interface of a hydrocarbon surface with these solvents is sufficiently stable that no bridging by an additive is required to lower the energy.

1.2.3 Atom transfer reactions

In the above cases hydrophobic binding involves reactants that add to each other, and both appear in the product. It seemed to us likely that we would also see such effects in water with reactions such as reductions or oxidations in which one of the reagents simply transfers an atom, rather than itself adding to a substrate. This turned out to be true.

Our first studies involved hydride reductions of carbonyl groups.[159–161] We compared the reduction of a phenyl ketone group with a methyl ketone group in water with lithium borohydride, lithium phenylborohydride, lithium pentafluorophenyl borohydride, and lithium 2-naphthylborohydride, and saw that there was a preference for the reaction of the arylborohydrides with the aryl ketone in water, diminished when the antihydrophobic materials methanol or sodium perchlorate were added, and increased with the prohydrophobic additive sodium chloride.[159] The effects were even larger when the ketone carried a 2-naphthyl ring rather than a phenyl ring.

Then we examined the reduction of a steroidal diketone **34** in which one carbonyl group is next to a phenyl ring and the other is not (Fig. 1.20).[160] With lithium borohydride in water there was an 87:13 preference for reducing the ketone group that was not deactivated by conjugation with the phenyl group. However, with lithium pentafluorophenyl borohydride in water with 4 M added lithium chloride, a prohydrophobic additive, there was now an 85:15 preference for reduction of the carbonyl group next to the phenyl ring. The hydrophobic preferences for packing the aryl groups next to each other completely overcame the greater intrinsic reactivity of the other carbonyl group. In a related study, we have seen that similar results can be obtained with amineboranes, in which a hydrophobic amine complexed with borane shows such preferences for reduction of carbonyl groups next to hydrophobic substrate segments.[161]

We examined the epoxidation of carbon–carbon double bonds by oxygen atom transfer reactants.[162] We used the competition between cinnamic acid derivatives and crotonic acid in water (D_2O) and water with added 2-propanol. There was no selectivity when the oxidant

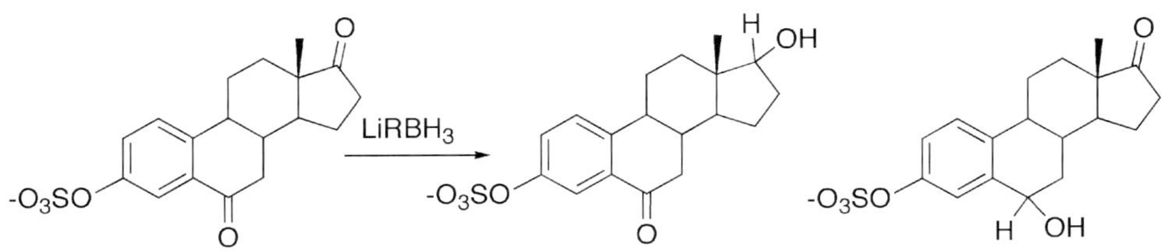

R = H	87	:	13
R = C_6F_5 [4 M LiCl]	15	:	85

Figure 1.20 Hydrophobic binding of a reagent to a substrate in water reverses the selectivity of reduction of a diketone.

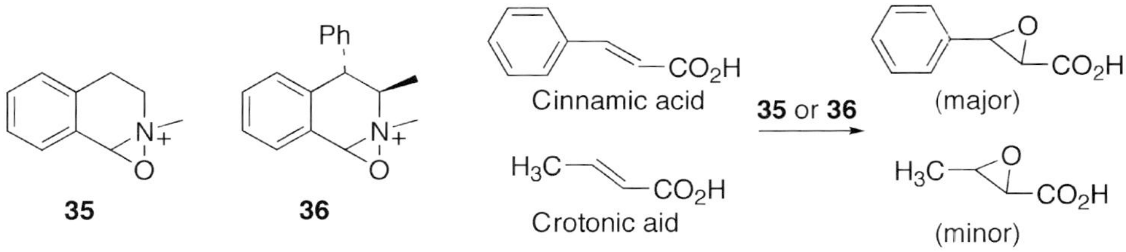

Figure 1.21 Oxaziridinium ions in water bind to the cinnamic acid selectively and reverse the higher reactivity of crotonic acid in the absence of such binding.

was perbenzoic acid. This was consistent with the geometry of epoxidation by peracids, in which the phenyl ring of the peracid cannot overlap the phenyl rings of the substrates. However, with oxaziridinium cations **35** and **36** carrying phenyl rings there was significant preference for epoxidizing the cinnamic acid derivatives in water, which was suppressed by the added alcohol (Fig. 1.21). This was consistent with hydrophobic packing in the transition state with these oxidants and substrates, in one case leading to a Gibbs energy advantage in the transition state of 3.23 kcal/mol and a selectivity increase for the hydrophobic reactions of as much as 240-fold.

1.3 Quantitative antihydrophobic effects in water and the geometries of transition states

Throughout this account we have been using antihydrophobic additives such as guanidinium cation or alcohols to show that there was some hydrophobic overlap in the transition states of various reactions. We realized that this could be made quantitative – the magnitude of the effect of the antihydrophobic additive on a rate could be related to the amount of hydrophobic surface that becomes hidden from the water in the transition state. This work was reviewed recently.[163]

In our first approach,[164] we simply examined the quantitative effects of antihydrophobic agents on binding constants and on solubilities in water. In particular, we saw that the effects of urea and of guanidinium chloride on water solubilities of *t*-butylphenyl rings were parallel to their effects on diminishing the binding of such groups into cyclodextrins. However, we then took up the use of antihydrophobic agents as probes for transition state structures.[165,166]

In our first study,[165] we examined the effect of added alcohols on the solubility of benzaldehyde in water and on the rate of the benzoin condensation in water. We saw – in a plot with 15 points – that there was a good parallel between the log of the solubility and the log of the rate constant effect (increased solubility and decreased rate with added alcohols in water). Since the effect of added alcohols on the solubility of benzaldehyde simply reflects the antihydrophobic role of the additives, this must also be true of the rate effects. That is, apparently there was no extra effect relating to solvation of charges in the rate changes with additives; they simply reflected the fact that the transition state for the benzoin condensation – under these conditions the transition state occurs during the addition of the cyanohydrin anion to the aldehyde – is less stabilized by additives than is the starting material. Some of the hydrophobic surface of the two phenyl groups is inaccessible to solvent, since they are packed on each other in the transition state. In the rate-determining step a cyanohydrin carbanion is converted to an oxyanion, so apparently differential charge solvation essentially cancels.

From the slope of the plot, we could see that about 40% of one face of each phenyl group is inaccessible to water solvent in the transition state. We will return to such quantitative calculations later. In that paper,[165] we also examined the effect of antihydrophobic additives on the rates of some displacement reactions, such as the reaction of *N*-methylaniline with the sodium salt of *p*-carboxybenzyl chloride (the carboxylate group was added to achieve water solubility). We must return later to such displacement reactions to see how much the additives affect the solvation of charges in these cases.

In the second early paper in this series, we examined the quantitative effect of antihydrophobic additives on some Diels–Alder reactions.[166] We plotted the effect of three different concentrations of added ethanol in water on the solubility of cyclopentadiene and on its rate constant for dimerization. As with the benzoin case, we saw that a log/log plot was linear and indicated that in the transition state for dimerization the two cyclopentadienes essentially covered one face of each. Thus the transition state was less well stabilized by the additive than was the starting material, so the added ethanol slowed the rate.

In the paper we showed that the effects of various alcohol additives on the solubilities of a number of hydrocarbons correlated with their calculated solvent-accessible surface areas. We also showed that the geometry of the transition state for Diels–Alder dimerization of cyclopentadiene that we had determined, from cosolvent effects in water, agreed with the geometry calculated by AM1 and Macromodel computations. Finally, the paper reports the effect of added antihydrophobic alcohols on the rate, in water, of the Diels–Alder reaction of various *N*-alkylmaleimides with anthracene-9-carbinol, a reaction we had studied previously in a qualitative way. We saw that the maleimide covered a little over 50% of one face of the anthracene in the transition state, a reasonable picture (Fig. 1.22). In the reaction product the structure opens up, and now only 10% of one face of the anthracene is inaccessible to solvent. The Diels–Alder transition state geometries we determined are not surprising, but our results indicate that this novel method for determining transition state geometries is reliable, at least in these cases.

In an extension of the earlier work using this method to examine displacement reactions, we also described a striking result.[167] The geometry of displacement on the sodium salt

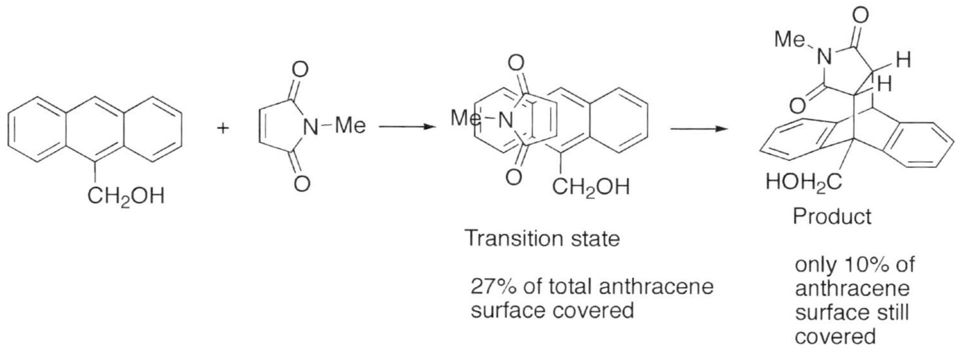

Figure 1.22 The transition state for this Diels–Alder reaction is in sensible agreement with expectations.

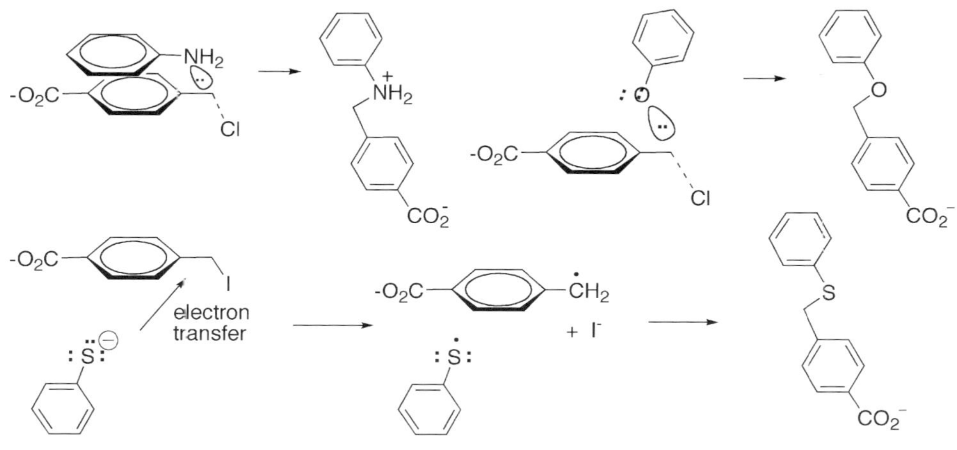

Figure 1.23 Displacement on a benzylic chloride by aniline has a transition state with significant hydrophobic overlap of the phenyls, but there is no overlap in the transition state for displacement by phenoxide ion. Thiophenoxide ion can perform a single-electron-transfer displacement mechanism with a benzylic iodide, but not with a benzylic mesylate.

of *p*-carboxybenzyl chloride with *N*-methylaniline involved significant overlap of the two phenyl groups in the transition state, as we had indicated previously,[165] but displacement by phenoxide ion involved no phenyl overlap at all in the transition state.

This difference can be understood. With aniline the single unshared pair of electrons on the nitrogen is conjugated with the phenyl group, forcing a geometry that requires overlap with the phenyl of the substrate during nucleophilic approach (Fig. 1.23). However, phenoxide ion has three unshared pairs of electrons on the oxygen, and only the use of the pair conjugated with the phenyl ring would require phenyl overlap in the transition state for reaction. If instead one of the electron pairs not so conjugated is used for displacement, the predicted geometry has the phenyl nowhere near each other, certainly not overlapping (Fig. 1.23). With this choice, the phenoxide ion retains the stabilization by the conjugated electron pair. We will return to these displacement reactions, as we examine the cosolvent effects on such ionic reactions. This early work was summarized in a review.[168]

Displacements by thiophenoxide ion have an interesting possibility – the nucleophile can attack one electron at a time, transferring an electron to produce a thiophenoxy radical while reductively cleaving the electrophile to form an alkyl radical. Then the two radicals, in a solvent cage, can couple (Fig. 1.23). In an exploration of this process, called the SET mechanism, we used thiophenoxide with the sodium salt of *p*-carboxybenzyl iodide, and with the corresponding mesylate.[169] We saw that there was a large *acceleration* by added ethanol in the iodide case, but not with the mesylate. We proposed that in the iodide displacement this reflected the conversion of thiophenoxide ion, with its delocalized charge, into the much more hydrophobic thiophenoxy radical at the transition state. Other evidence as well supported the SET mechanism. The carbon–iodine bond is more easily reductively cleaved than is the carbon–mesylate bond.

Further studies of displacement reactions correlated with detailed quantum mechanical calculations and showed that concerns about the solvation of charges needed to be addressed.[170] One approach to this was to examine the reaction of 2,6-dimethylphenoxide ion with substituted benzyl chlorides.[171] This reaction follows two paths, one with direct

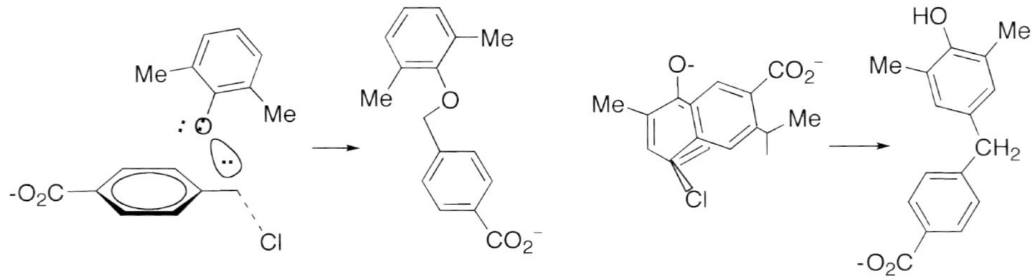

Figure 1.24 Reaction of 2,6-dimethylphenoxide ion with a benzylic chloride involves again no overlap of the phenyls in the reaction at oxygen, but there is significant hydrophobic overlap in the reaction at the para position of the phenoxide ring.

alkylation of the phenoxide ion and the other with alkylation of the phenoxide species on the para position of the phenyl ring (Fig. 1.24). We saw that the path involving oxygen alkylation showed no evidence of hydrophobic packing by the two aryl rings, while the phenyl alkylation showed such evidence. Since the nucleophile and leaving group, which carry charge, are the same in both paths, the difference in the effect of ethanol reflects the difference in hydrophobic packing, supporting our conclusion that direct phenoxide O-alkylation involves a path without aryl overlap. This work clarified some earlier results by others with very different interpretations.

In a major examination of the alkylation of phenoxides, we showed that the C-alkylation path was not promoted by steric hindrance of the oxygen by the two methyl groups, but by increased hydrophobicity added by the methyl groups on the phenoxide ring.[172] The result of all this was support of our original conclusion: Aniline acts as a nucleophile to benzyl chloride with a transition state involving phenyl overlap, but direct attack by phenoxide oxygen does not. Thus even with the complexities of cosolvent effects on charges, as well as on hydrophobic surfaces, it was possible to sort out the effects and determine the geometries of the transition states for the reactions.

This technique promises to be a unique tool for determining an essential detail of chemical reactions – the geometries of their transition states. It remains to be seen how this approach, critically dependent on the special hydrophobic effect in water solution, can be applied generally to such questions. It should be added that the conclusions we have reached so far are consistent with detailed quantum mechanical calculations that are also reported in the papers. Thus these experimental approaches can be used to test and validate theoretical conclusions. We can look forward to a future in which the detailed paths of chemical reactions are calculated with theoretical methods and validated by techniques such as the one developed in our work.

1.4 The importance of water as a reaction solvent

As this chapter indicates water is not simply a cheap and environmentally benign solvent, important though that is. Instead, the hydrophobic effect seen in water permits chemistry otherwise not accessible. It permits the construction of enzyme mimics that use hydrophobic forces to bind the substrates to the catalysts. It permits those reactions, and others, to achieve high positional selectivities in many cases, as well as stereoselectivities in particular systems. It even permits us to learn the detailed geometry of transition states for many reactions – information that answers the most fundamental questions of chemistry while giving us the insight to permit the design of novel reactions and novel catalysts. The remaining chapters of this book will illustrate these and other features of water as a solvent, so important to biological chemistry and yet so neglected for many years by synthetic chemists.

References

1. Breslow, R., Biomimetic chemistry. Centenary lecture, *Chem. Soc. Rev.*, 1972, **1**, 553–580.
2. Breslow, R., Biomimetic chemistry in oriented systems, *Isr. J. Chem.*, 1979, **18**, 187–191.
3. Breslow, R., Biomimetic control of chemical selectivity, *Acc. Chem. Res.*, 1980, **13**, 170–177.
4. Breslow, R., Adjusting the lock and adjusting the key in cyclodextrin chemistry, In: Dolphin, D., McKenna, C., Murakami, Y. and Tabushi, I. (Eds.), *Biomimetic Chemistry*, American Chemical Society, Washington, DC, 1980, pp. 1–15.
5. Breslow, R., Artificial enzymes, *Science*, 1982, **218**, 532–537.
6. Breslow, R., Artificial enzymes, *Chem. Br.*, 1983, **19**, 126–131.
7. Breslow, R., Approaches to artificial enzymes, In: Yoshida, Z.-I. and Ise, N. (Eds.), *Biomimetic Chemistry*, Kodansha Ltd. Publishing Co., Tokyo, Japan, 1983, pp. 1–20.
8. Breslow, R., Chin, J. and Hilvert, D., Studies on zinc enzymes and models, *Inorg. Chim. Acta*, 1983, **79**, 61.
9. Breslow, R., In: Atwood, J.L. and Davies, J.E. (Eds.), *Enzyme Models Related to Inclusion Compounds in Inclusion Compounds*, Vol. 3, Academic Press, Orlando, FL, 1984, pp. 473–508.
10. Breslow, R., Attempts to mimic enzymes – report from the battle front, In: Setton, R. (Ed.), *Chemical Reactions in Organic and Inorganic Constrained Systems*, D. Reidel Publishing Co., Dordrecht, 1986, pp. 17–28.
11. Breslow, R., Artificial enzymes and enzyme models, In: Meister, A. (Ed.), *Advances in Enzymology and Related Areas of Molecular Biology*, Vol. 58, John Wiley & Sons, New York, 1986, pp. 1–60.
12. Breslow, R., Approaches to artificial enzymes, *Ann. N. Y. Acad. Sci.*, 1986, **471**, 60–69.
13. Breslow, R., Biomimetic control of chemical selectivity, In: van Binst, G. (Ed.), *Proceedings of the XVIII Solvay Conference on Chemistry*, Springer-Verlag, Berlin, 1986, pp. 185–197.
14. Breslow, R., Artificial enzymes, *Cold Spring Harb. Symp. Quant. Biol.*, 1987, **52**, 75–81.
15. Breslow, R., Selective reactions catalyzed by enzyme mimics, *Chem. Scr.*, 1987, **27**, 555–556.
16. Breslow, R., Artificial enzymes: studies relevant to reaction mechanisms of enzymes and enzymes models, In: *Proc. Robert A. Welch Foundation Conference on Chemical Research*, 1988, Vol. 31, pp. 73–89.
17. Breslow, R., Enzyme mimics, *Pure Appl. Chem.*, 1990, **62**, 1859–1866.
18. Breslow, R., Enzyme mimics, *Ciba Found. Symp.*, 1991, **158**, 115–127.
19. Breslow, R., Hydrophobic effects on simple organic reactions in water, *Acc. Chem. Res.*, 1991, **24**, 159–164.
20. Breslow, R., Binding and catalysis in water, *Supramol. Chem.*, 1992, 411–428.
21. Breslow, R., Selective binding and catalysis by cyclodextrins, In: *Minutes of the Sixth International Symposium on Cyclodextrins*, Editions de Sante, Paris, 1992, pp. 625–630.
22. Breslow, R., Molecular recognition, *Proc. Natl. Acad. Sci. U. S. A.*, 1993, **90**, 1139.
23. Breslow, R., Biomimetic chemistry, *Pure Appl. Chem.*, 1994, **66**, 1573–1582.
24. Breslow, R., Biomimetic chemistry and artificial enzymes: catalysis by design, *Acc. Chem. Res.*, 1995, **28**, 146–153.
25. Breslow, R., Imitating antibodies and enzymes, In: *Proc. Robert A Welch Foundation 40th Conference on Chemical Research*, 1997, Vol. 40, pp. 1–11.
26. Breslow, R. and Dong, S.D., Biomimetic reactions catalyzed by cyclodextrins and their derivatives, *Chem. Rev.*, 1998, **98**, 1997–2011.
27. Breslow, R., Biomimetic reactions directed by templates and removable tethers, In: Diederich, F. and Stang, P. (Eds.), *Templated Organic Synthesis*, Wiley-VCH, Weinheim, Germany, 1999, Chap. 6.
28. Breslow, R., Belvedere, S., Gershell, L. and Leung, D., The chelate effect in binding, catalysis, and chemotherapy, *Pure Appl. Chem.*, 2000, **72**, 333–342.
29. Breslow, R. (Ed.), *Artificial Enzymes*, Wiley-VCH, Weinheim, Germany, 2005.
30. Breslow, R., Water as a solvent for chemical reactions, In: Anastas, P.T. and Williamson, T.C. (Eds.), *Green Chemistry*, Oxford Press, New York, 1998, Chap. 13.
31. Breslow, R., Artificial enzymes, In: Breslow, R. (Ed.), *Artificial Enzymes*, Wiley-VCH, Weinheim, Germany, 2005, pp. 1–35.

32. Liu, L. and Breslow, R., Vitamin B6 enzyme models, In: Breslow, R. (Ed.), *Artificial Enzymes*, Wiley-VCH, Weinheim, Germany, 2005, pp. 37–62.
33. Breslow, R., Rapid deuterium exchange in thiazolium salts, *J. Am. Chem. Soc.*, 1957, **79**, 1762.
34. Breslow, R., Mechanism of thiamine action: participation of a thiazolium zwitterion, *Chem. Ind.*, 1957, 893–894.
35. Breslow, R., On the mechanism of thiamine action. IV. Evidence from studies on model systems, *J. Am. Chem. Soc.*, 1958, **80**, 3719–3726.
36. Breslow, R. and McNelis, E., Studies on model systems for thiamine action. Synthesis of reactive intermediates, and evidence on the function of the pyrimidine ring, *J. Am. Chem. Soc.*, 1959, **81**, 3080–3082.
37. Breslow, R. and McNelis, E., On the mechanism of thiamine action. VI. 2-Acetylthiazolium salts as 'active acetate', *J. Am. Chem. Soc.*, 1960, **82**, 2394.
38. Breslow, R., The mechanism of thiamine action: evidence from studies on model systems, *CIBA Foundation Study Group II*, J.A. Churchill Ltd., London, 1961, p. 65.
39. Lopez-Calahorra, F., Castro, E., Ochoa, A. and Marti, J., Further evidences about the role of bis(thiazolin-2-ylidene)s as the actual catalytic species in the generalized benzoin condensation, *Tetrahedron Lett.*, 1996, **37**, 5019.
40. Breslow, R. and Kim, R., The thiazolium catalyzed benzoin condensation with mild base does not involve a dimer intermediate, *Tetrahedron Lett.*, 1994, **35**, 699–702.
41. Breslow, R. and Schmuck, C., The mechanism of thiazolium catalysis, *Tetrahedron Lett.*, 1996, **37**, 8241–8242.
42. Cramer, F. and Mackensen, G., Inclusion compounds. XXII. Cyclodextrin-imidazole compounds, *Chem. Ber.*, 1970, **103**, 2138–2147.
43. D'Souza, V.T. and Bender, M.L., Miniature organic models of enzymes, *Acc. Chem. Res.*, 1987, **20**, 146–152.
44. Siegel, B. and Breslow, R., Lyophobic binding of substrates by cyclodextrins in nonaqueous solvents, *J. Am. Chem. Soc.*, 1975, **97**, 6869–6870.
45. VanEtten, R.L., Sebastian, J.F., Clowes, G.A. and Bender, M.L., Acceleration of phenyl ester cleavage by cycloamyloses. A model for enzymic specificity, *J. Am. Chem. Soc.*, 1967, **89**, 3242–3253.
46. Breslow, R. and Campbell, P., Selective aromatic substitution within a cyclodextrin mixed complex, *J. Am. Chem. Soc.*, 1969, **91**, 3085.
47. Breslow, R. and Campbell, P., Selective aromatic substitution by hydrophobic binding of a substrate to a simple cyclodextrin catalyst, *Bioorg. Chem.*, 1971, **1**, 140–156.
48. Breslow, R., Kohn, H. and Siegel, B., Methylated cyclodextrin and a cyclodextrin polymer as catalysts in selective anisole chlorination, *Tetrahedron Lett.*, 1976, **17**, 1645–1646.
49. Czarniecki, M.F. and Breslow, R., Very fast acylation of beta-cyclodextrin by bound *p*-nitrophenyl ferrocinnamate, *J. Am. Chem. Soc.*, 1978, **100**, 7771–7772.
50. Thiem, H.-J., Brandl, M. and Breslow, R., Molecular modeling calculations on the acylation of cyclodextrin by ferroceneacrylate esters, *J. Am. Chem. Soc.*, 1988, **110**, 8612–8616.
51. Breslow, R., Czarniecki, M.F., Emert, J. and Hamaguchi, H., Improved acylation rates within cyclodextrin complexes from flexible capping of the cyclodextrin and from adjustment of the substrate geometry, *J. Am. Chem. Soc.*, 1980, **102**, 762–770.
52. Trainor, G. and Breslow, R., High acylation rates and enantioselectivity with cyclodextrin complexes of rigid substrates, *J. Am. Chem. Soc.*, 1981, **103**, 154–158.
53. Breslow, R., Trainor, G. and Ueno, A., Optimization of metallocene substrates for cyclodextrin reactions, *J. Am. Chem. Soc.*, 1983, **105**, 2739–2744.
54. le Noble, W.J., Srivastava, S., Breslow, R. and Trainor, G., Effect of pressure on two cyclodextrin-promoted ester hydrolyses, *J. Am. Chem. Soc.*, 1983, **105**, 2745–2748.
55. Breslow, R. and Chung, S., Additional flexibility solves the leaving group problem in cyclodextrin acylation, *Tetrahedron Lett.*, 1990, **31**, 631–634.
56. Breslow, R. and Nesnas, N., Burst kinetics and turnover in an esterase mimic, *Tetrahedron Lett.*, 1999, **40**, 3335–3338.
57. Nesnas, N., Lou, J. and Breslow, R., The binding of cocaine to cyclodextrins, *Bioorg. Med. Chem. Lett.*, 2000, **10**, 1931–1933.
58. Breslow, R. and Chung, S., A novel synthesis of substituted imidazoles, and a reexamination of a purported chymotrypsin model, *Tetrahedron Lett.*, 1989, **30**, 4345–4346.
59. Breslow, R. and Chipman, D., Mixed metal complexes as enzyme models. I. Intracomplex nucleophilic catalysis by an oxime anion, *J. Am. Chem. Soc.*, 1965, **87**, 4195–4196.
60. Breslow, R. and Overman, L.E., An 'artificial enzyme' combining a metal catalytic group and a hydrophobic binding cavity, *J. Am. Chem. Soc.*, 1970, **92**, 1075–1077.
61. Breslow, R. and Zhang, B., Very fast ester hydrolysis by a cyclodextrin dimer with a catalytic linking group, *J. Am. Chem. Soc.*, 1992, **114**, 5882–5883.
62. Breslow, R. and Schmir, M., Strongly catalyzed ligand–ligand reaction within a mixed metal complex, *J. Am. Chem. Soc.*, 1971, **93**, 4960–4961.

63. Breslow, R. and Mcallister, C., Intramolecular bifunctional catalysis of ester hydrolysis by metal ion and carboxylate in a carboxypeptidase model, *J. Am. Chem. Soc.*, 1971, **93**, 7096–7097.
64. Tang, C.C., Davalian, D., Huang, P. and Breslow, R., Models for metal binding sites in zinc enzymes. Syntheses of Tris[4(5)-imidazoyl]carbinol (4-TIC), Tris(2-imidazolyl)carbinol (2-TIC), and related ligands, and studies on metal complex binding constants and spectra, *J. Am. Chem. Soc.*, 1978, **100**, 3918–3922.
65. Chin, J. and Breslow, R., Some reactions of phenolic maleamic acids relevant to carboxypeptidase catalysis, *Tetrahedron Lett.*, 1982, **23**, 4221–4224.
66. Breslow, R., Hunt, J., Smiley, R. and Tarnowski, T., Synthesis of some polyimidazole ligands related to zinc enzymes, *J. Am. Chem. Soc.*, 1983, **105**, 5337–5342.
67. Gellman, S., Petter, R. and Breslow, R., Catalytic hydrolysis of a phosphate triester by tetracoordinated zinc complexes, *J. Am. Chem. Soc.*, 1986, **108**, 2388–2394.
68. Schepartz, A. and Breslow, R., Hydrolysis of an amide in a carboxypeptidase model using CoIII and bifunctional catalysts, *J. Am. Chem. Soc.*, 1987, **109**, 1814–1826.
69. Breslow, R. and Singh, S., Phosphate ester cleavage catalyzed by bifunctional zinc complexes: comments on the '*p*-nitrophenyl ester syndrome', *Bioorg. Chem.*, 1988, **16**, 408–417.
70. Light, J. and Breslow, R., The hydrolysis of a CoIII chelated amide catalyzed by an internal phosphonic acid group, *Bioorg. Chem.*, 1990, **18**, 63–77.
71. Breslow, R., Berger, D. and Huang, D.-L., Bifunctional zinc-imidazole and zinc-thiophenol catalysts, *J. Am. Chem. Soc.*, 1990, **112**, 3686–3687.
72. Breslow, R., Duggan, P.J. and Light, J.P., Cyclodextrin-B_{12}, a potential enzyme–coenzyme mimic, *J. Am. Chem. Soc.*, 1992, **114**, 3982–3983.
73. Breslow, R. and Huang, D.-L., Effects of metal ions, including Mg^{2+} and lanthanides, on the cleavage of ribonucleotides and RNA model compounds, *Proc. Natl. Acad. Sci. U. S. A.*, 1991, **88**, 4080–4083.
74. Breslow, R. and Zhang, B., Cleavage of phosphate esters by a cyclodextrin dimer catalyst that binds the substrates together with La^{3+} and hydrogen peroxide, *J. Am. Chem. Soc.*, 1994, **116**, 7893–7894.
75. Rezac, M. and Breslow, R., A mutase mimic with cobalamin linked to cyclodextrin, *Tetrahedron Lett.*, 1997, **38**, 5763–5766.
76. Zhang, B. and Breslow, R., Ester hydrolysis by a catalytic cyclodextrin dimer enzyme mimic with a metallobipyridyl linking group, *J. Am. Chem. Soc.*, 1997, **119**, 1676–1681.
77. Dong, S.D. and Breslow, R., Bifunctional cyclodextrin metalloenzyme mimics, *Tetrahedron Lett.*, 1998, **39**, 9343–9346.
78. Yan, J. and Breslow, R., An enzyme mimic that hydrolyzes an unactivated ester with catalytic turnover, *Tetrahedron Lett.*, 2000, **41**, 2059–2062.
79. Breslow, R., Greenspoon, N., Guo, T. and Zarzycki, R., Very strong binding of appropriate substrates by cyclodextrin dimers, *J. Am. Chem. Soc.*, 1989, **111**, 8296–8297.
80. Zhang, B. and Breslow, R., Enthalpic domination of the chelate effect in cyclodextrin dimers, *J. Am. Chem. Soc.*, 1993, **115**, 9353–9354.
81. Breslow, R. and Zhang, B., Cholesterol recognition and binding by cyclodextrin dimers, *J. Am. Chem. Soc.*, 1996, **118**, 8495–8496.
82. Breslow, R. and Chung, S., Strong binding of ditopic substrates by a doubly linked occlusive C_1 clamshell as distinguished from an aversive C_2 loveseat cyclodextrin dimer, *J. Am. Chem. Soc.*, 1990, **112**, 9659–9660.
83. Breslow, R., Halfon, S. and Zhang, B., Molecular recognition by cyclodextrin dimers, *Tetrahedron*, 1995, **51**, 377–388.
84. Leung, D.K., Atkins, J.H. and Breslow, R., Synthesis and binding properties of cyclodextrin trimers, *Tetrahedron Lett.*, 2001, **42**, 6255–6258.
85. Breslow, R., Yang, Z., Ching, R., Trojandt, G. and Odobel, F., Sequence selective binding of peptides by artificial receptors in aqueous solution, *J. Am. Chem. Soc.*, 1998, **120**, 3536–3537.
86. Wilson, D., Perlson, L. and Breslow, R., Helical templating of oligopeptides by cyclodextrin dimers, *Bioorg. Med. Chem.*, 2003, **11**, 2649–2653.
87. Maletic, M., Wennemers, H., McDonald, D.Q., Breslow, R. and Still, W.C., Selective binding of the dipeptides L-Phe-D-Pro and D-Phe-L-Pro to β-cyclodextrin, *Angew. Chem., Int. Ed. Engl.*, 1996, **35**, 1490–1492.
88. Leung, D.K., Yang, Z. and Breslow, R., Selective disruption of protein aggregation by cyclodextrin dimers, *Proc. Natl. Acad. Sci. U. S. A.*, 2000, **97**, 5050–5053.
89. Moore, S., Askew, J.A., Gibson, G.L., Aborgrein, A., El-Agnaf, O., Allsop, D., Leung, D.K. and Breslow, R., Novel cyclodextrin dimers and trimers as inhibitors of amyloid peptide aggregation, *Neurobiol. Aging*, 2002, **23**, S105.
90. Breslow, R., Duggan, P.J., Wiedenfeld, D. and Waddell, S.T., The binding properties of cyclophane dimers, *Tetrahedron Lett.*, 1995, **36**, 2707–2710.
91. Siegel, B., Pinter, A. and Breslow, R., Synthesis of cycloheptaamylose 2-,3-, and 6-phosphoric acids, and a comparative study of their effectiveness as general acid or general base catalysts with bound substrates, *J. Am. Chem. Soc.*, 1977, **99**, 2309–2312.

92. Ueno, A. and Breslow, R., Selective sulfonation of a secondary hydroxyl group of beta-cyclodextrin, *Tetrahedron Lett.*, 1982, **23**, 3451–3454.
93. Breslow, R. and Labelle, M., Sequential general base–acid catalysis in the hydrolysis of RNA by imidazole, *J. Am. Chem. Soc.*, 1986, **108**, 2655–2659.
94. Anslyn, E. and Breslow, R., On the mechanism of catalysis by ribonuclease: cleavage and isomerization of the dinucleotide UpU by imidazole buffers, *J. Am. Chem. Soc.*, 1989, **111**, 4473–4482.
95. Breslow, R., How do imidazole groups catalyze the cleavage of RNA in enzyme models and in enzymes? Evidence from negative catalysis, *Acc. Chem. Res.*, 1991, **24**, 317–324.
96. Breslow, R. and Xu, R., Quantitative evidence for the mechanism of RNA cleavage by enzyme mimics. Cleavage and isomerization of UpU by morpholine buffers, *J. Am. Chem. Soc.*, 1993, **115**, 10705–10713.
97. Breslow, R., Bifunctional acid–base catalysis by imidazole groups in enzyme mimics, *J. Mol. Cat.*, 1994, **91**, 161–174.
98. Breslow, R., Dong, S.D., Webb, Y. and Xu, R., Further studies on the buffer-catalyzed cleavage and isomerization of uridyluridine. Medium and ionic strength effects on catalysis by morpholine, imidazole, and acetate buffers help clarify the mechanisms involved and their relationship to the mechanism used by the enzyme ribonuclease and by a ribonuclease mimic, *J. Am. Chem. Soc.*, 1996, **118**, 6588–6600.
99. Breslow, R., Huang, D.-L. and Anslyn, E., On the mechanism of action of ribonucleases: dinucleotide cleavage catalyzed by imidazole and Zn^{2+}, *Proc. Natl. Acad. Sci. U. S. A.*, 1989, **86**, 1746–1750.
100. Breslow, R., Doherty, J., Guillot, G. and Lipsey, C., Beta-cyclodextrinyl-bisimidazole, a model for ribonuclease, *J. Am. Chem. Soc.*, 1978, **100**, 3227–3229.
101. Breslow, R., Bovy, P. and Lipsey Hersh, C., Reversing the selectivity of cyclodextrin bisimidazole ribonuclease mimics by changing the catalyst geometry, *J. Am. Chem. Soc.*, 1980, **102**, 2115–2117.
102. Anslyn, E. and Breslow, R., Geometric evidence on the ribonuclease model mechanism, *J. Am. Chem. Soc.*, 1989, **111**, 5972–5973.
103. Matta, M.S. and Vo, D.T., Proton inventory of the second step of ribonuclease catalysis, *J. Am. Chem. Soc.*, 1986, **108**, 5316.
104. Anslyn, E. and Breslow, R., Proton inventory of a bifunctional ribonuclease model, *J. Am. Chem. Soc.*, 1989, **111**, 8931–8932.
105. Breslow, R., Anslyn, E. and Huang, D.-L., Ribonuclease mimics, *Tetrahedron*, 1991, **47**, 2365–2376.
106. Breslow, R. and Schmuck, C., Goodness of fit in complexes between substrates and ribonuclease mimics: effects on binding, catalytic rate constants, and regiochemistry, *J. Am. Chem. Soc.*, 1996, **118**, 6601–6605.
107. Breslow, R., Kinetics and mechanism in RNA cleavage, *Proc. Natl. Acad. Sci. U. S. A.*, 1993, **90**, 1208–1211.
108. Breslow, R. and Graff, A., Geometry of enolization using a bifunctional cyclodextrin-based catalyst, *J. Am. Chem. Soc.*, 1993, **115**, 10988–10989.
109. Koh, J.T., Delaude, L. and Breslow, R., Geometric control of a pyridoxal-catalyzed aldol condensation, *J. Am. Chem. Soc.*, 1994, **116**, 11234–11240.
110. Desper, J.M. and Breslow, R., Catalysis of an intramolecular aldol condensation by imidazole-bearing cyclodextrins, *J. Am. Chem. Soc.*, 1994, **116**, 12081–12082.
111. Breslow, R., Desper, J. and Huang, Y., A selective intramolecular aldol condensation directed by a bifunctional enzyme mimic, *Tetrahedron Lett.*, 1996, **37**, 2541–2544.
112. Yuan, D.-Q., Dong, S.D. and Breslow, R., Cyclodextrin-based class I aldolase enzyme mimics to catalyze crossed aldol condensations, *Tetrahedron Lett.*, 1998, **39**, 7673–7676.
113. Breslow, R., Hammond, M. and Lauer, M., Selective transamination and optical induction by a beta-cyclodextrin-pyridoxamine artificial enzyme, *J. Am. Chem. Soc.*, 1980, **102**, 421–422.
114. Breslow, R. and Czarnik, A.W., Transaminations by pyridoxamine selectively attached at C-3 in beta-cyclodextrin, *J. Am. Chem. Soc.*, 1983, **105**, 1390–1391.
115. Winkler, J., Coutouli-Argyropoulou, E., Leppkes, R. and Breslow, R., Artificial transaminase carrying a synthetic macrocyclic binding group, *J. Am. Chem. Soc.*, 1983, **105**, 7198–7199.
116. Czarnik, A.W. and Breslow, R., Hexadeoxycycloheptaamylose-pyridoxamine, an artificial transaminase with a deeper binding pocket, *Carbohydr. Res.*, 1984, **128**, 133–139.
117. Breslow, R., Czarnik, A.W., Lauer, M., Leppkes, R., Winkler, J. and Zimmerman, S., Mimics of transaminase enzymes, *J. Am. Chem. Soc.*, 1986, **108**, 1969–1979.
118. Breslow, R., Canary, J.W., Varney, M., Waddell, S.T. and Yang, D., Artificial transaminases linking pyridoxamine to binding cavities: controlling the geometry, *J. Am. Chem. Soc.*, 1990, **112**, 5212–5219.
119. Zimmerman, S.C., Czarnik, A.W. and Breslow, R., Intramolecular general base–acid catalysis in transaminations catalyzed by pyridoxamine enzyme analogues, *J. Am. Chem. Soc.*, 1983, **105**, 1694–1695.
120. Zimmerman, S.C. and Breslow, R., Asymmetric synthesis of amino acids by pyridoxamine enzyme analogues utilizing general base–acid catalysis, *J. Am. Chem. Soc.*, 1984, **106**, 1490–1491.
121. Chmielewski, J. and Breslow, R., Increasing the racemase activity versus transaminase activity of a pyridoxal enzyme model by the attachment of a rigid base, *Heterocycles*, 1987, **25**, 533.
122. Liu, L., Rozenman, M. and Breslow, R., Stereoselectivity in reactions of amino acids catalyzed by pyridoxal derivatives carrying rigidly-attached chirally-mounted basic groups – transamination, racemization,

decarboxylation, retro-aldol reaction, and aldol condensations, *Bioorg. Med. Chem.*, 2002, **10**, 3973–3979.
123. Breslow, R., Chmielewski, J., Foley, D., Johnson, B., Kumabe, N., Varney, M. and Mehra, R., Optically active amino acid synthesis by artificial transaminase enzymes, *Tetrahedron*, 1988, **44**, 5515–5524.
124. Klotz, I.M. and Suh, J., Evolution of synthetic polymers with enzyme-like catalytic activities, In: Breslow, R. (Ed.), *Artificial Enzymes*, Wiley-VCH, Weinheim, Germany, 2005, pp. 63–88.
125. Klotz, I.M., Royer, G.P. and Sloniewsky, A.R., Macromolecule – small molecule interactions. Strong binding and cooperativity in a model synthetic polymer, *Biochemistry*, 1969, **8**, 4752.
126. Liu, L. and Breslow, R., A potent polymer/pyridoxamine enzyme mimic, *J. Am. Chem. Soc.*, 2002, **124**, 4978–4979.
127. Zhou, W., Liu, L. and Breslow, R., Transamination by polymeric enzyme mimics, *Helv. Chim. Acta*, 2003, **86**, 3560–3567.
128. Liu, L., Zhou, W., Chruma, J.J. and Breslow, R., Transamination reactions with multiple turnovers catalyzed by hydrophobic pyridoxamine cofactors in the presence of polyethylenimine polymers, *J. Am. Chem. Soc.*, 2004, **126**, 8136–8137.
129. Liu, L. and Breslow, R., Polymeric and dendrimeric pyridoxal enzyme mimics, *Bioorg. Med. Chem.*, 2004, **12**, 3277–3287.
130. Liu, L. and Breslow, R., Dendrimeric pyridoxamine enzyme mimics, *J. Am. Chem. Soc.*, 2003, **125**, 12110–12111.
131. Liu, L., Rozenman, M. and Breslow, R., Hydrophobic effects on rates and substrate selectivities in polymeric transaminase mimics, *J. Am. Chem. Soc.*, 2002, **124**, 12660–12661.
132. Chruma, J.J., Liu, L., Zhou, W. and Breslow, R., Hydrophobic and electronic factors in the design of dialkylglycine decarboxylase mimics, *Bioorg. Med. Chem.*, 2005, **13**, 5873–5883.
133. Zhou, W., Yerkes, N., Chruma, J., Liu, L. and Breslow, R., Chiral polyamines from reduction of polypeptides: asymmetric pyridoxamine-mediated transaminations, *Bioorg. Med. Chem. Lett.*, 2005, **15**, 1351–1355.
134. Weiner, W., Winkler, J., Zimmerman, S.C., Czarnik, A.W. and Breslow, R., Mimics of tryptophan synthetase and of biochemical dehydroalanine formation, *J. Am. Chem. Soc.*, 1985, **107**, 4093–4094.
135. Koh, J.T., Delaude, L. and Breslow, R., Geometric control of a pyridoxal-catalyzed aldol condensation, *J. Am. Chem. Soc.*, 1994, **116**, 11234–11240.
136. Fasella, E., Dong, S.D. and Breslow, R., Reversal of optical induction in transamination by regioisomeric bifunctionalized cyclodextrins, *Bioorg. Med. Chem.*, 1999, **7**, 709–714.
137. Liu, L. and Breslow, R., Selective racemization in preference to transamination catalyzed by pyridoxal enzyme analogs, *Tetrahedron Lett.*, 2001, **42**, 2775–2777.
138. Breslow, R., Brown, A.B., McCullough, R.D. and White, P.W., Substrate selectivity in epoxidation by metalloporphyrin and metallosalen catalysts carrying binding groups, *J. Am. Chem. Soc.*, 1989, **111**, 4517–4518.
139. Breslow, R., Zhang, X., Xu, R., Maletic, M. and Merger, R., Selective catalytic oxidation of substrates that bind to metalloporphyrin enzyme mimics carrying two or four cyclodextrin groups and related metallosalens, *J. Am. Chem. Soc.*, 1996, **118**, 11678–11679.
140. Breslow, R., Zhang, X. and Huang, Y., Selective catalytic hydroxylation of a steroid by an artificial cytochrome P-450 enzyme, *J. Am. Chem. Soc.*, 1997, **119**, 4535–4536.
141. Breslow, R., Huang, Y., Zhang, X. and Yang, J., An artificial cytochrome P450 that hydroxylates unactivated carbons with regio- and stereoselectivity and useful catalytic turnovers, *Proc. Natl. Acad. Sci. U.S.A.*, 1997, **94**, 11156–11158.
142. Breslow, R., Gabriele, B. and Yang, J., Geometrically directed selective steroid hydroxylation with high turnover by a fluorinated artificial cytochrome P-450, *Tetrahedron Lett.*, 1998, **39**, 2887–2890.
143. Yang, J. and Breslow, R., Selective hydroxylation of a steroid at C-9 by an artificial cytochrome P-450, *Angew. Chem. Int. Ed.*, 2000, **39**, 2692–2694.
144. Breslow, R., Yang, J. and Yan, J., Biomimetic hydroxylation of saturated carbons with artificial cytochrome P-450 enzymes – liberating chemistry from the tyranny of functional groups, *Tetrahedron*, 2002, **58**, 653–659.
145. Breslow, R., Yan, J. and Belvedere, S., Catalytic hydroxylation of steroids by cytochrome P-450 mimics. Hydroxylation at C-9 with novel catalysts and steroid substrates, *Tetrahedron Lett.*, 2002, **43**, 363–365.
146. Yang, J., Gabriele, B., Belvedere, S., Huang, Y. and Breslow, R., Catalytic oxidations of steroid substrates by artificial cytochrome P-450 enzymes, *J. Org. Chem.*, 2002, **67**, 5057–5067.
147. Breslow, R. and Fang, Z., Hydroxylation of steroids with an artificial P-450 catalyst bearing synthetic cyclophanes as binding groups, *Tetrahedron Lett.*, 2002, **43**, 5197–5200.
148. Breslow, R. and Gellman, S.H., Tosylamidation of cyclohexane by a cytochrome P-450 model, *J. Chem. Soc., Chem. Commun.*, 1982, 1400–1401.
149. Breslow, R. and Gellman, S.H., Intramolecular nitrene C–H insertions mediated by transition-metal complexes as nitrogen analogues of cytochrome P-450 reactions, *J. Am. Chem. Soc.*, 1983, **105**, 6728–6729.
150. Breslow, R., Hydrophobic and antihydrophobic effects on organic reactions in aqueous solution, In: Cramer, C.J. and Truhlar, D.G. (Eds.), *Structure and Reactivity in Aqueous Solution*, American Chemical Society, Washington, DC, 1994, pp. 291–302.

151. Rideout, D. and Breslow, R., Hydrophobic acceleration of Diels–Alder reactions, *J. Am. Chem. Soc.*, 1980, **102**, 7816–7817.
152. Breslow, R. and Rizzo, C.J., Chaotropic salt effects in a hydrophobically accelerated Diels–Alder reaction, *J. Am. Chem. Soc.*, 1991, **113**, 4340–4341.
153. Breslow, R., Hydrophobic effects on simple organic reactions in water, *Acc. Chem. Res.*, 1991, **24**, 159–164.
154. Breslow, R., Maitra, U. and Rideout, D., Selective Diels–Alder reactions in aqueous solutions and suspensions, *Tetrahedron Lett.*, 1983, **24**, 1901–1904.
155. Breslow, R. and Maitra, U., On the origin of product selectivity in aqueous Diels–Alder reactions, *Tetrahedron Lett.*, 1984, **25**, 1239–1240.
156. Kool, E.T. and Breslow, R., Dichotomous salt effects in the hydrophobic acceleration of the benzoin condensation, *J. Am. Chem. Soc.*, 1988, **110**, 1596–1597.
157. Breslow, R. and Guo, T., Surface tension measurements show that chaotropic salting-in denaturants are not just water-structure breakers, *Proc. Natl. Acad. Sci. U. S. A.*, 1990, **87**, 167–169.
158. Breslow, R. and Guo, T., Diels–Alder reactions in nonaqueous polar solvents. Kinetic effects of chaotropic and antichaotropic agents and of beta-cyclodextrin, *J. Am. Chem. Soc.*, 1990, **110**, 5613–5617.
159. Biscoe, M. and Breslow, R., Hydrophobically directed selective reduction of ketones, *J. Am. Chem. Soc.*, 2003, **125**, 12718–12719.
160. Biscoe, M., Uyeda, C. and Breslow, R., Requirements for selective hydrophobic accelerations in the reduction of ketones, *Org. Lett.*, 2004, **6**, 4331–4334.
161. Uyeda, C., Biscoe, M., Le Plae, P. and Breslow, R., Hydrophobically-directed selective reduction of ketones using amine boranes, *Tetrahedron Lett.*, 2005, **47**, 127–130.
162. Biscoe, M. and Breslow, R., Oxaziridinium salts as hydrophobic epoxidation reagents: remarkable hydrophobically-directed selectivity in olefin epoxidation, *J. Am. Chem. Soc.*, 2005, **127**, 10812–10813.
163. Breslow, R., Determining the geometries of transition states by use of antihydrophobic additives in water, *Acc. Chem. Res.*, 2004, **3**, 471–478.
164. Breslow, R. and Halfon, S., Quantitative effects of antihydrophobic agents on binding constants and solubilities in water, *Proc. Natl. Acad. Sci. U. S. A.*, 1992, **89**, 6916–6918.
165. Breslow, R. and Connors, R.V., Quantitative antihydrophobic effects as probes for transition state structures. 1. Benzoin condensation and displacement reactions, *J. Am. Chem. Soc.*, 1995, **117**, 6601–6602.
166. Breslow, R. and Zhu, Z., Quantitative antihydrophobic effects as probes for transition state structures. 2. Diels–alder reactions, *J. Am. Chem. Soc.*, 1995, **117**, 9923–9924.
167. Breslow, R. and Connors, R., Antihydrophobic cosolvent effects detect two different geometries for an S_N2 displacement and the change to a single-electron transfer mechanisms in related cases, *J. Am. Chem. Soc.*, 1996, **118**, 6323–6324.
168. Breslow, R., Connors, R. and Zhu, Z., Mechanistic studies using antihydrophobic agents, *Pure Appl. Chem.*, 1996, **68**, 1527–1533.
169. Mayer, M.U. and Breslow, R., Antihydrophobic evidence for the single electron transfer mechanism of nucleophilic substitution, *J. Am. Chem. Soc.*, 1998, **120**, 9098–9099.
170. Breslow, R., Groves, K. and Mayer, M.U., The hydrophobic effect as a mechanistic tool, *Pure Appl. Chem.*, 1998, **70**, 1933–1938.
171. Breslow, R., Groves, K. and Mayer, M.U., Antihydrophobic cosolvent effects in organic displacement reactions, *Org. Lett.*, 1999, **1**, 117–120.
172. Breslow, R., Groves, K. and Mayer, M.U., Antihydrophobic cosolvent effects for alkylation reactions in water solution, particularly oxygen versus carbon alkylations of phenoxide ions, *J. Am. Chem. Soc.*, 2002, **124**, 3622–3635.

Chapter 2

Structure and Properties of Water

Jan B.F.N. Engberts

Life processes on this planet are crucially dependent on the presence of water. A recent meeting[1] organized by the Royal Society of Chemistry in London was focused on the question 'Is life possible without water?' In the final discussion section an attempt was made to formulate an answer. The dominant opinion was 'No, we cannot imagine life in the absence of water'. One participant even said: 'You cannot take away *any* feature of water without destroying life'.

The individual water molecule, apart from being very small (hard sphere diameter 2.75 Å, compare neon: 2.79 Å), has no peculiar properties.[2] But liquid water, containing huge three-dimensional (3-D) networks of hydrogen bonds (H-bonds), has a large number of unusual properties. Any of them can perhaps be found in another liquid, but it is this combination of so many anomalous properties that makes liquid water unique. The importance of these properties for maintaining life processes is now well recognized, although our understanding of the relations between biochemical processes and the medium effects of water is often far from complete.

In physical organic chemistry, and particularly in reaction mechanistic studies, water has always been the solvent of choice.[3] The high dielectric constant of water (78 at 25°C) avoids kinetic complexities due to ion pair formation and many important mechanistic parameters (pK_a values, substituent constants, etc.) are abundantly available for water as the solvent. But also its transparency above 200 nm and its lack of toxicity and relatively low volatility make water a convenient solvent for quantitative measurements. This strong preference for water in physical organic chemistry contrasts sharply with the lack of popularity of water in synthetic organic chemistry. There are two main reasons for this situation: (1) a number of chemical functionalities are not stable in water (for example many organometallic compounds) and, still more importantly, (2) most organic molecules have limited solubilities in water, making it difficult to perform synthetic conversions at useful reactant concentrations.

In the past 25 years there have been two major developments that have led to a rather complete revision of the traditional views regarding aqueous reaction media in organic chemistry.

First of all, the 1980 communication by Breslow and Rideout[4] revealed surprisingly high-rate enhancements of Diels–Alder (DA) cycloadditions in water. The DA reaction was always considered to be only weakly sensitive to solvent effects because of the small changes in charge during the activation process. In fact, the small observed solvent effects were used as support for the concertedness of the DA reactions. Therefore, it was clear from the beginning that the peculiar nature of the aqueous reaction medium had to be invoked to explain the large rate enhancements in water. Many groups have contributed to the solution of the question why water was beneficial for the DA reaction. Sophisticated mechanistic studies, hand in hand with advanced computational studies, finally indicated that a combination of enforced hydrophobic interactions and H-bond stabilization of the activated complex were responsible for the aqueous rate accelerations.[5] This theory was extensively tested by variation of the hydrophobicity and H-bonding ability of diene and dienophile.[6] And it also rationalized other observations as, for example, the increased preference for the formation of the endo products. Much further work showed that the aqueous rate enhancements were not specific for the DA reaction but also occurred for a large variety of other organic reactions, including multicomponent reactions.[7] An enormous amount of work has been performed to probe the possibilities of organic synthesis in aqueous media, or, at least, to examine which synthetic procedures tolerate substantial concentrations of water in the reaction medium.[8–10] The field has been recently summarized by Li in a review with 919 references.[11]

It was natural that extensive attempts were made to further increase reaction rates in aqueous media by applying Lewis acid catalysis[12] and, more specifically, (transition) metal ion and rare earth metal ion catalysis. Also combinations of Lewis acid catalysis with micellar or vesicular catalysis were sometimes found to be highly effective.[13] And, of course, the possibility of stereospecific catalytic processes in water was examined in detail.[14]

The second breakthrough came in 2005 when Sharpless and colleagues[15] published an exciting communication in which they showed that heterogeneous aqueous reaction conditions can be turned into an advantage. The traditional view was that heterogeneity was unfavorable both for reaction rates and for yields of the desired product(s). Sharpless' results revealed that reaction rates, product yields, reaction specificity, and workup benefit from stirring a two-layer mixture of water and two liquid (or one liquid and one solid) reaction partners to give an aqueous suspension.[16] After reaction, the product can usually be separated either from the upper layer or from the bottom of the flask. This so-called 'on water' protocol was successful for a variety of reactions and further testing of the scope of this procedure is going on.[17] The exact mechanism of these reactions is also under investigation.

These recent developments indicate that biochemical and organic reactions have been brought closer together in the sense that for both types of conversions aqueous reaction media provide the best circumstances for highly efficient kinetic processes with excellent stereochemistry of the reaction products.

In the following sections an attempt will be made to summarize a selection of those properties of liquid water that are most adequate and useful for rationalizing the kinetics and stereochemistry of organic reactions in water.

2.1 Water, the molecule and the liquid

2.1.1 The single water molecule

The individual water molecule (H_2O) contains three nuclei, one modestly heavy (oxygen) and two light ones (hydrogen) (Fig. 2.1). The O—H bond lengths are 0.9572 Å and the average HOH bond angle is 104.52°, slightly smaller than the tetrahedral angle (109.5°). The molecule is very small with a hard sphere diameter of 2.75 Å, similar to that of neon (2.79 Å). The smallness of the water molecule has important consequences for the hydration of solutes (*vide infra*).

The contours of the *total* electron density in the HOH plane of the water molecule, as obtained from sophisticated quantum chemical calculations, are shown in Fig. 2.2 and indicate a shape not far from spherical. The water molecule has a dipole moment of approximately

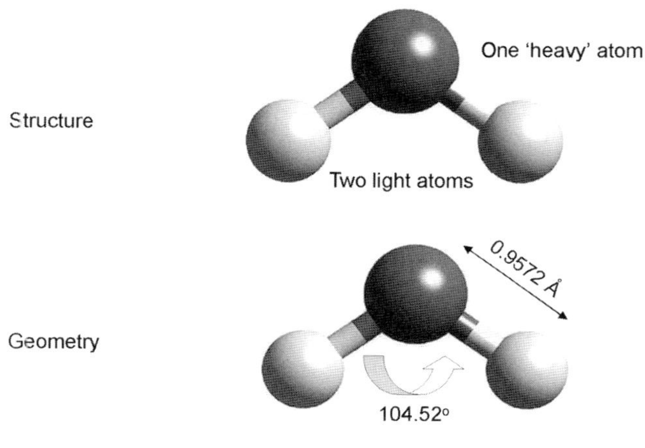

Figure 2.1 The single water molecule and its average geometry. Reproduced from Finney (2004)[2] with kind permission of The Royal Society.

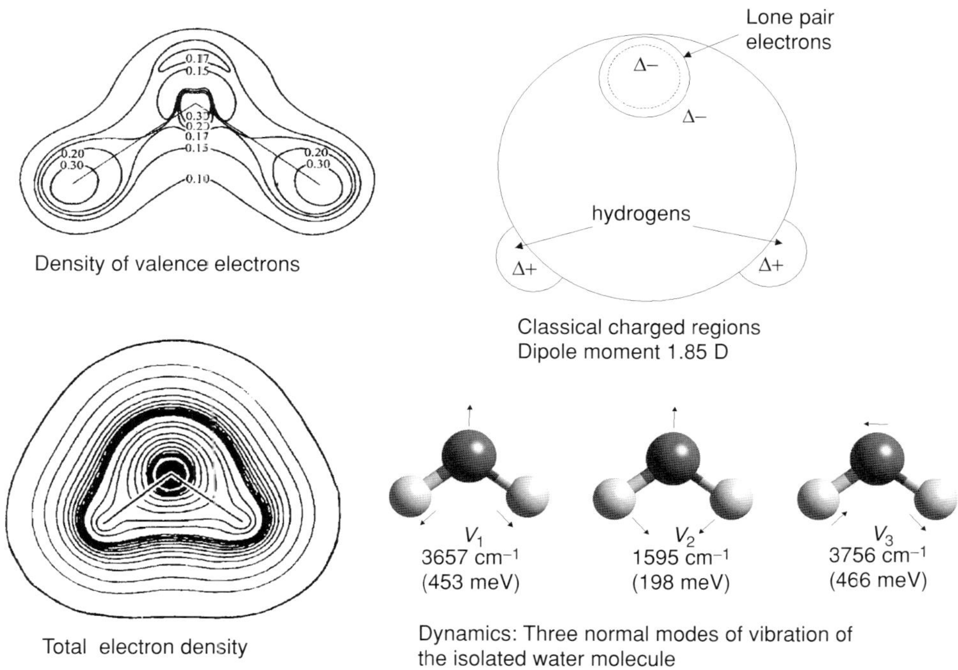

Figure 2.2 Contours of the density of the valence electrons and the total electron density in the HOH plane of the water molecule. The shape of the molecule is not far from spherical. The charges on the hydrogen and oxygen atoms and the three normal modes of vibration are also shown. Reproduced from Finney (2004)[2] with kind permission of The Royal Society.

1.85 D with two partial positive charges on the hydrogen atoms and a single zone of negative charge on the oxygen atom.[2] The popular view of a tetrahedral symmetry of the charges with two negative charges residing in two lobes on oxygen (the lone pair electrons associated with a definite quadrupole moment) is probably an oversimplification as revealed by advanced quantum mechanics.[18] There are no rabbit ears on water![19] The molecule has a significant dipole polarizability leading to an enhanced dipole moment in liquid water.

As discussed by Finney,[2] the near trigonality of the charge distribution in the water molecule has also been taken into account in attempts to construct more realistic potential functions in molecular dynamics (MD) computer simulations of water and aqueous solutions.

In sum, the isolated water molecule is by no means exceptional, only the smallness of the molecule has to be particularly recognized in our thinking about aqueous reaction media.

2.1.2 Liquid water[20]

It is well known that water molecules are strongly associated in the liquid phase by H-bonding. Water can act twice as a H-bond donor (pointing one of the hydrogen atoms to an oxygen atom of a second molecule) and twice as a H-bond acceptor (with its oxygen atom interacting with a hydrogen atom of a second molecule). The double acceptor ability in the lone pair region is determined more by space limitations for more than two waters interacting with the negative charge than by the presence of two lobes carrying negative charges (Section 2.1.1). Hydrogen bonds result dominantly from electrostatic interactions with a smaller contribution of covalent bond formation. There is still some controversy here (see Section 2.1.1). It could be argued[21] that the fact that different ice phases retain their tetrahedral H-bonding even under high external pressures could be better consistent with dominant sp^3 hybridization of the oxygen atom because ice with this geometry has not the smallest volume.

The H-bond interactions in liquid water constitute a 3-D H-bond network with localized and structured clustering. Therefore the liquid structure is not determined by hard-core repulsions between the individual water molecules but rather by intermolecular and directional H-bonding. The water–water H-bonds have average strengths of approximately 20 kJ mol^{-1}, which is about 10 times kT, with kT being the typical thermal fluctuation at 25°C. The rather strong intermolecular H-bonds explain the high heat of vaporization (2447 kJ kg^{-1}). The H-bond network is highly dynamic with water molecule reorientation times[22] of about 2 ps and times to move over one molecular distance[23] of about 7 ps.

In this respect the absolute molar entropies (273 K, 1 atm) also reflect the H-bond association: ice, 41 J K^{-1}; liquid water, 63.2 J K^{-1}; and water in the gas phase, 188 J K^{-1}.

The properties of the 3-D H-bond network of water are at the heart of the many water anomalies, summarized by Finney[2] and by Chaplin.[24] These include the fact that ice floats on water, the temperature of highest density at 4°C, the unusually high specific heat capacity, and the high proton and hydroxide mobilities. For organic reactivity in water, particular attention will be focused on hydrophobic interactions, which also originate from the tendency of water to form extensive H-bond networks.

There has been and still is much attention focused on the nature and detailed structure of the aqueous H-bond network and how it is affected by the presence of polar and apolar solutes. There is a lot of controversy in this field and several scientists favor their own ideas about the preference for certain types of water clusters in the liquid at 25°C. For example, in the recent literature the formation of large clusters (with 14, 17, 21, and even 196 and 280 waters) has been proposed and it was suggested that these intricate structures play a key role in aqueous (bio)chemistry.

Although water has sometimes been considered as rather 'liquid ice', it is more realistic to view the 3-D H-bond structure as quite disordered as shown particularly by advanced neutron diffraction measurements.[25] Of course, despite the lack of long-range order, there is a definite directionality in the interaction between waters and the four-coordinated motif is certainly preferred.[25] But there is also trigonal coordination, reflecting the continuous zone of electron density on the oxygen atom as discussed above. The broad variation of H-bond angles further demonstrates the noncrystalline arrangement of the water molecules and appears to indicate that some bond angle distortion in the H-bonds is rather easily allowed in the liquid. The significant presence of five-, six-, and sevenfold ring structures has been stressed in the literature,[2] although smaller and larger rings also occur to a smaller extent.

Finally the presence of bifurcated, threefold H-bond interactions has to be noted. Water molecules with five neighbors have already been proposed in early studies but in 1989 their presence has been confirmed by Raman scattering measurements. These special H-bonds have often been considered as 'defects' with one hydrogen atom of a water molecule interacting with oxygen atoms of two neighboring waters (Fig. 2.3).

This formal fivefold coordination, with two bifurcated H-bonds being about as strong as one normal H-bond, leads to a situation that the average coordination in water is larger than 4 as shown by spatial distribution functions, obtained both experimentally and computationally. A graph of the various most frequently occurring coordination numbers vs the fraction of O—O coordination 2.0–3.2 Å is shown in Fig. 2.4.

It is important to recognize that intermolecular interaction between water molecules is aided by the substantial dipole polarizability of water. Whereas the dipole moment of a single water molecule is 1.85 D, the dipole moment in the liquid is significantly larger, ranging from about 2 to 4 D, depending on the nature of the water–water interaction.[26] The average

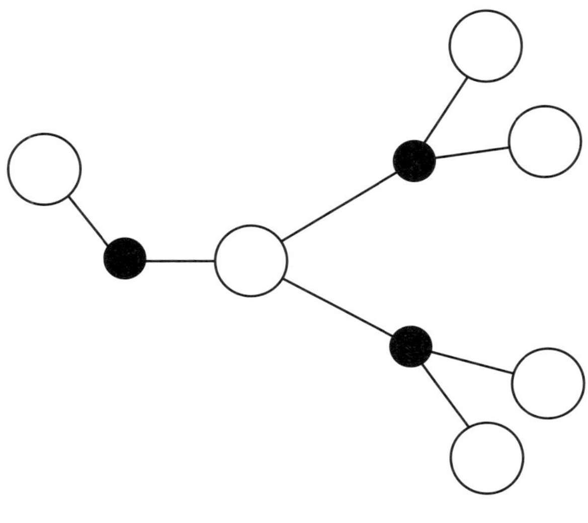

Figure 2.3 The bifurcated hydrogen bond.

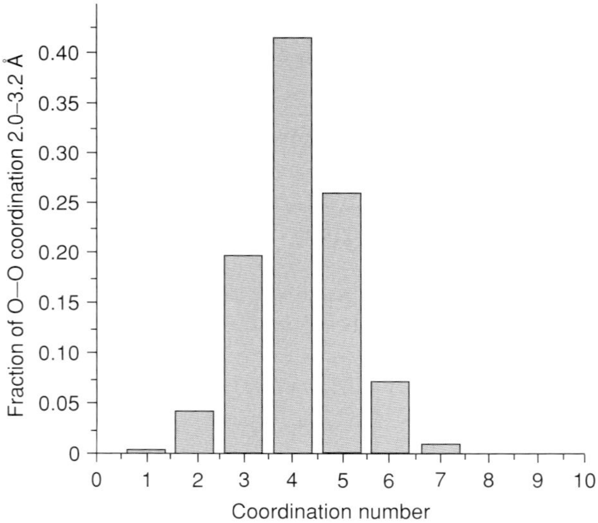

Figure 2.4 Graph of the coordination number vs the fraction of O—O coordination 2.0–3.2 Å. Reproduced from Finney (2004)[2] with kind permission of The Royal Society.

value is probably about 2.8 D. Cooperativity in water–water H-bonding interactions due to mutual polarization effects is an important issue and may enhance the strength of H-bonds of the participating waters by about 20%.[27]

Given the relatively strong association of liquid water, the highly dynamic character of the H-bond structure (*vide supra*) is surprising. Several mechanisms have been proposed to rationalize this phenomenon.[26,28,29]

It will be clear that the water's ability to reside in a large variety of locally different orientational order provides the solvent with ample opportunities to adapt its H-bond structure to the presence of a solute, either polar or apolar. However, the waters will always tend to retain as many water–water H-bonds as possible in their attempt to have the lowest Gibbs energy.

The extremely rapid dynamics of the H-bonds in water should be viewed as an enormous number of interrelated processes representing, at fixed temperature and pressure, an equilibrium without changing the surroundings. The most stable state of the system represents a compromise between counteracting enthalpic and entropic contributions, together leading to the lowest Gibbs energy. There are many metaphors that refer to the dynamics of water,[30] one of the most beautiful being the 'dance of water molecules' as described by Ball.[31]

During the past 25 years Monte Carlo and MD computer simulations have contributed a great deal to our understanding of the properties of liquid water. Of course, high-quality force fields and adequate configurational sampling are essential prerequisites for reliable results and the enormous progress in these fields has recently been reviewed and tested by Jorgensen and Tirado-Rives[32] and Csonka et al.[33] No attempt will be made to review this field within the context of this chapter.

Interestingly, water can also form hydrodynamic dissipative structures (Bénard cells) above a critical temperature when a layer of water is strongly heated from below.[34] Under these conditions fluctuations become reinforced and macroscopic convection streams lead to a macroscopic order stabilized by energy exchange with the environment. The order manifests itself as regular hexagonal cells.

Very recently, 'hydrophobic water' has been identified.[35] A monolayer of water formed on single crystals of Pt(III) (at $T > 60$ K) has no dangling OH bonds or free lone pair electrons (one bond to Pt(III), three to neighboring waters), resulting in a 'hydrophobic' surface on which water diffusion is facile and which is a poor template for 3-D crystalline ice growth.

Finally, supercritical water should be mentioned. Its properties and propensities as a reaction medium are briefly introduced below in Section 2.2.1.

For *organic reactions in water*, the tendency of water molecules to find the most favorable arrangement of the H-bonds around the solute will apply for both the *reactant(s)* as the *activated complex*. Therefore, our understanding of aqueous reactivity will hinge strongly on a quantitative assessment of these hydration processes.

2.2 Properties of water

2.2.1 Solvent properties and parameters

It has been common practice for a long time to analyze solvent effects on organic reactions by correlating kinetic data with solvent parameters. In these attempts solvent effects on reactants and activated complex are usually not distinguished. Although this approach may lead to helpful insights into the factors that primarily determine kinetic effects, a really quantitative analysis of solvation effects on reactants and activated complex is usually not feasible. In Table 2.1 some of the most popular solvent parameters are summarized for water and, for the sake of comparison, for four organic solvents ranging from highly apolar to polar. A brief discussion of these and some other parameters and properties of water follows.

Dielectric constant

The macroscopic dielectric constant of a solvent (ε_r) is the most popular parameter to characterize the polarity of the medium and controls the ionic dissociation of salts. In better terminology, ε_r is the relative permittivity and amounts to $\varepsilon/\varepsilon_0$, with ε_0 being the permittivity of the vacuum. Values of ε_r range from about 2 (apolar, hydrocarbons) to high values such as 180 for secondary amides.[36] Electrostatic interactions (ion–ion, ion–dipole,

Table 2.1 Selected solvent properties of water and four organic solvents

Solvent	ε_r	E_T^N	α_p	μ	ced	π	α	ß	π^*
n-Hexane	1.88	0.009	11.9	0.00	225	239	0.00	0.00	−0.08
Diethyl ether	4.20	0.117	8.9	1.14	251	264	0.00	0.47	0.27
Acetonitrile	35.94	0.460	4.4	3.54	590	379	0.19	0.31	0.75
Ethanol	24.55	0.654	5.1	1.74	703	291	0.83	(0.77)[a]	0.54
Water	78.30	1.000	1.4,[b] 2–4[c]	1.8,[b] 2.8[c]	2302	151	1.17	0.47[d]	1.09

[a] Less certain value.
[b] For the single water molecule.
[c] In the bulk liquid.
[d] For the bulk liquid.

dipole–dipole, dipole-induced dipole) vary with $1/\varepsilon_r$. Thus half of the effect is found for changing ε_r from 1 to 2, the other half from 2 to ∞. The high dielectric constant of water (78.3), related to the dipole orientations in the H-bond network, makes the liquid a highly polar solvent. Consequently, electrostatic interactions are greatly depressed compared to apolar solvents with $\varepsilon_r < 10$. Unfortunately the dielectric constant is not a reliable parameter for electrostatic effects on organic reactions in aqueous media in terms of solute–water interactions. This failure is primarily caused by the fact that the local permittivity in the solvation shell close to a solute differs from that of the bulk solvent because of the solute–solvent interactions.

It is noted that the polarity of water falls off at high temperatures and pressures as anticipated since the H-bond network breaks down. For example, synthesis of benzimidazoles at 358°C and 20.0 MPa was found to be successful,[36,37] at least in part due to the improved solubility of the organic reactants.

Solvatochromic solvent parameters

Solvent polarity is one of the most common solvent characteristics that have been used for correlations of rate constants with the nature of the reaction medium. Since relative permittivities and dipole moments did not give satisfactory results, there was a strong need for a *microscopic* parameter and many attempts have been made to develop empirical solvent parameters that are based on a physical parameter which is sensitive to the solvent polarity. Solvatochromic dyes have been the most successful and particularly Reichardt's $E_T(30)$ solvent polarity parameter.[36] The latter is based on the transition energy of the longest wavelength solvatochromic absorption band of the betaine dye pyridinium-*N*-phenoxide (in fact, this is dye number 30 in the first publication[38]). Its value is given by

$$E_T(30) = hc\nu\, N_A = 2.859 \times 10^{-3} \times \nu \text{ kcal mol}^{-1} = 1.196 \times 10^{-2} \times \nu \text{ kJ mol}^{-1}$$

in which ν is the wave number (cm^{-1}) of the photon necessary for the electronic excitation, h is Planck's constant, c is the velocity of light, and N_A is Avogadro's number. Solutions of the probe in methanol are red, in ethanol violet, and in acetone green.

The solubility of the dye in water is limited, but the $E_T(30)$ value (63.1) can be determined and again demonstrates the high polarity of water. In Table 2.1 the normalized E_T^N values[39] are listed:

$$E_T^N = [E_T(\text{solvent}) - E_T(\text{TMS})]/[E_T(\text{water}) - E_T(\text{TMS})] = [E_T(\text{solvent}) - 30]/32.4$$

Herein water ($E_T^N = 1$) and tetramethylsilane (TMS; $E_T^N = 0$) are employed as reference solvents.

Polarizability

An electric field can induce a dipole (μ_{ind}) in an apolar molecule which is given by

$$\mu_{ind} = 4\pi\varepsilon_0\alpha_p E$$

in which ε_0 is the permittivity of the vacuum, E the electric field strength, and α_p the polarizability of the molecule. The highly distance-dependent London dispersion forces between two molecules depend on the molecule's polarizabilities α_1 and α_2 (defined in volume units, cm^{-3} or m^{-3}). For many molecules, even highly polar ones like HCl, the pairwise interaction is primarily of the London dispersion type.[40] However, water is an exception to this rule: London dispersion contributes only about 24% to the total pairwise interaction energy.

Table 2.1 records the small α_p value for water as anticipated for a small molecule lacking π electrons. This does not necessarily mean that London dispersion interactions with liquid water are negligible since water is such a small molecule and hydration of a solute usually involves a number of water molecules.

Dipole moment

The permanent dipole moment for the individual water molecule and for water molecules in bulk water have already been discussed (Section 2.1.1). For a neutral molecule, containing an asymmetric charge distribution, the dipole moment is given by

$$\mu = ql$$

in which q is the magnitude of the two equal and opposite charges and l is the distance of charge separation. Traditionally the unit of the dipole moment is Debije (D; 1 D = 10^{-18} esu cm). Comparison of the μ values for water and ethanol (Table 2.1) shows how little the ethyl group contributes to the charge separation.

Cohesive energy density

The cohesive energy density (ced, in MPa = 1 J cm^{-3}) is a measure of the total molecular cohesion per unit volume:

$$\text{ced} = \Delta U_v / V_m = [\Delta H_v - RT] / V_m$$

where ΔU_v and ΔH_v are the energy and enthalpy of vaporization, respectively, of the solvent, R is the gas constant, T the absolute temperature, and V_m the molar volume of the solvent. The extremely high ced for water compared to the organic solvents is consistent with the strong 3-D intermolecular H-bond interactions in the liquid and the small value of V_m.

In the literature, sometimes the high ced of water has been taken as the ultimate cause of the hydrophobicity of apolar solutes. However, it has been argued that this is not correct and that probably the difference between the ced and the internal pressure is a better rational for the low solubility of apolar solutes in water.

Internal pressure

The internal pressure of a solvent (π, in MPa) represents the change in internal energy upon a very small isothermal volume expansion[40,41]:

$$\pi = [\delta U / \delta V_m]_T$$

in which U is the molar internal energy. Approximate values of π are obtained from

$$\pi = \alpha T / ß$$

where α is the thermal expansion coefficient and ß is the isothermal compressibility of the liquid.[42] In contrast to the ced, the π value is not representative for a complete disruption of all intermolecular interactions in the solvent. In the literature there is sometimes confusion about this issue and fallacious conclusions have been drawn in attempts to correlate kinetic aqueous solvent effects with internal pressure.[43] Recently, Graziano has clarified these matters.[44]

The internal pressure primarily responds to rupture of London dispersion and dipole–dipole interactions, whereas the ced also includes breaking of the less distance- and orientation-dependent H-bond interactions. For water the ced is extremely high but the

π is low (151 MPa). Dack[45] has proposed that for H-bonding liquids [ced $-\pi$] may be taken as an indication of the strength of the H-bonds. The π value of water may be compared to those of the organic solvents shown in Table 2.1. For water ced/π amounts to 15.2, and for n-hexane this ratio is 0.94, confirming the overwhelming importance of H-bonding in water.

Parameters in the linear solvation energy relationships (α, ß, π)*

Abraham,[46] in collaboration with Taft, Kamlet, and Abboud, has made an interesting attempt to characterize the nature of a solvent in terms of its H-bond donor ability (HBD; α), H-bond acceptor ability (HBA; ß), and its specific dipolarity/polarizability (π^*). He has used these parameters in multiparameter relationships to analyze medium effects on a variety of chemical processes. Briefly, the HBD propensities (α) were obtained from the enhanced solvatochromism of the $E_T(30)$ probe relative to 4-nitroanisole, the HBA propensities (ß) from the enhanced solvatochromism of 4-nitroaniline relative to N,N-diethyl-4-nitroaniline in HBA solvents, and the π^* values from solvent effects on the $\pi \rightarrow \pi^*$ transition of nitro-substituted aromatic compounds. For a more detailed discussion of these parameters and their applications, the reader is referred to the original literature.[46]

The α and ß values for water (Table 2.1) are not surprising. The π^* value (normalized by using $\pi^* = 0$ for cyclohexane and 1.00 for dimethyl sulfoxide) illustrates the ability of water to stabilize charges or dipoles.

Several scientists have tried to further improve the method by employing other systems for the definition of the three parameters. Of course, the π^* values contain a blend of different interactions and more complex approaches can be used for their definition. For water the definition of α, ß, and π^* is difficult because of the special and complex nature of the hydration process and the importance of dipole polarizability.

Thermodynamic activity of water

The thermodynamic activity of water (a_1) in an aqueous solution[47] is not often considered in relation with aqueous rate effects on organic reactions. It is given by the difference between the chemical potential of water in an aqueous mixture (μ_1(aq)) and the chemical potential of pure water (μ_1^*(l)):

$$\mu(\mathrm{aq}) = \mu_1^*(l) + RT\ \ln a_1 = \mu_1^* - \varphi RTM_1 m_j$$

where φ is the practical osmotic coefficient, M_1 is the molar mass of water, and m_j is the molality of solute j. If the thermodynamic properties of the aqueous solution are ideal, φ is unity.

In case of an aqueous liquid mixture we also have

$$\ln a_1 = \ln\ x_1 f_1$$

in which x_1 is the mole fraction of water and f_1 is the rational activity coefficient. Values for a_1 can, for example, be obtained from vapor pressure measurements:

$$a_1 = p_1(\mathrm{aq})/p_1^*(\mathrm{l})$$

where p_1^* is the water vapor pressure of pure water and p_1(aq) that of the mixture. However, many more techniques are available.

Table 2.2 Selected properties of bulk and cellular water[a]

Property	Bulk	Vicinal
Density (g cm^{-3})	1.00	0.97
Specific heat (cal kg^{-1})	1.00	1.25 ± 0.05
Thermal expansion coefficient (°C^{-1})	250 × 10^{-6} (25°C)	(300–700) × 10^{-6}
(Adiabatic) compressibility coefficient (atm^{-1})	45 × 10^{-6}	(60–100) × 10^{-6}
Excess sound absorption ($cm^{-1}s^2$)	7 × 10^{-17}	~35 × 10^{-17}
Heat conductivity (cal s^{-1}) cm^{-2} °C^{-1} cm^{-1})	0.0014	~0.01–0.05
Viscosity (cP)	0.089	2–10
Energy of activation ionic conduction (kcal mol^{-1})	~4	5–8
Dielectric relaxation frequency (Hz)	19 × 10^9	2 × 10^9

[a]Reproduced from Pollack (2001)[48] with kind permission of Ebner and Sons.

Water activities are relevant in dealing with the fact that the aqueous medium within the biological cell (cytosol) is a highly concentrated solution of various biomolecules, making it far from ideal in a thermodynamic sense (Table 2.2). It has been argued[48] that the number of water molecules in between biomolecules in the cytosol is very limited (approximately 5–6 waters between protein surfaces).

The consequences for enzymic reactions have not yet been examined in much detail, but interest in this problem is increasing.[49,50] There is little doubt that both the translational and rotational freedom of water molecules is seriously restricted as evidenced by quasi-elastic neutron scattering.

Evidence has been presented that the charged cytoplasmic macromolecules provide electrochemical gradients that determine the flow of biochemical ions through the cytoplasm. This makes the cellular metabolism vectorial, not only across the membrane but also throughout the contents of the cell.[51] Needless to say that hydration effects play a major role in these processes.

Recently it has even been argued that the cytosol is, in fact, a hydrogel, providing better opportunities for repression of leakage pathways.[52]

Normal water and heavy water

Primary, secondary, and solvent kinetic deuterium isotope effects (KDIEs; k_H/k_D) provide important mechanistic information about organic reactions in aqueous reaction media, particularly acid–base-catalyzed processes. Measurements are usually restricted to H_2O and D_2O because of the inconvenience of handling reactions in radioactive T_2O (tritium has a half-life of 12.5 years).

Apart from these KDIEs, which depend on the relatively large change in mass going from H to D thereby affecting the occupancies of kinetically relevant vibrational energy levels, the two liquids have a number of different physical properties (Table 2.3). Sometimes these differences are modest, but the differences in temperature of maximal density (H_2O, 4°C; D_2O, 11.2°C) and in viscosity (H_2O, 0.7975 cP; D_2O, 0.969 cP, both at 30°C) are certainly not negligible.

The data in Table 2.3 suggest that intermolecular association is somewhat stronger in D_2O than in H_2O, in accordance with the higher value of [ced – π] for heavy water than for normal water. The fact that some noble gases and hydrocarbons are slightly better soluble in

Table 2.3 Selected solvent properties of water and heavy water[a]

Property	H_2O	D_2O
Molecular mass (g mol^{-1})	18.015	20.028
Melting point (°C)	0.00	3.81
Boiling point (°C)	100.00	101.42
Temperature of maximum density (°C)	3.98	11.23
Density (g cm^{-3})	0.9970	1.1045
Molar volume (cm^3 mol^{-1})	18.069	18.133
Viscosity (Pa·s)	8.9×10^{-4}	11.0×10^{-4}
Heat of vaporization (kJ mol^{-1})	44.04	45.46
Dielectric constant	78.46	77.94
Dipole moment (D)	1.83	1.84
Ionization constant (mol L^{-1})	1.81×10^{-16}	0.354×10^{-16}

[a]Data taken from Ref. 36.

D_2O than in H_2O is most likely caused[53] by the slightly larger void volume in D_2O, making cavity formation less unfavorable than in H_2O.

The self-dissociation of water and the proton/hydroxide mobilities in water

Water is amphoteric; at 298.15 K and 1 atm, the ionization constant (pK_w) of water is 14.004. The proton and hydroxide concentrations are so small that the water activity is almost unity. The standard enthalpy of self-dissociation is 55.81 kJ mol^{-1}, the heat capacity –215 J K^{-1}, and the standard volume of self-dissociation is approximately –20 cm^3 mol^{-1}. Ultrafast mid-infrared spectroscopic measurements[54] have suggested that the first step of the autodissociation of water proceeds through an excited vibrational state of the OH bond, probably with $\nu = 2$.

Ionic mobilities (unit: 10^{-4} cm^2 s^{-1} V^{-1}) have been measured in water. They are high for the proton (36.23) and for the hydroxide ion (20.64) relative to those for other cations and anions (for example, Na^+, 5.19; Cl^{-1}, 7.91). For the proton this has been explained in terms of the Grotthuss mechanism, which implies that protons are hopping along arrays of H-bonds instead of single, strongly hydrated protons moving through the water. However, the exact mechanism of Grotthuss prototropic mobility for a long time remained a matter of speculation. In a thorough discussion, Agmon[55] has considered the previous mechanisms and argued that cleavage of one of the three strong H-bonds in the first hydration shell of H_3O^+ cannot be involved in the abnormally rapid proton mobility. Instead, a periodic series of isomerizations between $H_9O_4{}^+$ and $H_5O_2{}^+$ is proposed with cleavage of a H-bond in the second hydration shell as the rate-limiting step. Proton mobility is, of course, important in acid–base-catalyzed organic reactions in water. Enzymes also benefit from the high prototropic mobility in water in carrying protons to their active sites.

Hydroxide-ion mobility has received much less attention. On the basis of symmetry arguments, viewing OH^- as water with a missing proton, one could envisage similar mechanisms for proton and hydroxide transport in water. Recent computational studies, however, do not support this notion and stress the formation of specific hydration complexes influenced by nuclear quantum effects.[56]

A recent MD simulation study considered the hydration of the OH radical and these results are of immediate relevance for an understanding of the rapid diffusion of this radical in aqueous solution.[57]

Viscosity

It is well known, of course, that water pours rather freely despite the rather strong intermolecular H-bonding. The viscosity of water at 298.15 K is 0.8903 cP, much smaller than that of glycerol (954 cP). The difference in viscosity between 'normal' and heavy water is noticeable. Overall, the bulk transport properties appear to look rather normal, but, as noted by Franks,[58] the simple numbers hide a complex behavior that is revealed by the temperature and pressure dependence of the transport properties.

Supercritical water

At elevated temperatures and pressures, water forms dense supercritical states that possess a definite potential for being used as solvents and as reaction media for organic reactions. Above the critical temperature of 647 K, water can be compressed from gas-like to liquid-like densities. The properties of these 'hydrothermal' fluids can be tuned by variation of the temperature and pressure (density), and these systems find important applications in industrial chemistry.[59] IR and NMR spectral data in combination with neutron scattering experiments have shown that H-bonding interactions to form dimers and small clusters have been retained in these fluids. But the extensive 3-D H-bond network has disappeared, and so the peculiar and unique solvent properties of water under normal conditions have been lost. Since the relative dielectric permittivity ε_r decreases with increasing temperature and increases with increasing density, useful conditions emerge for organic reactions. For example, ε_r values in the range 10–25 can be created, allowing convenient solubilities of both salts and relatively apolar reactants. Also the autodissociation of water is greatly affected. At 1273 K and liquid density, K_w is increased by a factor of about 10^6 with immediate consequences for acid–base reactions. Spontaneous acid-catalyzed hydrolysis of carboxylic esters can be performed in such media.[60] Other useful reactions in these hydrothermal fluids are the partial oxidation of unfunctionalized hydrocarbons and the rapid uncatalyzed Diels–Alder reactions of a variety of dienes and dienophiles. No increased endo/exo ratios were found in the latter case. The Diels–Alder reactions in supercritical water have also been studied computationally, emphasizing the increased solubilities of the reaction components.[61] See Chapter 9 for an extensive treatise of reactions in high-temperature water.

2.2.2 Thermodynamics of hydration

The properties of water are beautifully reflected in the thermodynamics of hydration of ionic, polar, and particularly apolar molecules. The smallness of the water molecule, making it possible for more than one water to approach the interaction site, water's greediness to interact with negative and positive charges by dipole interactions (H-bonding), its subtle handling of apolar solutes not amenable with H-bonding so that water tries to retain its own H-bonding interactions as efficiently as possible, all these properties have fascinated scientists for decades. But in the absence of other data, the thermodynamic data are notoriously difficult to rationalize and several controversies have emerged in the long history of all these endeavors.

Since organic reactivity in water depends crucially on the hydration of reactants and the activated complex of the rate-determining step, some important aspects will be discussed in more detail below.

Ion hydration

The hydration of ions has been studied in great detail and excellent reviews are available.[62,63] As anticipated, water interacts strongly with both cations and anions and thereby becomes 'electrostricted'. A useful table showing heats of solution of a series of ions and salts is given in a recent textbook on physical organic chemistry.[64]

The strong hydration of ions has immediate and great consequences for their reactivity. For example, hydration water bound to anionic nucleophiles has to be largely removed before a new covalent bond can be formed in an S_N2 process, and the Gibbs energy of activation will be strongly determined by this dehydration effect (see Section 2.3).

At least two hydration spheres (A and B) have been considered and, of course, waters in the innermost sphere interact most strongly. The final Gibbs energy of hydration results from counteracting enthalpic and entropic contributions. Neutron scattering is one of the most useful experimental techniques to assess the structure of the hydration shells. Hydrophilic anions (i.e. anions not carrying hydrophobic groups) try to form strong linear H-bonds with water, and cations interact strongly with the negative charge on the oxygen atom of water. The inner zone A has strong and well-oriented H-bonds as demonstrated, *inter alia*, by X-ray studies. The second hydration layer (zone B), undergoing dipole polarization by the first one and increasing in size with the size of the anion, is largely responsible for effects extended to waters still further away from the central ionic species.

As anticipated, ion-pair formation (both contact and water-separated) does not readily occur in water and ion-pair association constants are usually small (about 1–10). Two types of ions are often distinguished: (1) salting-out ions, with ion–water interactions stronger than water–water interactions and which decrease the aqueous solubility of nonpolar solutes and (2) salting-in ions, with ion–water interactions weaker than the water–water interactions and which increase the aqueous solubility of nonpolar solutes. The first type ions are small ions of high charge density (kosmotropes), whereas second type ions are large monovalent ions of low charge density (chaotropes). The ranking of ions according to the extent to which they solubilize or precipitate proteins is called the Hofmeister series.[3] Hydrophobic interactions are weakened by salting-in ions and strengthened by salting-out ions.

As argued by Collins,[65] inner sphere ion pairs are preferentially formed between oppositely charged ions with matching absolute enthalpies of hydration.

The hydration spheres of 'hydrophobic ions' like R_4N^+ (with R being an alkyl group) have also received a lot of attention, and the presence of the apolar groups shields the charge from direct interaction with water. Therefore, the hydration of these ions is much weaker than that of hydrophilic ions. Computational studies[66] have shown that waters in the hydration shell of the Me_4N^+ cation form H-bonds with other water molecules whereas the interaction with the cation is mainly via London dispersion. In contrast to the chloride anion, the Me_4N^+ ion behaves as a hydrophobic solute.

Ion–ion interactions in water are largely governed by the volumes of the overlapping hydration spheres. This 'Gurney overlap' term[67] for many cation–anion interactions depends strongly on entropic contributions due to release of hydration water.

Hydration of nonionic solutes

Traditionally, apolar compounds have been called hydrophobic ('water-hating'). But, as noted by Haymet,[68] no SI unit of 'hate' is available! It is well known that oil and water do not mix. But this concept is a bit confusing: the solubility of water in hydrocarbons is significant. Rather the reverse is true: water hates oil. The solubility of hydrocarbons in water is very

Table 2.4 Hard sphere diameters and thermodynamic hydration parameters of noble gases and aliphatic hydrocarbons at 298 K

	σ^{a} (Å)	$\Delta C_{p,tr}$ (J K^{-1} mol^{-1})	ΔH_{tr} (kJ mol^{-1})	$\Delta S_{p,tr}$ (J K^{-1} mol^{-1})	$\Delta G_{p,tr}$ (kJ mol^{-1})
He	2.63	122	1.8	−32.5	11.5
Ne	2.79	143	−1.3	−41.9	11.2
Ar	3.41	195	−9.6	−60.4	8.4
Kr	3.67	218	−13.0	−66.8	6.9
Xe	3.96	250	−16.8	−74.8	5.5
Rn	4.23	—	−19.1	−76.2	3.7
CH_4	3.70	218	−10.9	−64.4	8.3
C_2H_6	4.38	284	−17.2	−83.2	7.6
C_3H_8	5.06	332	−21.0	−97.9	8.2
i-C_4H_{10}	5.55	377	−21.9	−106.0	9.7
n-C_4H_{10}	5.65	390	−23.6	−108.3	8.7
$C(CH_3)_4$	5.89	486	−22.8	−111.7	10.5
n-C_5H_{12}	6.16	—	−26.0	−120.4	9.9
n-C_6H_{14}	6.30	—	−29.1	−135.5	10.7
n-C_7H_{16}	6.74	—	−31.6	−142.9	11.0
n-C_8H_{18}	7.22	—	−37.4	−166.0	12.1

[a]Hard sphere diameter (water: 2.75 Å).

limited. The reason for this situation is beautifully demonstrated by the thermodynamics of hydration of hydrocarbons and noble gases, the prototypes of apolar molecules.[43,69] The interpretation of these often capricious data in terms of solute–water interactions has been and still is a significant intellectual challenge. But these adventures in hydration theories have led to greatly improved insights into the way water can handle dissolution of apolar solutes and into the driving force for hydrophobic interaction.

Gibbs energies, enthalpies, and entropies of hydration of noble gases and of unbranched and some branched hydrocarbons upto 18 carbon atoms are listed in Table 2.4. Following Lee[70] and Graziano,[71] these data are given in terms of Ben–Naim standard quantities which describe the transfer of an apolar molecule to a fixed position in water. This means that the standard hydration Gibbs energy is defined by

$$\Delta G_{tr} = RT \ln(C_g/C_l)$$

where C_g and C_l are the molar concentrations of the solute in the gas and liquid phases at equilibrium. Now the solubility of the solute in water is directly given by ΔG_{tr}. Table 2.4 also shows the heat capacity change resulting from hydration of the solute. Since we are interested in the question, how these thermodynamic data depend on the size of the solute, hard sphere diameters for the solutes are also provided in Table 2.4.

We note that all Gibbs energies of hydration are positive (unfavorable). Interestingly, ΔG_{tr} becomes less unfavorable with increasing size of the noble gas, which contrasts with the opposite behavior for the hydrocarbons (*vide infra*). Except for helium, all other ΔH_{tr} values are negative as expected, for solute–solvent interactions now play a role. The entropies of transfer are negative and unfavorable as anticipated for going from the gas to the liquid phase. The available heat capacities are large and positive because the enthalpies of hydration become less favorable upon increasing temperature.

If we compare these data with similar quantities for solvation of the noble gases and hydrocarbons in organic solvents at 25°C, three important conclusions can be drawn. First, ΔG_{tr} for water is strongly unfavorable, characterizing the pronounced hydrophobicity of these solutes. Second, ΔH_{tr} for water is *more* favorable than for any organic solvent. This is at first sight counterintuitive since the solutes cannot participate in H-bonding with water. In fact, the only interaction mechanism is London dispersion interaction and this is expected to be weak taking into account the low polarizability of the water molecule. And third, the favorable ΔH_{tr} is strongly overcompensated by a highly unfavorable ΔS_{tr}, again much more unfavorable than for any organic solvent.

In a very influential paper, Frank and Evans[72] in 1945 suggested a molecular interpretation of the data for water which was clear and elegant and which survived for many years. In their microscopic 'iceberg model' they argued that the hydration of apolar solutes is characterized by increased H-bonding (either stronger or more H-bonds per volume unit, i.e., greater tetrahedrality and a trend toward greater 'crystallinity') in the so-called hydrophobic hydration shell, characterized by the iceberg metaphor. The icebergs were supposed to be rather rigid and 'clathrate-like'. These ideas were often expressed by saying that there exists enhanced 'water structure' in the hydrophobic hydration shell. These 'water-structure' arguments have been used in the past in numerous explanations for aqueous kinetic and equilibrium processes in water.

The 'iceberg theory' remained in a dogmatic slumber for more than 40 years. During the past 10–15 years, however, it has become clear that the proposed increased H-bonding in the hydrophobic hydration shells has been much overestimated. Neutron scattering experiments,[73,74] NMR measurements,[75] and vibrational studies,[76] in combination with advanced computational studies (MD simulations[77] and quantum chemical calculations[78]) and thermodynamic analyses,[79] have indicated that hardly any increased structuring is involved. In fact, most recently it was concluded on the basis of neutron diffraction experiments[80] that the H-bond network in an aqueous methane solution is marginally less tetrahedral than in pure water. Because of the absence of solute–water H-bonding, the waters keep their 3-D H-bonding network as much as possible intact, which, for not too big solutes, results in a preference for tangential orientation of the water OH-bonds relative to the apolar surface. Of course, this orientational preference of the water molecules decreases with increasing temperature (Fig. 2.5) because of the breakdown of the H-bonds. The heat capacity changes (Table 2.4) are consistent with this picture and are large compared with those for solvation of the same solutes in organic solvents. In fact, large heat capacity changes are characteristic for water and find their origin in large fluctuations of the H-bond network, associated with compensating enthalpy and entropy effects. A recent theoretical model supports the view that the heat capacity of hydrophobic hydration is due to the H-bonding properties of water.[81] It is found quite generally that solute hydrophobicity causes an upfield shift in the anomalies of water, as, *inter alia*, illustrated by an increase in the temperature of maximum density.[82]

Upon increasing temperature the ΔG_{tr} of transfer becomes dominated at a critical temperature by the enthalpic rather than the entropic contribution. As noted by Haymet et al.,[78] 'hydrophobicity is entropic in cold water and enthalpic in hot water'.

Two important temperatures can be defined. The first is T_s above which ΔS_{tr} becomes favorable. The other is T_h, the temperature above which ΔH_{tr} becomes unfavorable. For relatively small solutes T_h is found around room temperature, whereas T_s is about 113°C.

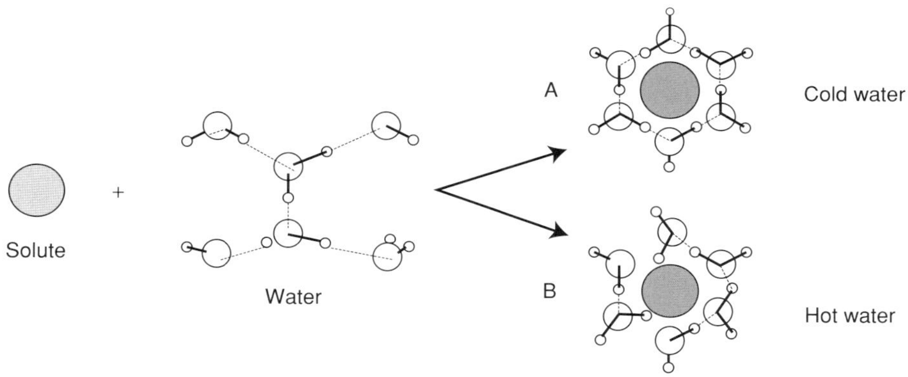

Figure 2.5 Preferred tangential orientation of the O—H bonds of water relative to an apolar molecular surface. At higher temperatures the hydration shell becomes less organized. Reproduced with permission from Southall et al. (2002).[78] Copyright 2002 American Chemical Society.

The ΔG_{tr} is then given by

$$\Delta G_{tr}(T) = \Delta C_p[(T - T_h) - T\ln(T/T_s)]$$

assuming that ΔC_p is temperature independent. This behavior is highly suggestive for the occurrence of hydrophobic hydration of liquid hydrocarbons.

Interestingly, Bowron et al.[83] have made a direct comparison of the hydrophobic hydration spheres of krypton dissolved in water and of its solid clathrate using extended X-ray absorption fine structure spectroscopy. It was found that the average orientational configurations of waters in contact with krypton dissolved in water deviate from the largely tangential orientations in the clathrate cage and are clearly more loosely organized.

For large, inert hydrophobic surfaces with small curvature, it becomes, for geometric reasons, increasingly more difficult to maintain the 3-D H-bond network and then the thermodynamics are dominated by positive enthalpic contributions with small changes in heat capacity. The temperature dependence of hydrophobic hydration of small (<1 nm) and larger (>1 nm) hydrophobic surfaces is different. Chandler et al.[84] contend that this is caused by water depletion for the larger hydrophobic length scales. For other complexities encountered in the hydration of extended apolar surfaces the reader is referred to the literature.[85]

Looking for a possible relation between the size of the solute and ΔG_{tr} for the hydrocarbons, it is remarkable that these Gibbs energies change only modestly with size. The different size dependence of the ΔG_{tr} parameters for noble gases and hydrocarbons has already been noted. This difference has been explained[70] by viewing the hydration of an apolar solute as resulting from two steps: (1) the creation of a suitable cavity in water to host the solute and which needs the exclusion of water molecules from the cavity space and (2) the insertion of the solute from the gas phase into the cavity and turning on the attractive solute–water potential. This two-step description of the dissolution process is convenient for application of scaled particle theory[86] and is useful although the definition of the separate steps remains rather arbitrary. The cavity term (ΔG_c) is strongly unfavorable for water because of the small molecular size of the water molecules. However, the low packing density of water (due to the H-bond network) in part counteracts this effect.

Graziano[69] has argued that for the noble gases the second favorable term increases with size more rapidly than the second unfavorable one, leading to an increased solubility with increasing hard sphere diameter (Table 2.4). The opposite situation applies for the hydrocarbons and now the ΔG_{tr} term (and their aqueous solubility) becomes more unfavorable with increasing size of the solute.

The rather strongly favorable ΔH_{tr} values for dissolution of apolar molecules in water (Table 2.4) are surprising. They probably arise from London dispersion interactions involving a number of (small) water molecules covering the total apolar surface area. Tentative suggestions have been made that these interactions might benefit from rapid proton exchange between the waters in direct contact with the apolar solute.[87]

The strongly negative ΔS_{tr} values are likely dominated by the excluded volume effect and are geometric in nature.[69] Again the smallness of the water molecule plays an important role. The chance of finding a naturally occurring cavity of the size of a liquid hydrocarbon in water is lower than for any other solvent because water molecules are so small. The reduced motional freedom of the waters in the hydrophobic hydration shell will also contribute to the unfavorable entropic term.

A thermodynamic analysis[71] has shown that the enthalpic and entropic contributions to the reorganization of the H-bonds in the hydration shell compensate each other:

$$\Delta H_h = T\Delta S_h$$

so that reorganization of the H-bonds does not contribute to the Gibbs energy of hydration. Graziano[71] then argues that

$$\Delta G_{tr} = E_a - T\Delta S_x$$

in which E_a is the average value of the solute–water interaction energy and ΔS_x is the entropy change due to the excluded volume effect resulting from cavity creation. The latter term is geometric in nature, independent of H-bonding and much dependent on the small size of the water molecule.

However, many organic molecules possess one or more polar functional groups that can favorably interact with the waters in the overall hydration shell. The result is a decrease in overall hydrophobicity. Recent evidence has been obtained that for a mix of hydrophilic/hydrophobic segments in the solute leading to completely water-soluble compounds like methanol, incomplete mixing occurs.[88]

Thus different parts of the molecule may carry hydration shells that have significantly different properties. This leads to mutually destructive 'intramolecular' overlap of water shells and the resulting overall hydration characteristics will definitely represent these complexities. This makes it difficult to define 'hydrophobic parameters' for organic functional groups because they will depend on their position in the molecule. Only long alkyl functionalities do not suffer from these overlap effects. Recent kinetic studies have demonstrated these effects and rough estimates have been made of the extent of the 'overlap region' in the total hydration shell (*vide supra*).

Aromatic molecules are hydrophobic but their thermodynamic hydration parameters are essentially different from those of aliphatic hydrocarbons (Table 2.5).[89] The data in this table reveal that for benzene and toluene the ΔG_{tr} values are *negative*, primarily resulting from less unfavorable ΔS_{tr} values. Advanced MD simulations[90] of the hydration of benzene and cyclohexane have shown that there is a definite tendency for H-bond interactions of the waters with the π-electron system of the aromatics. This is confirmed by the solubility

Table 2.5 Thermodynamic hydration parameters of benzene and toluene[a]

	ΔH° (kJ mol^{-1})	ΔS° (J K^{-1} mol^{-1})	ΔG° (kJ mol^{-1})
Benzene	−29.6	−87.2	−3.6
Toluene	−33.9	−101.4	−3.7

[a]Data taken from Ref. 89.

of water in binary benzene–cyclohexane mixtures, which increases with increasing benzene concentration. A recent theoretical study[91] emphasizes the importance of the quadrupole moment of benzene in explaining this observation.

Finally, in this section, the enormous importance of natural gas water hydrates should be briefly mentioned. These hydrates are polyhedral cells consisting of H-bonded waters and stabilized by encaged guest molecules such as methane and carbon dioxide. These solid hydrates constitute a great problem in gas transmission lines and inhibitors for their formation have been developed.[92]

2.2.3 Hydrophobic interactions

In the previous section we have seen that the formation of hydrophobic hydration shells aids the dissolution of apolar solutes in water. Upon increasing concentration and/or size of the solute it is inevitable that, at a critical concentration, the large hydrophobic hydration shells start to overlap, leading to mutually destructive breakdown of these water arrangements (Fig. 2.6). This sacrifice of H-bonding interactions results in a solvent-induced sticking

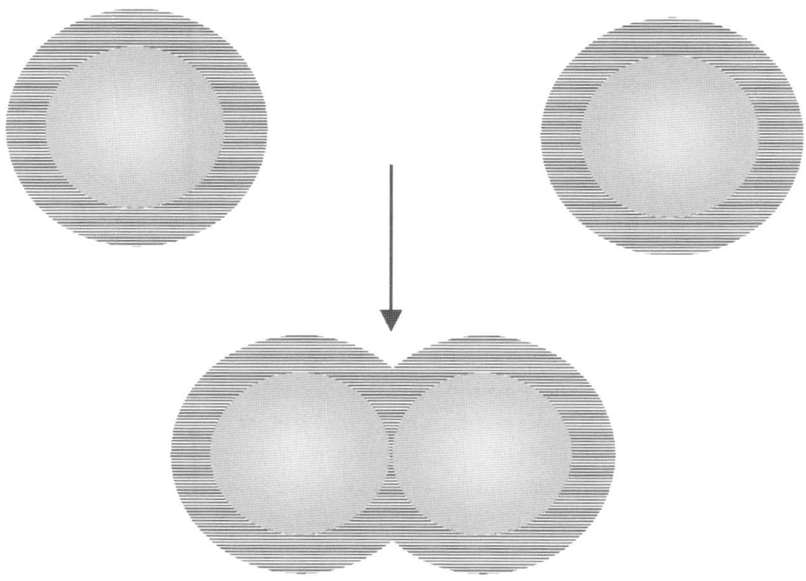

Figure 2.6 Hydrophobic interactions. Cartoon of the destructive overlap of two hydrophobic hydration shells (striped areas) resulting in a release of water to the bulk solution.

together of hydrophobic surface areas of the two solutes, primarily driven by the gain in entropy that comes from the release of water molecules from the hydrophobic hydration shells into bulk water. These *hydrophobic interactions* (HI),[78,87] introduced by Kauzmann[93] in 1954, are, however, the results of a complex mix of factors that cannot be all discussed in the context of this chapter. In brief, one can imagine attraction between the hydrophobic entities around room temperature by the entropy gain just mentioned and by the van der Waals interactions due to direct contact between the apolar molecules. But entropy will be lost due to a smaller mixing entropy. HI are further characterized by large negative changes in the heat capacity and, in fact, these $\Delta C_p{}^*$ effects have been taken by Herman[94] as indications of the size of the solvent accessible surface area involved in the HI between the apolar entities. Although this theory has been criticized since heat capacity effects may also originate from other types of interactions, the $\Delta\Delta C_p{}^*$ values provide at least a reasonable estimate of the hydrophobic contact areas and are dependent on molecular structure and geometry.[95] Interestingly, heat capacity changes due to pairwise hydrophobic interactions have recently been related to the structure of the joint hydration shells.[96]

It should be emphasized here that hydrophobic hydration shells are quite voluminous. Computer simulations can be used for estimating hydration numbers. For example, Jorgensen[97] has reported hydration numbers of 20 and 34 for, respectively, methane and *n*-pentane. A ^{13}C-NMR study[98] gave a hydration number of 20 for aqueous methane.

Depending on the nature of the hydrophobic molecules, one can distinguish pairwise (1:1) interactions, the formation of small aggregates ('moving units', for example in the case of hydrotropes) and the formation of larger aggregates (bulk hydrophobic interactions) as in the case of surfactant aggregates like micelles and vesicles. The ultimate state of aggregation will, of course, involve phase separation. It should be noted here that recent Monte Carlo simulations[99] provided evidence for two types of HI: (1) direct contact HI and (2) pairwise interactions to form a complex separated by a single water layer. Large solutes prefer contact HI whereas smaller solutes tend to form water-separated dimers, particularly in cold water.

HI between aromatic molecules have a different thermodynamic signature and are driven by favorable enthalpy effects rather than entropy effects ('nonclassical HI').[89]

A thermodynamic study has demonstrated the difference in binding of the inhibitors benzamidinium chloride and cyclohexylamidinium chloride to the binding site of trypsin.[100] The enthalpies of binding are rather similar, but the weaker binding of the cyclohexyl compound is caused by a less favorable entropy of binding, possibly caused by the restricted flexibility of the cyclohexyl ring in the binding pocket. At this point it should be stressed that the thermodynamic signature of ligand–protein association depends on the degree of hydration and architecture of the binding pocket. If the cleft is not strongly hydrated, the Gibbs energy of binding might be dominated by the enthalpic contribution.[101] The potential of NMR methods for the derivation of per-residue thermodynamic parameters for biomolecular association has recently been discussed.[102]

For HI between two π-systems the heat capacity changes are relatively small and negative. For aromatics London dispersion interactions involving the polarizable π electrons contribute more significantly than in the case of aliphatic solutes.[89]

Hydrophobic interactions, still the least understood representative of noncovalent interactions, are of great importance in (bio)chemistry. They contribute to protein folding, the formation of enzyme–substrate and enzyme–inhibitor complexes, the formation and functioning of cell membranes, just to mention processes of crucial biological importance. But HI also occur between ordinary apolar organic molecules and operate in many molecular recognition processes in water. Intramolecular hydrophobic interactions in molecules carrying long alkyl tails may lead to extensive coiling of the molecule (Section 2.3). Finally, HI are important in many industrial applications, such as detergency, coagulation, and mineral flotation.

HI also affect organic reactivity in water. This topic has been reviewed recently.[103] The thermodynamics of hydration and of HI are of immediate relevance for the interpretation of aqueous kinetic rate effects using transition state theory.

For not too big organic molecules these HI are relatively weak, but significant. Using statistical perturbation theory, Jorgensen et al.[104] have estimated the Gibbs energy of interaction of two methane molecules to be around -1.5 kJ mol^{-1} of which -1.2 kJ mol^{-1} originates from simple Lennard–Jones attraction. Similar values were also found in subsequent MD simulations and their dominating entropic origin was confirmed.

Significant effects due to HI were found in kinetic studies of organic reactions. For example, hydrolysis reactions in dilute (<5 mol%) aqueous solutions of simple alcohols have been examined in detail.[105,106] The reactants are usually better stabilized by HI than the activated complexes leading to rate retardations. The kinetic effects increase with increasing hydrophobicity of the alcohols (EtOH < n-PrOH < t-BuOH).[106] The cosolvent-induced changes in the isobaric activation parameters reflect entropy-driven reactant–alcohol interactions.

Further examples of the effects of HI on organic reactions will be presented in other chapters in this book Let me only briefly mention 'enforced hydrophobic interactions'.[107,108] Reactivity in the liquid phase is enormously slowed down relative to the gas phase because close contact between molecules, allowing sufficiently effective orbital overlap, needs partial desolvation of the reaction partners. In terms of Gibbs energies, this desolvation is costly in any solvent. However, in water, favorable hydrophobic contacts are a natural part of the activation processs, leading to rate enhancements as compared with nonaqueous reaction media. Diels–Alder reactions are a good example. These cycloadditions can be much accelerated in water although no diene–dienophile aggregation is involved. These *enforced* HI have been studied experimentally in detail and the kinetic effects are in good accord with sophisticated computer simulations carried out by Jorgensen et al.[109]

HI also influence the conformational preferences of organic solutes in water. The tendency is to reduce the solvent-accessible surface area as exemplified by MD simulations of *n*-butane.[110] If *n*-butane is transferred from the gas phase to water, there is an entropy-driven increased population of the more compact gauche conformation compared to the extended all-trans conformation. Other interesting cases of folding of small molecules in water have been recorded in the literature.[111] The relevance for organic reactivity is exemplified by Diels–Alder reactions in water, which occur with a significantly enhanced endo/exo isomer ratio.[11]

Many attempts have been made to develop a quantitative, experimental measure for the strength of hydrophobic interactions. This has been accomplished for hydrophobic gases by measuring the coalescence rate of gas bubbles under well-defined conditions.[112,113] But, as already mentioned (Section 2.2.2), many organic molecules carry polar functional groups which will be stabilized by H-bonding in water. The 'overall' hydrophobicity of the molecule, determining its propensity to participate in HI, is difficult to predict and assess. This problem is, for example, demonstrated by the large number of hydrophobicity scales that can be found in the literature for α-amino acids.[78,114] Recently a study has been made of hydrophobic regions at the binding site of proteins considering not only atom type but also nonadditive effects arising from the shape and extent of a nonpolar region.[115] In general, the MD study

Table 2.6 Gibbs energies and enthalpies of pairwise group interactions in aqueous solution at 298 K[119,120]

Group		G_{ij}[a](J kg mol^{-2})	H_{ij}[a](J kg mol^{-2})
i	j		
CH_2	CH_2	−34	33
OH	OH	−24	−19
OH	CH_2	27	9
CONH	CONH	−115	−252
CONH	OH	−34	—
CONH	CH_2	55	66
CHOH	CHOH	−3	−14
CHOH	CONH	−6	−47
CHOH	CH_2	10	44

[a]Standard errors are large, often 20–40%.

has shown that in biological systems HI effects exhibit nonadditivity, modulability, context-dependence, and long-range action.

Systematic kinetic studies have been made aimed at quantifying the effects of overlapping hydrophilic and hydrophobic hydration shells.[116–118] It could be shown that the presence of an anionic substituent strongly diminishes the hydrophobicity of the first three methylene units of an alkyl chain attached to the functional group.

For organic compounds, attempts have been made to assess the Gibbs energies and enthalpies of solute–solute interactions in water by summation of pairwise functional group interactions ('Savage–Wood additivity').[119,120] Three assumptions are made: (1) each group i of solute A interacts with each group j of solute B, (2) each of these interactions makes a characteristic contribution to the overall interaction between A and B, and (3) each group interaction is independent of other groups and their relative position in the solute. It will be clear on the basis of the previous discussions that this approach is an oversimplification because no neighboring group and stereochemical effects are taken into account. Nevertheless, the analysis has merit and is relatively simple to apply also for kinetic rate effects in water in the presence of a series of structurally related solutes.[121,122] Selected Gibbs energies and enthalpies of pairwise functional group interactions[123,124] are listed in Table 2.6.

In line with HI, the methylene–methylene interaction is characterized by $g_{AB} < 0$ and $Ts_{AB} > h_{AB} > 0$. The rather strong interaction between two CONH groups is remarkable because these functionalities are quite heavily hydrated. Still more surprising perhaps is the observation that small carbohydrate molecules (n-hexoses) demonstrate a weakly hydrophobic character which for each compound has been ascribed to the quality of the fit of the OH groups into the H-bond network of water.[125]

2.3 Kinetic solvent effects in aqueous solution

Water is unique among solvents because it has only an exceptionally small molecular volume but also the capacity to form a 3-D H-bond network with a dominant tetrahedral coordination. Kinetic solvent effects in water and in highly aqueous binary solvent mixtures respond to these properties. Organic reactivity can be best analyzed using transition state theory. This means that, in relation with reactivity in other solvents, the hydration has to be considered of the reactant(s) (or initial state, IS) and of the activated complex (AC, residing in the transition state, TS). The Gibbs energy of activation, and thus the rate constant, is given by

$$\Delta^{\ddagger} G^{\circ} = G^{\circ}_{(AC)} - G^{\circ}{}_{(IS)}$$

This difference in Gibbs energy is strongly influenced by hydration effects. For a bimolecular reaction, dehydration plays an important role since water has to be removed from the hydration shells of both reaction partners to allow sufficiently close contact for overlap of those molecular orbitals responsible for formation of the AC. Jorgensen[126] has shown in his Monte Carlo computer simulations of the S_N2 process

$$CH_3 - Cl + Cl^- \rightarrow CH_3 - Cl + Cl^-$$

how overwhelmingly important these dehydration phenomena are (Fig. 2.7).

Following a tradition pioneered by Ingold,[127] solvent effects are often intuitively assigned to dominating effects on the AC. Indeed, usually the AC has an increased polarity compared to the IS. But for reactions in water, in many cases the specific medium effect of water most strongly influences the IS. Therefore it is important to make a distinction between hydration effects on the IS and AC. Let us see how this can be done.[46,128]

Going from solvent A to (reference) solvent B normally leads to a change in $\Delta^{\ddagger} G^{\circ}$. This change originates from different changes in $G^{\circ}{}_{(IS)}$ and $G^{\circ}{}_{(TS)}$ because both species possess different solvation characteristics. In Fig. 2.8 an example is given in which going from solvent A to solvent B the difference in the Gibbs energy of activation ($\Delta^{\ddagger} G^{\circ}{}_{B} - \Delta^{\ddagger} G^{\circ}{}_{A}$) is expressed as the Gibbs energies of transfer from solvent A to solvent B of the activated complex and of the reactant(s) R. The first and last terms are experimentally accessible, the first from the

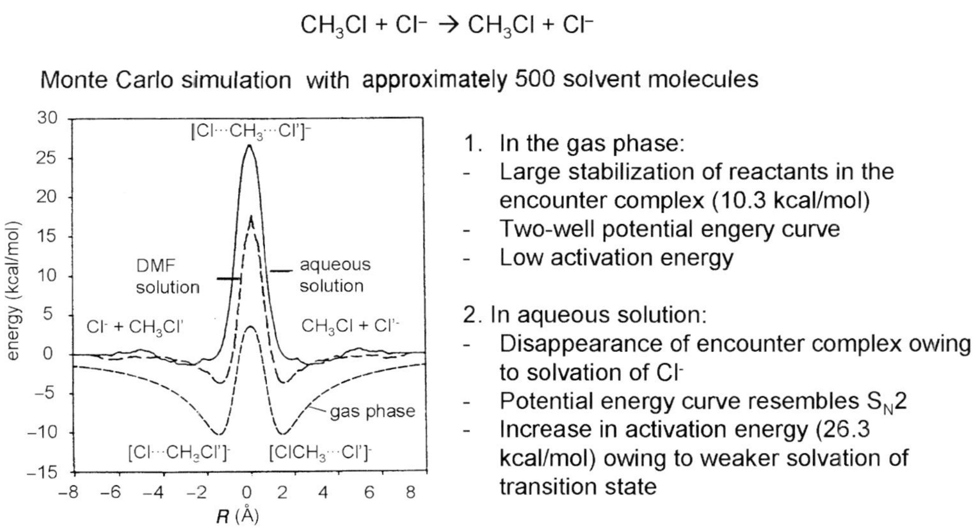

Figure 2.7 Monte Carlo reaction profile for an S_N2 reaction in the gas phase, dimethylformamide, and water. Reproduced with permission from Jorgensen (1989).[126] Copyright 1989 American Chemical Society.

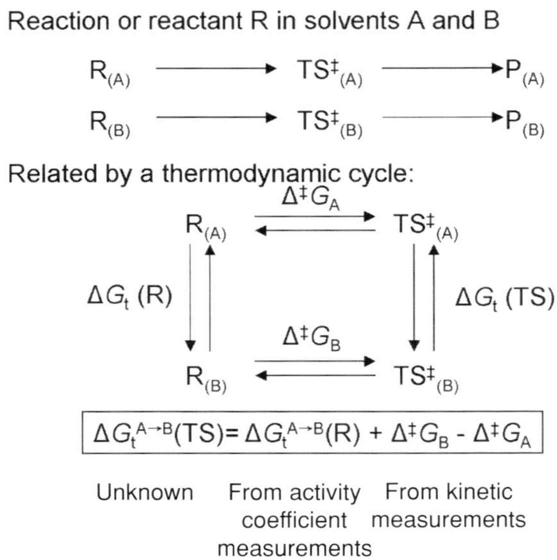

Figure 2.8 Thermodynamic (or unified) approach for distinguishing solvent effects on the reactants (R) and on the activated complex (TS).

difference in rate constants, the other from differences in solubility, vapor pressure, etc. In case the reactant(s) reacts with water, one or two model solutes have to be selected with solvation characteristics as closely as possible related to those of the reactant(s). One can then calculate the transfer Gibbs energy of the activated complex. This makes it possible to conclude whether the solvent change primarily affects the reactants or the activated complex. The next step then is, of course, to interpret the Gibbs energies of transfer for reactants and activated complex in terms of solute–solvent interactions, which is far from a trivial task. In the case of Diels–Alder reactions, combined quantum chemical calculations and MD/Monte Carlo computer simulations can be of great help as shown by Jorgensen et al.[109] and Gao and Furlani.[129] But the sophistication of the thermodynamic (or unified) approach is obvious and warrants the necessary careful experimentation.

Interestingly, an extensive kinetic study of cycloaddition reactions in mixed aqueous solvents[130] (cosolvents: methanol, acetonitrile, and poly(ethylene glycol)) showed that there are often small maxima in rate constants around 40 M water. Most likely they originate from different dependencies of the Gibbs energies of transfer of the IS and AC as a function of solvent composition with the largest effect operating on the IS. The dominating effects of H-bond interactions on 1,3-dipolar cycloadditions were also emphasized in another kinetic study.[131]

In a detailed study, Abraham et al.[46] have shown that the method of multiple linear regression can be successfully applied to the transfer Gibbs energies of the IS and AC. Finally, it should be noted here that the method can also be employed for enthalpies and entropies of transfer of IS and AC, but this needs careful temperature-dependent measurements of rate constants as well as transfer parameters and there are few examples of such demanding studies in the literature.[132]

Cosolute effects on hydrolysis reactions in water-rich aqueous media can be examined quantitatively by using a combined thermodynamic and kinetic analysis. The background theory[133] shows that

$$\ln[k_{(mc)}/k_{(mc=0)}] = [2/RTm_0{}^2][g_{(c\text{-IS})} - g_{(c\text{-TS})}] - N\varphi M_I m_c$$

in which $k_{(mc)}$ is the (pseudo-)first-order rate constant for hydrolysis in an m_c molal aqueous solution of cosolute c, $k_{(mc=0)}$ is the rate constant for hydrolysis in water, m_0 is the hypothetical, ideal reference state (i.e. 1 mol kg^{-1}), R is the gas constant, and T is the absolute temperature (in K). The term $[g_{(c\text{-IS})} - g_{(c\text{-TS})}]$ is denoted as $G(c)$ and represents the difference in Gibbs energy of interaction between the cosolute and the reactants and between the cosolute and the activated complex, respectively. N is the number of water molecules involved in the rate-determining step, φ is the practical osmotic coefficient for the solute solution of molality m_c, and M_I is the molal mass of water. For relatively dilute cosolvent solutions, φ can be taken as unity.

For hydrophobic cosolutes the interaction with the IS will usually be stronger than that with the more polar AC, leading to negative $G(c)$ values. The $G(c)$ values are easily obtained from a plot of $\ln[k_{(mc)}/k_{(mc=0)}]$ vs m_c and have been measured for a large variety of cosolutes.[134] Attempts have been made to analyze these $G(c)$ values in terms of additivity of functional group interactions but the results are not yet very satisfactory because of the simplifications inherent in this approach (Section 2.2.3). However, for the water-catalyzed hydrolysis of acyl-activated esters and 1-acyl-3-phenyl (or alkyl)-1,2,4-triazoles ($N = 2$) the $G(c)$ values respond nicely to variation in the hydrophobicity of both the substrate and the cosolute.[117,135] For example, for the neutral hydrolysis of 4-methoxyphenyl dichloroacetate the $G(c)$ values for the cosolutes EtOH, 1-PrOH, and 1-BuOH are –304, –474, and –709 J kg mol^{-2}, respectively. The errors in $G(c)$ are about 1–2%. These data primarily reflect stabilization of the ester by hydrophobic interaction with the alcohol in encounter complexes, thereby frustrating nucleophilic water attack on the ester carbonyl group. Isobaric activation parameters are in accord with this interpretation. MD computer simulations confirmed the linear relationship between $\ln k_{(mc)}$ and m_c and gave highly useful insights into the structure of the encounter complexes.[136]

Hydrogen-bonding interactions can also effectively decelerate organic reactions in water by dominant stabilization of the IS. This is well documented for, *inter alia*, unimolecular decarboxylation reactions.[137]

Self-coiling of elongated apolar molecules may exert significant effects on organic reactivity. As demonstrated by Jiang et al.[138–140] these effects occur only in water-rich media and originate from intramolecular hydrophobic interactions.

Interestingly, water has special potential for performing C—C bond forming reactions with unprotected sugars under mildly alkaline conditions.[141,142]

Beneficial reaction conditions for biphasic catalytic systems can be created using mixtures of water with supercritical CO_2 that have rather unique and controllable phase properties.[143,144]

Finally, it is noted that water can also have a beneficial effects on photochemical reactions. An example is the photodimerization of stilbenes, which is greatly accelerated in water relative to benzene, most likely due to HI between the reacting molecules.[145]

It is clear that the next decade will witness many further developments in the understanding and application of organic reactions in water. To appreciate the peculiar nature of the aqueous reaction medium, a useful approach will remain the comparison of organic reactivity in water with that in organic solvents like those shown in Table 2.1. To investigate the effects of HI, perhaps the most useful reference solvents are pure ethanol and 1-propanol and their binary mixtures with water below and above the concentration where the cosolvents start to aggregate. In most cases so far, H-bonding effects in water dominate over the kinetic effects due to HI.[109] However, (enforced) HI may be highly effective in inducing stereochemical preferences.

Experimental kinetic studies will go hand in hand with computational approaches, Monte Carlo and, particularly, MD computer simulations. The recent literature has already shown advanced and reliable computational results in which data have been obtained which are not, or with great difficulty, obtainable by experimental studies. In the near future still more realistic force fields will emerge and the speed of computational approaches will increase. I think at the moment a combination of kinetics, thermodynamics, and MD simulations is the most powerful approach to arrive at detailed and quantitative understanding of aqueous medium effects.

But also synthetic organic chemistry in homogeneous and, most likely, particularly in heterogeneous aqueous media, is anticipated to be flourishing in the coming years. The recent work by Sharpless et al.[15] has set the stage for efficient synthesis in heterogeneous media, in fact suspensions of hydrophobic reaction partners in water. For a brief review, see Ref. 16. Also supramolecular chemistry 'on water' is being developed.[146] These studies call for further detailed mechanistic analysis of organic reactions at such interfaces.[17,147]

Micellar and vesicular interfaces in homogeneous reaction mixtures have already been explored extensively and both the water concentration and the water–water H-bond interactions are different from those in bulk aqueous solution. An important technique to examine directly the properties of interfacial water is vibrational sum frequency spectroscopy.[148–150] These studies will now be followed up by detailed work to establish the benefits of such nanoparticles for organic reactions in heterogeneous media. Oil–water interfaces have already shown counterintuitive effects, such as the specific physiosorption of hydroxide ions even around neutral pH.[151,152] A recent MD study has suggested a mechanism for hydroxide-ion binding,[153] which may explain the negative zeta potential of small oil droplets under specific conditions. Similar effects may explain the negative zeta potential of small gas bubbles in water. Evidence suggests that the electrical properties of the bubbles are determined by the H-bond network at the gas–water interface.[154]

Hydrophobic interactions combined with H-bonding as well as specific molecular orientational effects at interfaces are probably the crucial topics for further exploration at the moment.

I have little doubt that water will continue to provide inspiration for creative study and thinking, not only for natural scientists but also for philosophers, poets, and all civilized human beings fascinated by the wonders this planet has to offer.

References

1. Daniel, R.M., Finney, J.L. and Stoneham, M. (Eds.), The molecular basis of life: is life possible without water? *Philos. Trans. R. Soc. Lond. B*, 2004, 1141–1328.
2. Finney, J.L., Water? What's so special about it? *Philos. Trans. R. Soc. Lond. B*, 2004, **359**, 1145–1165.
3. Jencks, W.P., *Catalysis in Chemistry and Enzymology*, McGraw-Hill, New York, 1969.
4. Breslow, R. and Rideout, D., Hydrophobic acceleration of Diels–Alder reactions, *J. Am. Chem. Soc.*, 1980, **102**, 7816–7817.

5. Blokzijl, W. and Engberts, J.B.F.N., Enforced hydrophobic interactions and hydrogen bonding in the acceleration of Diels–Alder reactions in water, In: Cramer, C.J. and Truhlar, D.G. (Eds.), *Structure and Reactivity in Aqueous Solution*, American Chemical Society Symposium Series, American Chemical Society, Washington, DC, 1994, pp. 303–317.
6. Wijnen, J.W. and Engberts, J.B.F.N., Retro Diels–Alder reaction in aqueous solution: toward a better understanding of organic reactivity in water, *J. Org. Chem.*, 1997, **62**, 2039–2044.
7. Pirrung, M.C. and Das Sarma, K., Multicomponent reactions are accelerated in water, *J. Am. Chem. Soc.*, 2004, **126**, 444–445.
8. Grieco, P.A. (Ed.), *Organic Synthesis in Water*, Blackie, London, 1998.
9. Li, C.-J. and Chan, T.-H., *Organic Reactions in Aqueous Media*, Wiley, New York, 1997.
10. Lindstrom, U.M. and Andersson, F., Hydrophobically directed organic synthesis, *Angew. Chem. Int. Ed.*, 2006, **45**, 548–551.
11. Li, C.-J., Organic reactions in aqueous media with a focus on carbon–carbon bond formations: a decade update, *Chem. Rev.*, 2005, **105**, 3095–3165.
12. Otto, S., Bertoncin, F. and Engberts, J.B.F.N., Lewis acid catalysis of a Diels–Alder reaction in water, *J. Am. Chem. Soc.*, 1996, **118**, 7702–7707.
13. Otto, S., Engberts, J.B.F.N. and Kwak, J.C.T., Million-fold acceleration of a Diels–Alder reaction due to combined Lewis acid and micellar catalysis in water, *J. Am. Chem. Soc.*, 1998, **120**, 9517–9525.
14. Lindström, U.M., Stereoselective organic reactions in water, *Chem. Rev.*, 2002, **102**, 2751–2772.
15. Narayan, S., Muldoon, J., Finn, M.G., Fokin, V.V., Kolb, H.C. and Sharpless, K.B., 'On water': unique reactivity of organic compounds in aqueous suspension, *Angew. Chem. Int. Ed.*, 2005, **44**, 3275–3279.
16. Klijn, J.E. and Engberts, J.B.F.N., Fast reactions 'on water', *Nature*, 2005, **435**, 746–747.
17. Dambacher, J., Zhao, W., El-Batta, A., Anness, R., Jiang, C. and Bergdahl, M., Water is an efficient medium for Wittig reactions employing stabilized ylides and aldehydes, *Tetrahedron Lett.*, 2005, **46**, 4473–4477.
18. Buckingham, A D., The structure and properties of a water molecule, In: Neilson, G.W.E.J.E. (Ed.), *Water and Aqueous Solutions*, Adam Hilger, Bristol, 1986, pp. 1–13.
19. Laing, M., No rabbit ears on water – the structure of the water molecule – what should we tell the students, *J. Chem. Educ.*, 1987, **64**, 124–128.
20. Franks, F. (Ed.), *Water: A Comprehensive Treatise*, Vols. 1–7, Plenum Press, New York, 1973–1982.
21. Zangi, R., personal communication, 2005.
22. Teixeiro, J., Bellisent-Funel, J.C.M.-C. and Dianoux, A.J., Experimental determination of the nature of diffusive motions of water molecules at low temperatures, *Phys. Rev. A*, 1985, **31**, 1913–1917.
23. Denisov, V.P. and Halle, B., Protein hydration dynamics in aqueous solution, *Faraday Discuss.*, 1996, **103**, 227–244.
24. Chaplin, M., www.sbu.ac.uk?water/index.html, 2000.
25. Soper, A.K., Tests of the empirical potential structure refinement method and a new method of application to neutron diffraction data on water, *Mol. Phys.*, 2001, **99**, 1503–1516.
26. Silvestrelli, P.L. and Parrinello, M., Water molecule dipole in the gas and in the liquid phase, *Phys. Rev. Lett.*, 1999, **82**, 3308–3311.
27. Jeffrey, G.A. and Saenger, W., *Hydrogen Bonding in Biological Structures*, Springer, Berlin, 1991.
28. Baum, J.O. and Finney, J.L., An SCF-CI study of the water dimer potential surface and the effects of including the correlation energy, the basis set superposition error and the Davidson correction, *Mol. Phys.*, 1985, **55**, 1097–1108.
29. Sciortino, F., Geiger, A. and Stanley, H.E., Effect of defects on molecular mobility in liquid water, *Nature*, 1991, **354**, 218–221.
30. Engberts, J.B.F.N., Water as a metaphoric model in process thought, In: Kelly, T. and Dibben, M. (Eds.), *Applied Process Thought: Frontiers of Theory and Research*, 1st edn., Ontos Verlag, Frankfurt/Lancaster, in Press.
31. Ball, P., *A Biography of Water*, Weidenfeld & Nicolson, London, 1999.
32. Jorgensen, W.L. and Tirado-Rives, J., Potential energy functions for atomic-level simulations of water and organic and biomolecular systems, *Proc. Nat. Acad. Sci. U. S. A.*, 2005, **102**, 6665–6670.
33. Csonka, G.I., Ruzsinszky, A. and Perdew, J.P., Proper Gaussian basis sets for density functional studies of water dimers and trimers, *J. Phys. Chem. B*, 2005, **109**, 21471–21475.
34. Jantsch, E., *The Self-Organizing Universe*, Pergamon Press, Oxford, 1980.
35. Kimmel, G.A., Petrik, N.G., Dohnalek, Z. and Kay, B.D., Crystalline ice growth on Pt(III): observation of a hydrophobic water monolayer, *Phys. Rev. Lett.*, 2005, **95**, 166102-1–166102-4.
36. Reichardt, Ch., *Solvents and Solvent Effects in Organic Chemistry*, 2nd edn., VCH, Weinheim, 1988.
37. Dudd, L.M., Venardou, E., Garcia-Verdugo, E., Licence, A.J., Blake, A.J., Wilson, C. and Poliakoff, M., Synthesis of benzimidazoles in high-temperature water, *Green Chem.*, 2003, **5**, 187–192.
38. Dimroth, K., Reichardt, Ch., Siepmann, T. and Bohlmann, F., Uber Pyridinium-*n*-phenol-betaine und ihre Verwendung zur Charakterisierung der Polarität von Lösungsmittel, *Liebigs Ann. Chem.*, 1963, **661**, 1–37.
39. Reichardt, Ch. and Schafer, G., Determination of new and corrections of old $E(T)(30)$ values as empirical measures of solvent polarity for 40 organic solvents, *Liebigs Ann. Chem.*, 1995, 1579–1582.

40. Israelachvili, J., *Intermolecular and Surface Forces*, 2nd edn., Academic Press, London, 1992.
41. Dack, M.R.J., Importance of solvent internal-pressure and cohesion to solution phenomena, *Chem. Soc. Rev.*, 1975, **4**, 211–229.
42. Allen, G., Gee, G. and Wilson, G.J., Intermolecular forces and chain flexibilities in polymers. I. Internal pressures and cohesive energy densities of simple liquids, *Polymer*, 1960, **1**, 456–466.
43. Graziano, G., Comment on 'reevaluation in interpretation of hydrophobicity by scaled particle theory', *J. Phys. Chem. B*, 2002, **106**, 7713–7716.
44. Graziano, G., Rate enhancement of Diels–Alder reactions in aqueous solutions, *J. Phys. Org. Chem.*, 2004, **17**, 100–101.
45. Dack, M.R.J., Solvent structure – use of internal-pressure and cohesive energy density to examine contributions to solvent–solvent interactions, *Aust. J. Chem.*, 1975, **28**, 1643–1648.
46. Abraham, M.H., Grellier, P.L., Nasehzadeh, A. and Walker, R.A.C., Substitution at saturated carbon. Part 26. A complete analysis of solvent effects on initial states and transition states for the solvolysis of the *t*-butyl halides in terms of G, H, and S using the unified method, *J. Chem. Soc. Perkin Trans. 2*, 1988, 1717–1725.
47. Blandamer, M.J., Engberts, J.B.F.N., Gleeson, P.T. and Reis, J.C.R., Activity of water in aqueous systems: a frequently neglected property, *Chem. Soc. Rev.*, 2005, **34**, 440–458.
48. Pollack, G.H, *Cells, Gels and the Engines of Life*, Ebner Publishers, Seattle, 2001.
49. Asaad, N. and Engberts, J.B.F.N., Cytosol-mimetic chemistry: kinetics of the trypsin-catalyzed hydrolysis of *p*-nitrophenyl acetate upon addition of polyethylene glycol and *n-tert*-butyl acetoacetamide, *J. Am. Chem. Soc.*, 2003, **125**, 6874–6875.
50. Asaad, N., Den Otter, M.J. and Engberts, J.B.F.N., Aqueous solutions that model the cytosol: studies on polarity, chemical reactivity and enzyme kinetics, *Org. Biomol. Chem.*, 2005, **2**, 1404–1412.
51. Spitzer, J.J. and Poolman, B., Electrochemical structure of the crowded cytoplasm, *Trends Biochem. Sci.*, 2005, **30**, 536–541.
52. Trevors, J.T. and Pollack, G.H., Hypothesis: the origin of life in a hydrogel environment, *Prog. Biophys. Mol. Biol.*, 2005, **89**, 1–8.
53. Graziano, G., On the isotope effect in hydrophobic hydration, *J. Phys. Chem. B*, 2000, **104**, 9249–9254.
54. Bakker, H.J. and Nienhuys, H.-K., Delocalization of protons in liquid water, *Science*, 2002, **297**, 587–590.
55. Agmon, N., The Grotthuss mechanism, *Chem. Phys. Lett.*, 1995, **244**, 456–462.
56. Tuckerman, M.E., Marx, D. and Parrinello, M., The nature and transport mechanism of hydrated hydroxide ions in aqueous solution, *Nature*, 2002, **417**, 925–929.
57. Vassilev, P., Louwerse, M.J. and Baerends, E.J., Hydroxyl radical and hydroxide ion in liquid water: a comparative electron density functional theory study, *J. Phys. Chem. B*, 2006, **109**, 23605–23610.
58. Franks, F., *Water: A Matrix of Life*, 2nd edn., Royal Society of Chemistry, Cambridge, UK, 2000.
59. Weingartner, H. and Franck, E.U., Supercritical water as a solvent, *Angew. Chem. Int. Ed.*, 2005, **44**, 2–22.
60. Krammer, P. and Vogel, H., Hydrolysis of esters in subcritical and supercritical water, *J. Supercrit. Fluids*, 2000, **16**, 189–206.
61. Harano, Y., Sato, H. and Hirata, F., Solvent effects on a Diels–Alder reaction in supercritical water: RISM-SCF study, *J. Am. Chem. Soc.*, 2000, **122**, 2289–2293.
62. Friedman, H.L. and Krishnan, C.V., Thermodynamics of ion hydration, In: Franks, F. (Ed.), *Water, A Comprehensive Treatise*, Vol. 3, 1st edn., Plenum Press, New York, 1973, pp. 1–118.
63. Enderby, J.E. and Neilson, G.W., X-ray and neutron scattering by aqueous solutions of electrolytes, In: Franks, F. (Ed.), *Water: A Comprehensive Treatise*, Vol. 6, 1st edn., Plenum Press, New York, 1979, pp. 1–46.
64. Anslyn, E.V. and Dougherty, D.A., *Modern Physical Organic Chemistry*, University Science Books, Sausalito, CA, 2006.
65. Collins, K.D., Charge density-dependent strength of hydration and biological structure, *Biophys. J.*, 1997, **72**, 65–76.
66. Saielli, G., Scorrano, G., Bagno, A. and Wakisaka, A., Solvation of tetraalkylammonium chlorides in acetonitrile–water mixtures: mass spectrometry and molecular dynamics simulations, *Chem. Phys. Chem.* 2005, **6**, 1307–1315.
67. Gurney, R.W., *Ionic Processes in Solution*, McGraw-Hill, New York, 1953.
68. Haymet, A.D.J., Hydrophobicity, *Curr. Biol.*, 1999, **9**, R81–R82.
69. Graziano, G., Comment on 'the mechanism of hydrophobic solvation depends on solute radius', *J. Phys. Chem. B*, 2001, **105**, 2079–2081.
70. Lee, B., The physical origin of the low solubility of nonpolar solutes in water, *Biopolymers*, 1985, **24**, 813–823.
71. Graziano, G., Cavity thermodynamics and hydrophobicity, *J. Phys. Soc. Jpn.*, 2000, **69**, 1566–1569.
72. Frank, H.S. and Evans, M.W., Free volume and entropy in condensed systems. III. Entropy in binary liquid mixtures; partial molar entropy in dilute solutions; structure and thermodynamics in aqueous electrolytes, *J. Chem. Phys.*, 1945, **13**, 507–532.
73. Soper, A.K., Orientational correlation-function for molecular liquids – the case of liquid water, *J. Chem. Phys.*, 1994, **101**, 6888–6901.

74. Finney, J.L. and Soper, A.K., Solvent structure and perturbations in solutions of chemical and biological importance, *Chem. Soc. Rev.*, 1994, 1–10.
75. Nakahara, M., Wakai, C., Yoshimoto, Y. and Matubayasi, N., Dynamics of hydrophobic hydration of benzene, *J. Phys. Chem.*, 1996, **100**, 1345–1349.
76. Scatena, L.F., Brown, M.G. and Richmond, G.L., Water at hydrophobic surfaces: weak hydrogen bonding and strong orientational effects, *Science*, 2001, **292**, 908–912.
77. Guillot, B., Guissani, Y. and Bratos, S., A computer-simulation study of hydrophobic hydration of rare gases and methane. I. Thermodynamic and structural properties, *J. Chem. Phys.*, 1991, **95**, 3643–3648.
78. Southall, N.T., Dill, K.A. and Haymet, A.D.J., A view of the hydrophobic effect, *J. Phys. Chem. B*, 2002, **106**, 521–533.
79. Soper, A.K., Dougan, L., Crain, J. and Finney, J.L., Excess entropy in alcohol–water solutions: a simple clustering explanation, *J. Phys. Chem. B*, 2006, **110**, 3472–3476.
80. Buchanan, P., Aldiwan, N., Soper, A.K., Creek, J.L. and Koh, C.A., Decreased structure on dissolving methane in water, *Chem. Phys. Lett.*, 2005, **415**, 89–93.
81. Xu, H. and Dill, K.A., Water's hydrogen bonds in the hydrophobic effect: a simple model, *J. Phys. Chem. B*, 2006, **110**, 4963–4970.
82. Chatterjee, S., Ashbaugh, H.S. and Debenedetti, P.G., Effects of nonpolar solutes on the thermodynamic response functions of aqueous mixtures, *J. Chem. Phys.*, 2005, **123**, 164503-1–164503-8.
83. Bowron, D.T., Filipponi, A., Roberts, M.A. and Finney, J.L., Hydrophobic hydration and the formation of a clathrate hydrate, *Phys. Rev. Lett.*, 1998, **81**, 4164–4167.
84. Lum, K., Chandler, D. and Weeks, J.D., Hydrophobicity at small and large length scales, *J. Phys. Chem. B*, 1999, **103**, 4570–4577.
85. Cheng, Y.-K. and Rossky, P.J., Surface topography dependence of biomolecular hydrophobic hydration, *Nature*, 1998, **392**, 696–699.
86. Pierotti, R.A., Aqueous solutions of nonpolar gases, *J. Phys. Chem.*, 1965, **69**, 281–288.
87. Blokzijl, W. and Engberts, J.B.F.N., Hydrophobic effects. Opinions and facts, *Angew. Chem. Int. Ed.*, 1993, **32**, 1545–1579.
88. Guo, J.-H., Luo, Y., Augustsson, A., Kashtanov, S., Rubensson, J.-E., Shuh, D.K., Agren, H. and Nordgren, J., Molecular structure of alcohol–water mixtures, *Phys. Rev. Lett.*, 2003, **91**, 157401-1–157401-4.
89. Meyer, E.A., Castellano, R.K. and Diederich, F., Interactions with aromatic rings in chemical and biological recognition, *Angew. Chem. Int. Ed.*, 2003, **42**, 1210–1250.
90. Raschke, T.M. and Levitt, M., Nonpolar solutes enhance water structure within hydration shells while reducing interactions between them, *Proc. Natl. Acad. Sci. U. S. A.*, 2005, **102**, 6777–6782.
91. Dorairaj, S., Jeon, J. and Kim, H.J., Solubility of water in a benzene–cyclohexane mixture, *J. Phys. Chem. B*, 2006, **110**, 429–435.
92. Anderson, B.J., Tester, J.W., Borghi, G.P. and Trout, B.L., Properties of inhibitors of methane hydrate formation via molecular dynamics simulations, *J. Am. Chem. Soc.*, 2005, **127**, 17852–17862.
93. Kauzmann, W., In: McElroy, W. and Glass, B. (Eds.), *The Mechanism of Enzyme Action*, 1st edn., The Johns Hopkins University Press, Baltimore, 1954, pp. 70–110.
94. Herman, R.B., Theory of hydrophobic bonding. II. The correlation of hydrocarbon solubility in water with solvent cavity surface area, *J. Phys. Chem.*, 1972, **76**, 2754–2759.
95. Lazardis, T. and Karplus, M., Effective energy functions for protein structure prediction, *Curr. Opin. Struct. Biol.*, 2000, **10**, 139–145.
96. Graziano, G., On the heat capacity change of pairwise hydrophobic interactions, *J. Chem. Phys.*, 2005, **123**, 34509-1–34509-4.
97. Jorgensen, W.L., Gao, J. and Ravimohan, C., Monte Carlo simulations of alkanes in water: hydration numbers and the hydrophobic effect, *J. Phys. Chem.*, 1985, **89**, 3470–3473.
98. Dec, S.F., Bowler, K.E., Stadterman, L.L., Koh, C.A. and Sloan, E.D., Direct measure of the hydration number of aqueous methane, *J. Am. Chem. Soc.*, 2006, **128**, 414–415.
99. Southall, N.T. and Dill, K.A., Potential of mean force between two hydrophobic solutes in water, *Biophys. Chem.*, 2002, **101–102**, 295–307.
100. Talhout, R. and Engberts, J.B.F.N., Probing the effect of the amidinium group and the phenyl ring on the thermodynamics of binding of benzamidinium chloride to trypsin, *Org. Biomol. Chem.*, 2004, **2**, 3071–3074.
101. Barrat, E., Bingham, R.J., Warner, D.J., Laughton, C.A., Phillips, S.E.V. and Homans, S.W., Van der Waals interactions dominate ligand–protein association in a protein binding site occluded from solvent water, *J. Am. Chem. Soc.*, 2005, **127**, 11827–11834.
102. Homans, S.W., Probing the binding entropy of ligand–protein interactions by NMR, *Chem. Biol. Chem.*, 2006, **6**, 1585–1591.
103. Otto, S. and Engberts, J.B.F.N., Hydrophobic interactions and chemical reactivity, *Org. Biomol. Chem.*, 2003, **1**, 2809–2820.

104. Jorgensen, W.L., Buckner, J.K., Boudon, S. and Tirado-Rives, J., Efficient computation of absolute free energies of binding by computer simulations. Application to the methane dimer in water, *J. Chem. Phys.*, 1988, **89**, 3742–3746.
105. Engberts, J.B.F.N. and Blandamer, M.J., Organic reactions in water, *Chem. Commun.*, 2001, 1701–1708.
106. Buurma, N.J., Pastorello, L., Blandamer, M.J. and Engberts, J.B.F.N., Kinetic evidence for hydrophobically stabilized encounter complexes formed by hydrophobic esters in aqueous solutions containing monohydric alcohols, *J. Am. Chem. Soc.*, 2001, **123**, 11848–11853.
107. Blokzijl, W. and Engberts, J.B.F.N., Diels–Alder reactions in aqueous solution. Enforced hydrophobic interactions between diene and dienophile, *J. Am. Chem. Soc.*, 1991, **113**, 4241–4246.
108. Blokzijl, W. and Engberts, J.B.F.N., Initial-state and transition-state effects on Diels–Alder reactions in water and mixed aqueous solvents, *J. Am. Chem. Soc.*, 1992, **114**, 5440–5442.
109. Chandrasekhar, J., Shariffskul, S. and Jorgensen, W.L., QM/MM simulations for Diels–Alder reactions in water: contribution of enhanced hydrogen bonding at the transition state to the solvent effect, *J. Phys. Chem. B*, 2002, **106**, 8078–8085.
110. Jorgensen, W.L. and Buckner, J.K., Use of statistical perturbation theory for computing solvent effects on molecular conformation: butane in water, *J. Phys. Chem.*, 1987, **91**, 6083–6085.
111. Gardner, R.R., Christianson, L.A. and Gellman, S.H., Quantitative analysis of hydrophobically induced folding in a minimal model system, *J. Am. Chem. Soc.*, 1997, **119**, 5041–5042.
112. Deschenes, L.A., Zilaro, P., Muller, L.J., Fourkas, J.T. and Mohanty, U., Quantitative measure of hydrophobicity: experiment and theory, *J. Phys. Chem. B*, 1997, **101**, 5777–5779.
113. Deschenes, L.A., Barrett, J., Muller, L.J., Fourkas, J.T. and Mohanty, U., Inhibition of bubble coalescence in aqueous solutions. 1. Electrolytes, *J. Phys. Chem. B*, 1998, **102**, 5115–5119.
114. Palecz, B., Enthalpic pair interaction coefficient between zwitterions of L-alpha-amino acids and urea molecule as a hydrophobicity parameter of amino acid side chains, *J. Am. Chem. Soc.*, 2005, **127**, 17768–17771.
115. San Biagio, P.L., Bulone, D., Martorana, V., Palma-Vittorelli, M.B. and Palma, M.U., Physics and biophysics of solvent induced forces: hydrophobic interactions and context-dependent hydration, *Eur. Biophys. J.*, 1998, **27**, 183–196.
116. Apperloo, J., Streefland, L., Engberts, J.B.F.N. and Blandamer, M.J., Pairwise Gibbs energy of interaction involving *N*-alkyl-2-pyrrolidinones and related compounds in aqueous solution obtained from kinetic medium effects, *J. Org. Chem.* 2000, **65**, 411–418.
117. Hol, P., Streefland, L., Blandamer, M.J. and Engberts, J.B.F.N., Kinetic medium effects of cationic cosolutes in aqueous solution: the effects of alkylammonium bromides on the neutral hydrolysis of 1-benzoyl-1,2,4-triazole, *J. Chem. Soc. Perkin Trans. 2*, 1997, 485–488.
118. Noordman, W., Blokzijl, W., Engberts, J.B.F.N. and Blandamer, M.J., Kinetic medium effects of amphiphilic cosolutes below their critical micelle concentration: the effect of sodium *n*-alkyl sulfates on the neutral hydrolysis of 1-benzoyl-1,2,4-triazole, *J. Org. Chem.*, 1993, **58**, 7111–7114.
119. Tasker, I.R. and Wood, R.H., Enthalpies of dilution of aqueous systems containing hexamethylenetetramine and other nonelectrolytes, *J. Solut. Chem.*, 1982, **11**, 729–747.
120. Spitzer, J.J., Suri, S.K. and Wood, R.H., Freezing temperatures of aqueous solutions of mixtures of amides and polyols. Gibbs energy of interaction of the amide group with the hydroxyl group in aqueous solution, *J. Solut. Chem.*, 1985, **14**, 561–569.
121. Blokzijl, W., Engberts, J.B.F.N. and Blandamer, M.J., Quantitative analysis of solvent effects in highly aqueous media. Application of the SWAG procedures and a critical appraisal of the additivity principle, *J. Am. Chem. Soc.*, 1990, **112**, 1197–1201.
122. Kerstholt, R.P.V., Engberts, J.B.F.N. and Blandamer, M.J., Kinetics of uncatalysed hydrolysis of 1-benzoyl-3-phenyl-1,2,4-triazole and *p*-methoxyphenyl dichloroethanoate in aqueous solutions containing ureas, carboxamides, sulfonamides, sulfones and sulfoxides, *J. Chem. Soc. Perkin Trans. 2*, 1993, 49–51.
123. Savage, J.J. and Wood, R.H., Enthalpy of dilution of aqueous mixtures of amides, sugars, ureas, ethylene glycol, and pentaerythritol at 25°C: enthalpy of interaction of the hydrocarbon, amide, and hydroxyl functional groups in dilute aqueous solutions, *J. Solut. Chem.*, 1976, **5**, 733–750.
124. Suri, S.K. and Wood, R.H., Freezing temperatures of aqueous solutions of methyl formate and ethyl acetate with a variety of nonelectrolytes. Gibbs energy of interaction of the ester group with other functional groups, *J. Solut. Chem.*, 1986, **15**, 705–715.
125. Galema, S.A., Howard, E., Engberts, J.B.F.N. and Grigera, J.R., The effect of stereochemistry upon carbohydrate hydration. A molecular dynamics simulation of beta-D-galactopyranose and (alpha,beta)-D-talopyranose, *Carbohydr. Res.*, 1994, **265**, 215–225.
126. Jorgensen, W.L., Free energy calculations: a breakthrough for modelling organic chemistry in solution, *Acc. Chem. Res.*, 1989, **22**, 184–189.
127. Ingold, C.K., *Structure and Mechanism in Organic Chemistry*, Cornell University Press, Ithaca, NY, 1953.
128. Abraham, M.H., Solvent effects on reaction rates, *Pure Appl. Chem.*, 1985, **57**, 1055–1064.

129. Furlani, T.R. and Gao, J., Hydrophobic and hydrogen-bonding effects on the rate of Diels–Alder reactions in aqueous solution, *J. Org. Chem.*, 1996, **61**, 5492–5497.
130. Rispens, T. and Engberts, J.B.F.N., Cycloadditions in mixed aqueous solvents: the role of the water concentration, *J. Phys. Org. Chem.*, 2005, **18**, 725–736.
131. Butler, R.N., Cunningham, W.J., Coyne, A.G. and Burke, L.A., The influence of water on the rates of 1,3-dipolar cycloaddition reactions: trigger points for exponential rate increases in water–organic solvent mixtures. Water-super versus water-normal dipolarophiles, *J. Am. Chem. Soc.*, 2004, **126**, 11923–11929.
132. Holterman, H.A.J. and Engberts, J.B.F.N., Heat capacities of activation for the neutral hydrolysis of two acyl-activated esters in water-rich 2-*n*-butoxyethanol–water mixtures. Analysis in terms of the pseudo-phase-separation model, *J. Org. Chem.*, 1983, **48**, 4025–4030.
133. Blokzijl, W., Engberts, J.B.F.N. Jager, J.B. and Blandamer, M.J., Description of solvent dependence of rate constants in terms of pairwise group Gibbs function interaction parameters. Medium effects for hydrolysis of *p*-methoxyphenyl dichloroacetate in aqueous solutions containing urea and alkyl-substituted ureas, *J. Phys. Chem.*, 1987, **91**, 6022–6027.
134. Engberts, J.B.F.N. and Blandamer, M.J., Reactant–solute encounters in aqueous solutions studied by kinetic methods: hydration cosphere overlap and camouflage effects, *J. Phys. Org. Chem.*, 1998, **11**, 841–846.
135. Benak, H., Engberts, J.B.F.N. and Blandamer, M.J., Kinetics of neutral hydrolysis of 1-benzoyl-3-phenyl-1,2,4-triazole in highly aqueous media: analysis of effects of added ethanol and propan-1-ol in terms of pairwise group interaction parameters, *J. Chem. Soc. Perkin Trans. 2*, 1992, 2035–2038.
136. Rispens, T., Lensink, M.F., Berendsen, H.J.C. and Engberts, J.B.F.N., Molecular dynamics simulation of the effect of hydrophobic cosolutes on the neutral hydrolysis of an activated ester, *J. Phys. Chem. B*, 2004, **108**, 5483–5488.
137. Grate, J.W., McGill, R.A. and Hilvert, D., Analysis of solvent effects on the decarboxylation of benzisoxazole-3-carboxylate ions using linear solvation energy relationships: relevance to catalysis in an antibody binding site, *J. Am. Chem. Soc.*, 1993, **115**, 8577–8584.
138. Jiang, X.-K., Fan, W.-Q. and Hui, Y.-Z., The effect of hydrophobic–lipophilic interactions on chemical reactivity. 4. A case of 17-membered 'neighboring group' participation: compelling evidence for self-coiling, *J. Am. Chem. Soc.*, 1984, **106**, 7202–7205.
139. Fan, W.-Q. and Jiang, X.-K., Effect of hydrophobic–lipophilic interactions on chemical reactivity. 6. The first example of correlation of hydrophobic–lipophilic substituent constants with chemical reactivity for a simple substrate–solvent system, *J. Am. Chem. Soc.*, 1985, **107**, 7680–7684.
140. Jiang, X.-K., Hui, Y.-Z. and Fei, Z.-X., A successful application of the concept of making use of hydrophobic forces to prepare large-ring compounds, *J. Chem. Soc., Chem. Commun.*, 1988, 689–690.
141. Rodrigues, F., Canac, Y. and Lubineau, A., A convenient, one-step, synthesis of beta-*C*-glycoside ketones in aqueous media, *Chem. Commun.*, 2000, 2049–2050.
142. Peters, S., Lichtenthaler, F. and Lindner, H., A 2-*C*-fructosyl-propanone locked in a 2,7-dioxabicyclo [3.2.1]octane framework, *Tetrahedron: Asymmetry*, 2003, **14**, 2475–2479.
143. Bhanage, B.M., Ikushima, Y., Shirai, M. and Arai, M., Multiphase catalysis using water-soluble metal complexes in supercritical carbon dioxide, *Chem. Commun.*, 1999, 1277–1278.
144. Jacobson, G.B., Lee, C.T., Johnston, K.P. and Tumas, W., Enhanced catalyst reactivity and separations using water/carbon dioxide emulsions, *J. Am. Chem. Soc.*, 1999, **121**, 11902–11903.
145. Syamala, M.S. and Ramamurthy, V., Consequences of hydrophobic association in photoreactions: photodimerization of stilbenes in water, *J. Org. Chem.*, 1986, **51**, 3712–3715.
146. Norgaard, K. and Bjornholm, T., Supramolecular chemistry on water – towards self-assembling molecular electronic circuitry, *Chem. Commun.*, 2005, 1812–1823.
147. Pratt, L.R. and Pohorille, A., Hydrophobic effects and modeling of biophysical aqueous solution interfaces, *Chem. Rev.*, 2002, **102**, 2671–2692.
148. Richmond, G.L., Molecular bonding and interactions at aqueous surfaces as probed by vibrational sum frequency spectroscopy, *Chem. Rev.*, 2002, **102**, 2693–2724.
149. Watry, M.R., Tarbuck, T.L. and Richmond, G.L., Vibrational sum-frequency studies of a series of phospholipid monolayers and the associated water structure at the vapor/water interface, *J. Phys. Chem. B*, 2003, **107**, 512–518.
150. Raymond, E.A. and Richmond, G.L., Probing the molecular structure and bonding of the surface of aqueous salt solutions, *J. Phys. Chem. B*, 2004, **108**, 5051–5059.
151. Pashley, R.M., Rzechowicz, M., Pashley, L.R. and Francis, M.J., De-gassed water is a better cleaning agent, *J. Phys. Chem. B*, 2005, **109**, 1231–1238.
152. Beattie, J.K. and Djerdjev, A.M., The pristine oil/water interface: surfactant-free hydroxide-charged emulsions, *Angew. Chem. Int. Ed.*, 2004, **43**, 3568–3571.
153. Zangi, R. and Engberts, J.B.F.N., Physisorption of hydroxide ions from aqueous solution to a hydrophobic surface, *J. Am. Chem. Soc.*, 2005, **127**, 2272–2276.
154. Takahashi, M., Zeta potential of microbubbles in aqueous solutions: electrical properties of the gas–water interface, *J. Phys. Chem. B*, 2005, **109**, 21858–21864.

Chapter 3

Acid Catalysis in Water

Chikako Ogawa and Shū Kobayashi

Water is a beautiful solvent in many ways, and performing organic reactions in this medium is now of great interest. It is no doubt the most inexpensive among various solvents used in organic synthesis. The lack of inflammable, explosive, mutagenic, and carcinogenic properties is a favorable aspect of water in academic laboratories as well as in industry. Furthermore, water is now regarded as one of the most suitable solvents from an environmental point of view.

In addition to these advantages, there are several synthetic uses of water as a solvent. First, simple operation systems can be attained by the use of water. That is, workup procedures can be simplified because many organic compounds are lipophilic and may be separated easily from the aqueous phase. Second, control of reaction temperature is easier because the heat capacity of water is extremely high compared with those of most organic solvents. Third, the need for protective groups is reduced in water as amino acids, carbohydrates, and other water-soluble materials can be used as they are. Finally, the unique solvent effects of water are expected to influence the course of many reactions in water.

In this chapter, we focus on acid catalysis in water.[1] While there are numerous examples of catalytic reactions in water, the main body of these involves acid catalysis. Homogeneous catalysis, heterogeneous catalysis, and micellar catalysis, including catalytic enantioselective reactions, are discussed in detail. Acid-catalyzed reactions using a small amount of water may not be included unless they are crucial for the further development of the field.

3.1 Homogeneous catalysis

3.1.1 Brønsted acid catalysis

While hundreds of Lewis acid catalysts have been developed for organic reactions, Brønsted acid catalysts have been paid less attention until recently. Conventional Lewis acids, such as titanium chloride and aluminium chloride, are known to be incompatible with aqueous media. On the other hand, Brønsted acids are stable toward water and oxygen. Thus, they are potential candidates as activators of electrophilic substrates in water. It was found that among various Brønsted acids, HBF_4 efficiently catalyzed Mannich-type reactions of silyl enol ethers with aromatic aldehydes derived from activated imines to afford the corresponding β-amino ketones (Scheme 3.1).[2]

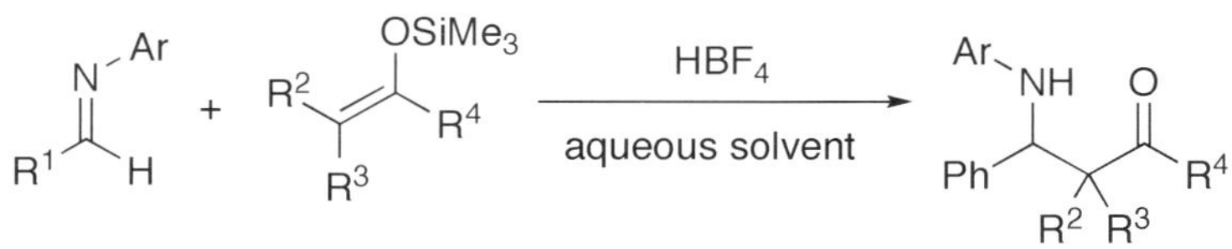

Scheme 3.1

Interesting chemoselectivity was observed in this addition reaction. HBF_4-catalyzed addition reaction selectively proceeded toward an aldimine in the presence of an aldehyde (Scheme 3.2). In general, common Lewis acids except for some lanthanide triflates[3] or transition metals[4] activate aldehydes rather than aldimines preferentially. The high chemoselectivity was realized because the more basic nitrogen was activated more effectively by HBF_4 than the carbonyl oxygen.

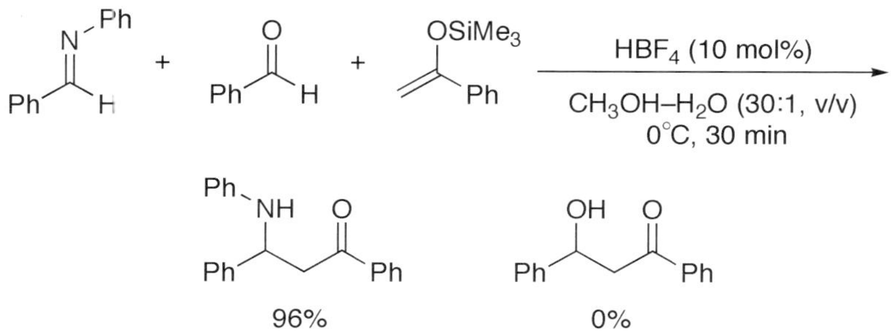

Scheme 3.2

Moreover, the use of a surfactant enabled the reaction to proceed smoothly in water (Scheme 3.3).

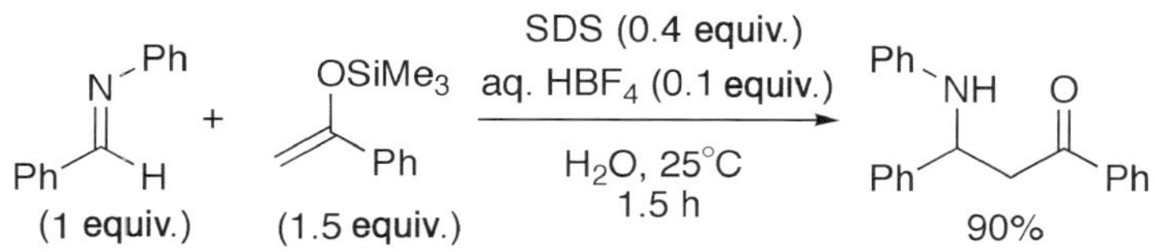

Scheme 3.3

By *in situ* preparation of imines, even aliphatic aldehyde-derived imines, which are generally known to be difficult to isolate, reacted with a ketene silyl acetal in water in the presence of HBF_4 and sodium dodecyl sulfate (SDS) to afford the corresponding adducts in good yields (Scheme 3.4).

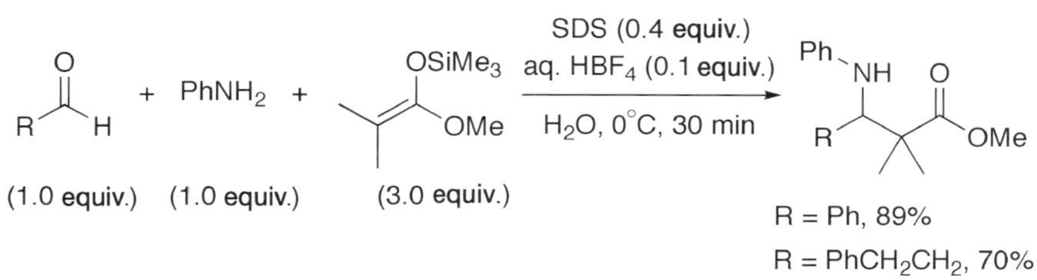

Scheme 3.4

Hetero Diels–Alder reaction of Danishefsky's diene with imines is a powerful method for the preparation of dihydropyridone derivatives. HBF_4 also catalyzed this reaction using aromatic and aliphatic imines that were prepared *in situ* from the corresponding aromatic or aliphatic aldehydes and aromatic amines (Scheme 3.5).

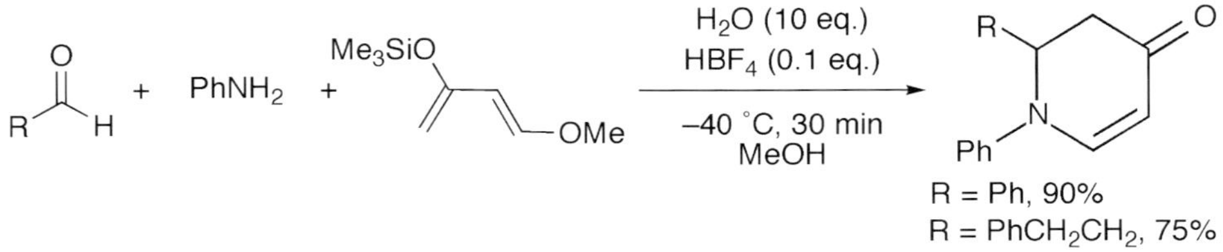

Scheme 3.5

Besides, this reaction could be conducted without any cosolvent in the presence of SDS. Further reports related to Brønsted acid and surfactant systems are summarized in Section 3.3.2.

3.1.2 Lewis acid catalysis

Lewis acid catalysis is one of the most useful methods in modern organic synthesis. However, many of the common Lewis acids are highly water-labile and have been used in organic synthesis under strictly anhydrous conditions. Contrary to this, it was found that lanthanide triflates catalyzed aldol reactions of formaldehyde (Scheme 3.6).[5] Formaldehyde is one of the most highly reactive C1 electrophiles. In this reaction, not gaseous formaldehyde but a commercially available aqueous solution was used as the formaldehyde source. This invaluable find introduced the concept of Lewis acid catalysis in aqueous media to many chemists. Later, it was also reported that other aldol reactions, with a variety of aldehydes and silyl enol ethers, as well as allylation reactions, proceeded smoothly in aqueous media to afford the desired compounds in high yields.[6]

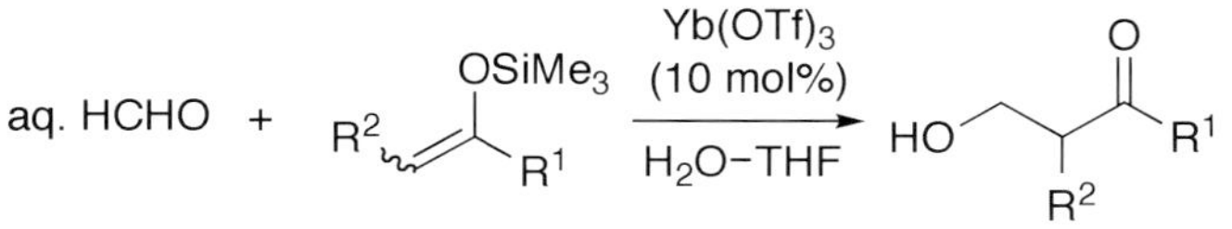

Scheme 3.6

Ytterbium triflate was shown to catalyze Diels–Alder reactions (Scheme 3.7). After the reaction, the catalyst was quantitatively recovered from the aqueous layer and reused several times without loss of activity.[7]

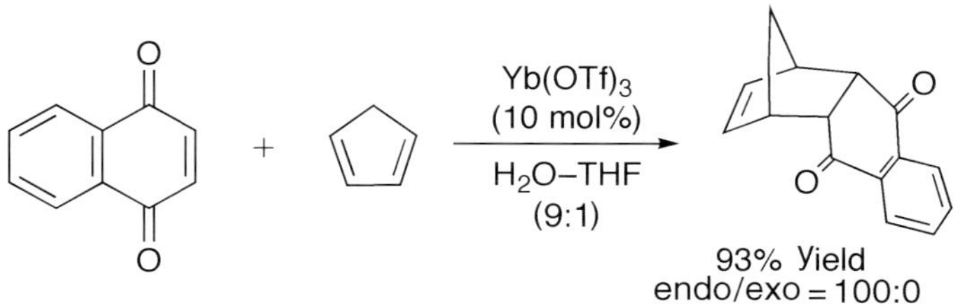

Scheme 3.7

Mannich-type reactions are among the most difficult reactions in terms of activation with a Lewis acid catalyst because of the basicity of the reactant and a product that will coordinate to the catalyst, leading to deactivation. Nevertheless, in the presence of a catalytic amount of $Yb(OTf)_3$, three component reactions with aldehydes, amines, and vinyl ethers reacted efficiently in H_2O–THF (1:9) to afford the desired Mannich-type adducts (Scheme 3.8).

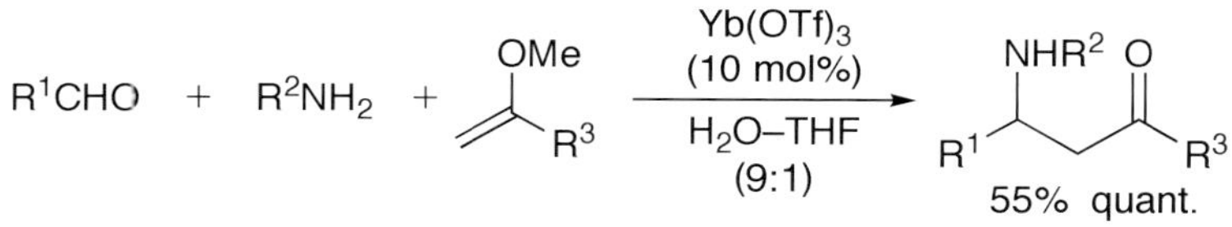

Scheme 3.8

In addition to rare earth triflates, copper triflate was also found to be a stable Lewis acid in aqueous media.[8] In a mixed aqueous solvent system (H_2O–EtOH–toluene = 1:7:4), allylation of various aldehydes with tetraallyltin and aldol reactions with silyl enol ethers proceeded smoothly in the presence of $Cu(OTf)_2$ (20 mol%) to give homoallylic alcohols and aldol adducts, respectively, in high yields (Schemes 3.9 and 3.10).

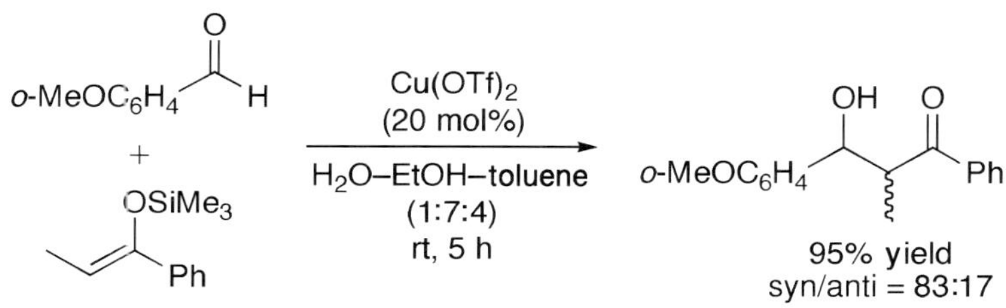

Scheme 3.9

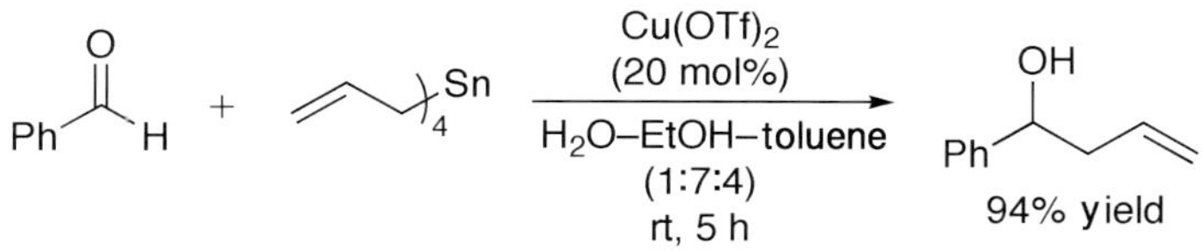

Scheme 3.10

It was also demonstrated that nitrogen-containing compounds such as acyl hydrazones and imines were activated by $Sc(OTf)_3$ and AgOTf in aqueous media (Schemes 3.11 and 3.12).[9,10]

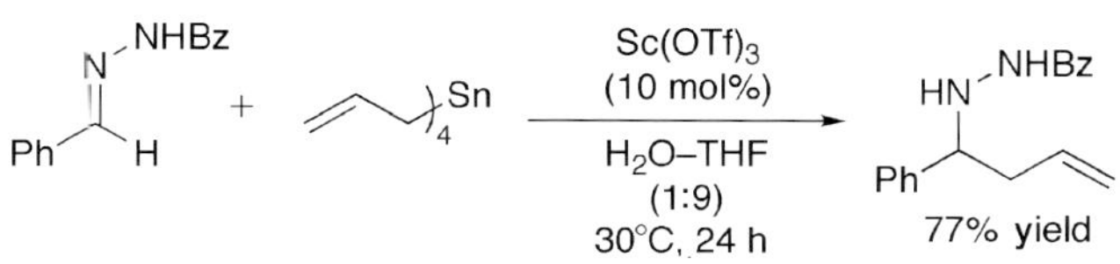

Scheme 3.11

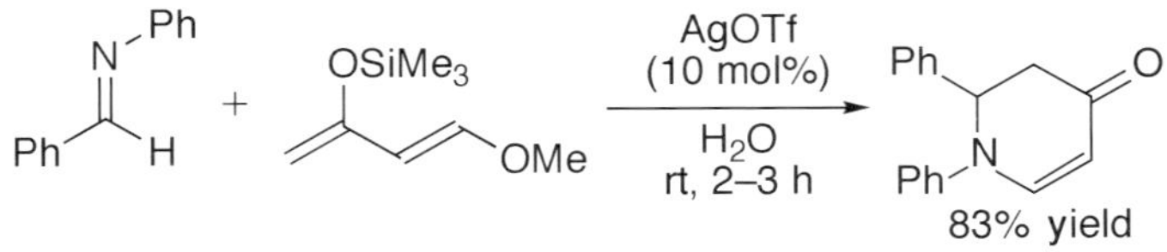

Scheme 3.12

As mentioned above, rare earth triflates, $Cu(OTf)_2$, and AgOTf were found to act as Lewis acids in aqueous media. It would be of great importance to know which factors are key to the success of organic reactions in water. To address this issue, 1–15 metal chlorides, perchlorates, and triflates were screened in the aldol reaction of benzaldehyde with the silyl enol ether in water–THF (1:9) (Scheme 3.13).[11] This screening revealed that not only Sc(III), Y(III), and Ln(III) but also Fe(II), Cu(II), Zn(II), Cd(II), and Pb(II) worked as Lewis acids in this medium to afford the desired aldol adducts in high yields.

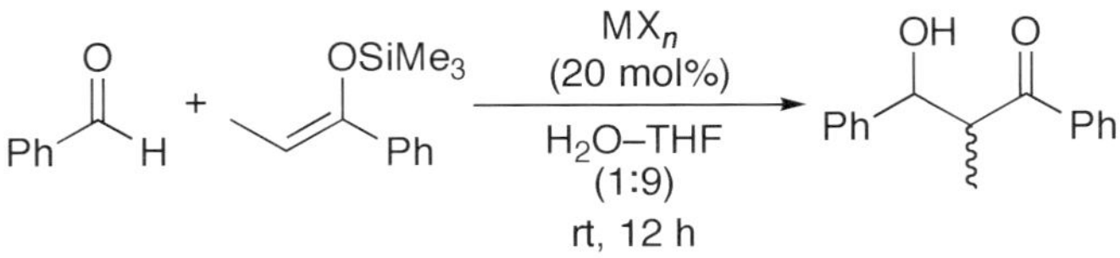

Scheme 3.13

From these results, a correlation was found between the catalytic activity of the metal cations and two of their property constants: the hydrolysis constant (K_h) and the exchange rate constant for substitution of inner-sphere water ligands (water-exchange rate constant (WERC)). Table 3.1 lists these constants for each metal cation. Metals that exhibited good catalytic activity in the screening are surrounded by squares outlined in black. These active metal compounds were found to have pK_h values between about 4 (4.3 for Sc(III)) and 10 (10.08 for Cd(II)) and WERC values greater than $3.2 \times 10^6\ \text{M}^{-1}\text{s}^{-1}$. Cations are generally difficult to hydrolyze when their pK_h values are large. In the case that pK_h values are less than 4, cations are readily hydrolyzed to produce certain amounts of protons. Under these conditions, the silyl enol ether decomposes rapidly. On the other hand, in the case that pK_h values are larger than 10, the Lewis acidities of the cations are too weak to catalyze the aldol reaction. Large WERC values may be necessary to secure fast exchange between hydrating water molecules and an aldehyde that must coordinate to the metal cation to be activated. 'Borderline' elements such as Mn(II), Ag(I), and In(III), whose pK_h and WERC values are close to the criteria limits, gave the aldol adducts in moderate yields. Whereas the precise activity as Lewis acids in aqueous media cannot be quantitatively predicted by pK_h and WERC values, these results have shown the possibility of using several promising metal compounds as water-compatible Lewis acid catalysts.

3.1.3 Asymmetric catalysis

Diels–Alder reaction

Compared to studies of enantioselective Diels–Alder reactions in organic solvents, there are few reports in water. The first reported enantioselective Diels–Alder reaction in water used

Table 3.1 Hydrolysis constants[a] and exchange rate constants for subsitution of innex-sphere water ligands[b]

Key: M^{+n} / pK_h [a] / WERC[b]

1	2	3	4	5	6	7	8	9	10	11	12	13	14	15
Li^{+1} 13.64 **4.7×10^7**	Be — —											**B^{+3}** — —	C — —	N — —
Na^{+1} 14.18 **1.9×10^8**	**Mg^{+2}** 11.44 5.3×10^5											**Al^{+3}** 4.97 1.6×10^0	**Si^{+4}** — —	**P^{+5}** — —
K^{+1} 14.46 **1.5×10^8**	**Ca^{+2}** 12.85 **5×10^7**	**Sc^{+3}** **4.3** **4.8×10^7**	**Ti^{+4}** ≤2.3 —	**V^{+3}** 2.26 1×10^3	**Cr^{+3}** 4.0 5.8×10^{-7}	**Mn^{+2}** **10.59** **3.1×10^7**	**Fe^{+2}** **9.5** **3.2×10^6**	**Co^{+2}** **9.65** 2×10^5	**Ni^{+2}** **9.86** 2.7×10^4	**Cu^{+2}** **7.53** **2×10^8**	**Zn^{+2}** **8.96** **5×10^8**	**Ga^{+3}** 2.6 7.6×10^2	**Ge^{+4}** — —	As — —
Rb — —	Sr — —	**Y^{+3}** **7.7** **1.3×10^7**	**Zr^{+4}** 0.22 —	**Nb^{+5}** (0.6) —	**Mo^{+5}** — —	Tc — —	**Ru^{+3}** — —	**Rh^{+3}** 3.4 3×10^{-8}	**Pd^{+2}** 2.3 —	**Ag^{+1}** 12 **$>5\times10^6$**	**Cd^{+2}** **10.08** **$>1\times10^8$**	**In^{+3}** 4.00 4.0×10^4	**Sn^{+4}** — —	**Sb^{+5}** — —
Cs — —	**Ba^{+2}** 13.47 **$>6\times10^7$**	**Ln^{+3}** **7.6-8.5** **10^6–10^8**	**Hf^{+4}** 0.25 —	**Ta^{+5}** (−1) —	**W^{+6}** — —	**Re^{+5}** — —	**Os^{+3}** — —	**Ir^{+3}** — —	**Pt^{+2}** 4.8 —	**Au^{+1}** — —	**Hg^{+2}** 3.40 **2×10^9**	**Tl^{+3}** 0.62 7×10^5	**Pb^{+2}** **7.71** **7.5×10^9**	**Bi^{+3}** 1.09 —

La^{+3} **8.5** **2.1×10^8**	**Ce^{+3}** **8.3** **2.7×10^8**	**Pr^{+3}** **8.1** **3.1×10^8**	**Nd^{+3}** **8.0** **3.9×10^8**	Pm — —	**Sm^{+3}** **7.9** **5.9×10^8**	**Eu^{+3}** **7.8** **6.5×10^8**	**Gd^{+3}** **8.0** **6.3×10^7**	**Tb^{+3}** **7.9** **7.8×10^7**	**Dy^{+3}** **8.0** **6.3×10^7**	**Ho^{+3}** **8.0** **6.1×10^7**	**Er^{+3}** **7.9** **1.4×10^8**	**Tm^{+3}** **7.7** **6.4×10^6**	**Yb^{+3}** **7.7** **8×10^7**	**Lu^{+3}** **7.6** **6×10^7**

[a] $pK_h = -\log K_h$.

[b] Exchange rate constants for substitution of inner-sphere water ligands.

the combination of a copper salt and an amino acid, especially *N*-α-methyl-L-tryptophan (L-abrine) (Scheme 3.14).[12]

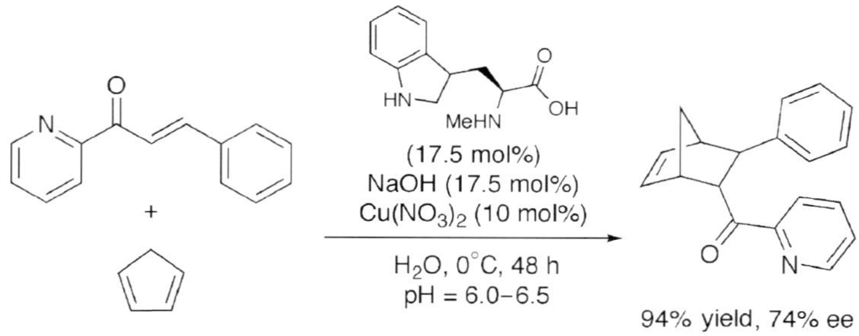

Scheme 3.14

The pyridine and the carbonyl moieties of the substrate are crucial for chiral induction. It was assumed that the copper cation coordinates strongly to the metal in a bidentate fashion.

Arene–arene interactions between the pyridine part of the substrate and L-abrine, which are enforced in water through the hydrophobic effect, also have beneficial effects on reaction rate and enantioselectivity (Table 3.2).

Michael reaction

Michael reactions provide synthetically useful 1,5-carbonyl and related compounds. While several excellent chiral catalysts for these reactions have been developed in organic solvents, examples in water are limited. It was revealed that the combination of BINAP derivatives and silver salts made good catalysts for asymmetric Michael reactions in water (Scheme 3.15). It should be noted that high selectivity was attained without using any organic cosolvents.[13]

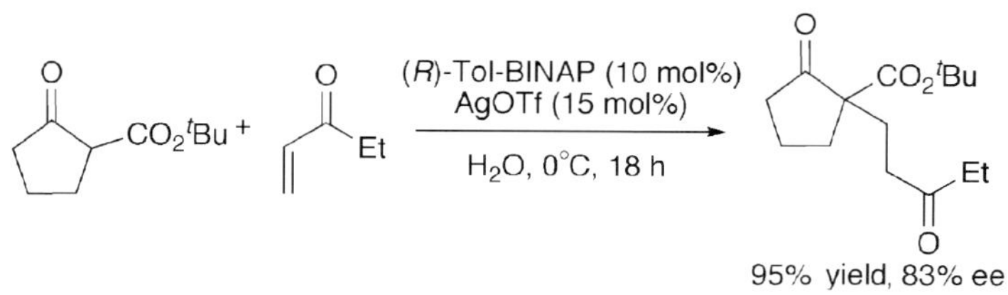

Scheme 3.15

Table 3.2 Effect of solvents

Medium	ee (%)
Acetonitrile	17
THF	24
Ethanol	39
Chloroform	44
Water	74

Addition to carbonyl compounds

Catalytic asymmetric aldol reactions have emerged as one of the most powerful carbon–carbon-bond-forming processes affording synthetically useful β-hydroxy ketones and esters in optically active form.[14] Although several successful examples of chiral Lewis acid-catalyzed reactions of silyl enol ethers with aldehydes have been developed in the past decade, most of them require aprotic, strict anhydrous solvents such as dichloromethane, toluene, and propionitrile at low temperature. In 1999, asymmetric aldol reactions in aqueous media using a copper triflate–bis(oxazoline) complex were reported. It should be noted that the reaction proceeded under mild conditions in this system (Scheme 3.16).[15] Moreover, asymmetric aldol reactions in water without using organic cosolvents were conducted with the aid of a Lewis acid surfactant-combined catalyst (Scheme 3.17).[16]

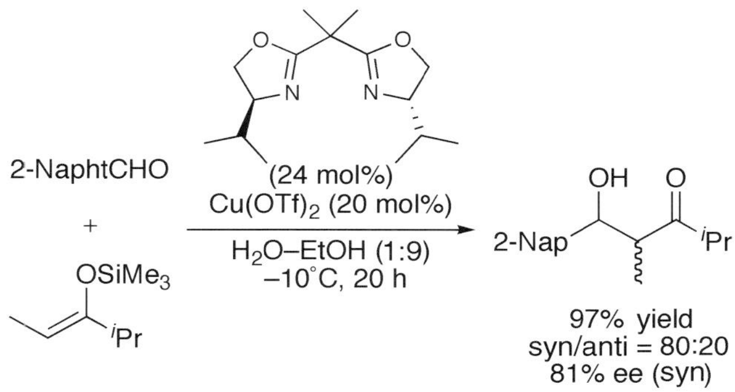

Scheme 3.16

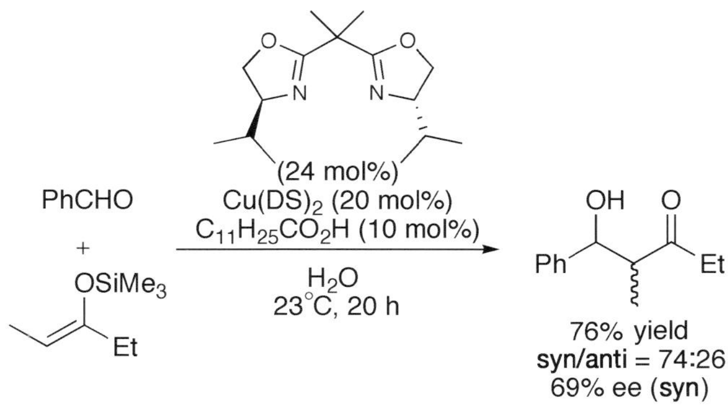

Scheme 3.17

As noted in Table 3.1, various cations were screened in an aqueous aldol reaction, and it was found that some cations such as Fe^{2+}, Cu^{2+}, Zn^{2+}, Cd^{2+}, Pb^{2+} as well as rare earth metal cations were water-compatible and worked as Lewis acid catalysts in water. On the other hand, even if metal cations are water-compatible, chiral ligand-coordinated metal complexes are decomposed in water, and this decomposition must be suppressed to realize asymmetric reactions successfully in aqueous media. An effective strategy to utilize multicoordination systems was demonstrated in order to address this issue. Combinations of metal cations and

Table 3.3 Effect of metals and ligands

MX_n	Ligand	Yield (%)	syn/anti	ee (%)
$Zn(OTf)_2$	**1**	88	69/31	2
$Cu(OTf)_2$	**1**	86	87/13	0
$Sc(OTf)_3$	**2**	75	52/48	1
$Yb(OTf)_3$	**2**	74	63/37	1
AgOTf	**3**	61	75/25	5
$Pb(OTf)_2$	**3**	62	90/10	55
$Pb(OTf)_2$	**1**	78	89/11	0
$Pb(OTf)_2$	**2**	92	89/11	0

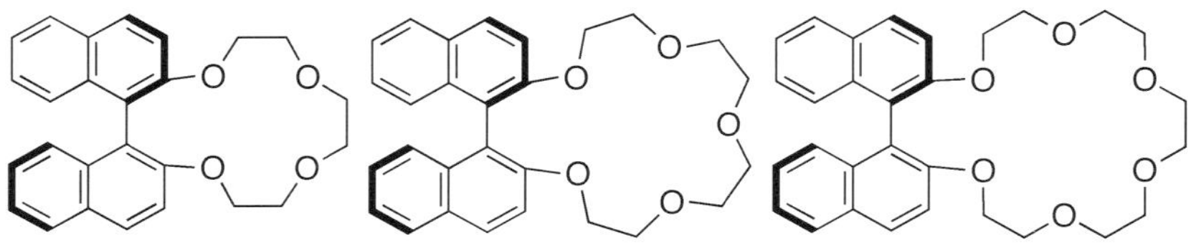

1 **2** **3**

chiral crown ethers were used as chiral Lewis acids in aldol reactions of silyl enol ethers with aldehydes. As shown in Scheme 3.18 and Table 3.3, it is noted that the fitting of the hole size of chiral crown ethers and metal cations is essential to obtain successful results. It was found that the combination of ligand **3** and $Pb(OTf)_2$ gave a promising result effective for the aldol reactions.[17]

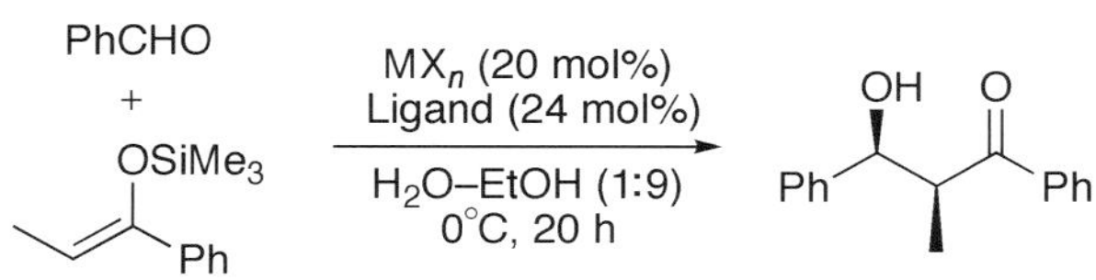

Scheme 3.18

Further investigations on Lewis acid–chiral ligand complexes revealed that the complexes of $La(OTf)_3$, $Ce(OTf)_3$, and $Pr(OTf)_3$ with ligand **4** gave high enantioselectivity (78–86% ee) (Scheme 3.19 and Fig. 3.1).[18]

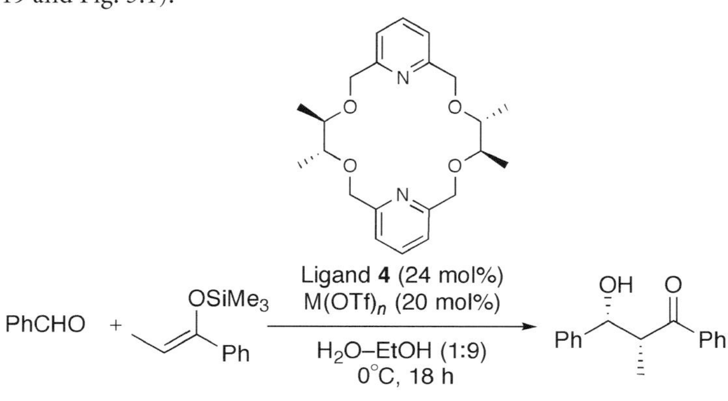

Scheme 3.19

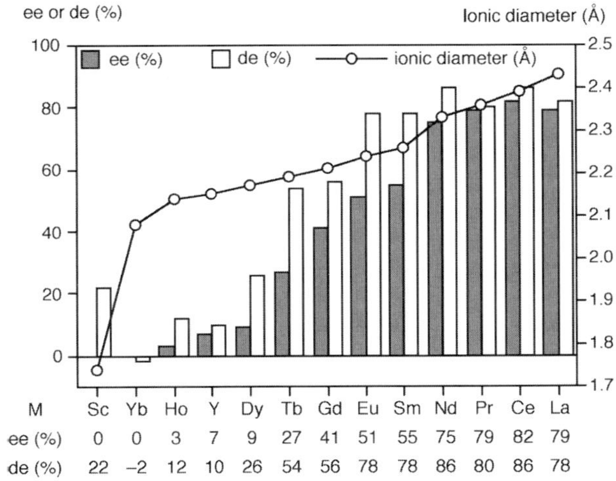

Figure 3.1 Enantio- and diastereoselectivities in the aldol reaction using rare earth metal triflates and ionic diameters (8-coordination for Sc, 9-coordination for other metals) of the metal cations (M^{3+}).

This study, based on the size-fitting strategy, showed that a slight change of ionic diameters of lanthanide cations greatly affected the diastereoselectivity as well as enantioselectivity of the reaction.

Applying the results of this study, the catalyst prepared from $Ga(OTf)_3$ and chiral semi-crown **5** was found to be effective for asymmetric aldol reactions in aqueous media. Interestingly, in the presence of the chiral ligand, hydrolysis of a silyl enol ether was suppressed, while rapid hydrolysis occurred in the absence of the ligand (Scheme 3.20).[19] In addition, water was necessary to give satisfactory yield and enantioselectivity of the aldol adduct. A similar system using pybox-type ligand **6** and $Zn(OTf)_2$ for aqueous asymmetric aldol reactions was also reported (Scheme 3.21).

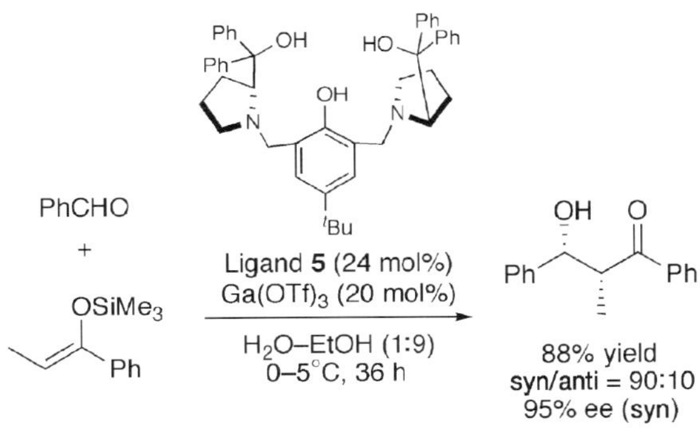

Scheme 3.20

It has been quite difficult to realize asymmetric hydroxymethylation in aqueous media.[20] After many trials, it was found that the combination of $Sc(OTf)_3$ and ligand **7** worked effectively in the reaction of a commercially available aqueous formaldehyde solution with several types of silyl enol ethers (Scheme 3.22).[21]

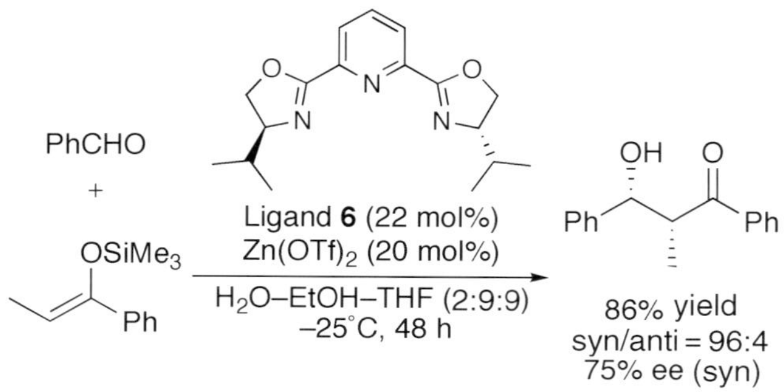

Scheme 3.21

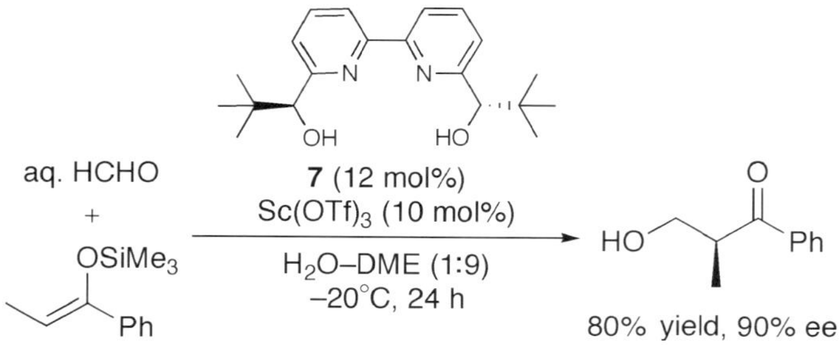

Scheme 3.22

As an extension of this work, other metal salts (10 mol%) and chiral bipyridine **7** (12 mol%) were tested, and $Bi(OTf)_3$[22] gave the most promising results. This result was unexpected because (1) the ionic diameter of bismuth (2.34 Å for 8-coordination) is much larger than that of scandium (1.74 Å for 8-coordination) and (2) $Bi(OTf)_3$ is known to be easily hydrolyzed.[23] Indeed, only a trace amount of the hydroxymethylated adduct was obtained using $Bi(OTf)_3$ without the chiral bipyridine. Several substrates were subjected to this catalyst system and the hydroxymethylation proceeded smoothly using an aqueous formaldehyde solution to afford the desired adducts in high yields and high enantioselectivities (Scheme 3.23).[24] It is to be noted that asymmetric quaternary carbons were constructed with high selectivities.

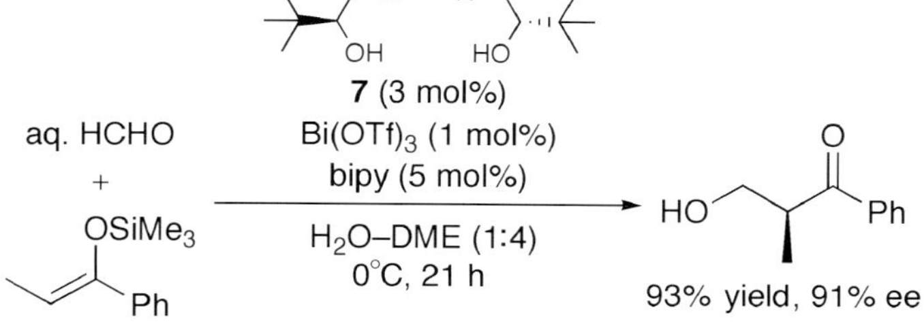

Scheme 3.23

For a long time, Lewis acids were believed to hydrolyze rapidly in the presence of water. Contrary to this, it was found that rare earth and other metal complexes were water-compatible Lewis acids. There are many strong Lewis acids that are unstable in water; however, these may be useful in water if they are combined with suitable basic ligands. As shown above, for example, $Bi(OTf)_3$ and $Ga(OTf)_3$ are unstable in the presence of water but are stabilized through coordination to a basic ligand, and may thus also be considered as water-compatible Lewis acids. In particular, the use of chiral ligands leading to new types of water-compatible chiral Lewis acids may enable a wide range of catalytic asymmetric reactions in aqueous media.

Asymmetric allylation reactions of carbonyl compounds using allylmetals have been the subject of extensive investigation. Asymmetric allylation in aqueous media was attained using combinations of BINAP derivatives and silver salts, a method originally developed for organic solvents.[25] In the aqueous system silver nitrate gave the best result (Scheme 3.24).[26]

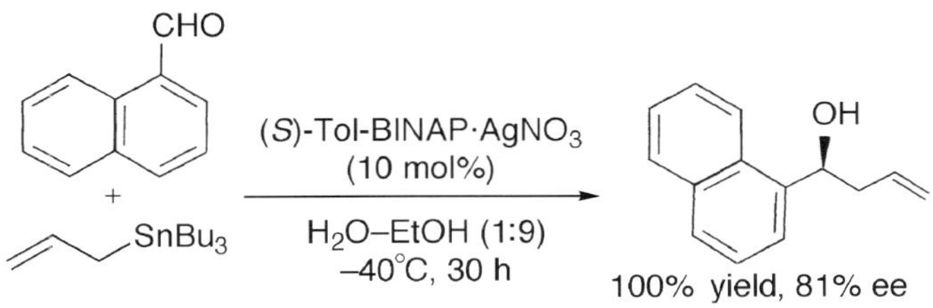

Scheme 3.24

In Lewis acid-catalyzed reactions, it is often observed that Lewis acids are deactivated by the coordination of ligands. Therefore, to prevent achiral pathways, ligand acceleration is a key factor to realize efficient asymmetric catalysis. It was reported that cadmium-catalyzed allylation of carbonyl compounds with allyltributyltin was significantly accelerated by several achiral ligands in aqueous media.[27] As an extension of this work, cadmium-catalyzed asymmetric allylation was reported (Scheme 3.25).[28] Although the present system needs further optimization, this ligand-accelerated reaction may provide a new insight into designing efficient catalytic systems that work in aqueous media.

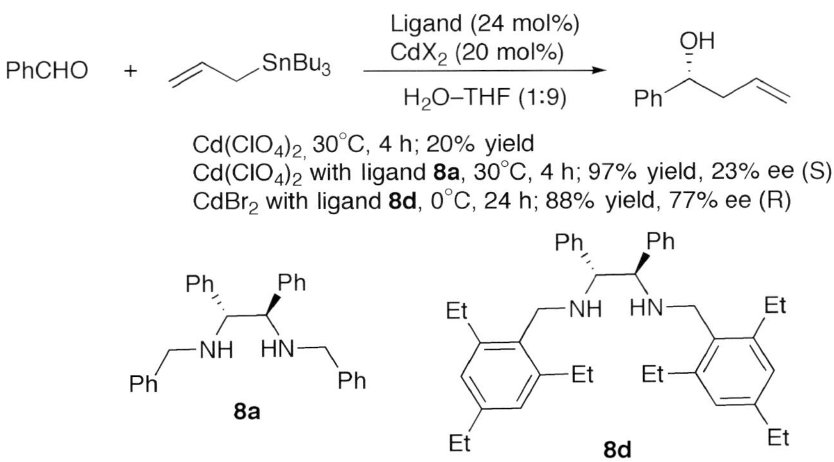

Scheme 3.25

Table 3.4 Improvement of asymmetric Mannich-type reaction

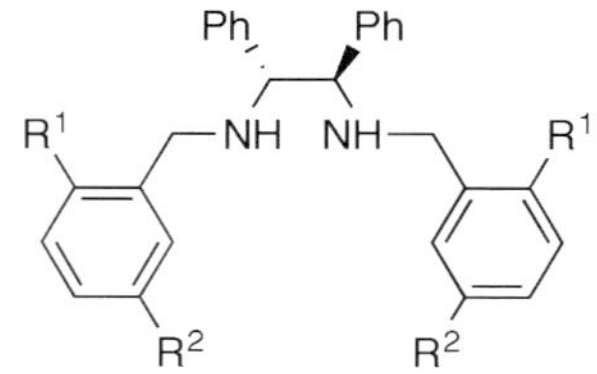

8a: R^1 = H, R^2 = H
8b: R^1 = OMe, R^2 = H
8c: R^1 = OMe, R^2 = ^{t}Bu

Entry	Diamine	Time (h)	Yield (%)	ee (%)
1[a]	**8a**	72	93	92
2	**8a**	72	55	95
3	**8b**	20	86	93
4	**8c**	20	95	96
5[b]	**8c**	20	91	95

[a] H_2O—THF = 1:9 was used as a solvent instead of H_2O.
[b] Without THF.

Additions to imine derivatives

Lewis acid catalysts have been developed for enantioselective addition reactions of imines.[29] Among them, asymmetric Mannich-type reactions provide useful routes for the synthesis of optically active α-amino ketones or esters, which are versatile chiral building blocks for the preparation of many nitrogen-containing, biologically important compounds.[30] Recently, the first example of catalytic asymmetric Mannich-type reactions in water has been developed (Table 3.4, Scheme 3.26).[31]

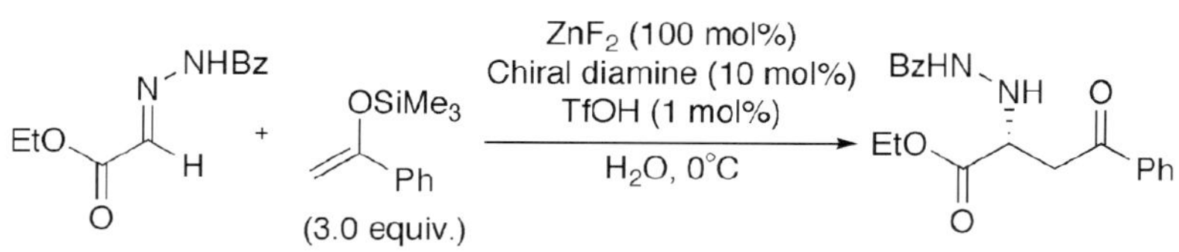

Scheme 3.26

In this reaction, acylhydrazones were chosen as imine surrogates because they could be more easily handled than imines.[32] Products were obtained as hydrazines, which could be used as unique building blocks as well as be transformed to the corresponding free amines after N—N bond cleavage under reductive conditons.[33] The reaction also proceeded smoothly in pure water with excellent enantioselectivity (entry 5).

In general, Lewis acid-catalyzed addition reactions of imines require highly sophisticated reaction systems because basicity of the starting imine and the product (amine) may lead to deactivation of the Lewis acid even in organic solvents, needless to say in water. It is noted that this study has addressed this issue for the first time in water.

This asymmetric Mannich-type reaction was successfully applied to the synthesis of (1*R*,3*R*)-*N*-(3-hydroxy-1-hydroxymethyl-3-phenylpropyl) dodecanamide (HPA-12), which is the first compound that specifically inhibits sphingomyelin synthesis in mammalian cells, and is also expected to act as a drug that inhibits intracellular trafficking of sphingolipids (Scheme 3.27).[34]

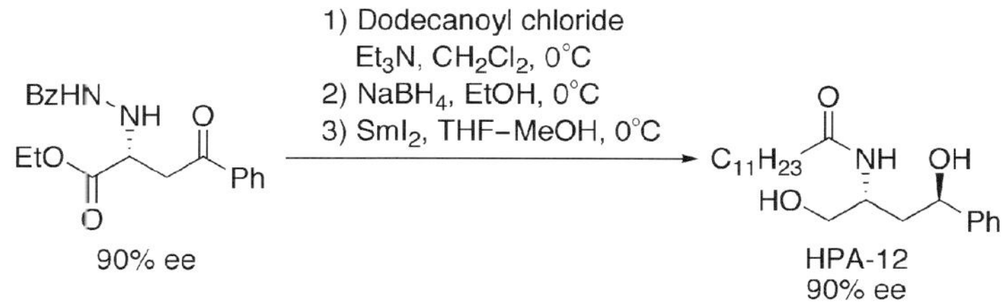

Scheme 3.27

Asymmetric allylation of acylhydrazono esters in aqueous media has been achieved using a catalytic amount of ZnF_2 and a chiral diamine ligand (Table 3.5, Scheme 3.28). While catalytic asymmetric allylation of imine derivatives remains one of the most difficult reactions to perform, the present reaction has elegantly achieved high efficiency in aqueous media.[35]

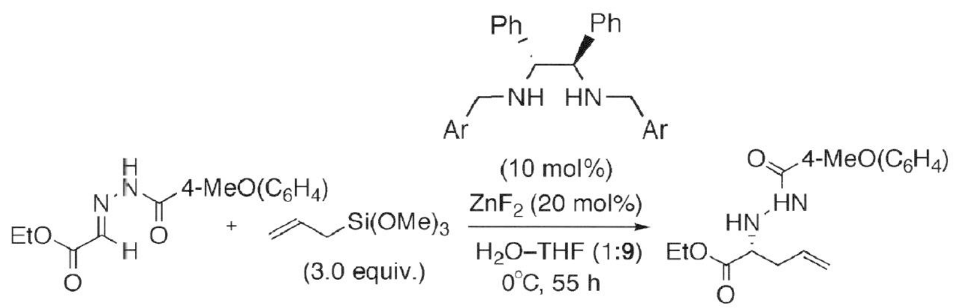

Scheme 3.28

Furthermore, the allylated products were easily converted into the synthetically important *N*-Boc α-amino esters as well as *N*-Boc α-amino ester hydrazine analogs without any loss of optical purity (Scheme 3.29).

Table 3.5 Effect of chiral diamines

Entry	Ar	Yield (%)	ee (%)
1	2-MeO-C_6H_4	84	81
2	C_6H_5	29	48
3	2-Me-C_6H_4	26	83
4	2,5-$(MeO)_2$-C_6H_3	81	76
5	3,5-$(MeO)_2$-C_6H_3	27	31
6	2-MeO-5-^{t}Bu-C_6H_3	99	64
7	1-MeO-2-naphthyl	19	70
8	8-MeO-2-naphthyl	59	84

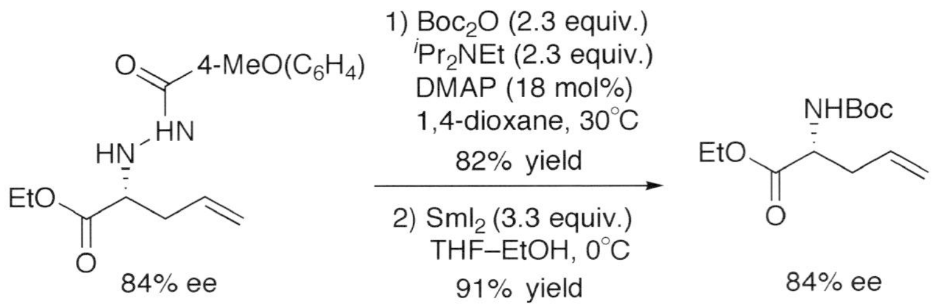

Scheme 3.29

Optically active propargylamines are important synthetic intermediates for various nitrogen-containing compounds, and a structural feature of many biologically active compounds and natural products. The most reliable and efficient methods for the preparation of optically active propargylic amines still depend on the addition of organometallic reagents to chiral imine derivatives. While catalytic methods for the preparation of optically active propargylic amines are limited, highly enantioselective and direct alkyne to imine addition was recently reported using a chiral Cu(I) complex.[36] The process is simple and provides a diverse range of propargylic amines in high enantiomeric excess (Scheme 3.30).

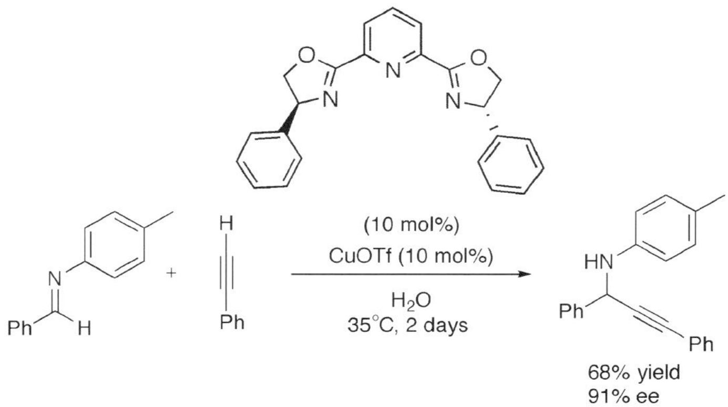

Scheme 3.30

3.2 Heterogeneous catalysis

In many cases, homogeneous catalysts are attached to organic or inorganic supports such as polymer, silica, and layered clay to form heterogeneous catalysts. The major advantages of 'heterogenized' catalysts are practical and economical since they may be readily recovered from reaction mixtures and reused multiple times without loss of catalytic activity.

3.2.1 Polymer-supported Brønsted catalysis

It was found that dehydrative esterification in water was effectively catalyzed by hydrophobic polystyrene-supported sulfonic acids as recoverable and reusable catalysts. As shown in

Table 3.6 Effect of catalysts on dehydrative esterification in water

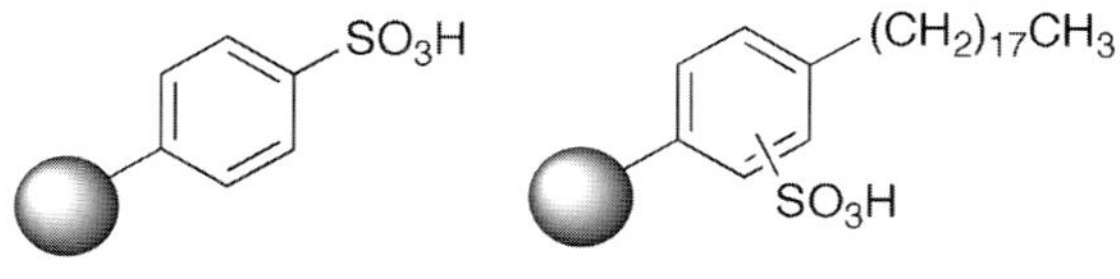

10 (0.879 mmol/g)
11 (0.352 mmol/g)
12 (1.01 mmol/g)

Entry	Catalyst (equiv.)	Yield (%)[a]
1	DBSA (0.10)	60 (83, 85[b])[c]
2	**9** (0.10)	0
3	**9** (0.10)	2
4	**10** (0.10)	41
5	**11** (0.10)	74 (81, 84[b])[c]
6	**12** (0.10)	72

[a]Isolated yield.
[b]NMR yield.
[c]For 120 h.

Table 3.6 and Scheme 3.31, in the model reaction of lauric acid with 3-phenyl-1-propanol, commercially available DOWEX 500W-X2 (H^+-form, **9**) did not promote esterification. The result indicates that a highly hydrophobic nature of the polymer-supported catalysts is important for activity in the dehydration reaction in water. It was found that resin **9** swelled significantly in water due to its high sulfonic acid content. On the other hand, both **11** and **12** scarcely swelled in water but worked as efficient catalysts. Resin **11** was easily recovered by simple filtration after the esterification was complete, and the catalyst could be continuously reused at least four times without loss of catalytic activity (Scheme 3.32).

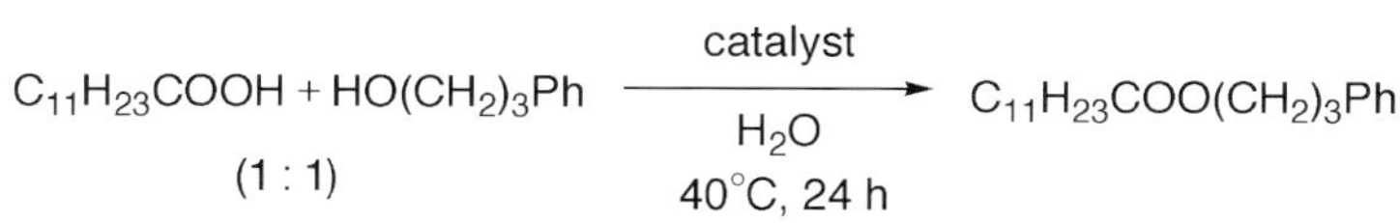

Scheme 3.31

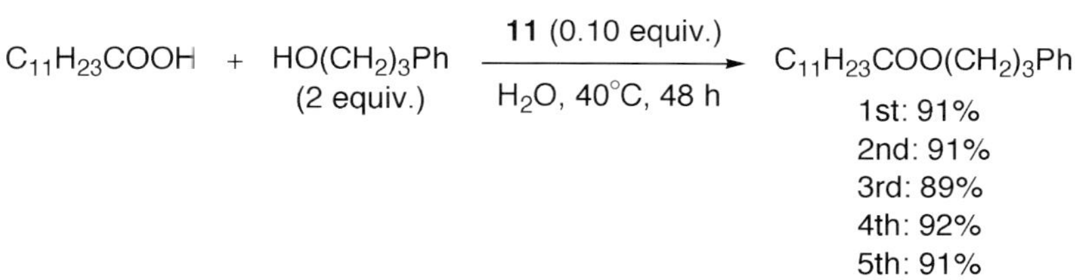

Scheme 3.32

Thioesters are biologically important compounds and synthetically useful as versatile protecting groups for thiols or carboxylic acids. In general, hydrolysis of thioesters to thiols and carboxylic acids is carried out under basic conditions. However, potential problems with this method are oxidation of thiols to disulfides and that more than equimolar amounts of reagents are needed. Although acid-promoted hydrolysis of thioesters provides an alternative important method, it is not common because of high activation energies for hydrolysis requiring the use of excess amounts of strong acids such as conc. HCl, etc. Furthermore, acid-catalyzed hydrolysis of water-insoluble thioesters in water without using organic co-solvents is extremely difficult to realize. Therefore, it was exciting to find that a hydrophobic polystyrene-supported sulfonic acid (PS-SO_3H, 0.462 mmol/g) prepared by sulfonation of 1% divinylbenzene-cross-linked polystyrene (200–400 mesh) was effective for the hydrolysis (Scheme 3.33). In addition, the catalyst was readily recovered and reused without loss of catalytic activity.

$$C_{12}H_{25}SAc \xrightarrow[\text{H}_2\text{O, reflux, 24 h}]{\text{PS-SO}_3\text{H (10 mol\%)}} C_{12}H_{25}SH$$

1st: quant.
2nd: quant.
3rd: quant.

Scheme 3.33

Racemization is an unavoidable problem when hydrolysis of thioesters is carried out under basic conditions. PS-SO_3H-catalyzed hydrolysis of an optically active thioester proceeded with only a slight loss in enantiomeric excess under water-reflux conditions (Scheme 3.34). This is one of the advantages of the acid-catalyzed hydrolysis of thioesters.

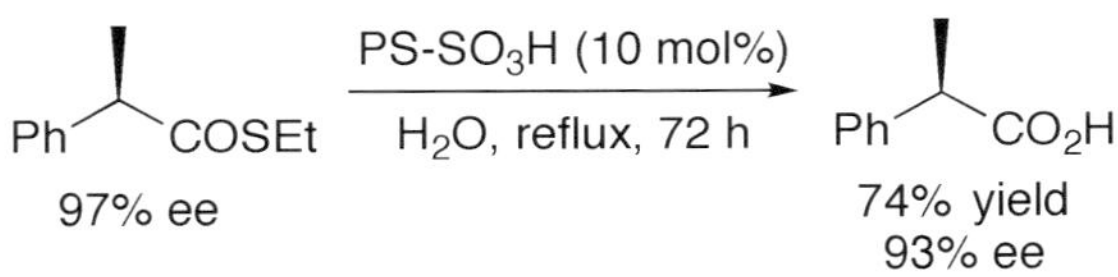

Scheme 3.34

By performing the hydrolysis of thioesters in the presence of benzylic alcohols, transprotection of thiols could be successfully carried out in water. The reactions proceeded smoothly to give benzylic thioethers from thioesters in good yields (Scheme 3.35).

$$C_{12}H_{25}SAc + Ph_3COH \xrightarrow[\text{H}_2\text{O, reflux, 48 h}]{\text{PS-SO}_3\text{H (10 mol\%)}} C_{12}H_{25}SCPh_3$$

91% yield

Scheme 3.35

It should be mentioned that, under the reaction conditions, opposite types of reactions (hydrolysis and dehydration) occur in the same pot. Thus, this system provides an effective method for conversion of thioesters to benzylic thioethers without isolating the thiol intermediates. In addition, an almost odorless dithioester instead of a dithiol, which has an unpleasant odor, can be used directly for dithioacetalization of a carbonyl compound (Scheme 3.36).

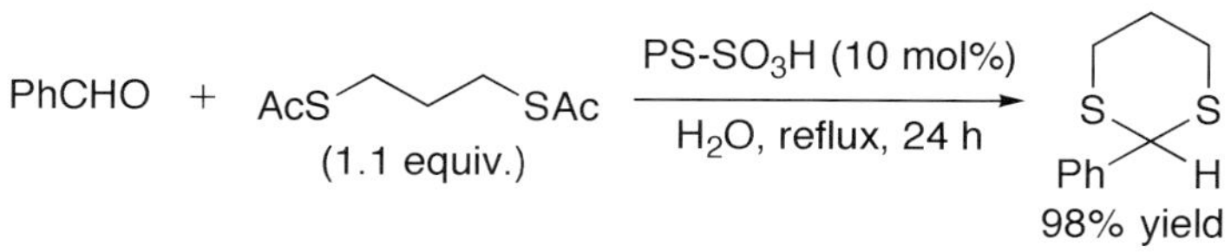

Scheme 3.36

Three-component Mannich-type reactions of aldehydes, amines, and silicon enolates also proceeded smoothly using PS-SO_3H in water (Scheme 3.37). In general, ketene silyl acetals are known to be easily hydrolyzed in the presence of water; however, such water-labile compounds could be successfully used in this reaction. Moreover, a remarkable effect of the loading levels of the polystyrene-supported sulfonic acid on yields was observed. It was suggested that the hydrophobic environment created by the catalyst might suppress hydrolysis of ketene silyl acetals.

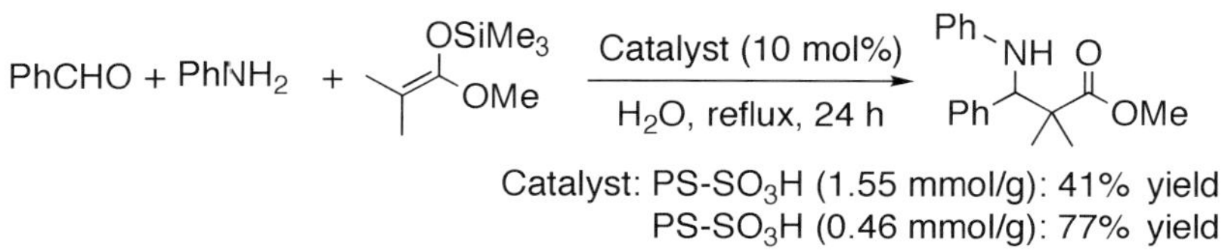

Scheme 3.37

Furthermore, low-loading (e.g. 0.17 mmol/g) and alkylated polystyrene-supported sulfonic acid (LL-ALPSSO_3H) such as **13** successfully calatyzed deprotection of *tert*-butyldimethylsilyl- (TBS) protected alcohols in water without using organic cosolvents (Scheme 3.38).

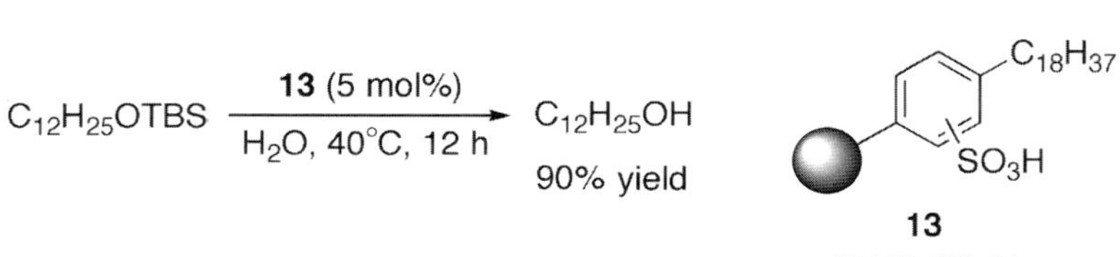

Scheme 3.38

3.2.2 Polymer-supported metal catalysis

Since the discovery of scandium triflate as a water-compatible Lewis acid, several supported scandium catalysts that work efficiently in water have been developed. Polymer-supported scandium-based Lewis acid (**14**) worked well in several carbon–carbon-bond-forming reactions (Schemes 3.39–3.42). It was suggested that the spacer could help form hydrophobic reaction environments in water. The reaction of 4-phenyl-2-butanone with tetraallyltin in water proceeded much faster than in other media including organic solvents (Scheme 3.39). As expected, **14** was easily recovered and reused.

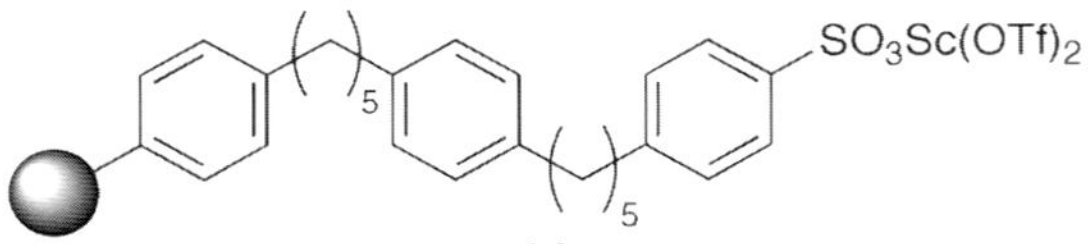

14

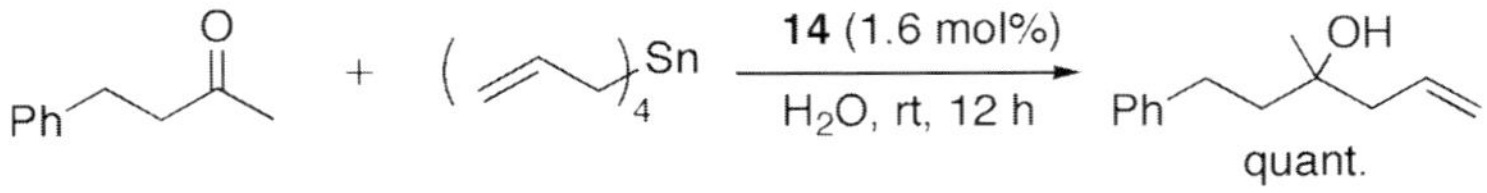

Scheme 3.39

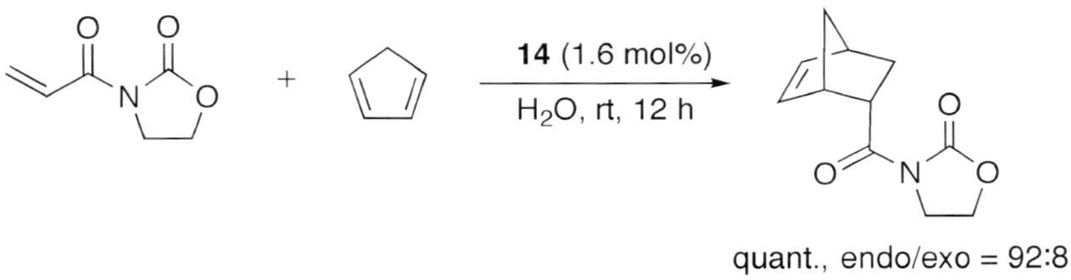

Scheme 3.40

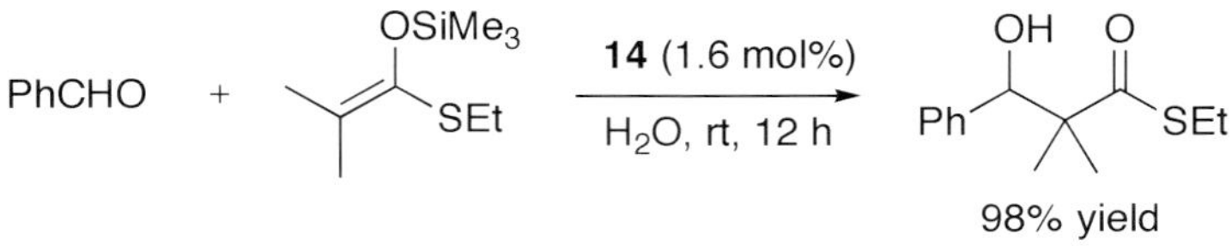

Scheme 3.41

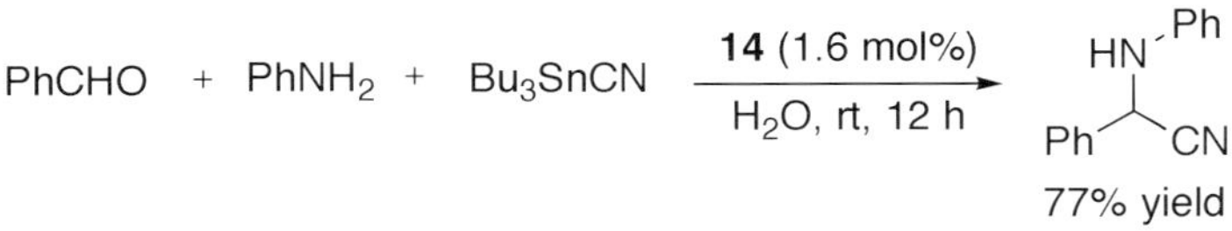

Scheme 3.42

Moreover, new PS-Sc **15** simply prepared from PS-SO_3H and $Sc(OTf)_3$ was also found to be an effective catalyst in several useful reactions as shown in Schemes 3.43–3.46.[38]

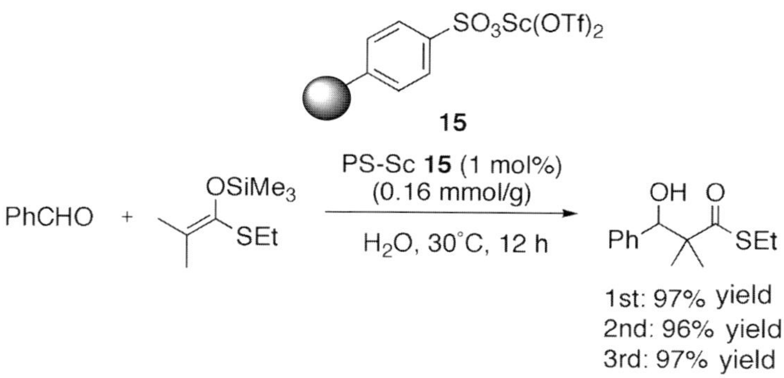

Scheme 3.43

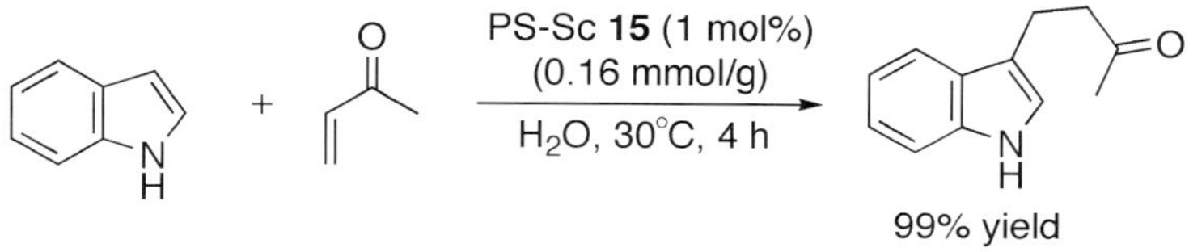

Scheme 3.44

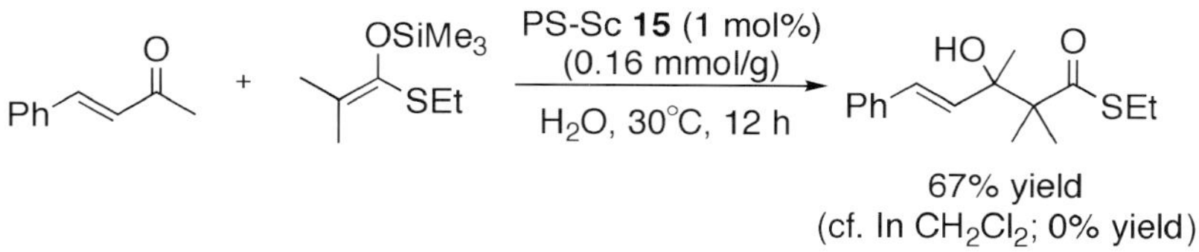

Scheme 3.45

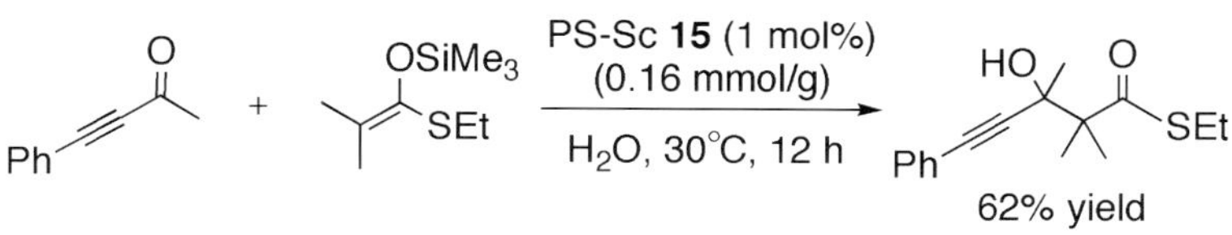

Scheme 3.46

3.3 Micellar catalysis

3.3.1 LASC (Lewis acid-surfactant-combined catalysts)

In an early stage of developing organic reactions in aqueous systems, rare earth triflates were used for aldol reactions in THF–water or ethanol–water,[39] giving successful results. On the other hand, when the reactions were carried out in pure water, the corresponding aldol products were obtained only in low yields.[40–42] This was probably because solubility of organic substrates was low and decomposition of silyl enol ethers occurred faster than the desired aldol reactions in water. To address this issue, micellar catalysis in water was

Table 3.7 Effect of surfactants

Surfactant	Time (h)	Yield (%)
None	4	3
SDS	4	88
Triton X-100	60	89
CTAB	4	Trace

invented.[43] The concept is based on the idea that by using a surfactant in water an emulsion in which organic compounds can reside is formed. A surfactant-aided Lewis acid was first demonstrated in the aldol reaction shown in Table 3.7 and Scheme 3.47.

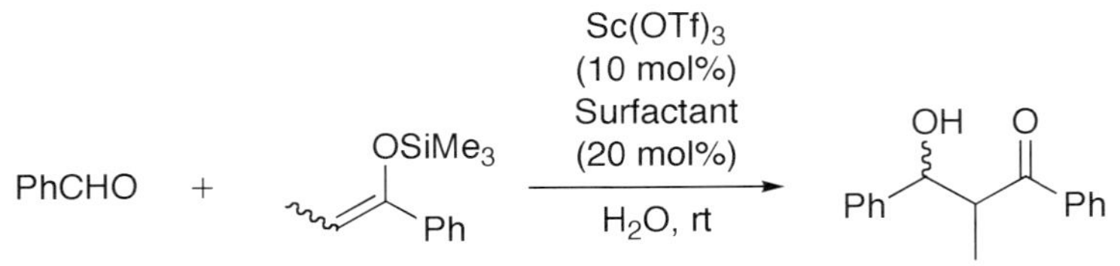

Scheme 3.47

It was found that the choice of surfactant had a significant influence on the yield. SDS gave a good yield, while Triton X required longer reaction time to give the same yield. Cetyltrimethylammonium bromide (CTAB) did not work well.

In the presence of catalytic amounts of $Sc(OTf)_3$ and SDS, three-component Mannich-type reactions of aldehydes, amines, and silyl enol ethers proceeded smoothly in micellar systems to afford the corresponding β-amino ketones or esters in high yields (Scheme 3.48). $Yb(OTf)_3$ and $Cu(OTf)_2$ also worked well in this system.[44] Furthermore, three-component reactions of aldehydes, amines, and allyltributylstannane proceeded smoothly in water to afford the corresponding homoallylic amines in high yields (Scheme 3.49).[45]

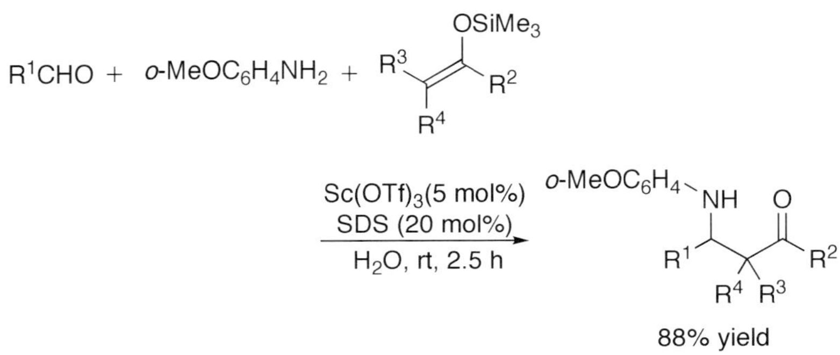

Scheme 3.48

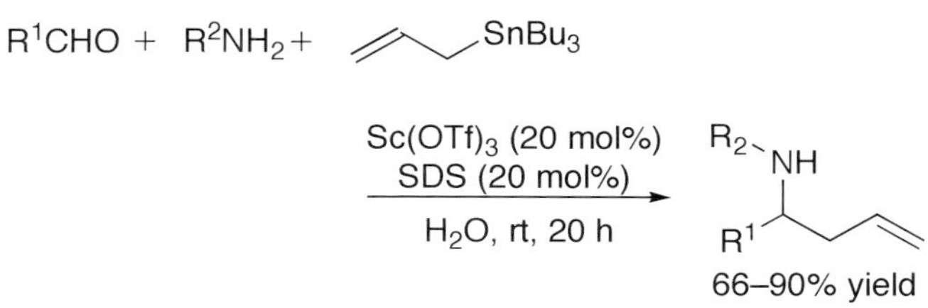

Scheme 3.49

In further studies, the simplified catalyst scandium tris(dodecyl sulfate) (Sc(DS)$_3$) was developed (Scheme 3.50).[46]

$$ScC_3 \cdot 6H_2O + 3\ NaOSO_3C_{12}H_{25} \longrightarrow Sc(OSO_3C_{12}H_{25})_3 \equiv Sc(DS)_3$$

Scheme 3.50

This catalyst was the first example of a LASC, and was expected to possess the characteristics of both a Lewis acid and a surfactant. Sc(DS)$_3$ showed quite high activity in an aldol reaction in water (Table 3.8, Scheme 3.51). A kinetic study on an initial rate of this reaction revealed that the reaction in water was approximately 100 times faster than that in dichloromethane. Besides, under neat conditions the reaction proceeded much slower and gave a low yield. The advantageous effect of water is attributed to the following factors:

1. Increased concentration of substrates and catalyst
2. Reduced hydrolysis of silyl enol ethers in the hydrophobic environment formed by LASC in water
3. High catalytic turnover caused by hydrolysis of scandium aldolates

Sc(DS)$_3$ worked well in aldol reactions of various substrates such as α,β-unsaturated, aliphatic, and heterocyclic aldehydes. As for nucleophiles, silyl enol ethers derived from ketones as well as ketene silyl acetals derived from thioesters and esters also reacted well to give the corresponding products in good yields. A key to the success in this system was assumed to be formation of stable emulsions. The size and the shape of emulsion droplets was examined by transmission electron microscopy.

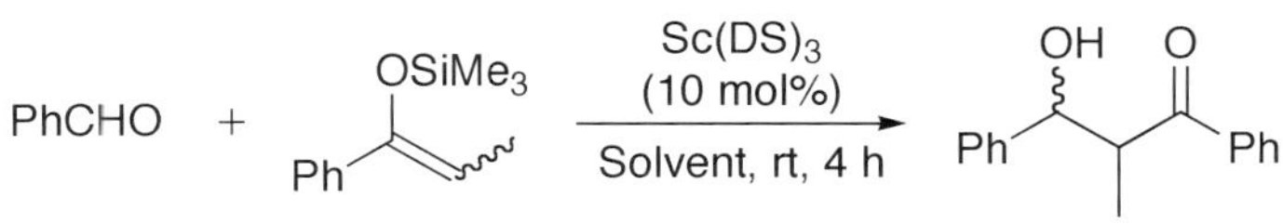

Scheme 3.51

As mentioned above, Lewis acids such as Cu(OTf)$_2$ and Sc(OTf)$_3$ with SDS were shown as effective catalysts in three-component Mannich-type reactions. Based on these results,

Table 3.8 Effect of solvents

Solvent	Yield (%)
H_2O	92
MeOH	4
DMF	14
DMSO	9
MeCN	3
CH_2Cl_2	3
THF	Trace
Et_2O	Trace
Toluene	Trace
Hexane	4
(Neat)	31

$Cu(DS)_2$ and $Sc(DS)_3$ were also developed for three-component Mannich-type reactions (Scheme 3.52).[47]

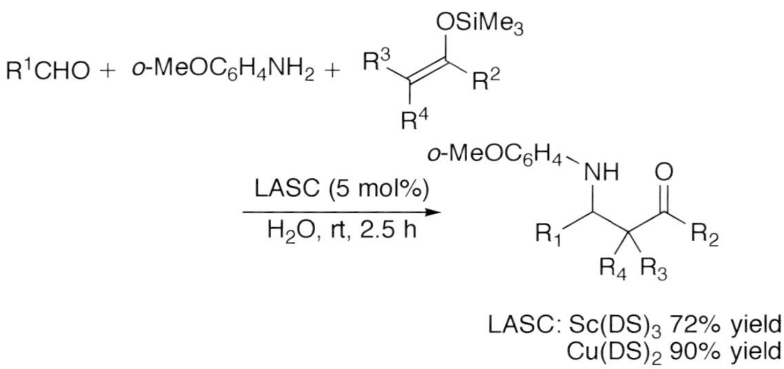

Scheme 3.52

Synthesis of α-aminophosphates was also conducted with LASC in water (Scheme 3.53).[48] It is noted that the catalysis occurred rapidly in water. In the cases of the aniline-type amines such as aniline and *o*-anisidine, the reactions proceeded rapidly and yields greater than 80% were obtained within 20–30 min. Thus, the turnover frequencies (TOFs) for these reactions are 17–26 h^{-1}, in contrast to reported procedures in organic solvents in which TOFs are less than 1 h^{-1}.

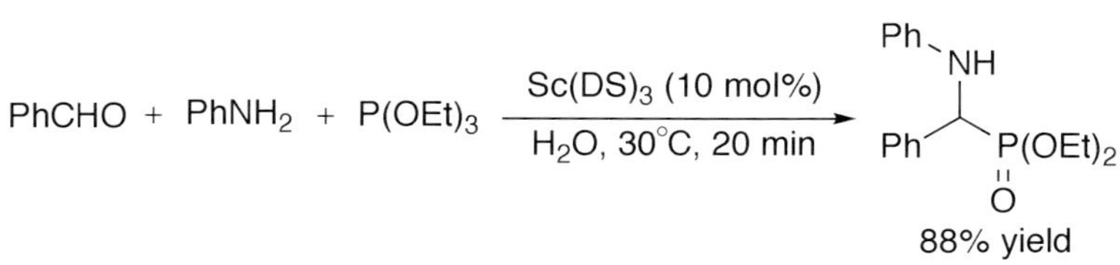

Scheme 3.53

LASC-catalyzed Michael reaction in water was also reported (Scheme 3.54).[49] Michael reactions are generally performed under basic conditions that sometimes cause serious side reactions. Using LASC, these problems could be solved. $Sc(DS)_3$ was shown to be the most active catalyst among the several types of Lewis acids or LASCs.[50] Similarly, Friedel–Crafts-type reactions of indoles with electron-deficient olefins were catalyzed by $Sc(DS)_3$ (Scheme 3.55).[51]

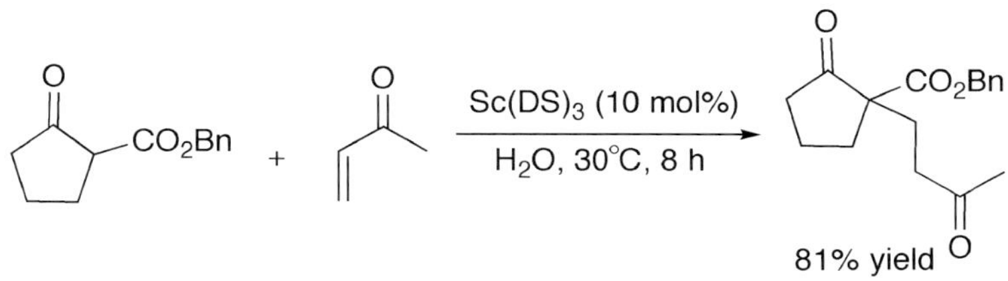

Scheme 3.54

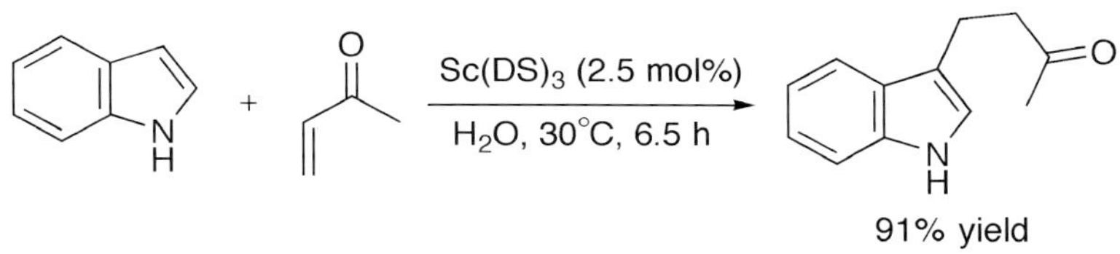

Scheme 3.55

The LACS system has been successfully applied to asymmetric catalysis. Catalytic asymmetric ring-opening reactions of *meso*-epoxides with aromatic amines proceeded smoothly in the presence of 1 mol% of $Sc(DS)_3$ and 1.2 mol% of chiral bipyridine ligand **7** in water to afford β-amino alcohols in high yields with excellent enantioselectivities (Scheme 3.56).[52] It is noted that a hydrophobic, excellent asymmetric environment has been created in water.

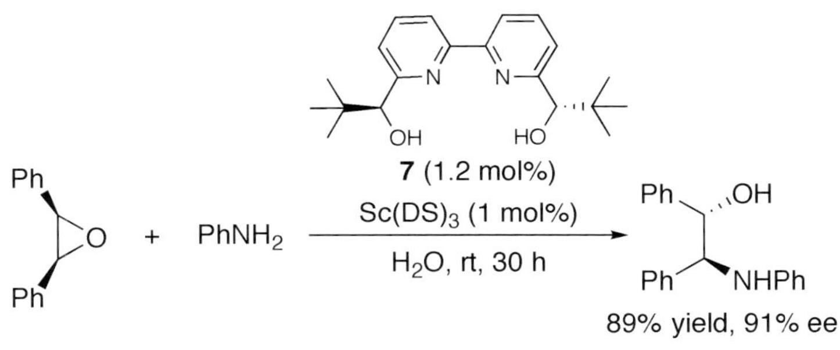

Scheme 3.56

Moreover, catalytic asymmetric ring-opening reactions of *meso*-epoxides with aromatic amines also proceeded in the presence of a catalytic amount of bismuth triflate ($Bi(OTf)_3$), chiral bipyridine ligand **7**, and sodium dodecylbenzene sulfonate (SDBS) in pure water. The corresponding β-amino alcohols were obtained in good yields with high enantioselectivities (Scheme 3.57).[53]

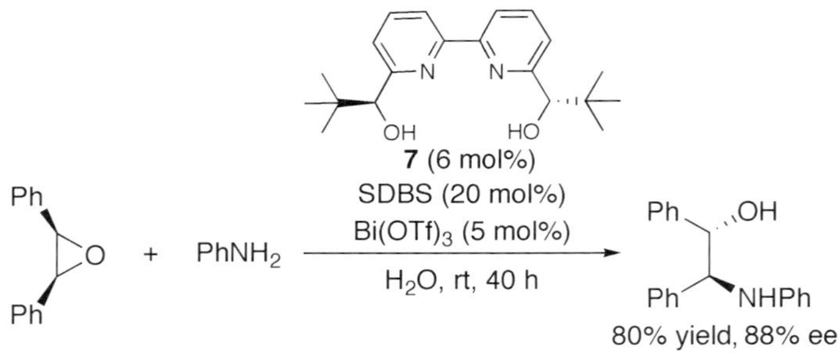

Scheme 3.57

3.3.2 BASC (Brønsted acid-surfactant-combined catalyst)

In the progress of investigations on the LASC systems, it was found that a Brønsted acid dramatically increased the activity of LASC (Table 3.9, Scheme 3.58).[54]

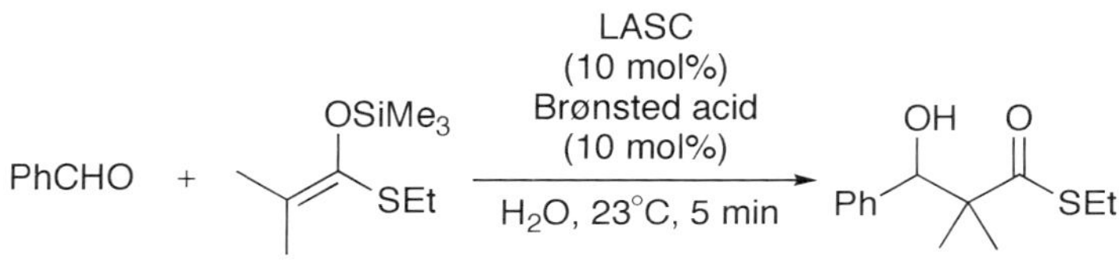

Scheme 3.58

The reaction using a stoichiometric amount of $Sc(O_3SC_{12}H_{25})_3$ was also performed (Scheme 3.59).

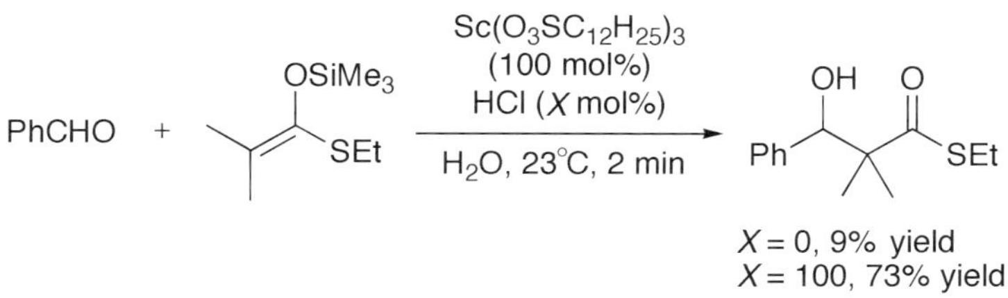

Scheme 3.59

Table 3.9 Effect of catalysts and Brønsted acid

Catalyst	Brønsted acid	Yield (%)
$Sc(DS)_3$	None	16
$Sc(DS)_3$	TsOH	52
$Sc(DS)_3$	TfOH	51
$Sc(DS)_3$	HCl	67
$Sc(O_3SC_{12}H_{25})_3$	HCl	67

Under these conditions, HCl was also found to promote the reaction, indicating that the rate-determining step, accelerated by HCl, was not the catalyst turnover step (Scheme 3.60). These results would suggest that the scandium cation and proton cooperatively act as a combined catalyst in the nucleophilic addition of the silyl enol ether to benzaldehyde.[55,56]

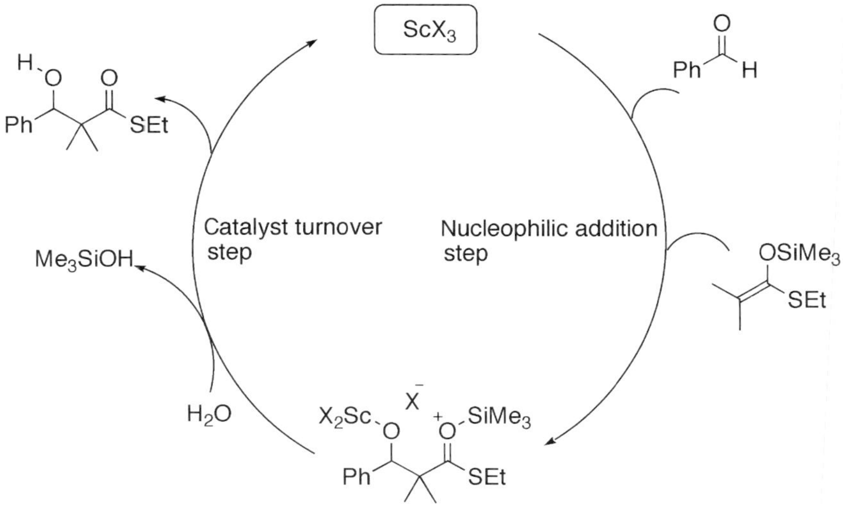

Scheme 3.60

In allylation reactions of carbonyl compounds in water, the reaction is greatly accelerated in the presence of both $Sc(DS)_3$ and HCl (Scheme 3.61).[57,58]

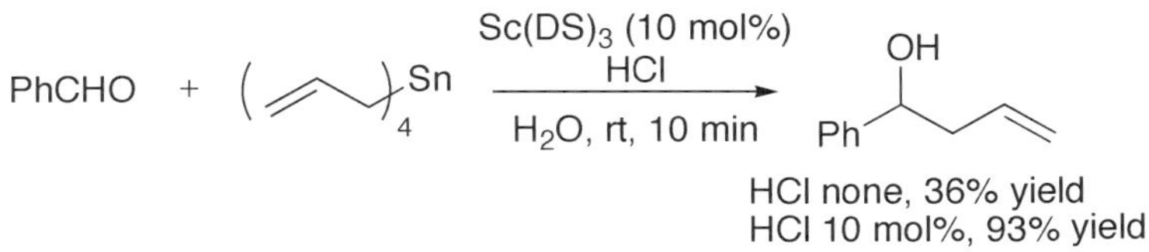

Scheme 3.61

In Mannich-type reactions, too, a certain acceleration was shown by a Brønsted acid, although a dramatic effect such as that in aldol and allylation reactions was not observed (Table 3.10, Scheme 3.62). More interestingly, however, a combination of HCl and SDS catalyzed Mannich-type reactions. In view of the above results, it is clear that BASCs can create a hydrophobic reaction environment and provide efficient acid catalysis, simultaneously, by acting as both a surfactant and a Brønsted acid.

Several Brønsted acids as catalysts were tested in a model Mannich-type reaction in water, and it was found that dodecylbenzenesulfonic acid (DBSA) afforded the desired product in

Table 3.10 Effect of LASCs and HCl

LASC	HCl (mol%)	Yield (%)
$Sc(DS)_3$	0	38
$Sc(DS)_3$	10	50
$Na(O_3SOC_{12}H_{25})_3$	10	29

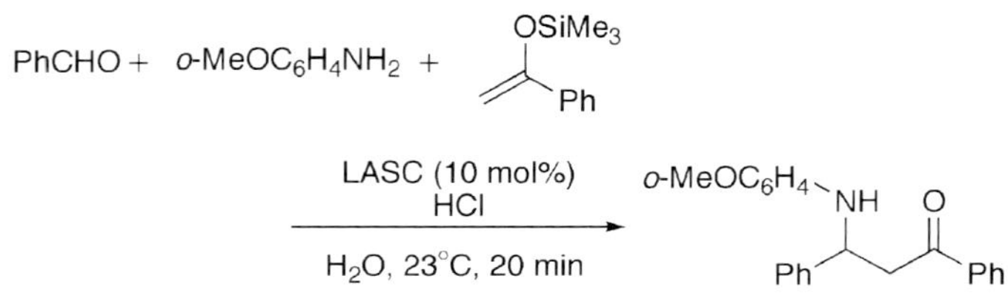

Scheme 3.62

high yield (Table 3.11, Scheme 3.63). It should be noted that *p*-toluenesulfonic acid, which has a much shorter alkyl chain than DBSA does, gave only a trace amount of the product. On the basis of this result, it was suggested that the long alkyl chain of the acid was indispensable for efficient catalysis probably due to formation of a hydrophobic reaction environment, and that the strong acidity of DBSA was essential for catalysis because a carboxylic acid having a long alkyl chain, lauric acid, was much less effective than DBSA.

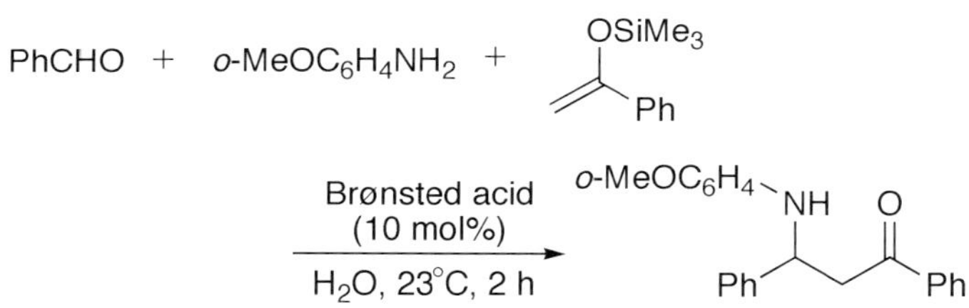

Scheme 3.63

From atom economical and environmental points of view, it is desirable to develop a new efficient system for Mannich-type reactions in which the parent carbonyl compounds are directly used.[59] Remarkably, DBSA catalyzes Mannich-type reactions in a colloidal dispersion system using ketones as nucleophilic components (Scheme 3.64).[60]

Table 3.11 Effect of Brønsted acid

Brønsted acid	Yield (%)
DBSA	83
TsOH	Trace
$C_{11}H_{25}COOH$	6

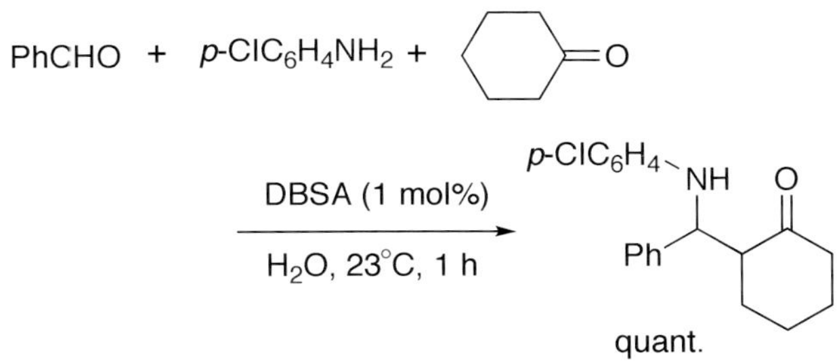

Scheme 3.64

DBSA also catalyzed direct esterification of carboxylic acids with alcohols.[61] Generally, direct esterification is carried out in organic solvents and needs either of two methods to shift the equilibrium between reactants and products. One is removal of water (azeotropically or using dehydrating agents) generated as reactions proceed, and the other is the use of a large excess of one of the reactants. On the other hand, a new concept realized that the esterification could be carried out even in water without using a large excess of reactants (Scheme 3.65). The direct esterification is a dehydration step, and it is really remarkable that the described reactions proceeded smoothly in water.

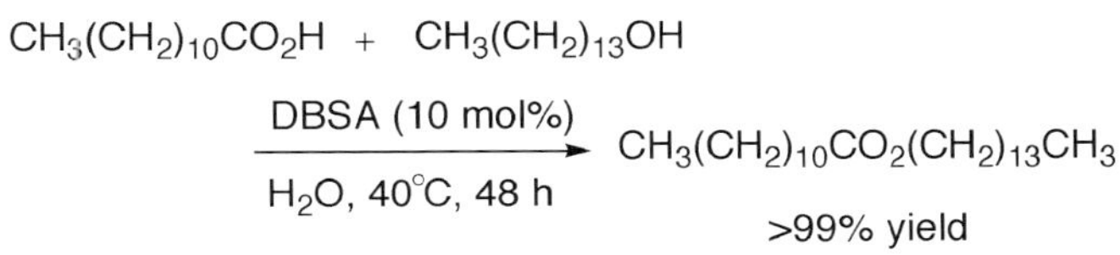

Scheme 3.65

The success of this esterification can be attributed to the fact that the surfactant-type catalysts and organic substrates (carboxylic acids and alcohols) in water form droplets with a hydrophobic interior. Catalytic species, such as a proton, concentrate at the polar surface of the droplets, where reaction takes place.

When a 1:1 mixture of lauric acid and acetic acid was esterified with dodecanol in the presence of DBSA in water, the laurate ester was predominantly obtained in 81% yield, while in neat conditions the laurate ester was obtained in 63% yield (Scheme 3.66). This

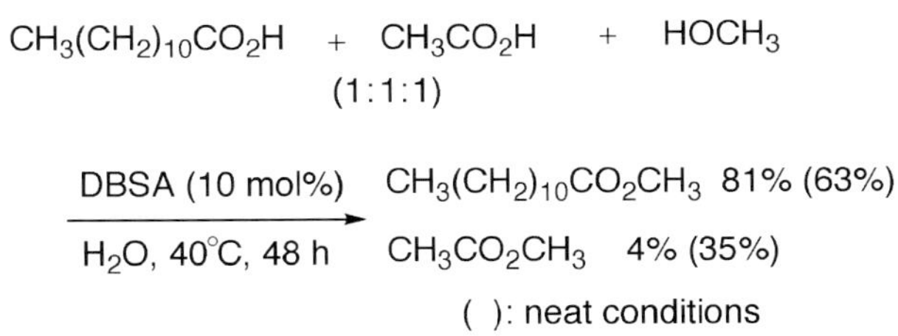

Scheme 3.66

unique selectivity is attributed to the hydrophobic nature of lauric acid and to the high hydrophilicity of acetic acid. This result led to investigations of transesterification (Schemes 3.67 and 3.68), etherification (Scheme 3.69), and thioacetalization (Scheme 3.70) in water.[62] It is noted again that dehydration proceeded smoothly in water, as shown in Schemes 3.69 and 3.70.

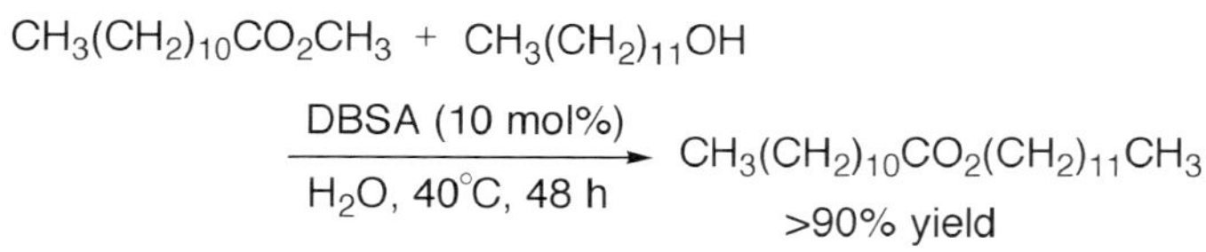

Scheme 3.67

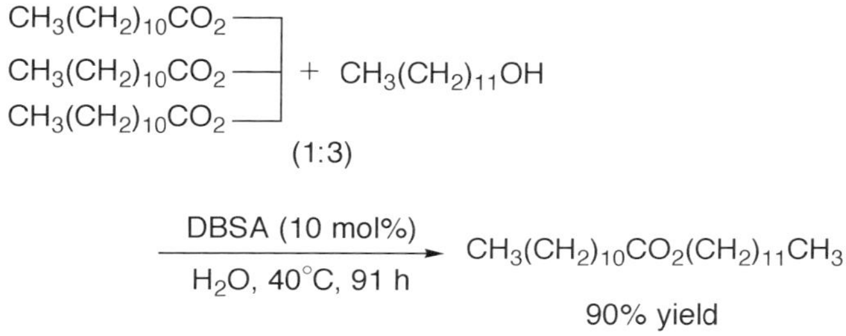

Scheme 3.68

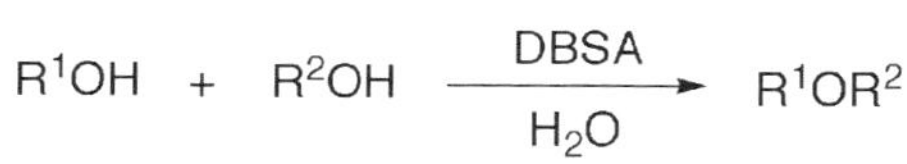

Scheme 3.69

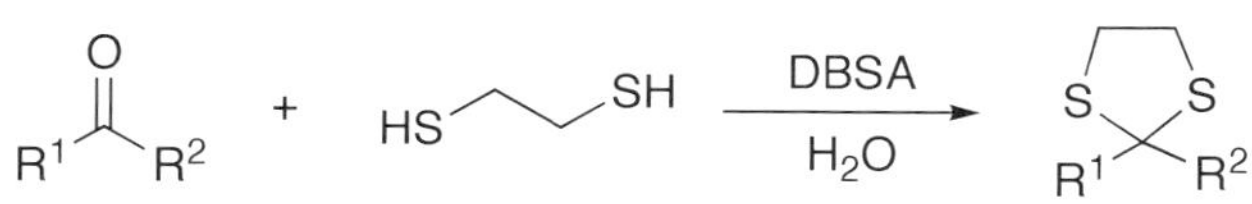

Scheme 3.70

The combination of SDS and a Brønsted acid was also investigated in aldol reaction using diphenylboronic acid (Scheme 3.71).[63] In this reaction, it should be noted that boron enolates were formed in water, and that high syn selectivity of the product was observed. Boron enolates are known to be water-sensitive, but such species can be successfully used in water under these reaction conditions.

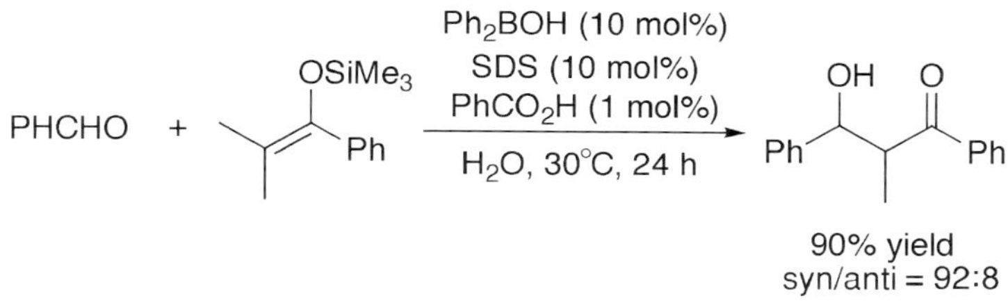

Scheme 3.71

3.4 Conclusion

Acid catalysis in water has been surveyed. Remarkable progress in this area has been achieved through the discovery and development of water-compatible Lewis acids. Most Lewis acids were for a long time believed to be unstable in water, but the initial discovery of water-compatible rare earth metal compounds was soon followed by several other metal compounds. While Lewis acid-promoted reactions have been developed in aqueous media using catalysts based on these metals, another exciting advance is asymmetric catalysis in aqueous media. Although some metal compounds are stable in water, most chiral catalysts based on these metals, that is, combinations of such metal compounds and chiral ligands, are not stable in the presence of water. Chiral crown ethers and other half-crowns as chiral ligands have provided some solution to this tremendously important issue. Furthermore, the discovery that water-sensitive Lewis acids are stabilized by basic ligands will expand the use of Lewis acids in aqueous media, especially in the area of asymmetric catalysis. On the other hand, Brønsted acid catalysis is also promising because Brønsted acids are stable in water and their acidic characters clearly appear in water rather than in organic solvents in many cases. This area is now developing. Heterogeneous catalysis in water is also a very important field that has been investigated rather recently. The use of recoverable and reusable catalysts in water can provide an ultimate solution to many environmental issues. Finally, micellar catalysis is now expanding rapidly. The idea of LASC and BASC will provide many possibilities for their use in organic synthesis. Asymmetric catalysis based on LASC and a chiral ligand in water without using any organic solvents is really remarkable and promising for the future.

In another aspect, enzymes organize very efficient reactions in the presence of a large amount of water in living cells. Approaching the efficiency of enzymatic reactions, which proceed in high yields with high selectivity under mild conditions, is indeed the goal for many organic chemists when developing new reactions. Focusing on the media in which enzymes work, we notice that it is exactly water. With this point of view, many new and exciting aspects of water may appear. We believe work in this field will promote the study, learning, and real understanding of life and nature.

References

1. Recent reviews on organic reactions in water: (a) Lindström, U.M., *Chem. Rev.*, 2002, **102**, 2751; (b) Sinou, D., *Adv. Synth. Catal.*, 2002, **344**, 237; (c) Manabe, K. and Kobayashi, S., *Chem. Eur. J.*, 2002, **8**, 4094; (d) Li, C.-J., *Chem. Rev.*, 2005, **105**, 3095.
2. (a) Akiyama, T., Takaya, J. and Kagoshima, H., *Synlett*, 1999, **1045**; (b) Akiyama, T., Takaya, J. and Kagoshima, H., *Adv. Synth. Catal.*, 2002, **344**, 338.

3. (a) Kobayashi, S., *Synlett*, 1994, **689**; (b) Kobayashi, S., *Eur. J. Org. Chem.*, 1999, **15**; (c) Kobayashi, S., Sugiura, M., Kitagawa, H. and Lam, W.W.-L. *Chem. Rev.*, 2002, **102**, 2227.
4. (a) Nakamura, H., Iwama, H. and Yamamoto, Y., *J. Am. Chem. Soc.*, 1996, **118**, 6641; (b) Akiyama, T., Takaya, J. and Kagoshima, H., *Chem. Lett.*, 1999, 947.
5. Kobayashi, S., *Chem. Lett.*, 1991, 2187.
6. (a) Kobayashi, S. and Hachiya, I., *Tetrahedron Lett.*, 1992, **33**, 1625; (b) Hachiya, I. and Kobayashi, S., *J. Org. Chem.*, 1993, **58**, 6958; (c) Kobayashi, S. and Hachiya, I., *J. Org. Chem.*, 1994, **59**, 3590.
7. Kobayashi, S., Hachiya, I., Takahori, T., Araki, M. and Ishitani, H., *Tetrahedron Lett.*, 1992, **33**, 6815.
8. Kobayashi, S., Nagayama, S. and Busujima, T., *Chem. Lett.*, 1997, 959.
9. Kobayashi, S., Hamada, T. and Manabe, K., *Synlett*, 2001, 1140.
10. Loncaric, C., Manabe, K. and Kobayashi, S., *Adv. Synth. Catal.*, 2003, **345**, 475.
11. Kobayashi, S., Nagayama, S. and Busujima, T., *J. Am. Chem. Soc.*, 1998, **120**, 8287.
12. (a) Otto, S., Boccaletti, G. and Engberts, J.B.F.N., *J. Am. Chem. Soc.*, 1998, **120**, 4238; (b) Otto, S. and Engberts, J.B.F.N., *J. Am. Chem. Soc.*, 1999, **121**, 6798.
13. Kobayashi, S., Kakumoto, K., Mori, Y. and Manabe, K., *Isr. J. Chem.*, 2001, **41**, 247.
14. Mahrwald, R. (Ed.), *Modern Aldol Reactions*, Wiley-VCH, Weinheim, Germany, 2004.
15. (a) Kobayashi, S., Nagayama, S. and Busujima, T., *Chem. Lett.*, 1999, 71; (b) Kobayashi, S., Nagayama, S. and Busujima, T., *Tetrahedron*, 1999, **55**, 8739.
16. Kobayashi, S., Mori, Y., Nagayama, S. and Manabe, K., *Green Chem.*, 1999, **1**, 175; see also Yang, B.-Y., Chen, X.-M., Deng, G.-J., Zhang, Y.-L. and Fan, Q.-H., *Tetrahedron Lett.*, 2003, **44**, 3535.
17. (a) Nagayama, S. and Kobayashi, S., *J. Am. Chem. Soc.*, 2000, **122**, 11531; (b) Kobayashi, S., Hamada, T., Nagayama, S. and Manabe, K., *J. Braz. Chem. Soc.*, 2001, **12**, 677.
18. (a) Kobayashi, S., Hamada, T., Nagayama, S. and Manabe, K., *Org. Lett.*, 2001, **3**, 165; (b) Hamada, T., Manabe, K., Ishikawa, S., Nagayama, S., Shiro, M. and Kobayashi, S., *J. Am. Chem. Soc.*, 2003, **125**, 2989.
19. (a) Li, H.-J., Tian, H.-Y., Chen, Y.-J., Wang, D. and Li, C.-J., *Chem. Commun.*, 2002, 2994; (b) Li, H.-J., Tian, H.-Y., Wu, Y.-C., Chen, Y.-J., Liu, L., Wang, D. and Li, C.-J., *Adv. Synth. Catal.*, 2005, **347**, 1247.
20. (a) Ozawa, N., Wadamoto, M., Ishihara, K. and Yamamoto, H., *Synlett*, 2003, 2219; (b) Manabe, K., Ishikawa, S., Hamada, T. and Kobayashi, S., *Tetrahedron*, 2003, **59**, 10439.
21. Ishikawa, S., Hamada, T., Manabe, K. and Kobayashi, S., *J. Am. Chem. Soc.*, 2004, **126**, 12237.
22. Gaspard-Iloughmane, H. and Le Roux, C., *Eur. J. Org. Chem.*, 2004, 2517.
23. Répichet, S., Zwick, A., Vendier, L., Le Roux, C. and Dubac, J., *Tetrahedron Lett.*, 2002, **43**, 993.
24. Kobayashi, S., Ogino, T., Shimizu, H., Ishikawa, S., Hamada, T. and Manabe, K., *Org. Lett.*, 2005, **7**, 4729.
25. Yanagisawa, A., Nakashima, H., Ishiba, A. and Yamamoto, H., *J. Am. Chem. Soc.*, 1996, **118**, 4723.
26. Loh, T.-P. and Zhou, J.-R., *Tetrahedron Lett.*, 2000, **41**, 5261.
27. Kobayashi, S., Aoyama, N. and Manabe, K., *Synlett*, 2002, 483.
28. (a) Kobayashi, S., Aoyama, N. and Manabe, K., *Chirality*, 2003, **15**, 124; (b) Aoyama, N., Hamada, T., Manabe, K. and Kobayashi, S., *J. Org. Chem.*, 2003, **68**, 7329.
29. Kobayashi, S. and Ishitani, H., *Chem Rev.*, 1999, **99**, 1069.
30. For reviews on asymmetric Mannich reactions, see Ref. 23; (a) Taggi, A.E., Hafez, A.M. and Lectka, T., *Acc. Chem. Res.*, 2003, **36**, 10; (b) Kobayashi, S. and Ueno, M., In: Jacobsen, E.N., Pfaltz, A. and Yamamoto, H. (Eds.), *Comprehensive Asymmetric Catalysis, Supplement 1*, Springer, Berlin, 2003, Chap. 29.5, p. 143; (c) Arend, M., Westermann, B. and Risch, N., *Angew. Chem., Int. Ed. Engl.*, 1998, **37**, 1044.
31. (a) Kobayashi, S., Hamada, T. and Manabe, K., *J. Am. Chem. Soc.*, 2002, **124**, 5640; (b) Hamada, T., Manabe, K. and Kobayashi, S., *J. Am. Chem. Soc.*, 2004, **126**, 7768; (c) Wolfgang, N., Tanaka, F., Watanabe, S.-I., Chowdari, N.S., Turner, J.M., Thayumanavan, R. and Barbas, C. F., III, *J. Org. Chem.*, 2003, **68**, 9624; (d) Ibrahem, I., Casas, J. and Córdova, A., *Angew. Chem. Int. Ed.*, 2004, **43**, 6528.
32. For review on acylhydrazone, see Sugiura, M. and Kobayashi, S., *Angew. Chem. Int. Ed.*, 2005, **44**, 5176.
33. Burk, M. J. and Feaster, J., *J. Am. Soc. Chem.*, 1992, **114**, 6266.
34. (a) Kobayashi, S., Matsubara, R. and Kitagawa, H., *Org. Lett.*, 2003, **4**, 143; (b) Kobayashi, S., Matsubara, R., Nakamura, Y., Kitagawa, H. and Sugiura, M., *J. Am. Chem. Soc.*, 2003, **125**, 2507.
35. Hamada, T., Manabe, K. and Kobayashi, S., *Angew. Chem. Int. Ed.*, 2003, **42**, 3927.
36. (a) Wei, C. and Li, C.-J., *J. Am. Chem. Soc.*, 2002, **124**, 5638; (b) Wei, C., Mague, J.T. and Li, C.-J., *Proc. Natl. Acad. Sci. U.S.A.*, 2004, **101**, 5749.
37. Nagayama, S. and Kobayashi, S., *Angew. Chem. Int. Ed.*, 2000, **39**, 567.
38. Iimura, S., Manabe, K. and Kobayashi, S., *Tetrahedron*, 2004, **60**, 7673.
39. Kobayashi, S., Hachiya, I. and Oyamada, H., *Bull. Chem. Soc. Jpn.*, 1994, **67**, 2342.
40. Kobayashi, S. and Hachiya, I., *J. Org. Chem.*, 1994, **59**, 3590.
41. Lubineau, A., *J. Org. Chem.*, 1986, **51**, 2142.
42. Kobayashi, S., Nagayama, S. and Busujima, T., *Chem. Lett.*, 1997, 959.
43. (a) Kobayashi, S., Wakabayashi, T. and Oyamada, H., *Chem. Lett.*, 1997, 831; (b) Kobayashi, S.,Wakabayashi, T., Nagayama, S. and Oyamada, H.,*Tetrahedron Lett.*, 1997, **38**, 4559.
44. Kobayashi, S., Busujima, T. and Nagayama, S., *Synlett*, 1999, 545.

45. Kobayashi, S., Busujima, T. and Nagayama, S., *J. Chem. Soc., Chem. Commun.*, 1998, 19.
46. (a) Kobayashi, S. and Wakabayashi, T., *Tetrahedron Lett.*, 1998, **39**, 5389; (b) Manabe, K. and Kobayashi, S., *Synlett*, 1999, 547.
47. Manabe, K., Mori, Y., Wakabayashi, T., Nagayama, S. and Kobayashi, S., *J. Am. Chem. Soc.*, 2000, **122**, 7202.
48. Manabe, K. and Kobayashi, S., *Chem. Commun.*, 2000, 669.
49. Mori, Y., Kakumoto, K., Manabe, K. and Kobayashi, S., *Tetrahedron Lett.*, 2000, **41**, 3107.
50. (a) Keller, E. and Feringa, B.L., *Tetrahedron Lett.*, 1996, **37**, 1879; (b) Kobayashi, S., Hachiya, I., Takahori, T., Araki, M. and Ishitani, H., *Tetrahedron Lett.*, 1992, **33**, 6815.
51. Manabe, K., Aoyama, N. and Kobayashi, S., *Adv. Synth. Catal.*, 2001, **343**, 174.
52. Azoulay, S., Manabe, K. and Kobayashi, S., *Org. Lett.*, 2005, **7**, 4593.
53. (a) Ogawa, C., Azoulay, S. and Kobayashi, S., *Heterocycles*, 2006, **66**, 201; see also (b) Ollevier, T. and Lavie-Compin, G., *Tetrahedron Lett.*, 2004, **45**, 49.
54. Manabe, K., Mori, Y., Odashima, K. and Kobayashi, S., *Inorg. Chim. Acta*, 1999, **296**, 158.
55. Kobayashi, S. and Hachiya, I., *J. Org. Chem.*, 1994, **59**, 3590.
56. Kobayashi, S., Nagayama, S. and Busujima, T., *Chem. Lett.*, 1997, 959.
57. (a) Yanagisawa. A., Inoue, H., Morodome, M. and Yamamoto, H., *J. Am. Chem. Soc.*, 1993, **115**, 10356; (b) Loh, T.-P. and Xu, J., *Tetrahedron Lett.*, 1999, **40**, 2431.
58. Aspinall, H.C., Greeves, N. and McIver, E.G., *Tetrahedron Lett.*, 1998, **39**, 9283.
59. (a) Blatt, A.H. and Gross, N., *J. Org. Chem.*, 1964, **29**, 3306; (b) Yi, L., Zou, J., Lei, H., Lin, X. and Zhang, M., *Org. Prep. Proc. Int.*, 1991, **23**, 673.
60. Manabe, K. and Kobayashi, S., *Org. Lett.*, 1999, **1**, 1965.
61. (a) Manabe, K., Sun, X.-M. and Kobayashi, S., *J. Am. Chem. Soc.*, 2001, **123**, 10101; (b) Manabe, K., Iimura, S., Sun, X.-M. and Kobayashi, S., *J. Am. Chem. Soc.*, 2002, **124**, 11971, and references therein.
62. Kobayashi, S., Iimura, S. and Manabe, K., *Chem. Lett.*, 2002, 10.
63. Mori, Y., Manabe, K. and Kobayashi, S., *Angew. Chem. Int. Ed.*, 2001, **40**, 2815.

Chapter 4
Metal-Mediated C—C Bond Formations in Aqueous Media

Chao-Jun Li

4.1 Introduction

In the middle of the nineteenth century, alkylzinc compounds were discovered by Frankland, from the reaction of iodoalkanes with metallic zinc, and were then added onto aldehydes.[1] A quarter of a century later, the zinc-mediated coupling of aldehydes and ketones with α-halo carbonyl compounds – the Reformatsky reaction[2] – was developed. However, it was the discovery of magnesium-mediated addition of organic halides to aldehydes and ketones by Barbier[3] and Grignard[4] at the turn of the nineteenth and twentieth centuries that established the foundation of using organometallic reagents for organic synthesis purposes.[5] More recently, organometallic reactions have been extended to the use of lithium, copper, boron, silicon, aluminum, tin, and other metals (including low-valent transition metals). In their most general form, these reagents serve as nucleophiles and react with various electrophiles, forming carbon–carbon bonds. When the electrophile is a carbonyl compound, the transformations are now frequently referred to as the Barbier–Grignard-type reactions (Eq. 4.1), which has been the most important C—C bond formation reaction in organic chemistry textbooks and one of the most utilized reactions in the synthesis of fine chemicals and pharmaceutical compounds (both academically and industrially). The generation of the organometallic reagents can be *in situ* (Barbier) or stepwise (Grignard). Conventionally, such reactions are carried out under the strict exclusion of moisture and air. Thus, conventional wisdom tells us that 'Grignard-type' reactions are highly sensitive toward water and air. Such a restriction can impose limitations on synthetic design as various acidic hydrogens in the substrates will have to be replaced with unreactive protective groups. The use of various protection–deprotection sequences prolongs synthetic routes and decreases synthetic efficiency.

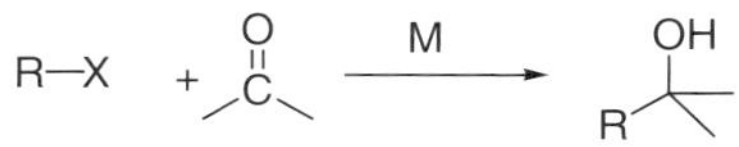

(Eq. 1)

Recently, however, extensive studies by us and others have shown that various metal-mediated C—C bond formations, including 'Grignard-type' reactions, can be carried out well in aqueous media and sometimes these reactions are even more effective than those in organic solvents in terms of both product yields and chemo- (as well as stereo-) selectivities.[6] This chapter will provide an understanding of such reactions and summarize the current research progress on this subject.

$C^{\ominus}$ $M^{\oplus}$ C—M

ionic covalent

Figure 4.1 Bonding of C—M.

4.2 Reactivity of organometallic compounds with water

In order to understand why the notoriously 'water-sensitive' metal-mediated reactions can proceed in water at all, it is necessary to look at three basic aspects: (1) the fundamental bonding in organometallic compounds, (2) the kinetics of the hydrolysis of C—M bonds, and (3) the kinetics of carbonyl addition (and other nucleophilic reactions) by C—M bonds.

4.2.1 C—M bonding

A carbon–metal bond can be more ionic or more covalent, depending on the relative electronegativities between carbon and the corresponding metal (Fig. 4.1). Due to the large electronegativity difference between carbon and alkali (and alkali-earth) metals, these metals primarily form ionic bonds with carbon in their corresponding organometallic reagents. On the other hand, there is significant covalent bonding character with post-transition metals because the electronegativities of these metals and carbon are close. Furthermore, the formation of a C—M intermediate from an organic halide is greatly affected by the structure of the organic halides. Li and coworkers explored the relative difficulty in forming organoindium reagent under solventless conditions and found that indium can react with allyl halides and propargyl halides readily at room temperature, whereas the corresponding reaction between indium and α-halo carbonyl compounds is more difficult and no reaction was observed between indium and benzyl bromide even at an elevated temperature. On the basis of these results, they proposed a five-membered cyclic transition state (Fig. 4.2) for the formation of organometallic compounds from allyl halides, propargyl halides, and α-halo carbonyls (to explain the great differences in relative reactivities of these compounds, which seemingly possess similar reductive potentials).[7]

4.2.2 C—M hydrolysis

The overall result of hydrolysis of a C—M bond is the formation of a C—H bond and an M—OH bond and might be considered as one of the simplest reactions. Generally speaking, the overall transformation is thermodynamically favorable; if C—M is a pure ionic bond,

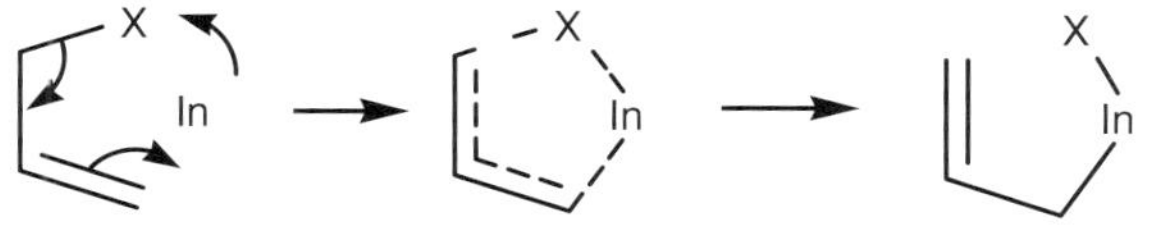

Figure 4.2 Li's proposed mechanism for the formation of allylindium from allyl halide.

then a simple combination of the carbanion with the proton in water (protonolysis) will form the hydrolyzed product and the metal ions will be dispersed in water (Fig. 4.3(a)). However, besides alkali metals, a C—M bond is rarely entirely ionic and often has some covalent character. Sometimes, the covalent C—M bonds are quite stable. For example, many compounds involving C—Si, C—B, C—Sn, and C—Hg bonds are stable in water. The relative stability of methylmercury in water has been known for a century[8] and tribenzyltin chloride has been prepared by reacting benzyl bromide and metallic tin in water in the early 1960s.[9] Thus, how to proceed from the C—M bond (of the organometallic compound) to the hydrolyzed product is an interesting question as it would require a considerable amount of energy to break a covalent C—M bond in order to form a carbanion, which can combine with a proton. Alternatively, the hydrolysis can proceed in a 'concerted' fashion through a six-membered cyclic transition state, so that the formation of 'pure' ions will not be required (Fig. 4.3(b)). Such a mechanism would require less energy. Indeed, even the seemingly very reactive Grignard reagent may undergo hydrolysis (mostly) via a six-membered cyclic transition state.[10] It is likely that the hydrolysis of most C—M bonds of post-transition metals has a similar mechanism.

4.2.3 C—M reactions

A C—M bond can react with nucleophiles other than a proton in several different ways. Take a carbonyl group (the Barbier–Grignard-type reactions) as an example: a pure carbanion can undergo direct nucleophilic addition to the carbonyl group (Scheme 4.1, route A). Alternatively, a concerted reaction that involves a six-membered cyclic transition state would be favorable for C—M bonds with a significant covalent character (Scheme 4.1, route B). However, in order to form a six-membered cyclic transition state (enthalpically favored), three reactants are frequently required to come together, which has a 'cost' in entropy (Scheme 4.1, route C). An exception exists when a six-membered ring can be formed by only two components, as the cases are with allylmetals, propargylmetals, and metal enolates. Because of the entropic cost associated with three components coming together, a two-component, four-membered cyclic transition state is possible for certain C—M-based reagents (such as in the cases of vinyl-, aryl-, propargyl-, and alkylmetal reagents) (Scheme 4.1, route D), with some cost in enthalpy (four- vs six-membered rings). Finally, the reaction of a C—M bond may

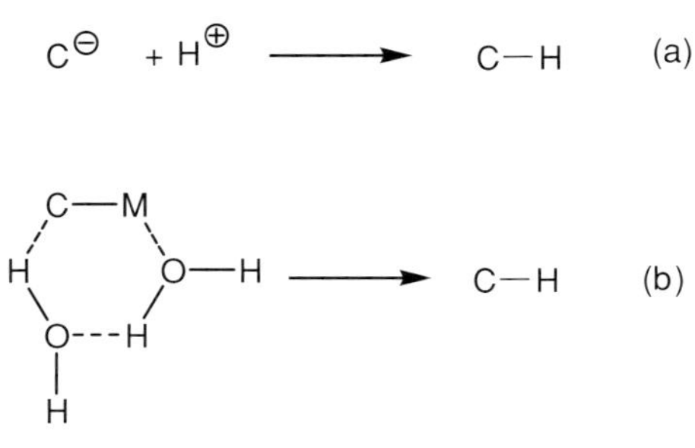

Figure 4.3 Hydrolysis of a C—M bond.

not need to form a discrete C—M-based reagent; a free radical or a surface-bound radical anion can also react with a carbonyl or another electrophile to generate the corresponding C—C bond (Scheme 4.1, route E).

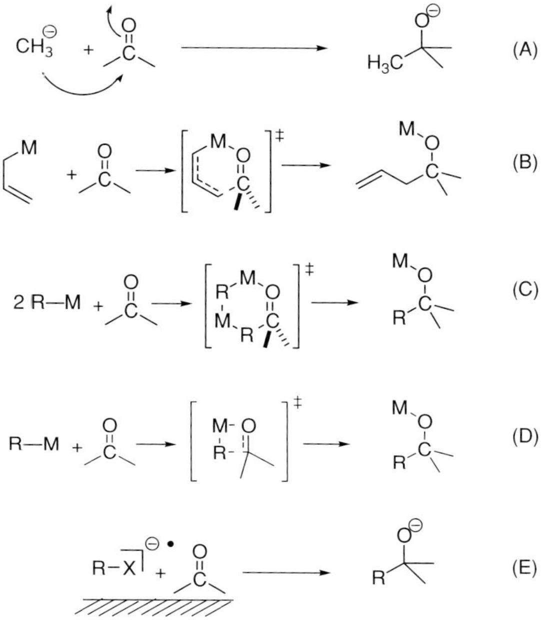

Scheme 4.1 Various pathways for 'Barbier–Grignard-type' reactions.

4.2.4 C—C bond formations via C—M reactions in water

First of all, in order to form a C—C bond by using metal-mediated reactions in water the C—M bond needs to be reactive enough under any one of the five possible routes. Second, it is required that the reaction between the C—M bond and the corresponding electrophile runs much faster than the corresponding reaction between the C—M bond and the (solvent) water molecule. We will use the 'Grignard-type' reaction as an example for analysis.

Based on the Klopman–Salem equation for estimating chemical reactivities,[11] a free carbanion will normally be protonated much faster (a charge–charge interaction) than it reacts with a neutral carbonyl (an orbital–orbital interaction). However, with the decrease in polarization and the increase in covalent-bond character, the orbital–orbital interaction becomes more important. When all other factors are equal, a predominantly 'covalent' C—M bond will be more reactive toward the carbonyl than toward hydrolysis because of the favorable interactions of the 'Frontier molecular orbitals'. Thus, carbonyl allylation can proceed well in water unless the allylmetal is predominantly carbanion (in which case hydrolysis product will be formed). Scheme 4.2 illustrates the allylation reaction of an aldehyde with an allylmetal reagent vs the hydrolysis of the allylmetal reagent. Indeed, the classical Barbier–Grignard-type reactions involving organomagnesium reagents and organolithium reagents are highly sensitive toward water. A similar reasoning of relative reactivity in product formation vs hydrolysis can be applied to reactions involving a four-membered transition state as well as to reactions involving radicals and radical anions on a metal surface (surface effect also plays an important role in this case). There are two potential benefits of performing reactions involving organometallic reagents in water: (1) the stabilization of more polarized transition states and (2) the dissolution (cleaning) of metal salts formed on the metal surface (when elemental metals are used as the reagents). Indeed, in the last decade or so, various 'Barbier–Grignard-type' reactions and transition-metal catalyzed C—C bond forming reactions have been developed in water, which is in agreement with the above reasoning. Chan and coworkers carried out quantum mechanics MP2/6-31+G* calculations for the reactions of a series of monomeric allylmetals with water and carbonyl compounds in the gas phase.[12] It has been shown that allyl complexes of groups IA and IIA and low-valent group IIIA and IVA metals are π-complexes or reactive σ-complexes; they show high reactivities toward hydrolysis. Group IIB, trivalent group IIIA, tetravalent group IVA, and both tri- and pentavalent group VA metals form σ-complexes with allyl, which are less reactive toward hydrolysis than toward allylation. The calculated intrinsic kinetic preference of allylation over hydrolysis is found to correlate well with the rate of hydrolysis, the nucleophilicity of the allylmetals, and the lateness of hydrolysis transition structures. Both the nucleophilicity of the allylmetal complexes and the thermodynamic driving force are important factors determining the rate of hydrolysis.

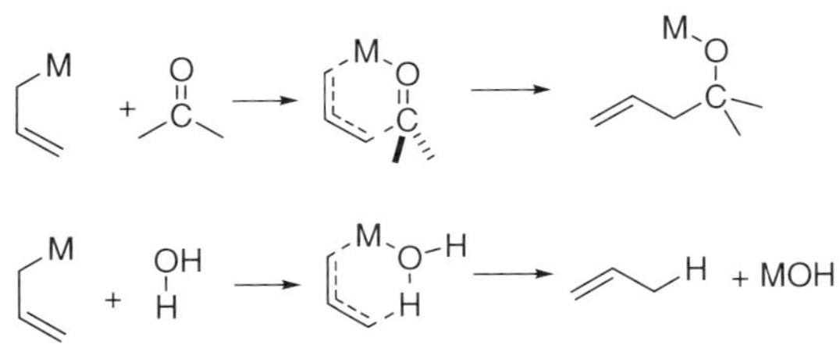

Scheme 4.2 Carbonyl allylation vs hydrolysis of allylmetals.

4.3 Allylation of carbonyls and imines

Because of the high reactivity of allyl halides toward metals (to generate allylmetal species) and the favorable entropy effect (requires only two components to form a six-membered cyclic transition state) as well as favorable 'frontier orbital' interactions between the allylmetal reagents and carbonyl compounds (and imines), the 'Barbier–Grignard-type' allylation reactions are highly successful in water. Many metals such Zn, Sn, In, Bi, Mn, Ga, Al, Tl, Ge, Pb, Fe, and even Mg, as well as preformed allylsilicon, allyltin, allylgermanium, and allylboron reagents, can be used for allylation reactions under aqueous conditions (Scheme 4.3). Based on the above analysis of C—M reactivities, the most effective metal for mediating such reactions would be one that has the most favorable reductive potential, while still having a strong covalent character of the allylmetal reagent (generated from the metal). By considering these two criteria, indium appears to have the lowest first ionization potential among its neighboring elements and, in fact, it is similar to alkali metals; yet, unlike alkali C—M bonds, a C—In bond is primarily a covalent bond.[13] Indeed, indium has become the most effective metal for various allylation reactions.[14]

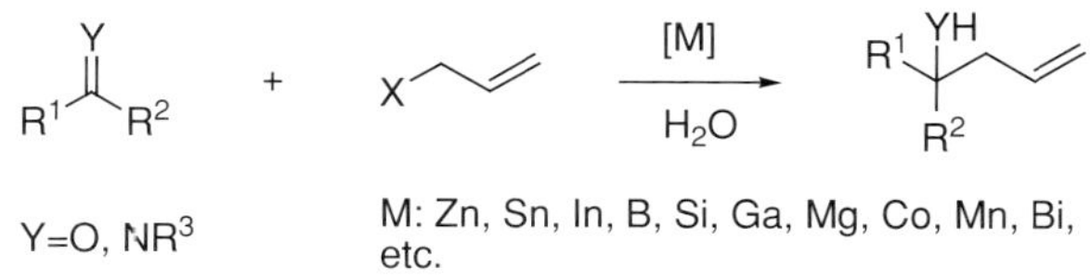

Scheme 4.3 Allylation of carbonyl compounds and imines.

4.3.1 Alyllation of carbonyl compounds

Mediated by zinc

In the late 1970s, Wolinsky and coworkers[15] studied the Barbier-type allylation of aldehydes and ketones with allyl bromide mediated by 'activated' zinc dust in 95% ethanol at 78°C. The reaction yields are comparable to those obtained in aprotic solvents (Eq. 4.2). Then, in 1985, Luche and coworkers reported the allylation of aldehydes and ketones in aqueous media using zinc as the metal and THF as a cosolvent under magnetic stirring or sonication conditions (Eq. 4.3).[16,17] The replacement of water by aqueous saturated ammonium chloride solution enhanced the efficiency. In this case, comparable results were obtained with or without the use of sonication. In the same year, Mattes and Benezra reported[18] that ethyl (2-bromomethyl)acrylate can couple with carbonyl compounds mediated by metallic zinc in refluxing saturated aqueous NH_4Cl/THF to give α-methylene-γ-butyrolactones (Eq. 4.4). The same reaction in THF alone gives only a low yield (15%) of the product within the same time range and under the same conditions. Since then, the zinc-mediated allylation reactions have been studied under a variety of conditions. The use of a solid organic support instead of the cosolvent THF[19] proceeded at about the same rate as reactions with THF as a cosolvent. Both allyl bromide and allyl chloride can be used. Zinc-mediated allylation of methyl γ-oxocarboxylates in a mixture of saturated aqueous ammonium chloride and THF provides a convenient synthesis of 5-allyl-substituted γ-lactone.[20]

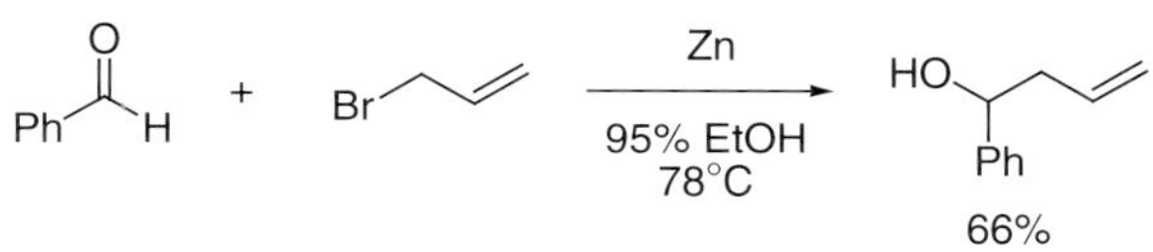

(Eq. 2)

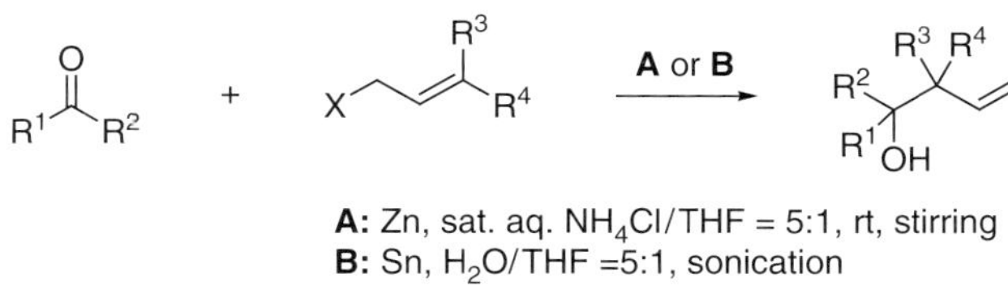

(Eq. 3)

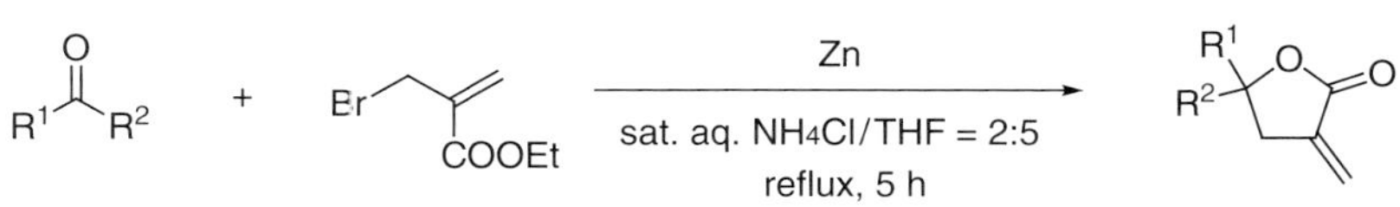

(Eq. 4)

Conjugated 1,3-butadienes were produced in moderate yields when carbonyl compounds reacted with 1,3-dichloropropene or, more effectively, 3-iodo-1-chloropropene and zinc in water (Eq. 4.5).[21] When the reactions were interrupted after their initial allylations, subsequent base treatment of the intermediate compounds produced vinyloxiranes. Similarly, reactions of carbonyl compounds with 3-iodo-2-chloromethyl-1-propene followed by base treatment produced 2-methylenetetrahydrofurans (Eq. 4.6).[22] The reaction of 2,3-dichloro-1-propene with aldehydes and ketones in a two-phase system of water and toluene containing a small amount of acetic acid gave 2-chloroallylation products (Eq. 4.7).[23] No reaction was observed in the absence of water. Interestingly, the reaction of 2,3-dichloropropene plus zinc powder in aqueous ethanol gave the dechlorinated allene product.[24]

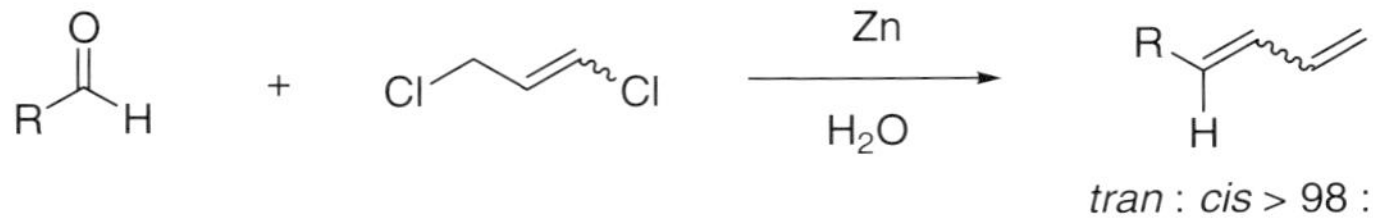

(Eq. 5)

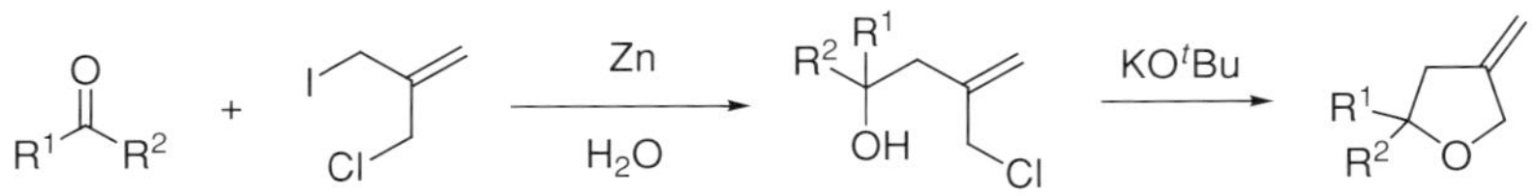

(Eq. 6)

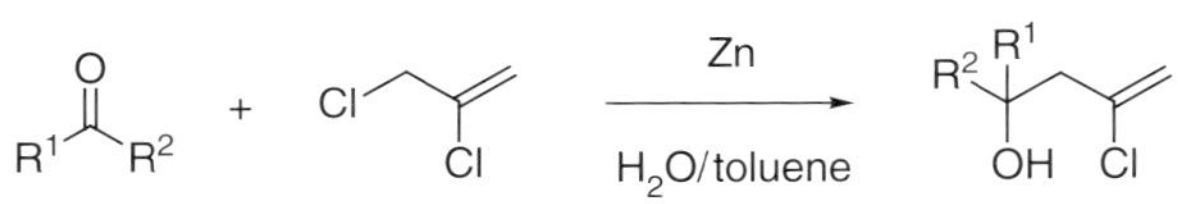

(Eq. 7)

The reactivity of zinc-mediated reactions can be increased by using 'activated' submicromic zinc powder[25] produced by pulsed sonoelectro reduction. The stereochemical course of the allylation (and propargylation) of several aldehydes with crotyl (and propargyl halides) using zinc powder as the condensing agent in cosolvent/water (salt) media has been extensively studied.[26] The Zn-mediated reactions of cinnamyl chlorides with aldehydes and ketones in THF/NH_4Cl(aq) give α- and γ-addition products, as well as phenylpropenes and dicinnamyls. Radical intermediates were proposed to be involved in the reaction.[27] Enolizable 1,3-dicarbonyl compounds can be allylated by zinc and allyl halide.[28]

The zinc-mediated allylation of carbonyl compounds have been used in a number of applications. For example, the allylation reaction in water could be used to prepare α,α-difluorohomoallylic alcohols from *gem*-difluoro allyl halides.[29] An efficient route for the synthesis of the Phe–Phe hydroxyethylene dipeptide isostere precursors utilized for the design of potential inhibitors of renin and HIV-protease was developed. The key step is the zinc-mediated stereoselective allylation of N-protected α-aminoaldehydes in aqueous solution.[30] Chan and Li used the zinc-mediated allylation as a key step in a total synthesis of (+)-muscarine.[31] The strategy was based on the observation that the diastereoselectivity of the allylation reaction in water can be reversed through protection of the α-hydroxyl group. Diastereoselective allylation of α-aminoaldehydes with the chiral building block (*Ss*)-3-chloro-2-(*p*-tolylsulfinyl)-1-propene gave enantiomerically pure sulfinyl aminoalcohols in good yields and diastereoselectivities (Eq. 4.8).[32]

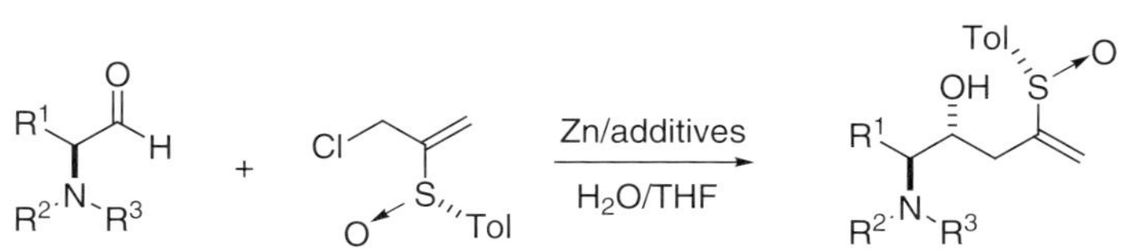

(Eq. 8)

Mediated by tin

In 1983, Nokami et al. observed that water had an accelerating effect on the allylation of carbonyl compounds with diallyltin dibromide in ether.[33] For example, benzaldehyde was allylated in 75% yield in 1.5 h in a 1:1 mixture of ether/water, while the same reaction gave less than 50% yield in a variety of organic solvents such as ether, benzene, or ethyl acetate, even after a reaction time of 10 h. The tin reaction can be activated by using a combination of tin metal and a catalytic amount of hydrobromic acid. In this case, the addition of metallic aluminum powder or foil to the reaction mixture dramatically improved the yield of the product. The use of allyl chloride for such a reaction, however, was not successful. The reaction can also proceed intramolecularly and ketones having allylic halide functionality were cyclized to form five- and six-membered rings.[34] Similar intramolecular reactions occurred stereoselectively with aldehydes (Eq. 4.9).[35]

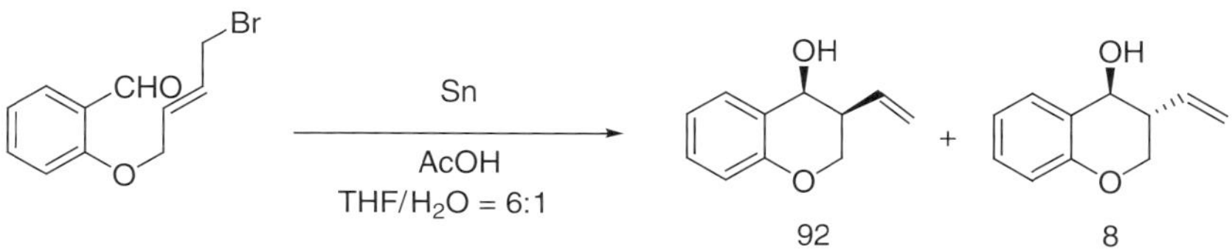

(Eq. 9)

Torii and coworkers found that when a mixture of alcohol/water/acetic acid was used as the solvent the tin/aluminum-mediated allylation can be carried out with the less expensive allyl chloride, instead of allyl bromide.[36] When combined with stoichiometric amounts of aluminum powder, both stoichiometric and catalytic amounts of tin are effective. The tin-mediated allylations can be activated through ultrasonic radiation,[37,38] instead of using aluminum powder and hydrobromic acid. When a mixture of aldehyde and ketone was subjected to the reaction, highly selective allylation of the aldehyde was achieved. A higher temperature can be used instead of aluminum powder.[39] Under such conditions, after oxidation of the product with ferric chloride, allyl quinones were obtained from 1,4-quinones. A phase transfer catalyst is found to help the allylation mediated by tin.[40] Recently, nanometer tin was found to provide enhanced reactivity in mediating allylation of aldehydes or ketones in distilled or tap water without any other assistance such as heat or ultrasonic or acidic media.[41] Tin-mediated allylation reactions in water/organic solvent mixtures were also carried out electrochemically, with the advantage that the allyltin reagent could be recycled.[42]

When 2-bromo- and 2-acetoxy-3-bromo-1-propene were used, the allylation with tin produced the corresponding functionalized coupling products (Eq. 4.10).[43] In the case of 2,3-dibromopropene, the reaction occurred exclusively through allylation in the presence of the vinylic bromide. The reaction can tolerate other electrophiles such as a nitrile (—CN) or an ester (—COOR).

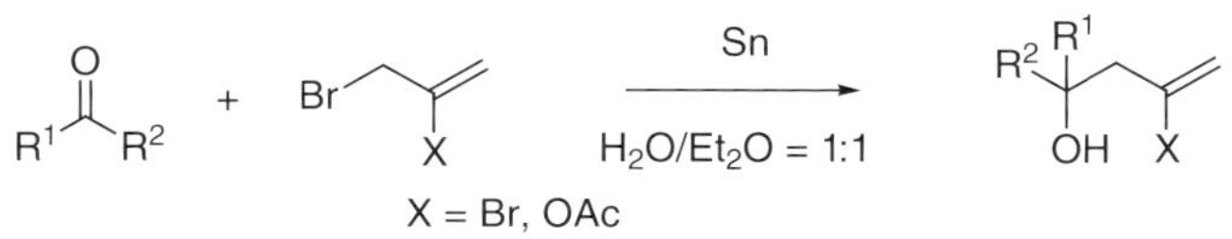

(Eq. 10)

Besides metallic tin, Sn(II) is also a good reducing reagent and can thus mediate allylation of carbonyl compounds in aqueous media. In the presence of a palladium catalyst, allylic alcohols[44] or carboxylates[45] can be used instead of allyl halides (Eq. 4.11). The diastereoselectivity of this palladium-catalyzed allylation with substituted crotyl alcohols was solvent dependent. A mixture of water and THF or DMSO, instead of the organic solvent alone, led to an increased diastereoselectivity, possibly due to coordination of water onto palladium or tin intermediates. The reaction of 1-bromobut-2-ene with aldehydes in a dichloromethane/water biphasic system at 25°C led to α-regioselective addition with $SnBr_2$ to produce 1-substituted pent-3-en-1-ols, and γ-regioselective addition to with $SnBr_2/Bu_4NBr$ to produce 1-substituted 2-methylbut-3-en-1-ols (Eq. 4.12).[46] Transition metals such as cupric chloride or copper,[47] hydrophilic palladium complex[48,49] and $TiCl_3$[50] can also catalyze the allylation of aldehydes using stannous chloride and allyl halides. As in the case of tin metal, sonication is also effective for the Sn(II) allylation.[51] Bis-allylation of dialdehydes or their acetals occurred with allyl bromide, tin(II) chloride and potassium iodide in water or water/THF (Eq. 4.13).[52]

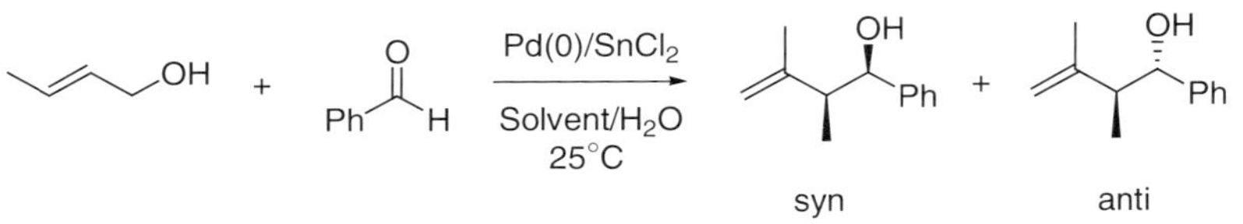

	syn	anti
THF/H_2O (10 equiv.)	16	84
DMSO/H_2O (3 equiv.)	86	14

(Eq. 11)

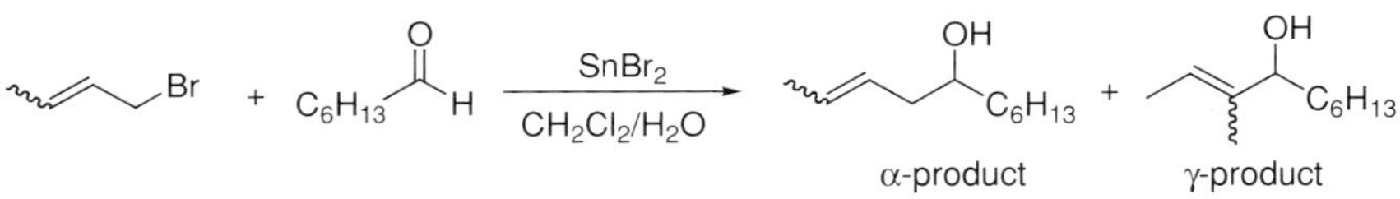

	α-product	γ-product
With Bu_4NBr	91	9
Without Bu_4NBr	91	9

(Eq. 12)

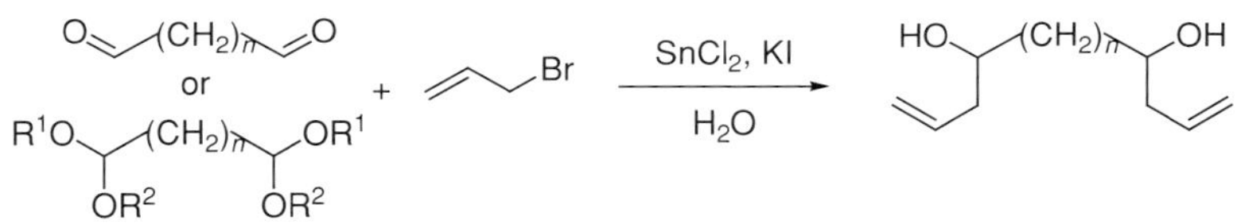

(Eq. 13)

In the presence of $PdCl_2(PPh_3)_2$, HCl, and $SnCl_2$, allylation of aldehydes can be performed by allenes in aqueous/organic media with good regio- and stereoselectivity. The reaction of 1,1′-dimethylallene and $SnCl_2$ with PhCHO gave the corresponding carbonyl allylation product in 95% isolated yield (Eq. 4.14). The reaction likely occurs via allylation of the aldehyde by the allyltrichlorotin generated *in situ* from hydrostannylation of the starting allene.[53]

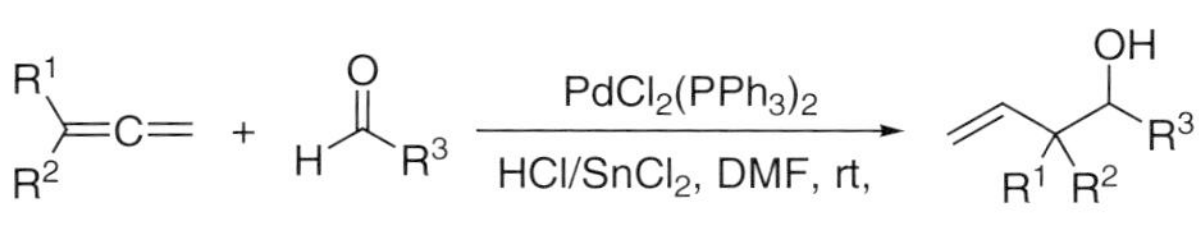

(Eq. 14)

Allylations (as well as allenylations and propargylations) of carbonyl compounds in aqueous media can also be carried out with preformed organic tin reagents, rather than using tin metal.[54–56] Using scandium triflate as a catalyst, the allylation reaction of a wide variety of carbonyl compounds with tetraallyltin was successfully carried out in aqueous media.[57] Exclusive aldehyde selectivity was observed when both aldehydes and ketones were present in a mixture of aqueous HCl and THF.[58] Other Lewis and protonic acids have also been able to promote the allylation with preformed allyltin reagents. Methyltin trichloride and In(III) chloride promote the addition of aldehydes to cyclic allylic stannanes providing good yields of the corresponding homoallylic alcohols.[59] Bu_4NBr/PbI_2 acts as an effective catalyst for the allylation of aldehydes with allylic tin reagents in water with a high syn-selectivity irrespective of their E/Z geometry.[60] A Lewis acid-surfactant-combined catalyst[61] and a divinylbenzene-cross-linked polystyrene-supported $PhSO_3Sc(O_3SCF_3)_2$[62] catalyzed allylation by allyltin reagents in water. Cadmium perchlorate was found to catalyze allylation reactions using allyltributyltin in aqueous media very efficiently in the presence of N,N,N',N'',N''-pentamethyldiethylenetriamine or 2,9-dimethylphenanthroline as a ligand.[63]

For synthetic applications, Whitesides and coworkers reported the first application of aqueous medium Barbier–Grignard allylation to carbohydrate synthesis using tin in an aqueous/organic solvent mixture (Eq. 4.15).[64] Higher carbon aldoses were obtained by the ozonolysis of the deprotected allylation product, followed by suitable derivatizations. A higher diastereoselectivity was observed when a hydroxyl group was present at C-2. However, no reaction was observed under the reaction conditions when there was an *N*-acetyl group present at C-2 position.

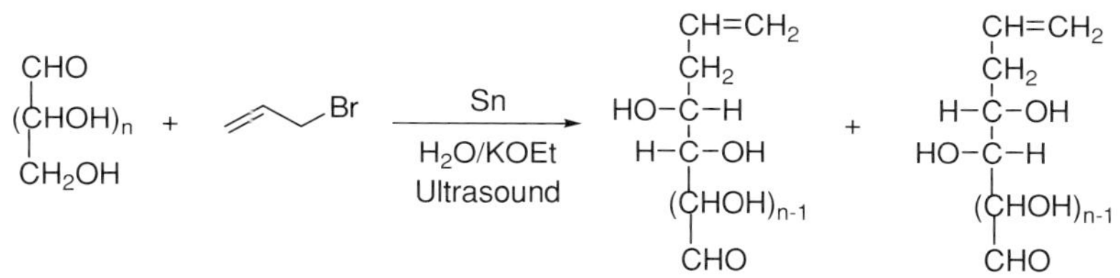

(Eq. 15)

Mediated by indium

In 1991, Li and Chan reported the first use of indium to mediate Barbier–Grignard-type reactions in water.[65] The work was designed on the basis of the first ionization potentials of different elements.[66] Indium has the lowest first ionization potential relative to the other metal elements near it in the periodic table, yet it is not sensitive to boiling water or alkali and does not form oxides readily in air. Such special properties are highly desirable for aqueous Barbier–Grignard-type reactions. Indeed, indium has been shown to be the most effective metal for such reactions (Eq. 4.16). When allylations were mediated by indium in water, the reactions went smoothly at room temperature without any promoter, whereas zinc- and tin-mediated allylations usually require acid catalysis, heat, or sonication. Furthermore, many functional groups can tolerate the indium-mediated allylations due to its mild conditions. The coupling of ethyl 2-(bromomethyl)acrylate with carbonyl compounds proceeds equally well under the same reaction conditions, which makes the synthesis of various sialic acids possible. Gordon and Whitesides found that indium-mediated allylations at room temperature gave results comparable to the tin-mediated allylations at reflux.[67] Replacement of the aqueous phase with 0.1 N HCl further increased the rate of the reaction most likely due to the generation of fresh indium surface. The transformation can also be carried out with preformed allylindium chloride.

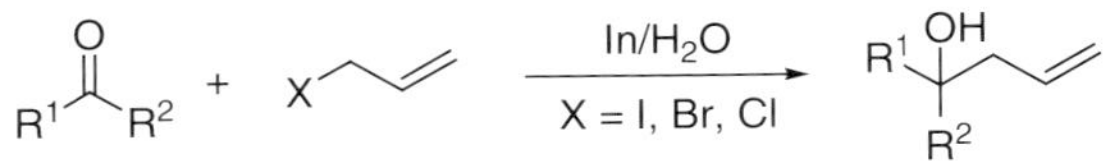

(Eq. 16)

Instead of using a stoichiometric amount of indium, the allylation of aldehydes and ketones can be carried out by using catalytic amounts of indium(III) chloride in combination with aluminum or zinc metals in a THF/water (5:2) mixture at room temperature.[68] The reaction is much slower relative to the same reaction mediated by a stoichiometric amount of indium. When the reaction was carried out in anhydrous THF, the yield dropped considerably, while side reactions (such as reducing to alcohol) increased. The combinations of Al/$InCl_3$ or Zn/$InCl_3$ gave comparable results. The indium-mediated allylation carried out with allylstannanes in combination with indium chloride in aqueous medium was reported by Marshall and Hinkle.[69] Allyl indium was proposed as the reaction intermediate. Various allylic compounds undergo palladium-catalyzed allylation of carbonyls via the transmetalation of π-allylpalladium(II) intermediates with indium in the presence of indium trichloride in aqueous media.[70]

Because of its superior reactivity, the indium-mediated reaction in water has found wider applications in natural product synthesis. Chan and Li[71] reported an efficient synthesis of (+)-3-deoxy-D-*glycero*-D-*galacto*-nonulosonic acid (KDN) (Scheme 4.4) using the indium-mediated alkylation reaction in water. A similar synthesis of 3-deoxy-D-manno-octulosonic acid (KDO) led primarily to the undesired diastereomer. However, through the disruption of the newly generated stereogenic center,[72] a formal synthesis[73] of KDO was completed (Scheme 4.5). In contrast to the tin-mediated reactions, the indium-mediated reaction also occurred on a substrate with an *N*-acetyl group present at C-2. Whitesides et al. reported the synthesis of *N*-acetyl-neuraminic acid (Scheme 4.6),[75] as well as other sialic acid derivatives based on this strategy and the use of indium is essential. KDO was synthesized via indium-mediated allylation of 2,3:4,5-di-*O*-isopropylidene-D-arabinose (Scheme 4.7).[74] In this case, the desired product became the major product due to the protection of the α-hydroxyl group. Since carboxylic acid functionality on allyl halides is compatible with indium-mediated allylations,[76,77] the reaction of 2-(bromomethyl)acrylic acid with carbonyl compounds and indium in water gives the corresponding γ-hydroxyl-α-methylenecarboxylic acids in good yields. The indium-mediated reaction of α-(bromomethyl)acrylic acid with sugars further shortened the already concise sialic acids' synthesis to only two steps. Both KDN and *N*-acetyl-neuraminic acid have been synthesized in such a way (Scheme 4.8). The indium-mediated allylation reaction has also been applied to the elongation of the carbon chain of carbohydrates in forming higher analogs (Scheme 4.9)[78] and to the synthesis of deoxy-sugars (Fig. 4.4).[79] The reaction has also been applied to erythrose.[80] C-branched sugars

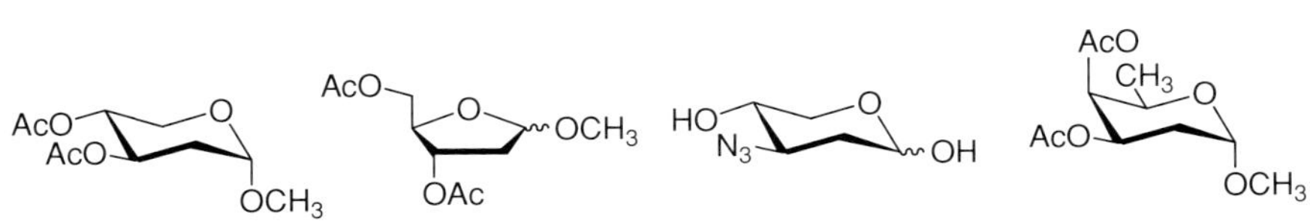

Figure 4.4 Deoxy-sugars synthesized via indium-mediated allylation in water.

or C-oligosaccharides are obtainable through indium-promoted Barbier-type allylations in aqueous media.[81] The chemistry has been extended to six-carbon sialic acid derivatives by Chappell and Halcomb[82] and the protocol has been further improved by Warwel and Fessner.[83]

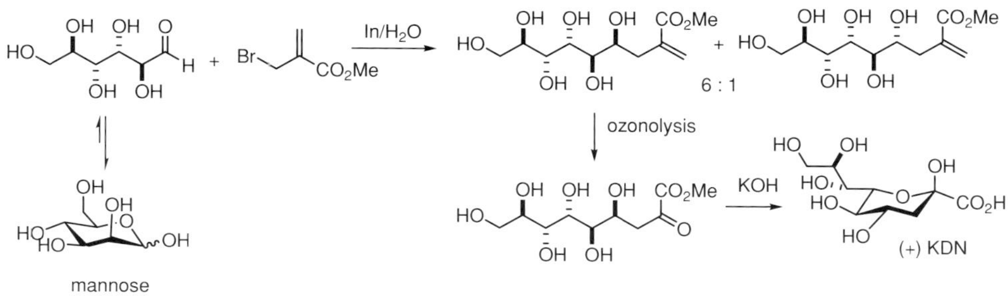

Scheme 4.4 Synthesis of (+) KDN.

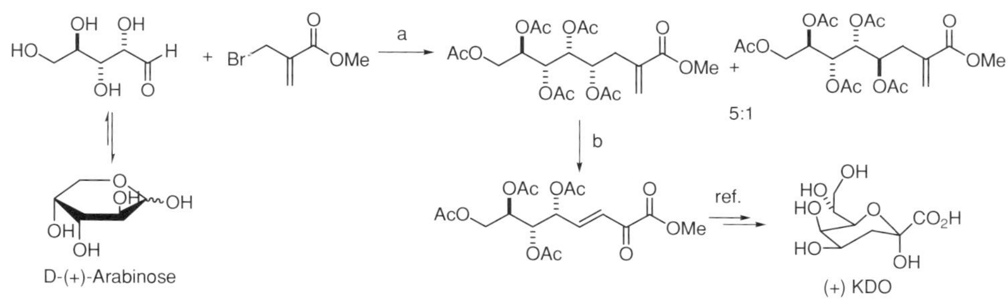

Scheme 4.5 Synthesis of (+) KDO: (a) In/H_2O, vigorous stirring, then Ac_2O/pyridine/4-dimethylaminopyridine, 79%; (b) O_3/methylene chloride $-78\,^\circ C$ to rt, then column chromatography; 67%.

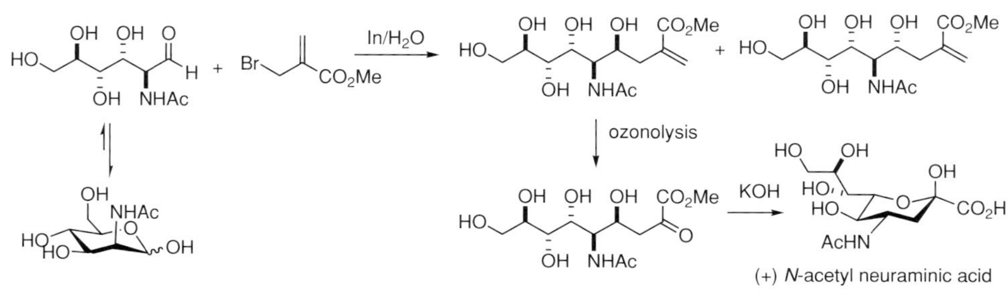

Scheme 4.6 Synthesis of (+) *N*-acetyl neuraminic acid.

The coupling of lower carbohydrates with dimethyl 3-bromopropenyl-2-phosphonate in water generated phosphonic acid analogs of both KDN and *N*-acetyl neuraminic acid by using the indium-mediated allylation (Scheme 4.10).[84]

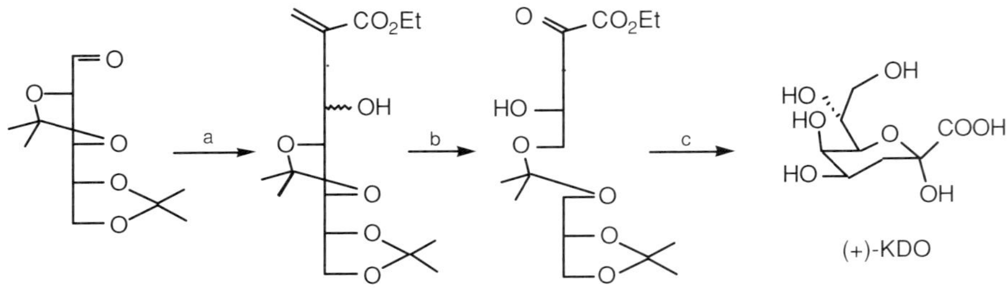

Scheme 4.7 Synthesis of KDO: (a) In/ethyl (bromomethyl)acrylate, 10% formic acid, aqueous MeCN, 61%; (b) O_3, MeOH, -78°C to rt 92%, (c) Trifluoroacetic acid, NH_4OH, 55%.

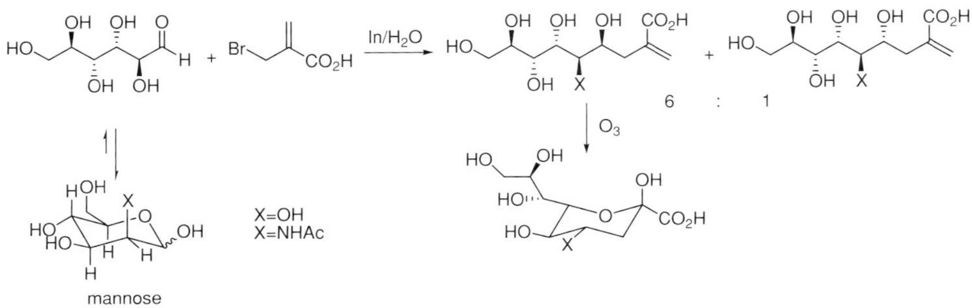

Scheme 4.8 Synthesis of sialic acids using bromomethacrylic acid.

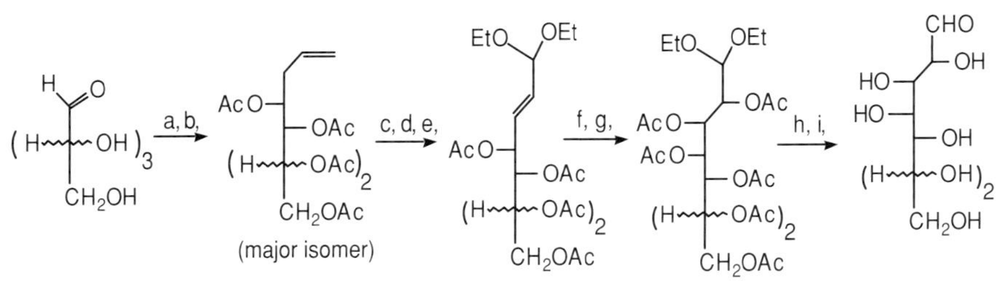

Scheme 4.9 Synthesis of higher carbon sugars: (a) In/allyl bromide, ultrasound; (b) Ac_2O/pyridine/4-dimethylaminopyridine; (c) OsO_4, KIO_4; (d) TBAF; (e) H^+/$HC(OEt)_3$; (f) OsO_4, *N*-methylmorpholine *N*-oxide; (g) Ac_2O/pyridine/4-dimethylaminopyridine; (h) NaOMe/MeOH; (i) H^+.

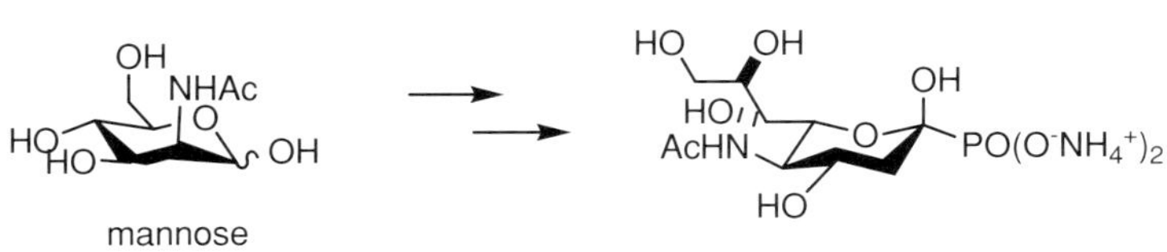

Scheme 4.10 Synthesis of sialic acid analogs.

Besides being useful in the synthesis of carbohydrates, the indium-mediated allylation has also been applied to other targets. The combination of 2-halomethyl-3-halo-1-propene with carbonyl compounds mediated by indium in water generates bis-allylation products (Eq. 4.17).[85] Reaction of 1,3-dihaloalkene with carbonyl compounds mediated by indium in water gave predominately 1,1-bis-allylation product.[86] Indium efficiently mediates the alkylation of various aldehydes with 3-bromo-2-chloro-1-propene in water at room temperature. Subsequent treatment of the compound with ozone in methanol followed by workup with sodium sulfite provided the desired hydroxyl ester in high yield.[87] The addition of ytterbium trifluoromethanesulfonate [$Yb(OTf)_3$] enhances both the reactivity and the diastereoselectivity of the allylation reaction of the glucose-derived aldehyde with allyl bromide mediated by indium to give a nonchelation product as the major diastereomer.[88] Enolizable 1,3-dicarbonyl compounds undergo efficient carbonyl allylation reactions in aqueous medium (Eq. 4.18).[89] A variety of 1,3-dicarbonyl compounds have been alkylated successfully using allyl bromide or allyl chloride in conjunction with either tin or indium. The reaction can be used readily for the synthesis of cyclopentane derivatives (Eq. 4.19).[90] Application of aqueous Barbier-type reaction in a carbocycle ring enlargement methodology was developed by Li et al. (Eq. 4.20).[91] By using the indium mediated Barbier-type reaction in water, 5-, 6-, 7-, 8-, and 12-membered rings are enlarged by two carbon atoms into 7-, 8-, 9-, 10-, and 14-membered ring derivatives respectively. The use of water as a solvent was found to be critical for the success of the reaction. Similar ring expansion in organic solvents was not successful. The ring expansion has also been applied to the synthesis of a heterocyclic medium ring (C.J. Li, D.L. Chen, Synlett.1999, 735). Similar one carbon-ring expansion was also reported.[92]

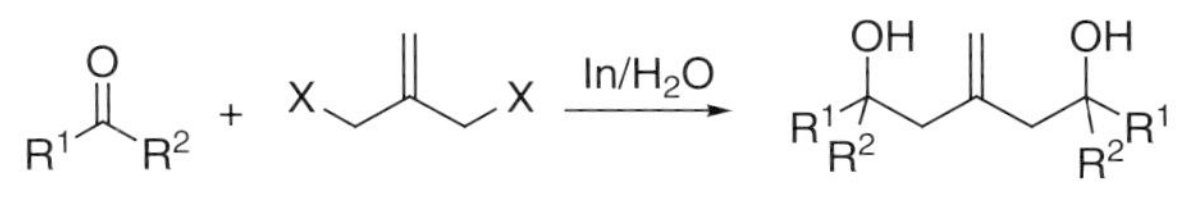

(Eq. 17)

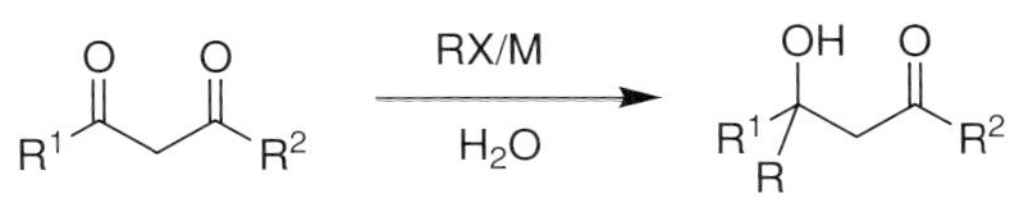

(Eq. 18)

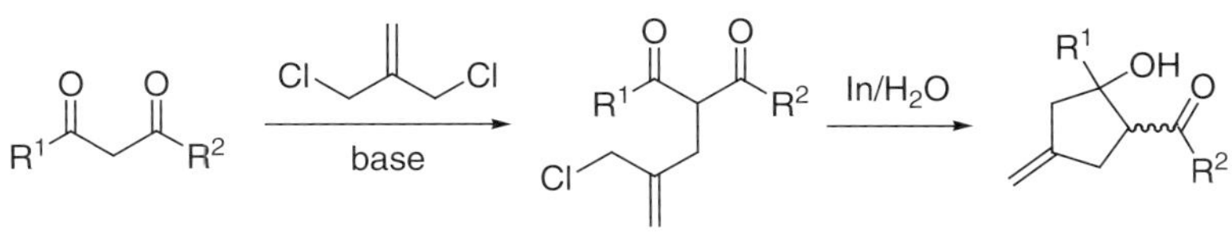

(Eq. 19)

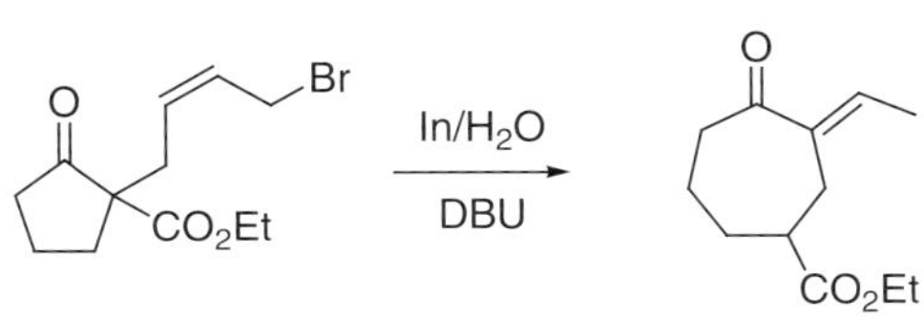

(Eq. 20)

The indium-mediated allylation has also been used to reach an advanced intermediate in the synthesis of antillatoxin.[93] In the presence of a lanthanide triflate, the indium-mediated allylation of Z-2-bromocrotyl chloride and aldehyde in saturated NH_4Cl under sonication yielded the desired advanced intermediate in a 1:1 mixture of diastereomers in 70% yield.

On the other hand, the coupling of methyl (*Z*)-2-(bromomethyl)-2-butenoate with the aldehyde under the same conditions generated the desired homoallylic alcohol in 80% yield with a high 93:7 syn/anti selectivity (Eq. 4.21).[94] The indium-mediated allylation of trifluoroacetaldehyde hydrate (R = H) or trifluoroacetaldehyde ethyl hemiacetal (R = Et) with an allyl bromide in water yielded α-trifluoromethylated alcohols.[95] Lanthanum triflate-promoted indium-mediated allylations of aminoaldehydes in aqueous media generated β-aminoalcohols stereoselectively.[96] Indium-mediated intramolecular carbocyclization in aqueous media generated fused α-methylene-γ-butyrolactones.[97] An advanced intermediate for azaspiracids was only accessible by the indium-mediated allylation.[98] Other potentially reactive functionalities, such as azide, enone, and ketone, did not interfere with the reaction.

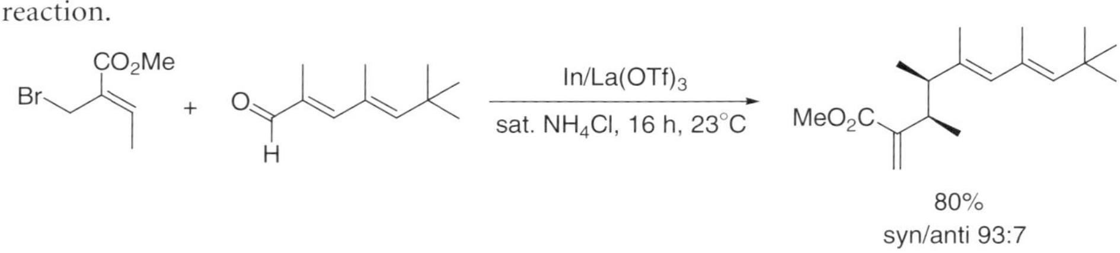

(Eq. 21)

Indium-mediated allylation of gem-diacetates gave excellent yields of the corresponding homoallylic acetates in aqueous media.[99] Allylation of α-diketones by treatment with allyl bromide and indium in water/THF 4:1 gives diallyl diols with moderate stereoselectivities. Ring-closing metathesis of the diallyl diols with Grubbs' ruthenium olefin metathesis catalyst, followed by catalytic hydrogeneration and diol cleavage with lead tetraacetate, gave *cis*-alkenediones (Eq. 4.22).[100] The allylation was applied in the total asymmetric synthesis of the putative structure of the cytotoxic diterpenoid (−)-sclerophytin and of the authentic natural sclerophytins.[101] An aqueous indium chemistry/ring-closing metathesis tandem reaction was reported by Mendez-Andino and Paquette as a general route to fused-ring α-methylene-γ-butyrolactones.[102] Vogel et al. reported the synthesis of moenomycin analogs with modified lipid side chains from indium-mediated Barbier-type reactions.[103]

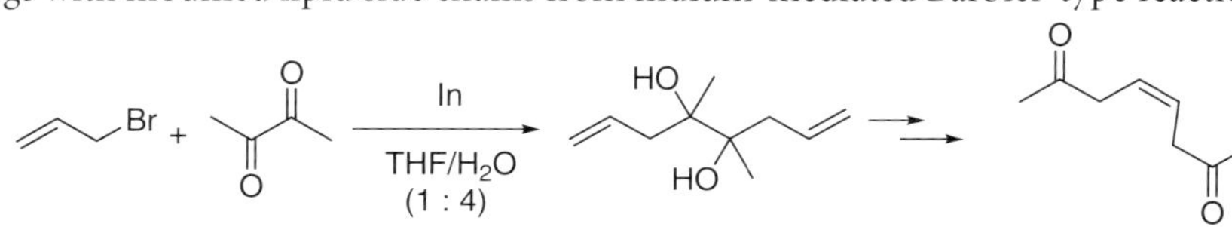

(Eq. 22)

Allylation of the C-3 position of the cephem nucleus was accomplished by either indium mediated or indium trichloride promoted tin-mediated allylation reactions in aqueous media. Both methods gave 3-allyl-3-hydroxycephams in moderate to excellent yields.[104] The synthesis of (*E*)-β-methyl Baylis–Hillman adducts was obtained with high E/Z selectivity (>93%) in modest to good yields by using an indium-mediated allylation reaction followed by a simple base-catalyzed isomerization step (Eq. 4.23).[105]

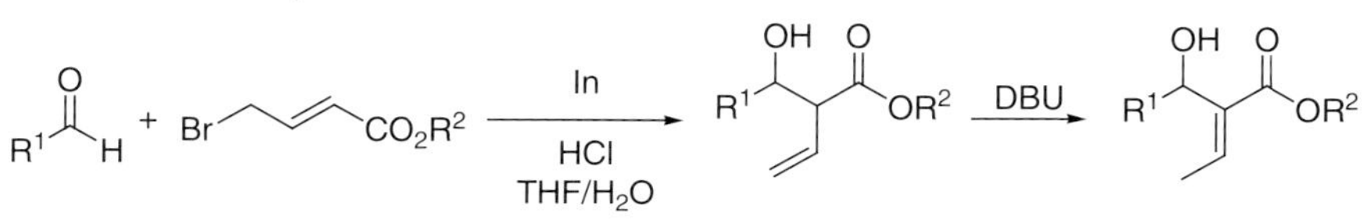

(Eq. 23)

Indium-mediated allylation of α-chlorocarbonyl compounds with various allyl bromides in aqueous media gave the corresponding homoallylic chlorohydrins, which could be transformed into the corresponding epoxides in the presence of a base (Eq. 4.24).[106] The reaction of cyclopentadienylindium(I) with aldehydes gives isomeric mixtures in aqueous media.[107] Indium-promoted reaction of 1,4-dibromo-2-butyne with carbonyl compounds gave 1,3-butadien-2-ylmethanols via the allenic intermediates.[108]

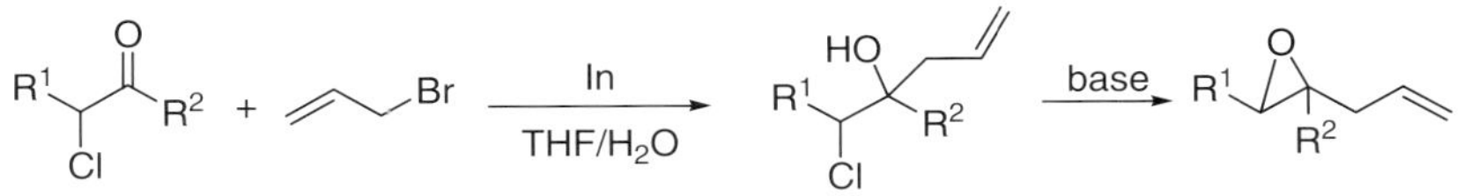

(Eq. 24)

Linear α-homoallylic alcohol adducts were obtained with high regioselectivities in moderate to good yields using allylic indium reagents in the presence of 10 M water.[109] A new mechanism is proposed for the α-regioselective indium-mediated allylation reaction in water. It is suggested that the initially formed γ-adduct undergoes a retro-ene reaction followed by a 2-oxonia [3,3]-sigmatropic rearrangement to furnish the α-adduct.[110] Indium trichloride-catalyzed indium-mediated allylations of dihydropyrans and dihydrofurans in water afforded the allylated diols in moderate to high yields.[111] Indium-mediated allylation reaction in THF/water 1:1 was used as the key step in the total synthesis of dysiherbaine (Scheme 4.11).[112] A chemoenzymatic methodology has been developed using indium-mediated allylation of heterocyclic aldehydes under aqueous conditions followed by *Pseudomonas cepacia* lipase-catalyzed enantioselective acylation of racemic homoallylic and homopropargylic alcohols in organic media.[113]

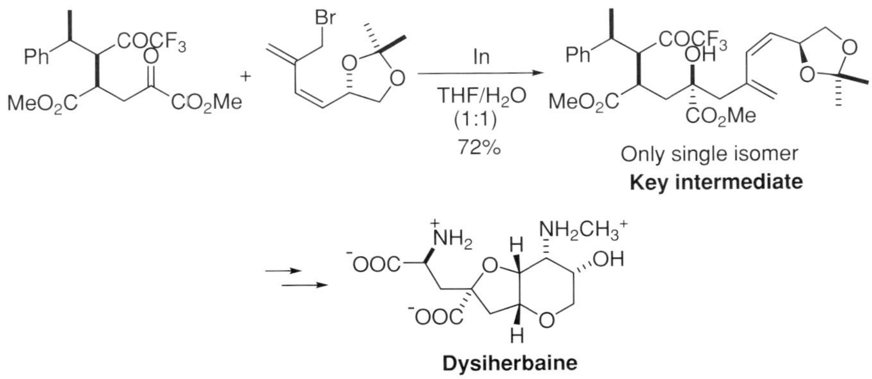

Scheme 4.11

Mediated by other metals

As C—M bonds of most post-transition metals have a strong covalent character and because many reactions can occur also via radical and radical anion processes on metal surfaces, it is not surprising that many other metals have been found to mediate the Barbier–Grignard-type reactions in water.

Covalent C—B-based reagents have been applied with some success. For example, the reaction of aldehydes with potassium allyl- and crotyltrifluoroborates, in biphasic media or water alone, provided homoallylic alcohols in high yields (>94%) and excellent diastereoselectivities (dr > 98:2). The presence of a phase transfer catalyst (e.g. Bu_4NI) significantly accelerates the rate of reaction, whereas added fluoride ion retards the reaction (Eq. 4.25).[114] These results are consistent with the six-membered cyclic transition state argument in the discussion of C—M reactivities. The method was applied to the total synthesis of the antiobesity agent tetrahydrolipstatin (orlistat).[115]

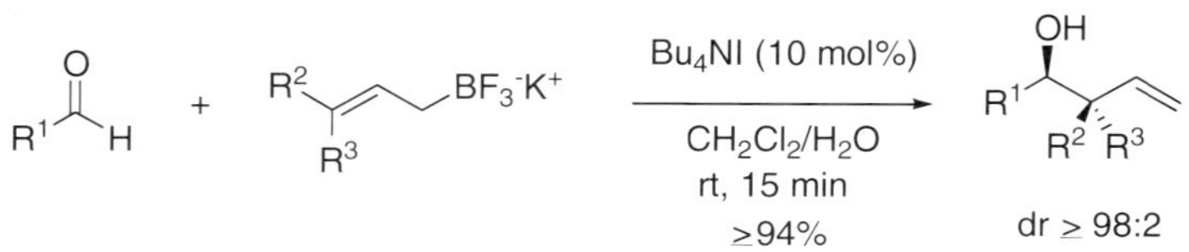

(Eq. 25)

Metal-mediated allylation of difluoroacetyltrialkylsilanes with various allyl bromides in aqueous media formed homoallylic alcohols exclusively. The common Brook rearrangement, carbon to oxygen silyl migration, was totally suppressed with no detectable formation of silyl enol ether.[116] The reaction afforded high syn-selectivity regardless of the allylic bromide geometry. $Sc(OTf)_3$ catalyzes allylation of hydrates of α-keto aldehydes and glyoxylates and activated aromatic aldehydes with allyltrimethylsilane in H_2O/CH_3CN to give α-keto and α-ester homoallylic alcohols in good to excellent yields.[117]

Metallic gallium was used to mediate the allylation of carbonyl compounds in water (Eq. 4.26).[118] The reaction can also be carried out by using preformed allylgallium reagents.[119] The corresponding homoallyl alcohols were obtained in high yields without the assistance of acidic media or sonication. Li and Zhang reported that even magnesium can mediate Barbier–Grignard allylation of benzaldehyde in water in low yield together with pinacol-coupling products (Eq. 4.27).[120]

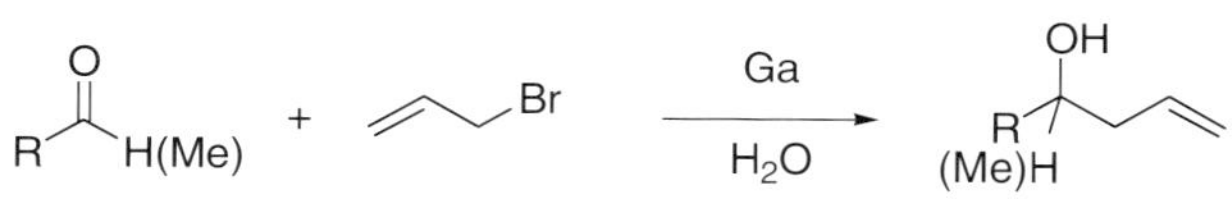

(Eq. 26)

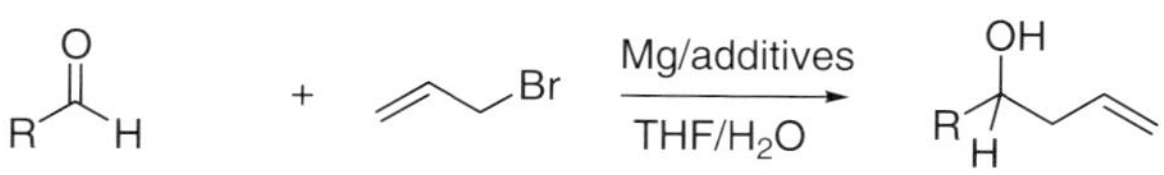

(Eq. 27)

Homoallylic alcohols can also be obtained from allylation of aldehydes and ketones with allyl bromide promoted by metallic lead in aqueous media.[121] Manganese in combination with a catalytic amount of copper or NH_4Cl is also effective for mediating such allylations and pinacol-coupling reactions in aqueous medium. Manganese offers a higher reactivity than other metals and a complete chemoselectivity toward allylation of aromatic aldehydes was observed.[122] Antimony metal, in aqueous 1 M HCl or DCl solution, reacts with allyl bromide and aldehydes to give the corresponding homoallylic alcohols in good yield. The reaction proceeds through the formation of allylstibine intermediates.[123] Wada et al.[124] reported that metallic bismuth, in combination with aluminum powder and hydrobromic acid, can also be used for allylation reaction. Again, the reaction is more effective than the same one conducted in an organic solvent. As a comparison, the allylation of phenylacetaldehyde carried out in a mixture of THF/water at room temperature gave the corresponding alcohol in 90% yield, whereas under otherwise same conditions, the use of THF as solvent led to decreased yields and irreproducible results. Other effective combinations include

Al(0)/$BiCl_3$, Zn(0)/$BiCl_3$, Fe(0)/$BiCl_3$, and Mg/$BiCl_3$.[125] Bismuth-mediated allylation was found to be promoted by the presence of fluoride ion[126] or sonication.[127] Allylation of aldehydes carried out by electrochemically regenerated bismuth metal in an aqueous two-phase system was reported by Minato and Tsuji.[128]

Regio- and stereoselectivity

Qualitatively, most metal-mediated allylation reactions in aqueous medium display similar regio- and stereoselectivities. Such studies have been carried out most extensively on the indium-mediated allylations. For regioselectivities of the allylic moiety, both electronic and steric effects are important. Usually, the carbon–carbon bond formation occurs at the more substituted carbon of the allyl halide, irrespective of the position of halogen in the starting material (Eq. 4.28). However, the carbon–carbon bond forms at the less substituted carbon when the γ-substituents of allyl halides are large enough (e.g. trimethylsilyl or *tert*-butyl) as shown by Chan and Isaac.[129]

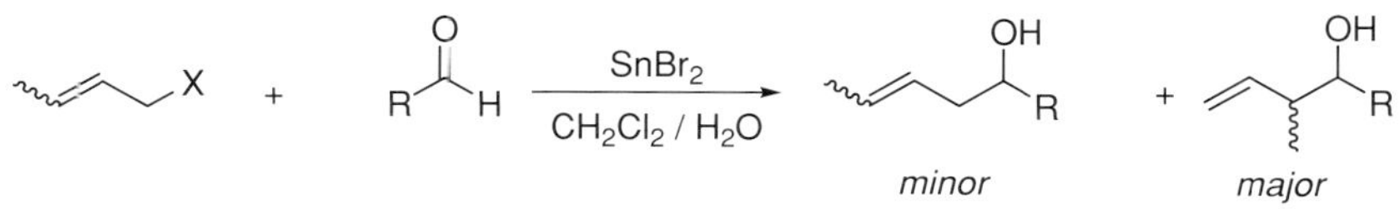

(Eq. 28)

The stereochemistry of the allylation reaction in aqueous medium is somewhat analogous to that in organic medium. In terms of diastereoselectivity, two types of situation prevail (Type A and Type B) (Scheme 4.12). The Type A situation usually gives an anti-diastereoselectivity that is independent of the stereochemistry of the double bond in the allyl bromide moiety. The diastereoselectivity (anti/syn ratio) is governed by the steric size of the substituent on the aldehydes. The anti/syn ratio increases as the size of the aldehyde R group increases (Scheme 4.13).

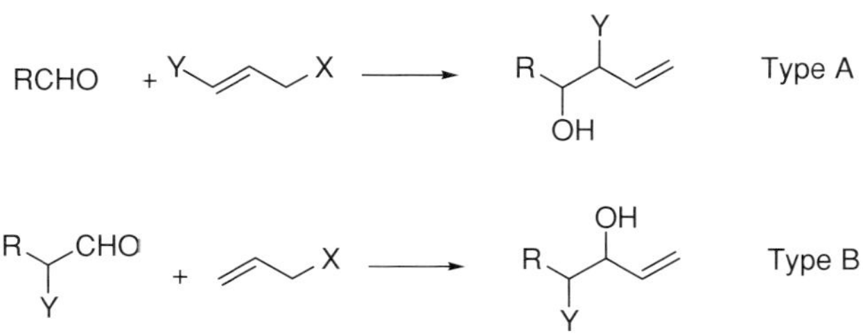

Scheme 4.12 Two types of diastereoselectivities in allylation reactions.

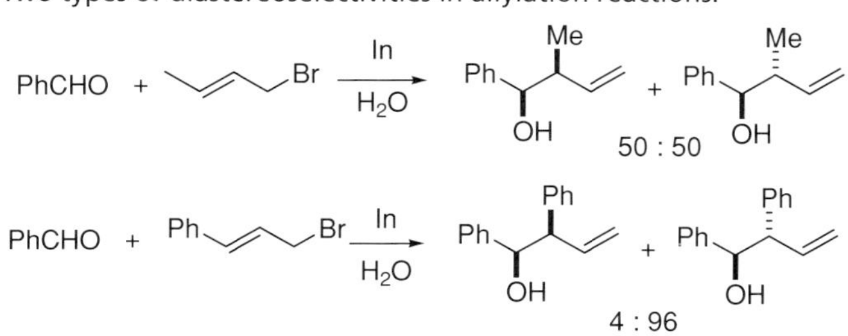

Scheme 4.13 Effect of substituents on the diastereoselectivity of indium mediated allylation of aldehydes.

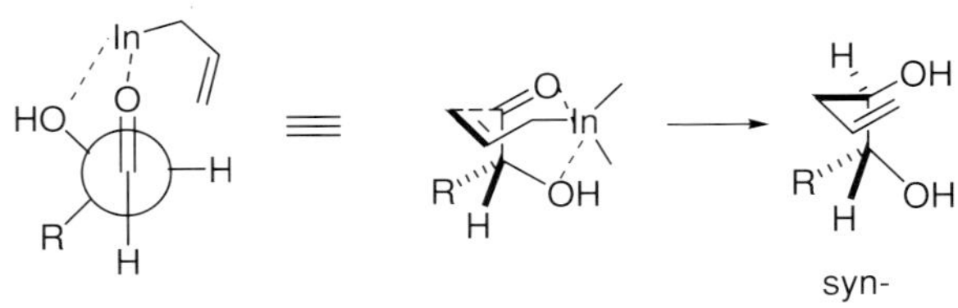

Figure 4.5 Chelation model in indium-mediated allylations of aldehyde inaqueous media.

Within the Type B situation, the reaction can favor either syn- or anti-diastereoselectivity, depending on the properties of the α-substituents. Normally, the addition of C-nucleophiles to chiral α-alkoxyaldehydes in organic solvents is opposite to the Cram's rule (Scheme 4.14). The anti-Cram selectivity has been rationalized on the basis of chelation control.[130] In water, it was shown that the presence of a strong α-chelating group, such as a hydroxyl, leads to syn-product, whereas a non-α-chelating group, such as a methyl, produces anti-product (Scheme 4.15). The regio- and diastereoselectivity of metal-mediated reactions have been studied extensively by both Chan and Paquette.[131] For example, when a weak chelating alkoxyl is present, allylation in an organic solvent usually favors a chelation-controlled product, whereas the nonchelating anti-preference was observed in the reactions of α-alkoxyaldehydes with allyl bromide/indium in water (Scheme 4.16). On the other hand, in the allylation of α-hydroxyaldehydes with allyl bromide/indium, the syn-isomer is the major product. The syn-selectivity can be as high as 10:1 (syn/anti) in the reaction of arabinose. It is argued that in this case, the allylindium intermediate coordinates with both the hydroxy and the carbonyl function as in Fig. 4.5, leading to the syn-adduct. Thus, it is possible to reverse the diastereoselectivity of an allylation simply either by use of a free hydroxyl group or by protecting it as an alkoxyl. An application of such a selectivity change is utilized in the synthesis of muscarine, carried out by Chan and Li.[132] The stereogenic center can also be further away from the carbonyl group. Such an example can be found in Waldmann's studies of the diastereoselectivity of allylations using proline benzyl ester as a chiral auxiliary to produce α-hydroxylamides. The diastereoselectivities were around 4–5:1 (Scheme 4.17).[133]

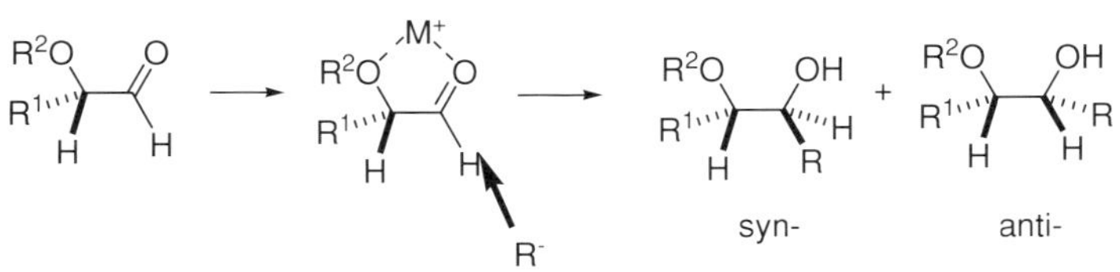

Scheme 4.14 Chelation model for the nucleophilic addition of aldehydes.

Coordination control is also used to account for the observed anti-preference in the allylation of β-hydroxybutanal with allyl bromide/indium in water (Scheme 4.18). The reaction leads to the anti-product. In support of the intramolecular chelation model, it was found that if the hydroxy group is converted to the corresponding benzyl or *tert*-butyldimethylsilyl ether, the reaction gives nearly equal amounts of syn- and anti-products.

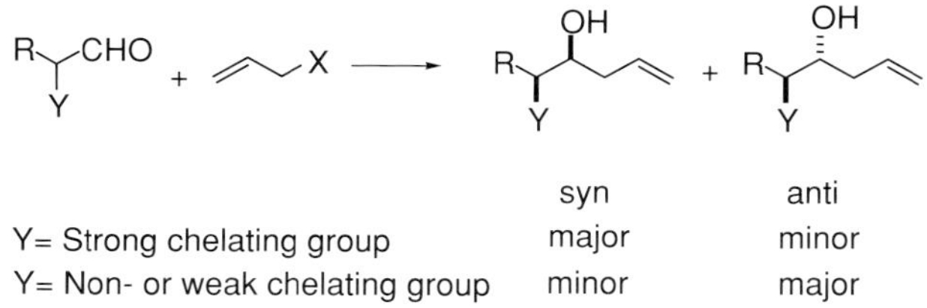

Scheme 4.15 Effect of chelating substituents on the diastereoselectivity.

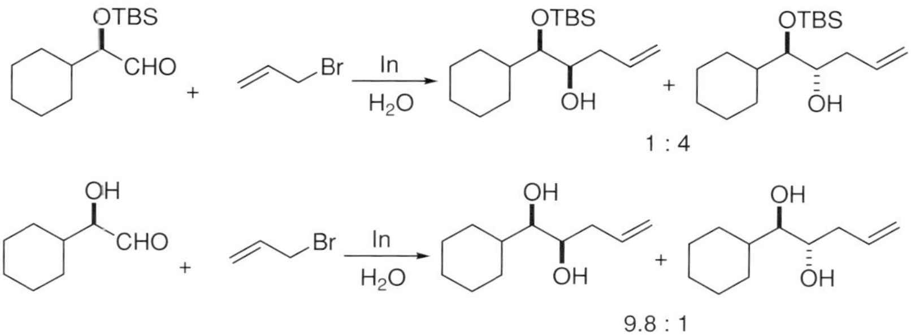

Scheme 4.16 Diastereoselectivity switch with and without protecion of the α-hydroxyl group.

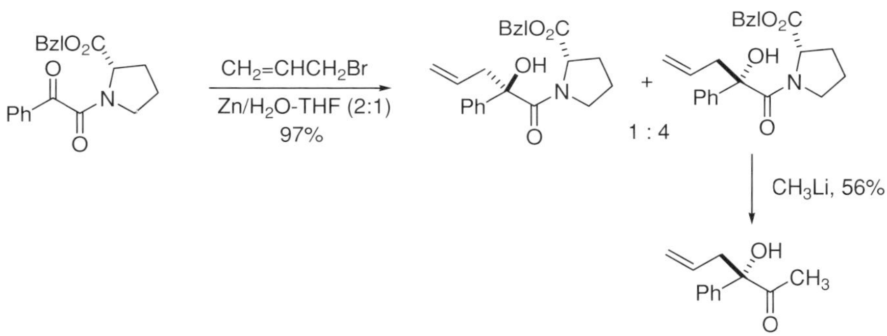

Scheme 4.17 Synthesis of an optically active α-hydroxyl ketone.

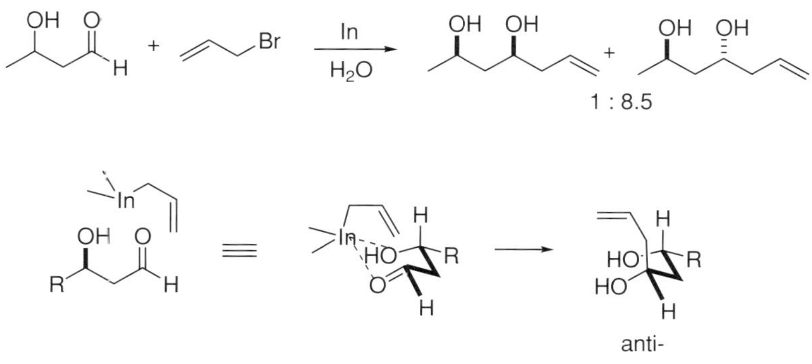

Scheme 4.18 Diastereoselectivity in indium-mediated allylation of β-hydroxyl aldehydes.

It is possible to combine both the Type A and the Type B situations in the coupling of a chiral aldehyde with a substituted allylic halide. Such is the case in the coupling of unprotected aldoses (e.g. glyceraldehyde) with cinnamyl bromide (Scheme 4.19).[134] In such a coupling, two new stereogenic centers are created. It has been found that the syn,syn-isomer is formed preferentially. To account for this, chelation of the allylindium species with the hydroxyaldehyde function and intramolecular attack through a cyclic transition state is postulated. The stereochemistry of the adduct is then dependent on the geometry of the attacking allylindium species.

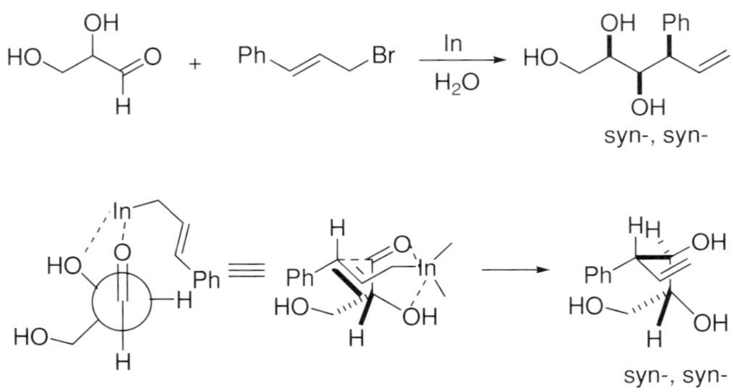

Scheme 4.19 Diastereoselectivity in indium-mediated allylatin of glyceraldehyde in water.

Reactions of racemic as well as optically pure 2-carbonyl-β-lactams with indium and allyl halides offer a convenient asymmetric entry into densely functionalized hydroxy-β-lactams, with good regio- and stereocontrol (Eq. 4.29).[135] A highly stereoselective indium allylation was used for asymmetric synthesis of a highly functionalized THF derivative (Eq. 4.30).[136]

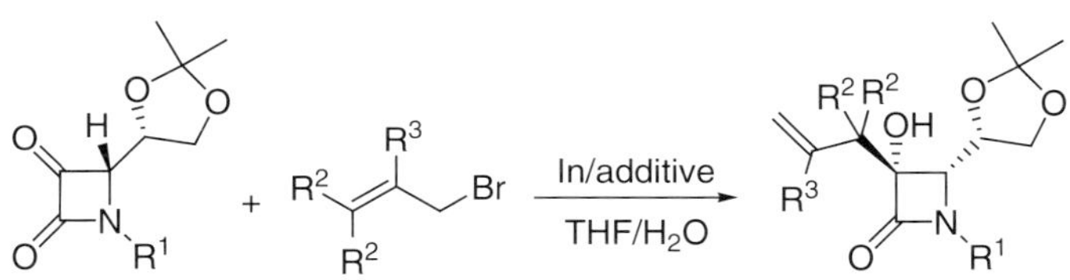

(Eq. 29)

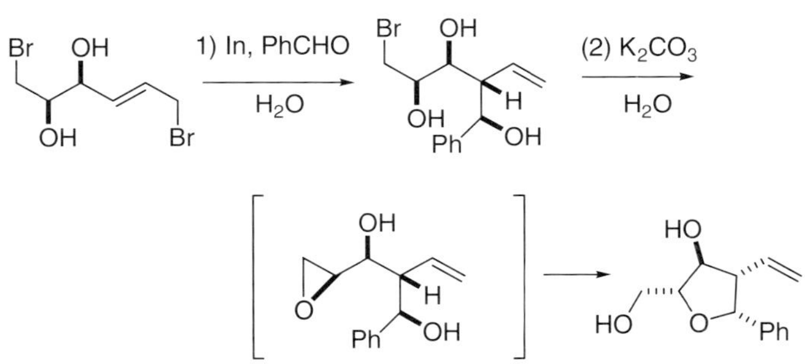

(Eq. 30)

Asymmetric allylation

Recently, asymmetric allylation reactions in aqueous media have been investigated. Native and trimethylated cyclodextrins were found to promote enantioselective allylation of 2-cyclohexenone and aldehyde by using Zn dust and alkyl halides in water/THF 5:1. The reactions proceeded with moderate enantioselectivities and enantiomeric excesses (ee) up to 50% were obtained.[137] The results can be rationalized in terms of formation of inclusion complexes between the substrates and the cyclodextrins and their interaction with the metal surface. In the presence of the monothiobinaphthol (MTB) ligand, aryl ketones were allylated by a mixture of $Sn(CH_2CH{=}CH_2)_4/RSn(CH_2CH{=}CH_2)_3$ (R = Et, Bu), with high enantioselectivity. The presence of water was shown to suppress racemic background allylation. Allylation reactions using the pure components alone were rather ineffective. The (1*R*)-2-mercapto[1,1′-binaphthalen]-2-ol-mediated allylation of acetophenone using a mixture of tetra(2-propenyl)stannane/ethyl tri(2-propenyl)stannane/butyl tri(2-propenyl)stannane gave (αR)-α-methyl-α-(2-propenyl)benzenemethanol in >98% yield and in 86–89% ee (Eq. 4.31). Aliphatic ketones gave a complex mixture of products, whereas only 59% ee was observed with cyclohexyl methyl ketone.[138] A combination of cadmium bromide with chiral diamine ligands was also effective. The diamine ligands were found to accelerate the reactions significantly.[139]

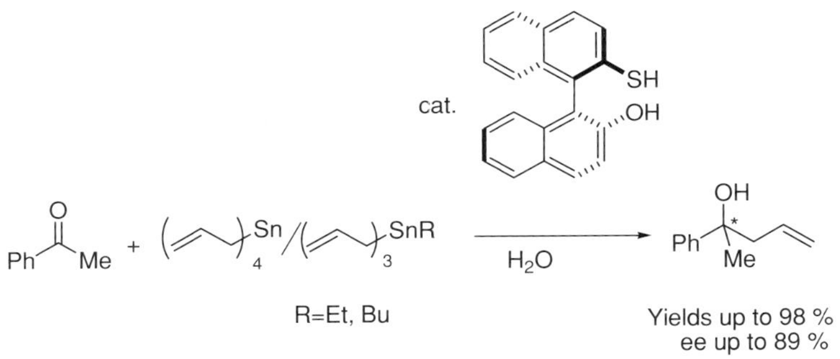

(Eq. 31)

Mechanistic studies

For the mechanism of the metal-mediated Grignard-type reactions in water, Li previously proposed a carbanion/allylmetal/radical triad, in which the specific mechanism of the reaction is dependent on the metal being used (Fig. 4.6).[6] Recently, mechanistic studies of the allylation detected secondary deuterium kinetic isotope effects in the metal-mediated allylation of benzaldehyde in aqueous media.[140] The inverse SDKIE observed for the indium and tin cases are consistent with the polar addition mechanism. For magnesium and antimony, normal SDKIE were observed. These were interpreted as single electron transfer processes on metal surface in the magnesium case, or between the allylmetal and the carbonyl compound in the antimony case.

4.3.2 Allylation of imines and related compounds

Grieco and Bahsas[141] reported that iminium salts generated *in situ* from primary amines and formaldehyde can be allylated with preformed allylstannane under aqueous conditions. Katritzky et al. found that the bismuth(III)/aluminum system also mediated the allylation of iminium cations to give amines.[142] In this case, even methylation with iodomethane

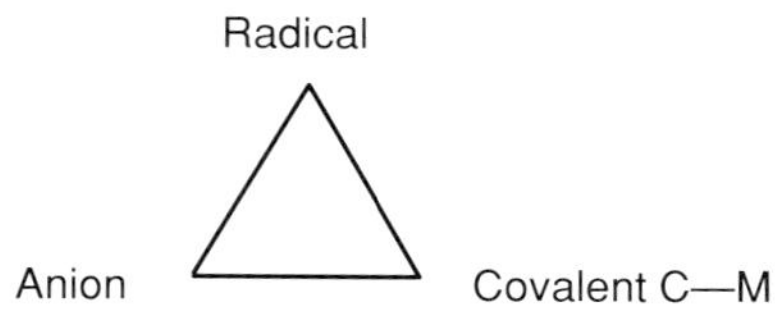

Figure 4.6 Li's Triad for metal-mediated allylations in water.

took place smoothly. Allylation of β-keto aldehydes and functionalized imines by diallyltin dibromide was carried out to generate skipped and conjugated dienes.[143] Chan and Lu reported that the allylation of sulfonimines in water can be mediated by either indium[144] or zinc (Eq. 4.32).[145]

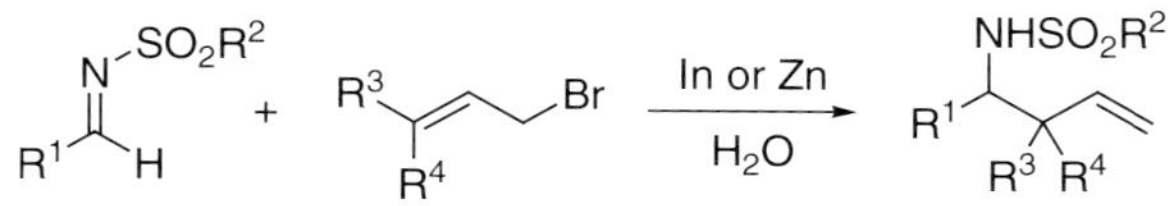

(Eq. 32)

Allylation of acyloyl-imidazoles and pyrazoles[146] with allyl halide mediated by indium in aqueous media provides a facile regioselective synthesis of β,γ-unsaturated ketones. The reaction has been applied to the synthesis of the monoterpene artemesia ketone. Similar products can be obtained by indium-mediated allylation of acyl cyanide.[147] Samarium, gallium, and bismuth can be used as a mediators for the allylation of hydroxylamine and hydrazides in aqueous media in the presence of Bu_4NBr.[148] The reaction with gallium and bismuth can be improved dramatically under microwave activation. Allylation of the nitro group on nitrobenzene derivatives proceeded under similar reaction conditions.[149] Allylation reactions of various benzoylhydrazones with tetraallyltin were carried out in the presence of scandium triflate as a Lewis acid catalyst in aqueous media.[150] A three-component reaction of aldehydes, benzoylhydrazine, and tetraallyltin was catalyzed by scandium triflate in aqueous media, and a three-component synthesis of homoallylic amines starting from aldehyde, amine, and allyltributylstannane were realized with the use of Sn(II) chloride dihydrate in H_2O in the presence of sodium dodecyl sulfate surfactant.[151] Iminium ions, generated in aqueous solution from secondary amines and formaldehyde, undergo a Barbier-type allylation mediated by tin, aluminum, and zinc (Eq. 4.33). The reaction is catalyzed by copper and produces tertiary homoallylamines in up to 85% yield.[152] Imines generated *in situ* from 2-pyridinecarboxaldehyde/2-quinolinecarboxaldehyde and aryl amines undergo indium-mediated Barbier allylation in aqueous media.[153] Crotyl and cinnamyl bromides lead to diastereoselective allylation, with diastereomeric ratios of up to 98:2 by this method.

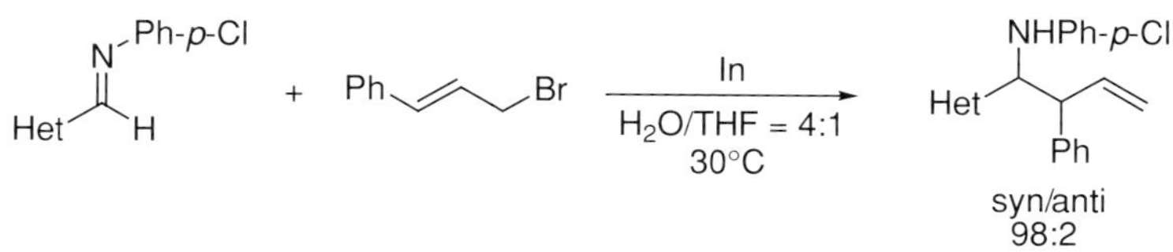

(Eq. 33)

Allylation reactions of electron-deficient imines with allylic alcohol derivatives were effective in the presence of a catalytic amount of palladium(0) complex and indium(I) iodide in the presence of water.[154] Homoallylic *o*-methylhydroxylamines are prepared by indium-mediated addition of allylic bromides to oxime ethers derived from 2-pyridinecarboxaldehyde and glyoxylic acid (Eq. 4.34). γ-Substituted allylic bromides undergo bond formation at the most substituted termini; when the allylic bromide is γ-substituted, the syn stereoisomers of the hydroxylamine products predominate. The reaction does not occur if the oxime ether does not possess a chelating group in close proximity. Water as a solvent was found to accelerate the indium-mediated Barbier-type allylation and benzylation of β,γ-unsaturated piperidinium ion which was generated from β,γ-unsaturated α-methoxy-*N*-methoxycarbonylpiperidine, while ring-opened allylated product was obtained in a case using β,γ-saturated α-methoxy-*N*-methoxycarbonylpiperidine. Other solvents than water resulted in low yield of the allylated and benzylated products, suggesting that water is essential to generate the piperidinium ion intermediate from β,γ-saturated α-methoxy-*N*-methoxycarbonylpiperidine (Scheme 4.20).[155]

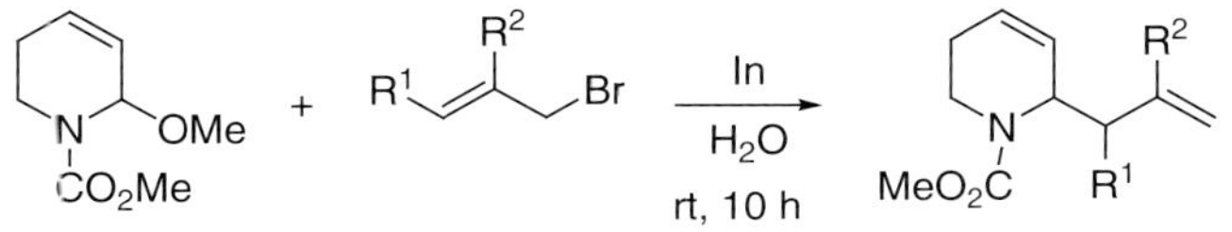

(Eq. 34)

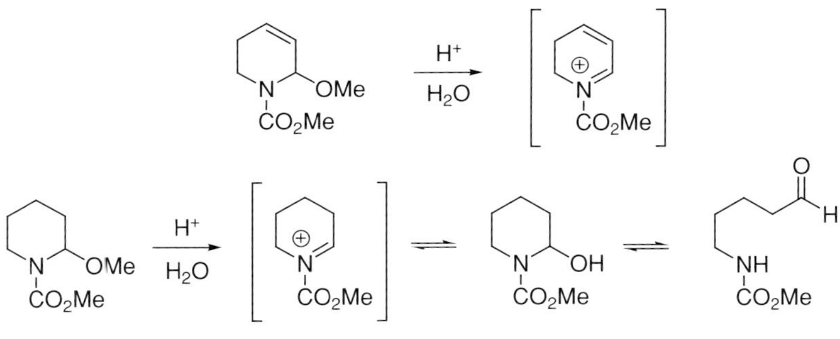

Scheme 4.20 Generation o the piperidinium ion intermediate from β,γ-unsaturated α-methoxy-*N*-methoxycarbonylpiperidine.

The stereoselective allylation reactions of carbon–nitrogen multiple bonds have also been studied. The addition of allylzinc bromide to aromatic imines derived from (*S*)-valine esters was affected by reversibility, which caused the lowering of the diastereoisomeric ratio with increasing reaction time. The retroallylation reaction could be avoided by performing the reaction in the presence of trace amount of water, or by using $CeCl_3 \cdot 7H_2O$ as the catalyst with a decreased reaction rate.[156] Hanessian and Yang reported the synthesis of enantiomerically pure or highly enriched allylglycine and its chain-substituted analogs from the reaction of the sultam derivatives of *O*-benzyl glyoxylic acid oxime with allylic bromides in the presence of powdered zinc in aqueous ammonium chloride.[157]

Indium-mediated allylation reactions of α-keto imides derived from Oppolzer's sultam proceeded in aqueous THF in good yields and excellent diastereoselectivity.[158] The indium-mediated allylation of the Oppolzer camphorsultam derivatives of glyoxylic oxime ether proceeded with excellent diastereoselectivity in aqueous media, providing a variety of enantiomerically pure α-amino acids.[159]

More recently, catalytic asymmetric allylations of imines and imine derivatives in aqueous media have been studied. A N-spiro C_2-symmetrical chiral quaternary ammonium salt (S,S)-I-Br [$(S,S) = \beta$-Np-NAS-Br] has been evaluated in the allylation of glycine *tert*-butyl ester benzophenone Schiff base $Ph_2C{:}NCH_2COOCMe_3$ for synthesis of both natural and unnatural α-amino acids.[160] The asymmetric allylation of hydrazono esters with allylsilanes in the presence of a catalytic amount of ZnF_2/chiral diamines in aqueous media generate (benzoyl)hydrazino-4-pentenoates in high enantiomeric excess (Eq. 4.35).[161]

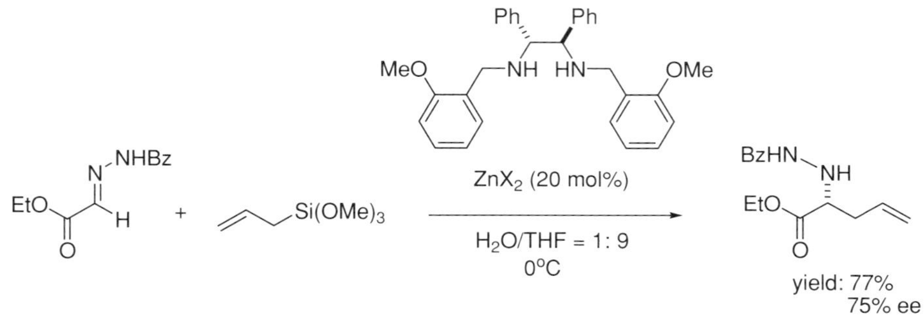

(Eq. 35)

4.4 Propargylation/allenylation of carbonyls, imines, and related compounds

Similar to allylmetal reagents, propargylmetal reagents can also form six-membered-ring transition state, although it is slightly less stable due to the linear geometry of the three carbons in the propargyl system. Furthermore, it is also possible to form a cyclic transition state during the reaction of propargyl halides with metals. Thus, it can be expected that propargylation/allenylation is also quite facile. However, there will be a selectivity issue in forming either the allene product (via the propargylmetal intermediate) or the homopropargyl product (via the allenylmetal intermediate) if a six-membered cyclic transition state is involved in the product formation. Indeed, the reaction of propargyl bromide with aldehydes mediated by tin in water generated a mixture of propargylation and allenylation products with very low product selectivity.[162] Allenylations and propargylations of carbonyl compounds in aqueous media could also be carried out with preformed organic tin reagents, instead of using metals.[163–166] The combination of $SnCl_2$ and KI[167] (or $SnCl_2/NiCl_2$–KI)[168] was found to be more effective for the reaction (Eq. 4.36).

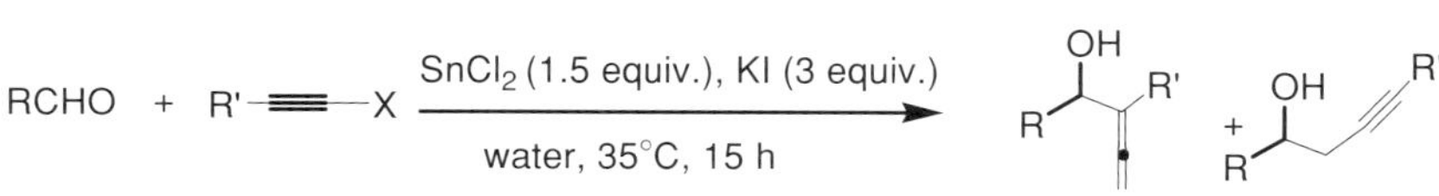

(Eq. 36)

The zinc-mediated propargylation of 3-formylcephalosporins was also studied in aqueous media.[169] Issac and Chan studied the behavior of aldehydes with propargyl bromides in aqueous medium mediated by indium.[170] They found that simple prop-2-yn-1-yl bromide reacted with both aliphatic and aromatic aldehydes in water to give mainly the homopropargyl alcohols. In contrast, when propargyl bromide was γ-substituted the coupling products were predominantly, or exclusively, the allenylic alcohols. Such couplings also proceed with α-chloropropargyl phenyl sulfide.[171]

In synthetic applications, Li et al. examined the propargylation–allenylation of carbonyl compounds by using a variety of metals including Sn, Zn, Bi, Cd, and In.[172] By using the indium-mediated allenylation reaction, Li and coworkers developed the synthesis of the antiviral, antitumor compound (+)-goniofufurone,[173] a key component isolated from the Asian trees of the genus *Goniothalamus*,[174] and other styryl lactone derivatives (Eqs. 4.37 and 4.38).[175]

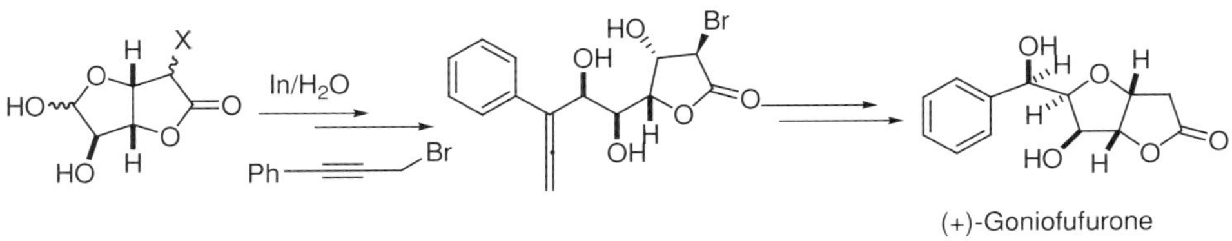

(Eq. 37)

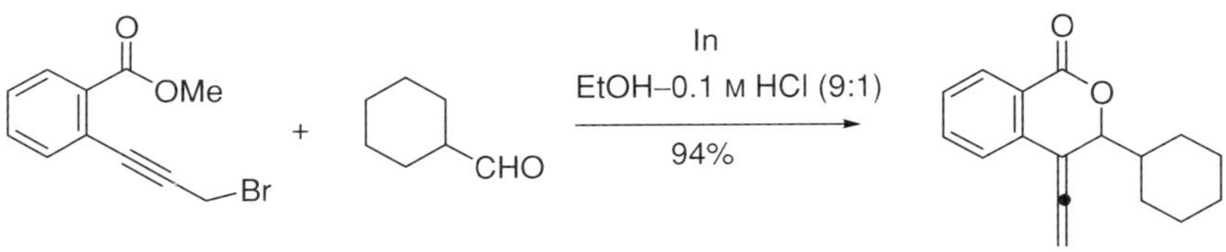

(Eq. 38)

Propargylations (allylations) of diphenylmethyl 6-oxopenicillanate and 7-oxocephalosporanate were accomplished in moderate yields by reaction with the corresponding bromides in the presence of indium or zinc in aqueous conditions.[176] In-mediated propargylation of acetals and ketals with various allyl or propargyl bromides in aqueous media successfully provided the corresponding homopropargylic (and allenylic) alcohols (Eq. 4.39).[177]

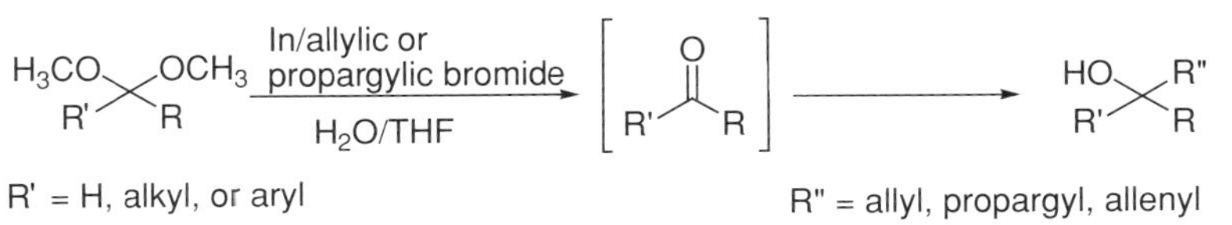

(Eq. 39)

Metal-mediated carbonyl allylation, allenylation, and propargylation of optically pure azetidine-2,3-diones were investigated in aqueous environments.[178] The regioselectivity of these reactions depended on the type of metal promoters that were used. The configuration of the new C_3-substituted C_3-hydroxy quaternary center was controlled by placing a chiral auxiliary at C_4.

The indium-mediated coupling of propargyl bromide with a variety of imines and imine oxides afforded homopropargylamine derivatives in aqueous media under mild conditions.[179] Propargylation of glyoxylic oxime ether in the presence of a catalytic amount of palladium(0) complex and indium(I) iodide in aqueous media was also studied (Eq. 4.40).[180]

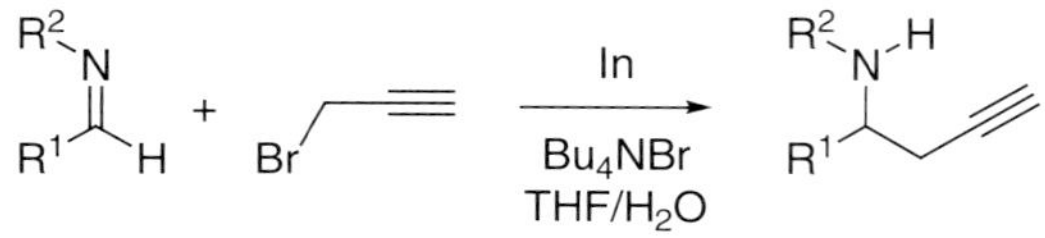

(Eq. 40)

4.5 Metal-mediated benzylation of carbonyls and imines

Benzyl halides and allyl (propargyl) halides are structurally similar but have drastically different chemical reactivities in the aqueous Barbier–Grignard-type reactions. Although tribenzyl and dibenzyltin derivatives have been prepared in aqueous conditions since the 1960s, they do not add onto carbonyls,[66] most likely because it is not possible to form a six-membered cyclic transition state with the carbonyl group in a 'two-component' fashion. Still, zinc-mediated benzylation of carbonyl compounds in aqueous media was reported by Bieber et al. recently.[181] The benzylation of 4-nitrobenzaldehyde could be controlled chemoselectively by using various phase transfer catalysts and metal reductants in water (Eq. 4.41).[182]

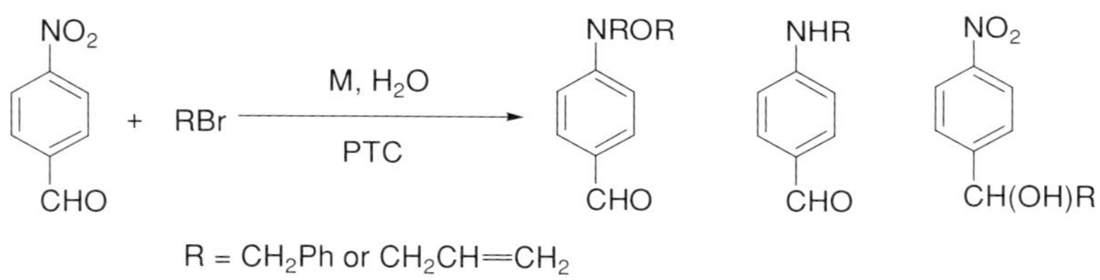

(Eq. 41)

4.6 Arylation and vinylation of carbonyls and imines

For arylation of carbonyls and imines to occur bimolecularly, a four-membered cyclic transition state is needed. Although this reaction pathway cannot be ruled out, it will require a considerably strained transition state (Fig. 4.7(a)). Alternatively, a η^2-coordination of the carbonyl or imine to the metal, followed by 'ligand' migration, will lead to the vinylation or arylation product (Fig. 4.7(b)). The latter is typical of transition metal-catalyzed reactions.[183] Thus, it can be expected that for a Barbier–Grignard-type reaction involving vinyl/arylmetals to occur, a transition metal-catalyzed process will be required. In order to develop a transition metal catalyst for such a process, Li and coworkers examined the use of palladium-catalyzed and metal-mediated reactions between aryl halides and aldehydes.[184] Unfortunately, although the arylmetal intermediate was formed successfully in water, the corresponding Barbier–Grignard reaction was not observed. Instead, an Ullmann-type reaction product was obtained. Interestingly, when the reaction was carried out in water, the reaction was successful in open air, whereas no reaction was observed in air if the reaction was carried out in anhydrous organic solvent.[185]

Figure 4.7 Two possible mechanism for the vinylation and arylation of carbonyl compounds in water.

4.6.1 Arylation and vinylation of aldehydes

In 1998, Miyaura and coworkers reported a Rh(acac)$(CO)_2$/dppp-catalyzed addition of aryl or alkenylboronic acids to aldehydes in aqueous organic mixtures under an inert atmosphere (Eq. 4.42).[186] The use of electron-rich tri(*tert*-butyl)phosphine as ligand was found to be beneficial for obtaining good yields of the corresponding aldehyde addition products.[187]

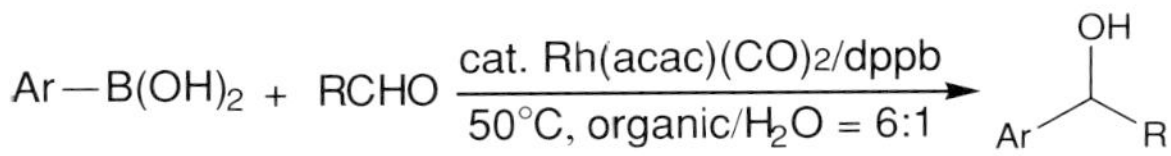

(Eq. 42)

On the other hand, after being unable to carry out a direct 'Grignard-type' reaction of aryl halide with aldehyde, Li and coworkers attributed the failure to the addition of arylmetal onto the carbonyl. Subsequently, they studied the addition of various aryl- and vinylmetal reagents to aldehydes in air and water by using a variety of catalyst. It was found that aryl- and vinyltin compounds added to aldehydes smoothly when catalyzed by either $Rh_2(COD)_2Cl_2$ or $Rh(COD)_2BF_4$ (Eq. 4.43).[188] The carbonyl addition was found to be highly sensitive to the choice both of metal and of groups attached to the metal. Except for organoarsen and organoantimony reagents, aryl or vinyl derivatives of all other metals (and metalloids) examined were able to generate the desired carbonyl addition and conjugate addition products with variable efficiency. Among them, aryl and vinyltin, silicon, boron, lead, and bismuth derivatives were found to be the most effective. The corresponding indium and germanium reagents provided only low yields of the products. Taking the organotin reagents as an example, in the presence of a catalytic amount of $Rh(COD)_2BF_4$ at refluxing temperature in air and water, benzaldehyde underwent nucleophilic addition with trimethylphenyltin and dibutyldiphenyltin to give the corresponding nucleophilic addition product smoothly. On the other hand, under the same reaction conditions, no reaction was observed between benzaldehyde and phenyltin trichloride even after several days. When the reaction was carried out in the presence of potassium hydroxide, a smooth reaction again occurred to give the desired product overnight. A more dramatic effect was observed among triphenyltin chloride, triphenyltin hydroxide, and butyltriphenyltin. No reaction was observed with the chloride derivative, but the reaction with either hydroxide or butyl derivatives proceeded smoothly. The use of different bases also affects the reaction progress. Various bases such as lithium hydroxide, sodium hydroxide, and potassium hydroxide were tested (showing the same trend as the relative basicity), and potassium hydroxide appeared to be the most effective for this reaction. A similar electronic effect was also observed with organobismuth, organolead, organoindium, and organoboron compounds. Li and coworkers explained the phenomena by d–p–π bonding between these metals and the substituent and developed analogous chemistry using aromatic compounds (Fig. 4.8).[189] Toward this end, aryltriethoxysilanes

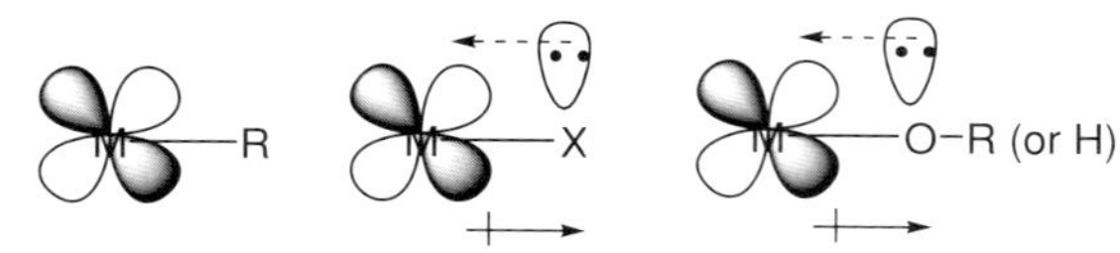

Figure 4.8 Li's postulated electronic effect on tin and other metals.

added to aldehydes in high yield in the presence of a Rh(I) catalyst and aq. NaOH.[190] On the other hand, treatment of α,β-acetylenic ketones with chromium(II) in the presence of aldehydes, Me_3SiCl, and water in THF gives 2,5-disubstituted furans in good to excellent yields.[191] Under air and water conditions, as developed by Li and coworkers, carbonyl hydrates can also react. A highly diastereoselective rhodium-catalyzed addition of arylbismuth and aryllead reagents to chiral glyoxylate hydrate in air and water was reported (Eq. 4.44).[192]

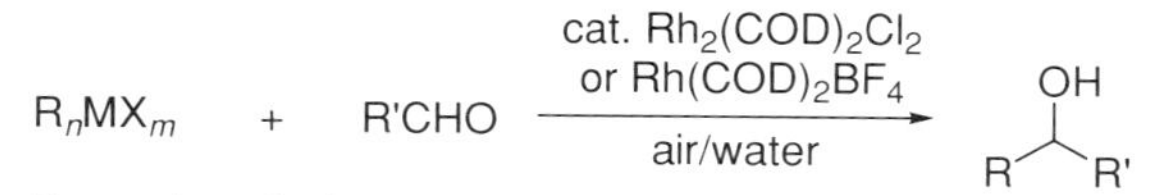

R = aryl or vinyl
M = B, In, Si, Sn, Ge, Pb, As, Sb, Bi
X = alkyl, aryl, halogen, hydroxyl, and alkoxyl

(Eq. 43)

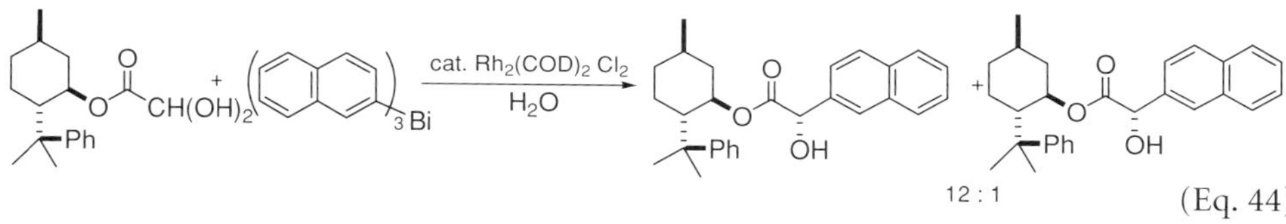

(Eq. 44)

4.6.2 Arylation and vinylation of imines and related compounds

The reaction of aromatic radicals, generated by decomposition of diazonium salts, with iminium salts in the present of $TiCl_3$ in aqueous media produced secondary amines.[193] The iminium salts are formed *in situ* from aromatic amines and aldehydes. Petasis and Zavialov reported an efficient addition of vinyl boronic acid to iminium salts.[194] While no reaction was observed when acetonitrile was used as solvent, the reaction went smoothly in water to give allyl amines (Eq. 4.45). The reaction of the boron reagent with iminium ions, generated from glyoxylic acid and amines, affords novel α-amino acids (Eq. 4.46). Carboalumination of alkynes in the presence of catalytic Cp_2ZrCl_2 and H_2O affords vinylalane intermediates, which serve as nucleophiles in the subsequent addition to enantiomerically enriched (*tert*-butyl)- and (*p*-tolyl)sulfinimines. Chiral allylic sulfinamides are obtained in high diastereomeric excesses and good yields.[195]

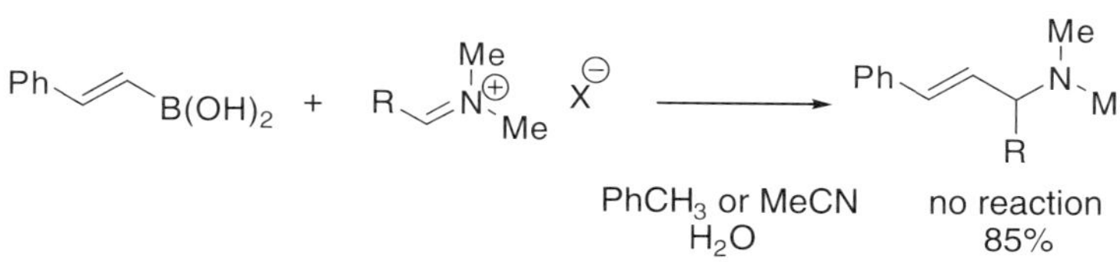

(Eq. 45)

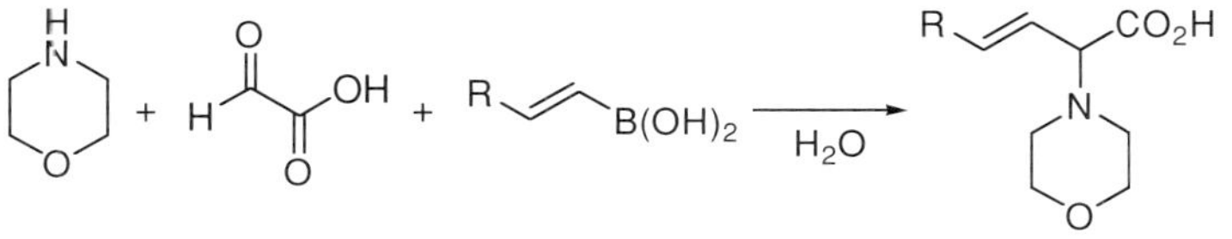

(Eq. 46)

Miyaura and coworkers reported the rhodium-catalyzed reaction of arylboronic esters with *N*-sulfonylaldimines under aqueous conditions.[196] Recently, Wang et al.[197] reported that in the presence of a rhodium catalyst, imines react with phenyltrimethyltin or phenyltrimethyllead in water and air under ultrasonic irradiation at 35°C to give corresponding diarylmethylamines in good yields.

4.7 Alkynylation of carbonyls, imines, and related compounds

For alkynylation reactions, the first challenge is the generation of alkynylmetal intermediates in water. The use of alkynyl halides to form alkynylmetals appears quite difficult in water. On the other hand, direct formations of alkynylmetals from acetylenes are well known for silver, copper, and late transition metals in water. Once the alkynylmetal is generated, another challenge is the subsequent reaction with electrophiles. The required six-membered cyclic transition state is not possible using only the alkynylmetal and the aldehyde (or ketone). In fact, the silver and copper acetylides are quite stable and often resist further reactions. Li and coworkers reasoned that the 'Barbier–Grignard-type' alkynylation reaction can occur if (1) the C—M bond is activated or (2) the corresponding electrophile (such as aldehyde or imine) is activated. Furthermore, if the catalyst also activates the terminal alkynyl C—H bond in water, then only a catalytic amount of the metal will be required.

4.7.1 Alkynylation of aldehydes

Li and coworkers found that the direct addition of terminal alkynes to aldehydes in water can proceed when a $RuCl_3/In(OAc)_3$ bicatalytic system was used (Eq. 4.47).[198] In this combination, the $In(OAc)_3$ presumably played the role of a Lewis acid and activated the carbonyl, whereas the ruthenium chloride converted the alkyne to an alkynylmetal intermediate. As the metal ions can be regenerated, only substoichiometric amounts of the catalysts are necessary (Scheme 4.21). The addition of a base improved the yield of the reaction. The simple alkynylsilver generated from acetylene and silver halides is unreactive toward aldehydes. Nevertheless, Li and coworkers reported a highly efficient alkynylation of aldehydes by using silver/phosphine complexes as catalysts in water (Eq. 4.48).[199] The reaction is dually promoted by the electron-donating phospine ligand and water to give yields of 63–98% (Scheme 4.22). An aldehyde containing a hydroxyl group has been alkynylated without the need for protection (Eq. 4.49).

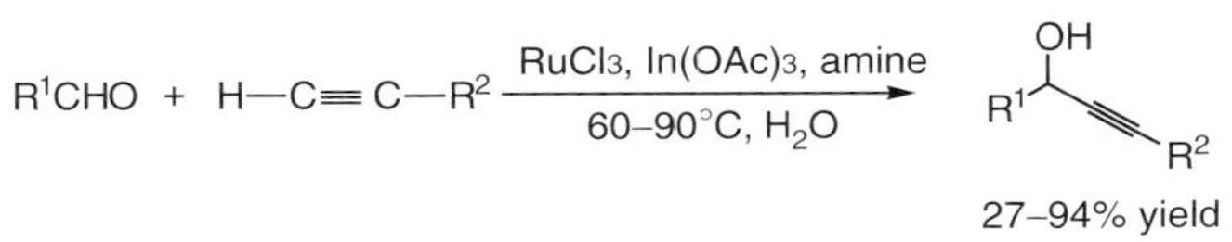

(Eq. 47)

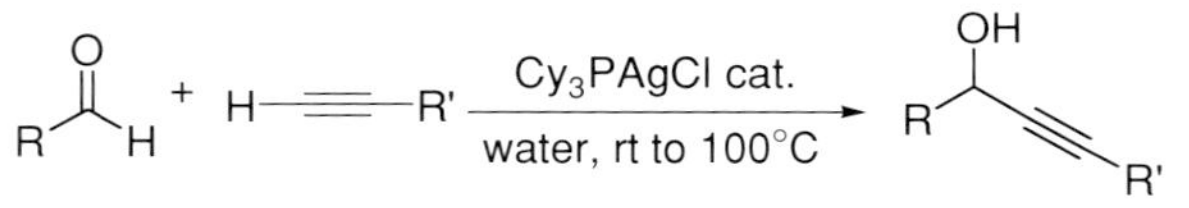

(Eq. 48)

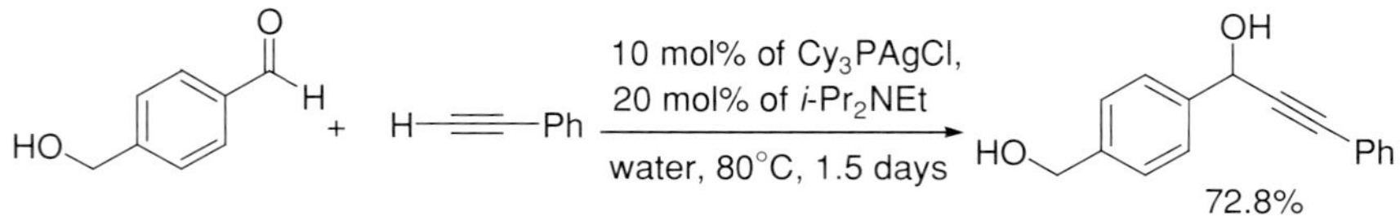

3s

(Eq. 49)

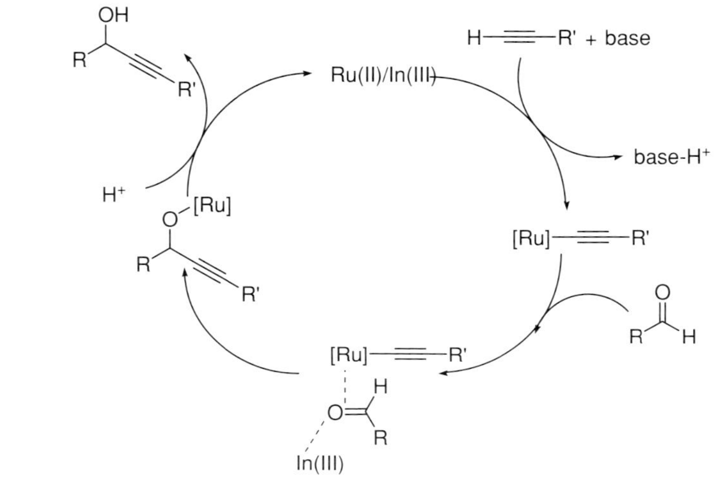

Scheme 4.21 Proposed mechanism of the alkynylation catalyzed by Ru/In in water.

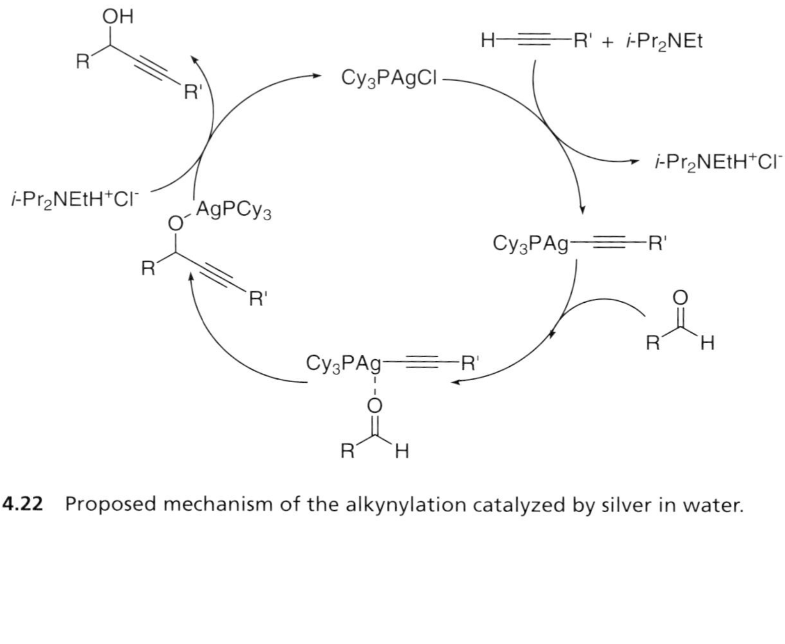

Scheme 4.22 Proposed mechanism of the alkynylation catalyzed by silver in water.

In addition to the formation of propargyl alcohols, indium chloride catalyzed the coupling of alkynes to aldehydes to give α,β-unsaturated carbonyl compounds in water in low yields (Eq. 4.50).[200] The product may be formed via the further reaction of the propargyl alcohol.

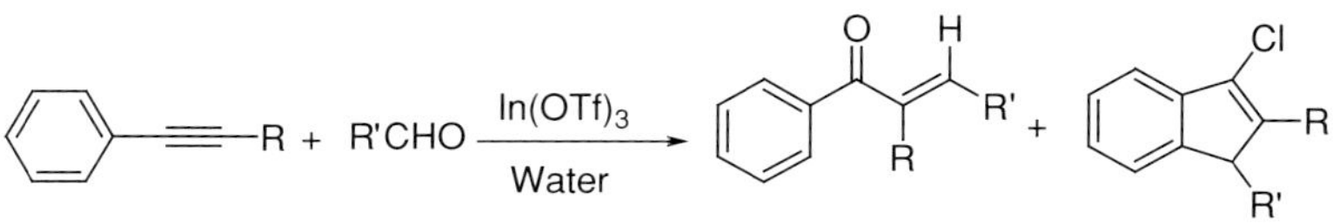

(Eq. 50)

A highly effective direct coupling of acid chlorides with terminal alkynes catalyzed by $PdCl_2(PPh_3)_2$/CuI together with a catalytic amount of sodium laurylsulfate as surfactant and K_2CO_3 as base provided ynones in high yields in water (Eq. 4.51).[201] The use of $PdCl_2(PPh_3)_2$/CuI as cocatalysts together with a catalytic amount of sodium laurylsulfate as the surfactant and K_2CO_3 as the base provided the desired product in 98% isolated yield. No reaction was observed when either Cu(I) alone or Pd(II) alone was used as the catalyst. The use of surfactant is also critical for the success of the reaction; without a surfactant/phase transfer reagent the yield dropped from 98% to 9%.

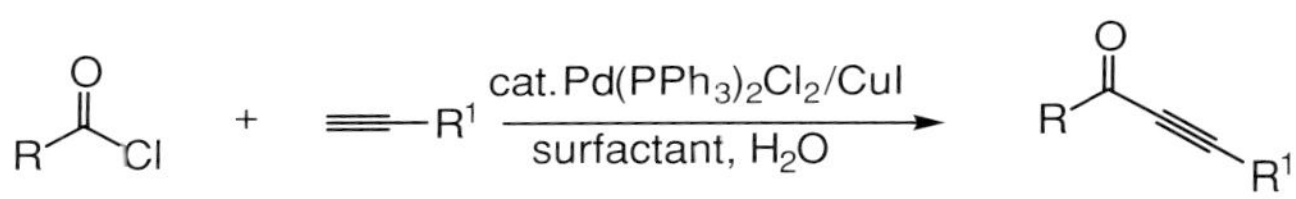

(Eq. 51)

4.7.2 Alkynylation of imines and related compounds

Direct 1,2-addition of terminal alkynes to the C—N double bond in imines and their derivatives via activation of the C—H bond in terminal alkynes is a convenient route to synthesize propargyl amines. As in the case of aldehydes, it may be assumed that such addition reactions can occur with either (1) the activation of C—M bond or (2) the activation of the imine by a nitrogenophilic Lewis acid. Indeed, Li and coworkers reported a highly efficient A^3-coupling (aldehyde–alkyne–amine) in water or without solvent by using a combination of $RuCl_3$ and Cu(I) in catalytic amount.[202] In the presence of catalytic Cu(I) alone, only a very small amount of the corresponding desired product was obtained in aqueous media. Likewise, no desired product was found with only $RuCl_3$. A broad range of substituted aromatic imines and aliphatic imines (Eq. 4.52) was converted to propargyl amines by this method. This simple process provides a convenient and efficient method for the preparation of propargyl amines. This is also the first general method for the catalytic addition of alkynes to imines reported in the literature.

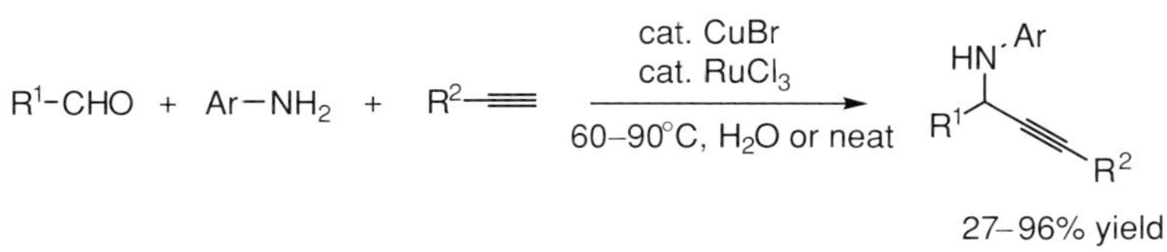

(Eq. 52)

Li and coworkers also reported a copper-mediated coupling of alkynes with N-acylimines and N-acyliminium ions in water to generate propargyl amide derivatives (Eq. 4.53).[203] N-Acylimines or N-acyliminium ions can be generated *in situ* from amines containing a good leaving group at α-position – for example, α-phenylsulfonyl N-acylamine and

α-methoxy *N*-(alkoxycarbonyl)pyrrolidine – and the products can be easily modified for various synthetic purposes. However, an excess of CuBr is required. In these cases, the imino C—N bonds are activated by the electron-withdrawing acyl group. Au(III) is also effective for the acylimine alkynylations.[204] Li and coworkers also found that the A^3-coupling reaction is highly efficient and general with gold as the catalyst (Eq. 4.54).[205] In the latter case, no cocatalyst or activator is needed for the reaction. Less than 1 mol% of catalyst is enough to generate an excellent yield of the corresponding propargyl amine products. Both aromatic and aliphatic aldehydes were able to undergo this three-component coupling with alkyne and amine. Dialkylamines are good for the reaction, whereas anilines gave the corresponding products in lower yields. *N*-Alkylanilines did not form the desired products. Aromatic aldehydes reacted more efficiently, and nearly quantitative yields were obtained in most cases. Aliphatic aldehydes can also be used; however, some trimerizations of aldehydes were observed that decreased the yields of the propargyl amine products. The properties of solvents significantly affect the reaction. Water is the best solvent and the reaction process is very clear with almost quantitative yield; the use of organic solvents such as THF, toluene, and DMF resulted in low conversions and more by-products.

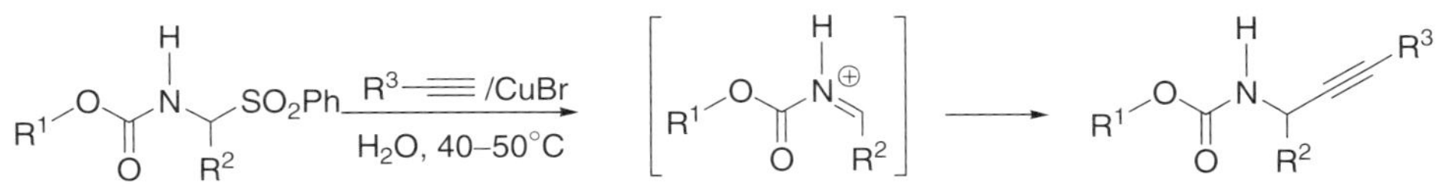

(Eq. 53)

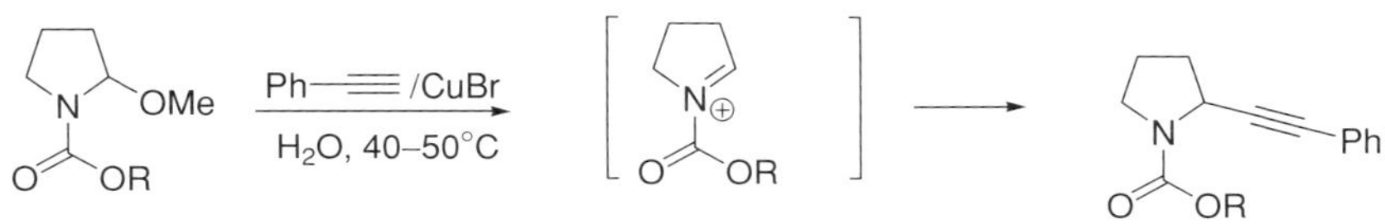

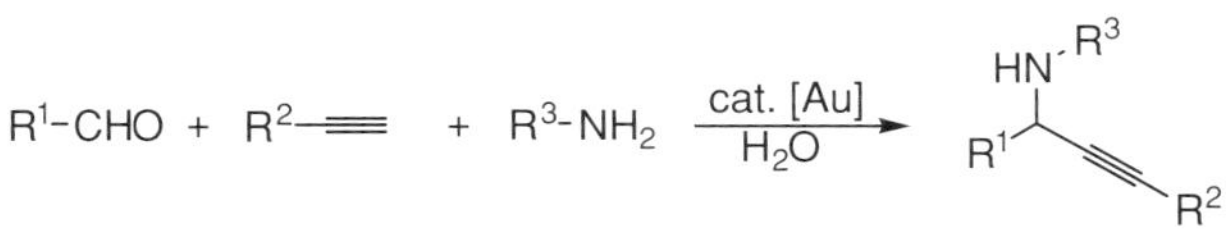

(Eq. 54)

Following the success of copper and gold, it was found that AgCl, AgBr, and AgI showed good catalytic activity for the three-component coupling in water (Eq. 4.55).[206] No other additive was needed for this reaction either. Aromatic aldehydes decreased the rate of the reaction, whereas aliphatic aldehydes displayed higher reactivity and cleaner reactions than reactions catalyzed by copper and gold.

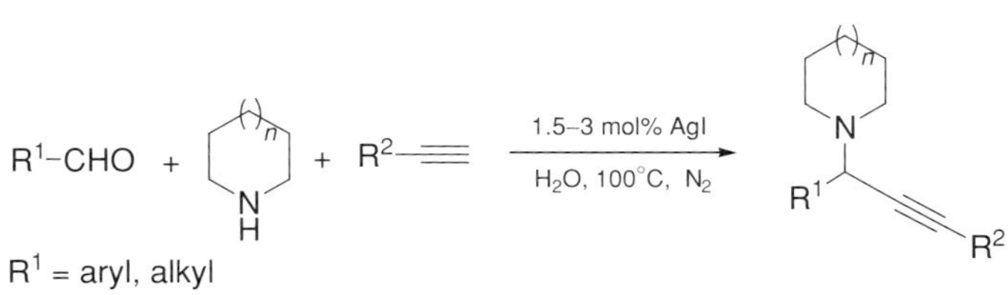

(Eq. 55)

R^1 = aryl, alkyl
n = 0, 1, 2
R^2 = aryl

4.7.3 Asymmetric alkynylation

The relatively low reactivity of the C—M bond alkynylmetals and the potential of activating C—M bond with a ligand provide opportunities for asymmetric alkynylations. Indeed, Li and coworkers reported a highly efficient asymmetric coupling of AA^3 (asymmetric aldehyde–alkyne–amine) in water (Eq. 4.56).[207] The use of the tridentate bis(oxazolinyl)pyridines (pybox) with Cu(OTf) afforded the product in high yield and up to 99.6% ee in organic solvent and 84% ee in water.[208] In most cases, imines were formed *in situ* and the addition was very simple consisting of simply mixing the aldehyde, aniline, and an alkyne with the catalyst. Recently, Shi et al. reported that a three-component coupling of aldehyde, alkyne, and amine catalyzed by CuI in water can be greatly accelerated by using microwave irradiation.[209]

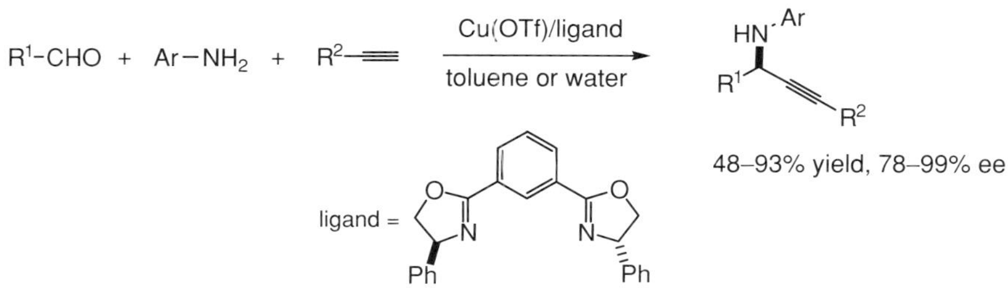

(Eq. 56)

4.8 Metal-mediated aldol and Reformatsky-type reactions

Recently, there has been a resurgence of the classical direct aldol reactions catalyzed by acids and bases. These reactions involve the condensation reactions of active methylene compounds such as acetophenone or cyclohexanone with aryl aldehydes under basic or acidic conditions and afford aldols along with dehydration compounds in water.[210] The presence of surfactants led mainly toward the dehydration reactions. The most common solvents for aldol reactions are ethanol, aqueous ethanol, and water.[211] Recently, various Lewis acids have also been examined as catalysts for aldol reactions. For example, zinc-aminoester or aminoalcohol complexes catalyzed a quantitative aldol reaction without concomitant dehydration.[212] Zn complexes of proline, lysine, and arginine catalyzed the aldol addition of *p*-nitrobenzaldehyde with acetone in aqueous medium to give quantitative yields and enantiomeric excesses up to 56% at room temperature.[213] Microporous polymer-supported $La(OiPr)_3$ catalyzed the aldol reaction between ketones and aldehydes in pure water at neutral pH.[214] The La network was stable against hydrolysis and maintained microporosity and reversible substrate binding, which mimics the function of an enzyme.[215] Recently, Janda and coworkers reported the aqueous asymmetric direct aldol reaction by using a nicotine metabolite.[216] Another recent development is the uncatalyzed aldol reaction; high-intensity ultrasound was employed to reinvestigate the aldol reaction in water without the use of a catalyst.[217] These reactions, by and large, are outside the scope of the current chapter.

The crossed aldol reaction of silyl enol ethers with carbonyl compounds (Mukaiyama aldol) was first studied by Lubineau and coworkers in aqueous solvents. Without any acid catalyst, these reactions took several days to complete.[218] A major development was the use of water-tolerant Lewis acids for such reactions, pioneered by Kobayashi and coworkers.[219]

Adding a catalytic amount of lanthanide triflate greatly improved the rate and the yield of such reactions.[220] Recently, asymmetrical aqueous Mukaiyama aldol reaction has been successfully performed in aqueous media.[221] In aqueous ethanol, Kobayashi and coworkers achieved asymmetric inductions by using $Cu(OTf)_2$/chiral bis(oxazoline) ligand,[222] $Pb(OTf)_2$/chiral crown ether,[223] or $Ln(OTf)_3$/chiral bis-pyridino-18-crown-6 (Eq. 4.57).[224] On the other hand, Li and coworkers recently developed a highly efficient asymmetric Mukaiyama reaction by using chiral gallium catalysts with Trost's chiral semi-crown ligands (Eq. 4.58).[225] Such a system can achieve high enantioselectivity even in pure water, as evidenced by an aqueous reaction that proceeded smoothly with good yield (89%), diastereoselectivity (syn/anti 89:11), and enantioselectivity of the syn-product (ee 87%). However, these reactions are again beyond the scope of this chapter and are covered in greater detail in Chapter 3.[226]

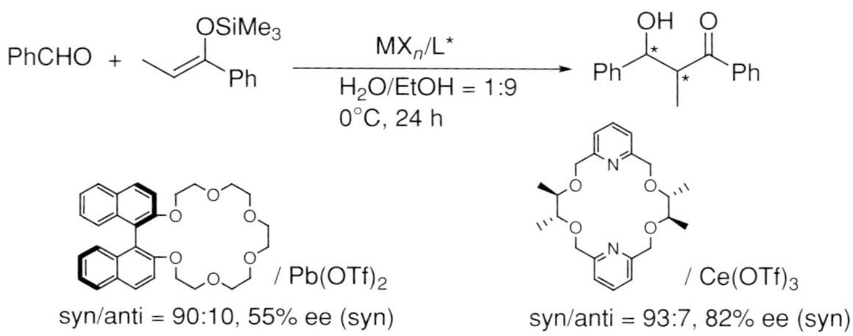

(Eq. 57)

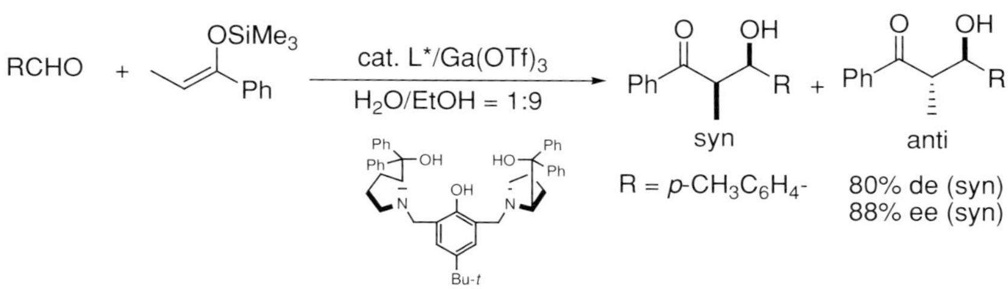

(Eq. 58)

The reaction of a metal with an α-halogen carbonyl compound will generate an organometallic intermediate that can exist in both the carbanion form and the enolate form, which are often in equilibrium. As in the case of allyl and propargyl, the enolate form will allow a six-membered cyclic transition state and are thus energetically favorable. Indeed, the reaction of an α-halogen carbonyl compound with zinc, tin, or indium together with an aldehyde in water gave a direct cross-aldol reaction product (Eq. 4.59).[227] A direct Reformatsky-type reaction occurred in low yields when an aromatic aldehyde reacted with a α-bromo ester in water mediated by zinc.[228] Recently, it was found that under sonication, such a reaction mediated by indium is successful.[229] The combination catalysts $BiCl_3$/Al,[230] $CdCl_2$/Sm,[231] and $Zn/Et_3B/Et_2O$[232] are also effective mediators. Bismuth metal, upon activation by zinc fluoride, effected the crossed aldol reaction between α-bromo carbonyl compounds and aldehydes in aqueous media. The reaction was found to be regiospecific and syn-diastereoselective.[233] Bieber et al. reported that the reaction of bromoacetates is greatly enhanced by catalytic amounts of benzoyl peroxide or peracids and gives satisfactory yields with aromatic aldehydes. A radical chain mechanism, initiated by electron abstraction from the organometallic Reformatsky reagent, is proposed (Eq. 4.60).[234] However, an alternative process of reacting aldehydes with 2,3-dichloro-1-propene and indium in water followed by ozonolysis provided the Reformatsky product in practical yields.[235] An electrochemical Reformatsky reaction in aqueous medium and in the absence of metal mediator has also been reported.[236] The indium-mediated aqueous Reformatsky reaction was successfully used in the synthesis of α,α-difluoro-β-hydroxy ketones.[237]

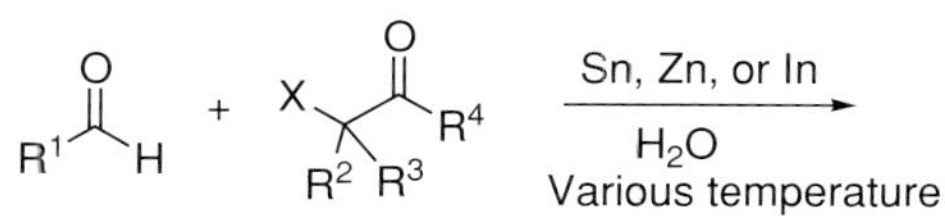

(Eq. 59)

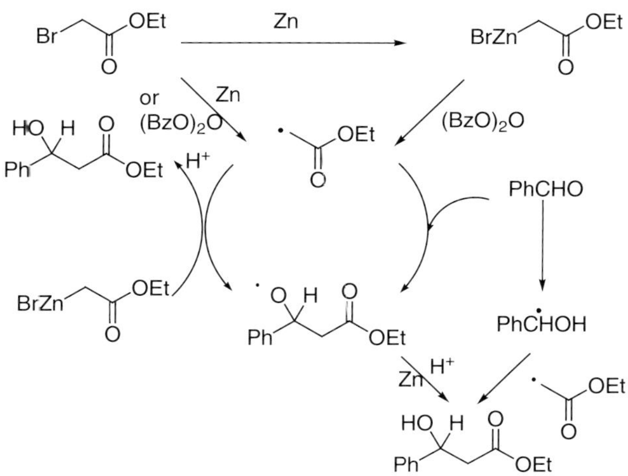

(Eq. 60)

Alternatively, Li et al. reported a $RuCl_2(PPh_3)_3$-catalyzed reshuffling of functional groups of homoallylic alcohols in water (Eq. 4.61).[238] The reaction led to an aldol-type reaction by reacting allyl alcohols with aldehyde (Scheme 4.23).[239] The presence of $In(OAc)_3$ promoted the aldol reaction with α-vinylbenzyl alcohol and aldehyde.[240] An indium hydride-promoted reductive aldol reaction of unsaturated ketones in aqueous media was developed.[241] The use of water/methanol as a solvent dramatically reverses stereochemistry from anti to syn. Boron enolates have been used for aldol reactions in water using catalytic amounts of boron reagents.[242]

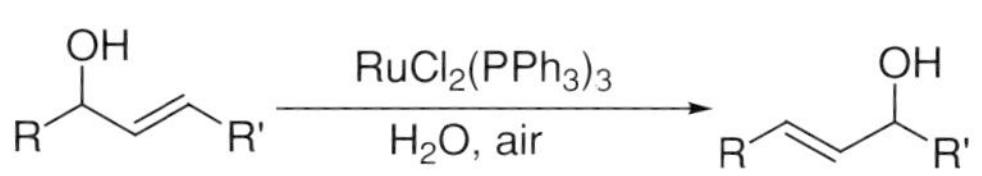

(Eq. 61)

4.9 Metal-mediated alkylation of carbonyls and imines

4.9.1 Alkylation of aldehydes

The direct addition of simple alkyl groups to aldehydes is a challenging reaction to perform in water. Mitzel and coworkers reported the indium-mediated alkylation of carbonyl compounds with α-sulfur stabilized systems.[243] Recently, Li and coworkers reported the first

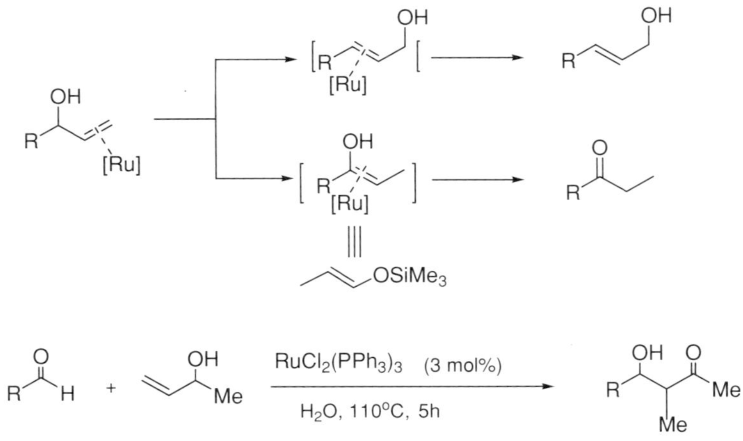

Scheme 4.23 Aldol reaction via the ruthenium-catalyzed olefin migration of allyl alcohols in water.

efficient addition of simple alkyl halides to aldehydes in water (Eq. 62).[244] A tentative mechanism for the reaction was proposed in which the alkyl iodide reacts with zinc (activated by copper as in the case of zinc–copper couple) to form a radical anion, which then reacts with the aldehyde. However, the reaction is still very limited in scope at the present stage.

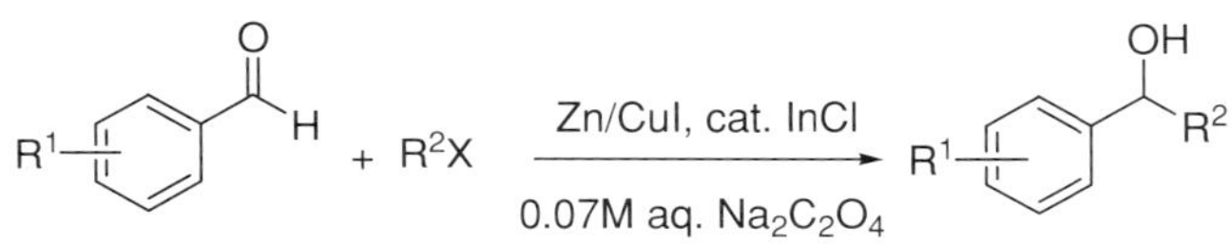

(Eq. 62)

4.9.2 Alkylation of imines

Katritzky et al. reported that in the presence of bismuth(III) chloride-metallic aluminum, alkyl (as well as allyl) halides react with *N*-(alkylamino)benzotriazoles at 20°C in THF/water to give the corresponding homoalkylated amines in high yields (Eq. 4.63).[245] Clerici and Porta reported that phenyl, acetyl, and methyl radicals add to the C_α atom of the iminium ion, $PhN^+Me{:}CHMe$, formed *in situ* by the titanium-catalyzed condensation of *N*-methylaniline with acetaldehyde to give PhNMeCHMePh and PhNMeCHMeAc in 80% overall yield.[246] Recently, Miyabe et al. studied the addition of various alkyl radicals to imine derivatives. Alkyl radicals generated from alkyl iodide and triethylborane added to imine derivatives, such as oxime ethers, hydrazones and nitrones, in aqueous medium.[247] The reaction also proceeds on solid support.[248] *N*-Sulfonylimines are also effective under such reaction conditions.[249] Indium is also effective as the mediator (Eq. 4.64).[250] A tandem radical addition–cyclization reaction of oxime ether and hydrazone was also developed (Eq. 4.65).[251] The zinc-mediated radical reaction of the hydrazone bearing a chiral camphorsultam provided with good diastereoselectivities the corresponding alkylated products, which could be converted into enantiomerically pure α-amino acids (Eq. 4.66).[252] Competitive addition of $^-CCl_3$ anions to *N*-alkyl-pyridinium salts was studied in a two-phase system (chloroform/conc. aq. NaOH) and in homogeneous medium.[253] Li and coworkers reported the synthesis of α-amino acid derivatives and amines via the addition of simple alkyl halides to imines and enamides mediated by zinc in water (Eq. 4.67).[254]

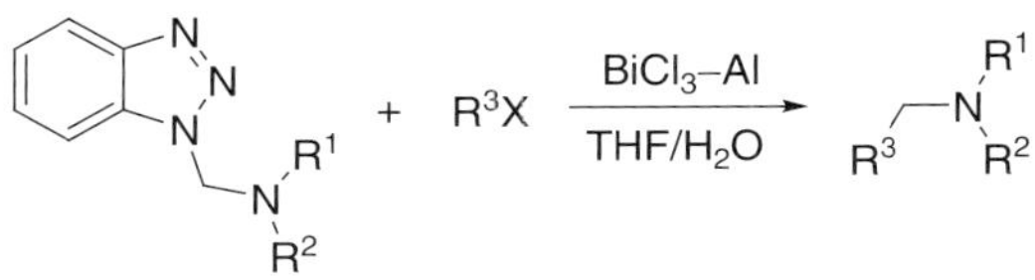

(Eq. 63)

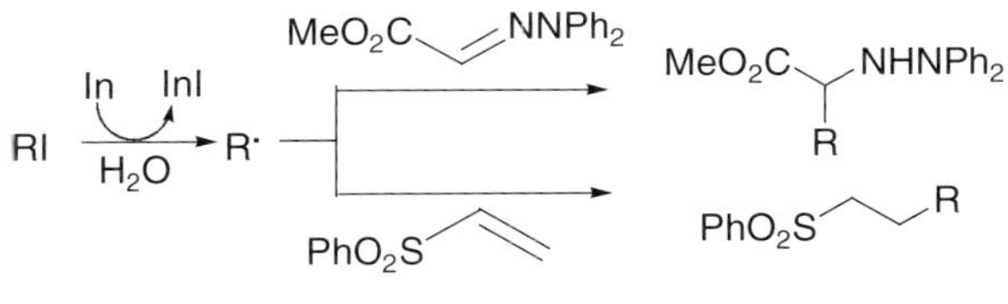

(Eq. 64)

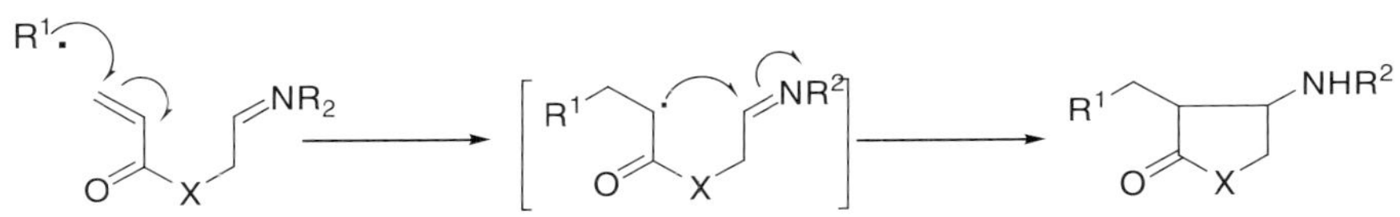

(Eq. 65)

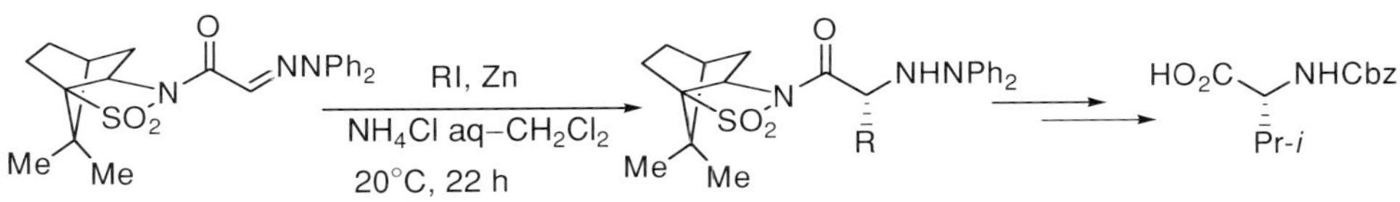

(Eq. 66)

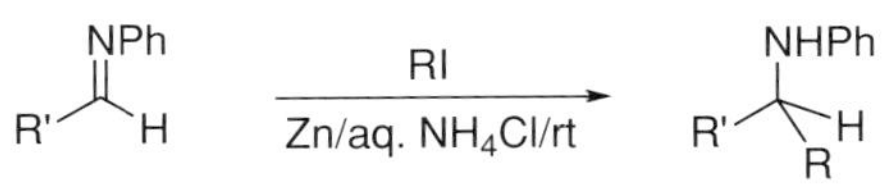

(Eq. 67)

R = *t*-Bu	R' = Ph	70%
R = *i*-Pr	R' = Ph	60%
R = *i*-Pr	R' = *o*-HOPh	55%

4.10 Metal-mediated conjugate addition reactions

In conjugate addition reactions, unlike in additions of organometallic reagents to carbonyls and imines, a six-membered cyclic transition state can be formed readily by two components for any type of organometallic reagents. Furthermore, a conjugated system is also an excellent acceptor for carbanions and radicals. This simple analysis suggests that metal-mediated conjugate addition should be quite feasible in water.

In the 1970s, Hajos and Parrish[255] and Wiechert and coworkers[256] independently reported that the Michael addition of 2-methylcyclopentane-1,3-dione to vinyl ketone in water gives the corresponding conjugated addition product without the use of a basic catalyst. A similar enhancement of reactivity was found in the Michael addition of 2-methyl-cyclohexane-1,3-dione to vinyl ketone, which eventually led to optically pure Wieland–Miescher ketone.[257] The reaction however proceeds under more drastic conditions. Since then, many reports have appeared on base- or acid- (Lewis acid-) catalyzed conjugate addition reactions in aqueous media, which is beyond the scope of this chapter.

4.10.1 Addition of alkyl groups

In metal-mediated conjugate additions involving alkyl groups, Luche and coworkers reported that when zinc–copper couple was used, alkyl halides reacted with conjugated carbonyl compounds and nitriles to give 1,4-addition products in good yields under sonication conditions (Eq. 4.68).[258] A moderate diastereoselectivity was observed in those reactions where a mixture of diastereomers could be generated.[259] The reactivity of the halides followed the order of tertiary > secondary ≫ primary, and iodide > bromide (chlorides did not react). The preferred solvent system was aqueous ethanol. The process was suggested to proceed by a free radical mechanism occurring on the metal surface under sonochemical conditions. Efforts to trap the intermediate intramolecularly only gave a very low yield of the cyclization product (Eq. 4.69).[260] Similar additions also occurred on vinylphosphine oxides. When optically active vinylphosphine oxide was used, P-chiral alkylphosphine oxide was obtained with retention of configuration (Eq. 4.70).[261]

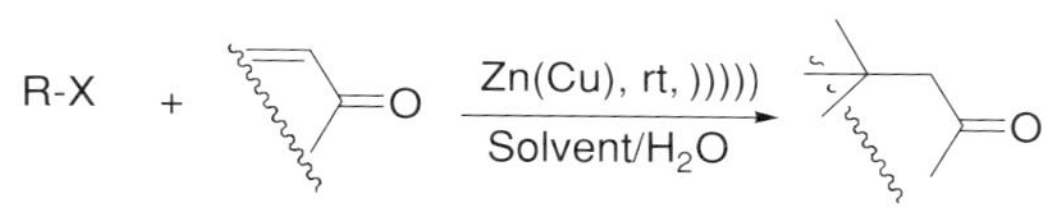

(Eq. 68)

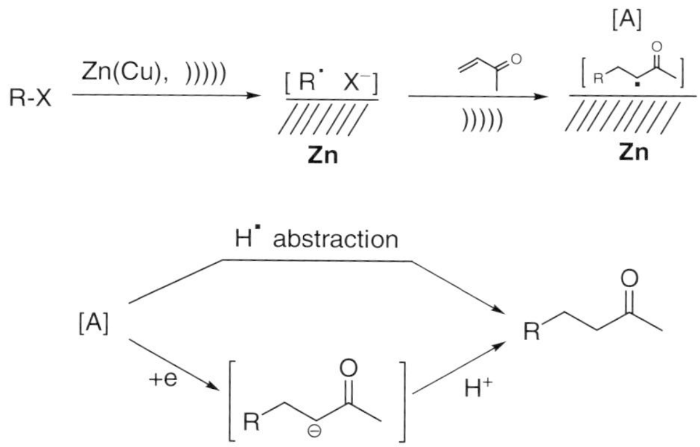

(Eq. 69)

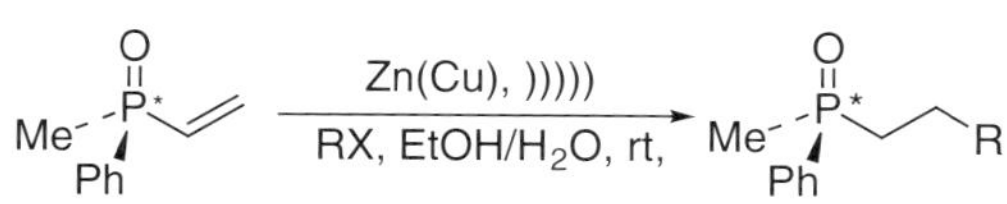

(Eq. 70)

Giese and coworkers studied the diastereoselectivity associated with a related addition in water (Eq. 4.71)[262] and found that the anti-isomer was the major product if the attacking radical is bulky. The results were explained that the more stable 'A-strain' conformer of the alkene reacts slower with bulky alkyl radical than the less stable 'Felkin-Anh' conformer.

The diastereoselective ultrasonically induced zinc–copper 1,4-addition of alkyl iodides to chiral α,β-unsaturated systems in aqueous media was studied by Suarez et al. and good diastereoselectivities were observed with the Z-isomer while reactions with the E-isomer were nonselective (Eq. 4.72).[263]

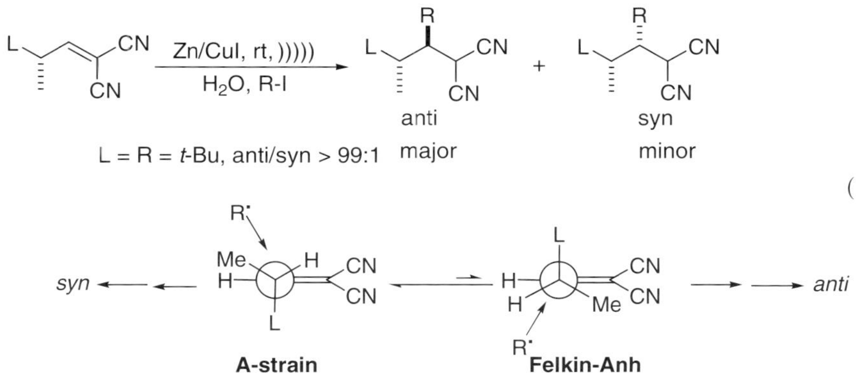

(Eq. 71)

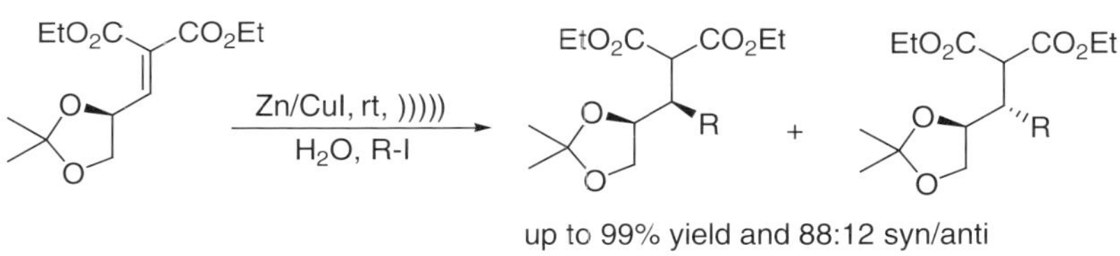

(Eq. 72)

Li and coworkers reported the conjugate addition of alkyl groups to enamides mediated by zinc in aq. NH_4Cl to generate α-amino acid derivatives (Eq. 4.73).[264] No reaction was observed in the absence of water. Both secondary and tertiary alkyl groups such as linear (2-butyl, 2-propyl, 2-pentyl), cyclic (cyclohexyl, cyclopentyl, cycloheptyl), and bulky ones (*tert*-butyl) were all transferred to the substrate successfully. Even simple primary iodides and methyl iodide provided the desired products in good yields. Miyabe et al.[265] as well as Jang and Cho[266] reported the addition of alkyl radicals from alkyl iodide to α,β-unsaturated ketones, esters, and nitriles mediated by indium in aqueous media. Indium-mediated Michael addition of allyl bromide to 1,1-dicyano-2-arylethenes also proceeded well in aqueous medium.[267]

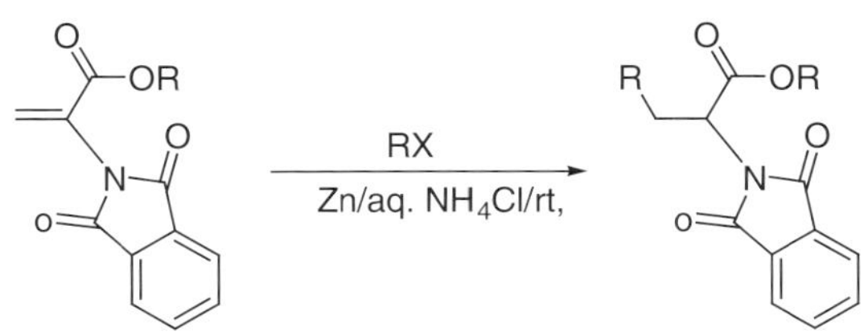

(Eq. 73)

4.10.2 Addition of vinyl and aryl groups

Miyaura and coworkers[268] reported the Rh(I)-catalyzed conjugate addition of aryl- or 1-alkenylboronic acids, $RB(OH)_2$,to enones in high yields at 50°C in an aqueous solvent. A combination of (acac)Rh$(CO)_2$ and dppb was found to be highly effective for the addition to acyclic and cyclic enones. For example, a 96% yield of 2-phenyl-4-octanone was obtained from $PhB(OH)_2$ and 2-octen-4-one in aq. MeOH in the presence of (acac)Rh$(CO)_2$ and dppb. Since then, extensive studies have been carried out on the boronic acid chemistry largely related to conjugated additions, including asymmetric conjugate additions, most noticeably by Hayashi and coworkers.[269] Reactions of α,β-unsaturated ketones with excess arylboronic acids in the presence of a rhodium catalyst generated in situ from Rh(acac)(C_2H_4) and (3*S*)-4,4′-bis(diphenylphosphino)-2,2′,6,6′-tetramethoxy-3,3′-bipyridine ((S)-P-Phos) in dioxane/water at 100°C gave high yields of the corresponding products in up to 99% ee.[270] For example, 2-cyclohexenone reacted with phenylboronic acid in the presence of Rh(acac)(C_2H_4) and (*S*)-P-Phos giving 3-phenylcyclohexanone in >99% yield and 99% ee (Eq. 4.74). Arylboronate esters bearing a pendant Michael-acceptor alkene can add to norbornene and cyclize to give indane systems in yields ranging from 62 to 95% and in high diastereomeric excess (>20:1).[271] The reaction is accelerated by bases and ligands.[272]

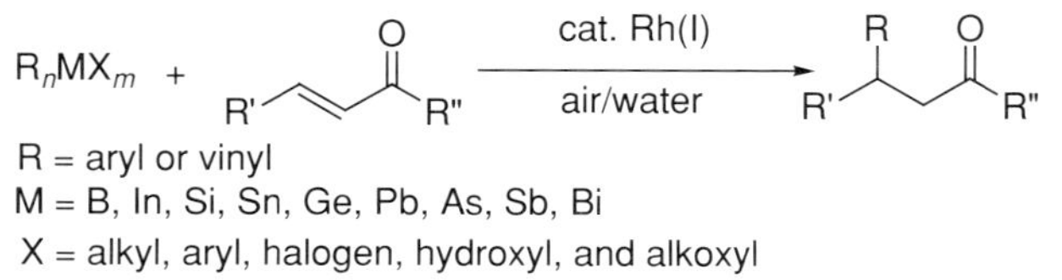

(Eq. 74)

On the other hand, Li and coworkers examined the addition of various aryl and vinyl *organometallic reagents* to α,β-unsaturated carbonyl compounds *in air and water*. It was found that both $Rh_2(COD)_2Cl_2$ and $Rh(COD)_2BF_4$ are effective.[273] The organometallic reagents include organotin,[274] organoindium, organobismuth,[275] organolead,[276] and organosilicon compounds (arylhalosilanes and aryl silanols) (Eq. 4.75)[277] in addition to organoboron compounds. For conjugate additions, both ketones (linear and cyclic) and esters were effective as electron-withdrawing functional groups. When either a mono- or disubstituted unsaturated C—C was involved, the reaction proceeded rapidly. A mixture of several products including both the conjugate addition and Heck-type reaction products was observed for the monosubstituted derivatives. Either no reaction was observed or very low yields of the products were obtained with trisubstituted derivatives. A novel synthesis of α-amino acids was developed by using the method in air and water.[278]

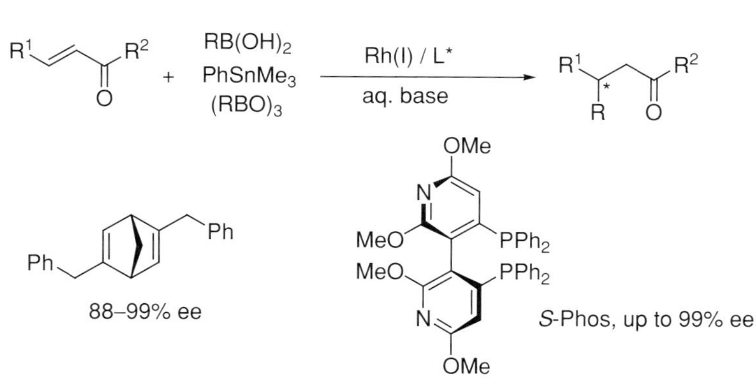

(Eq. 75)

In order to generate aryl and vinylmetal intermediates in situ, Li and coworkers reported a highly effective and stereoselective hydrosilylation of terminal alkynes under the ambient conditions of air, water, and room temperature by using Pt(DVDS)-P as the catalyst (Eq. 4.76).[279] Employing $Rh_2(COD)_2Cl_2$ as the catalyst and tributyltin hydride as the hydrometalating reagent, a one-pot alkyne-hydrostannylation and conjugate addition to unsaturated carbonyl compounds was developed in aqueous media (Eq. 4.77).[280]

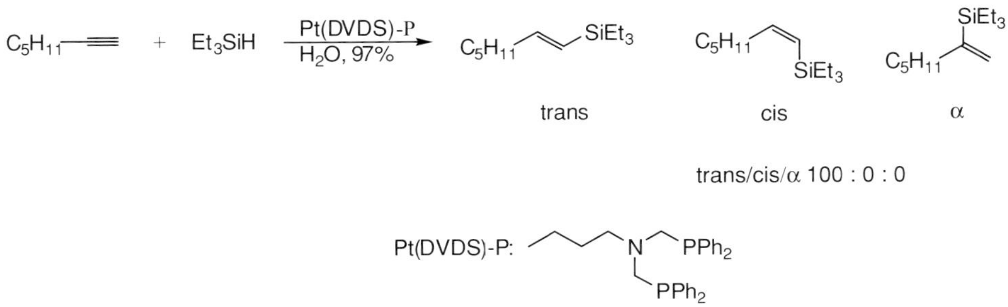

(Eq. 76)

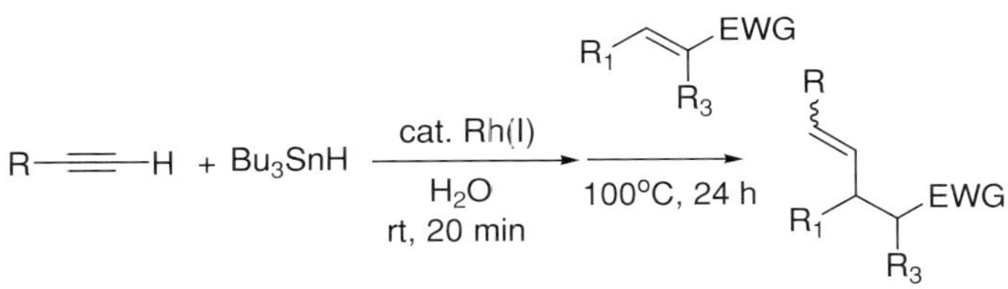

(Eq. 77)

4.10.3 Addition of alkynes

For the conjugate addition of terminal alkynes to unsaturated carbonyl compounds, Knoepfel and Carreira reported that alkynyl copper reagents, generated from terminal alkynes and catalytic $Cu(OAc)_2$ in the presence of sodium ascorbate, undergo additions to alkylidene Meldrum's acids at room temperature in aqueous media to give the corresponding adducts (Eq. 4.78).[281] The limitation to this methodology is that it seems to work only with highly activated alkylidene Meldrum's acids.

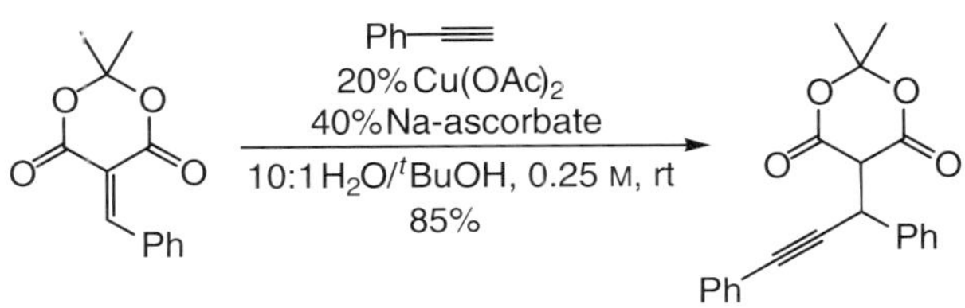

(Eq. 78)

On the other hand, Chen and Li reported a facile and selective copper–palladium-catalyzed addition of terminal alkynes to activated alkynes in water without the competition of the homocoupling of the terminal alkynes (Eq. 4.79).[282] Subsequently, a simple and highly efficient Pd-catalyzed addition of a terminal alkyne to a C—C double bond, such as a conjugated enone, either in water or in acetone, was also developed (Eq. 4.80). A variety of conjugated enones are effective in this coupling.

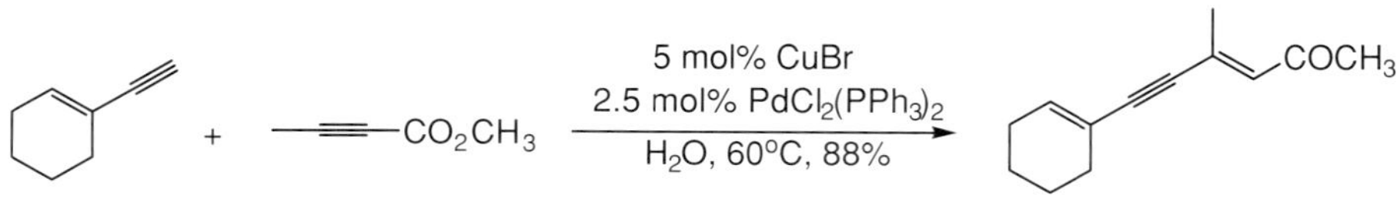

(Eq. 79)

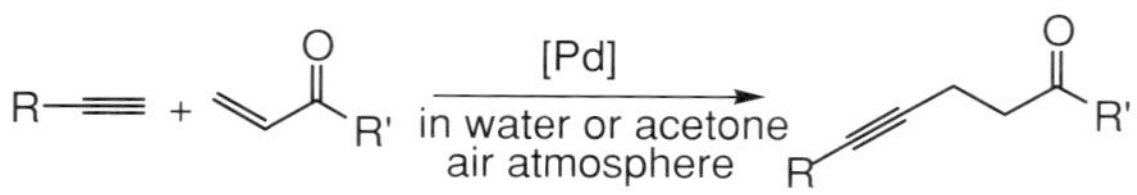

(Eq. 80)

4.11 Metal-mediated coupling reactions

4.11.1 Pinacol coupling

The pinacol coupling of carbonyl compounds[283] to give 1,2-diols has been carried out in aqueous media. Clerici and Porta extensively studied the aqueous pinacol coupling reactions mediated by Ti(III).[284] A stereoselective pinacol coupling with a cyclopentadienetitanium complex was reported by Barden and Schwartz.[285] Pinacol-type couplings were also developed by using a Zn–Cu couple,[286] Mg,[287] Mn,[288] Zn,[289] In,[290] Sm,[291] Al/NaOH,[292] Al/F^-,[293] Ga,[294] Cd,[295] and other metals (Eq. 4.81). As in the case of Barbier–Grignard-type reactions, sonication is beneficial to pinacol coupling of benzaldehyde[296] in aqueous medium and many metals including magnesium, zinc, iron, nickel, and tin have been studied under the effect of ultrasound on these reactions, with magnesium providing the best results.[297] In the absence of sonication, the reaction was much slower and the yield of the product was decreased by a factor of 2–3. Interestingly, *the reaction did not proceed under nitrogen protection*! Water alone or a 1:1 mixture of water and *t*-BuOH was used in these reactions. Aliphatic aldehydes and ketones are inert under the reaction conditions. Solid aldehydes resulted in poor yields or sometimes no product at all.

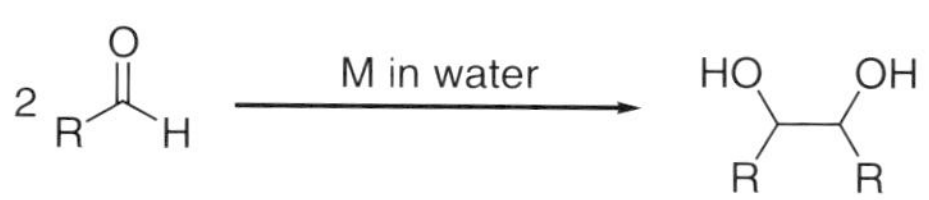

M = Zn–Cu, Mg, Mn, Zn, In, Sm, Al, Ga, Cd

(Eq. 81)

It is also possible to carry out cross-pinacol couplings under certain conditions. A pinacol-type cross-coupling reaction of aldehydes with α-diketones proceeded in the presence of water to give the corresponding adducts in moderate to good yield. It is possible to use the substrates such as phenylglyoxal monohydrate, aqueous methylglyoxal, formalin, and aqueous α-chloroacetaldehyde for this reaction.[298] Cross-coupling reactions between α,β-unsaturated carbonyl compounds and acetone were carried out by using a Zn–Cu couple and ultrasonic radiation in an aqueous acetone suspension.[299] The large excess of acetone was intended to alleviate the self-coupling of the α,β-unsaturated substrates.

4.11.2 Other reductive couplings

Reductive indium-mediated coupling of aldimines obtained from aromatic aldehydes and aromatic amines generates vicinal diamines in aqueous ethanol (Eq. 4.82).[300] Small indium rods were used in this study. The presence of NH_4Cl was found to accelerate the reaction and no side product due to unimolecular reduction was observed. The reaction fails completely in CH_3CN, DMF, or wet DMF. The use of nonaromatic substrates also resulted in the failure of the reaction.

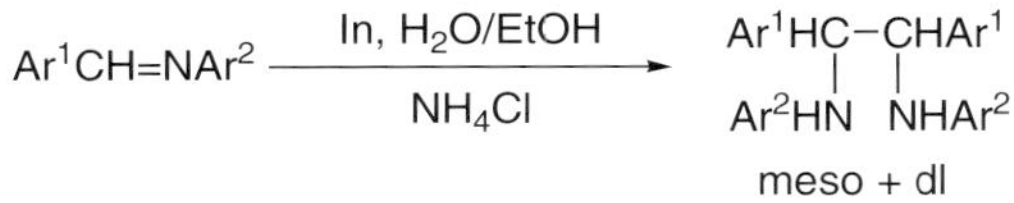

(Eq. 82)

Reductive coupling of aldimines into vicinal diamines has also been mediated by zinc powder in 10% NaOH solution at ambient temperature in high yields.[301] Additives such as NH_4Cl and L-tyrosine can be used in lieu of 10% NaOH.[302] Vicinal disulfonamides were generated by reductive coupling of *N*-sulfonylimines in Sm/HCl/THF,[303] whereas reductive coupling of aldimines and ketimine was examined by Sm(II)-based reagents (SmI_2, SmI_2-HMPA, $SmBr_2$, $Sm\{N[Si(CH_3)_3]_2\}_2$, and SmI_2, triethylamine, and water).[304] Nitrones, e.g. 2-HOC_6H_4CH:N(O)Ph, undergo deoxygenative reductive coupling and subsequent cyclization to 3-arylamino-2,3-dihydrobenzofuran derivatives, in good yields, in the presence of indium under aqueous conditions at ambient temperature (Eq. 4.83).[305] The reductive coupling cyclization reactions of 1,1-dicyanoalkenes were performed with metallic samarium in saturated aq. NH_4Cl/THF solution at room temperature (Eq. 4.84).[306] A similar reaction occurred by using zinc as the mediator.[307] The trans isomers are the major products under these conditions.

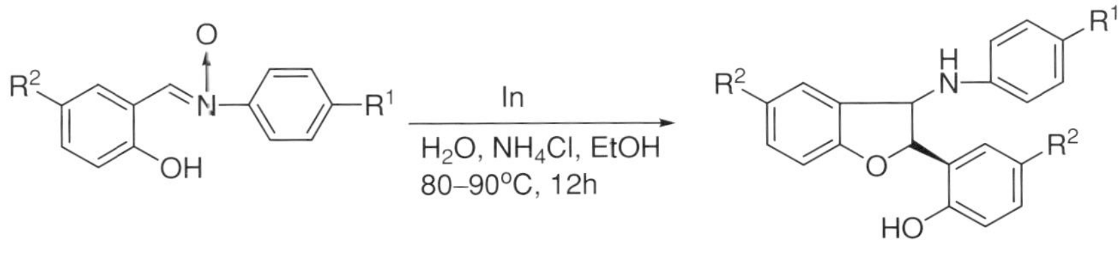

(Eq. 83)

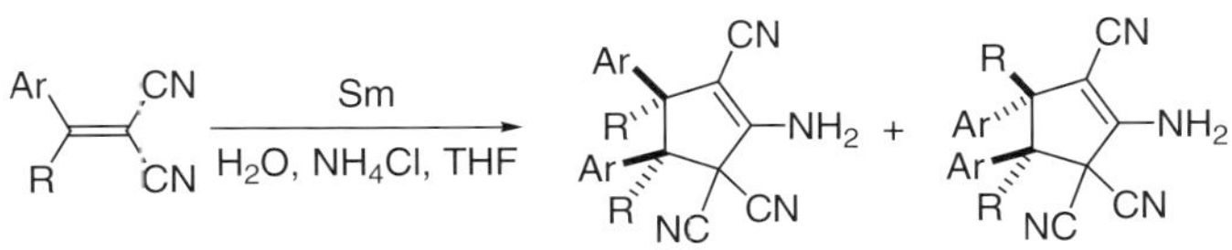

(Eq. 84)

Wurtz-type homocoupling of alkyl halides in aqueous media can be mediated by manganese/cupric chloride to give the dimerization products in good yield. Cross-coupling can also be controlled to give the desired product.[308] Wurtz-type coupling of allyl halides was (in low yields) the normal outcome in refluxing alcohol.[309] Reductive coupling reaction of benzyl, allyl, and alkyl halides in aqueous medium was promoted by zinc.[310] The reaction yields are strongly enhanced by copper catalysis. An allylgallium reagent was found to be effective for radical allylation of α-iodo or α-bromo carbonyl compounds. Treatment of benzyl bromoacetate with allylgallium, prepared from allylmagnesium chloride and gallium trichloride, in the presence of triethylborane in THF, provided benzyl 4-pentenoate in good yield. The addition of water as cosolvent improved the yields of allylated products.[311] Wurtz-type reductive coupling reaction of primary alkyl iodides or allyl halides and haloorganotins in cosolvent/$H_2O(NH_4Cl)$/Zn media provides a route to mixed alkyl and allylstannanes.[312]

A related aryl halide-aryl halide coupling is the so-called Ullmann-type coupling. The homocoupling of aryl halide to diaryl compounds has been studied in aqueous conditions. In 1970, arylsulfinic acids were coupled with Pd(II) in aqueous solvents to give biaryls (Eq. 4.85).[313] However, the reaction requires the use of a stoichiometric amount of palladium. In the presence of hydrogen gas, aryl halides homocoupled to give biaryl compounds in moderate yields (30–50%) in an aqueous/organic microemulsion (Eq. 4.86).[314]

$$2\ ArSO_2Na + Na_2PdCl_4 \xrightarrow[1\text{–}36\%]{H_2O} Ar\text{—}Ar + 2\ SO_2 + Pd + 4\ NaCl \qquad \text{(Eq. 85)}$$

$$2\ ArX \xrightarrow[H_2O/BuOH]{H_2,\ \text{cat. } PdCl_2,\ K_2CO_3} Ar\text{—}Ar \qquad \text{(Eq. 86)}$$

X = I, Br; Ar = Ph, *p*-MePh, *p*-MePPh, *p*-ClPh, *m*-CF_3Ph

In 1999, Venkatraman and Li reported a facile coupling of aryl halides catalyzed by palladium and mediated by zinc in air and aqueous acetone at room temperature by using Pd/C as catalyst (Eq. 4.87).[315] Various aryl iodides and aryl bromides coupled effectively under such conditions. Subsequently, they found that the addition of surfactant or a crown ether in water alone provided better isolated yields of the product.[316] Sasson et al. further developed this reaction by using polyethylene glycol as an additive and higher reaction temperature. In this case, aryl chloride also worked effectively.[317] Reductive homocoupling of chlorobenzenes affords high yields (93–95%) of biphenyls in the presence of catalytic PEG-400 and 0.4 mol% of a recyclable, heterogeneous trimetallic catalyst (4% Pd, 1% Pt, and 5% Bi on carbon). The competing reduction process is minimized.[318] It was assumed that dihydrogen is generated *in situ*. In addition to Pd/C, Rh/C is also effective as the catalyst.[319] Carbon dioxide was found to promote the palladium-catalyzed zinc-mediated reductive Ullmann coupling of aryl halides. In the presence of carbon dioxide, Pd/C, and zinc, various aromatic halides including less reactive aromatic chlorides were coupled to give the corresponding homocoupling products in good yields.[320] The Ullmann–Goldberg condensation of aryl halides with phenols and anilines worked efficiently in the presence of copper in water.[321] For example, coupling of 2-chlorobenzoic acid with 4-chlorophenol (K_2CO_3/pyridine/copper powder) gave 2-(4-chlorophenoxy)carboxylic acid.[322] The CuI-catalyzed transformation of 2-bromobenzoic acid into salicylic acid has also been studied in aqueous media.[323]

$$2\ ArX \xrightarrow[\substack{Zn,\ H_2O/18\text{-crown-6} \\ \text{air atmosphere, rt}}]{\text{Pd/C cat.}} Ar\text{—}Ar \qquad \text{(Eq. 87)}$$

Many transition metal-catalyzed cross-couplings such as the Stille coupling, the Heck reaction, the Suzuki reaction, and the Trost–Tsuji reactions have been carried out in aqueous conditions.[324] In our own studies, we recently discovered that various palladium-catalyzed couplings of the above types can be performed in air and water. For example, a highly efficient palladium-catalyzed coupling of aryl halides with arylhalosilanes was developed for water/open air in the presence of base or fluoride. Both $Pd(OAc)_2$ and Pd/C were effective catalysts (Eq. 4.88).[325] A Stille-type coupling and a Suzuki-type coupling in water and under an atmosphere of air were also discovered.[326]

$$\mathrm{ArBr} + \mathrm{Ph_2SiF_2} \xrightarrow[\mathrm{DMF/H_2O,\ 120^\circ C} \\ \mathrm{KF}]{\mathrm{2\ mol\%\ PdCl_2}} \mathrm{Ar{-}Ph} + [\mathrm{PhSiF_3}] \qquad \text{(Eq. 88)}$$

4.11.3 Cross-dehydrogenative coupling (CDC)

More recently, a new type of coupling, which was termed the cross-dehydrogenative coupling, was discovered by Li and coworkers.[327] The reactions involve the selective removal of two hydrogen atoms across two C—H bonds to form a new C—C bond (Scheme 4.24). Asymmetric reactions based on cross-dehydrogenative coupling are also possible.[328] Most reactions of this type developed so far either are tolerant of water or can be run in water with decreased yield. Of particular value is the report on the synthesis of mono- and diaryl-substituted 1,4-quinones via the $In(OTf)_3$-catalyzed reaction of electron-rich aromatic compounds and 1,4-quinone derivatives in water (Eq. 4.89).[329] It was observed that water was beneficial to the reaction. Furthermore, it was found that when indoles were used as the electron-rich aromatic compounds, the reaction proceeded in the absence of any catalyst, organic cosolvent, or additives. The use of water as solvent provided the best yields of the corresponding products and was the only system to produce bis-coupling products.[330]

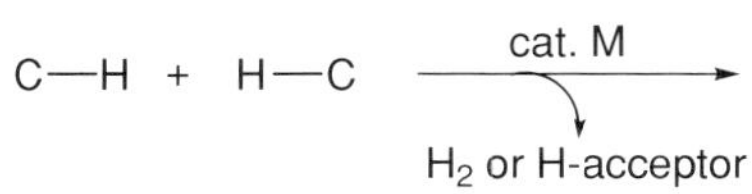

Scheme 4.24 Cross-dehydrogenative coupling for the formation of C—C bonds.

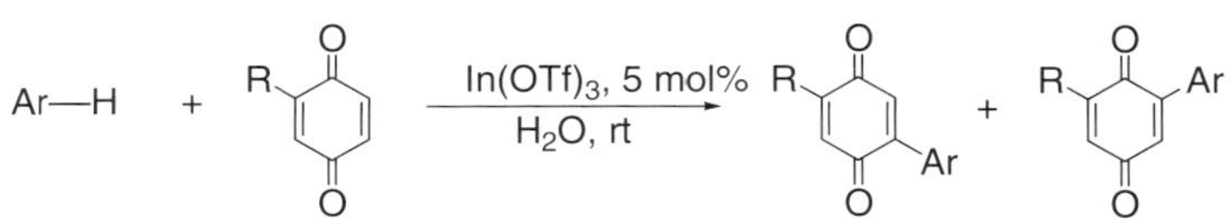

(Eq. 89)

4.12 Conclusion

The use of metals for mediating carbon–carbon bond formations, in particular the Grignard reaction, has been an inherent part of the history of organic chemistry. It has made the synthesis of numerous important organic compounds possible. Kharasch stated over half a century ago that '*he who knows and understands the Grignard reactions has a fair grasp of organic chemistry, for most fundamental processes have prototypes or analogous phenomena observable in Grignard systems.*'[10] Ironically, it is exactly because of the fundamental importance of 'Grignard reactions' in organic chemistry that the notion of such reactions as being *highly sensitive toward air and water* is so deeply imprinted in our mind, and a primary reason for the design of many protection–deprotection methods and inert atmosphere techniques in synthesis. The development of metal-mediated carbon–carbon bond formation reactions in water (and air) will open new avenues in chemistry and fundamentally change 'organometallic reactions' as we know it in the chemistry textbooks of today, and the way we carry out organic synthesis.

Acknowledgments

The author thank all the people (especially Prof. T.H. Chan) whose names are cited in the references, and who have contributed to this subject. He is deeply indebted to the past and present members of C. J. Li's laboratory for their dedication, and to the Canada Research Chair Foundation, NSERC, CFI, (US) NSF, (US) EPA, American Chemical Society (PRF), FQRNT, Merck Frosst, CIC (Merck Frosst/Bohringer Ingelheim/AstraZeneca), Louisiana Board of Regents for their support over the years related to this subject.

References

1. Frankland, E., *Annalen*, 1849, **71**, 171.
2. Reformatsky, A., *Ber.*, 1887, **20**, 1210.
3. Barbier P., *Compt. Rend.*, 1899**, 128**, 110.
4. Grignard, V., *Compt. Rend.*, 1890, **130**, 1322.
5. For a recent monograph, see: Blomberg, C., In: Hafner, K., Lehn, J.M., Rees, C.W., von Ragué Schleyer, P., Trost, B.M. and Zahradnîk, R. (Eds.), *Reactivity and Structure: Concepts in Organic Chemistry*, Springer-Verlag, Berlin, 1993; for a review, see: Lai, Y.H., *Synthesis*, 1981, 585.
6. Li, C.J., *Chem. Rev.*, 1993, **93**, 2023; Li, C.J., *Tetrahedron*, 1996, **52**, 5643; Li, C.J., In: Anastas, P.T. and Williamson, T.C. (Eds.), *Green Chemistry: Recent Advances in Chemical Syntheses*, ACS Symposium Series, Oxford University Press, Cambridge, UK, 1999; Zhang, W.C., Hua, X.G., Meng, Y., Venkatraman, S., Keh, C.C.K. and Li, C.J., In: Pandalai, S.G. (Ed.), *Recent Research Developments in Organic Chemistry*, Vol. 4, Part II, Transworld Research Network, Trivandrum, India, 2000, p. 397; Li, C.J., *Green Chem.*, 2002, **4**, 1; Li, C.J. and Wang, D., *Chemtract*, 2003, **16**, 59; Li, C.J., *Angew. Chem., Int. Ed. Engl.*, 2003, **42**, 4856; Li, C.J., *Chem. Rev.*, 2005, **105**, 3095; Li, C.J. and Chan, T.H., *Organic Reactions in Aqueous Media*, John Wiley & Sons, New York, 1997; Chen, L., Li, C.J. Chem. Sec. Rev. 2006, 5, 68.
7. Li, C.J., Haberman, J.X., Keh, C.C.K., Yi, X.H., Meng, Y., Hua, X.G., Venkatraman, S., Zhang, W.C., Nguyen, T., Wang, D., Huang, T.S. and Zhang, J.H., In: Moen, L. and Abraham, M. (Eds.), *Cleaner Media*, ACS Symposium Series, Oxford University Press, Cambridge, UK, 2001.
8. Peters, W., *Ber.*, 1905, **38**, 2567.
9. Sisido, K., Takeda,Y. and Kinugawa, Z., *J. Am. Chem. Soc.*, 1961, **83**, 538; Sisido, K., Kozima, S. and Hanada, T., *J. Organomet. Chem.*, 1967, **9**, 99; Sisido, K. and Kozima, S., *J. Organomet. Chem.*, 1968, **11**, 503.
10. In a classical book, Kharasch and Reinmuth frequently used the six-membered cyclic transition state to explain most Grignard reactions; see: Kharasch, M.S. and Reinmuth, O., *Grignard Reactions of Nonmetallic Substances*, Prentice-Hall, New York, 1954.
11. Fleming, I., *Frontier Orbitals and Organic Chemical Reactions*, John Wiley & Sons, Chichester, England, 1976.
12. Chuang, L.W., Chan, T.H. and Wu, Y.D., *Organometallics*, 2005, **24**, 1598.
13. Chan, T.H. and Yang, Y., *J. Am. Chem. Soc.*, 1999, **121**, 3228.
14. Li, C.J. and Chan, T.H., *Tetrahedron*, 1999, **55**, 11149.
15. Killinger, T.A., Boughton, N.A., Runge, T.A. and Wolinsky, J., *J. Organomet. Chem.*, 1977, **124**, 131.
16. Petrier, C. and Luche, J.L., *J. Org. Chem.*, 1985, **50**, 910.
17. Einhorn, C. and Luche, J.L., *J. Organomet. Chem.*, 1987, **322**, 177; Petrier, C., Einhorn, J. and Luche, J.L., *Tetrahedron Lett.*, 1985, **26**, 1449.
18. Mattes, H. and Benezra, C., *Tetrahedron Lett.*, 1985, **26**, 5697; Zhou, J.Y., Lu, G.D. and Wu, S.H., *Synth. Commun.*, 1992, **22**, 481.
19. Wilson, S.R. and Guazzaroni, M.E., *J. Org. Chem.*, 1989, **54**, 3087.

20. Kunz, T. and Reißig, H.U., *Liebigs Ann. Chem.*, 1989, 891.
21. Chan, T.H. and Li, C.J., *Organometallics*, 1990, **9**, 2649.
22. Li, C.J. and Chan, T.H., *Organometallics*, 1991, **10**, 2548.
23. Oda, Y., Matsuo, S. and Saito, K., *Tetrahedron Lett.*, 1992, **33**, 97.
24. Cripps, H.N. and Kiefer, E.F., *Org. Synth.*, 1962, **42**, 12.
25. Durant, A., Delplancke, J.L., Winand, R. and Reisse, J., *Tetrahedron Lett.*, 1995, **36**, 4257.
26. Marton, D., Stivanello, D. and Tagliavini, G., *J. Org. Chem.*, 1996, **61**, 2731.
27. Sjoholm, R., Rairama, R. and Ahonen, M., *Chem. Commun.*, 1994, 1217.
28. Ahonen, M. and Sjöholm, R., *Chem. Lett.*, **1995**, 341.
29. Yang, Z.Y. and Burton, D.J., *J. Org. Chem.*, 1991, **56**, 1037.
30. Hanessian, S., Park, H. and Yang, R.Y., *Synlett.*, 1997, 351; Hanessian, S., Park, H. and Yang, R.Y., *Synlett*, 1997, 353.
31. Chan, T.H., Li, C.J., Lee, M.C. and Wei, Z.Y., *Can. J. Chem.*, 1994, **72**, 1181.
32. Marquez, F., Montoro, R., Llebaria, A., Lago, E., Molins, E. and Delgado, A., *J. Org. Chem.*, 2002, **67**, 308.
33. Nokami, J., Otera, J., Sudo, T. and Okawara, R., *Organometallics*, 1983, **2**, 191.
34. Nokami, J., Wakabayashi, S. and Okawara, R., *Chem. Lett.*, 1984, 869.
35. Zhou, J.Y., Chen, Z. G. and Wu, S.H., *J. Chem. Soc., Chem Commun.*, 1994, 2783.
36. Uneyama, K., Kamaki, N., Moriya, A. and Torii, S., *J. Org. Chem.*, 1985, **50**, 5396.
37. Petrier, C. and Luche, J.L., *J. Org. Chem.*, 1985, **50**, 910; Petrier, C., Einhorn, J. and Luche, J.L., *Tetrahedron Lett.*, 1985, **26**, 1449.
38. Einhorn, C. and Luche, J.L., *J. Organomet. Chem.*, 1987, **322**, 177.
39. Wu, S.H., Huang, B.Z., Zhu, T.M., Yiao, D.Z. and Chu, Y.L., *Acta Chim. Sin.*, 1990, **48**, 372; Wu, S.H., Huang, B.Z., Zhu, T.M., Yiao, D.Z. and Chu, Y.L., *Acta Chim. Sin.*, 1987, **45**, 1135.
40. Zha, Z., Wang, Y., Yang, G., Zhang, L. and Wang, Z., *Green Chem.*, 2002, **4**, 578.
41. Wang, Z., Zha, Z. and Zhou, C., *Org. Lett.*, 2002, **4**, 1683.
42. Uneyama, K., Matsuda, H. and Torii, S., *Tetrahedron Lett.*, 1984, **25**, 6017.
43. Mandai, T., Nokami, J. and Yano, T., *J. Org. Chem.*, 1984, **49**, 172.
44. Masuyama, Y., Takahara, T.P. and Kurusu, Y., *Tetrahedron Lett.*, 1989, **30**, 3437; Masuyama, Y., Nimura, Y. and Kurusu, Y., *Tetrahedron Lett.*, 1991, **32**, 225.
45. Sati, M. and Sincu, D., *Tetrahedron Lett.*, 1991, **32**, 2025.
46. Masuyama, Y., Kishida, M. and Kurusu, Y., *Chem. Commun.*, 1995, 1405; Ito, A., Kishida, M., Kurusu, Y. and Masuyama, Y., *J. Org. Chem.*, 2000, **65**, 494.
47. Kundu, A., Prabhakar, S., Vairamani, M. and Roy, S., *Organometallics*, 1997, **16**, 4796; Tan, X.-H., Shen, B., Liu, L. and Guo, Q.-X., *Tetrahedron Lett.*, 2002, **43**, 9373; Debroy, P. and Roy, S., *J. Organomet. Chem.*, 2003, **675**, 105.
48. Okano, T., Kiji, J. and Doi, T., *Chem. Lett.*, 1998, 5.
49. Carde, L., Llebaria, A. and Delgado, A., *Tetrahedron Lett.*, 2001, **42**, 3299.
50. Tan, X.-H., Shen, B., Deng, W., Zhao, H., Liu, L. and Guo, Q.X., *Org. Lett.*, 2003, **5**, 1833.
51. Wang, J., Yuan, G. and Dong, C.-Q., *Chem. Lett.*, 2004, **33**, 286.
52. Samoshin, V.V., Gremyachinskiy, D.E., Smith, L.L., Bliznets, I.V. and Gross, P.H., *Tetrahedron Lett.*, 2002, **43**, 6329; Gremyachinskiy, D.E., Smith, L.L., Gross, P.H. and Samoshin, V.V., *Green Chem.*, 2002, **4**, 317.
53. Chang, H.-M. and Cheng, C.-H., *Org. Lett.*, 2000, **2**, 3439.
54. Boaretto, A., Marton, D., Tagliavini, G. and Gambaro, A., *J. Organomet. Chem.*, 1985, **286**, 9.
55. Boaretto, A., Marton, D. and Tagliavini, G., *J. Organomet. Chem.*, 1985, **297**, 149.
56. Furlani, D., Marton, D., Tagliavini, G. and Zordan, M., *J. Organomet. Chem.*, 1988, **341**, 345.
57. Hachiya, I. and Kobayashi, S., *J. Org. Chem.*, 1993, **58**, 6958; Kobayashi, S., Wakabayashi, T. and Oyamada, H., *Chem Lett.*, 1997, 831; McCluskey, A., *Green Chem.*, 1999, **1**, 167.
58. Yanagisawa, A., Morodome, M., Nakashima, H. and Yamamoto, H., *Synlett*, 1997, 1309.
59. Marshall, R.L., Muderawan, I.W. and Young, D.J., *J. Chem. Soc., Perkin Trans. 2*, 2000,(5), 957.
60. Shibata, I., Yoshimura, N., Yabu, M. and Baba, A., *Eur. J. Org. Chem.*, 2001, 3207.
61. Manabe, K., Mori, Y., Wakabayashi, T., Nagayama, S. and Kobayashi, S., *J. Am. Chem. Soc.*, 2000, **122**, 7202.
62. Nagayama, S. and Kobayashi, S., *Angew. Chem. Int. Ed.*, 2000, **39**, 567.
63. Kobayashi, S. and Aoyama, N., Manabe, K., *Synlett*, 2002, 483.
64. Schmid, W. and Whitesides, G.M., *J. Am. Chem. Soc.*, 1991, **113**, 6674; Kim, E., Gordon, D.M., Schmid, W. and Whitesides, G.M., *J. Org. Chem.*, 1993, **58**, 5500.
65. Li, C.J. and Chan, T.H., *Tetrahedron Lett.*, 1991, **32**, 7017.
66. Li, C.J., Ph.D. Thesis, McGill University, 1992.
67. Gordon, D.M. and Whitesides, G.M., *J. Org. Chem.*, 1993, **58**, 7937.
68. Araki, S., Jin, S.J., Idou, Y. and Butsugan, Y., *Bull. Chem. Soc. Jpn.*, 1992, **65**, 1736.
69. Marshall, J.A. and Hinkle, K.W., *J. Org. Chem.*, 1995, **60**, 1920.
70. Jang, T.-S., Keum, G., Kang, S.B., Chung, B.Y. and Kim, Y., *Synthesis*, 2003, 775.
71. Chan, T.H. and Li, C.J., *J. Chem. Soc., Chem. Commun.*, 1992, 747.

72. Chan, T.H. and Li, C.J., In: *203rd National Meeting of the American Chemical Society*, San Francisco, CA, April 1992, Abstract ORGN435.
73. Dondoni, A., Merino, P. and Orduna, J., *Tetrahedron Lett.*, 1991, **32**, 3247.
74. Gao, J., Harter, R., Gordon, D.M. and Whitesides, G.M., *J. Org. Chem.*, 1994, **59**, 3714.
75. Casiraghi, G. and Rassu, G., *Chemtracts: Org. Chem.*, 1993, **6**, 336.
76. Archibald, S.C. and Hoffmann, R.W., *Chemtracts: Org. Chem.*, 1993, **6**, 194.
77. Chan, T.H. and Lee, M.C., *J. Org. Chem.*, 1995, **60**, 4228.
78. Prenner, R.H., Binder, W.H. and Schmid, W., *Liebigs Ann. Chem.*, 1994, 73.
79. Binder, W.H., Prenner, R.H. and Schmid, W., *Tetrahedron*, 1994, **50**, 749.
80. Carda, M., Castillo, E. and Marco, J.A., *Tetrahedron: Asymmetry*, 1998, **9**, 1117.
81. Canac, Y., Levoirier, E. and Lubineau, A., *J. Org. Chem.*, 2001, **66**, 3206; Lubineau, A., Canac, Y. and Le Goff, N., *Adv. Synth. Catal.*, 2002, **344**, 319.
82. Chappell, M.D. and Halcomb, R.L., *Org. Lett.*, 2000, **2**, 2003.
83. Warwel, M. and Fessner, W.D., *Synlett*, 2000, 865.
84. Gao, J., Martichonok, V. and Whitesides, G.M., *J. Org. Chem.*, 1996, **61**, 9538.
85. Li, C.J., *Tetrahedron Lett.*, 1995, **36**, 517.
86. Chen, D.L. and Li, C.J., *Tetrahedron Lett.*, 1996, **37**, 295.
87. Yi, X.H., Meng, Y. and Li, C.J., *Tetrahedron Lett.*, 1997, **38**, 4731.
88. Wang, R., Lim, C.M., Tan, C.H., Lim, B.K., Sim, K.Y. and Loh, T.P., *Tetrahedron: Asymmetry*, 1995, **6**, 1825.
89. Li, C.J. and Lu, Y.Q., *Tetrahedron Lett.*, 1995, **36**, 2721.
90. Li, C.J. and Lu, Y.Q., *Tetrahedron Lett.*, 1996, **37**, 471.
91. Li, C.J., Chen, D.L., Lu, Y.Q., Haberman, J.X. and Mague, J.T., *J. Am. Chem. Soc.*, 1996, **118**, 4216; Li, C.-J., Chen, D.-L., Lu, Y.-Q., Haberman, J.X. and Mague, J.T., *Tetrahedron*, 1998, **54**, 2347.
92. Haberman, J.X. and Li, C.J., *Tetrahedron Lett.*, 1997, **38**, 4735.
93. Loh, T.-P., Cao, G.-Q. and Pei, J., *Tetrahedron Lett.*, 1998, **39**, 1453.
94. Loh, T.P., Cao, G.Q. and Pei, J., *Tetrahedron Lett.*, 1998, **39**, 1457; Loh, T.-P. and Song, H.-Y., *Synlett*, 2002, 2119.
95. Loh, T.P. and Li, X.R., *J. Chem. Soc., Chem. Commun.*, 1996, 1929.
96. Loh, T.P., Wang, R.B. and Sim, K.Y., *Main Group Met. Chem.*, 1997, **20**, 237.
97. Bryan, V.J. and Chan, T.H., *Tetrahedron Lett.*, 1996, **37**, 5341.
98. Hao, J., Aiguade, J. and Forsyth, C.J., *Tetrahedron Lett.*, 2001, **42**, 821.
99. Yadav, J.S., Reddy, B.V.S., Reddy, G.S. and Kiran, K., *Tetrahedron Lett.*, 2000, **41**, 2695.
100. Mendez-Andino, J. and Paquette, L.A., *Org. Lett.*, 2000, **2**, 1263.
101. Bernardelli, P., Moradei, O.M., Friedrich, D., Yang, J., Gallou, F., Dyck, B.P., Doskotch, R.W., Lange, T. and Paquette, L.A., *J. Am. Chem. Soc.*, 2001, **123**, 9021.
102. Mendez-Andino, J. and Paquette, L.A., *Adv. Synth. Catal.*, 2002, **344**, 303.
103. Vogel, S., Stembera, K., Hennig, L., Findeisen, M., Giesa, S., Welzel, P. and Lampilas, M., *Tetrahedron*, 2001, **57**, 4139.
104. Lee, J.E., Cha, J.H., Pae, A.N., Choi, K.I., Koh, H.Y., Kim, Y. and Cho, Y.S., *Synth. Commun.*, 2000, **30**, 4299.
105. Cha, J.H., Pae, A.N., Choi, K.I., Cho, Y.S., Koh, H.Y. and Lee, E., *J. Chem. Soc., Perkin Trans 1*, 2001, 2079.
106. Shin, J.A., Choi, K.I., Pae, A.N., Koh, H.Y., Kang, H.-Y. and Cho, Y.S., *J. Chem. Soc., Perkin Trans. 1*, 2001, 946.
107. Yang, Y. and Chan, T.H., *J. Am. Chem. Soc.*, 2000, **122**, 402.
108. Lu, W., Ma, J. and Yang, Y., *Org. Lett.*, 2000, **2**, 3469.
109. Loh, T.-P., Tan, K.-T., Yang, J.-Y. and Xiang, C.-L., *Tetrahedron Lett.*, 2001, **42**, 8701.
110. Loh, T.-P., Tan, K.-T. and Hu, Q.-Y., *Tetrahedron Lett.*, 2001, **42**, 8705; Tan, K.-T., Chng, S.-S., Cheng, H.-S. and Loh, T.-P., *J. Am. Chem. Soc.*, 2003, **125**, 2958.
111. Juan, S., Hua, Z.-H., Qi, S., Ji, S.-J. and Loh, T.-P., *Synlett*, 2004, 829.
112. Huang, J.-M., Xu, K.-C. and Loh, T.-P., *Synthesis*, 2003, 755.
113. Singh, S., Kumar, S. and Chimni, S.S., *Tetrahedron: Asymmetry*, 2002, **13**, 2679.
114. Thadani, A.N. and Batey, R.A., *Org. Lett.*, 2002, **4**, 3827.
115. Thadani, A.N. and Batey, R.A., *Tetrahedron Lett.*, 2003, **44**, 8051.
116. Chung, W.J., Higashiya, S., Oba, Y. and Welch, J.T., *Tetrahedron*, 2003, **59**, 10031.
117. Wang, M., Chen, Y.-J., Liu, L., Wang, D. and Liu, X.-L., *J. Chem. Res. (S)*, 2000, 80.
118. Wang, Z., Yuan, S. and Li, C.-J., *Tetrahedron Lett.*, 2002, **43**, 5097.
119. Tsuji, T., Usugi, S.-I., Yorimitsu, H., Shinokubo, H., Matsubara, S. and Oshima, K., *Chem. Lett.*, 2002, 2.
120. Li, C.-J. and Zhang, W.-C., *J. Am. Chem. Soc.*, 1998, **120**, 9102; Zhang, W.-C. and Li, C.-J., *J. Org. Chem.*, 1999, **64**, 3230.
121. Zhou, J.-Y., Jia, Y., Sun, G.-F. and Wu, S.-H., *Synth. Commun.*, 1997, **27**, 1899.
122. Li, C.-J., Meng, Y., Yi, X.-H., Ma, J. and Chan, T.H., *J. Org. Chem.*, 1997, **62**, 8632.
123. Li, L.-H. and Chan, T.H., *Can. J. Chem.*, 2001, **79**, 1536.

124. Wada, M., Ohki, H. and Akiba, K.Y., *Bull. Chem. Soc. Jpn.*, 1990, **63**, 1738; Wada, M., Ohki, H. and Akiba, K.Y., *J. Chem. Soc., Chem. Commun.*, 1987, 708; Fukuma, T., Lock, S., Miyoshi, N. and Wada, M., *Chem. Lett.*, 2002, 376.
125. Wada, M., Fukuma, T., Morioka, M., Takahashi, T. and Miyoshi, N., *Tetrahedron Lett.*, 1997, **38**, 8045.
126. Smith, K., Lock, S., El-Hiti, G.A., Wada, M. and Miyoshi, N., *Org. Biomol. Chem.*, 2004, **2**, 935.
127. Matsumura, N., Doi, T., Mishima, K., Kitagawa, Y., Okumura, Y. and Mizuno, K., *ITE Lett. Batt. New Technol. Med.*, 2003, **4**, 473.
128. Minato, M. and Tsuji, J., *Chem. Lett.*, 1988, 2049.
129. Chan, T.H. and Isaac, M., *Tetrahedron Lett.*, 1995, **36**, 8957.
130. Cram, D.J. and Kopecky, K.R., *J. Am. Chem. Soc.*, 1959, **81**, 2748; Reetz, M.T., *Angew. Chem., Int. Ed. Engl.*, 1984, **23**, 556; Midland, M.M. and Koops, R.W., *J. Org. Chem.*, 1990, **55**, 5058.
131. Paquette, L.A., In: Anastas, P.A. and Williamson, T.C. (Eds.), *Green Chemistry: Frontiers in Benign Chemical Syntheses and Processes*, Oxford University Press, Cambridge, UK, 1998; Isaac, M.B. and Chan, T.H., *Tetrahedron Lett.*, 1995, **36**, 8957; Chan, T.H. and Li, C.J., *Can. J. Chem.*, 1992, **70**, 2726; Paquette, L. and Mitzel, T.M., *J. Am. Chem. Soc.*, 1996, **118**, 1931.
132. Chan, T.H. and Li, C.J., *Can. J. Chem.*, 1992, **70**, 2726.
133. Waldmann, H., *Synlett*, 1990, 627.
134. Paquette, L.A., Bennett, G.D., Isaac, M.B. and Chhatriwalla, A., *J. Org. Chem.*, 1998, **63**, 1836; Maguire, R.J., Mulzer, J. and Bats, J.W., *J. Org. Chem.*, 1996, **61**, 6936; Paquette, L.A., Rothhaar, R.R., Isaac, M., Rogers, L.M. and Rogers, R.D., *J. Org. Chem.*, 1998, **63**, 5463; Lloyd-Jones, G.C. and Russell, T., *Synlett*, 1998, 903.
135. Alcaide, B., Almendros, P., Aragoncillo, C. and Rodriguez-Acebes, R., *Synthesis*, 2003, 1163.
136. Hidestål, O., Ding, R., Almesåker, A. and Lindström, U.M., *Green Chem.*, 2005, **7**, 259.
137. Fornasier, R., Marcuzzi, F., Piva, M. and Tonellato, U., *Gazz. Chim. Ital.*, 1996, **126**, 633; Fornasier, R., Marcuzzi, F. and Marton, D., *Main Group Met. Chem.*, 1998, **21**, 65; Abele, E. and Lukevics, E., *Latv. Kim. Zur.*, 1998, 73.
138. Cunningham, A. and Woodward, S., *Synlett*, 2002, 43.
139. Kobayashi, S., Aoyama, N. and Manabe, K., *Chirality*, 2003, **15**, 124.
140. Gajewski, J.J., Bocian, W., Brichford, N.L. and Henderson, J.L., *J. Org. Chem.*, 2002, **67**, 4236; Lucas, P., Gajewski, J.J. and Chan, T.H., *Can. J. Anal. Sci. Spectrosc.*, 2003, **48**, 1.
141. Grieco, P.A. and Bahsas, A., *J. Org. Chem.*, 1987, **52**, 1378.
142. Katritzky, A.R., Shobana, N. and Harris, P.A., *Organometallics*, 1992, **11**, 1381.
143. Kumaraswamy, S., Nagabrahmanandachari, S. and Kumara Swamy, K.C., *Synth. Commun.*, 1996, **24**, 729.
144. Chan, T.H. and Lu, W., *Tetrahedron Lett.*, 1998, **39**, 8605; Sampath Kumar, H.M., Anjaneyulu, S., Jagan Reddy, E. and Yadav, J.S., *Tetrahedron Lett.*, 2000, **41**, 9311.
145. Lu, W. and Chan, T.H., *J. Org. Chem.*, 2000, **65**, 8589.
146. Bryan, V.J. and Chan, T.H., *Tetrahedron Lett.*, 1997, **38**, 6493.
147. Yoo, B.W., Choi, K.H., Lee, S.J., Nam, G.S., Chang, K.Y., Kim, S.H. and Kim, J.H., *Synth. Commun.*, 2002, **32**, 839.
148. Laskar, D.D., Prajapati, D. and Sandhu, J.S., *Tetrahedron Lett.*, 2001, **42**, 7883; Laskar, D.D., Gohain, M., Prajapati, D. and Sandhu, J.S., *New J. Chem.*, 2002, **26**, 193.
149. Kang, K.H., Choi, K.I., Koh, H.Y., Kim, Y., Chung, B.Y. and Cho, Y.S., *Synth. Commun.*, 2001, **31**, 2277.
150. Kobayashi, S., Hamada, T. and Manabe, K., *Synlett*, 2001, 1140.
151. Akiyama, T. and Onuma, Y., *J. Chem. Soc., Perkin Trans. 1*, 2002, 1157.
152. Estevam, I.H.S. and Bieber, L.W., *Tetrahedron Lett.*, 2003, **44**, 667.
153. Kumar, S. and Kaur, P., *Tetrahedron Lett.*, 2004, **45**, 3413.
154. Miyabe, H., Yamaoka, Y., Naito, T. and Takemoto, Y., *J. Org. Chem.*, 2003, **68**, 6745.
155. Matsumura, Y., Onomura, O., Suzuki, H., Furukubo, S., Maki, T. and Li, C.-J., *Tetrahedron Lett.*, 2003, **44**, 5519.
156. Basile, T., Bocoum, A., Savoia, D. and Umani-Ronchi, A., *J. Org. Chem.*, 1994, **59**, 7766.
157. Hanessian, S. and Yang, R.-Y., *Tetrahedron Lett.*, 1996, **37**, 5273.
158. Shin, J.A., Cha, J.H., Pae, A.N., Choi, K.I., Koh, H.Y., Kang, H.-Y. and Cho, Y.S., *Tetrahedron Lett.*, 2001, **42**, 5489.
159. Miyabe, H., Nishimura, A., Ueda, M. and Naito, T., *Chem. Commun.*, 2002, 1454.
160. Ooi, T., Uematsu, Y. and Maruoka, K., *Adv. Synth. Catal.*, 2002, **344**, 288.
161. Hamada, T., Manabe, K. and Kobayashi, S., *Angew. Chem. Int. Ed.*, 2003, **42**, 3927.
162. Wu, S.H., Huang, B.Z. and Gao, X., *Synth. Commun.*, 1990, **20**, 1279.
163. Boaretto, A., Marton, D., Tagliavini, G. and Gambaro, A., *J. Organomet. Chem.*, 1985, **286**, 9.
164. Boaretto, A., Marton, D. and Tagliavini, G., *J. Organomet. Chem.*, 1985, **297**, 149.
165. Furlani, D., Marton, D., Tagliavini, G. and Zordan, M., *J. Organomet. Chem.*, 1988, **341**, 345.
166. Hachiya, I. and Kobayashi, S., *J. Org. Chem.*, 1993, **58**, 6958.
167. Houllemare, D., Outurquin, F. and Paulmier, C., *J. Chem. Soc., Perkin Trans. 1*, 1997, 1629.
168. Wang, J., Maio, X.P. and Yuan, G., *Aust. J. Chem.*, 2005, **58**, 611.

169. Chattopadhyay, A., *J. Org. Chem.*, 1996, **61**, 6104; Keltjens, R., Vadivel, S.K., de Gelder, R., Bieber, L.W., da Silva, M.F., da Costa, R.C. and Silva, L.O.S., *Tetrahedron Lett.*, 1998, **39**, 3655; Klunder, A.J.H. and Zwanenburg, B., *Eur. J. Org. Chem.*, 2003, 1749.
170. Issac, M.B. and Chan, T.H., *J. Chem. Soc., Chem. Commun.*, 1995, 1003.
171. Mitzel, T.M., Palomo, C. and Jendza, K., *J. Org. Chem.*, 2002, **67**, 136.
172. Yi, X.H., Meng, Y., Hua, X.G. and Li, C.J., *J. Org. Chem.*, 1998, **63**, 7472.
173. Yi, X.H., Meng, Y. and Li, C.J., *Chem. Commun.*, 1998, 449.
174. Fang, X.P., Anderson, J.E., Chang, C.J., Fanwick, P.E. and McLaughlin, J.L., *J. Chem. Soc., Perkin Trans. 1*, 1990, 1655.
175. Hua, X.G. and Li, C.J., *Main Group Met. Chem.*, 1999, **22**, 533; Hua, X.G., Mague, J.T. and Li, C.J., *Tetrahedron Lett.*, 1998, **39**, 6837; Mague, J.T., Hua, X.G. and Li, C.J., *Acta Crystallogr. Sect. C Cryst. Struct. Commun.*, 1998, **54**, 1934.
176. Cho, Y.S., Lee, J.E., Pae, A.N., Choi, K.I. and Koh, H.Y., *Tetrahedron Lett.*, 1999, **40**, 1725.
177. Kwon, J.S., Pae, A.N., Choi, K.I., Koh, H.Y., Kim, Y. and Cho, Y.S., *Tetrahedron Lett.*, 2001, **42**, 1957.
178. Alcaide, B., Almendros, P., Aragoncillo, C. and Rodriguez-Acebes, R., *J. Org. Chem.*, 2001, **66**, 5208; Alcaide B., Almendros P. and Aragoncillo C., *Chemistry*, 2002, **8**, 1719; *Chem. Eur. J.*, 2002, **8**, 1719; Alcaide, B., Almendros, P., Aragoncillo, C. and Rodriguez-Acebes, R., *Synthesis*, 2003, 1163.
179. Prajapati, D., Laskar, D.D., Gogoi, B.J. and Devi, G., *Tetrahedron Lett.*, 2003, **44**, 6755.
180. Miyabe, H., Yamaoka, Y., Naito, T. and Takemoto, Y., *J. Org. Chem.*, 2004, **69**, 1415.
181. Bieber, L.W., Storch, E.C., Malvestiti, I. and Sila, M.F., *Tetrahedron Lett.*, 1998, **39**, 9393.
182. Zha, Z.-G., Xie, Z., Zhou, C.-L., Wang, Z.-Y. and Wang, Y.-S., *Chin. J. Chem.*, 2002, **20**, 1477.
183. Crabtree, R.H., *The Organometallic Chemistry of Transition Metals*, 3rd edn. Wiley-Interscience, New York, 2005.
184. Venkatraman, S. and Li, C.J., *Org. Lett.*, 1999, **1**, 1133; Venkatraman, S. and Li, C.J., *Tetrahedron Lett.*, 2000, **41**, 4831; Venkatraman, S., Huang, T.S. and Li, C.J., *Adv. Synth. Catal.*, 2002, **344**, 399; Li, C.J., Slaven, W.T., John, V.T. and Banerjee, S., *J. Chem. Soc., Chem. Commun.*, 1997, 1569.
185. Li, C.J., *Acc. Chem. Res.*, 2002, **35**, 533.
186. Sakai, M., Ueda, M. and Miyaura, N., *Angew. Chem., Int. Ed. Engl.*, 1998, **37**, 3274.
187. Ueda, M. and Miyaura, N., *J. Org. Chem.*, 2000, **65**, 4450.
188. Meng, Y. and Li, C.J., *J. Am. Chem. Soc.*, 2000, **120**, 9538; Li, C.J., Huang, T.S., Venkatraman, S., Meng Y., Kort, D., Ding, R., Wang, D. and Nguyen, T., *Pure Appl. Chem.*, 2001, **73**, 1315.
189. Huang, T.-S., Meng, Y., Venkatraman, S., Wang, D. and Li, C.-J., *J. Am. Chem. Soc.*, 2001, **123**, 7451.
190. Murata, M., Shimazaki, R., Ishikura, M., Watanabe, S. and Masuda, Y., *Synthesis*, 2002, 717.
191. Takai, K., Morita, R. and Sakamoto, S., *Synlett*, 2001, **10**, 1614.
192. Ding, R., Ge, C.S., Chen, Y.J., Wang, D. and Li, C.J., *Tetrahedron Lett.*, 2002, **43**, 7789.
193. Clerici, A. and Porta, O., *Tetrahedron Lett.*, 1990, **31**, 2069.
194. Petasis, N.A. and Zavialov, I.A., *J. Am. Chem. Soc.*, 1997, **119**, 445.
195. Wipf, P., Nunes, R.L. and Ribe, S., *Helv. Chim. Acta*, 2002, **85**, 3478.
196. Ueda, M. and Miyaura, N., *J. Organomet. Chem.*, 2000, **595**, 31.
197. Ding, R., Zhao, C.H., Chen, Y.J., Liu, L., Wang, D. and Li, C.J., *Tetrahedron Lett.*, 2004, **45**, 2995.
198. Wei, C.M. and Li, C.-J., *Green Chem.*, 2002, **4**, 39.
199. Yao, X. and Li, C.-J., *Org. Lett.*, 2005, **7**, 4395.
200. Viswanathan, G.S. and Li, C.-J., *Tetrahedron Lett.*, 2002, **43**, 1613.
201. Chen, L. and Li, C.-J., *Org. Lett.*, 2004, **6**, 3151.
202. Li, C.-J. and Wei, C., *Chem. Commun.*, 2002, 268.
203. Zhang, J., Wei, C. and Li, C.-J., *Tetrahedron Lett.*, 2002, **43**, 5731.
204. Wei, C. and Li, C.J., *Lett. Org. Chem.*, 2005, **2**, 410.
205. Wei, C. and Li, C.-J., *J. Am. Chem. Soc.*, 2003, **125**, 9584.
206. Wei, C., Li, Z.G. and Li, C.-J., *Org. Lett.*, 2003, **5**, 4473.
207. Wei, C. and Li, C.-J., *J. Am. Chem. Soc.*, 2002, **124**, 5638.
208. Wei, C., Mague, J.T. and Li, C.-J., *Proc. Natl. Acad. Sci. U.S.A.*, 2004, **101**, 5749.
209. Shi, L., Tu Y.-Q., Wang, M., Zhang, F.-M. and Fan, C.-A., *Org. Lett.*, 2004, **6**, 1001; Ju, Y., Li, C.-J. and Varma, R.S., *QSAR Comb. Sci.*, 2004, **23**, 891.
210. For a recent example, see: Ayed, T.B. and Amri, H. *Synth. Commun.*, 1995, **25**, 3813.
211. For an excellent review on classical Aldol reactions, see: Nielsen, A.T. and Houlihan, W.J., *Org. React.*, 1968, **16**, 1.
212. Bounora, P.T., Rosauer, K.G. and Dai, L., *Tetrahedron Lett.*, 1995, **26**, 4009.
213. Darbre, T. and Machuqueiro, M., *Chem. Commun.*, 2003, 1090.
214. Dewa, T., Saiki, T. and Aoyama, Y., *J. Am. Chem. Soc.*, 2001, **123**, 502.
215. For recent reviews, see: Machajewski, T.D. and Wong, C.-H., *Angew. Chem. Int. Ed.*, 2000, **39**, 1352; Fessner, W.-D. and Helaine, V., *Curr. Opin. Biotechnol.*, 2001, **12**, 574; Breuer, M. and Hauer, B., *Curr. Opin. Biotechnol.*, 2003, **14**, 570.
216. Dickerson, T.J. and Janda, K.D., *J. Am. Chem. Soc.*, 2002, **124**, 3220.

217. Cravotto, G., Demetri, A., Nano, G.M., Palmisano, G., Penoni, A. and Tagliapietra, S., *Eur. J. Org. Chem.*, 2003, 4438.
218. Lubineau, A., *J. Org. Chem.*, 1986, **51**, 2143; Lubineau, A. and Meyer, E., *Tetrahedron*, 1988, **44**, 6065.
219. Kobayashi, S. and Hachiya, I., *Tetrahedron Lett.*, 1992, **33**, 1625; Hachiya, I. and Kobayashi, S., *J. Org. Chem.*, 1993, **58**, 6958; Kobayashi, S., Sugiura, M., Kitagawa, H. and Lam, W.L., *Chem. Rev.*, 2002, **102**, 2227, and references cited therein.
220. Kobayashi, S. and Hachiya, I., *J. Org. Chem.*, 1994, **59**, 3590; for a review on lanthanides-catalyzed organic reactions in aqueous medium, see: Kobayashi, S., *Synlett*, 1994, 689.
221. For a review, see: Correa, I.R., Jr. and Pilli, R.A., *Quim. Nova*, 2003, **26**, 531; Kobayashi, S., Hamada, T., Nagayama, S. and Manabe, K., *J. Braz. Chem. Soc.*, 2001, **12**, 627.
222. Kobayashi, S., Nagayama, S. and Busujima, T., *Chem. Lett.*, 1999, 71; Kobayashi, S., Mori, Y., Nagayama, S. and Manabe, K., *Green Chem.*, 1999, **1**, 175; Kobayashi, S., Nagayama, S. and Busujima, T., *Tetrahedron*, 1999, **55**, 8739.
223. Nagayama, S. and Kobayashi, S., *J. Am. Chem. Soc.*, 2000, **122**, 11531.
224. Kobayashi, S., Hamada, T., Nagayama, S. and Manabe, K., *Org. Lett.*, 2001, **3**, 165.
225. Li, H.J., Tian, H.Y., Chen, Y.J., Wang, D. and Li, C.J., *Chem. Commun.*, 2002, 2994; Li, H.J., Tian, H.Y., Chen, Y.J., Wang, D. and Li, C.J., *J. Chem. Res. (S)*, 2003, 153; Li, H.-J., Tian, H.-Y., Wu, Y.-C., Chen, Y.-J., Liu, L., Wang, D. and Li, C.-J., *Adv. Synth. Catal.*, 2005, **347**, 1247. For other related studies by the authors, see: Tian, H.Y., Chen, Y.J., Wang, D., Zeng, C.C. and Li, C.J., *Tetrahedron Lett.*, 2000, **41**, 2001, **42**, 1803; Tian, H.Y., Chen, Y.J., Wang, D. and Li, C.J., *Ind. Eng. Chem. Res.*, 2002, **41**, 4523.
226. For other examples of asymmetric aqueous Mukaiyama aldol reaction, see: Yamashita, Y., Ishitani, H., Shimizu, H. and Kobayashi, S., *J. Am. Chem. Soc.*, 2002, **124**, 3292; Ozasa, N., Wadamoto, M., Ishihara, K. and Yamamoto, H., *Synlett*, 2003, 2219; Yang, B.-Y., Chen, X.-M., Deng, G.-J., Zhang, Y.-L. and Fan, Q.-H., *Tetrahedron Lett.*, 2003, **44**, 3535; Ohkouchi, M., Yamaguchi, M. and Yamagishi, T., *Enantioselectivity*, 2000, **5**, 71; Kobayashi, S., Ueno, M., Suzuki, R. and Ishitani, H., *Tetrahedron Lett.*, 1999, **40**, 2175.
227. Chan, T.H., Li, C.J. and Wei, Z.Y., *J. Chem. Soc., Chem. Commun.*, 1990, 505; Zhou, J.Y., Jia, Y., Sha, Q.Y. and Wu, S.H., *Synth. Commun.*, 1996, **26**, 769.
228. Chan, T.H., Li, C.J., Lee, M.C. and Wei, Z.Y., *Can. J. Chem.*, 1994, **72**, 1181.
229. Chung, W.J., Higashiya, S. and Welch, J.T., *J. Fluorine Chem.*, 2001, **112**, 343; Lee, P.H., Bang, K., Lee, K., Sung, S.Y. and Chang, S.B., *Synth. Commun.*, 2001, **31**, 3781.
230. Shen, Z., Zhang, J., Zou, H. and Yang, M., *Tetrahedron Lett.*, 1997, **38**, 2733.
231. Xu, X.L., Lu, P. and Zhang, Y.M., *Chin. Chem. Lett.*, 1999, **10**, 729.
232. Chattopadhyay, A. and Salaskar, A., *Synthesis*, 2000, 561.
233. Lee, Y.J. and Chan, T.H., *Can. J. Chem.*, 2003, **81**, 1406.
234. Bieber, L.W., Malvestiti, I. and Storch, E.C., *J. Org. Chem.*, 1997, **62**, 9061.
235. Yi, X.H., Meng, Y. and Li, C.J., *Tetrahedron Lett.*, 1997, **38**, 4731.
236. Areias, M.C.C., Bieber, L.W., Navarro, M. and Diniz, F.B., *J. Electroanal. Chem.*, 2003, **558**, 125.
237. Chung, W.J., Higashiya, S. and Welch, J.T., *J. Fluorine Chem.*, 2001, **112**, 343.
238. Li, C.J., Wang, D. and Chen, D.L., *J. Am. Chem. Soc.*, 1995, **117**, 12867.
239. Wang, M.W. and Li, C.J., *Tetrahedron Lett.*, 2002, **43**, 3589.
240. Wang, M., Yang, X.-F. and Li, C.-J., *Eur. J. Org. Chem.*, 2003, 998.
241. Inoue, K., Ishida, T., Shibata, I. and Baba, A., *Adv. Synth. Catal.*, 2002, **344**, 283.
242. Mori, Y., Kobayashi, J., Manabe, K. and Kobayashi, S., *Tetrahedron*, 2002, **58**, 8263.
243. Engstrom, G., Morelli, M., Palomo, C. and Mitzel, T., *Tetrahedron Lett.*, 1999, **40**, 5967.
244. Keh, C.C.K., Wei, C. and Li, C.J., *J. Am. Chem. Soc.*, 2003, **125**, 4062.
245. Katritzky, A.R., Shobana, N. and Harris, P.A., *Tetrahedron Lett.*, 1991, **32**, 4247; Katritzky, A.R., Shobana, N. and Harris, P.A., *Organometallics*, 1992, **11**, 1381.
246. Clerici, A. and Porta, O., *Gazz. Chim. Ital.*, 1992, **122**, 165.
247. Miyabe, H., Ueda, M. and Naito, T., *J. Org. Chem.*, 2000, **65**, 5043.
248. Miyabe, H., Nishimura, A., Fujishima, Y. and Naito, T., *Tetrahedron*, 2003, **59**, 1901.
249. Miyabe, H., Ueda, M. and Naito, T., *Chem. Commun.*, 2000, 2059.
250. Miyabe, H., Ueda, M., Nishimura, A. and Naito, T., *Org. Lett.*, 2002, **4**, 131.
251. Miyabe, H., Ueda, M., Fujii, K., Nishimura, A. and Naito, T., *J. Org. Chem.*, 2003, **68**, 5618.
252. Ueda, M., Miyabe, H., Nishimura, A., Sugino, H. and Naito, T., *Tetrahedron: Asymmetry*, 2003, **14**, 2857.
253. Makosza, M. and Krylowa, I., *Tetrahedron*, 1999, **55**, 6395.
254. Huang, T., Keh, C.C.K. and Li, C.-J., *Chem. Commun.*, 2002, 2440.
255. Hajos, Z.G. and Parrish, D.R., *J. Org. Chem.*, 1974, **39**, 1612.
256. Eder, U., Sauer, G. and Wiechert, R., *Angew. Chem., Int. Ed. Engl.*, 1971, **10**, 496.
257. Harada, N., Sugioka, T., Uda, H. and Kuriki, T., *Synthesis*, 1990, 53.
258. (a) Petrier, C., Dupuy, C. and Luche, J.L., *Tetrahedron Lett.*, 1986, **27**, 3149; (b) Luche, J.L. and Allavena, C., *Tetrahedron Lett.*, 1988, **29**, 5369; (c) Dupuy, C., Petrier, C., Sarandeses, L.A. and Luche, J.L., *Synth. Commun.*, 1991, **21**, 643.

259. Giese, B., Damm, W., Roth, M. and Zehnder, M., *Synlett*, 1992, 441; Erdmann, P., Schafer, J., Springer, R., Zeitz, H.G. and Giese, B., *Helv. Chim. Acta*, 1992, **75**, 638.
260. Luche, J.L., Allavena, C., Petrier, C. and Dupuy, C., *Tetrahedron Lett.*, 1988, **29**, 5373.
261. Pietrusiewicz, K.M. and Zablocka, M., *Tetrahedron Lett.*, 1988, **29**, 937.
262. Roth, M., Damm, W. and Giese, B., *Tetrahedron Lett.*, 1996, **37**, 351.
263. Suarez, R.M., Sestelo, J.P. and Sarandeses, L.A., *Synlett*, 2002, 1435; Suarez, R.M., Sestelo, J.P. and Sarandeses, L.A., *Chem. Eur. J.*, 2003, **9**, 4179.
264. Huang, T., Keh, C.C.K. and Li, C.-J., *Chem. Commun.*, 2002, 2440.
265. (a) Miyabe, H., Ueda, M., Nishimura, A. and Naito, T., *Org. Lett.*, 2002, **4**, 131; (b) Itooka, R., Iguchi, Y. and Miyaura, N., *Chem. Lett.*, 2001, 722.
266. Jang, D.O. and Cho, D.H., *Synlett*, 2002, **4**, 631.
267. Wang, L., Sun, X. and Zhang, Y., *Synth. Commun.*, 1998, **28**, 3263.
268. Sakai, M., Hayashi, H. and Miyaura, N., *Organometallics*, 1997, **16**, 4229.
269. For an excellent account review on this subject, see: Hayashi, T., *Synlett*, 2001, 879; for additional information, see (a) Hayashi, T., Ueyama, K., Tokunaga, N. and Yoshida, K., *J. Am. Chem. Soc.*, 2003, **125**, 11508; (b) Shintani, R., Tokunage, N., Doi, H. and Hayashi, T., *J. Am. Chem. Soc.*, 2004, **126**, 6240; PS-supported BINAP as ligand: (c) Otomaru, Y., Senda, T. and Hayashi, T., *Org. Lett.*, 2004, **6**, 3357; (d) Uozumi, Y. and Kobayashi, Y., *Heterocycles*, 2003, **59**, 71; (e) Uozumi, Y., Tanaka, H. and Shibatomi, K., *Org. Lett.*, 2004, **6**, 281.
270. Shi, Q., Xu, L., Li, X., Jia, X., Wang, R., Au-Yeung, T.T.-L., Chan, A.S.C., Hayashi, T., Cao, R. and Hong, M., *Tetrahedron Lett.*, 2003, **44**, 6505.
271. Lautens, M. and Mancuso, J., *Org. Lett.*, 2002, **4**, 2105.
272. Itooka, R., Iguchi, Y. and Miyaura, N., *J. Org. Chem.*, 2003, **68**, 6000.
273. Huang, T.S., Venkatraman, S., Meng, Y., Kort, D., Wang, D., Ding, R., Nguyen, T. and Li, C.J., *Pure Appl. Chem.*, 2001, **73**, 1315.
274. (a) Huang, T.-S., Meng, Y., Venkatraman, S., Wang, D. and Li, C.-J., *J. Am. Chem. Soc.*, 2001, **123**, 7451; (b) Venkatraman, S., Meng, Y. and Li, C.-J., *Tetrahedron Lett.*, 2001, **42**, 4459; (c) Oi, S., Moro, M., Ito, H., Honma, Y., Miyano, S. and Inoue, Y., *Tetrahedron*, 2002, **58**, 91.
275. Venkatraman, S. and Li, C.J., *Tetrahedron Lett.*, 2001, **42**, 781.
276. Ding, R., Chen, Y.-J., Wang, D. and Li, C.-J., *Synlett*, 2001, 1470.
277. Huang, T.-S. and Li, C.-J., *Chem. Commun.*, 2001, 2348.
278. Huang, T. and Li, C.J., *Org. Lett.*, 2001, **3**, 2037.
279. Wu, W. and Li, C.-J., *Chem. Commun.*, 2003, 1668.
280. Wu, W. and Li, C.-J., *Lett. Org. Chem.*, 2004, **1**, 122.
281. Knoepfel, T.F. and Carreira, E.M., *J. Am. Chem. Soc.*, 2003, **125**, 6054.
282. Chen, L. and Li, C.J., *Tetrahedron Lett.*, 2004, **45**, 2771.
283. For representative reviews, see: Kahn, B.E. and Rieke, R.D., *Chem. Rev.*, 1988, **88**, 733; Pons, J. M. and Santelli, M., *Tetrahedron*, 1988, **44**, 4295.
284. Clerici, A. and Porta, O., *J. Org. Chem.*, 1989, **54**, 3872.
285. Barden, M.C. and Schwartz, J., *J. Am. Chem. Soc.*, 1996, **118**, 5484.
286. Delair, P. and Luche, J.L., *J. Chem. Soc., Chem. Commun.*, 1989, 398.
287. Zhang, W.C. and Li, C.J., *J. Chem. Soc., Perkin Trans. 1*, 1998, 3131; Zhang, W.-C. and Li, C.-J., *J. Org. Chem.*, 1999, **64**, 3230; Li, J.-T., Bian, Y.-J., Zang, H.-J. and Li, T.-S., *Synth Commun.*, 2002, **32**, 547.
288. Li, C.-J., Meng, Y., Yi, X.-H., Ma, J. and Chan, T.-H., *J. Org. Chem.*, 1997, **62**, 8632; Li, C.-J., Meng, Y., Yi, X.-H., Ma, J. and Chan, T.-H., *J. Org. Chem.*, 1998, **63**, 7498; Li, J.-T., Bian, Y.-J., Liu, S.-M. and Li, T.-S., *Youji Huaxue*, 2003, **23**, 479.
289. Tsukinoki, T., Kawaji, T., Hashimoto, I., Mataka, S. and Tashiro, M., *Chem. Lett.*, 1997, 235; Wang, L., Sun, X. and Zhang, Y., *J. Chem. Res. (S)*, 1998, 336; Hekmatshoar, R., Yavari, I., Beheshtiha, Y.S. and Heravi, M.M., *Monatsh. Chem.*, 2001, **132**, 689; Yang, J.-H., Li, J.-T., Zhao, J.-L., Li, T.-S., *Synth. Commun.*, 2004, **34**, 993.
290. Lim, H.J., Keum, G. and Kim, Y., *Tetrahedron Lett.*, 1998, **39**, 4367; Nair, V., Ros, S., Jayan, C. N. and Rath, N.P., *Tetrahedron Lett.*, 2002, **43**, 8967.
291. Wang, L. and Zhang, Y., *Tetrahedron*, 1998, **54**, 11129; Wang, L. and Zhang, Y.-M., *Chin. J. Chem.*, 1999, **17**, 550; Wang, L. and Zhang, Y., *Tetrahedron*, 1998, **54**, 11129.
292. Bhar, S. and Panja, C., *Green Chem.*, 1999, **1**, 253; Bian, Y.-J., Liu, S.-M., Li, J.-T. and Li, T.-S., *Synth. Commun.*, 2002, **32**, 1169.
293. Li, L.-H. and Chan, T.H., *Org. Lett.*, 2000, **2**, 1129.
294. Wang, Z.-Y., Yuan, S.-Z., Zha, Z.-G. and Zhang, Z.-D., *Chin. J. Chem.*, 2003, **21**, 1231.
295. Zheng, Y., Bao, W. and Zhang, Y., *Synth. Commun.*, 2000, **30**, 3517.
296. Lim, H.J., Keum, G. and Kim, Y., *Tetrahedron Lett.*, 1998, **39**, 4367.
297. Meciarova, M., Toma, S. and Babiak, P., *Chem. Pap.*, 2001, **55**, 302.
298. Miyoshi, N., Takeuchi, S. and Ohgo, Y., *Chem. Lett.*, 1993, 2129.
299. Delair, P. and Luche, J.L., *J. Chem. Soc., Chem. Commun.*, 1989, 398.

300. Kalyanam, N. and Venkateswara Rao, G., *Tetrahedron Lett.*, 1993, **34**, 1647.
301. Dutta, M.P., Baruah, B., Boruah, A., Prajapati, D. and Sandhu, J.S., *Synlett*, 1998, 857.
302. Tsukinoki, T., Nagashima, S., Mitoma, Y. and Tashiro, M., *Green Chem.*, 2000, **2**, 117.
303. Liu, X., Liu, Y. and Zhang, Y., *Tetrahedron Lett.*, 2002, **43**, 6787.
304. Kim, M., Knettle, B.W., Dahlen, A., Hilmersson, G. and Flowers, R.A., II, *Tetrahedron*, 2003, **59**, 10397.
305. Jeevanandam, A. and Ling, Y.-C., *Tetrahedron Lett.*, 2001, **42**, 4361.
306. Wang, L. and Zhang, Y., *Tetrahedron Lett.*, 1998, **39**, 5257.
307. Wang, L. and Zhang, Y., *Synth. Commun.*, 1998, **28**, 3269.
308. Ma, J. and Chan, T.-H., *Tetrahedron Lett.*, 1998, **39**, 2499.
309. Nosek, J., *Collect. Czech. Chem. Commun.*, 1964, **29**, 597.
310. De Sa, A.C.P.F., Pontes, G.M.A., Dos Anjos, J.A.L., Santana, S.R., Bieber, L.W. and Malvestiti, I., *J. Braz. Chem. Soc.*, 2003, **14**, 429.
311. Usugi, S., Yorimitsu, H. and Oshima, K., *Tetrahedron Lett.*, 2001, **42**, 4535.
312. Marton, D. and Tari, M., *J. Organomet. Chem.*, 2000, **612**, 78; Furlani, D., Marton, D., Tagliavini, G. and Zordan, M., *J. Organomet. Chem.*, 1988, **341**, 345.
313. Garves, K., *J. Org. Chem.*, 1970, **35**, 3273.
314. Davydov, D.V. and Beletskaya, I.P., *Russ. Chem. Bull.*, 1995, **44**, 1139.
315. Venkatraman, S. and Li, C.-J., *Org. Lett.*, 1999, **1**, 1133.
316. Venkatraman, S. and Li, C.-J. *Tetrahedron Lett.*, 2000, **41**, 4831; Venkatraman, S., Huang, T. and Li, C.-J., *Adv. Synth. Catal.*, 2002, **344**, 399; Mukhopadhyay, S., Yaghmur, A., Baidossi, M., Kundu, B. and Sasson, Y., *Org. Proc. Res. Dev.*, 2003, **7**, 641.
317. Mukhopadhyay, S., Rothenberg, G., Gitis, D. and Sasson, Y., *Org. Lett.*, 2000, **2**, 211.
318. Mukhopadhyay, S., Rothenberg, G. and Sasson, Y., *Adv. Synth. Catal.*, 2001, **343**, 274.
319. Mukhopadhyay, S., Joshi, A.V., Peleg, L. and Sasson, Y., *Org. Proc. Res. Dev.*, 2003, **7**, 44.
320. Li, J.-H., Xie, Y.-X. and Yin, D.-L., *J. Org. Chem.*, 2003, **68**, 9867.
321. Pellon, R.F., Carrasco, R. and Rodes, L., *Synth. Commun.*, 1993, **23**, 1447; Pellon, R.F., Mamposo, T., Carrasco, R. and Rodes, L., *Synth. Commun.*, 1996, **26**, 3877.
322. Pellon, R.F., Carrasco, R., Millian, V. and Rodes, L., *Synth. Commun.*, 1995, **25**, 1077.
323. Saphier, M., Masarwa, A., Cohen, H. and Meyerstein, D., *Eur. J. Inorg. Chem.*, 2002, 1226.
324. Papadogianakis, G. and Sheldon, R.A., *New J. Chem.*, 1996, **20**, 175; Cornils, B., *J. Mol. Catal. A*, 1999, **143**, 1; Genet, J.P. and Savignac, M., *J. Organomet. Chem.*, 1999, **576**, 305; Oehme, G., Grassert, I., Paetzold, E., Meisel, R., Drexler, K. and Fuhrmann, H., *Coord. Chem. Rev.*, 1999, **185–186**, 585.
325. Huang, T.S. and Li, C.J., *Tetrahedron Lett.*, 2002, **43**, 403.
326. Venkatraman, S. and Li, C.J., *Org. Lett.*, 1999, **1**, 1133; Venkatraman, S., Huang, T.S. and Li, C.J., *Adv. Synth Catal.*, 2002, **344**, 399.
327. (a) Li, Z. and Li, C.-J., *Eur. J. Org. Chem.*, 2005, 3173; (b) Li, Z. and Li, C.-J., *J. Am. Chem. Soc.*, 2005, **127**, 6968; (c) Li, Z. and Li, C.-J., *J. Am. Chem. Soc.*, 2005, **127**, 3672; (d) Li, Z. and Li, C.-J., *J. Am. Chem. Soc.*, 2004, **126**, 11810; (e) Li, Z. and Li, C.J., *J. Am. Chem. Soc.*, 2006, **128**, 56; (f) Zhang, Y. and Li, C.J., *Angew. Chem., Int. Ed. Engl.*, 2006, **45**, 1949.
328. Li, Z. and Li, C.-J., *Org. Lett.*, 2004, **6**, 4997.
329. Zhang, H.B., Liu, L., Chen, Y. J., Wang, D. and Li, C.J., *Adv. Catal. Synth.*, 2006, **348**, 229.
330. Zhang, H.-B., Liu, L., Chen, Y. J., Wang, D. and Li, C.J., *Eur. J. Org. Chem.*, 2006, 869.

Chapter 5
Pericyclic Reactions in Aqueous Media

Francesco Fringuelli, Oriana Piermatti, Ferdinando Pizzo, and Luigi Vaccaro

Pericyclic reactions make up a large class of organic reactions that are fundamental tools in organic synthesis because they are versatile, occur with high regio- and stereoselectivity, and generally with high yields.[1] Some of them are related to famous scientists such as the cycloadditions of Diels–Alder[2] and Paternò–Büchi[3] and the sigmatropic transpositions of Cope[4] and Claisen.[5] These reactions proceed with a concerted reaction mechanism in which six electrons (σ, π, ω), but sometimes some other number, move in a cyclic transition state. There are no intermediates and the chemical bonds are made or broken in a synchronous or asynchronous way. This allows a very high regio- and stereochemical control of the reaction.[6]

Pericyclic reactions are characterized by a large negative activation volume,[7] and the reactivity and selectivity are influenced by high pressure.

Woodward and Hoffman have introduced a general method to characterize a pericyclic process based on the number and type of electrons of breaking bonds and they have interpreted the formation and stereochemistry of the product by the principle of conservation of orbital symmetry.[8]

The polarity difference between the reagents and the transition state of the pericyclic reaction is generally minimal and therefore, in principle, the reaction rate is not greatly influenced by the solvent polarity[9] except when an aqueous medium is used.[10]

The Diels–Alder cycloadditions were the first organic reactions for which it was shown that the aqueous medium, when compared to organic solvents, has a beneficial effect on both the reactivity and selectivity of the reaction.[11]

Among the pericyclic reactions investigated in aqueous medium, the Diels–Alder cycloadditions and the 1,3-dipolar cycloadditions have been the most widely studied. Beneficial effects of aqueous medium were also observed for [2 + 2] and [4 + 4] cycloaddition reactions and sigmatropic rearrangements.

5.1 Diels–Alder cycloaddition reactions

Before 1980 only a few examples of Diels–Alder reactions in aqueous medium had been reported because little consideration was given to water as a reaction medium for organic synthesis.[12]

In 1931 Diels and Alder used water as reaction medium for the cycloaddition of furan and maleic anhydride.[12a] This work was revised in 1948 by Woodward and Baer[12b] and some years later by De Koning.[12c]

In the 1940s two patents appeared concerning practical applications of beneficial effects of the aqueous medium for Diels–Alder cycloadditions of simple dienes with activated dienophiles.[13]

In 1975 Carlson performed the cycloadditions of aromatic diazonium salts with methyl-substituted 1,3-butadienes in aqueous medium.[14]

In 1980 Breslow[15] was the first to report the kinetic data of the effect of water as reaction medium on the reactivity of Diels–Alder reactions of cyclopentadiene with methyl vinyl ketone and acrylonitrile and the cycloaddition of anthracene-9-carbinol with *N*-ethylmaleimide. The results were surprising. The cycloaddition in water of cyclopentadiene with methyl vinyl ketone was 740 times faster than in isooctane and the endo/exo ratio was approximately 10 times higher[16]. In addition the reaction was 2.5 times faster in the presence of prohydrophobic agent lithium chloride (salting-out agent) and was slowed down by adding an antihydrophobic agent such as guanidinium chloride (salting-in agent).

Breslow's investigation stimulated research in this field and today many other examples are known. Important contributions were reported in the 1980s and 1990s by J.B.F.N. Engberts, P.A. Grieco, S. Kobayashi, A. Lubineau, and other distinguished scientists and appeared in books and reviews.[7,10,11,17]

In the last decade of the last century there was great interest toward Lewis acid-catalyzed Diels–Alder reaction in aqueous medium. Many catalysts have been discovered (Cu(II), Ni(II), Zn(II), $Ln(OTf)_3$, $Sc(OTf)_3$, $Bi(OTf)_3$, $InCl_3$, $MeReO_3$, $Cu(DS)_2$, $Zn(DS)_2$) that have also contributed to the development of environmental-friendly organic synthesis; this topic was recently reviewed.[7,11c]

More recently the research on the Diels–Alder reaction in the aqueous medium has been focused mainly on the diastereoselectivity of the cycloaddition, particularly the enantioselectivity, and on the utilization of biomolecules as highly stereoselective catalysts.

5.1.1 Carbo Diels–Alder reactions

Bioconjugation (the covalent attachment of small molecules to biomolecules) of carbohydrates to proteins provides the foundation for generating anti-oligosaccharide antibodies that are reagents in glycobiology and are valuable glycol pharmaceuticals used as vaccines.[18] With the used protocol the yields are generally low[19] and the oligosaccharides cannot usually be recovered in their reactive form after the coupling procedure. A new approach[20a] to bioconjugation utilizes the Diels–Alder cycloaddition in pure water between a dienophile-equipped protein **1** and a diene-derivative saccharide (in Scheme 5.1 the ramnopyranosyl derivative **2** is reported) to give the neoglycoprotein **3** at room temperature with a reaction half-life of approximately 2 h. The uncoupled saccharide is recovered with complete conservation of the diene moiety and can be reused.

The Diels–Alder cycloaddition in aqueous medium was also successfully exploited in the synthesis of biotinylated oligonucleotides by bioconjugation of diene-modificated oligonucleotides and biotinmaleimide.[20b]

α,β-Unsaturated ketones activated by chiral amines have been recently used by McMillan[21a] to perform asymmetric Diels–Alder reactions in aqueous medium. The best level of enantiofacial discrimination, maintaining in the same time good reaction efficiency, was found by using the perchlorate salt of (2*S*,5*S*)-5-benzyl-3-methyl-2-(5-methyl-furan-2-yl)-imidazolidin-4-one (**4**). This organocatalytic strategy is quite general with respect to the diene and dienophile (cyclic and acyclic), allowing to perform highly enantio-, diastereo-, and regioselective cycloadditions. Some examples are illustrated in Table 5.1.

Optically active compounds (ee 11–77%) were recently prepared (yields 13–81%) by a one-pot procedure in a water suspension using inclusion complexes of *N*-alkylmaleimides

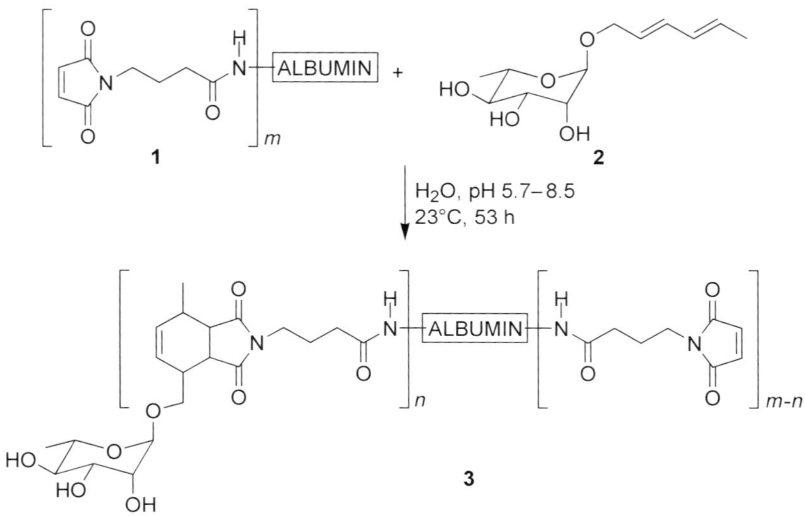

Scheme 5.1

or isoprene with an optically active host, and isoprene and *N*-alkylmaleimides as diene and dienophile, respectively.[21b]

5-Norbornene-2-carboxylic acid (**8**) was obtained in acceptable ee by the use of a chiral auxiliary. Optically active tetrahydropyrimidones **6** underwent Diels–Alder cycloadditions with cyclopentadiene (**5**) in water at room temperature. Subsequent removal of the auxiliary was achieved by boiling the carboxamide **7** in water (Scheme 5.2).[21c] The cycloaddition conversions were at least 90% and the endo adduct was the prevalent diastereoisomer. Using 70% aqueous ethanol as reaction medium, both the endo/exo ratio and ee were lower than in pure water. Since tetrahydropyrimidones **6** can be prepared in water from L-asparagine and a suitable aldehyde and subsequent acylation with acryloyl chloride, the entire synthesis of **8** was performed in water by one-pot procedure. The yield was fair, but the ee of norbornene carboxylic acid **8** was lower than that obtained by using a step-by-step procedure. This result was probably due to some acryloyl chloride being hydrolyzed to acrylic acid, which then reacted with **5** in a non-stereo-biased manner.

An extensive investigation was carried out on the cycloadditions of 3-nitrocoumarin (**9**) with a variety of 1,3-dienes **10–14** (Scheme 5.3) in water, organic solvent, and under solvent-free conditions.[22] In water, the reactions occurred under heterogeneous conditions and were faster (1–3 h; 20–90°C) than when carried out in homogeneous toluene solution (21–30 h; 90°C) and high yields were always observed (80–95%). 3-Nitrocoumarin (**9**) behaved as a 2π component in the cycloadditions in water with methylbutadienes **10–12** and gave the adducts **15–18** (Scheme 5.4). A good exo selectivity was obtained in the case of **15** and **16**. This was explained by considering that the secondary orbital interactions, originating from the exo approach, between the 1,3-diene and nitro groups lead to a greater stabilization than those originating from the 1,3-diene and carbonyl group. It is interesting to observe that **9**, α, β-unsaturated nitroalkenes, and (*E*)-3-diazenylbut-2-enes behave as 4π components in cycloadditions with electron-rich alkenes such as vinyl ethers, and the second orbital interactions between the oxygen of the dienophile and the nitrogen atom of nitro group favor the endo approach (see Section 5.1.3).[23]

Table 5.1 Aminocatalyzed Diels–Alder cycloadditions between cyclopentadiene (**5**) and α,β-unsaturated enones[a]

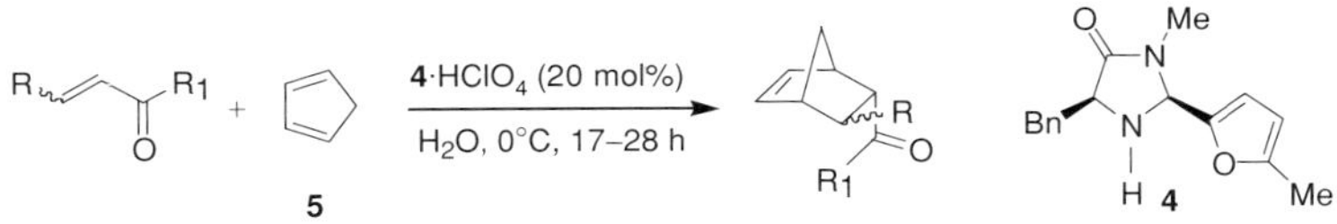

Enone	endo/exo	ee (%)	Yield (%)
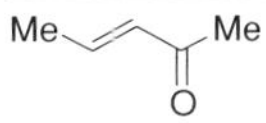	93:7	61	85
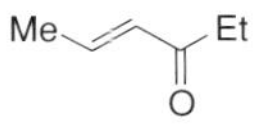	96:4	90	89
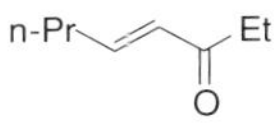	94:6	92	84
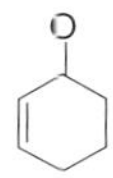	93:7	63	81
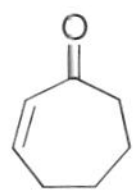	95:5	90	85

[a] Cycloadditions by using open dienes were carried out in EtOH.

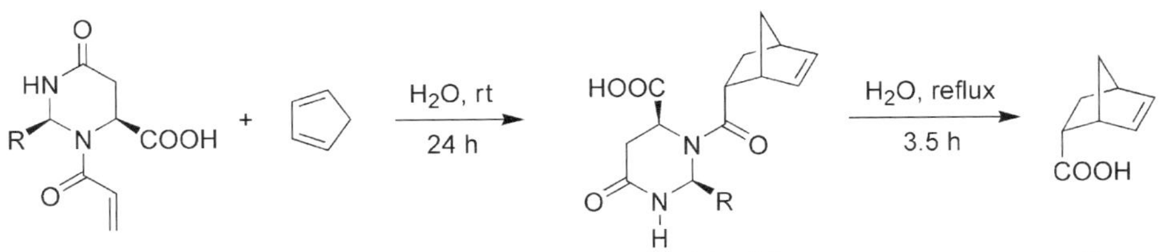

6 **5** **7** (endo) **8**

R	endo/exo[a]	ee (% endo)[b]
i-Pr	93:7	60
t-Bu	82:18	64
Ph	95:5	55

[a] Conversion >90%.
[b] Configuration at C-2 is S.

Scheme 5.2

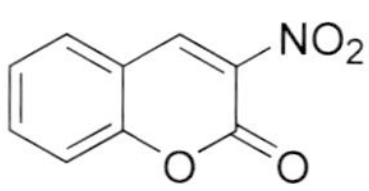

9

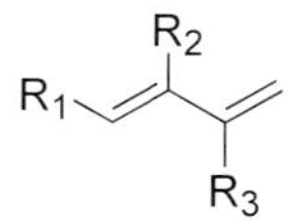

	R_1	R_2	R_3
10	Me	H	H
11	H	H	Me
12	H	Me	Me
13	H	H	OMe
14	H	OMe	OMe

Scheme 5.3

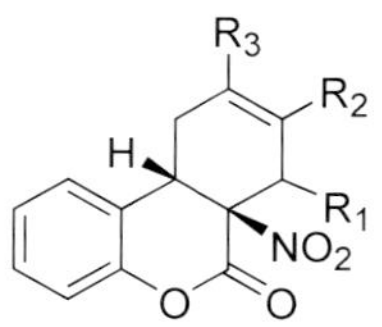

	R_1	R_2	R_3
15	β-Me	H	H
16	α-Me	H	H
17	H	H	Me
18	H	Me	Me
19	H	H	OMe
20	H	OMe	OMe

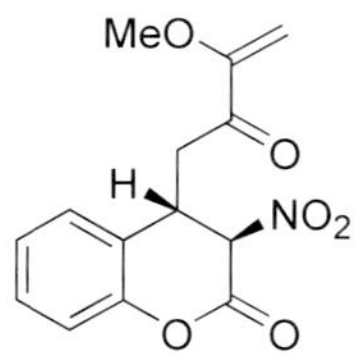

21a

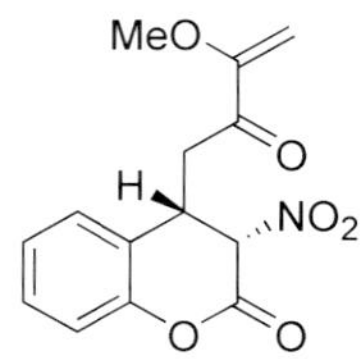

21b

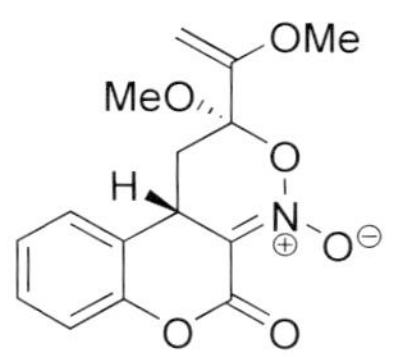

22

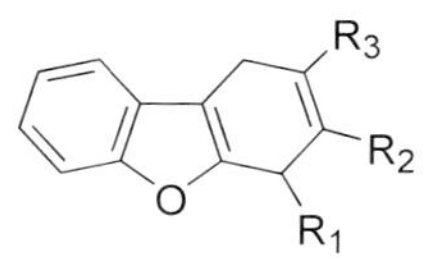

23 R_1 = Me, R_2 = R_3 = H
24 R_1 = R_2 = H, R_3 = Me
25 R_1 = H, R_2 = R_3 = Me

Scheme 5.4

The cycloaddition in water of **9** with 2,3-dimethoxy-1,3-butadiene (**14**) (20°C, pH 8.3, 1 h) gave the nitrochromanones **21a** and **21b** in 85:15 ratio and 85% yield. In toluene, the reaction does not work, while in dichloromethane (DCM) the expected Diels–Alder adduct **20** was isolated (20°C, 168 h) in 80% yield. The cycloaddition of **9** with 2-methoxy-1,3-butadiene (**13**) gave high yield of adduct **19** in both water and DCM. These results have been explained,[22] suggesting that **9** reacts with **14** as 4π component, producing selectively the endo adduct nitronate **22** which hydrolyzes in water to **21a** and **21b**. In water, **14** assumes an s-transoid conformation (or a nonplanar s-cisoid conformation) because of the solvation of the methoxy group which precludes the behavior of **14** as 1,3-diene and favors its reactivity as vinyl ether.[23] The solvation does not greatly influence the s-cisoid–s-transoid conformational equilibrium of either **14** in DCM or **13** in water, and therefore, **9** gives normal electron-demand Diels–Alder reactions with these dienes in DCM and water, respectively.

The Diels–Alder adducts **15–18**, subjected to strong basic conditions followed by *in situ* Nef-cyclodehydration reaction in acidic medium, gave by one-pot procedure 1,4-dihydrodibenzo[*b*,*d*]furans **23–25** in 48–50% overall yields.

The catalytic effect of thiourea derivative **26** (a bidentate hydrogen-bond donor) was investigated[24] in the Diels–Alder reaction of methyl vinyl ketone (**27**) with cyclopentadiene (**5**) in cyclohexane (that has negligible interactions with the solutes), chloroform (a weak hydrogen-bond donor), and in water (a high polar solvent) to give 5-norbornene-2-methylketone (**28**) (Table 5.2).

Table 5.2 Diels–Alder reactions of methyl vinyl ketone (**27**) with cyclopentadiene (**5**) catalyzed by thiourea derivative **26** in water and organic solvents

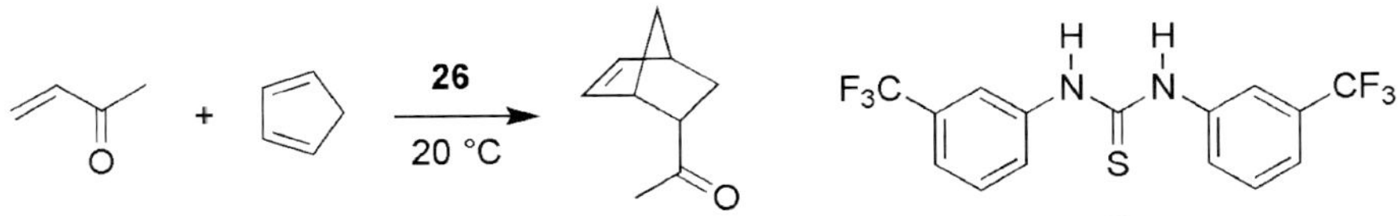

27 **5** **28** **26**

Medium	**26** (mol%)	Yield (%)[a]	c.e. (%)[b]
c-C_6H_{12}	—	18	
	1	42	51
$CHCl_3$	—	31	
	1	52	40
H_2O[c]	—	74	
	1	85	13

[a] After 1 h.
[b] Increase percent of the yield after 1 h in the presence of catalyst.
[c] With 10% of t-butanol.

The reaction rate was faster in water than in the organic solvents, but the catalytic effect (c.e., expressed in terms of percent increase of yield; Table 5.2) was less in the polar reaction medium because water does not favor the complexation of dienophile by the catalyst. An analogous result was obtained[17g,25] in the Diels–Alder reaction of 3-(4-nitrophenyl)-1-(2-pyridyl)-2-propen-1-one (a bidentate dienophile) with cyclopentadiene (**5**) in water and MeCN catalyzed by $Cu(NO_3)_2$. In sole water, the cycloaddition was 287 times faster than in MeCN but the catalytic effect of $Cu(NO_3)_2$ (expressed in terms of relative reaction rates)[11a] in water was 808, while in MeCN it was 158,000.

$Yb(OTf)_3$ (10 mol%) was recently used to catalyze the cycloaddition of alkylated cyclopentadienes, obtained as intermediates in the addition reaction of aldehydes to cyclopentadienyl indium(I) in aqueous medium, with dimethylacetyleneolicerboxylate.[26a]

Fluorous reverse-phase silica gel (FRPSG)-supported Lewis acids are new and effective catalysts of Baeyer–Villiger and Diels–Alder reactions in water. FRPSG-supported $Sc[C(SO_2C_4F_9)_3]_3$ (5 mol%) catalyzes the Diels–Alder cycloaddition of 2,3-dimethylbutadiene with methyl vinyl ketone in water at room temperature (16 h, 91%) and can be recycled by simple filtration after the reaction.[26b]

The easily synthesizable water-soluble organotungsten Lewis acid $[OP(2Py)_3W(CO)(NO)_2](BF_4)_2$ (**29**) was used[26c] to catalyze a series of Diels–Alder reactions in sole water and in water under microwave- (MW) -controlled irradiation. Some results are reported in Table 5.3. The combined effects of Lewis acid catalyst and MW irradiation strongly accelerated the cycloadditions without affecting the diastereoselectivity.

Other recent examples of water–MW synergy are the cycloadditions of dimethylfumarate and methyl vinyl ketone (**27**) with the air-stable and water-soluble aquocobaloxime 1,3-butadien-2-yl-[(H_2O)bis(dimethylglyoximato)cobalt(III)] **30** (Scheme 5.5). The reactions occurred quickly and with good yields.[27]

The Diels–Alder cycloaddition was also investigated in water under supercritical conditions (373.9°C, 220.5 bar) and in conditions of subcritical or super-heated water (200–350°C, at pressure of expansion of water). The density, viscosity, and dielectric constant of water in its supercritical state are very different from those of water under standard conditions.[28] Some examples of Diels–Alder cycloadditions investigated in supercritical water are illustrated in Scheme 5.6.[28a,29]

Table 5.3 Organotungsten [a]-catalyzed Diels–Alder reactions in sole water and in water under MW irradiation

		H_2O (50°C)			H_2O + MW (50°C)[b]	
Reagents	*t* (min)	endo/exo	Yield (%)	*t* (min)	endo/exo	Yield (%)
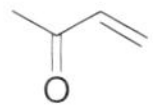	35	91:9	90	0.83	96:4	90
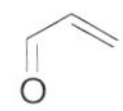	40	94:6	92	0.83	89:11	87
 5 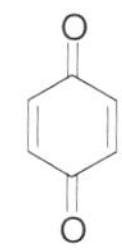	36	100:0	82	0.83	100:0	84
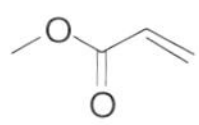	210	79:21	96	1.00	78:22	97

[a] $[OP(2Py)_3W(CO)(NO)_2](BF_4)_2$ (**29**), 3 mol%.
[b] 20% of 300 W maximum power, 50°C preselected maximum temperature.

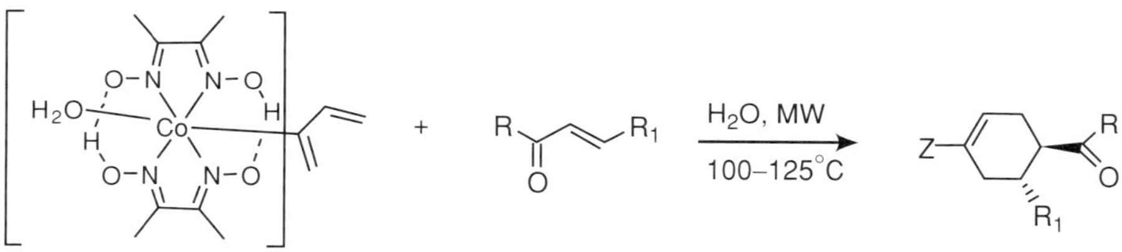

30

Z = $[(H_2O)(DMG)_2Co^{III}]$

R	R_1	*t* (min)	Yield (%)
OMe	CO_2Me	15	76
Me	H	20	89[a]

[a] para/meta = 84:16.

Scheme 5.5

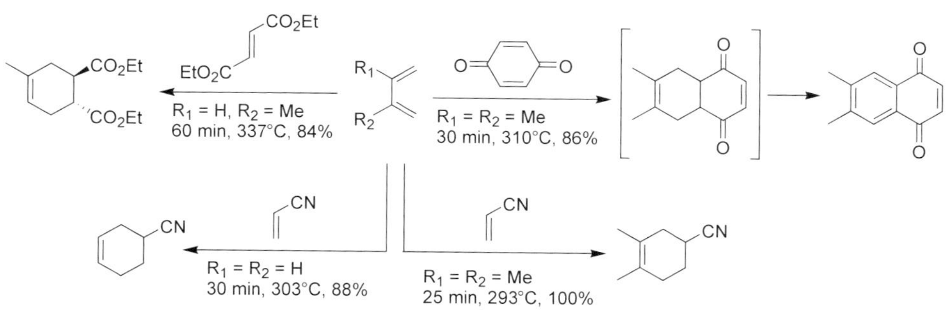

Scheme 5.6

5.1.2 Biocatalyzed carbo Diels–Alder reactions

Biocatalysts (enzymes and proteins) and conditions resembling biological systems (micelles, cyclodextrins [CDs], molecular cages, catalytic antibodies, RNA-based mixtures of metals) have received great attention in the last few years because of the ease with which regio- and stereoselective reactions can be performed under mild conditions.[30]

The influence of micelle and CDs on the reactivity and selectivity of Diels–Alder reaction in aqueous medium was mainly investigated in the 1990s and has been reviewed.[10c,11a] The effect of surfactants (cetyltrimethylammonium bromide [CTAB], sodium dodecyl sulfate [SDS]) at their critical micellar concentration significantly influences the yield and diastereoselectivity of Diels–Alder reactions of acrylates with cyclopentadiene performed in water at room temperature.[31a] A recent contribution on the effects of micelle on the Diels–Alder reaction was reported by Engberts[25c] who observed that in the cycloadditions of N-substituted maleimides with cyclopentadiene, sorbyl alcohol, and sorbyl trimethyl ammonium bromide, a micellar catalysis seems to be active, but if one considers the micellar rate constant, the neat effect on the rate constant of the reaction is remarkably small.

Organic molecules bearing the 2-pyridyldimethylsilyl group (2-$PyMe_2Si$) are miscible in water when HCl is added and dynamic light-scattering experiments reveal the presence of molecular aggregates. Diels–Alder reaction of 2-$PyMe_2Si$-substituted 1,3-dienes with *p*-benzoquinone occurs at room temperature, using 1.0 equiv. of HCl, with simultaneous desilylation and aromatization affording the corresponding naphthoquinones. It has been suggested that the cycloaddition occurs in the interior of aggregates where there are enhanced hydrophobic interactions and not in the aqueous bulky phase.[31b,c]

Recent examples of Diels–Alder reactions in aqueous medium in a self-assembled coordination cage are the cycloadditions of 1,4-naphthoquinone, enclathrated in the cavity of a cage compound, with isoprene and 1,3-cyclohexadiene: the reactions were accelerated 113 and 21 times, respectively,[31d] with respect to those carried out in the absence of cage.

Currently the Diels–Alder reactions catalyzed by enzymes or antibodies in aqueous-buffered medium are a very promising topic because neither is capable of emulating the extraordinary activity and specificity of these catalysts.

One of the first examples of protein-promoted Diels–Alder cycloadditions was the reaction of 1,4-naphthoquinones **31** with 1,3-carbodienes performed in the presence of bovine serum albumin (BSA).[31] The cycloaddition of methoxy juglone **31** (R = Me) with 1-methoxy-1,3-cyclohexadiene (**32**) proceeded with 76% yield giving the regioisomers **33** (R = Me) and **34** (*R* = Me) in a 1:6 ratio with low ee (Scheme 5.7).

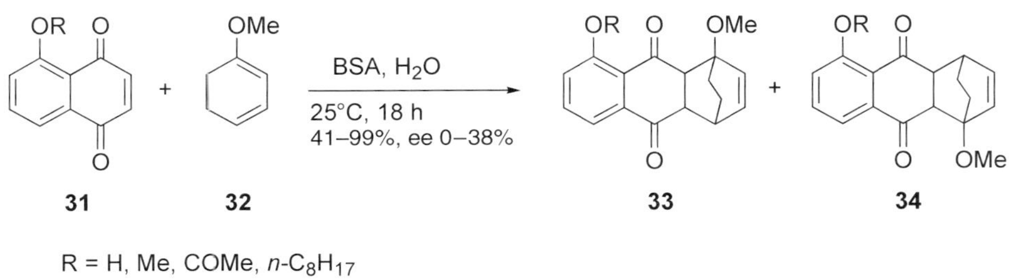

Scheme 5.7

Baker's yeast (By) catalyzed the Diels–Alder reactions of maleic acid and its ethyl ester with cyclopentadiene (**5**) (Scheme 5.8) at pH 7.2 in H_2O/EtOH 70:30 at 37°C.[32] After an incubation period of 48 h the cycloadditions gave exo adducts exclusively while, in the absence of biocatalyst, the endo adducts were prevalent in both water and organic solvent.

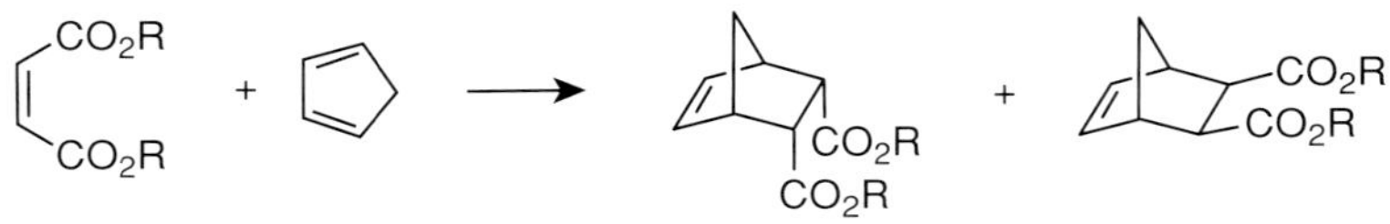

5 (endo) (exo)

R	Organic medium	H_2O	H_2O/EtOH + BY
H	80:20	98:2	0:100
Et	73:27	93:7	0:100

Scheme 5.8

A Diels-Alderase enzyme that catalyzes the Diels–Alder reaction in biosynthetic processes was isolated in 1995 from cell-free extracts of the fungus *Alternaria solani*.[33] The fungus produces toxins known as solanapyrones that are biosynthesized via Diels–Alder reaction exo selectively. In buffered aqueous medium at pH 7.0 the Diels-Alderase catalyzed the cycloaddition of prosolanapyrone (**35**) to (–)-solanapyrone A (**36**) with high exo diastereoselectivity (exo/endo 86:14) and excellent enantioselectivity (ee 99%; Scheme 5.9). In sole water the reaction occurred at 30°C and after 3 h a reversed diastereoselectivity (exo/endo 4:96) was observed.

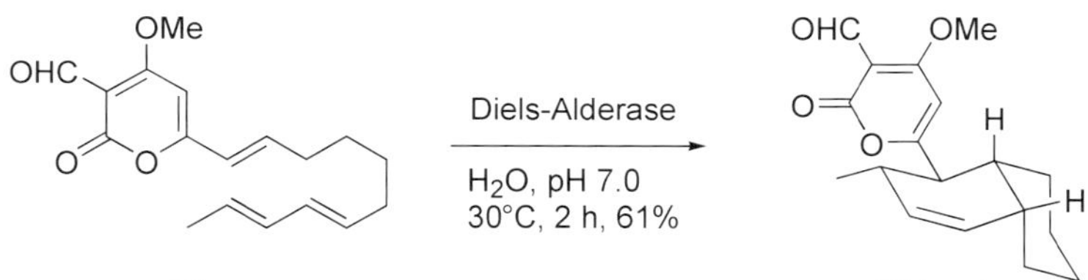

35 **36** (exo)

Scheme 5.9

In aqueous buffer medium at pH 7.7 the protein lavastatin nonaketide synthase catalyzed the intramolecular Diels–Alder reaction of triene **37** to bicyclic compounds **39–41** (**39/40/41** ratio 15:15:1) (Scheme 5.10).[34] The exo-syn adduct **38** was not detected. In the absence of enzyme, a 1:1 mixture of **39** and **40** was detected in aqueous media at 20°C.

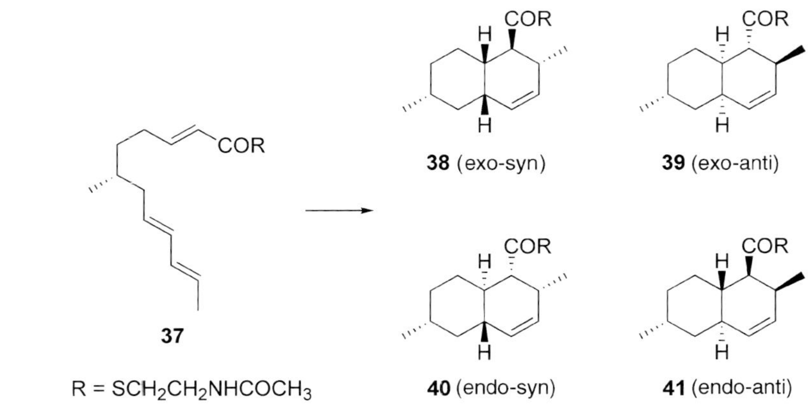

Scheme 5.10

Macrophomate synthase enzyme (MPHS), isolated from the fungus *Macrophoma commelinae*, catalyzes the Diels–Alder cycloaddition between 2-pyrones **42** and decarboxylated oxalacetic acid **43** in aqueous buffered medium at pH 7.0, giving the benzoates **44** (Scheme 5.11).[35] These types of aromatic compounds are commonly biosynthesized by either a shikimate or polyketide pathway and therefore the reaction depicted in Scheme 5.11 supports the fact that the Diels–Alder reaction takes place in biosynthesis.

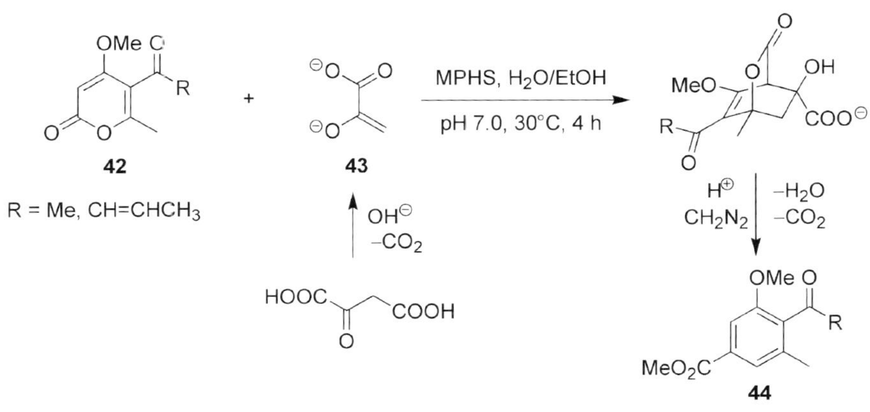

Scheme 5.11

Biosynthetic studies show that when (−)-pregaliellalactone (**45**) and (+)-desoxygaliellalactone (**46**) are fed the mycelium of fungus *Galiella rufa* they are converted to (−)-galiellalactone (**47**), showing that the biosynthesis of (−)-(**47**) presumably involves an enzymatic intramolecular Diels–Alder reaction of (−)-(**45**) to (+)-(**46**) (Scheme 5.12).[36a] Sterner and coworkers found that compound (−)-(**45**) cyclized to (+)-(**46**) stereoselectively at 24°C faster in a sterile saline (0.9% NaCl) suspension of pretreated mycelium of *G. rufa* than under various abiotic reaction conditions (in $CHCl_3$ at rt; in PhMe at 140°C; in sole water at rt; in water/LiCl at rt).[36b] This is the first example of an enzymatic intramolecular Diels–Alder reaction with inverse electron demand. No attempts were yet made to isolate the enzyme responsible for cyclization.

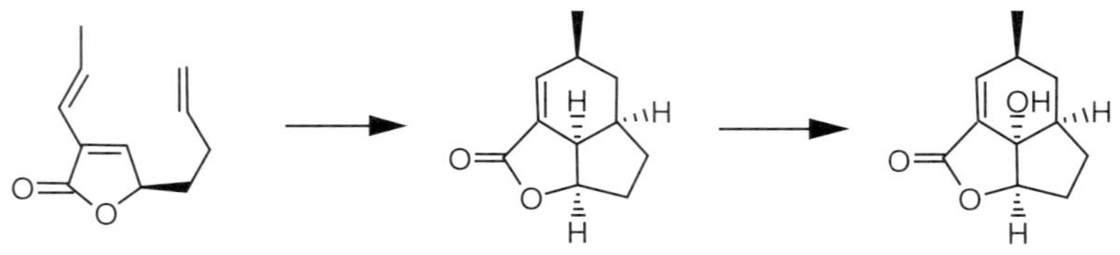

45 **46** **47**

Scheme 5.12

Jencks' postulate says that antibodies generated against an organic molecule resembling the transition state of a given reaction should catalyze this process.[37]

Monoclonal antibody catalysts are generally prepared by displaying a small organic molecule, called hapten and resembling the transition state of the reaction under study (usually a reaction product or analogs), on a carrier protein to be recognized by the immune system that elicits the antibodies.[38]

The Diels–Alder reaction between butadienyl carbamate **48** and *N,N*-dimethylacrylamide (**49**) is catalyzed by antibodies 22C8, 7D4, 13G5, and 4D5 elicited by haptens **52–55**, respectively (Scheme 5.13).[39] In the absence of antibodies, the endo adduct **50** is favored, while with antibodies 22C8 and 7D4 that mimic the exo and endo approach of reagents, only exo-**51** and endo-**50** adducts were obtained, respectively, with high enantioselectivity. The antibodies 13G5 and 4D5 behave analogously to 22C8 and 7D4, respectively, showing that, for reactions that proceed through a highly ordered transition state, it is not necessary that the haptens be conformationally restricted.[39c]

The transition state of Diels–Alder reaction of diene **56** with dienophile **57** to give the adduct **58** (Scheme 5.14) is a boat-like type. The antibody 39-A11, generated to the bicyclo[2.2.2]octane hapten **59** that mimics the transition state of the reaction, efficaciously catalyzed the cycloaddition, selectively giving **58** in buffered aqueous medium (Scheme 5.14).[40]

Ribozymes are RNA molecules with catalytic properties. Artificial ribozymes, isolated from synthetic combinatorial libraries, accelerate a broad range of reactions including the Diels–Alder cycloaddition.[41] These ribozymes generally require that at least one of the reactants be RNA or be covalently tethered to RNA.

Diels-Alderase ribozymes (DAR), isolated from a combinatorial RNA library, cause a (2×10^4)-fold acceleration of the Diels–Alder cycloaddition of anthracene covalently tethered to ribozyme and a biotinylated maleimide in aqueous-buffered medium (Scheme 5.15).[41c] Jäschke recently reported[41a] the action of Diels-Alderase ribozymes as true catalysts, in the sense that they catalyze the cycloaddition of anthracene that is not covalently tethered to RNA and biotin maleimide in aqueous-buffered medium.

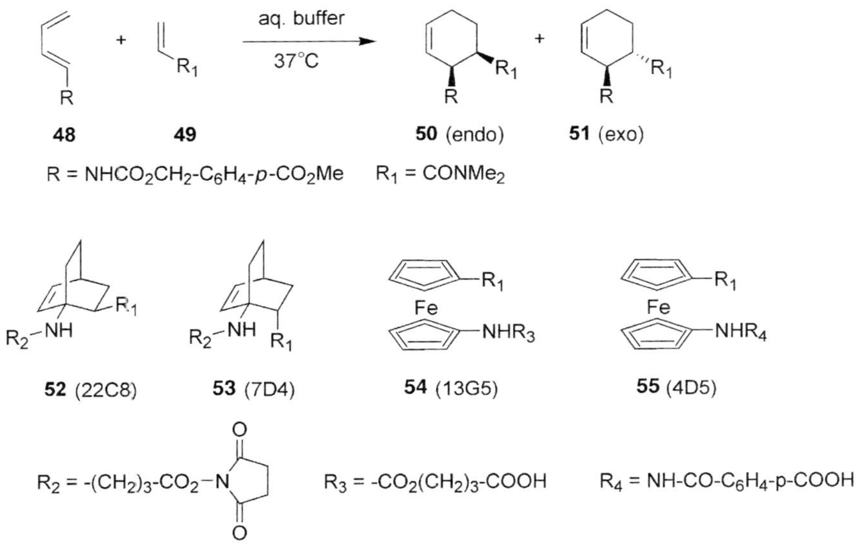

Hapten	Antibody	**50/51**	ee (%)
Uncatalyzed		85:15	
52	22C8	0:100	>98
53	7D4	100:0	>98
54	13G5	2:98	95
55	4D5	98:2	95

Scheme 5.13

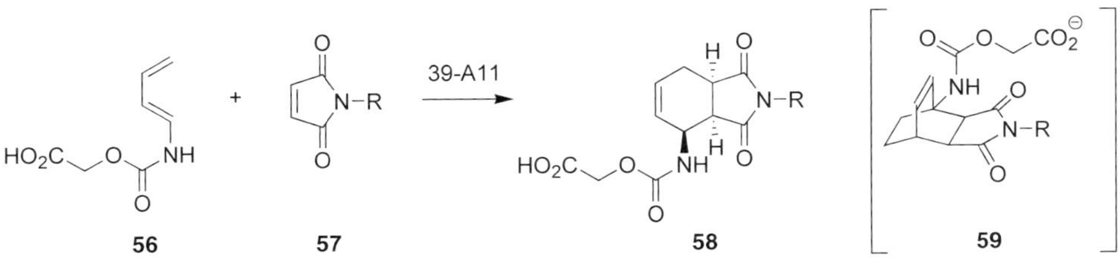

R = C_6H_4-*p*-$NHCOCH_3$

Scheme 5.14

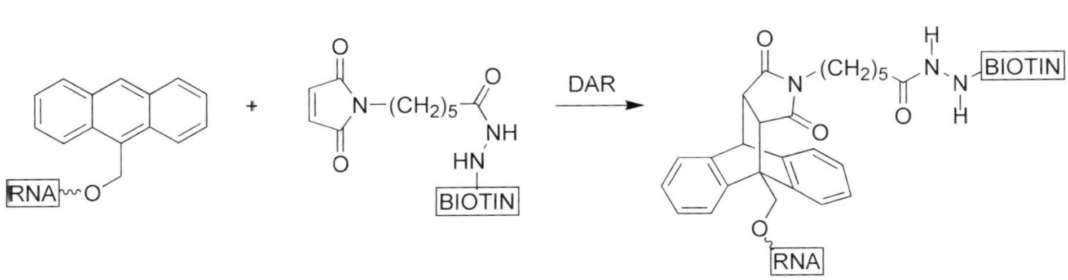

Scheme 5.15

5.1.3 Hetero Diels–Alder reactions

The nitroso functionality is a powerful dienophile and *N*-alkyl- and *N*-acylnitroso compounds give inter- and intrahetero Diels–Alder reactions easily. The cycloadditions also occur in aqueous medium although some nitroso compounds (i.e. *N*-acylnitroso derivative) are short-lived in the presence of water.[42] The NO functionality is generated *in situ* by periodate oxidation of the hydroxylamine group and the cycloaddition with butadienes gives a 1,2-oxazine ring. Scheme 5.16 illustrates the utility of the nitroso Diels–Alder cycloaddition for the synthesis of (−)-swainsonine,[42c] (−)-pumiliotoxin C,[42d] and BCX-1812 (RWJ-270201), a neuramidase inhibitor used as an anti-influenza agent.[42e–g]

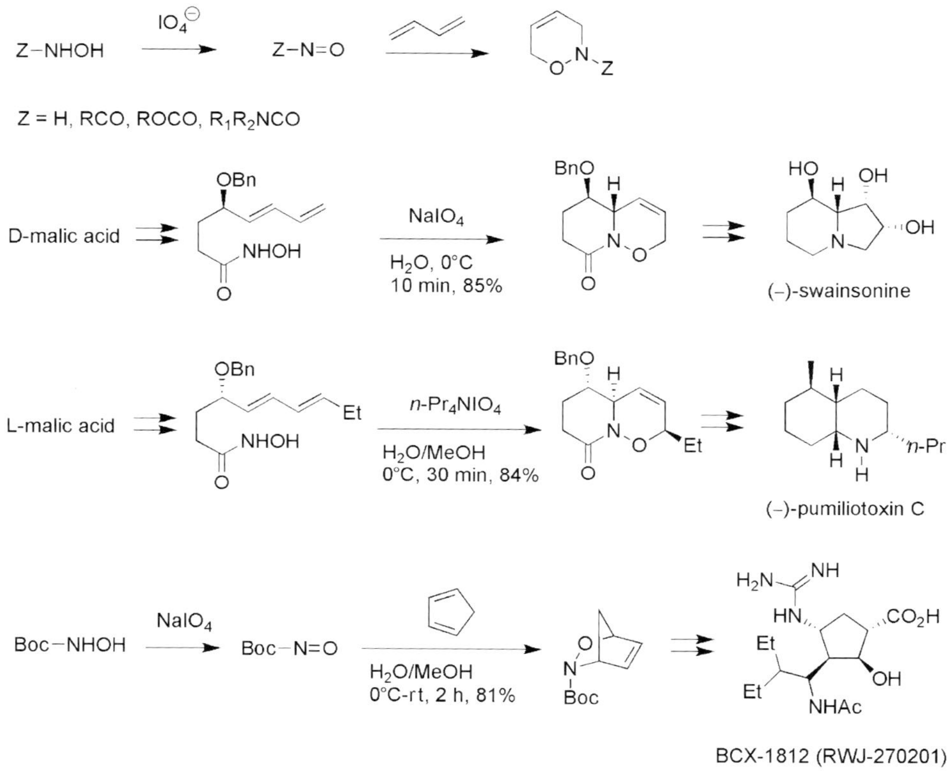

Scheme 5.16

The aza-Diels–Alder reaction of imines with diene of Danishefsky is an important route to 2,3-dihydro-4-pyridones.[43] A number of Lewis acids have been used to catalyze the reaction in organic solvents. In water the reaction was realized by acid catalysis via iminium salts[11a] or by Brønsted acids.[43a] The montmorillonite K-10 catalyzed this cycloaddition in water or in aqueous acetonitrile in excellent yield.[43b] Recently Kobayashi has performed the reaction in water at room temperature under neutral conditions in two (imine + diene) or three (aldehyde + amine + diene) component versions by using sodium triflate as catalyst.[43c]

Imine intermediates from the indium-mediated reaction, in aqueous medium at 50°C, between aromatic nitro compounds and 2,3-dihydrofuran undergo aza-Diels–Alder cycloadditions to give tetrahydroquinoline derivatives in good overall yields.[43d]

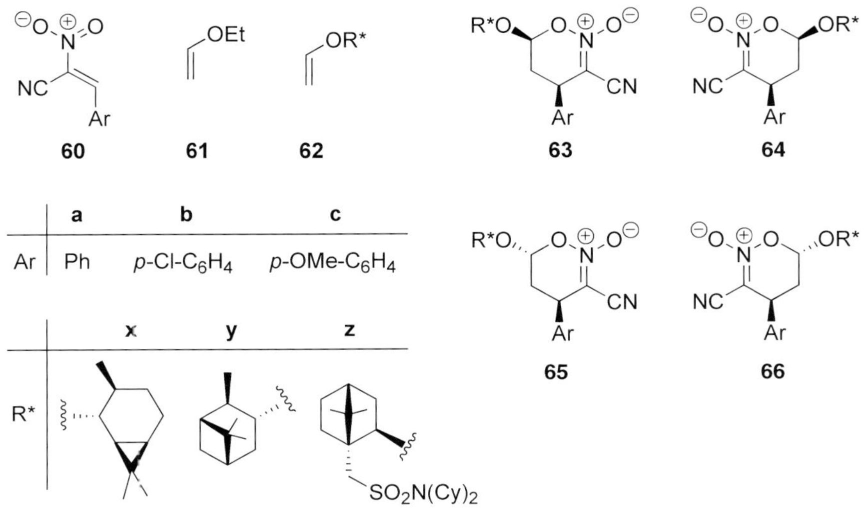

Scheme 5.17

Thermal and catalyzed [4 + 2] cycloadditions of α,β-unsaturated nitroalkenes in organic solvent have been investigated by using optically active vinyl ethers.[44] [Note: It is not certain if reactions of nitroalkenes as 4π components with electron-rich olefins are pericyclic Diels–Alder cycloadditions or involve a Michael additions to give betaines followed by a collapse to form the nitronate.[45]] [4 + 2] Cycloadditions between an electron-rich diene with an electron-deficient dienophile have been widely investigated in aqueous medium,[10b,17h] while examples of [4 + 2] cycloadditions in water between an electron-deficient diene and an electron-rich dienophile are rare.[23b,46] The cycloadditions of (*E*)-2-aryl-1-cyano-1-nitroetenes **60** with achiral and enantiopure vinyl ethers **61** and **62** performed in pure water are some recent examples of the latter[23c] (Scheme 5.17). All the reactions were totally regioselective and highly endo diastereoselective. The endo approach of reactants is due to strong secondary orbital interactions between the oxygen of electron-rich vinyl ether and the positively charged nitrogen atom of nitroalkene, which are present in the endo transition state and absent in the exo-mode orientation. Table 5.4 illustrates the results of asymmetric cycloadditions of **60** in sole water with enantiopure vinyl ethers **62**. When (−)-*N*,*N*-dicyclohexyl-(1*S*)-isoborneol-10-sulfonamide was used as chiral auxiliary, the cycloadditions (**60a–c** with **62z**) were totally asymmetric.

Table 5.4 Asymmetric [4 + 2] cycloadditions of nitroetenes **60** with vinyl ethers **62** in water

Reactants	*T* (°C)	*t* (min)	**63** (%)	**64** (%)	**65** (%)	**66** (%)	endo/exo	de (4R) (%)
60b + **62x**	0	30	54	35	5	6	89:11	18
60b + **62y**	0	30	60	30	5	5	90:10	30
60a + **62z**	rt	60	85		15		85:15	100
60b + **62z**	0	60	85		15		85:15	100
60c + **62z**	rt	60	83		17		83:17	100

Similarly, 3-nitrocoumarin (**9**) behaves as 4π component in the cycloaddition reactions with electron-rich dienophiles such as ethyl vinyl ether (**61**), 2,3-dihydrofuran, and 3,4-dihydro-2*H*-pyran.[23a] Under solvent-free conditions and in organic solvent (DCM, CH_3COCH_3, PhH), the reactions are totally regio- and endo-stereoselective. For example, the cycloaddition of **9** with **61** gives only the adduct **67** with 90% yield (Scheme 5.18).

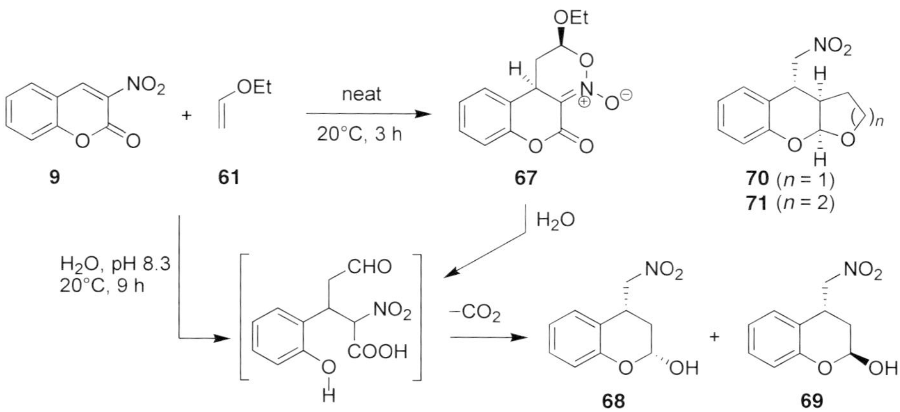

Scheme 5.18

In aqueous $NaHCO_3$ (pH 8.3), the cycloaddition occurred at 20°C under heterogeneous conditions and after 9 h gave a mixture of chromanols **68** and **69** in a 56:44 ratio and with 56% overall yield. The cycloadduct **67** in the presence of water (pH 8.3, 20°C, 8 h) gave **68** and **69** in 60% yield and again in a 56:44 ratio. The endo diastereoselectivity of cycloaddition of **9** with **61** was explained on the basis of strong secondary orbital interactions between the oxygen of **61** and the nitrogen atom of **9** that are absent in the exo transition state. These results allowed tetrahydrofuro- and tetrahydropyranochromenes **70** and **71** to be synthesized by a one-pot procedure starting from **9** and the suitable vinyl ether.

Other recent examples of inverse electron-demand Diels–Alder reactions in water are the cycloadditions of (*E*)-3-diazenylbut-2-enes **72** with a variety of vinyl ethers.[23b] The results of cycloaddition of **72** with ethyl vinyl ether (**61**) are reported in Table 5.5. The reactions were always faster in heterogeneous aqueous medium than in organic solvent and the endo adduct was the prevalent reaction product. Pyrrole derivatives such as ethyl-2-methyl-1-ureido-1*H*-pyrrole-3-carboxylates, derived from zwitterionic [3 + 2] cycloaddition reactions, were sometimes observed and a reaction mechanism of their formation has been proposed.[23b] In water, as well as in DCM, **72** (R = OEt, R_1 = H) behaves like an electron-acceptor heterodiene even with a highly reactive diene such as cyclopentadiene, giving quickly, at 15°C, only the endo adduct. The cycloaddition of **72** (R = OEt, R_1 = H) with the chiral vinyl ether (+)-2-(ethenyloxy)-3,7,7-trimethylbicyclo[4.1.0]heptane (**62**x) was complete in water in 68 h at 15°C and gave a mixture of 83:17 endo/exo adducts with modest enantioselectivity. This is the first example of an asymmetric inverse electron-demand Diels–Alder reaction performed in water.[23b]

Table 5.5 Diels–Alder cycloadditions of (*E*)-3-diazenylbut-2-enes **72** with ethyl vinyl ether (**61**) at 15°C

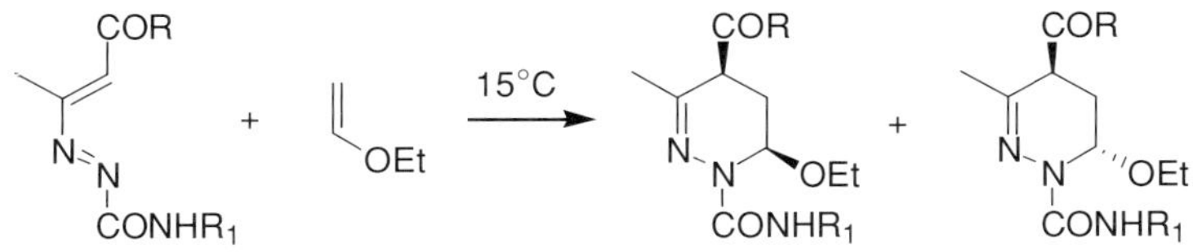

R	R_1	Medium	*t* (h)	*C* (%)[a]	**73/74**	Yield (%)
OEt	H	H_2O	16	100	92:8[b]	83
OEt	H	CH_2Cl_2	16	81	90:10	60
OEt	H	AcOEt	16	21	77:23	
OEt	H	PhMe	16	19	74:26	
OEt	H	THF	16	16	75:25	
OMe	Ph	H_2O	13	100	89:11	75
OMe	Ph	CH_2Cl_2	13	79	91:9	65
NMe_2	H	H_2O	24	82	85:15	62
NMe_2	H	CH_2Cl_2	24	31	84:16	

[a] Reaction conversion.
[b] 3:1 mixture of exo-**74** and ethyl-2-methyl-1-ureido-1*H*-pyrrole-3-carboxylate.

5.1.4 The role of water

On the basis of kinetic data and of the fact that the Diels–Alder reaction is accelerated in aqueous medium by adding salting-out salts (LiCl, NaCl) and retarded by adding salting-in salts ($LiClO_4$, guanidinium chloride), Breslow[15,47] suggested that the acceleration and the increased selectivity of the reaction are ascribable to hydrophobic packing of diene and dienophile.

Engberts[11c,17f,i,25,48] reported evidence that enforced hydrophobic interactions and hydrogen-bonding interactions are responsible for the rate enhancement of aqueous Diels–Alder reaction. The term 'enforced' has been used to emphasize that the rate enhancement is not the result of the hydrophobic packing of reactants but rather 'to stress that hydrophobic interactions occur simply because they are an integral part of the activation process'.

Hydrogen bonding has an important role[48d,49a–c] as shown in the relationship between the acceptor number of the solvent and the Diels–Alder reaction rate.[50] The micellar effect,[48a,51] the cohesive energy density of water,[49c,52] and the antihydrophobic cosolvent effect,[47,53] when mixtures of water and organic solvents are used, have also been suggested and discussed to explain the beneficial effects of water on both rate acceleration and the increased regio- and stereoselectivity of Diels–Alder reaction.

In contrast with previous hypotheses,[49d,e] it has been observed[49f,g] that the internal pressure of water cannot facilitate the hydrophobic packing of diene and dienophile and therefore it cannot be invoked to explain the strong rate enhancement of Diels–Alder cycloaddition when carried out in water with respect to organic solvent.

The first mechanistic study on the benefits of a combination of Lewis acid catalysis and water as reaction medium for a Diels–Alder reaction appeared in 1995.[25a] $Cu(DS)_2$ micelles (DS: dodecyl sulfate) catalyze the cycloaddition of 3-(*para*-substituted phenyl)-1-(2-pyridyl)-2-propen-1-ones with cyclopentadiene, with very high efficiency leading to a rate enhancement of up to 1.8×10^6 compared to the uncatalyzed reaction in acetonitrile.[48a]

An efficient coordination between Lewis acid and dienophile ensures an efficient catalysis provided that the coordination of adduct to the catalyst is not very strong. Water hampers the complexation because it is a hard solvent and interferes with the interactions between a hard Lewis acid and hard sites of the dienophile. This is the reason why the catalytic effect can be more marked in organic media than in aqueous media.[11a] By combining the aqueous medium and the Lewis acid catalyst, the Diels–Alder reaction can be markedly accelerated but it is the water that contributes much more to the rate enhancement because of the enforced hydrophobic interactions and hydrogen-bonding interactions.

Theoretical studies[9b,54] and computer simulations[49b,55] have also been reported regarding the role of water in Diels–Alder reaction in aqueous medium.

Sharpless[56] has recently observed that when water-insoluble reactants were vigorously stirred in pure water, the reaction was accelerated with respect to the use of solvent-free conditions or organic solvents that allowed homogeneous conditions. Sharpless coined the term 'on water' to indicate that the reaction proceeds in pure water without solvation but by simply stirring the reactants to generate a suspension.[57] Examples of the Diels–Alder reaction of water-insoluble *trans,trans*-2,4-hexadienyl acetate (**75**) with *N*-propylmaleimide (**76**) and the $[4\sigma + 2\pi]$ cycloaddition of quadricyclane (**77**) with dimethyl azodicarboxylate (**78**) are reported in Scheme 5.19. The acceleration of the reactions 'on water' did not depend on the amount of water used, provided that a minimum amount was present for clear phase separation. The cycloaddition of **77** with **78** carried out in protic solvent under homogeneous conditions (i.e. $MeOH/H_2O$ 3:1) was 24 times slowed down compared to the reaction performed under heterogeneous conditions either in sole water or in $MeOH/H_2O$ 1:1. When D_2O was used in the place of H_2O the reaction rate was 4.5 times slowed down.

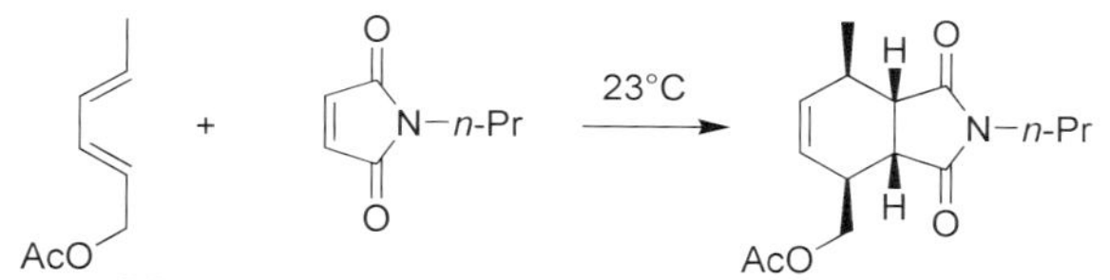

75 **76**

Medium	t (h)	Yield (%)
PhMe	144	79
SFC	10	82
on water	8	81

77 **78**

Medium	t (h)	Yield (%)
PhMe	24[a]	74
SFC	48	85
on water	0.15	82

[a] At 80°C.

Scheme 5.19

The reason for the rate acceleration of reactions carried out by 'on water' protocol is presently unclear. Supposing that a small amount of reactants is dissolved in water, the acceleration can, in principle, be explained arguing as above for reactions performed under homogeneous conditions or with reactants that are partially soluble in aqueous medium.

However, if one considers that the reactions performed under heterogeneous conditions are too fast for the acceleration to be solely due to solution-phase phenomena, perhaps it must be considered, as Sharpless writes that 'the unique properties of molecules at the macroscopic phase boundary between water and insoluble hydrophobic oil play a role'.[56] For an extensive discussion of reactions 'on water', see Chapter 11.

5.2 1,3-Dipolar cycloaddition reactions

In the area of cycloaddition reactions in aqueous medium much work has been done on the Diels–Alder reaction and, until 10 years ago, little attention had been paid to 1,3-dipolar cycloaddition although this reaction shows many similarities to Diels–Alder reaction: both are $[4\pi + 2\pi]$ reversible pericyclic processes with negative volume of activation and the dipolarophiles are the counterpart of dienophiles.

1,3-Dipolar cycloadditions give access to a variety of five-membered heterocycles.[58] Scheme 5.20 illustrates some reactions performed in aqueous medium before 1995; more recent studies based on the type of cycloadduct are reported below.

5.2.1 Pyrrole and pyrrolidine-ring formation

Kinetic studies of cycloadditions of phthalazinium-2-dicyanomethanide **79** and pyridazinium **80** with a range of dipolarophiles[65] showed that 'water super dipolarophiles' such as methyl and ethyl vinyl ketone and but-3-yn-2-one displayed rate enhancements of 45–59 times with **79** and 156–202 times with **80** (Scheme 5.21) in reactions carried out in water compared to MeCN. 'Water normal dipolarophiles,' such as methyl acrylate and methyl propiolate, accelerated the reactions in water by 7–12 times with **79** and 13–17 times with **80**. By adding LiCl or guanidinium chloride, the reaction rates in water increased or decreased respectively, paralleling the effects observed in Diels–Alder reactions[17k] and suggesting that the hydrophobic effect and the hydrogen-bonding effect are mainly responsible for the reaction rate enhancement.

5.2.2 Isoxazole and hydroderivative-ring formation

2-Methyl-4-nitro-3-isoxazolin-5-one (**81**) acts as a precursor of nitrile oxide **82**. When **81** is reacted with acetylenes and with both electron-rich and electron-deficient alkenes in MeCN/H_2O 3:1 at room temperature, isoxazole **83** and isoxazoline derivatives **84** are produced regioselectively and in good yield, respectively (Scheme 5.22).[66] In sole MeCN, the isoxazolinone **81** was recovered unchanged even under reflux.

The second-order rate constant of 1,3-dipolar cycloadditions of benzonitrile oxide (**85**) with various dipolarophiles **A–E** were determined in water and organic solvents.[67] The greatest accelerating effect was observed in water for the cycloaddition of **85** with electron-rich dipolarophiles **A** and **B** and in *n*-hexane for the cycloadditions with electron-poor dipolarophiles **C–E** (Table 5.6). The results were explained in terms of frontier molecular orbital theory and enforced hydrophobic interactions.

Engberts[68] extended the systematic study of the effects of micelles on Diels–Alder reaction[25c] to 1,3-dipolar cycloadditions of benzonitrile oxide (**85**) with N-substituted maleimides (ethyl, *n*-butyl, benzyl). The rate constants of these reactions in water and 1-propanol were only slightly different ($k_{H_2O}/k_{1-\text{propanol}} = 1.2$–$1.7$). When anionic SDS or

Ar = X-C_6H_4 (X = H, *p*-Me, *o*-Cl, *p*-Cl, *p*-NO_2, *m*-Cl, *m*-NO_2)

(A)[59]

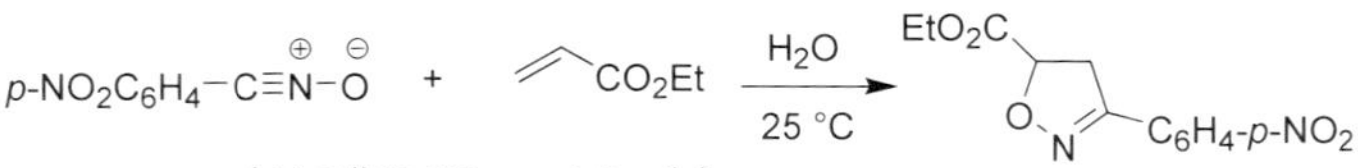

$k(H_2O)/k(H_2O/\text{dioxane } 1{:}1) = 1.9$

(B)[59]

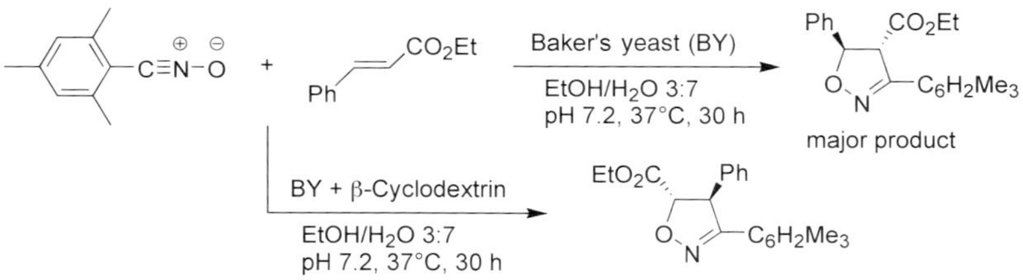

(C)[60, 61]

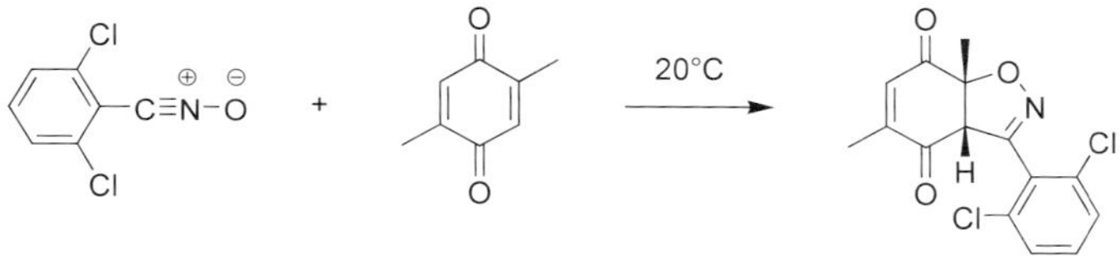

(D)[62]

Solvent	*c*-C_6H_{12}	PhH	$CHCl_3$	MeCN	DMSO	EtOH	EtOH/H_2O 4:1	EtOH/H_2O 1.5:1
$k_{solv}/k(c\text{-}C_6H_{12})$	1	0.32	0.23	0.42	0.82	0.86	1.93	3.26

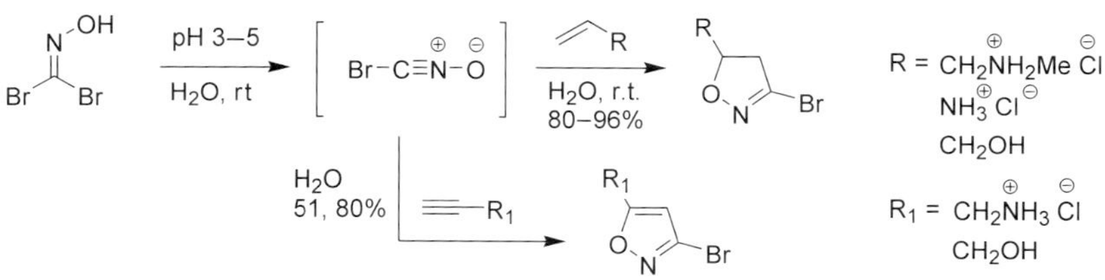

(E)[63]

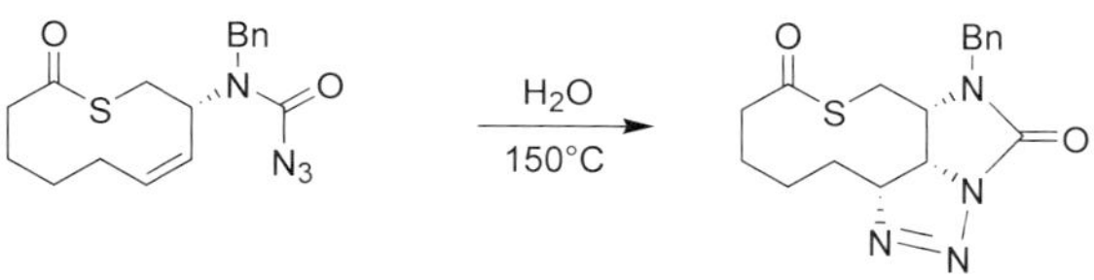

(F)[64]

Scheme 5.20

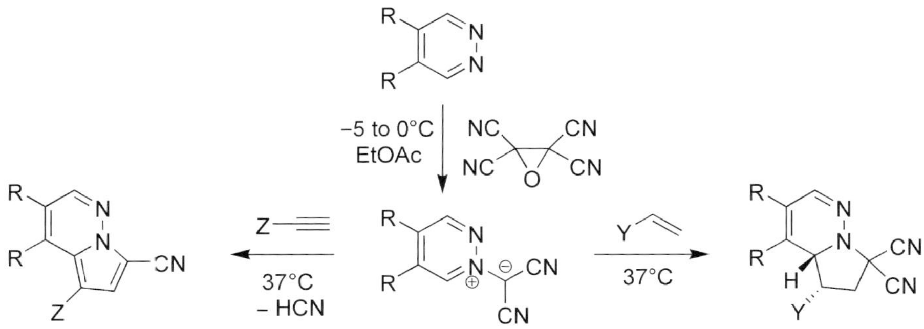

79 R–R = $(CH_2)_4$
80 R = R = H

Z	**79** (k_{rel})	**80** (k_{rel})
COMe	45	156
CO_2Me	7	12.7

Y	**79** (k_{rel})	**80** (k_{rel})
COMe	59	202
CO_2Me	12	16.6

$k_{rel} = kH_2O/kMeCN$

Scheme 5.21

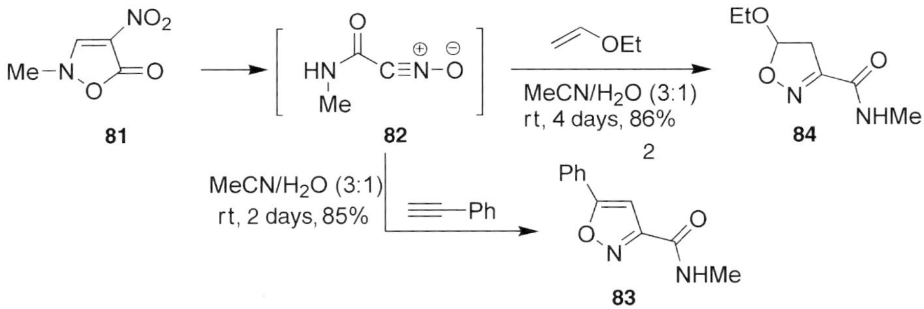

Scheme 5.22

cationic CTAB were added, the rate constants increased significantly (as much as a factor of 17), which showed that the micellar acceleration for 1,3-dipolar cycloadditions was greater than that found for Diels–Alder reactions that rarely exceed a factor of 2.[25c] This is because that 1,3-dipolar cycloadditions have only slightly reduced rate constants for the reactions in micellar phase with respect to water ($k_m/k_w \sim 0.25$–0.45) whereas the Diels–Alder reactions have much smaller relative rate constants ($k_m/k_w \sim 0.02$–0.05).[68]

Second-order reaction rate constants of reaction between *C,N*-diphenylnitrone **86** and dibutylfumarate (**87**) were determined in various solvents (Table 5.7).[69] The formation of isoxazolidine **88** was approximately 125 times faster in water than in ethanol and increased up to 518 times when 2 M LiCl was added and was decelerated with the addition of urea, paralleling the effects observed for Diels–Alder reactions.[15,46]

Table 5.6 Relative rate constants for 1,3-dipolar cycloadditions of benzonitrile oxide (**85**) with dipolarophiles **A–E** in various reaction media

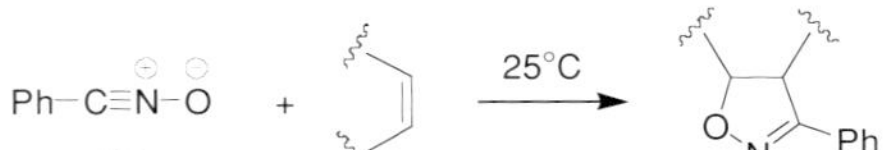

85

Medium	A	B	C	D	E
H_2O	7.1	11.1	5	5.5	4.3
EtOH	1.9	2.3	5.5	7	3.7
CF_3CH_2OH	2.7	3.1	1	1	1
DMSO	2.2	4.0	3.8	15.5	5.0
DCM	1	1.1	1.5	3	1.2
1,4-Dioxane	1.4	1.4	3.0	6.5	2.0
n-C_6H_{14}	2.2	1	8.0	7.5	6.0

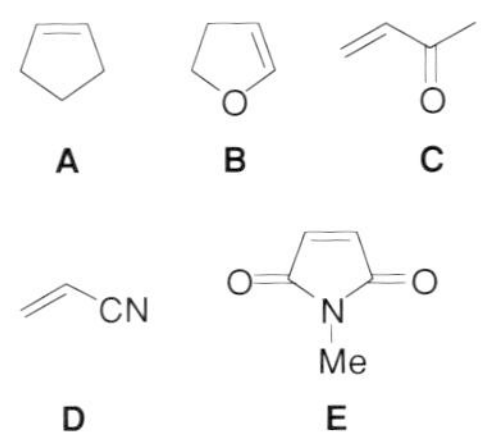

5.2.3 Pyrazole and pyrazoline-ring formation

The high HOMO–LUMO energy gap between diazocarbonyl compounds and alkynes[8] and the poor stability of alkynes in the presence of Lewis acids have prevented the use of diazocarbonyl compounds as dipoles for 1,3-dipolar cycloadditions. Li and Jiang[70] found that by using $InCl_3$ (20% mol) as catalyst, the cycloadditions of a series of aryldiazocarbonyl

Table 5.7 Second-order reaction rate constants of 1,3-cycloaddition of *C,N*-diphenylnitrone **86** with dibutylfumarate **87** in various reaction media relative to ethanol

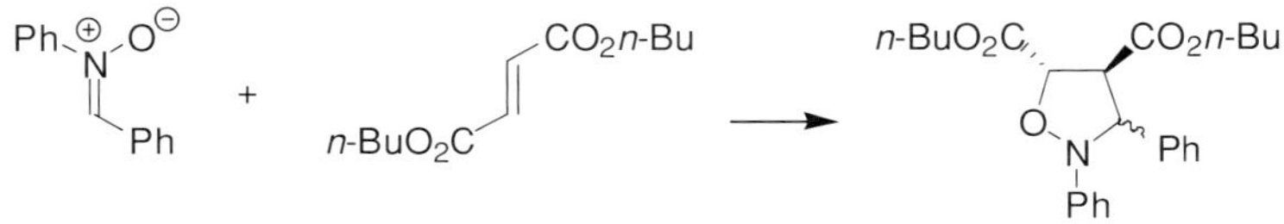

86 **87** **88**

Medium	k_{rel}
H_2O	125
H_2O/LiCl	518
H_2O/Urea	75
$(CH_2)_2(OH)_2$	18
n-C_6H_{14}	10
PhH	3
PhMe	3
THF	3
i-PrOH	2
n-PrOH	1
DMF	1
EtOH	1

compounds **89** and methyl propiolate occurred easily in water at room temperature with good yields (Scheme 5.23). The initially formed cycloadduct **90** easily isomerizes via hydrogen, aryl, and carboxylate shifts to pyrazoles **91** and **92**.

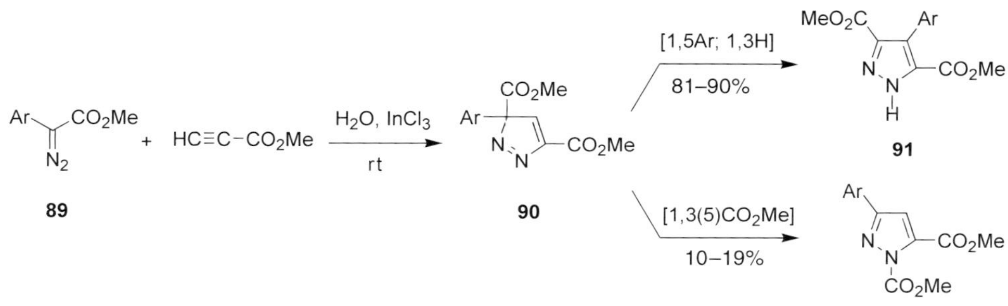

Scheme 5.23

Molteni[71] found that nitrilimine **94**, generated *in situ* from hydrazonoyl chloride **93**, reacted with electron-poor dipolarophiles (i.e. ethylacrylate in Scheme 5.24) to give pyrazolines in water under basic conditions and in the presence of tetrahexylammonium chloride. The presence of surfactant reduced the formation of the by-product 1,2,4,5-tetrazine derivative coming from the dimerization of **94**. On the basis of a computational study, the authors conclude that the water has a rather small effect on the cycloaddition with respect to that played by the surfactant.

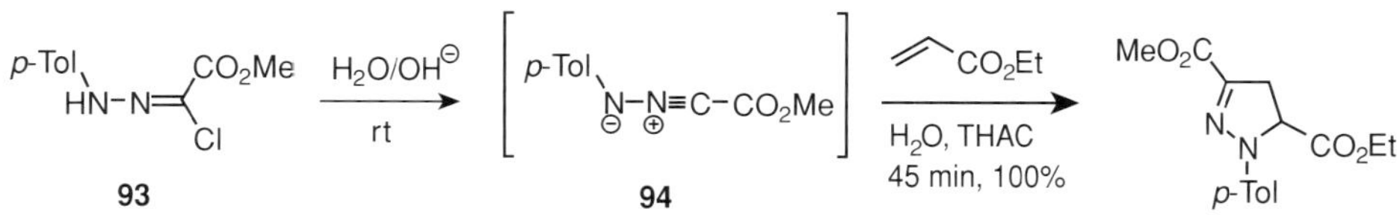

Scheme 5.24

5.2.4 Triazole and triazoline-ring formation

The high stability of azides toward water, oxygen, and high temperature make these compounds very useful dipoles for the synthesis of triazoles and triazolines by 1,3-dipolar cycloaddition reaction. When the reaction involves an unsymmetrical alkyne, a mixture of 1,4- and 1,5-regioisomer triazoles is generally obtained. Using Cu(I) salts (1 mol%), prepared *in situ* by reducing Cu(II) salts with ascorbic acid and/or sodium ascorbate, in H_2O/*t*-BuOH 2:1, allows 1,4-disubstituted 1,2,3-triazoles to be obtained (Scheme 5.25) regioselectively and with high yields.[72] A mechanistic proposal for the catalytic cycle was advanced.

In a study aimed at optimizing the 1,3-dipolar cycloaddition process for DNA analysis and immobilization, Ju found that the cycloaddition of non-bioconjugated azides can be carried

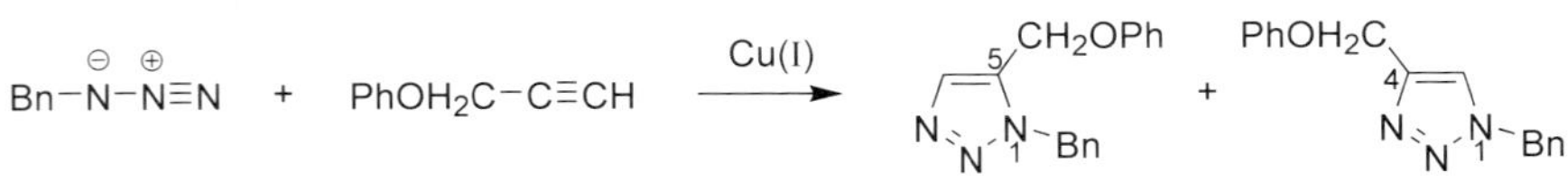

Medium	T(°C)	t(h)	1,4	1,5	Yield (%)
H_2O/*t*-BuOH (2:1)	rt	8	100	–	91
Neat	92	18	62	38	–

Scheme 5.25

out at room temperature and without catalyst if an electron-deficient alkyne is used (Scheme 5.26, A).[73] The cycloaddition of conjugated azido-DNA sometimes requires H_2O/MeCN 4:1 as reaction medium and Cu(I) salts as catalysts to achieve the best results (Scheme 5.26, B).

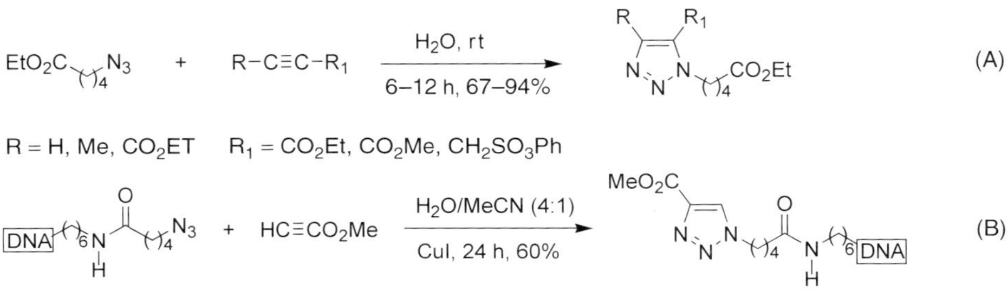

Scheme 5.26

The 1,3-dipolar cycloaddition between nonfluorescent compounds containing azido and alkyne moieties as pro-fluorophores has been used to prepare compounds that enhance fluorescent emission due to the formation of a triazolyl unit.[74] This strategy is of particular interest in the fields of bioconjugation and bioimaging. 3- and 4-Azidocoumarins, chosen as pro-fluorophores, react in aqueous ethanol at room temperature and in the presence of Cu(I), generated from $CuSO_4$ (5% mol) and ascorbic acid or sodium ascorbate (10%), to give intensely fluorescent 1,2,3-triazoles in high yields (Scheme 5.27).

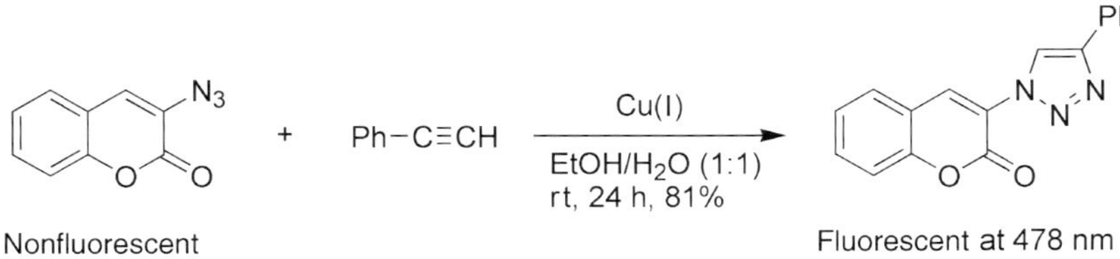

Scheme 5.27

Bitriazolyl unit **96** was prepared as a surrogate of biaryl system by 1,3-dipolar cycloaddition of methyl-5-azido-1,2,4-triazole-3-carboxylate (**95**) with a variety of alkynes in H_2O/THF 2:1 at 80°C in the presence of Cu(I) (Scheme 5.28).[75]

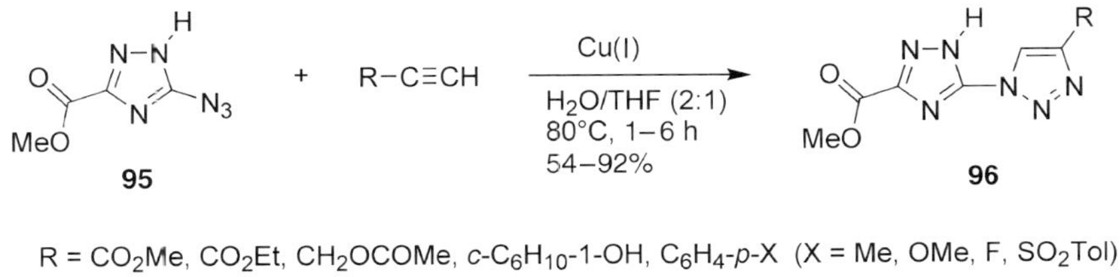

Scheme 5.28

Azide and acetylene functionalities are generally compatible with the enzymes under physiological conditions and can be easily incorporated into diverse organic building blocks. This allows an enzyme target to be used to selectively synthesize compounds of biological and pharmaceutical interest from suitable building blocks by 1,3-dipolar cycloaddition reaction. The enzyme acetylcholinesterase was used[76] to select and synthesize the triazol-linked **97** inhibitor by using building block inhibitors based on tacrine-azide and phenanthridinium acetylene, which vary in carbon chain length (2 and 6 in the Scheme 5.29).

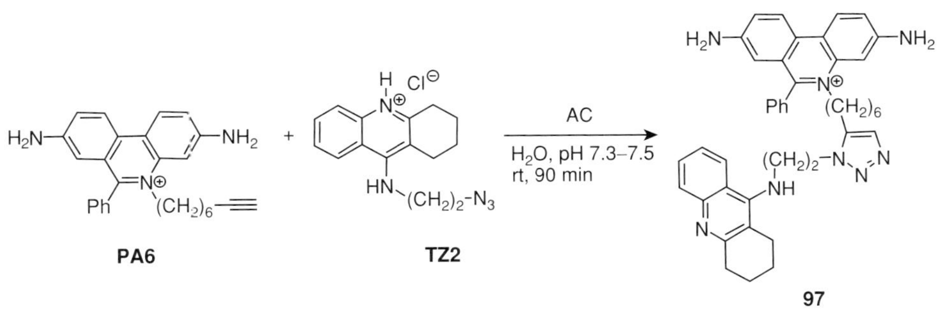

Scheme 5.29

Kinetic data regarding the 1,3-dipolar cycloaddition of phenyl azide with norbornene in various solvents to give triazoline **98** were reported by Engeberts[77] (Table 5.8). A high reaction rate enhancement (a factor of 53 relative to *n*-hexane) was observed when the reaction was performed in water containing 1% of 1-cyclohexyl-2-pyrrolidinone (NCP).

A simple one-pot, two-step procedure for the regioselective preparation of 1,2,3-triazoles **99** (Scheme 5.30) was recently reported by Kacprzak.[78] Water, alkyne, sodium ascorbate, and aqueous $CuSO_4$ were added to the DMSO solution of alkyl or aryl azides generated under anhydrous conditions. The product, obtained in good to excellent yields, usually precipitated and it was collected by simple filtration. Only 1,4-regioisomers were detected when monosubstituted alkynes were used.

Table 5.8 Reaction rate constants of cycloaddition of phenylazide with norbornene in various reaction media relative to *n*-hexane

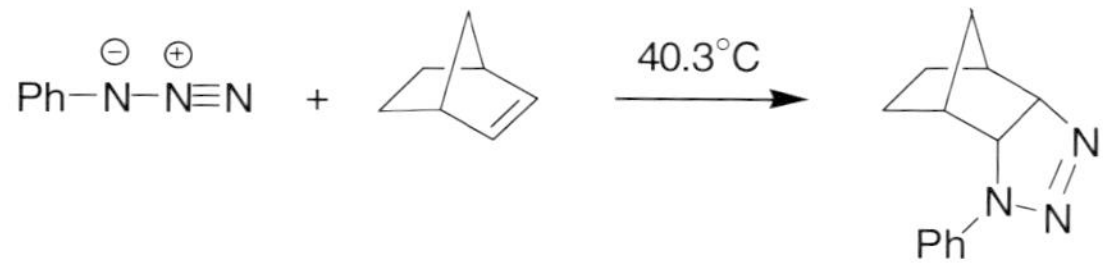

98

Medium	k_{rel}
H_2O/NCP (99:1)	53.2
H_2O/MeOH (75:25)	7.4
EtOH	1.6
DMSO	3.7
MeCN	1.6
THF	1.1
n-C_6H_{14}	1

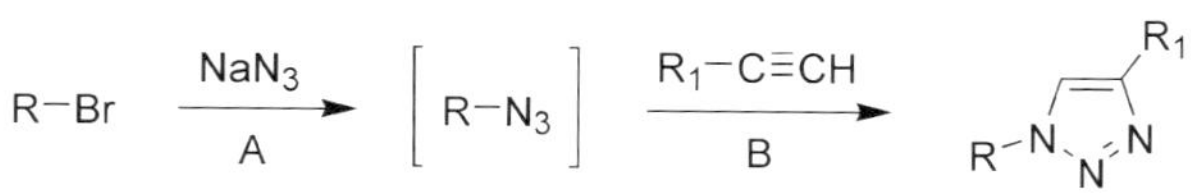

99

R = Benzyl, alkyl; R_1 = CO_2Me, Ph, OAc

A = Anhydrous DMSO, rt, 12–18 h

B = H_2O, sodium ascorbate, $CuSO_4$ aq. solution, rt, 6–18 h, 56–96%

Scheme 5.30

5.2.5 Tetrazole-ring formation

The tetrazole functionality has an important role in organic synthesis, medicinal chemistry, and various materials science applications. The simplest method to prepare this heterocyclic ring is the 1,3-dipolar cycloaddition of azides and nitriles. The reaction mechanism was recently revised.[79]

Sharpless[80] found that the formation of 1*H*-tetrazole was achieved with excellent yield by zinc salt-catalyzed cycloaddition of activated and inactivated nitriles with sodium azide in water (generally under heterogeneous conditions) at reflux (Scheme 5.31). The reaction occurs at pH 8, which minimizes the release of hydrazoic acid even at 100°C. Several zinc salts were effective. $ZnBr_2$ was chosen as the best compromise between cost, selectivity, and

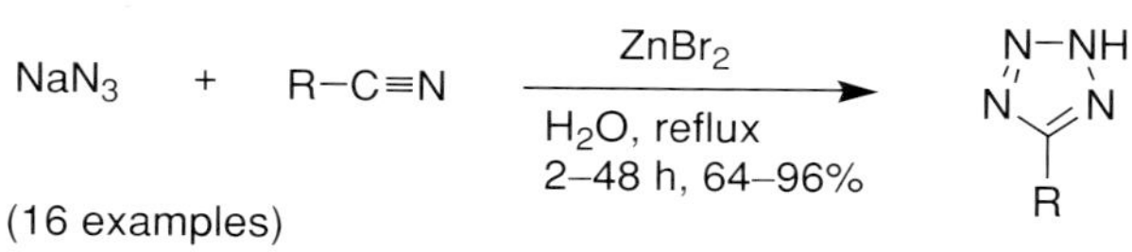

(16 examples)

Scheme 5.31

reactivity. The exact role of zinc salt is not clear but it is hypothesized that it does not act simply as a Lewis acid. Other Lewis acids were tested but little to no reaction rate acceleration was observed. This protocol is suited for the synthesis of α-amino tetrazoles.[81] Amino-protected tetrazoles were prepared in excellent yields (Scheme 5.32) by using protected α-amino nitriles, in H_2O/i-PrOH 1:2 at reflux and in the presence of $ZnBr_2$. By employing α-amino nitriles prepared from natural α-amino acids, optically active tetrazoles were synthesized.

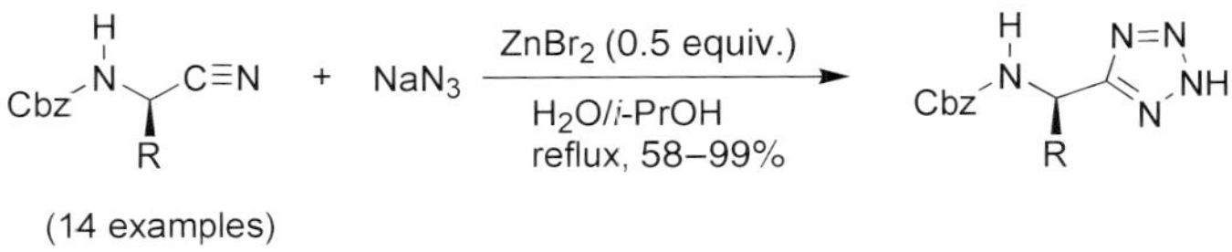

Scheme 5.32

5.3 [2 + 2] Photocycloaddition reactions

The reaction mechanism of photochemical cycloadditions is not always well known. A concerted pericyclic process is likely when the reaction is highly stereospecific, while a two-step mechanism involving a diradical or diion intermediate is more probable when a mixture of stereoisomers is obtained. This is not a rule and a recent example is the [2 + 2] photodimerization reaction of diethyl 1,2-benzoxaphosphorine-6-bromo-3-carboxylate (an analog to coumarin) performed in solvents of different polarities (H_2O, MeOH, PhH, CH_3CO_2H).[82] In water, the reaction is highly stereoselective in favor of a centrosymmetric anti-head-to-tail stereoisomer. Theoretical data, however, indicate that the process is not pericyclic, but that the reaction proceeds through a diradical or dipolar intermediate.

Photocycloadditions are currently carried out in organic solvent[11c]; the study in pure water is still at an infancy stage even if there is great interest from an environmental point of view. Most of the investigations carried out in aqueous medium concern [2 + 2] processes and have been performed in the presence of micelles, CD, or in inclusion complexes, with the scope of favoring the molecular aggregation and therefore the regio- and stereoselectivity of the reaction.[83]

Photodimerization of *trans*-stilbene (**100**) in water gave after 24 h a mixture of stilbenes (44%, cis/trans 1:3), phenanthrene (33%), and cycloadducts **101** and **102** (1:1 ratio; Scheme 5.33), while in benzene no cycloadduct was detected. In the presence of LiCl (a salting-out solute) the amount of cycloadducts increased to 42% and their ratio did not change.[84] In 3:7 benzene/methanol the cycloadducts **101** and **102** were only 9% of the reaction mixture in a 2:1 ratio.[85]

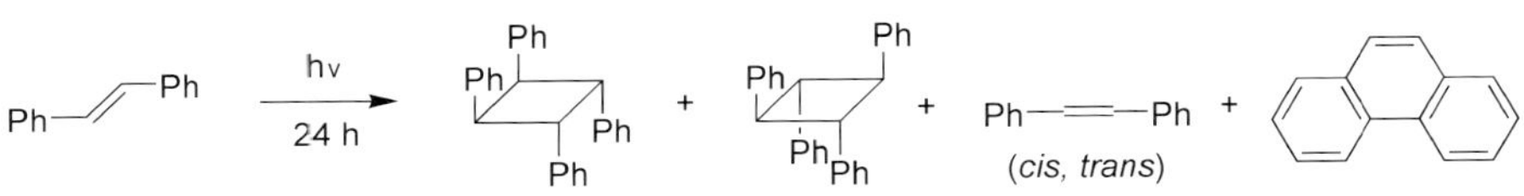

100 **101** **102**

Scheme 5.33

Coumarins have historically been the subject of intense photochemical interest. Photodimerization of coumarin **103** (0.01 M) in sole water or in aqueous micellar media (SDS, CTAB, Triton X-100) afforded only the adduct **104** (Scheme 5.34).[86] In MeOH at the same concentration a mixture of **104** and **105** in a 1:13 ratio was obtained, while in benzene (0.31 M) only the adduct **105** was detected.

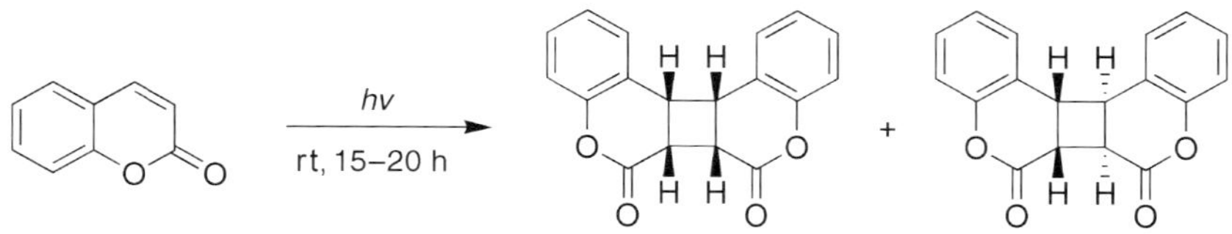

103 **104** **105**

Scheme 5.34

Aqueous medium containing micelles have a great influence on the photodimerization of 3-*n*-butylcyclopentenone (**106**)[87] and isophorone (**109**)[88] (Scheme 5.35). Reversed orientations in aqueous micellar medium, with respect to organic solvent, were also observed in the photodimerization of 2-substituted naphthalenes,[89] 2-pyridones[90] (a [4 + 4] cycloaddition), and in the photocycloaddition reaction of 1-heptenylacetate with 3-*n*-butylcyclopenten-2-one (**106**)[91] (Scheme 5.36).

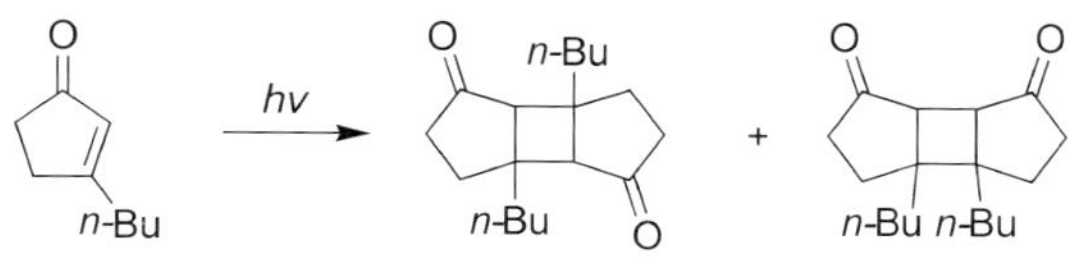

106 **107** **108**

Medium	**107** (%)	**108** (%)
PhH	91	9
MeOH	50	50
H_2O/KDC[a]	2	98

[a]KDC = potassium dodecanoate.

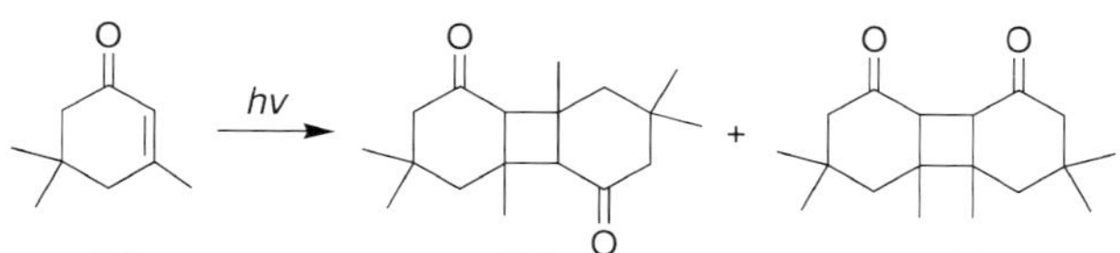

109 **110** **111**

Medium	**110** (%)	**111** (%)
c-C_6H_{12}	80	20
H_2O/SDS	5	95

Scheme 5.35

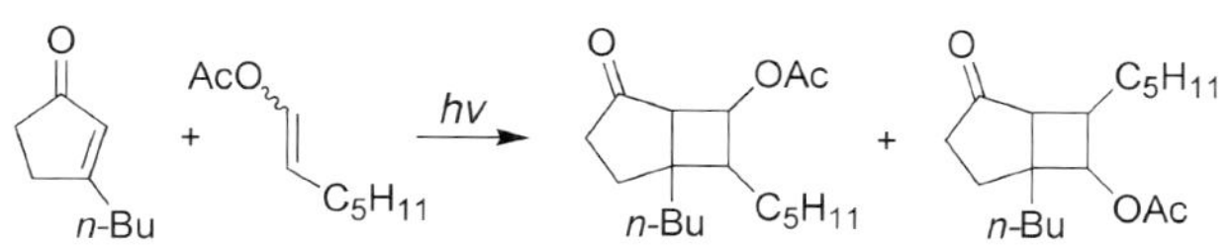

106 **112** **113**

Medium	**112** (%)	**113** (%)
MeOH	0	100
c-C_6H_{12}	0	100
H_2O/KDC[a]	70	30

[a]KDC = potassium dodecanoate.

Scheme 5.36

An example of the effect of CDs on the photocycloaddition reaction is the photodimerization of water-soluble 2-anthracene sulfonate **114** (Scheme 5.37).[92] In sole water, the cycloadducts **115, 116, 117,** and **118** were obtained in relative percentage of 45, 35, 18, and 2%, respectively. The presence of α- and γ-CD did not change the isomeric distribution, while the β-CD allowed only the dimer **115** to be obtained. The results were explained by the formation of different guest/host inclusion complexes.

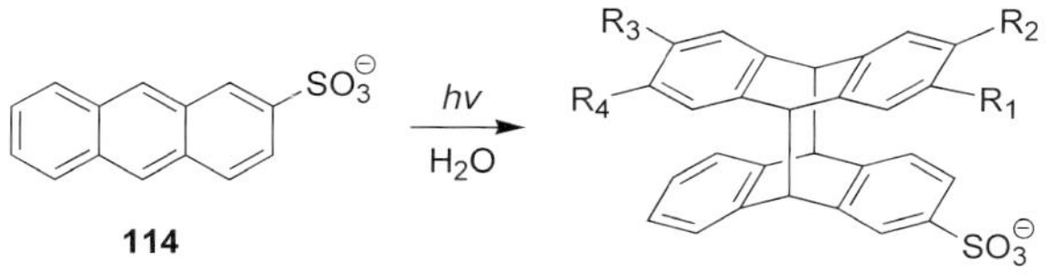

	R_1	R_2	R_3	R_4
115	H	H	SO_3^-	H
116	H	H	H	SO_3^-
117	H	SO_3^-	H	H
118	SO_3^-	H	H	H

Scheme 5.37

A similar effect of α-, β-, and γ-CD was observed in the photocycloaddition of various 5-substituted-(F, Cl, Br, OH, Ph, *t*-Bu)-adamantan-2-one with fumaronitrile[93] (Scheme 5.38; the case of 5-*t*-butyl-adamantan-2-one). All the compounds gave an analogous **119/120** ratio in various reaction media (MeCN, H_2O, α-CD, γ-CD), while a dramatic reversal in face selectivity was observed in aqueous solution containing β-CD. This result was interpreted in terms of a favored syn-face-attack by complexation with β-CD.

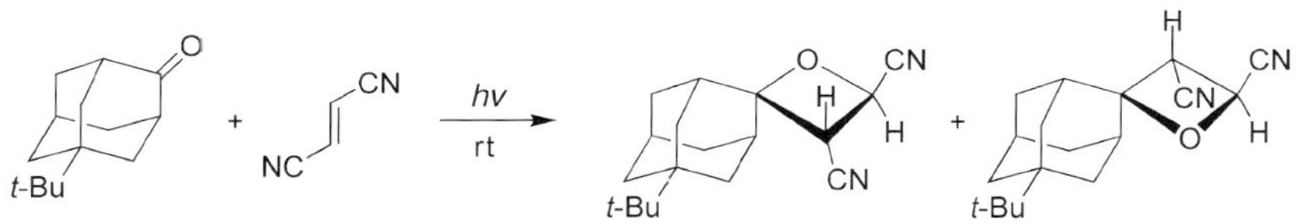

119 **120**

	MeCN	H_2O	α-CD	β-CD	γ-CD
119/120	64:36	62:38	60:40	14:86	55:45

Scheme 5.38

An example of a [2 + 2] photocycloaddition in water of inclusion complexes was reported by Toda.[94] Inclusion crystals of *N*-allyl-3-oxo-1-cyclohexenecarboxamides with chiral hosts derived from tartaric acid, suspended in water containing an alkyl sulfate as surfactant, irradiated at room temperature for 10 h, gave regio- and enantioselectively cycloadducts **121** (Scheme 5.39).

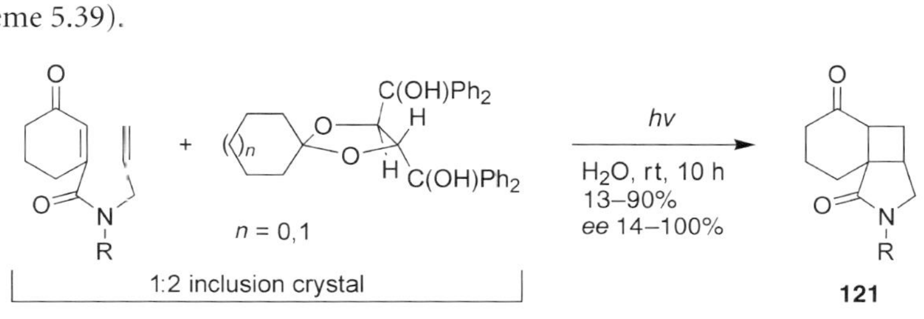

121

R = Me, Et, *n*-Bu, Bn

Scheme 5.39

5.4 Claisen rearrangement reactions

In 1912 Claisen described the transposition of allyl vinyl ethers to γ,δ-unsaturated carbonyl compounds under thermal conditions.[5,95] The first example of this type of [3,3]-sigmatropic shift performed in pure water appeared approximately 60 years later and concerned the conversion of chorismate **122** to prephenate **123** (Scheme 5.40).[96]

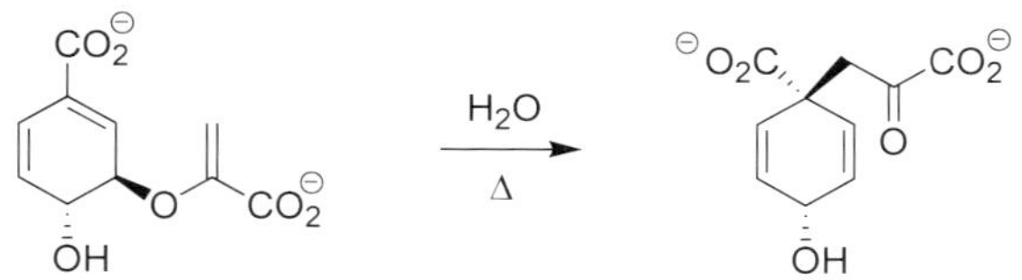

122 **123**

Scheme 5.40

The rearrangement of **122** to **123** is a key reaction along the shikimate biosynthetic pathway for generating aromatic amino acids in plant, fungal, and bacterial systems, and it is catalyzed by the enzyme chorismate mutase more than a millionfold. This has stimulated an in-depth investigation of the mechanism of the Claisen rearrangement.[97]

Extensive experimental and theoretical studies have been focused on the role of the solvent in the acceleration of the Claisen rearrangement.[97c,d,98] The finding of rate acceleration in aqueous medium has been of particular interest.

The rearrangement of allyl vinyl ether, for instance, is approximately 10^3 times faster in water than in gas phase at 75°C.[98c] Table 5.9 lists the relative reaction rate constants of allyl vinyl ether sodium- and methyl carboxylate **124** and **125** in various reaction media of different polarities.[99] The benefit of the water on the reaction rate is evident but the effect cannot be related simply to the polarity of the reaction medium. The reasons behind the water effect are still debated. The hydrogen bonds and the presence of water molecules in the transition state, the increased polarity of the transition state, and the destabilization of the ground state by hydrophobic effects seem to be the most important causes when the reactants are soluble or mostly soluble in aqueous medium.[98] When the reactants are insoluble or highly insoluble it is suggested[56] that the role of water is that of supporting medium, in the

Table 5.9 Relative reaction rate constants of Claisen rearrangement of allyl vinyl ethers 124 and 125 in various reaction media

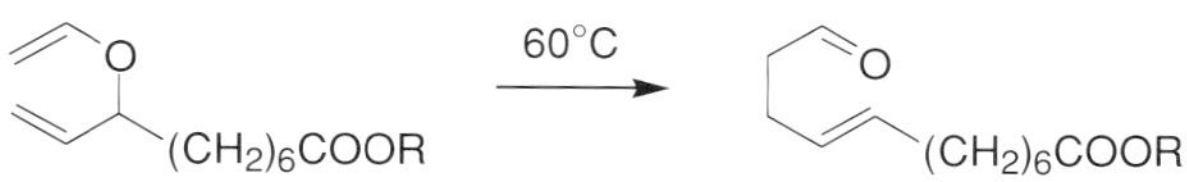

124 R = Na
125 R = Me

Medium	k_{rel} **124**	Medium	k_{rel} **125**
H_2O	66.6	$MeOH/H_2O$ (1:1)	42.8
CF_3CH_2OH	9.6	MeOH	8.6
MeOH	2.9	MeCN	3.0
$MeOH/H_2O$ (1:1)	17.0	Me_2CO	2.1
$DMSO/H_2O$ (1:1)	5.9	PhH	2.0
$DMSO/H_2O$ (9:1)	1	c-C_6H_{12}	1

sense that the water generates a suspension and a strong stirring promotes the reaction by increasing the area of surface contact between the organic and aqueous phase. An example is the rearrangement of the water-insoluble 4-chloro-1-naphthol **126** (Scheme 5.41). After 120 h in sole water, at room temperature, under vigorous stirring ('on water' protocol; see Section 5.1.4) the transposition is complete, while in PhMe, DMF, MeCN, MeOH, or under neat conditions only 16, 21, 27, 56, and 73% of naphthol **127** was found, respectively.[56]

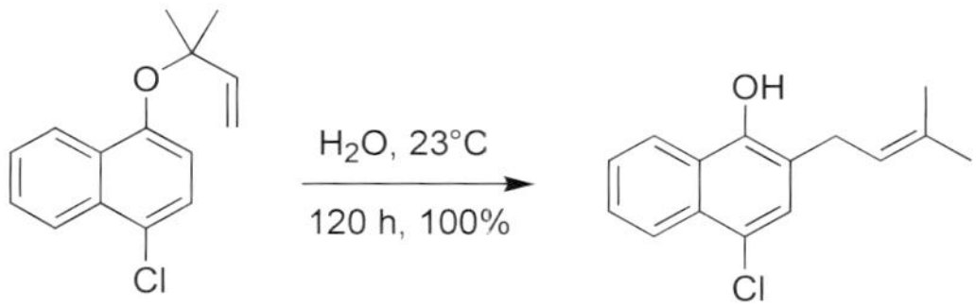

126 **127**

Scheme 5.41

The Claisen rearrangement was also investigated in aqueous medium at high temperature by using a pressurized microwave reactor.[100] The example in Scheme 5.42 is the first Claisen rearrangement of an allyl aryl ether in water. Under neat condition, the transposition of allyl phenyl ether (**128**) occurred at 290°C, while in water a temperature of 240°C was enough. The control of temperature was important: at 245°C a mixture of 2-allylphenol (46%), 2-(2-hydroxyprop-1-yl)-phenol (31%), and 2-methyl-2,3-dihydrobenzofuran (23%) was obtained.

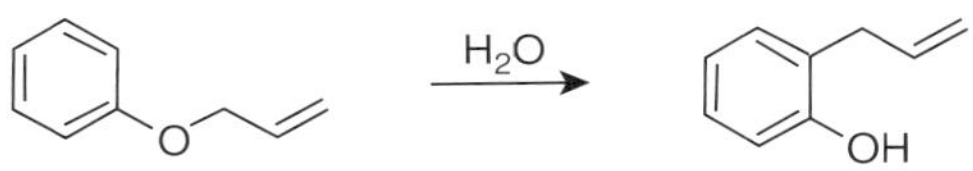

128

T (°C)	*t* (min)	Yield (%)
180	60	–
200	10	10
240	10	84

Scheme 5.42

The Claisen rearrangement is of great interest since it is often one of the key steps in total synthesis and the discovery of the beneficial effect of aqueous medium has increased its practical application. Compounds that are inaccessible because they decompose at the high temperatures necessary for the rearrangement are accessible by carrying out the reaction in aqueous medium.

Enantiomerically pure 1,3-diols have been prepared via the Claisen rearrangement in aqueous medium of α- and β-glucopyranosides (β-anomer in Scheme 5.43).[101] The glucoside **129** rearranged at 80°C in 1 h, while the rearrangement of peracetyl analog at 80°C required 13 days in toluene.

Grieco found (Scheme 5.44) that the aldehyde **131**, a key intermediate in the synthesis of aphidicolin, could be obtained by rearrangement of **130** in H_2O/MeOH 2.5:1 at 80°C in

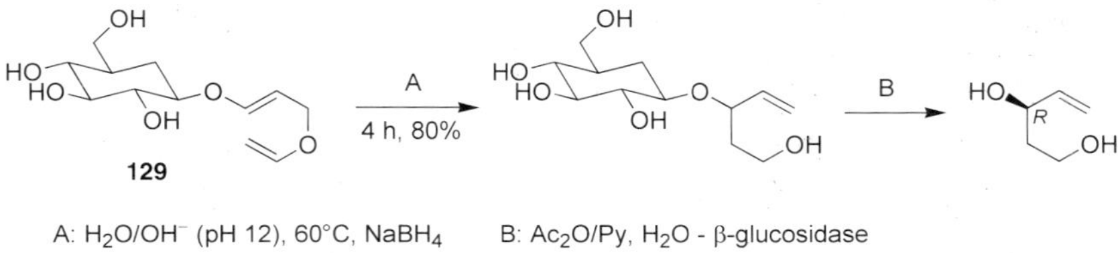

Scheme 5.43

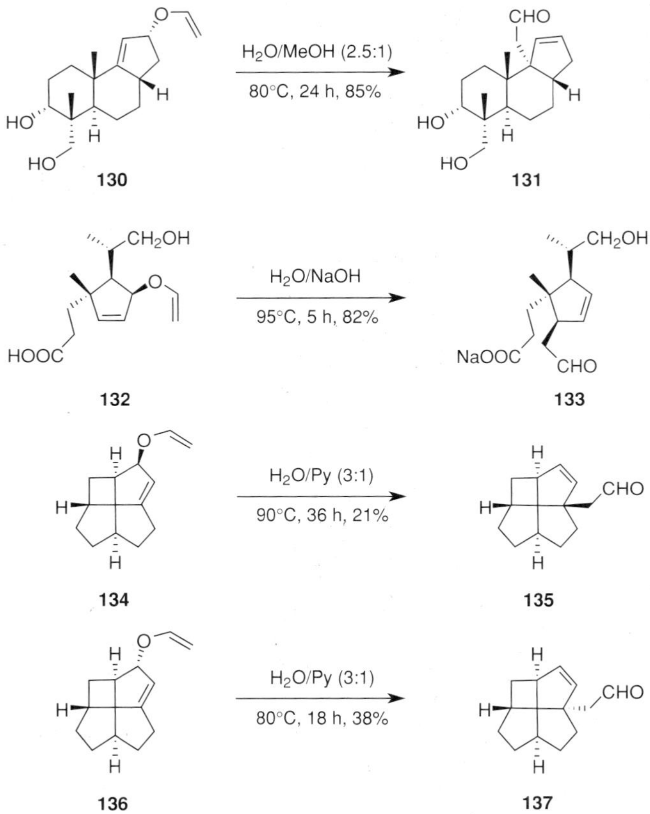

Scheme 5.44

85% yield.[102] Working under thermal conditions (220°C) and in a base-washed, silylated sealed glass tube the protected **130** gives the Claisen rearrangement product in 60% yield.[103] Similarly the vinyl ether **132**, a key intermediate in the synthesis of the Inhoffen-Lythgoe diol, rearranges to **133** in aqueous NaOH in 82% yield, while the corresponding methyl ester was recovered unreacted after prolonged heating in decalin at the same temperature. The rearrangements of **134** and **136** to **135** and **137** respectively, previously attempted without success, were also easy in aqueous medium (Scheme 5.44).

In the course of biomimetic total synthesis of gambogin, Nicolaou reported a dramatic rate acceleration in water of the Claisen rearrangement of **138** to **139** (Scheme 5.45).[104] In d_6-benzene at 100°C there was no detectable reaction, while in d_7-DMF and MeOH the rearrangement was complete in 30 min. The use of water (MeOH/H_2O 1:1) reduced the reaction time in half. The authors also report the first examples of a Claisen–Diels–Alder cascade sequence (**140** → **141** → **142**) accelerated by aqueous medium. In this case, the beneficial effect of water with respect to d_7-DMF was even greater.

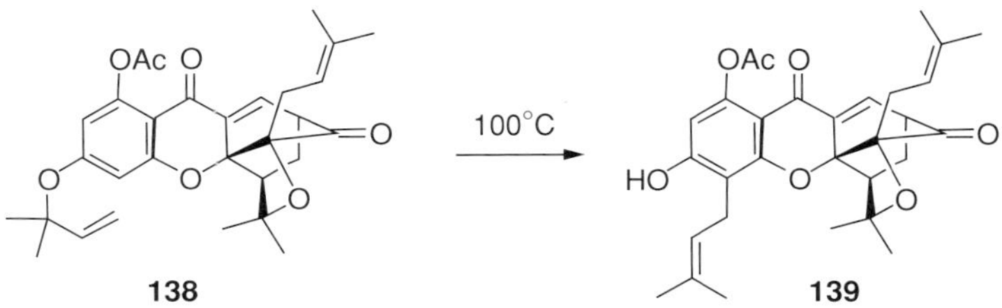

Medium	t (h)	C (%)
d_6–PhH	4	0
d_7–DMF	0.5	100
MeOH	0.5	100
MeOH/H_2O (1:1)	0.25	100

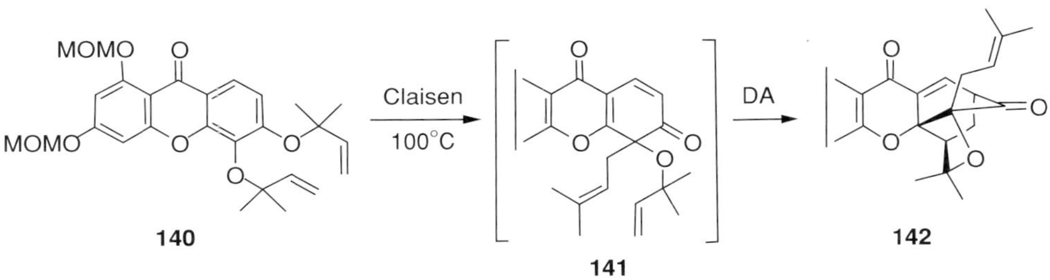

Medium	t (h)	C (%)
d_6–PhH	4	0
d_7–DMF	2	75
MeOH	1	100
MeOH/H_2O (1:1)	0.5	100

Scheme 5.45

References

1. (a) Gill, G.B. and Willis, M.R., *Perycíclic Reactions*, Chapman and Hall, London, 1974; (b) Houk, K.N., In: Scott, A.F. (Ed.), *Pericyclic Reactions and Orbital Symmetry in Survey of Progress in Chemistry*, Vol. 6, Academic Press, New York, 1973, pp. 113–208.
2. Diels, O. and Alder K., Synthesen in der hydroaromatischen reihe. I Mitteilung: anlagerungen von 'dien'-kohlen wasserstoffen,*Justus Liebigs Ann. Chem.*, 1928, **460**, 98–122.

3. (a) Paternò, E. and Chieffi, G., Synthesis in organic chemistry using light. Note II. Compounds of unsaturated hydrocarbons with aldehydes and ketones, *Gazz. Chim. Ital.*, 1909, **39**, 341–361; (b) Büchi, G., Inman, C.G. and Lipinsky, E.S., Light-catalyzed organic reactions. I. The reaction of carbonyl compounds with 2-methyl-2-butene in the presence of ultraviolet light, *J. Am. Chem. Soc.*, 1954, **76**, 4327–4331.
4. Cope, A.C. and Hardy, E.M., The introduction of substituted vinyl groups. V. A rearrangement involving the migration of an allyl group in a three-carbon system, *J. Am. Chem. Soc.*, 1940, **62**, 441–444.
5. Claisen, L., Üser umlangerung von phenol-allyläthern in c-allyl-phenole, *Berichte*, 1912, **45**, 3157–3166.
6. Fleming, I., *Frontier Orbitals and Organic Chemical Reactions*, Wiley, New York, 1976.
7. van Eldik, R., Asano, T. and Le Noble, W.J., Activation and reaction volumes in solution. 2, *Chem. Rev.*, 1989, **89**, 549–688.
8. Woodward, R.B. and Hoffmann, R., *The Conservation of Orbital Symmetry*, Academic Press, New York, 1970.
9. (a) Sauer, J. and Sustmann, R., Mechanistic aspects of Diels–Alder reactions: a critical survey, *Angew. Chem., Int. Ed. Engl.*, 1980, **19**, 779–807; (b) Cativiela, C., Garcia, J.I., Mayoral, J.A. and Salvatella, L., Modelling of solvent effects on the Diels–Alder reaction, *Chem. Soc. Rev.*, 1996, **25**, 209–218.
10. (a) Li, C.-J. and Chang, T.H., *Organic Reactions in Aqueous Media*, Wiley, New York, 1997; (b) Grieco, P.A. (Ed.),*Organic Synthesis in Water*, Blackie Academic Professional, London, 1998; (c) Fringuelli, F. and Taticchi, A., *The Diels–Alder Reaction: Selected Practical Methods*, Wiley, Chichester, UK, 2002.
11. (a) Fringuelli, F., Piermatti, O., Pizzo, F. and Vaccaro, L., Recent advances in Lewis acid catalyzed Diels–Alder reactions in aqueous media, *Eur. J. Org. Chem.*, 2001, **3**, 439–455; (b) Parker, D.T., In: Grieco, P.A. (Ed.),*Organic Synthesis in Water*, Blackie Academic Professional, London, 1998, p. 47; (c) Blokzijl, W. and Engberts, J.B.F.N., Hydrophobic effects: opinions and facts, *Angew. Chem., Int. Ed. Engl.*, 1993, **32**, 1545–1579; (d) Breslow, R., Hydrophobic effects on simple organic reactions in water, *Acc. Chem. Res.*, 1991, **24**, 159–164.
12. (a) Diels, O. and Alder, K., Synthesen der hydroaromatischen reihe. XII Mitteilung. ('dien-synthesen' sauerstoffhaltiger heteroringe. 2. dien-synthesen des furans), *Ann. Chem.*, 1931, **490**, 243–257; (b) Woodward, R.B. and Baer, H., The reaction of furan with maleic anhydride, *J. Am. Chem. Soc.*, 1948, **70**, 1161–1166; (c) Eggelte, T.A., De Koning, H. and Huisman, H.O., Diels–Alder reaction of furan with some dienophiles, *Tetrahedron*, 1973, **29**, 2491–2493.
13. (a) Hopff, H. and Rautenstrauch, C.W., Addition products of butadiene with vinyl methyl ketone, maleic acid, etc., US Patent 2262002, 1941; *Chem. Abstr.*, 1942, **36**, 10469; (b) Lane, L.C. and Parker, C.H., Adducts from fumaric acid and conjugated dienes, US Patent 2444263, 1948; *Chem. Abstr.*, 1948, **42**, 7102.
14. Carlson, B.A., Sheppard, W.A. and Webster, O.W., The reaction of diazonium salts with dienes. A route to pyridazines and pyridazinium salts, *J. Am. Chem. Soc.*, 1975, **97**, 5291–5293.
15. Rideout, D.C. and Breslow, R., Hydrophobic acceleration of Diels–Alder reactions, *J. Am. Chem. Soc.*, 1980, **102**, 7816–7817.
16. (a) Breslow, R., Maitra, U. and Rideout, D., Selective Diels–Alder reactions in aqueous solutions and suspensions, *Tetrahedron Lett.*, 1983, **24**, 1901–1904; (b) Ahamd-Zadeh Samii, A., de Savignac, A., Rico, I. and Lattes, A., Waterless microemulsions. IV. Diels–Alder reaction of cyclopentadiene and methylacrylate as a probe of formamide microemulsions,*Tetrahedron*, 1985, **41**, 3683–3688.
17. (a) Grieco, P.A., Organic chemistry in unconventional solvents, *Aldrichim. Acta*, 1991, **24**(3), 59–66; (b) Li, C.-J., Organic reactions in aqueous media – with a focus on carbon–carbon bond formation, *Chem. Rev.*, 1993, **93**, 2023–2035; (c) Lubineau, A., Augé, J. and Queneau, Y., Water-promoted organic reactions, *Synthesis*, 1994, 741–760; (d) Chan, T.H., Li, C.-J., Lee, M.C. and Wei, Z.Y., 1993 R.U. Lemieux Award Lecture. Organometallic-type reactions in aqueous media – a new challenge in organic synthesis, *Can. J. Chem.*, 1994, **72**, 1181–1192; (e) Kobayashi, S., Rare earth metal trifluoromethanesulfonates as water-tolerant Lewis acid catalysts in organic synthesis, *Synlett*, 1994, 689–701; (f) Engberts, J.B.F.N., Diels–Alder reactions in water: enforced hydrophobic interaction and hydrogen bonding, *Pure Appl. Chem.*, 1995, **67**, 823–828; (g) Engberts, J.B.F.N., Feringa, B.L., Keller, E. and Otto, S., Lewis acid catalysis of carbon–carbon bond forming reactions in water, *Recl. Trav. Chim. Pays-Bas*, 1996, **115**, 457–464; (h) Fringuelli, F., Piermatti, O. and Pizzo, F. In: Attanasi, A.O. and Spinelli, D. (Eds.), *Targets in Heterocyclic Systems*, Vol. 1, SCI, Rome, 1997, pp. 57–73; (i) Otto, S. and Engberts, J.B.F.N., Diels–Alder reactions in water, *Pure Appl. Chem.*, 2000, **72**, 1365–1372; (j) Kumar, A., Salt effects on Diels–Alder reaction kinetics, *Chem. Rev.*, 2001, **101**, 1–19; (k) Lindström, U.M., Stereoselective organic reactions in water, *Chem. Rev.*, 2002, **102**, 2751–2772.
18. (a) Robbins, J.B., Schneerson, R., Anderson, P. and Smith, D.H., The 1996 Albert Lasker Medical Research Awards. Prevention of systemic infections, especially meningitis, caused by *Haemophilus influenzae* type b. Impact on public health and implications for other polysaccharide-based vaccines, *J. Am. Med. Assoc.*, 1996, **276**, 1181–1185; (b) Hermanson, G.T., *Bioconjugate Techniques*, Academic Press, San Diego, CA, 1996.
19. (a) Pozsgay, V., Oligosaccharide–protein conjugates as vaccine candidates against bacteria, *Adv. Carbohydr. Chem. Biochem.*, 2001, **56**, 153–199; (b) Lee, Y.C. and Lee, R.T., *Methods in Enzimology*, Academic Press, San Diego, CA, 1994; (c) Davis, B.G., Recent developments in glycoconjugates, *J. Chem. Soc., Perkin Trans. 1*, 1999, 3215–3237; (d) Davis, B.G., Synthesis of glycoproteins, *Chem. Rev.*, 2002, **102**, 579–601.

20. (a) Pozsgay, V., Vieira, N.E. and Yergey, A., Method for bioconjugation of carbohydrates using Diels–Alder cycloaddition, *Org. Lett.*, 2002, **4**, 3191–3194; (b) Hill, K.W., Rigby, J.T., Carter, J.D., Kropp, E., Vagle, K., Pieken, W., McGee, D.P.C., Husar, G.M., Leuk, M., Anziano, D.J. and Sebesta, D.P., Diels–Alder bioconjugation of diene-modified oligonucleotides, *J. Org. Chem.*, 2001, **66**, 5352–5358.
21. (a) Northrup, A.B. and McMillan, D.W.C., The first general enantioselective catalytic Diels–Alder reaction with simple α,β-unsaturated ketones, *J. Am. Chem. Soc.*, 2002, **124**, 2458–2460; (b) Miyamoto, H., Kimura, T., Daikawa, N. and Tanaka, K., Preparation of optically active *cis*-4-methylcyclohex-4-ene-1,2-dicarboximides by a combination of Diels–Alder reaction and complexation with optically active host and enantioselective Diels–Alder reaction in inclusion crystals in a water suspension medium, *Green Chem.*, 2003, **5**, 57–59; (c) Lakner, F.J. and Negrete, G.R., A new and convenient chiral auxiliary for asymmetric Diels–Alder cycloadditions in environmentally benign solvents, *Synlett*, 2002, 643–645.
22. Amantini, D., Fringuelli, F., Piermatti, O., Pizzo, F. and Vaccaro, L., 3-Nitrocoumarins as dienophiles in the Diels–Alder reaction in water an approach to the synthesis of nitrotetrahydrobenzo[*c*]chromenones and dihydrodibenzo[*b,d*]furans, *J. Org. Chem.*, 2003, **68**, 9263–9268.
23. (a) Amantini, D., Fringuelli, F. and Pizzo, F., Uncatalyzed [4 + 2] cycloadditions of 3-nitrocoumarins with vinyl ethers in solventless conditions. A new entry to chromene derivatives, *J. Org. Chem.*, 2002, **67**, 7238–7243; (b) Attanasi, O.A., De Crescentini, L., Filippone, P., Fringuelli, F., Mantellini, F., Matteucci, M., Piermatti, O. and Pizzo, F., Inverse-electron-demand Diels–Alder reactions of (*E*)-3-diazenylbut-2-enes in water, *Helv. Chim. Acta*, 2001, **84**, 513–525; (c) Fringuelli, F., Matteucci, M., Piermatti, O., Pizzo, F. and Burla, M.C., [4 + 2] Cycloadditions of nitroalkenes in water. Highly asymmetric synthesis of functionalized nitronates, *J. Org. Chem.*, 2001, **66**, 4661–4666.
24. Wittkopp, A. and Schreiner, P.R., Metal-free, noncovalent catalysis of Diels–Alder reactions by neutral hydrogen bond donors in organic solvents and in water, *Chem. Eur. J.*, 2003, **9**, 407–414.
25. (a) Otto, S. and Engberts, J.B.F.N., Lewis-acid catalysis of a Diels–Alder reaction in water, *Tetrahedron Lett.*, 1995, **36**, 2645–2648; (b) Otto, S., Bertoncin, F. and Engberts, J.B.F.N., Lewis acid catalysis of a Diels–Alder reaction in water, *J. Am. Chem. Soc.*, 1996, **118**, 7702–7707; (c) Rispens, T. and Engberts, J.B.F.N., Micellar catalysis of Diels–Alder reactions: substrate positioning in the micelle, *J. Org. Chem.*, 2002, **67**, 7369–7377.
26. (a) Yang, Y. and Chan, T.H., Organic reactions in aqueous media. Cyclopentadienylindium(I) as the first example of organoindium(i) reagent for carbon–carbon bond formation and the demostration of one-pot tandem addition/intramolecular Diels–Alder reaction in aqueous media, *J. Am. Chem. Soc.*, 2000, **122**, 402–403; (b) Yamazaki, O., Hao, X., Yoshida, A. and Nishikido, J., Fluorous reverse-phase silica gel-supported Lewis acids as recyclabe catalysts in water, *Tetrahedron Lett.*, 2003, **44**, 8791–8795; (c) Chen, I.-H., Young, J.-N. and Yu, S.J., Recyclable organotungsten Lewis acid and microwave assisted Diels–Alder reactions in water and in ionic liquids, *Tetrahedron*, 2004, **60**, 11903–11909.
27. Tucker, C.J., Welker, M.E., Day, C.S. and Wright, M.W., Preparation of an aquocobaloxime 1,3-dienyl complex and its Diels–Alder reactions in water and organic solvents, *Organometallics*, 2004, **23**, 2257–2262.
28. (a) Shaw, R.W., Brill, T.B., Clifford, A.A., Eckert, C.A. and Franck, E.U., Supercritical water. A medium for chemistry, *Chem. Eng. News*, 1991, **69**(51), 26–39; (b) Clifford, A.A., *Fundamentals Supercritical Fluids*, Oxford University Press, Oxford, UK, 1998; (c) Ikushima, Y., Noncatalytic synthetic reactions using supercritical water: the implications of the unique reactivities based on the nature of supercritical water, *Rec. Res. Dev. Chem. Phys.*, 2000, **1**, 123–131; (d) Kuhlmann, B., Arnett, E.M. and Siskin, M., Classical organic reactions in pure superheated water, *J. Org. Chem.*, 1994, **59**, 3098–3101; (e) Katritzky, A.R., Allin, S.M. and Siskin, M., Aquathermolysis: reactions of organic compounds with superheated water, *Acc. Chem. Res.*, 1996, **29**, 399–406.
29. Korzenski, M.B. and Kolis, J.W., Diels–Alder reactions using supercritical water as an aqueous solvent medium, *Tetrahedron Lett.*, 1997, **38**, 5611–5614.
30. (a) Drauz, K. and Waldmann, H. (Eds.), *Enzyme Catalysis in Organic Synthesis*, 2nd edn., Wiley-VCH, Weinheim, Germany, 2002; (b) Santaniello, E., Ferraboschi, P., Grisenti, P. and Manzocchi, A., The biocatalytic approach to the preparation of enantiomerically pure chiral building blocks, *Chem. Rev.*, 1992, **92**, 1071–1140; (c) Grupta, M.N. and Roy, I., Enzymes in organic media. Forms, functions and applications, *Eur. J. Biochem.*, 2004, 2575–2583; (d) Kim, S.P., Leach, A.G. and Houk, K.N., The origins of noncovalent catalysis of intermolecular Diels-Alder reactions by cyclodextrins, self-assembling capsules, antibodies, and RNAses, *J. Org. Chem.*, 2002, **67**, 4250–4260.
31. (a) Diego-Castro, M.J. and Hailes, H.C., Studies on the use of surfactants in aqueous Diels–Alder reactions, *Tetrahedron Lett.*, 1998, **39**, 2211–2214; (b) Itami, K., Nokami, T. and Yoshida, J.-I., 2-Pyridylmethylsilyl as a removable hydrophilic group in aqueous Diels–Alder reactions, *Angew. Chem. Int. Ed.*, 2001, **40**, 1074–1076; (c) Itami, K., Nokami, T. and Yoshida, J.-I., 2-Pyridyldimethylsilyl group as a removable hydrophilic group in aqueous organic reactions: formation of molecular aggregates and dramatic rate enhancement in Diels–Alder reactions, *Adv. Synth. Catal.*, 2002, **344**, 441–451; (d) Kusukawa, T., Nakai, T., Okano, T. and Fujita, M., Remarkable acceleration of Diels–Alder reactions in a self-assembled coordination cage, *Chem. Lett.*, 2003, **32**, 284–285.

32. (a) Colonna, S., Manfredi, A. and Annunziata, R., Changes of stereoselectivity and rate in Diels–Alder reactions by hydrophobic solvent effects and by bovine serum albumin, *Tetrahedron Lett.*, 1988, **29**, 3347–3350; (b) Rama Rao, K., Srinivasan, T.N. and Bhanumathi, N., A stereoselective biocatalytic Diels–Alder reaction, *Tetrahedron Lett.*, 1990, **31**, 5959–5960.
33. (a) Oikawa, H., Katayama, K., Suzuki, Y. and Ichihara, A., Enzymatic activity catalysing exo-selective Diels–Alder reaction in solanapyrone biosynthesis, *J. Chem. Soc., Chem. Commun.*, 1995, 1321–1322; (b) Oikawa, H., Kobayashi, T., Katayama, K., Suzuki, Y. and Ichihara, A., Total synthesis of (−)-solanapyrone a via enzymatic Diels–Alder reaction of prosolanapyrone, *J. Org. Chem.*, 1998, **63**, 8748–8756; (c) Oikawa, H., Suzuchi, Y., Katayama, K., Naya, A., Sakano, C. and Ichihara, A., Involvement of Diels–Alder reactions in the biosynthesis of secondary natural products: the late stage of the biosynthesis of phytotoxins solanapyrones, *J. Chem. Soc., Perkin Trans.* 1, 1999, 1225–1232; (d) Katayama, K., Kobayashi, T., Oikawa, H., Honma, M. and Ichiara, A., Enzymatic activity and partial purification of solanapyrone synthase: first enzyme catalyzing Diels–Alder reaction, *Biochim. Biophys. Acta*, 1998, **1384**, 387–395.
34. Auclair, K., Sutherland, A., Kennedy, J., Witter, D.J., Van den Heever, J.P., Hutchinson, C.R. and Vederas, J.C., Lovastatin nonaketide synthase catalyzes an intramolecular Diels–Alder reaction of substrate analogue, *J. Am. Chem. Soc.*, 2000, **122**, 11519–11520.
35. Watanabe, K., Mie, T., Ichihara, A., Oikawa, H. and Honma, M., Detailed reaction mechanism of macrophomate synthase, *J. Biol. Chem.*, 2000, **275**, 38393–38401.
36. (a) Johansson, M., Köpcke, B., Anke, H. and Sterner, O., Synthesis of (−)-pregaliellalactone, conversion of (−)-pregaliellalactone to (−)-galiellalactone by mycelia of *Galiella rufa*, *Tetrahedron*, 2002, **58**, 2523–2528; (b) Johansson, M., Köpcke, B., Anke, H. and Sterner, O., Cyclization of (−)-pregaliellalactone in fungus *Galiella rufa*, *Angew. Chem. Int. Ed.*, 2002, **41**, 2158–2160.
37. Jencks, W.P., *Catalysis in Chemistry and Enzymology*, McGraw-Hill, New York, 1969.
38. (a) Avalle, B., Zanin, V., Thomas, D. and Friboulet, A., Antibody catalysis based on functional mimicry, *Appl. Biochem. Biotech.*, 1998, **75**, 3–11; (b) Hasserodt, J., Organic synthesis supported by antibody catalysis, *Synlett*, 1999, 2007–2022.
39. (a) Gouverneur, V.E., Houk, K.N., de Pascual-Teresa, B., Beno, B., Janda, K.D. and Lerner, R.A., Control of the exo and endo pathways of the Diels–Alder reaction by antibody catalysis, *Science*, 1993, **262**, 204–208; (b) Heine, A., Stura, E.A., Yli-Kauhaluoma, J.T., Gao, C., Deng, Q., Beno, B.R., Houk, K.N., Janda, K.D. and Wilson, I.A., An antibody exo Diels-Alderase inhibitor complex at 1.95 angstrom resolution, *Science*, 1998, **279**, 1934–1940; (c) Yli-Kauhaluoma, J.T., Ashley, J.A., Lo, C.-H., Tucker, L., Wolfe, M.M. and Janda, K.D., Anti-metallocene antibodies: a new approach to enantioselective catalysis of the Diels–Alder reaction, *J. Am. Chem. Soc.*, 1995, **117**, 7041–7047.
40. (a) Romesberg, F.E., Spiller, B., Schultz, P.G. and Stevens, R.C., Immunological origins of binding and catalysis in a Diels-Alderase antibody, *Science*, 1998, **279**, 1929–1933; (b) Braisted, A.C. and Schultz, P.G., An antibody-catalyzed bimolecular Diels–Alder reaction, *J. Am. Chem. Soc.*, 1990, **112**, 7430–7431.
41. (a) Seelig, B., Keiper, S., Stuhlmann, F. and Jäschke, A., Enantioselective ribozyme catalysis of a biomolecular cycloaddition reaction, *Angew. Chem. Int. Ed.*, 2000, **39**, 4576–4579; (b) Unrau, P.J. and Bartel, D.P., RNA-catalysed nucleotide synthesis, *Nature*, 1998, **395**, 260–263; (c) Seelig, B. and Jäschke, A., A small catalytic RNA motif with Diels-Alderase activity, *Chem. Biol.*, 1999, **6**, 167–176.
42. (a) Kibayashi, C. and Aoyagi, S., Nitrogenous natural products synthesis via *n*-acylnitroso Diels–Alder methodology, *Synlett*, 1995, 873–879; (b) Naruse, M., Aoyagi, S. and Kibayashi, C., Enhanced stereoselectivity in aqueous intramolecular hetero Diels–Alder cycloaddition of chiral acylnitroso compounds, *Tetrahedron Lett.*, 1994, **35**, 595–598; (c) Naruse, M., Aoyagi, S. and Kibayashi, C., New chiral route to (−)-swainsonine via an aqueous acylnitroso cycloaddition approach, *J. Org. Chem.*, 1994, **59**, 1358–1364; (d) Naruse, M., Aoyagi, S. and Kibayashi, C., Stereoselective total synthesis of (−)-pumiliotoxin C by an aqueous intramolecular acylnitroso Diels–Alder approach, *J. Chem. Soc., Perkin Trans. 1*, 1996, 1113–1124, corrigendum 2077; (e) Mineno, T. and Miller, M.J., Stereoselective total synthesis of racemic BCX-1812 (RWJ-270201) for the development of neuraminidase inhibitors as anti-influenza agents, *J. Org. Chem.*, 2003, **68**, 6591–6596; (f) Zhang, D., Süling, C. and Miller, M.J., The hetero Diels–Alder reactions between D-mannose-derived halonitroso compounds and cyclopentadiene: scope and limitations, *J. Org. Chem.*, 1998, **63**, 885–888; (g) Mulvihill, M.J., Gage, J.L. and Miller, M.J., Enzymatic resolution of aminocyclopentenols as precursors to D- and L-carbocyclic nucleosides, *J. Org. Chem.*, 1998, **63**, 3357–3363.
43. (a) Manabe, K., Mori, Y. and Kobayashi, S., Three-component carbon–carbon bond-forming reactions catalyzed by a Brønsted acid-surfactant-combined catalyst in water, *Tetrahedron*, 2001, **57**, 2537–2544; (b) Akiyama, T., Matsuda, K. and Fuchibe, K., Montmorillonite K10-catalyzed aza Diels–Alder reaction of Danishefsky's diene with aldimines, generated *in situ* from aliphatic aldehydes and amine, in aqueous media, *Synlett*, 2002, 1898–1900; (c) Loncaric, C., Manabe, K. and Kobayashi, S., Alkaline salt-catalyzed aza Diels-Alder reactions of danishefsky document s diene with imines in water under neutral conditions, *Chem. Commun.*, 2003, 574–575; (d) Chen, L., Li, Z. and Li, C.-J., Indium-mediated domino reaction of nitroarenes with 2,3-dihydrofuran in water: an efficient synthesis of 1,2,3,4-tetrahydroquinoline derivatives, *Synlett*, 2003, 732–734.

44. (a) Denmark, S.E. and Marcin, L.R., Asymmetric nitroalkene [4 + 2] cycloadditions: enantioselective synthesis of 3-substituted and 3,4-disubstituted pyrrolidines, *J. Org. Chem.*, 1995, **60**, 3221–3235; (b) Denmark, S.E. and Dixon, J.A., Tandem inter [4 + 2]/intra [3 + 2] cycloadditions of nitroalkenes. Asymmetric synthesis of highly functionalized aminocyclopentanes using the bridged mode (β-tether) process, *J. Org. Chem.*, 1998, **63**, 6178–6195; (c) Denmark, S.E. and Hurd, A.R., Tandem [4 + 2]/[3 + 2] cycloadditions with nitroethylene, *J. Org. Chem.*, 1998, **63**, 3045–3050.
45. Denmark, S.E., Dappen, M.S. and Cramer, C.J., Intramolecular [4 + 2] cycloadditions of nitroalkenes with olefins, *J. Am. Chem. Soc.*, 1986, **108**, 1306–1307.
46. Wijnen, J.W., Zavarise, S., Engberts, J.B.F.N. and Charton, M., Substituent effects on an inverse electron demand hetero Diels–Alder reaction in aqueous solution and organic solvents: cycloaddition of substituted styrenes to di(2-pyridyl)-1,2,4,5-tetrazine, *J. Org. Chem.*, 1996, **61**, 2001–2005.
47. (a) Breslow, R. and Connors, R.V., Quantitative antihydrophobic effects as probes for transition state structures. 1. Benzoin condensation and displacement reactions, *J. Am. Chem. Soc.*, 1995, **117**, 6601–6602; (b) Breslow, R. and Zhu, Z., Quantitative antihydrophobic effects as probes for transition state structures. 2. Diels–Alder reactions, *J. Am. Chem. Soc.*, 1995, **117**, 9923–9924; (c) Breslow, R., Connors, R. and Zhu, Z., Mechanistic studies using antihydrophobic agents, *Pure Appl. Chem.*, 1996, **68**, 1527–1534.
48. (a) Otto, S., Engberts, J.B.F.N. and Kwak, I.C.T., Million-fold acceleration of a Diels–Alder reaction due to combined Lewis acid and micellar catalysis in water, *J. Am. Chem. Soc.*, 1998, **120**, 9517–9525; (b) Otto, S., Boccaletti, G. and Engberts, J.B.F.N., A chiral Lewis-acid-catalyzed Diels–Alder reaction. Water-enhanced enantioselectivity, *J. Am. Chem. Soc.*, 1998, **120**, 4238–4239; (c) Otto, S. and Engberts, J.B.F.N., A systematic study of ligand effects on a Lewis-acid-catalyzed Diels–Alder reaction in water. Water-enhanced enantioselectivity, *J. Am. Chem. Soc.*, 1999, **121**, 6798–6806; (d) Meijer, A., Otto, S. and Engberts, J.B.F.N., Effects of the hydrophobicity of the reactants on Diels–Alder reactions in water, *J. Org. Chem.*, 1998, **63**, 8989–8994.
49. (a) Pagni, R.M., Kabalka, G.W., Hondrogiannis, G., Bains, S., Anosike, P. and Kurt, R., The cycloaddition reactions of unsaturated esters with cyclopentadiene on γ-alumina, *Tetrahedron*, 1993, **49**, 6743–6756; (b) Blake, J.F., Lim, D. and Jörgensen, W.L., Enhanced hydrogen bonding of water to Diels–Alder transition states. Ab initio evidence, *J. Org. Chem.*, 1994, **59**, 803–805; (c) Pawar, S.S., Phalgune, U. and Kumar, A., Ionic effects on the stereoselectivities of Diels–Alder reactions: the cyclopentadiene–methyl acrylate reaction in aqueous LiCl, NaCl, NaBr, $CaCl_2$, $LiClO_4$, and guanidinium chloride, *J. Org. Chem.*, 1999, **64**, 7055–7060; (d) Kumar, A. and Deshpande, S.S., Exploring the re-evaluated interpretation of hydrophobicity to explain rates of Diels–Alder reactions in aqueous salt solutions, *J. Phys. Org. Chem.*, 2002, **15**, 242–245; (e) Kodaka, M., Reevaluation in interpretation of hydrophobicity by scaled particle theory, *J. Phys. Chem. B*, 2001, **105**, 5592–5594; (f) Graziano, G., Rate enhancement of Diels–Alder reactions in aqueous solutions, *J. Phys. Org. Chem.*, 2004, **17**, 100–101; (g) Blokzijl, W. and Engberts, J.B.F.N., Initial-state and transition-state effects on Diels–Alder reactions in water and mixed aqueous solvents, *J. Am. Chem. Soc.*, 1992, **114**, 5440–5442.
50. (a) Desimoni, G., Faita, G., Righetti, P., Tornaletti, N. and Visigalli, M., Solvent effects resulting from frontier molecular-orbital interactions. Part 2. The Diels–Alder reaction between 5-substituted 1,4-naphthoquinones and 2,3-dimethylbutadiene, *J. Chem. Soc., Perkin Trans 2*, 1989, 437–441; (b) Desimoni, G., Faita, G., Pasini, D. and Righetti, P.P., Solvent effect as the result of frontier molecular orbital interaction. VII. The retro-Diels–Alder reaction, *Tetrahedron*, 1992, **48**, 1667–1674.
51. (a) Yoshida, K. and Grieco, P.A., Aqueous intermolecular Diels–Alder chemistry: vernolepin revisited, *J. Org. Chem.*, 1984, **49**, 5257–5260; (b) Grieco, P.A., Garner, P. and He, Z., 'Micellar' catalysis in the aqueous intermolecular Diels–Alder reaction: rate acceleration and enhanced selectivity, *Tetrahedron Lett.*, 1983, **24**, 1897–1900; (c) Grieco, P.A., Yoshida, K. and Garner, P., Aqueous intermolecular Diels–Alder chemistry: reactions of diene carboxylates with dienophiles in water at ambient temperature, *J. Org. Chem.*, 1983, **48**, 3137–3139; (d) Grieco, P.A., Galatsis, P. and Spohn, R.F., Aqueous intermolecular Diels–Alder chemistry: reaction of (*E*)-2,4-pentadienyl ammonium chloride and related ammonium salts with dienophiles in water, *Tetrahedron*, 1986, **42**, 2847–2853.
52. (a) Gajewsky, J.J., A semitheoretical multiparameter approach to correlate solvent effects on reactions and equilibria, *J. Org. Chem.*, 1992, **57**, 5500–5506; (b) Cativiela, C., García, J.I., Gil, J., Martínez, R.M., Mayoral, J.A., Salvatella, L., Urieta, J.S., Mainar, A.M. and Abraham, M.H., Solvent effects on Diels–Alder reactions. The use of aqueous mixtures of fluorinated alcohols and the study of reactions of acrylonitrile, *J. Chem. Soc., Perkin Trans. 2*, 1997, 653–660.
53. (a) Breslow, R., Groves, K. and Mayer, M.U., Antihydrophobic cosolvent effects in organic displacement reactions, *Org. Lett.*, 1999, **1**, 117–120; (b) Breslow, R., In: Cramer, C.J. and Truhlar, D.G. (Eds.), *Structure and Reactivity in Aqueous Solution*, Vol. 568, American Chemical Society, Washington, DC, 1994, p. 291.
54. (a) Kong, S. and Evanseck, J.D., Density functional theory study of aqueous-phase rate acceleration and endo/exo selectivity of butadiene and acrolein Diels–Alder reaction, *J. Am. Chem. Soc.*, 2000, **122**, 10418–10427; (b) Chandrasekhar, J., Shariffskul, S. and Jorgensen, W.L., QM/MM simulations for Diels–Alder reactions in water: contribution of enhanced hydrogen bonding at the transition state to the solvent effect, *J. Phys. Chem. B*, 2002, **106**, 8078–8085.

55. (a) Furlani, T.R. and Gao, J., Hydrophobic and hydrogen-bonding effects on the rate of Diels–Alder reactions in aqueous solution, *J. Org. Chem.*, 1996, **61**, 5492–5497; (b) Jörgensen, W.L., Blake, J.F., Lim, D. and Severance, D.L., Investigation of solvent effects on pericyclic reactions by computer simulations, *J. Chem. Soc., Faraday Trans.*, 1994, **90**, 1727–1732.
56. Narayan, S., Muldoon, J., Finn, M.G., Fokin, V.V., Kolb, H.C. and Sharpless, K.B., 'On water': unique reactivity of organic compounds in aqueous suspension, *Angew. Chem. Int. Ed.*, 2005, **44**, 3275–3279.
57. For a long time our group has been working on organic reactions carried out in pure water. Most of them occurred under heterogeneous conditions and some of them were carried out in 'on water conditions': see Refs 22 and 23 and (a) Brufola, G., Fringuelli, F., Piermatti, O. and Pizzo, F., Simple and efficient one-pot preparation of 3-substituted coumarins in water, *Heterocycles*, 1996, **43**, 1257–1266; (b) Brufola, G., Fringuelli, F., Piermatti, O. and Pizzo, F., Efficient one-pot synthesis of 7-azacoumarins by Knoevenagel reaction using water as reaction medium, *Heterocycles*, 1997, **45**, 1715–1721; (c) Fringuelli, F., Piermatti, O. and Pizzo, F., One-pot synthesis of 3-carboxycoumarins via consecutive Knoevenagel and Pinner reactions in water, *Synthesis*, 2003, **15**, 2331–2334; (d) Fringuelli, F., Piermatti, O. and Pizzo, F.,One-pot synthesis of 7-hydroxy-3-carboxycoumarin in water, *J. Chem. Educ.*, 2004, **81**, 874–876.
58. (a) Huisgen, R., In: Padwa, A. (Ed.), *1,3-Dipolar Cycloaddition Chemistry*, Wiley, New York, 1984; (b) Gothelf, K.V. and Jörgensen, K.A., Asymmetric 1,3-dipolar cycloaddition reactions, *Chem. Rev.*, 1998, **98**, 863–909; (c) Mulzer, J., *Organic Synthesis Highlights*, Wiley-VCH, Weinheim, Germany, 1991, pp. 77–95; (d) Kadaba, P.K., Role of protic and dipolar aprotic solvents in heterocyclic syntheses via 1,3-dipolar cycloaddition reactions, *Synthesis*, 1973, 71–84.
59. Dignam, K.J., Hegarty, A.F. and Quain, P.L., Reactivity of 1,3-dipoles in aqueous solution. 2. Stereospecific reactions of benzonitrile oxides with oxygen, carbon, and nitrogen nucleophiles, *J. Org. Chem.*, 1978, **43**, 388–393.
60. Rama Rao, K., Bhanumathi, N., Srinivasan, T.N. and Sattur, P.B., A regioselective enzyme catalyzed cycloaddition, *Tetrahedron Lett.*, 1990, **31**, 899–902.
61. Rama Rao, K., Bhanumathi, N. and Sattur, P.B., Baker's yeast catalyzed asymmetric cycloaddition of nitrileoxides to C=C bond: improved chiral recognition by using β-cyclodextrin, *Tetrahedron Lett.*, 1990, **31**, 3201–3204.
62. Inoue, Y., Araki, K. and Shiraishi, S., Unusual solvent effect on the cycloadditions of aromatic nitrile *n*-oxide with alkyl-substituted *p*-benzoquinones in ethanol–water systems, *Bull. Chem. Soc. Jpn.*, 1991, **64**, 3079–3083.
63. Rohloff, J.C., Robinson, J., III and Gardner, J.O., Bromonitrile oxide [3+2] cycloadditions in water, *Tetrahedron Lett.*, 1992, **33**, 3113–3116.
64. Deroose, F.D. and De Clercq, P.J., A novel enantioselective synthesis of (+)-biotin, *Tetrahedron Lett.*, 1994, **35**, 2615–2618.
65. (a) Butler, R.N., Farrell, D.M. and Pyne, C.S., Reactions of phthalazinium-2-(unsubstituted methanide) and -2-dicyanometamide 1,3-dipoles with alkene and alkyne dipolarophiles: stereochemistry and secondary reactions – substituted pyrrolo[2,1-*a*]phthalazines, *J. Chem. Res. (S)*, 1996, 418–419; (b) Butler, R.N., Coyne, A.G., Cunningham, W.J. and Burke, L.A., Kinetic and synthetic influences of water and solvent-free conditions on 1,3-dipolar cycloaddition reactions: the phthalazinium and pyridazinium dicyanomethanide 1,3-dipoles – surprisingly successful synthetic methods, *J. Chem. Soc., Perkin Trans. 2*, 2002, 1807–1815; (c) Butler, R.N., Cunningham, W.J., Coyne, A.G. and Burke, L.A., The influence of water on the rates of 1,3-dipolar cycloaddition reactions: trigger points for exponential rate increases in water–organic solvent mixtures. Water-super versus water-normal dipolarophiles, *J. Am. Chem. Soc.*, 2004, **126**, 11923–11929.
66. Nishiwaki, N., Uehara, T., Asaka, N., Tohda, Y., Ariga, M. and Kanemasa, S., Improved generation method for functionalized nitrile oxide, *Tetrahedron Lett.*, 1998, **39**, 4851–4852.
67. van Mersbergen, D., Wijnen, J.W. and Engberts, J.B.F.N., 1,3-Dipolar cycloadditions of benzonitrile oxide with various dipolarophiles in aqueous solutions: a kinetic study, *J. Org. Chem.*, 1998, **63**, 8801–8805.
68. Rispens, T. and Engberts J.B.F.N., A kinetic study of 1,3-dipolar cycloadditions in micellar media, *J. Org. Chem.*, 2003, **68**, 8520–8528.
69. Gholami, M.R. and Yangjeh, A.H., Hydrophobic effects in 1,3-dipolar cycloaddition of *c*,*n*-diphenylnitrone with dibutyl fumarate in aqueous solutions, *J. Chem. Res. (S)*, 1999, 226–227.
70. Jiang, N. and Li, C.-J., Novel 1,3-dipolar cycloaddition of diazocarbonyl compounds to alkynes catalyzed by $InCl_3$ in water, *Chem. Commun.*, 2004, 394–395.
71. (a) Molteni, G., Orlandi, M. and Broggini, G., Nitrilimine cycloadditions in aqueous media, *J. Chem. Soc., Perkin Trans. 1*, 2000, 3742–3745; (b) Molteni, G., Ponti, A. and Orlandi, M., Uncommon aqueous media for nitrilimine cycloadditions. I. Synthetic and mechanistic aspects in the formation of 1-aryl-5-substituted-4,5-dihydropyrazoles, *New. J. Chem.*, 2002, **26**, 1340–1345; (c) Ponti, A. and Molteni, G., Uncommon aqueous media for nitrilimine cycloadditions. II. Computational study of the effect of water on reaction rate, *New. J. Chem.*, 2002, **26**, 1346–1351.
72. Rostovtsev, V.V., Green, L.G., Fokin, V.V. and Sharpless, K.B., A stepwise Huisgen cycloaddition process: copper(I)-catalyzed regioselective 'ligation' of azides and terminal alkynes, *Angew. Chem. Int. Ed.*, 2002, **41**, 2596–2599.

73. Li, Z., Seo, T.S. and Ju, J., 1,3-Dipolar cycloaddition of azides with electron-deficient alkynes under mild condition in water, *Tetrahedron Lett.*, 2004, **45**, 3143–3146.
74. Sivakumar, K., Xie, F., Cash, B.M., Long, S., Barnhill, H.N. and Wang, Q., A fluorogenic 1,3-dipolar cycloaddition reaction of 3-azidocoumarins and acetylenes, *Org. Lett.*, 2004, **6**, 4603–4606.
75. Xia, Y., Qu, F., Li, W., Wu, Q. and Peng, L., Synthesis of bitriazolyl compounds via Huisgen reaction, *Heterocycles*, 2005, **65**, 345–352.
76. Lewis, W.G., Green, L.G., Grynszpan, F., Radić, Z., Carlier, P.R., Taylor, P., Finn, M.G. and Sharpless, K.B., Click chemistry *in situ*: acetylcholinesterase as a reaction vessel for the selective assembly of a femtomolar inhibitor from an array of building blocks, *Angew. Chem. Int. Ed.*, 2002, **41**, 1053–1057.
77. Wijnen, J.W., Steiner, R.A. and Engberts, J.B.F.N., 1,3-Dipolar cycloaddition of phenyl azide to norbornene in aqueous solutions, *Tetrahedron Lett.*, 1995, **36**, 5389–5392.
78. Kacprzak, K., Efficient one-pot synthesis of 1,2,3-triazoles from benzyl and alkyl halides, *Synlett*, 2005, 943–946.
79. Himo, F., Demko, Z.P., Noodleman, L. and Sharples, K.B., Mechanism of tetrazole formation by addition of azide to nitriles, *J. Am. Chem. Soc.*, 2002, **124**, 12210–12216.
80. Demko, Z.P. and Sharpless, K.B., Preparation of 5-substituted 1*H*-tetrazoles from nitriles in water, *J. Org. Chem.*, 2001, **66**, 7945–7950.
81. Demko, Z.P. and Sharpless, K.B., An expedient route to the tetrazole analogues of α-amino acids, *Org. Lett.*, 2002, **4**, 2525–2527.
82. Nikolava, R.D., Vayssilov, G.N., Rodios, N. and Bojilova, A., Regio- and stereoselective [2+2] photodimerization of 3-substituted 2-alkoxy-2-oxo-2*H*-1,2-benzoxaphosphorines, *Molecules*, 2002, **7**, 420–432.
83. Ramamurthy, V., Organic photochemistry in organized media, *Tetrahedron*, 1986, **42**, 5753–5839.
84. Syamala, M.S. and Ramamurthy, V., Consequences of hydrophobic association in photoreactions: photodimerization of stilbenes in water, *J. Org. Chem.*, 1986, **51**, 3712–3715.
85. Ito, Y., Kajita, T., Kunimoto, K. and Matsuura, T., Accelerated photodimerization of stilbenes in methanol and water, *J. Org. Chem.*, 1989, **54**, 587–591.
86. Muthuramu, K. and Ramamurthy, V., Photodimerization of coumarin in aqueous and micellar media, *J. Org. Chem.*, 1982, **47**, 3976–3979.
87. (a) Lee, K.H. and de Mayo, P., Biphasic photochemistry: photochemical regiospecificity and critical micelle concentration determination, *Photochem. Photobiol.*, 1980, **31**, 311–314; (b) Lee, K.H. and de Mayo, P., Biphasic photochemistry: micellar regioselectivity in enone dimerisation, *J. Chem. Soc., Chem. Commun.*, 1979, 493–495.
88. (a) Fargues, R., Maurette, M.-T., Oliveros, E., Rivière, M. and Lattes, A., Réactivité chimique et photochimique dans les milieux micellaires etles microemulsions. Photodimerisation de l'isophorone, *Nouv. J. Chem.*, 1979, **3**, 487–489; (b) Fargues, R., Maurette, M.T., Oliveros, E., Riviere, M. and Lattes, A., Chemical and photochemical reactivity in micellar media and microemulsions. IV. Concentration effects on isophorone dimerization, *J. Photochem.*, 1982, **18**, 101–107.
89. Ramesh, V. and Ramamurthy, V., Control of regiochemistry in photodimerization through micellar preorientational effect: 2-substituted naphthalenes, *J. Org. Chem.*, 1984, **49**, 536–539.
90. Nakamura, Y., Kato, T. and Morita, Y., A micellar alignment effect in the photodimerizations of *n*-ω-carboxyalkyl-2-pyridones and their 4-alkyl derivatives in micelle or reversed micelle, *Tetrahedron Lett.*, 1981, **22**, 1025–1028.
91. (a) de Mayo, P. and Sydnes, L.K., Biphasic photochemistry: micellar control of regioselectivity in enone photoannelations, *J. Chem. Soc., Chem. Commun.*, 1980, 994–995; (b) Berenjian, N., de Mayo, P., Sturgeon, M.-E., Sydnes, L.K. and Weedon, A.C., Biphasic photochemistry: micelle solutions as media for photochemical cycloadditions of enones, *Can. J. Chem.*, 1982, **60**, 425–436.
92. Tamaki, T. and Kokubu, T., Acceleration of the photodimerization of water-soluble anthracenes included by β- and γ-cyclodextrins, *J. Inclusion Phenom.*, 1984, **2**, 815–822.
93. Chung W.-S., Turro, N.J., Silver, J. and le Noble, W.J., Modification of face selectivity by inclusion in cyclodextrins, *J. Am. Chem. Soc.*, 1990, **112**, 1202–1205.
94. Toda, F., Miyamoto, H. and Kikuchi, S., Enantioselective intramolecular [2+2] photocycloaddition reaction of *n*-allyl-3-oxo-1-cyclohexenecarboxamides in inclusion crystals with chiral host compounds, *J. Chem. Soc., Chem. Commun.*, 1995, 621–622.
95. (a) Blechert, S., The hetero-cope rearrangement in organic synthesis, *Synthesis*, 1989, 71–82; (b) Bartlett, P.A., Stereocontrol in the synthesis of acyclic systems: applications to natural product synthesis, *Tetrahedron*, 1980, **36**, 2–72.
96. Andrews, P.R., Smith, G.D. and Young, I.G., Transition-state stabilization and enzymic catalysis. Kinetic and molecular orbital studies of the rearrangement of chorismate to prephenate, *Biochemistry*, 1973, **12**, 3492–3498.
97. (a) Gajewski, J.J., Jurayj, J., Kimbrough, D.R., Gande, M.E., Ganem, B. and Carpenter, B.K., On the mechanism of rearrangement of chorismic acid and related compounds, *J. Am. Chem. Soc.*, 1987, **109**, 1170–1186; (b) Ganem, B., The mechanism of the Claisen rearrangement: déjà vu all over again, *Angew. Chem., Int. Ed.*

Engl., 1996, **35**, 936–945; (c) Repasky, M.P., Guimarães, C.R.W., Chandrasekhar, J., Rives, J.T. and Jörgensen, W.L., Investigation of solvent effects for the Claisen rearrangement of chorismate to prephenate: mechanistic interpretation via near attack conformations, *J. Am. Chem. Soc.*, 2003, **125**, 6663–6672; (d) Hur, S. and Bruice, T.C., Comparison of formation of reactive conformers (NACS) for the Claisen rearrangement of chorismate to prephenate in water and in the *E. coli* mutase: the efficiency of the enzyme catalysis, *J. Am. Chem. Soc.*, 2003, **125**, 5964–5972.

98. (a) Coates, R.M., Rogers, B.D., Hobbs, S.J., Peck, D.R. and Curran, D.P., Synthesis and Claisen rearrangement of alkoxyallyl enol ethers. Evidence for a dipolar transition state, *J. Am. Chem. Soc.*, 1987, **109**, 1160–1170; (b) Copley, S.D. and Knowles, J.R., The conformational equilibrium of chorismate in solution: implications for the mechanism of the non-enzymic and the enzyme-catalyzed rearrangement of chorismate to prephenate, *J. Am. Chem. Soc.*, 1987, **109**, 5008–5013; (c) Severance, D.L. and Jörgensen, W.L., Effects of hydration on the Claisen rearrangement of allyl vinyl ether from computer simulations, *J. Am. Chem. Soc.*, 1992, **114**, 10966–10968; (d) Sehgal, A., Shao, L. and Gao, J., Transition structure and substituent effects on aqueous acceleration of the Claisen rearrangement, *J. Am. Chem. Soc.*, 1995, **117**, 11337–11340; (e) Guest, J.M., Craw, J.S., Vincent, M.A. and Hillier, I.H., The effect of water on the Claisen rearrangement of allyl vinyl ether: theoretical methods including explicit solvent and electron correlation, *J. Chem. Soc., Perkin Trans. 2*, 1997, 71–74; (f) Gajewski, J.J., The Claisen rearrangement. Response to solvents and substituents: the case for both hydrophobic and hydrogen bond acceleration in water and for a variable transition state, *Acc. Chem. Res.*, 1997, **30**, 219–225.
99. Brandes, E., Grieco, P.A. and Gajewski, J.J., Effect of polar solvents on the rates of Claisen rearrangements: assessment of ionic character, *J. Org. Chem.*, 1989, **54**, 515–616.
100. An, J., Bagnell, L., Cablewski, T., Strauss, C.R. and Trainor, R.W., Applications of high-temperature aqueous media for synthetic organic reactions, *J. Org. Chem.*, 1997, **62**, 2505–2511.
101. Lubineau, A., Augé, J., Bellanger, N. and Caillebourdin, S., Water-promoted organic synthesis using glyco-organic substrates: the Claisen rearrangement, *J. Chem. Soc., Perkin Trans. 1*, **1992**, 1631–1636.
102. Grieco, P.A., Brandes, E.B., McCann, S. and Clark, J.D., Water as a solvent for the Claisen rearrangement: practical implications for synthetic organic chemistry, *J. Org. Chem.*, 1989, **54**, 5849–5851.
103. McMurry, J.E., Andrus, A., Ksander, G.M., Musser, J.H. and Johnson, M.A., Total synthesis of aphidicolin, *Tetrahedron*, 1981, **37**(Suppl. 1), 319–327.
104. Nicolaou, K.C., Xu, H. and Wartmann, M., Biomimetic total synthesis of gambogin and rate acceleration of pericyclic reactions in aqueous media, *Angew. Chem. Int. Ed.*, 2005, **44**, 756–761.

Chapter 6

Catalyzed Reductions in Aqueous Media

T.V. RajanBabu and Seunghoon Shin

There has been an increasing interest in the use of water as a reaction medium for organic and organometallic reactions during the last three decades.[1] Water as a preferred reaction medium has several desirable characteristics: it is cheap, nontoxic, nonflammable, and widely available. In addition, the immiscibility of water with most organic solvents (and products) improves the prospect of easy separation of a water-soluble catalyst from the reaction mixture.[2] From a large-scale manufacturing perspective, poor recovery of homogeneous catalysts (often expensive) from the medium and contamination of products by metallic salts have often hindered their use in many practical applications involving fine chemical synthesis.[3] Thus the advantage of anchoring a homogeneous catalyst in an 'aqueous support' is apparent. Ideally a simple decantation should allow catalyst recovery and reuse. In addition to being a solvent and an immobilizing agent, water has also been shown to give higher rates (especially for hydrogen activation by Ru(II) complexes) and selectivities for several reactions. Phase separation can sometimes help with removal of components of the reaction, thereby preventing substrate/product inhibition of catalysis. However such advantages should be contrasted with other problems such as the high heat capacity, heat of evaporation, and reactivity of water. These properties could pose difficulties in separating and purifying a water-soluble product. Remediation of a contaminated aqueous stream is another factor to consider in designing a process in aqueous medium. Nonetheless, several hugely successful processes conducted in water, among them the Rhone-Poulenc/Ruhrchemie hydroformylation[4] and the Rhone-Poulenc aldehyde hydrogenation,[5] underscore the importance of continued research in this area (see Chapter 12). Another application that has been less explored is the *selective* functionalization of water-soluble biomolecules such as polypeptides, oligonucleotides, and oligosaccharides, where water-soluble organometallic catalysts may shorten the synthesis by avoiding multiple protection–deprotection steps currently needed to prepare some of the derivatives.[6]

All of the above considerations provide ample impetus for further research in homogeneous catalytic systems in aqueous media. Toward this goal, ligand modification that endows water solubility has been the most widely utilized approach to immobilize catalysts in the aqueous phase. In this chapter, we will focus on various water-soluble ligands developed for reductions including hydrogenations of C=C, C=O, and C=N bonds. The primary focus will be on reactions and processes that are not extensively dealt with in recent reviews.[1] Along these lines, we also plan to discuss how special solvent properties of water, such as its unique ligand properties and pH-dependent behavior, can be used to improve efficiency and selectivity of hydrogenation reactions. Literature up to the end of March 2006 is covered here, and the coverage is intended to be illustrative, rather than exhaustive. Many important topics such as the classical heterogeneous hydrogenations[7] using Ni, Pd, and Pt and enzyme-catalyzed reductions are not included (for a discussion on enzymatic reductions in water, see Chapter 10).

6.1 Special features of catalytic hydrogenation in water by organometallic complexes

Hydrogenations in water have several distinct, yet less well-recognized features than those occurring in organic solvents owing to the unique physical characteristics of water. Apart from the solubility of catalytic metal complexes and organic substrates, the solubility of gaseous hydrogen also has a pronounced effect on the selectivity and reactivity. Hydrogen, being a nonpolar molecule, has poor solubility in water, which is polar with a high dielectric constant. Recent studies by Blackmond have indicated that the kinetics of mass transfer of the gaseous reactant, as influenced by the agitation speed, is important for achieving high selectivity, including as in the cases reported, high enantioselectivity.[8] The key kinetic parameter affecting enantioselectivity was found to be concentration of molecular hydrogen in the liquid phase, which depends on the rate of gas–liquid mass transfer.

Additives such as surfactants or salts in aqueous solvent have marked influence on the solvation properties of reacting species. Selke, Oehme, and coworkers[1d,9] have reported that the use of micelle-forming additives has beneficial effect on the catalyst activity and selectivity. In several cases, micelle acts both as a solubilizing agent for poorly soluble organic substrates and as reaction medium for the catalytic reaction.

Börner and coworkers have carried out a systematic study on the effect of ionic strength on the enantioselectivity and activity of a catalyst.[10] In Na_2SO_4 solution (6.7 mM), they observed a significantly decreased activity and more or less similar enantioselectivity compared with the reactions in neat water. By increasing the ionic strength, the solubility of both catalyst and substrates decreases, resulting in a lowering of the reaction rate. Although hydrogenation reactions in water have many similarities to those in organic media in its operational details, the special properties of water have to be taken into consideration when developing water-soluble catalysts.

In aqueous organometallic catalysis, water may have other roles than simply acting as a reaction medium. For example, water plays a fundamental role in coordination chemistry and many metal complexes can bind water to fill their coordination sphere. Water, having a good ligand field, can act as ligand itself or it can exchange proton with metal hydrides according to deuterium labeling studies using D_2O or D_2 (Eqs. 6.1 and 6.2).

$$\mathrm{Rh(R)L_n} + \mathrm{D_2O} \longrightarrow \mathrm{RhL_n(OH)} + \mathrm{R{-}D} \qquad \text{(Eq. 1)}$$

$$\mathrm{Rh(D_2)L_n} + \mathrm{H_2O} \longrightarrow \mathrm{RhDHL_n(OH)} + \mathrm{HOD} \qquad \text{(Eq. 2)}$$

Thermodynamic stability of M—C bond in water has raised questions about lifetimes of intermediates in the aqueous organometallic catalytic cycles. From deuterium-labeling experiments (Eq. 6.3), Sinou and coworkers[11] have concluded that the protonation of the M—C bond can occur readily. When the reduction of (*Z*)-acetamidocinnamic acid methyl ester was performed under hydrogen in a two-phase system (EtOAc–D_2O) in the presence of the catalyst $[Rh(cod)Cl]_2$+ TPPTS [tris-(*m*-sulfonato-phenyl)phosphine], regioselective deuterium incorporation up to 75% at the α-position was observed by ^{1}H and ^{13}C NMR and mass spectrometry. Performing the same reaction under deuterium (D_2) atmosphere in a two-phase system (EtOAc–H_2O) led to hydrogen incorporation up to 94% at the α-position and deuterium exclusively at the β-position. While the deuterium content depends on the nature of cosolvent and phosphine ligands, these results clearly show that water can act as *reactant* under hydrogenation conditions. In spite of such limitations, however, there have been a number of recent reports of achieving high reaction rates and selectivities in aqueous media, where many of these factors are not detrimental, when optimal protocols are employed.[12]

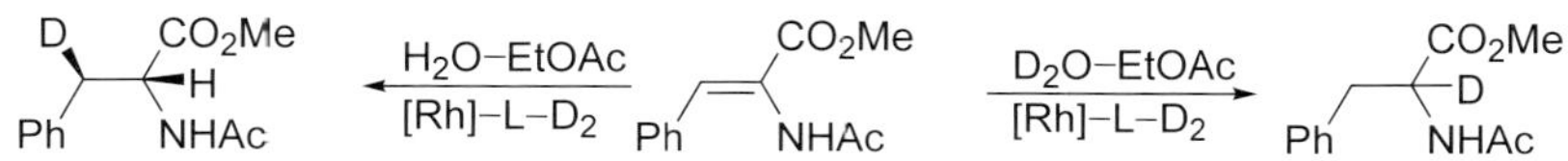

(Eq. 3)

6.2 Water-soluble complexes for aqueous hydrogenation

One of the most effective ways to prevent catalyst from leaching into organic phase is to incorporate highly water-soluble prosthetic groups into the ligand skeleton. Most often this has been accomplished by introduction of polar groups such as amines, carboxylic and sulfonic acids, multiple hydroxyl groups, and oligomeric chains with heteroatoms. However, modification of a given phosphine ligand to impart water solubility is not always straightforward. The following brief overview of preparative methods for each class of ligands will provide a guideline in designing a new water-soluble catalytic system and show how water-soluble ancillary groups can be incorporated into a ligand backbone containing phosphorus and nitrogen donor atoms. Recent reviews[13] on water-soluble ligands for aqueous systems should be consulted for a complete list of ligands and other methods for their preparation.

6.2.1 Sulfonated phosphine and other ligands

Historically, prototypical hydrogenation reactions in water dates back to 1973, when sulfonated aryl phosphine ligands **1a–c** were introduced by Joó and coworkers (Fig. 6.1).[1a,14] Most early chiral ligands also utilized aryl- and alkyl sulfonate groups for aqueous solubility.[15] Sulfonated phosphine ligands having C—S bond are stable toward hydrolysis and remain in their ionic forms in the entire pH range and are thus well soluble in water. Not surprisingly, many sulfonated phosphine ligands have been prepared and representative methods for the preparation of these ligands are summarized below. Standard sulfonation conditions that utilize $H_2SO_4{\cdot}SO_3$ (oleum) are quite severe and acid-sensitive functional groups are not compatible. Several novel and milder procedures for the preparation of sulfonated ligands have since been published.[16] Controlling the reaction conditions allows selective sulfonation of each diphenylphosphino group into mono-, di-, tri- or tetrasulfonated diarylphosphines. However, direct sulfonation strategy in the case of diarylphosphine ligands involve tedious separation of mono-, di-, tri- and tetrasulfonated reaction mixture through reversed phase-high performance liquid chromatography (RP-HPLC), and in the case of chiral phosphines, the possibility of formation of stereoisomeric mixtures.[17] Very attractive alternative methods of introducing sulfonate groups on chiral backbones have been developed by Williams and

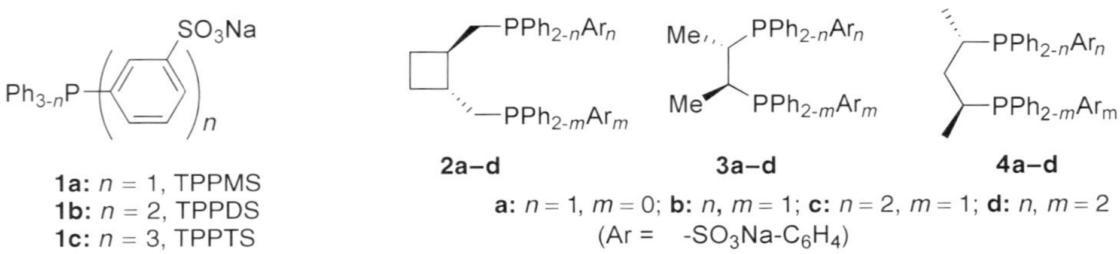

Figure 6.1 Selected examples of sulfonated phosphine ligands.

coworkers, which involve oxidation of a chiral disulfide precursor by H_2O_2.[18] In addition, this group reported derivatization through the reaction of an amine with benzene-1,2-disulfonic anhydride, which also provided *ortho*-sulfonated phenyl derivatives. Improved workup procedures for the isolation of sulfonated diphosphines should permit large-scale synthesis of these ligands.[19] Researchers at Roche used a lithiated sulfonamide derivative to install this group on OMe-BIPHEP ligands by displacement reactions on a chlorophosphine.[16e]

Standard methods to introduce sulfonate group into the ligand can be summarized as follows:

1. Direct sulfonation of aryl group by fuming sulfuric acid (oleum) followed by neutralization to give the sulfonyl group at meta position.[20]
2. Remote sulfonation at an arene not linked to P carried out under milder conditions (conc. H_2SO_4, rt).[16a]
3. Nucleophilic substitution of phosphide ion on fluoroarene sulfonic acid resulting in the sulfonyl group at ortho or para position or both.[21]
4. Reaction of diarylphosphide with haloalkyl sulfonic acid salt or sultone giving diaryl alkyl phosphine.[22]
5. Cross-coupling of substituted iodobenzenes with Ph_2PH catalyzed by $Pd(OAc)_2$.[23]
6. Conjugate addition of secondary phosphine.[24]
7. Displacement on a chlorophosphine by protected sulfone-containing organometallics.[16e]

6.2.2 Nitrogen-containing phosphine ligands

Quaternary ammonium ligands are widely utilized because of their high solubility in water at a wide range of pH.[25] More recently, other polar functional groups containing N-atom, such as guanidinium,[26] phosphonium,[27] 1,3,5-triaza-7-phosphaadamantane (PTA),[28] and tertiary amine have also been reported. The feasibility of catalyst recycling by pH-controlled extraction has been explored by several workers including, in one of the recent reports, by Uemura and coworkers, in the context of allylic alkylation catalyzed by a water-soluble Pd catalyst.[29] However, in many cases, coordination of tertiary amine to the metal center complicates the characterization of precatalyst, and the ligands bearing quaternary ammonium salt are the most widely used. Figure 6.2 lists representative ligands with a quaternary ammonium salt as the hydrophilic ancillary group.[30–35]

A common procedure for quaternization of an amine is the use of Meerwein's salt[31,32] or MeI (excess).[34,35] In the alkylation of amine-containing phosphine ligand, the nucleophilic P-atom could potentially compete with N-alkylation and therefore precomplexation with the

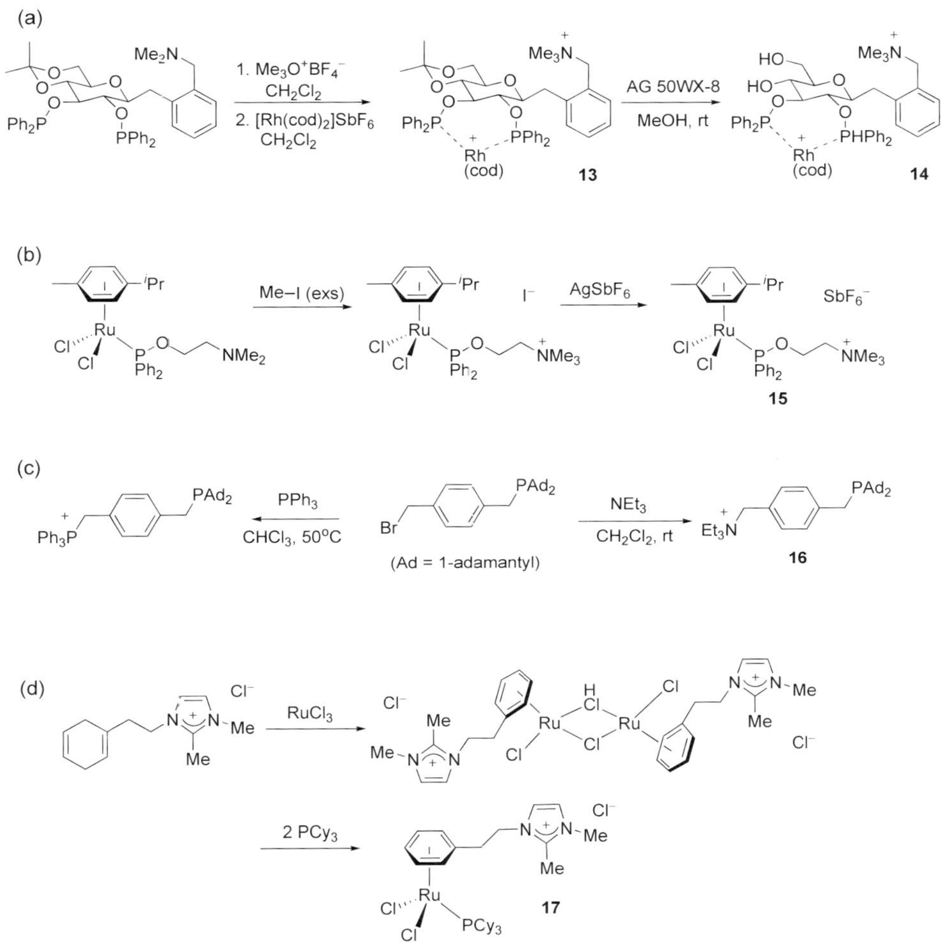

Scheme 6.1 Preparative methods for quaternary ammonium-containing phosphine ligands and their water-soluble metal complexes (**13–17**).

desired metal generally precedes the alkylation step.[30] But for diarylphosphinites this is not necessary. Thus, treating a dimethylamino-phosphinite with Meerwein's salt ($Me_3O^+BF_4^-$, 1.0 equiv.) in dichloromethane, followed by complexation with Rh precursor, results in a clean formation of Rh complex **13** (Scheme 6.1a). Further increase in water solubility could be achieved by deprotection of an isopropylidene acetal upon treatment with acidic resin in methanol. If necessary, the crude complex **14** can be purified further by reprecipitation with ether. Similarly, water-soluble ruthenium–arene complex **15** can be prepared by treating dimethylamino precursor with excess MeI, followed by anion exchange with AgX salt (Scheme 6.1b).[35] Quaternary ammonium salts **16** can be prepared by alkylation of triethylamine with a corresponding arylphosphino-alkylhalide (Scheme 6.1c).[27] Another interesting class of quaternary ammonium ligands is those containing imidazolium group, such as **17**.[33a] They are prepared by reduction of $RuCl_3$ with 1,4-cyclohexadiene bearing

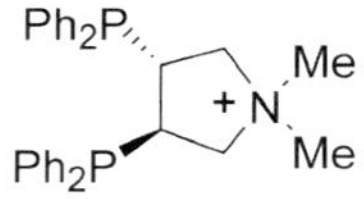

5[30]

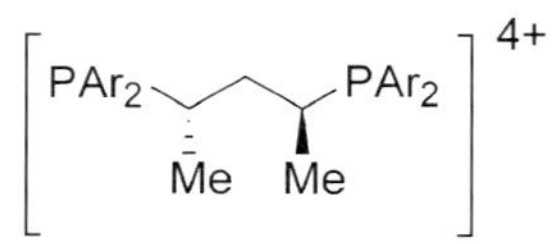

6[31]

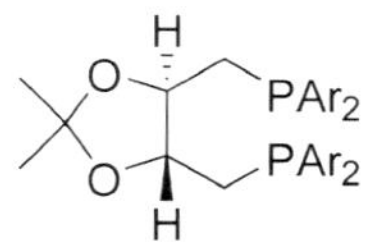

7[31]

Ar = $p\text{-}\overset{+}{\mathrm{Me_3N}}\text{-}C_6H_4$

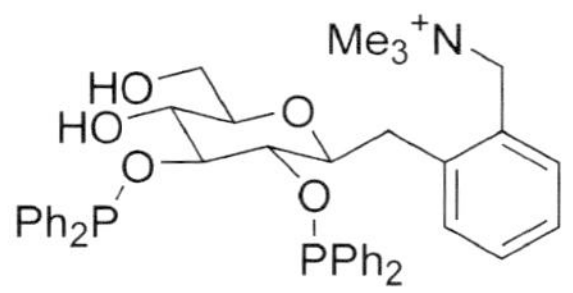

8[32]

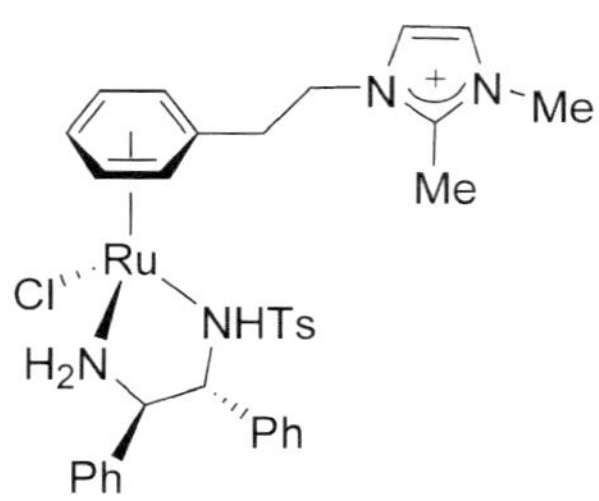

9[33]

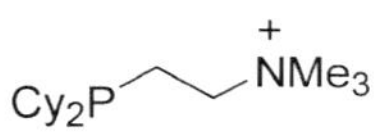

10[34]

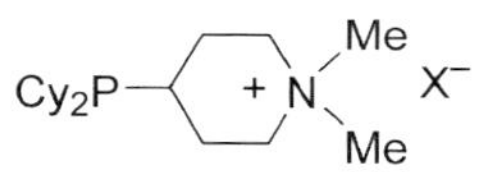

11[34]

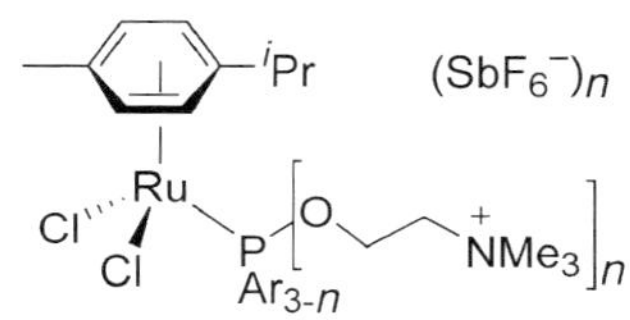

12[35]

Figure 6.2 Ligands containing quaternary ammonium salts.

imidazolium salt in methanol to yield ruthenium–arene complex dimer in an excellent yield (Scheme 6.1d). Subsequent treatment with 2 equiv. of phosphine leads to **17**. A new water-soluble N-heterocyclic carbene–Ru complex has also been reported recently.[36]

6.2.3 Hydroxyphosphine and other oxygen-containing ligands

Nonionic ligands are beginning to receive increasing attention principally because biphasic reactions of poorly soluble organic substrates can be expected to benefit from these ligands. In addition, such neutral ligands may find applications for metal-catalyzed reduction of membrane lipids where the catalyst transport into the bilayer may be precluded by highly charged ionic groups generally seen in many water-soluble ligands.[37] Notable advances have been made in the use of hydroxyphosphine ligands in asymmetric catalysis and selected examples of such ligands are listed in Fig. 6.3. In addition to the hydrophilicity, hydroxy groups might exert special effects in organometallic catalysis, such as acting as 'hemi-labile' ligand coordinating to the metal center,[38] or providing a hydrogen bond that may influence the energetics of catalyst–substrate interactions. For example, in mechanistic studies of

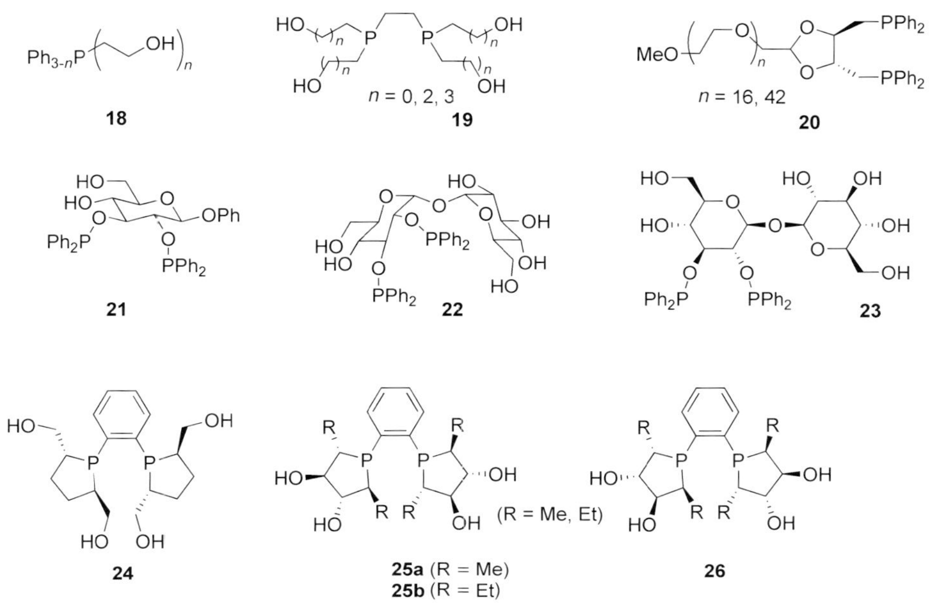

Figure 6.3 Selected examples of hydroxyphosphine ligands.

asymmetric hydrogenation done by Börner, hydroxyphosphine ligands were found to have a rate-reducing but enantioselectivity-enhancing effect.[39]

Early investigation of water-soluble hydroxyphosphine ligands led to polyhydroxyphosphine ligands such as **18–20**,[40] although few applications in aqueous organometallic catalysis were reported. Some of these ligands (**19**, $n = 2, 3$) were prepared by free-radical reaction between primary phosphine and olefinic alcohol,[41] or by reaction of metal phosphide with formaldehyde (**19**, $n = 0$).[42] Ethylene glycol unit could also be used as water-solubilizing group. For example, synthesis of water-soluble DIOP derivative **20** incorporates a hydrophilic ethylene glycol unit by hemiacetal condensation, which is depicted in Scheme 6.2.[43]

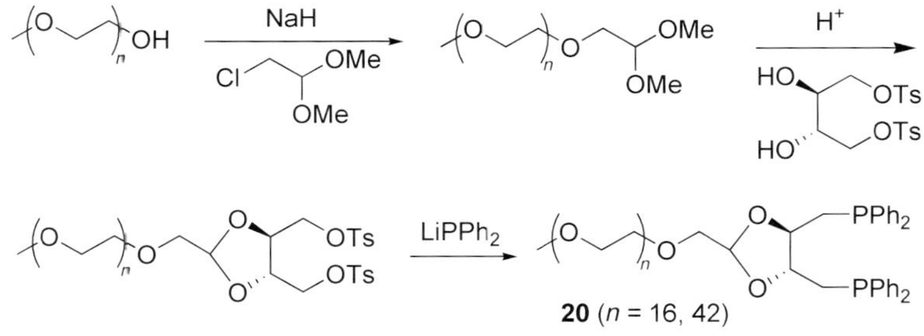

Scheme 6.2 Synthesis of a hydrophilic DIOP derivative.

Carbohydrates are among the most widely available natural products and their use as ligand scaffolding has attracted considerable attention.[44] Once a metal-chelating group is appropriately incorporated into the carbohydrate frame, the ligand would be water soluble by virtue of the polyhydroxylic nature of these molecules. An added benefit would be that the chirality of the ligand backbone (the sugar) could induce asymmetry in metal-catalyzed reactions. Since there are thousands of carbohydrates of different configurations and substitution patterns to choose from, the enantio-differentiating property of the catalyst could be optimized for efficiency and selectivity with relative ease.

Pioneering studies by Selke and Oehme[45] have demonstrated that monosaccharide ligand **21** has some solubility in water and they reported its use in asymmetric hydrogenation of dehydroamino acids in water. However, without the addition of micelle-forming surfactants, Rh(I) complexes derived from **21** seem to have only limited solubility as well as moderate activity and selectivity. A study of the influence of carbohydrate-derived amphiphiles on the reduction of (Z)-acetamidocinnamic acid in water using $\{[Rh(COD)_2]BF_4 + t\ BPPM\}$ catalyst shows that enantiomeric excesses (ee's) similar to or better than what is observed in MeOH can be obtained.[46] The chiral amphiphile alone gave a small (6%), but reproducible selectivity when used with an achiral Rh catalyst.

Uemura[47] and RajanBabu[48] groups independently reported the synthesis of Rh(I) complexes of **22** and its catalytic application in asymmetric hydrogenation in water. Synthesis of **22** relied upon the availability of α,α'-trehalose-derived diol **27** (Scheme 6.3),[49] from which diarylphosphinites were prepared in excellent yield. The corresponding β,β'-derivatives are also readily available. After complexation with $[Rh(NBD)_2]BF_4$, deprotection of cyclohexylidene ketal was achieved simply by treatment with acidic resin (AG 50 WX-8 from BioRad®) in alcohol, followed by filtration through Celite. In handling highly water-soluble and oxygen-sensitive organometallic complexes, this simplified workup protocol proved powerful for obtaining acceptable quality of catalyst precursors for organometallic reactions.

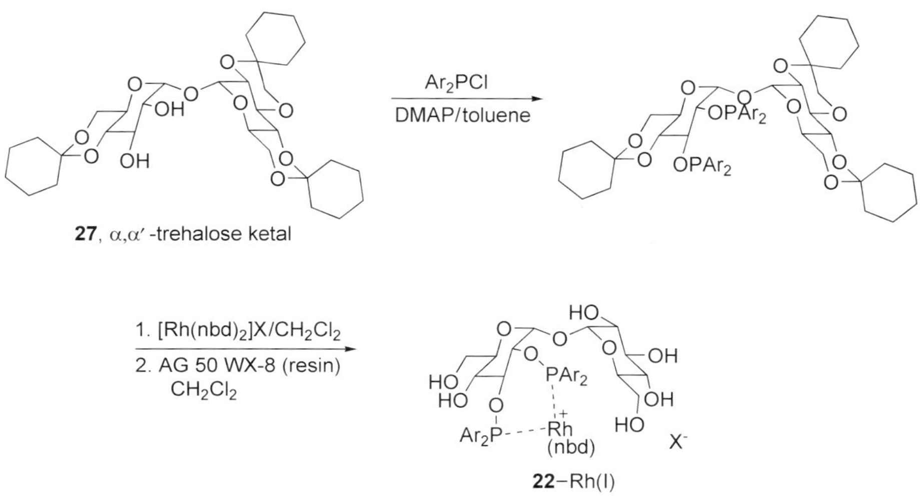

Scheme 6.3 Synthesis of **22**–Rh(I) complex.

Later improvements addressing both problems of limited solubility and hydrolytic stability came with the development of water-soluble ligands having the notable DuPHOS skeleton, for which excellent enantioselectivity in ketone, olefin and imine reductions have been reported.[50] Zhang[51] and RajanBabu[52,53] groups independently pursued the synthesis of **25a**, modifying the procedure reported by Burk and coworkers (Scheme 6.4). Also, RajanBabu[52,53] and Börner[54] have independently reported synthesis of **24** and **25**. These ligands showed much improved water solubility by reducing the number of the hydrophobic *P*-aryl groups.

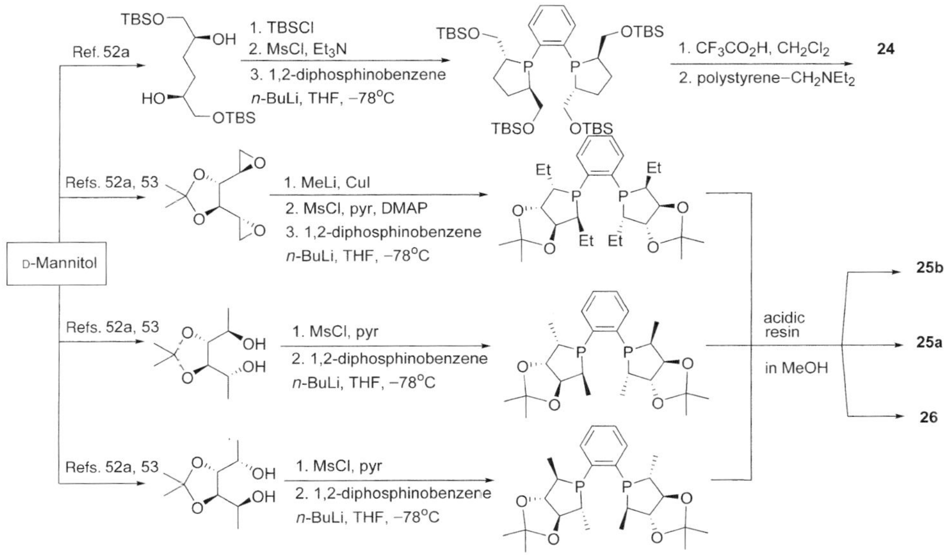

Scheme 6.4 Overview of the synthesis of polyhydroxylic phospholane **24–26**.

Successful preparation of these highly water-soluble ligands (**24–26**) depends critically on a simplified purification protocol for these oxygen-sensitive and water-soluble compounds. RajanBabu group devised a protocol using solid-phase reagents, which proved useful for the preparation of **22** and **24–26**. The key steps are outlined in Scheme 6.4.[52a] Börner and coworkers have noted that the deprotection and purification of **24** (from a Tetrahydropyranyl-protected derivative) could not be easily achieved owing to its high polarity. Instead they precomplexed the ligand with [Rh(cod)]BF_4-precursor and treated the resulting Rh(I) complex with HBF_4 in MeOH to obtain fully deprotected [Rh(I)**24**]BF_4 complex. This protocol, while perfectly acceptable for Rh catalysis and electron-deficient phosphines, may impose some limitations if other metal–ligand combinations are to be considered for catalysis in water. RajanBabu group desilylated the *t*-butyldimethylsilyl-protected derivative of **24** with CF_3CO_2H in MeOH, which resulted in a mixture of phosphonium salts as indicated by poorly defined ^{31}P NMR spectra. Removal of excess acid could be cleanly achieved to afford **24** by filtration through a short pad of diethylaminomethyl-polystyrene resin.[52a]

Zhang and coworkers reported the synthesis of 'free' ligand **26** by treating its isopropylidene acetal with CH_3SO_3H in methanol–water. The resulting product obtained had characteristic broad peaks at δ 15.0 ppm, suspiciously similar to a phosphonium salt. Instead, RajanBabu group took a solid-phase approach advantageously, i.e. treating isopropylidene acetal derivative of **26** with AG 50 WX-8 acidic resin in methanol. This procedure gave analytically pure sample of **26**, which showed a single sharp peak in ^{31}P NMR. The latter protocol could be similarly applied to the preparation of **25a** and **25b**.

6.3 Hydrogenation of C=C bond

Earlier work on this hydrogenation has been extensively reviewed[1g,1i,55] and will not be covered here. In this area, use of cyclodextrins and cyclodextrin-derived ligands in connection with $[HCo(CN)_5]^{3-}$ and Rh-catalyzed reductions are significant highlights that need further developmental efforts before this can be practiced on preparative scale.[56] Use of the water-soluble Ru complex **17** for aqueous phase hydrogenation of styrene (0.1 mol% catalyst in water, 40 bar H_2, 80°C, 2 h, >99% conversion) has been reported.[33a] Reduction of CO_2 using this and other related complexes (*vide infra*) is a much less satisfactory reaction.

Increased demand for enantiomerically pure pharmaceutical, agrochemical, flavor and other fine chemicals has provided added impetus for research in catalytic asymmetric synthesis. In the field of aqueous organometallic reductions, there have been significant advances in hydrogenations of alkenes, carbonyl compounds and imines, where enantioselectivities over 95% have been realized in selected cases. Since the earlier reviews mostly dealt with organometallic hydrogenation reactions in water using achiral ligands, we will focus our discussion on asymmetric versions of these reactions.

6.3.1 Reductions of dehydroamino acid and acrylic acid derivatives

The enantioselective hydrogenations of dehydroamino acid derivatives **28** and **29** (α-acetamidoacrylate and α-acetamidocinnamates, respectively) have been employed as prototypical catalytic asymmetric reactions in water (Fig. 6.4), because of the apparent utility of the resulting amino acid derivatives and the detailed knowledge of the mechanism of this reaction in organic media. Other substrates that have been successfully used in asymmetric hydrogenation include itaconic acid derivatives **30** and α-arylacrylic acids such as **31**. Although both sulfonated Ru-BINAP and Rh-BINAP complexes had been known to catalyze asymmetric hydrogenation in water,[57] water-soluble Rh(I) complexes have played a pivotal role in the development of several successful asymmetric hydrogenation reactions in water. Excellent level of enantioselectivity as well as activity that rival the organic counterpart has become possible. Alongside, continued efforts are being made to attain the facile recovery of catalyst and reuse.

In their pioneering work, Sinou and coworkers applied ligands **2–4**[58] and a PEG-attached-DIOP[43] in asymmetric hydrogenation of α-amino acid precursors. Both sulfonated phosphines and those bearing quaternary ammonium salts[31a,b,59] show good to excellent enantioselectivity in aqueous asymmetric hydrogenation (Table 6.1 and Fig. 6.5). In many cases, the enantioselectivity significantly deteriorated in going from an organic to an aqueous solvent. For example, while ligands **3d** (m-$SO_3^-Na^+$) and **32** (Ar = C_6H_4-4-$NMe_3^+BF_4^-$) gave enantioselectivities, both in organic solvent and in a two-phase system, close to the values obtained in organic media, ligands **2d** and **7** showed much lower enantioselectivities in water or biphasic media.

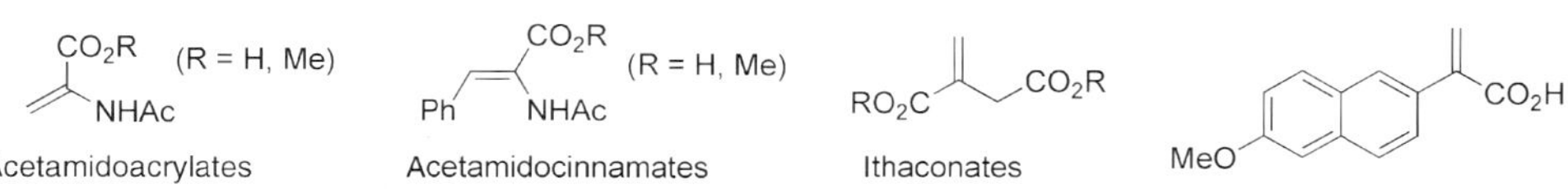

Acetamidoacrylates — **28a** (R = Me), **28b** (R = H)

Acetamidocinnamates — **29a** (R = Me), **29b** (R = H)

Ithaconates — **30a** (R = Me), **30b** (R = H)

31

Figure 6.4 Substrates used in hydrogenation.

Table 6.1 Asymmetric reduction of α-acetamidocinnamic acid **(29b)** by Rh(I)–L complexes (see Fig. 6.5)

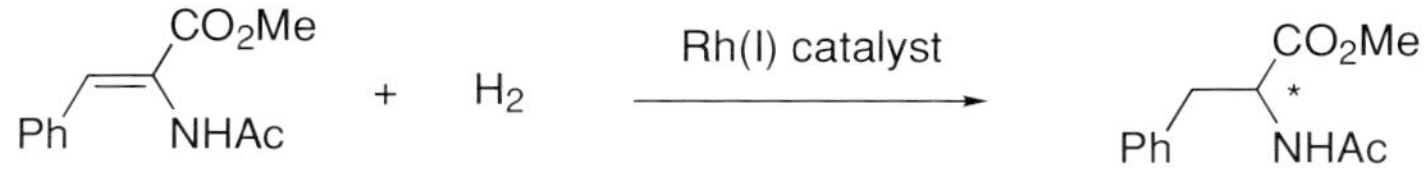

29b

Entry	Ligand	Solvent	P_{H2} (atm)	Product % ee	Product Reference
1	**2d** (S, S)	H_2O—EtOAc (1:1)	1	34 (S)	58a
2	**3d** (S, S)	H_2O—EtOAc (1:2)	10	87 (R)	58a
3	**4d** (S, S)	H_2O—EtOAc (1:1)	15	65 (R)	58a
4	**7** (R, R)	H_2O	14	25 (S)	31b
5	**32** (S, S)	H_2O	14	94 (R)	31b
6	**33** (S, S)	H_2O	14	67 (R)	31b

More recently recyclable Rh catalysts have been prepared by the coupling of (2*S*,4*S*)-4-diphenylphosphino-2-diphenylphosphino-methylpyrrolidine to polyacrylic acid.[60] The macromolecular ligand shows high water solubility and this property can be varied by changing the ratio of phosphine to carboxylate on the polymer. A catalyst prepared by treating the polymer with bis-norbornadienerhodium(I) triflate, $[Rh(NBD)_2]^+OTf^-$, is active in enantioselective hydrogenation in water or under biphasic conditions (H_2O–EtOAc). The best enantioselectivity (89% ee) is obtained for α-acetamidocinnamic acid, giving (*R*)-*N*-acetylphenylalanine. The enantioselectivity is high at low phosphorus loading on the polymer and is low at high phosphorus loading. The aqueous catalyst solutions are easily recovered and recycled by phase separation, with no apparent loss in enantioselectivity (Fig. 6.5).

Selke and coworkers have examined the application of carbohydrate-derived bisphosphinite–Rh(I) complexes from ligand **21** for the reduction of (*Z*)-acetamidocinnamic acid derivatives.[45a,61] While a beneficial effect of sodium dodecyl sulfate (SDS) additive on enantioselectivity was observed in pure water, the Rh(I) complexes derived from these dihydroxy derivative of glucopyranoside had only limited solubility in water and they showed only moderate enantioselectivity. The polyhydroxylic nature of carbohydrates was

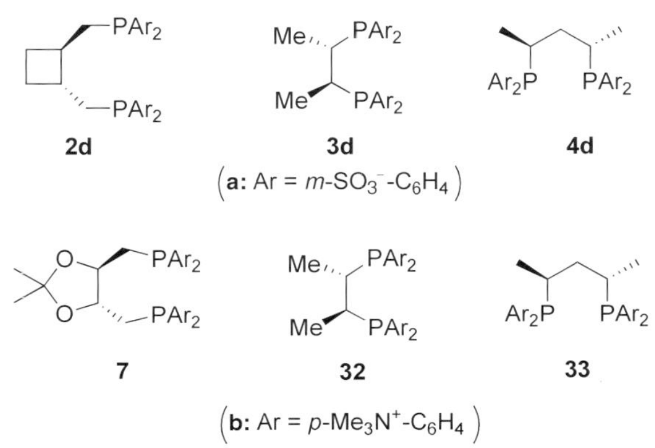

Figure 6.5 Chiral water-soluble ligands.

more advantageously utilized by the introduction of α,α'- and β,β'-trehalose-derived ligands **22**, reported by Uemura and RajanBabu groups.[47,48] Alternatively, ligand **8**, bearing a quaternary ammonium salt, was also used by the latter group.[32] In the hydrogenation of dehydrophenylalanine derivatives **29a** and **29b**, only modest selectivity (55% ee) was observed with **22** (α,α)–Rh(I) complex (Table 6.2). The enantioselectivity could be improved to 88% ee by modification of the glycosidic linkage by using the unnatural β,β-form **23**. However, in all cases, an increased amount of water in the solvent mixture resulted in decreased enantioselectivity and activity. The poor enantioselectivity in pure water could presumably result from insufficient solubility of catalytic Rh(I) species in water and/or from hydrolysis of phosphinite ligands in water, resulting in the formation of nonchiral Rh-containing species. The ligand **8** bearing quaternary ammonium salt gave 61% ee in asymmetric hydrogenation of methyl acetamidoacrylate **28a** in water after 1 h. The corresponding isopropylidene-protected derivative of **8** had sufficient solubility in THF and gave 86% ee in THF. One interesting observation made with ligand **8** is that the enantioselectivity depends on the reaction time. When the product is left in contact with water for longer periods of time under the reaction conditions, enantioselectivity drops off precipitously, giving another indication that some nonselective events are possible in water when using phosphinite ligands. One possible explanation for the deterioration of enantioselectivity in water is the intervention of protonolysis of the putative Rh—C bond before the final reductive elimination. Since significant α-deuterium incorporation has been observed in reactions run in D_2O-containing media, it has been suggested that Rh—H/Rh—D exchange could also take place in the penultimate intermediate in the catalytic cycle. Presumably, the racemization can also occur through the intermediacy of a Rh-bound enolate.

Effects of pH on reaction rates in prototypical rhodium-catalyzed hydrogenation in water have been examined carefully.[62] The complex $[Rh(DPPBTS)(NBD)][O_3SCF_3]$ (DPPBTS = tetrasulfonated 1,4-bis(diphenylphosphino)butane) upon reaction with hydrogen gives different complexes $[Rh(DPPBTS)(H_2O)_3(H)]^{2+}$, $[Rh(DPPBTS)(H_2O)_2]^+$,

Table 6.2 Asymmetric reduction of dehydroamino acid derivatives by Rh(I)–L complexes

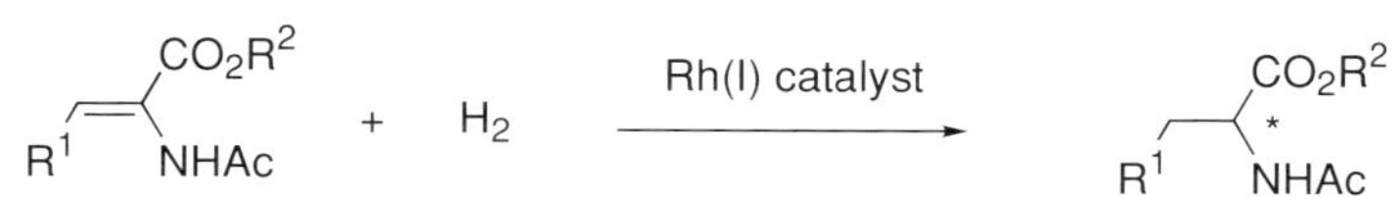

28a (R^1 = H; R^2 = Me)
29a (R^1 = Ph; R^2 = Me)
29b (R^1 = Ph; R^2 = H)

Entry	Substrate	Ligand	Solvent	P_{H2} (atm)	Product % ee	Reference
1	**29b**	**22** (α,α)	H_2O	5	55	47a
2	**29b**	**22** (α,α)	H_2O—EtOAc(1:1)	5	68	47a
3	**29a**	**22** (α,α)	H_2O—THF(1:1)	2.8	65	48
4	**29a**	**22** (α,α)	THF	2.8	70	48
5	**29b**	**23** (β,β)	H_2O	5	88	47a
6	**29b**	**23** (β,β)	H_2O—MeOH	5	94	47a
7	**28a**	**8**	H_2O	3	61	32
8	**28a**	**8**	H_2O[a]	3	8	32

[a] Second recycle experiment, where product was left in contact with catalytic aqueous phase for longer time.

or $[Rh(DPPBTS)(OH)]_2$ depending on the pH of the aqueous solution. Addition of α-acetamidoacrylic acid to an aqueous solution of $[Rh(DPPBTS)(H_2O)_2]^+$ gives a substrate complex in which the coordination mode is pH dependent. The amino acid coordinates via the double bond and the amide carbonyl at a pH below its pK_a. Coordination through the double bond and the carboxylate group takes place at a pH higher than the pK_a of the amino acid. This has a dramatic effect on the rate of hydrogenation. The reaction is extremely fast at a pH below the pK_a of the substrate (270,000 mol h^{-1}), but approximately 2000 times slower at a pH higher than the pK_a. A kinetic isotope effect k_H/k_D of 1.25 has been observed at pH 4.5, indicating that the oxidative addition of hydrogen is the rate-determining step in the hydrogenation using this catalyst. Thus the mechanism is not very different from that in organic medium.

One of the purported goals in developing water-soluble catalysts is the ease of recycling the catalyst 'immobilized' in the aqueous phase. Yet there are few reports in the literature where the actual distribution of the catalyst in biphasic medium has been measured under the reaction conditions. Viability of the catalyst recovery option is often demonstrated by carrying out the reaction in a biphasic medium in which the product is organic soluble, and the catalyst water soluble. At the end of the reaction, the organic phase is separated and the aqueous phase is used for subsequent reactions. Arguably, for highly efficient catalysts like the Rh(I)–bisphosphinites, unless careful quantitative rate studies are done with the recovered catalyst solution, this is *not* a satisfactory way of demonstrating the practicality of this approach, since residual amounts of the Rh complex can still facilitate the hydrogenation. Hence, RajanBabu and coworkers[48] sought to determine the distribution of the cationic Rh complexes between water and several common organic solvents by inductively coupled plasma mass spectrometry (ICP-MS) method, which gives more quantitative information on the Rh distribution under typical reaction conditions. The distribution of complex Rh(I)(**22**)(NBD)(BF_4) between water and organic solvents (CH_2Cl_2, EtOAc, Et_2O) are shown in Table 6.3. Surprisingly, this 'water-soluble' complex having six hydroxy groups had significant solubility in all these organic solvents, raising concern that catalyst leaching could be a significant problem. It became apparent that diaryl phosphinite ligands (and possibly many of the well-known diarylphosphines as well) could have significant problems when recovery of catalyst is desired.

The limited aqueous solubility of the hydrophobic diarylphosphinite ligands aside, hydrogenation study with ligand **8** showed that significant deterioration of catalytic activity could result when the aqueous solution was recycled in sequential reactions (entry 8, Table 6.2). One possible origin of such decrease in activity might be the hydrolytic degradation of the P—O bond, which was observed for Rh(I) complexes of **8** and **22** (Fig. 6.6). The dual problem of hydrophobicity and hydrolytic instability could simultaneously be addressed through the use of polyhydroxyphosphines bearing only P—C bonds and a reduced number of aryl substituents on the phosphorus atom. In this regard, analogs of Burk's DUPHOS ligand

Table 6.3 Distribution constant of Rh(I)–**22** complex between water and organic solvents[a]

Solvent	Total Rh (mg)	Ratio
CH_2Cl_2—H_2O	0.32	1.05
EtOAc—H_2O	0.17	0.63
Et_2O—H_2O	0.24	0.04

[a]In 9.0 mL each of solvent mixture; determined by ICP-MS, error limit ±4%.

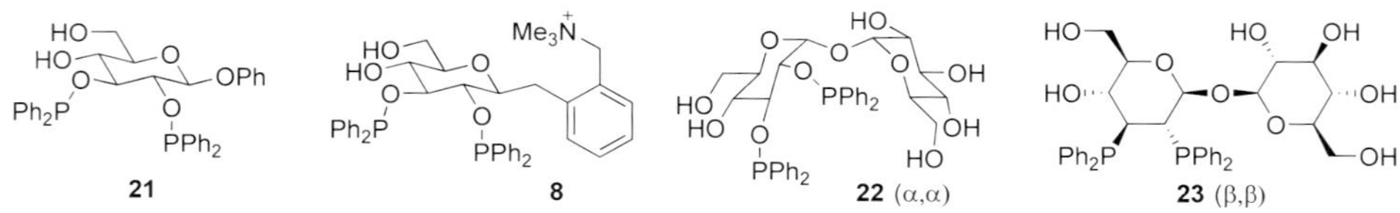

Figure 6.6 Chiral water-soluble, polyhydroxyphosphine ligands.

with polyhydroxylic substituents (e.g. **24–26**) offer an attractive option (Fig. 6.7).[63] These ligands show excellent enantioselectivity for asymmetric hydrogenation of methyl acetamidoacrylate as substrate as shown in Table 6.4. The enantioselectivity is highly dependent on the structure of the catalyst and the solvent. For example, while near 100% ee is observed reproducibly in neat methanol using **25a**, in MeOH–H_2O mixtures, often a capricious reaction is observed with enantioselectivity dropping off as the portion of MeOH increases. At the end of the reaction in several runs, precipitation of metallic Rh is noticed, indicating that nonselective Rh-mediated processes might be intervening. The problems of reproducibility and precipitation can be alleviated by the use of an extra equivalent of the ligand (entry 4, Table 6.4). Using this protocol, it is now possible to recycle the aqueous solution of the catalyst without loss of selectivity.

An often-stated, yet rarely achieved goal of developing water-soluble ligands is to find conditions where the aqueous layer containing the catalyst can be recycled with no loss of activity and selectivity. While a number of examples of such recovery of catalysts containing ionic ligands or polymeric support are known, ligands **24**, **25a/b**, and **26** (Fig. 6.7) are among the first nonionic ligands where this has been possible without apparent loss of selectivity.[52] The results for asymmetric hydrogenation of methyl acetamidoacrylate are shown in Table 6.5. The reaction for ligand **25b** was typically carried out in 1:1 MeOH–water with 1 mol% of the isolated cationic Rh complex as the precatalyst with 1 mol% of extra ligand, which is added to suppress precipitation of the metal. The product is separated at the end after each run by extraction into ether. The aqueous layer containing the catalyst was reused in subsequent hydrogenation. The results of hydrogenation using Rh(COD)[**25b**]BF_4 complex are shown in entries 1–3 in Table 6.5. The catalyst derived from ligand **25b** (a diethyl derivative) could be recycled three times with no apparent loss of enantioselectivity. It is tempting to speculate that the loss of reactivity upon repeated reactions is related to the increased hydrophobicity of the diethyl derivative and the attendant leaching of catalyst in ether, which is used to extract the product after each run. The catalyst from ligand **24** performed best for hydrogenation in neat water. The enantioselectivity and recyclability are outstanding, as the results shown in entries 4–7 of Table 6.5 indicate. The hydrogenation was

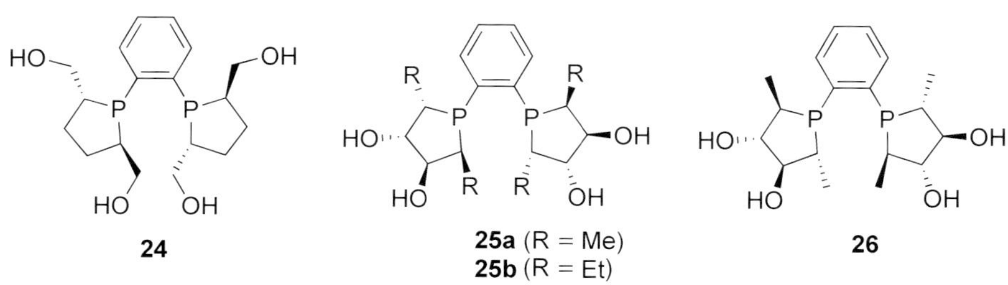

Figure 6.7 Water-soluble DuPHOS analogs.

Table 6.4 Asymmetric reduction of α-acetamidoacrylate **(28a)** by Rh(I)–L complexes

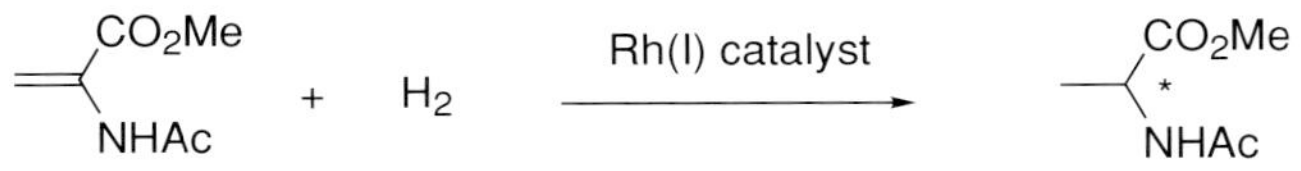

Entry	Ligand	Solvent	P_{H2} (atm)	Product:[a] % ee
1	**25a**	MeOH	3	99
2	**25a**	MeOH—H_2O(1:1)	3	69
3	**25a**	MeOH—H_2O(1:2)	3	21
4	**25a**	MeOH—H_2O(1:3)[b]	3	67
5	**25b**	MeOH	3	99
6	**25b**	H_2O	3	99
7	**24**	H_2O	3	99
8	**26**	MeOH—water(1:1)	3	83

[a]Ref. 52a.
[b]Extra 1 equiv. of ligand **25a** was added (Rh–L 1:2).

conducted in neat water, and at the end of the reduction, the product in aqueous phase was extracted with ether. Up to four sequential runs, ~99% ee and >90% isolated yield (100% conversion) were obtained in *neat* water. The recycling could be repeated up to seventh run with a small loss of selectivity. The catalytic efficiency suffers in runs beyond the fourth cycle. For example, in the seventh cycle, it is approximately 50%, as judged by the 12 h needed to complete the reaction instead of usual 6 h. Gratifyingly, high enantioselectivity (~95%) was still observed even for the seventh run.

The best example of supported aqueous phase catalysis,[64] in which a homogeneous catalyst is embedded in an aqueous layer over silica, is the use of tetrasulfonated BINAP ligand to reduce 2-(6′-methoxy-2′-naphthyl)acrylic acid (**31**; Fig. 6.4). The ee is dependent on the supported organic phase, and in this case, ethylene glycol on a porous glass gave up to 95% ee for the hydrogenation product (naproxen).

Enantiomerically pure sulfonated biphenylphosphine ligands [MeO-BIPHEP(SO_3Na)$_4$] have been synthesized by a novel nonsulfonation route and the corresponding ruthenium complexes used for aqueous and mixed phase reduction of several standard substrates such as geraniol (98% ee in ethyl acetate–water) and α-arylacrylic acid (up to 84% ee).[16e] This is one of a handful of systems where the reduction with the sulfonated ligand in aqueous medium gave the same enantiomeric excesses and turnover frequencies as the nonsulfonated ligand in

Table 6.5 Catalyst recycling experiments (see Fig. 6.7)

Entry	Ligand[a]	Run no.[b]	Conversion	% ee
1	**25b**	1	100	99
2	**25b**	3	100	99
3	**25b**	4	35	99
4	**24**	1	100	99
5	**24**	4	100	99
6	**24**	5	100	97
7	**24**	6	100	95

[a]1:1 MeOH–water with 1 % catalyst precursor and 1 % of extra ligand after each run.
[b]Intervening runs were omitted for clarity.

organic medium (methanol). A substrate/catalyst ratio up to 20,000 has been demonstrated. This area warrants further exploration.

Enantioselective hydrogenation of methyl 2-acetamidoacrylate has also been carried out with Rh(I)-MeDuPHOS occluded in a polydimethylsiloxane or polyvinyl alcohol (PVA) matrix.[65] The catalytic activity and, to some extent, the ee were lower than the homogeneous counterpart, even though for the reaction in water the catalyst can be recycled. In water, leaching of the catalyst was found to be higher from the PVA film. Significant improvements will be needed before this system can be used at a practical level.

Feringa prepared water-soluble analogs of his monodentate phosphoramidite ligands which served well in hydrogenation of dehydroamino acids.[66] Thus the attachment of a PEG residue to biscarbazole-derived phosphoramidite gave ligands whose Rh(I) complexes reduce acetamidoacrylate in water with ee up to 82%. Addition of surfactants increased the rate significantly, with a moderate improvement in enantioselectivity.

6.4 Hydrogenation of C=O bond

6.4.1 Chemoselectivity of C=C vs C=O bonds

Selective hydrogenation/reduction of α,β-unsaturated aldehydes by organometallic complexes in water leads to valuable intermediates for fragrance, flavor, and vitamin manufacture and has been developed into a bulk industrial process. A number of Ru(II)-based aqueous catalytic systems that reduce crotonaldehyde, cinnamaldehyde, methyl-3-buten-2-al, and citral into the corresponding alcohol have been developed, giving excellent yields (>95%) and selectivities (>97%).[67] Depending on which metal (Rh or Ru) is used with sulfonated triphenylphosphine, either the C=C or the C=O is reduced with excellent chemoselectivity.[68]

An interesting aspect of the hydrogenation of α,β-unsaturated aldehydes is the pH-dependent chemoselectivity change. In 1998, Joó and coworkers investigated pH-dependent protic equilibria of ruthenium hydride complexes having TPPMS ligands.[1g,69] The equilibrium distribution of hydride complexes indeed turned out to depend highly on the solution pH. As shown in Eq. 6.4, the major species in acidic solution is monohydrido complex **35**, while in neutral or basic media **36** predominates, as confirmed by ^{1}H and ^{31}P NMR spectroscopy. Hydrogenation of cinnamaldehyde at pH below 5 leads to a slow reduction of C=C bond to give 3-phenylpropanal selectively, while at pH values above 6, exclusive hydrogenation of the aldehyde functionality occurred, presumably catalyzed by **35**. The pH-dependent thermodynamics seems to be general phenomena and a number of similar studies have been reported for Rh(I) and Ir(I) in combination of variety of ligands.[70] Other recent studies have examined the pH-dependent hydrogenation of water-soluble carbonyl compounds by $[Cp^{*}Ir(III)(H_2O)_3]^{2+}$.[71] Further examples of this pH dependence will be discussed under asymmetric transfer hydrogenation in Section 6.5.2.

$$\underset{\mathbf{34}}{0.5\,[\{RuCl_2(TPPMS)_2\}_2]} + H_2 + \text{tppms} \rightleftharpoons \underset{\mathbf{35}}{[HRuCl(TPPMS)_3]} + H^+ + Cl^- \quad \text{(Eq. 6.4a)}$$

$$\underset{\mathbf{35}}{[HRuCl(TPPMS)_3]} + H_2 + (\text{tppms}) \rightleftharpoons \underset{\mathbf{36}}{[H_2RuCl(TPPMS)_4]} + H^+ + Cl^- \quad \text{(Eq. 6.4b)}$$

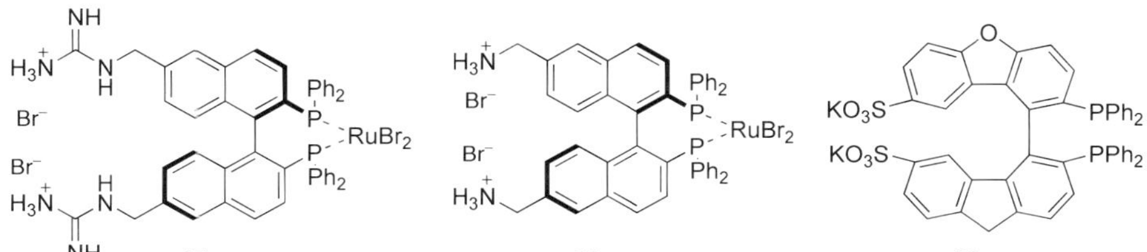

37 38 39

Figure 6.8 Water-soluble BINAP- or BIFAP-derivatives.

6.5 Asymmetric reduction of C=O bond in water

6.5.1 Asymmetric hydrogenation of C=O bond in water

Efficient asymmetric reduction of β-keto-carbonyl compounds and of alkyl aryl ketones have been extensively investigated. Genét and Lemaire groups reported water-soluble derivatives of BINAP (Fig. 6.8) for Ru(II)-catalyzed hydrogenation of methyl acetoacetate in aqueous solvent (Table 6.6). The parent catalyst of general formula RuX_2(BINAP) (X = Cl, Br, I) and $RuCl_2$(BINAP(dmf)$_2$ gives outstanding enantioselectivity for a wide variety of β-keto esters.[72] The enantioselectivities obtained with ligands **37** and **38**, water-soluble derivatives of BINAP, display marked dependence on the solvent. In general, enantioselectivity drops on going from MeOH to water.[73] For example, Genét reported that 62% ee was obtained with ammonium salt ligand **38** in asymmetric reduction of methyl acetoacetate in water, but an excellent enantioselectivity of 96% ee was obtained in methanol or in ethylene glycol (entries 1–3, Table 6.6). Diguanidinium salt ligand **37** similarly shows excellent enantioselectivity in ethylene glycol in the reduction of methyl acetoacetate. In contrast, Lemaire group claimed that ethyl acetoacetate is reduced in 94% ee with the ligand **38** in water under higher pressure of hydrogen.[74] The same group demonstrated that catalyst **38** could be recycled after extraction of product with pentane. The catalytic performance remained excellent without significant loss of activity until the third reuse, albeit with diminished enantioselectivity in the third run.

A novel skeleton having sulfonated benzofuran-based bisphosphine (BIFAPS) was reported by Hiemstra and coworkers.[75] They found that this ligand **39** in combination with

Table 6.6 Asymmetric hydrogenation of acetoacetate esters

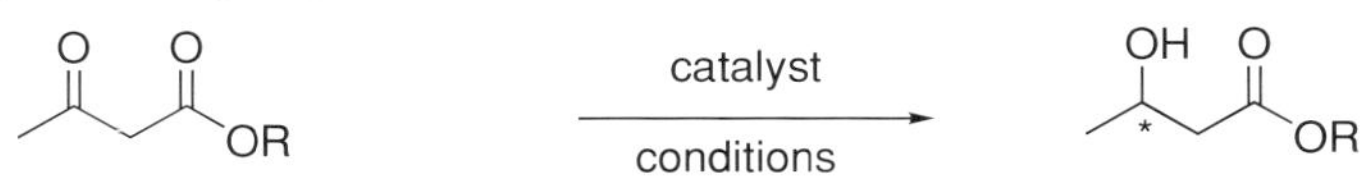

Entry	Ligand	R	Solvent	P_{H2} (atm)	Product % ee	Product Reference
1	**38**	Me	H_2O	20	62	73
2	**38**	Me	MeOH	20	80	73
3	**38**	Me	Ethylene glycol	20	96	73
4	**37**	Me	Ethylene glycol	20	96	73
5	**38**	Et	H_2O	40	94	74
6	**39**	Me	H_2O	100	22	75
7	**39**	Me	H_2O (1 % H_2SO_4)	100	86	75

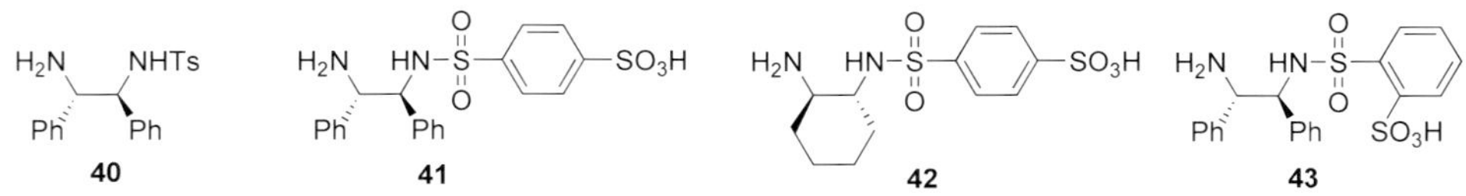

Figure 6.9 Sulfonated ligands for asymmetric transfer hydrogenation.

1% of H_2SO_4 gives an excellent enantioselectivity in the reduction of methyl acetoacetate in water, while poor conversion as well as mediocre enantioselectivity was observed without the addition of acid. While cases reporting beneficial effect of acid in combination with organometallic complexes have been known,[76] the origin of the effect of added acid remains unclear.

6.5.2 Asymmetric transfer hydrogenation of C=O bond in water

Asymmetric transfer hydrogenation with Ru(II) complexes derived from *N*-(*p*-toluenesulfonyl)-1,2-diphenylethylenediamine (**40**) (Ts-DPEN) shows outstanding performance in the reduction of a variety of aromatic and α,β-unsaturated ketones in organic media.[77] Transfer hydrogenation of ketones employs Ru, Rh, or Ir with various reductants for this reaction: formic acid or its salt, an azeotropic mixture of triethylamine and formic acid, or 2-propanol. A number of important design parameters have come to light in the development of catalysts for transfer hydrogenation, even though industrially this reaction is a less attractive option than the ones using hydrogen gas.

In 2001, Williams and coworkers reported sulfonated derivatives of Ts-DPEN **41–43** and examined asymmetric transfer hydrogenation in *i*-PrOH–water mixture (Fig. 6.9).[18] These ligands form active catalysts for asymmetric hydrogenation when combined with $[RuCl_2(p\text{-cymene})]_2$ or $[Cp^*MCl_2]_2$ (M = Rh and Ir). The resulting Ru(II), Rh(I), and Ir(I) complexes catalyze asymmetric hydrogenation of acetophenone derivatives in excellent enantioselectivities, with ligand **41** giving the highest selectivity (97% ee). An interesting observation they made in the application of these water-soluble ligands was the effect of increased proportion of water in solvent. In general, as in the hydrogenation of acetamidoacrylates or acetamidocinnamates, the lower enantioselectivity as well as activity is observed as the water content increases in solvent mixture. On the contrary, for the iridium-based catalyst derived from **41** and **42**, water significantly *enhanced activity and selectivity* as its content increases from 34 to 51% (entries 8–11, Table 6.7).

Chung and coworkers have investigated asymmetric transfer hydrogenation using Ru(II) complexes derived from L-proline-derived ligands **44** using azeotropic mixture of triethylamine–formic acid or sodium formate as reducing agent.[78] In the initial study it was shown that the Ru complexes efficiently catalyzed reduction of acetophenone derivatives in neat water. The observed enantioselectivity was highly dependent on the steric bulk of substrate, with ortho-substituted aryl ketones generally giving high enantioselectivity. In a subsequent study, the ligands **44a** and**44b** were found to be water soluble, and catalyst recovery and reuse was demonstrated by extraction of the reaction mixture with *n*-hexane. After each catalytic reaction, 1 equiv. of formic acid was replenished to maintain the same concentration of HCOONa. The results are shown in Table 6.8. The catalyst reuse has been demonstrated up to the sixth run (five recyclings) without significant decrease in yield or selectivity (entries 5 and 6). They further claimed that a surfactant, SDS, has a beneficial

Table 6.7 Asymmetric transfer hydrogenation of acetophenone using ligand **41–43**

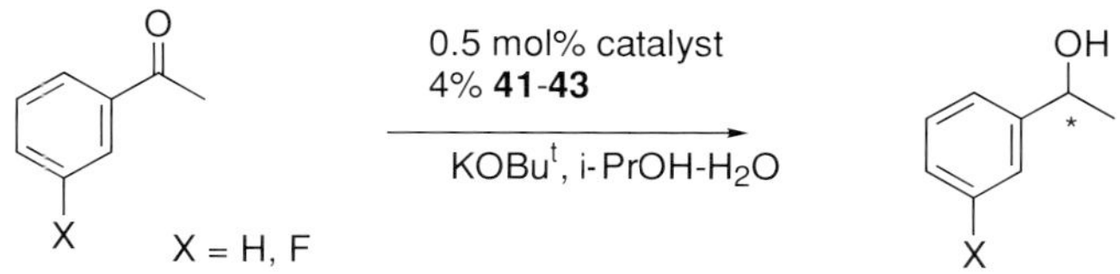

Entry	Ligand	Metal	Solvent	Time	Conversion	% ee	Reference
1	**41**	Ru	*i*-PrOH—H_2O (85:15)	48	96	94 (S)	18a
2	**42**	Ru	*i*-PrOH—H_2O (85:15)	48	91	88 (R)	18a
3	**43**	Ru	*i*-PrOH—H_2O (85:15)	48	11	91 (S)	18a
4	**41**(R, R)	Rh	*i*-PrOH—H_2O (85:15)	24	92	97 (R)	18b
5	**42**(R, R)	Rh	*i*-PrOH—H_2O (85:15)	18	94	95 (R)	18b
6	**41**(R, R)	Ir	*i*-PrOH—H_2O (85:15)	51[a]	83	85 (R)	18b
7	**42**(R, R)	Ir	*i*-PrOH—H_2O (85:15)	26[a]	99	94 (R)	18b
8	**41**(R, R)	Ir	*i*-PrOH—H_2O (2:1)	22[a]	74	92 (R)	18b
9	**41**(R, R)	Ir	*i*-PrOH—H_2O (1:1)	22[a]	90	92 (R)	18b
10	**42**(R, R)	Ir	*i*-PrOH—H_2O (2:1)	2.5[a]	82	94 (R)	18b
11	**42**(R, R)	Ir	*i*-PrOH—H_2O (1:1)	2.5[a]	94	93 (R)	18b

[a] *m*-fluoro acetophenone was used.

effect on the recoverability of the catalyst (entry 7).[78c] Under similar conditions, the sulfonated Ts-DPEN ligand **45** also showed good results in the reduction of acetophenone and 2-bromoacetophenone (entry 7).[79] Catalytic systems based on **45** displayed a more general substrate scope, giving high yields and enantioselectivities, albeit with slower turnover numbers. Heterogeneous catalytic systems based on ligands **46** and **47**, where Ts-DPEN group is anchored on to polymer (PEG),[80] or mesoporous silica gel,[81] provide another example of accelerating effect of water and enhanced recyclability in water in asymmetric transfer hydrogenation of acetophenones (Fig. 6.10).

Asymmetric transfer hydrogenation of insoluble ketones (e.g. α-bromomethyl aryl ketones) catalyzed by hydrophobic metal-amido complexes of Ir and Rh can be carried out

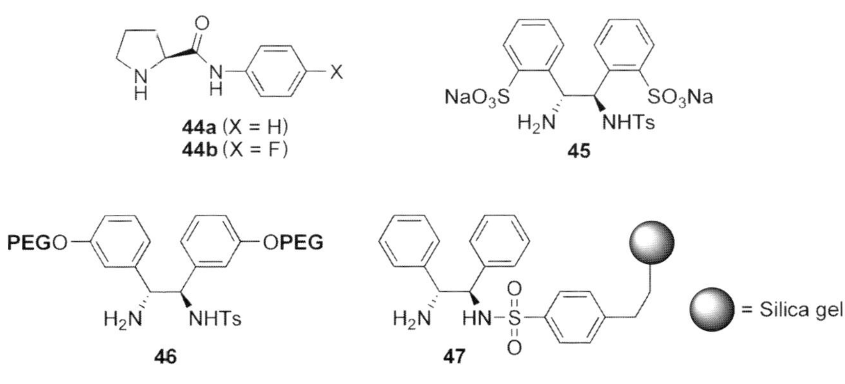

Figure 6.10 Chiral diamine ligands for asymmetric transfer hydrogenation in water.

Table 6.8 Asymmetric transfer hydrogenation of acetophenone (See Fig. 6.10)

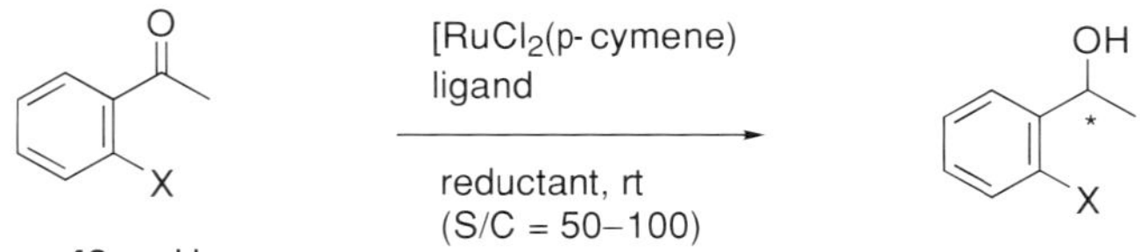

48a = H
48b = OMe

Entry	Ligand	Substrate[a]	Reductant[b]	Time(h)	Conversion	% ee	Reference
1	**44a**	**48a**[a]	HCOOH—Et_3N	7	97	79	78a
2	**44a**	**48b**[a]	HCOOH—Et_3N	6	98	95	78a
3	**44a**	**48a**[a]	HCOONa	4	99	61	78b
4	**44a**	**48b**[a]	HCOONa	4	99	95	78b
5	**44b**	**48b**[a]	HCOOH (2nd run)	5	99	95	78b
6	**44b**	**48b**[a]	HCOOH (6th run)	7	99	94	78b
7	**44b**	**48b**[c]	HCOONa, 40°C, SDS[d]	5.5	92	92	78c
8	**45**	**48a**[c]	HCOONa, 40°C, SDS[e]	24	95	93	79

[a] S/C = 50.
[b] 10 equiv. of reductant was used unless stated otherwise.
[c] S/C = 100.
[d] 2 mol% SDS was added and 5 equiv. of HCOONa was used.
[e] 4 mol% SDS was added and 5 equiv. of HCOONa was used.

in aqueous medium containing micelles, giving nearly perfect asymmetric induction.[82] Improved recyclability of micelle-embedded catalyst (up to six times) has also been claimed.

An intriguing discovery in asymmetric transfer hydrogenation in water was made by Xiao and coworkers in 2004.[83] They found that unmodified Noyori–Ikariya catalyst, [Ru(*p*-cymene)(**40**)Cl], showed superior activity as well as enantioselectivity both in HCOOH–Et_3N azeotrope and in water using HCOONa as a reductant compared to Ru(II)-catalytic system based on **41–43**. Surprisingly, for most acetophenone derivatives, reduction with HCOONa in water without any amphiphilic additives was complete in less than 2–3 h (Scheme 6.5) at S/C = 200. Corresponding reduction in HCOOH–Et_3N azeotrope proceeded approximately 10 times more slowly. This result indicated that the new protocol using 5 equiv. of HCOONa in water affords faster rates and comparable enantioselectivities, and thus water has a clearly beneficial effect in transfer hydrogenation. The results using S/C ratio of 1000 is depicted in Scheme 6.5.

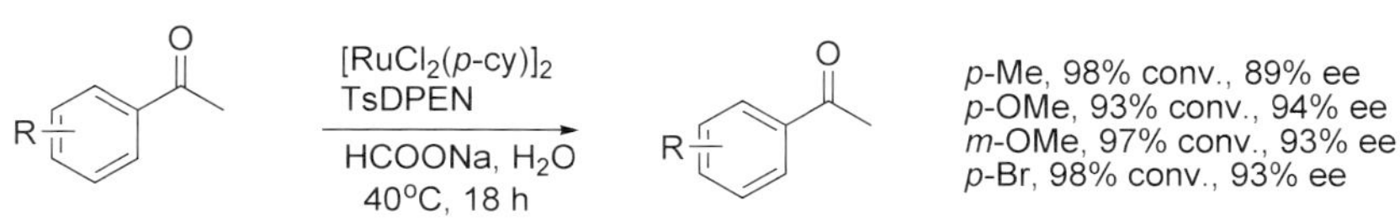

Scheme 6.5 Asymmetric transfer hydrogenation of acetophenone derivatives at S/C = 1000.

Above results inspired a kinetic study on the transfer hydrogenation as a function of pH.[84a] As in the case of $[(\eta^6\text{-}C_6M_6)Ru(bpy)(H_2O)]^{2+}$,[84b] and the previously mentioned Rh-catalyzed hydrogenations,[62] Ru-Ts-DPEN catalyst showed a strong pH dependence. The rate of reduction increases sharply when pH was changed from 4 to 5 in HCOOH–Et_3N

mixture. The authors ascribed the increased rate to the increased concentration of formate ion $[HCOO]^-$. As the reaction proceeds, conversion of HCOOH by Ru-Ts-DPEN into CO_2 and H_2 leads to increased basicity, thus increasing the concentration of $[HCOO]^-$. Another interesting observation is that enantioselectivity also varies with pH. Thus not only rate but also selectivity is significantly enhanced in basic media.

Recently Xiao and coworkers also reported[85] a Rh(III) catalyst prepared by a combination of $[Cp^*RhCl]_2$ and (1*R*,2*R*)-*N*-(*p*-toluenesulfonyl)1,2-cyclohexanediamine for the transfer hydrogenation of aryl alkyl ketones in air and water.

The most recent addition to the new protocols for transfer hydrogenation employs water-soluble cationic Ru complexes derived from *trans*-1,2-*trans*-diaminocyclohexane and $[(arene)Ru(Cl_2)]_2$ (arene = benzene, *p*-cymene, hexamethylbenzene).[86] The resulting complexes $[(arene)ClRu(H_2N\hat{}NH_2)]^+$ are four times more soluble in water than the corresponding, more familiar, neutral monotosyl amide complexes. However, the corresponding neutral monotosyl amides perform much better in transfer hydrogenation of acetophenone using an aqueous solution of sodium formate as the hydrogen source (up to 93% ee for $[(C_6Me_6)ClRu(H_2\hat{}NTs)]$. These authors also report crystal structures of a number of pre-catalysts. Along similar lines, a combination of $[(p\text{-cymene})Ru(Cl_2)]_2$ and a commercially available 1,2-aminoalcohol, (–)-ephedrine hydrochloride, has been found to be competent for transfer hydrogenation using sodium formate (2.5 mol% catalyst, ee up to 75% for acetophenone).[87]

Amphiphilic dendrimers derived from polyamidoamine and sugar lactones can also affect the enantioselectivity of $NaBH_4$ reduction of prochiral ketones including dialkyl ketones giving unprecedented enantioselectivity (up to 96% ee for 2-heptanone).[88]

6.5.3 Hydrogenation of C=N bond

Asymmetric hydrogenation of prochiral imines has been less systematically investigated than that of the C=C or C=O bond. However, notable examples have been reported for this reduction in water. Asymmetric hydrogenation of C=N bond in water was first examined by Sinou and coworkers in 1989. They prepared a mixture of mono-, di- and trisulfonated derivatives of DPBB ligand **4a–c** (Fig. 6.1) and tested hydrogenation of Schiff base derived from acetophenone using *in situ* generated catalyst from $[Rh(cod)Cl_2]_2$ and **4** in a biphasic mixture of EtOAc–water (Eq. 6.5). Although only moderate enantioselectivity (34–58% ee) was obtained at 70 bar of H_2, this marked the first example of asymmetric hydrogenation of imine in water.[58a]

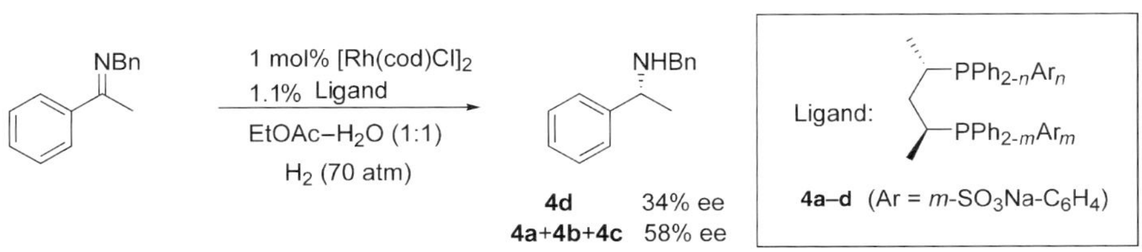

(Eq. 6.5)

Later, DeVries, Bakos, and their coworkers confirmed that the enantioselective catalyst for imine reduction is actually monosulfonated ligand **4a** prepared via RP-HPLC separation of mixtures of ligands **4a–c**.[58a,89] By using monosulfonated ligand **4a** as an inseparable mixture of diastereomers (chirality on phosphorus) under otherwise the same condition, the same ketimine was reduced with excellent conversion (100%) and enantioselectivity (94% ee). They further found that the disulfonated ligand **4b** gives almost racemic product. This interesting effect of degree of sulfonation on selectivity was presumed by the authors to be electronic in origin. Other imine reductions include rhodium/iridium-bimetallic-catalyzed hydroformylation/hydrogenation of terminal alkenes in the presence of ammonia. This reaction involves the *in situ* formation of a primary imine. Up to 91% selectivity for primary amines was achieved in a medium containing Rh and Ir sources and TPPTS.[90] Imine reduction in aqueous medium remains a largely unsolved problem in spite of several other effort.[91] Compared to other immobilization techniques, such as use of clay support[92] or *sc*-CO_2,[93] catalyst immobilization through the use of water-soluble ligands for hydrogenation of C=N bond has met with little success so far.

6.6 Miscellaneous reductions: reduction of epoxides, halides, and carbon dioxide

Hydrogenolysis of epoxides in water has been examined by Sinou and coworkers[94] using (*SS*)-BDDP-Rh(I) complexes. Even though the enantioselectivities obtained for hydrogenolysis of epoxysuccinate and for kinetic resolution of $(2S^*3R^*)$-phenylglycidate were <50%, these reactions deserve further attention. Hydrogenolysis of alkyl halides have been effected by $[RuCl_2(TPPMS)_2]_2$ and $[Ru(H_2O)_3(PTA)_3]^{2+}$ using sodium formate as the reductant.[95]

No discussion of hydrogenation in aqueous media will be complete without a mention of reduction of carbon dioxide to formate, methanol, and methane. This important topic has been reviewed recently and the reader is urged to consult the recent references to get a complete picture of this important area.[33a,96]

6.7 Summary and outlook

The success of several established industrial processes and many promising results from recent literature suggest that water has huge potential to be an outstanding solvent for hydrogenation reactions, especially when the substrate is at least partially soluble. Advantages by way of kinetic acceleration and higher selectivity have been corroborated in several cases, even though the fundamental rationale for some of these observations is only poorly developed. Likewise, catalyst tuning by manipulation of ligand structure and reaction parameters such as pH, temperature and solubility has been amply demonstrated. For large-scale synthesis, especially where expensive metals and ligands (for example, in asymmetric synthesis) are involved, catalyst recovery is crucial, and so is the use of gaseous hydrogen as a reductant. Use of macromolecular water-soluble ligands might aid catalyst recovery by membrane filtration.[97] Engineering the macromolecular scaffolding of a polymer to decrease its solubility under specific conditions such as pH or temperature (thermoregulated phase-transfer catalysis), once the reaction is complete, provides another attractive possibility for catalyst recovery.[98] For insoluble substrates, thus far, the best results are obtained with addition of surfactants and other micellar reagents, which is a major distraction. Lessons from 'micellar enzymology',[99] where it has been found that for some enzymes (e.g. pyrogallol peroxidase) reactions proceed much faster in reverse micelles than in water, could aid further developments in this area. Design of new ligands, especially neutral ones derived from the polyhydroxyphosphines, N-heterocyclic carbenes, or those derived from diamines and aminoalcohols will continue to play a major role in multiphase catalysis. An isolated example of a Cr(II) salt of *S*-valine giving up to 94% yield and 75% ee (*S*, *(1)-phenylethanol*) in the reduction of acetophenone bodes well for the use of simple readily available chiral sources in conjunction with appropriate metal ions.[100] Selective functionalization of biomolecules, which are often soluble in water, by catalytic reactions could be an area of potential value in connection with many biotechnological problems. Here water-soluble catalysts have several advantages. Replacement of enzymatic reactions, which are often the only alternatives currently available, by reactions catalyzed by water-soluble catalysts could improve volumetric productivity and broaden the substrate scope. In addition, there are many more reactions that homogeneous catalysts can promote vis-à-vis those aided by the commonly available enzymes. Combination of metal and enzyme catalysis[101] can also benefit from water-soluble homogeneous catalysts, and few examples have appeared since the early studies on the regeneration of NADH.[1i,102] Reduction of macromolecular biological membrane components[103] (especially under physiological conditions) and regeneration of biological cofactors are two other significant problems that await satisfactory solution. Another area where water-soluble catalysts could play an important role is in the conversion of biomass materials (e.g. ethanol, glycerol, carbohydrates) to other usable chemical feedstocks and intermediates currently derived from petroleum sources. In terms of disposal of side-products and other waste, an aqueous stream containing toxic organic materials is more burdensome than an organic stream containing the same materials, and innovative solutions must be found to avoid such situations. Even with such limitations, thanks to the vigorous research activity in the area, future prospects of finding useful processes that run in water are indeed excellent.

References

1. For recent monographs, see: (a) Horváth, I.T. and Joó, F. (Eds.), *Aqueous Organometallic Chemistry and Catalysis*, NATO ASI 3/5, Kluwer, Dordrecht, 1995; (b) Cornils, B. and Herrmann, W.A. (Eds.), *Aqueous-Phase Organometallic Catalysis*, 2nd edn., Wiley-VCH, Weinheim, Germany, 2004; (c) Cornils, B., Herrmann, W.A., Horváth, I.T., Leitner, W., Mecking, S., Olivier-Bourbigou, H. and Vogt, D. (Eds.), *Multiphase Homogeneous Catalysis*, Wiley-VCH, Weinheim, Germany, 2005. For recent reviews, see: (d) Dwars, T. and Oehme, G., Complex-catalyzed hydrogenation reactions in aqueous media, *Adv. Synth. Catal.*, 2002, **344**, 239; (e) Sinou, D., Asymmetric organometallic-catalyzed reactions in aqueous media, *Adv. Synth. Catal.*, 2002, **344**, 221; (f) Lindström, U.M., Stereoselective organic reactions in water, *Chem. Rev.*, 2002, **102**, 2751; (g) Joó, F., Aqueous biphasic hydrogenations, *Acc. Chem. Res.*, 2002, **35**, 738; (h) RajanBabu, T.V., Yan, Y.-Y. and Shin, S., Asymmetric catalysis in water: prospects and problems of using hydroxyphosphines and hydroxyphosphinites as ligands, *Curr. Org. Chem.*, 2003, **7**, 1759. The work before 1994 has been reviewed extensively; see: (i) Chaloner, P.A., Esteruelas, M.A., Joó, F. and Oro, L.A., Hydrogenation in aqueous systems, In: *Homogeneous Hydrogenation*, Kluwer, Dordrecht, 1994, pp. 183–239; (j) Joó, F. and Kathó A., Hydrogenation, In: Cornils, B. and Herrmann, W.A. (Eds.), *Aqueous-Phase Organometallic Catalysis*, Wiley-VCH, Weinheim, Germany, 1998, Chap. 6.2, pp. 340–372. For recent thematic issues of journals dedicated to applications of water-soluble organometallic compounds, see: (k) *J. Mol. Catal. A: Chem.*, 1997, **116**(1, 2), 1–309; *Catal. Today*, 1998, **42**(4), 371–478.
2. (a) Manassen, J., In: Basalo, F. and Burwell, R.L., Jr. (Eds.), *Catalysis: Progress in Research*, Plenum, London, 1973, pp. 177–183; (b) Dror, Y. and Manassen, J., Hydrogenation of olefins with rhodium–phosphine complexes, having substrate and catalyst in two different immiscible phases. An alternative method for the heterogenization of a homogeneous catalyst, *J. Mol. Catal.*, 1977, **2**, 219.
3. For a recent article on large-scale enantioselective catalysis, see: Blaser, H.-U., Enantioselective catalysis in fine chemicals production, *Chem. Commun.*, 2003, 293.

4. (a) Kuntz, E.G., Homogeneous catalysis... in water, *CHEMTECH*, 1987, **17**, 570; (b) Cornils, B. and Kuntz, E.G., Introducing TPPTS and related ligands for industrial biphasic processes, *J. Organomet. Chem.*, 1995, **502**, 177.
5. (a) Rhône-Poulenc, BP 362.037 (1987), EP 320.339; (b) Grosselin, J.M. and Mercier, C., Selective hydrogenation of α,β-unsaturated aldehydes in aqueous–organic two-phase solvent systems using ruthenium or rhodium complexes of sulfonated phosphines, *Organometallics*, 1991, **10**, 2126; (c) Grosselin, J.M. and Mercier, C., Selective hydrogenation of α,β-unsaturated aldehydes in aqueous–organic two phase solvent systems using ruthenium complexes of sulfonated phosphines, *J. Mol. Catal.*, 1990, **63**, L25–L27; For applications in vitamin synthesis, see: (d) Mercier, C. and Chabardes, P., Organometallic chemistry in industrial vitamin A and vitamin E synthesis, *Pure Appl. Chem.*, 1994, **66**, 1509. For earlier reports of use of water-soluble metal complexes, see: (e) Joó, F. and Bényei, A., A biphasic reduction of unsaturated aldehydes to saturated alcohols by ruthenium complex-catalyzed hydrogen transfer, *J. Organomet. Chem.*, 1989, **363**, C-19; (f) Bényei, A. and Joó, F., Organometallic catalysis in aqueous solutions: the biphasic transfer hydrogenation of aldehydes catalyzed by water-soluble complexes of ruthenuim, rhodium and iridium, *J. Mol. Catal.*, 1990, **58**, 151.
6. For an early outstanding example, see: Casalnuovo, A.L. and Calabrese, J.C., Palladium catalyzed alkylations in aqueous media, *J. Am. Chem. Soc.*, 1990, **112**, 4324.
7. (a) Keane, M.A., The importance of modifier pH in the generation of enantioselective nickel-catalysts, *Catal. Lett.*, 1993, **19**, 197; (b) Guo, H. and Ding, K.L., Reduction of 1,1′-binaphthyls to octahydro-1,1′-binaphthyl derivatives with Raney Ni–Al alloy in aqueous solution, *Tetrahedron Lett.*, 2000, **41**, 100061.
8. Sun, Y., Landau, R.N., Wang, J., LeBlond, C. and Blackmond, D.G., A reexamination of pressure effects on enantioselectivity in asymmetric hydrogenation, *J. Am. Chem. Soc.*, 1996, **118**, 1348.
9. Selke, R., Holz, J. and Riepe, A., Impressive enhancement of the enantioselectivity for a hydroxy-containing rhodium(I) bisphosphine catalyst in aqueous solution by micelle-forming amphiphiles, *Chem. Eur. J.*, 1998, **4**, 769, and references cited therein. For a recent application, see: Grassert, I., Schmidt, U., Ziegler, S., Fischer, C. and Oehme, G., Use of rhodium complexes with amphiphilic and nonamphiphilic ligands for the preparation of chiral α-aminophosphonic acid esters by hydrogenation in micellar media, *Tetrahedron: Asymmetry*, 1998, **9**, 4193.
10. (a) Trinkhaus, S., Kadyrov, R., Selke, R., Holz, J., Götze, L. and Börner, A., Internal vs. external ionic functionality – a comparative study in the asymmetric hydrogenation in water as solvent, *J. Mol. Catal. A*, 1999, **144**, 15; (b) Ludwig, M., Kadyrov, R., Fiedler, H., Haage, K. and Selke, R., Dependence of enantioselectivity on the distribution of a chiral hydrogenation catalyst between an aqueous and a micellar phase: investigation using pulsed field gradient spin echo NMR spectroscopy, *Chem. Eur. J.*, 2001, **7**, 3298. For other related effects, see: (c) Ding, H. and Hanson, B.E., Spectator cations and catalysis with sulfonated phosphines. The role of cations in determining reaction selectivity in the aqueous phase hydroformylation of olefins, *J. Mol. Catal. A*, 1995, **99**, 131. For an earlier report on salt effects, see: (d) Fache, E., Senocq, F., Santini, C. and Bassett, J.-M., Homogeneous catalysis in water: a remarkable salt effect in the hydrogenation of propionaldehyde with $RuCl_2(TPPTS)_3$ and related complexes [TPPTS = $(m\text{-}NaO_3SC_6H_4)_3P$], *J. Chem. Soc., Chem. Commun.*, 1990, 1776.
11. Bakos, J., Karaivanov, R., Laghmari, M. and Sinou, D., Chiral sulfonated phosphines. Role of water in the hydrogenation of dehydroamino acids, *Organometallics*, 1994, **13**, 2951.
12. For a critical, highly readable review of achievements, problems and prospects of this area for industrial applications, see: Cornils, B., Herrmann, W.A. and Eckl, R.W., Industrial aspects of aqueous catalysis, *J. Mol. Catal. A: Chem.*, 1997, **116**, 27.
13. (a) Dwars, T. and Ohme, G., Complex-catalyzed hydrogenation reactions in aqueous media, *Adv. Synth. Catal.*, 2002, **344**, 239; (b) Papadogianakis, G. and Sheldon, R.A., Solubilizing groups for phosphines, *Catalysis*, 1997, **13**, 114; (c) Cornils, B. and Herrmann, W.A. (Eds.), *Aqueous-Phase Organometallic Catalysis*, Wiley-VCH, Weinheim, Germany, 1998, Chap. 3.2, pp. 71–143; (d) Pinault, N. and Bruce, D.W., Homogeneous catalysts based on water-soluble phosphines, *Coord. Chem. Rev.*, 2003, **241**, 1.
14. (a) Joó, F. and Beck, M.T., Redox reactions of phosphine complexes in aqueous solution – new possibilities in homogeneous solution catalysis, *Magy. Kém. Folyóirat*, 1973, **79**, 189; *Chem. Abstr.*, 1973,**73**, 167966b; (b) Joó, F. and Beck, M.T., Formation and catalytic properties of water-soluble phosphine complexes, *React. Kinet. Catal. Lett.*, 1975, **2**, 257.
15. For a recent review, see: (a) Pinault, N. and Bruce, D.W., Homogeneous catalysts based on water-soluble phosphines, *Coord. Chem. Rev.*, 2003, **241**, 1. For history of the problem and an early review, see: (b) Sinou, D., Phosphines hydrosolubles. Synthèses et applications en catalyse, *Bull. Soc. Chim. Fr.*, 1987, 480. See also: (c) Alario, F., Amrani, Y., Colleuille, Y., Dang, T., Jenck, J., Morel, D. and Sinou, D., Asymmetric hydrogenation in aqueous organic 2-phase solvent systems using rhodium complexes of sulfonated phosphines, *J. Chem. Soc., Chem. Commun.*, 1986, **3**, 202; (d) Lecomte, L. and Sinou, D., Chiral sulfonated phosphines. Part III. Rh(I)-tetrasulfonated diphosphine complex: homogeneous or heterogeneous catalyst?, *J. Mol. Catal.*, 1989, **52**, L21–L24.

16. For sulfonation of other sensitive ligands, see: (a) Ding, H., Hanson, B.E. and Bakos, J., Preparation of a surface-active chiral diphosphane and its use in the hydrogenation of prochiral olefins, *Angew. Chem., Int. Ed. Engl.*, 1995, **34**, 1645. For biaryl derivatives, see: (b) Wan, K.-T. and Davis, M.E., Ruthenium (II)-sulfonated BINAP: a novel water-soluble asymmetric hydrogenation catalyst, *Tetrahedron: Asymmetry*, 1993, **4**, 2461. For use of boric acid to moderate reactivity of oleum, see: (c) Eckl, R.W., Priermeier, T. and Herrmann, W.A., Mono- and biphasic asymmetric hydroformylation with rhodium catalysts of diphosphine ligand NAPHOS and its sulfonated derivatives, *J. Organomet. Chem.*, 1997, **532**, 243; (d) Gelpke, A.E.S., Kooijman, H., Spek, A.L. and Hiemstra, H., Synthesis of the dibenzofuran-based diphosphine ligand BIFAP and its water-soluble derivative BIFAPS and their use in ruthenium-catalyzed asymmetric hydrogenation, *Chem. Eur. J.*, 1999, **5**, 2472; (e) Schmid, R., Broger, E.A., Cereghetti, M., Crameri, Y., Foricher, J., Lalonde, M., Müller, R.K., Scalone, M., Schoettel, G. and Zutter, U., Tetrasulfonated analogs of OMe-BIPHEP: new developments in enantioselective hydrogenation, *Pure Appl. Chem.*, 1996, **68**, 131; (f) Ma, Y., Liu, H., Chen, Li., Cui, X., Zhu, J. and Deng, J., Asymmetric transfer hydrogenation of prochiral ketones in aqueous media with new water-soluble chiral vicinal diamine as ligand, *Org. Lett.*, 2003, **5**, 2103.
17. Herrmann, W.A. and Kohlpaintner, C.W., Water-soluble ligands. Metal complexes, and catalysts: synergism of homogeneous and heterogeneous catalysts, *Angew. Chem., Int. Ed. Engl.*, 1993, **32**, 1524.
18. (a) Bubert, C., Blacker, J., Brown, S.M., Crosby, J., Fitzjohn, S., Muxworthy, J.P., Thorpe, T. and Williams, J.M.J., Synthesis of water-soluble aminosulfonamide ligands and their application in enantioselective transfer hydrogenation, *Tetrahedron Lett.*, 2001, **42**, 4037; (b) Thorpe, T., Blacker, J., Brown, S.M., Bubert, C., Crosby, J., Fitzjohn, S., Muxworthy, J.P. and Williams, J.M.J., Efficient rhodium and iridium-catalyzed asymmetric transfer hydrogenation using water-soluble aminosulfonamide ligands, *Tetrahedron Lett.*, 2001, **42**, 4041.
19. Mul, W.P., Ramkisoensing, K., Kamer, P.C.J., Reek, J.N.H., van der Liden, A.J., Marson, A. and van Leeuwen, P.W.N.M., New, highly efficient workup protocol for sulfonated diphosphines, *Ad. Synth. Catal.*, 2002, **344**, 293.
20. Joó, F., Kovács, J., Katho, Á., Bényei, A.C., Decuir, T. and Darensbourg, D.J., Ligands for water-solubilizing organometallic compounds, *Inorg. Synth.*, 1998, **32**, 1.
21. Bitterer, F., Herd, O., Hasseler, A., Kuhnel, M., Rettig, K., Stelzer, O., Sheldrick, W.S., Nagel, S. and Rosch, N., Water-soluble phosphines. 6. Tailor-made syntheses of chiral secondary and tertiary phosphines with sulfonated aromatic substituents: structural and quantum chemical studies, *Inorg. Chem.*, 1996, **35**, 4103.
22. Wedgwood, J.L., Hunter, A.P., Kresinski, R.A., Platt, A.W.G. and Stein, B.K., Water-soluble platinum(II) complexes of alkyl sulfonated phosphines, *Inorg. Chim. Acta*, 1999, **290**, 189.
23. (a) Herd, O., Hesseler, A., Tepper, M. and Stelzer, O., Palladium catalyzed P–C coupling – a powerful tool for the syntheses of hydrophilic phosphines, *J. Organometal. Chem.*, 1996, **522**, 69; (b) Herd, O., Hesseler, A., Hingst, M., Machnitzki, P., Tepper, M. and Stelzer, O., Palladium catalyzed P–C coupling – a powerful tool for the syntheses of hydrophilic phosphines, *Catal. Today*, 1998, **42**, 413.
24. Lavenot, L., Bortoletto, M.H., Roucoux, A., Larpent, C. and Patin, H., Synthesis of new hydrophilic phosphines by addition of diphenylphosphine in activated alkenes, *J. Organomet. Chem.*, 1996, **509**, 9. For a recent report, see: (b) Sadow, A.D., Isabelle Haller, I., Fadini, F. and Togni, A., Nickel(II)-catalyzed highly enantioselective hydrophosphination of methacrylonitrile, *J. Am. Chem. Soc.*, 2004, **126**, 14704.
25. See, for example, Nagel, U. and Albrecht, J., The enantioselective hydrogenation of *N*-acyl dehydroamino acids, *Top. Catal.*, 1998, **5**, 3.
26. (a) Hesseler, A., Stelzer, O., Dibowski, H., Worm, K. and Schmidtchen, F.P., Water-soluble cationic phosphine ligands containing *m*-guanidinium phenyl moieties. Syntheses and applications in aqueous Heck-type reactions, *J. Org. Chem.*, 1997, **62**, 2362; (b) Guerreiro, P., Ratovelomanana-Vodal, V., Genêt, J.-P. and Dellis, P., Recyclable diguanidinium-BINAP and PEG-BINAP supported catalysts: syntheses and use in Rh(I) and Ru(II) asymmetric hydrogenation reactions, *Tetrahedron Lett.*, 2001, **42**, 3423.
27. Remmele, H., Köllhofer, A. and Plenio, H., Recyclable catalyst with cationic phase tags for the Sonogashira coupling of aryl bromides and aryl chlorides, *Organometallics*, 2003, **22**, 4098.
28. (a) Darensbourg, D.J., Stafford, N.W., Joó, F. and Reibenspies, J.H., Water-soluble organometallic compounds. 5. The regioselective catalytic hydrogenation of unsaturated aldehydes to saturated aldehydes in an aqueous 2-phase solvent system using 1,3,5-triaza-7-phosphaadamantane complex of rhodium, *J. Organomet. Chem.*, 1995, **488**, 99; (b) Pruchnik, F.P., Smolenski, P., Galdecka, E. and Galdecki, Z., Structural, spectroscopic and catalytic properties of water-soluble hydride rhodium complexes $[RhH(rtpa^+)I^-]H_2O$ (R = Me, Et), *Inorg. Chim. Acta*, 1999, **293**, 110; (d) Joó, F., Kovács, J., Kathó, A., Bényei, A.C., Decuir, T. and Darensbourg, D.J., Ligands for water-solubilizing organometallic compounds, *Inorg. Synth.*, 1998, **32**, 1.
29. Hashizume, T., Yonehara, K., Ohe, K. and Uemura, S., A novel amphiphilic chiral ligand derived from D-glucosamine. Application to palladium-catalyzed asymmetric allylic substitution in an aqueous or organic medium, allowing catalyst recycling, *J. Org. Chem.*, 2000, **65**, 5197.
30. Nagel, U. and Kinzel, E., Enantioselective catalytic hydrogenation of α-(acetamido)cinnamic acid with a rhodium phosphane complex in aqueous solution, *Chem. Ber.*, 1986, **119**, 1731.

31. (a) Tóth, I. and Hanson, B.E., Novel chiral water-soluble phosphines. I. Functionalized DIOP, chiraphos and BDDP derivatives and quaternization of their rhodium complexes, *Tetrahedron Asymmetry*, 1990, **1**, 895; (b) Tóth, I., Hanson, B.E. and Davis, M.E., Novel chiral water-soluble phosphines II. Applications in catalytic asymmetric hydrogenation, *Tetrahedron Asymmetry*, 1990, **1**, 895.
32. Yong, Y-.Y. and RajanBabu, T.V., Water-soluble organometallic catalysts from carbohydrates. 2. A strategy for the preparation of catalysts with pendant quaternary ammonium groups using D-salicin, *J. Org. Chem.*, 2001, **66**, 3277.
33. (a) Geldbach, T.J., Laurenczy, G., Scopelliti, R. and Dyson, P.J., Synthesis of imidazolium-tethered ruthenium(II)–arene complexes and their application in biphasic catalysis, *Organometallics*, 2006, **25**, 733. Related catalysts with DPEN-Ts ligand are excellent for transfer hydrogenation in ionic media: (b) Geldbach, T.J. and Dyson, P.J., A versatile ruthenium precursor for biphasic catalysis and its application in ionic liquid biphasic transfer hydrogenation: conventional vs task-specific catalysts, *J. Am. Chem. Soc.*, 2004, **126**, 8114.
34. Mohr, B., Lynn, D.M. and Grubbs, R.H., Synthesis of water-soluble, aliphatic phosphines and their application to well-defined ruthenium olefin metathesis catalysts, *Organometallics*, 1996, **15**, 4317.
35. Crochet, P., Díez, J., Fernández-Zúmel, M.A. and Gimeno, J., Catalytic isomerization of allylic alcohols by (η^6-*p*-cymene)–ruthenium(II) complexes in organic and aqueous media: a new recyclable and highly efficient catalyst in water-containing ammonium-functionalized ligands, *Adv. Synth. Catal.*, 2006, **348**, 93–100.
36. Csabai, P. and Joó, F., Synthesis and catalytic properties of new water-soluble ruthenium(II)–N-heterocyclic carbene complexes, *Organometallics*, 2004, **23**, 5640.
37. (a) Quinn, P.J., In: Cornils, B. and Herrmann, W.A. (Eds.), *Aqueous-Phase Organometallic Catalysis*, Wiley-VCH, Weinheim, Germany, 1998, Chap. 3.2, pp. 418–431; (b) Madden, T.D., Peel, W.E., Quinn, P.J. and Chapman, D., *J. Biochem. Biophys. Methods*, 1980, **2**, 19. For biomedical applications of hydroxyphosphines, see: (c) Katti, K.V., Gali, H., Smith, C.J. and Berning, D.E., Design and development of functionalized water-soluble phosphines: catalytic and biomedical implication, *Acc. Chem. Res.*, 1999, **32**, 9.
38. Börner, A., Kless, A., Kempe, R., Heller, D., Holz, J.L. and Braumann, W., β-Alkoxyalkylphosphanes and β-hydroxyalkylphosphanes as ligands in the stereoselective hydrogenation – a comparison, *Chem. Ber.*, 1995, **128**, 767; (b) Boms, S., Kadyrov, R., Heller, D., Braumann, W., Spannenberg, A., Kempe, R., Holz, J. and Börner, A., Catalytic and structural features of hydroxy and methoxy groups as hemilabile coordinating ligands in chiral (diphosphane)-rhodium(I) hydrogenation catalysts, *Eur. J. Inorg. Chem.*, 1998, 1291; (c) Carmichael, D., Doucet, H. and Brown, J.M., Hybrid *P*-chiral diphosphines for asymmetric hydrogenation, *Chem. Commun.*, 1999, 261.; (d) Borns, S., Kadyrov, R., Heller, D., Braumann, W., Holz, J. and Börner, A., Counter-ligand and solvent dependent oxygen–metal interactions of hemilabile coordinating hydroxy groups in chiral diphosphine rhodium(I) hydrogenation catalysts, *Tetrahedron: Asymmetry*, 1999, **10**, 1425.
39. Börner, A., The effect of internal hydroxy groups in chiral diphosphine rhodium(I) catalysts on the asymmetric hydrogenation of functionalized olefins, *Eur. J. Inorg. Chem.*, 2001, 327.
40. (a) Klotzer, D., Mading, P. and Munze, R., 1,2-Bis[bis(hydroxylmethyl)phosphine]ethane – preparation, formation of complexes and characterization, *Z. Chem.*, 1984, **24**, 224; (b) Baxley, G.T., Miller, W.K., Lyon, D.K., Miller, B.E., Nieckarz, G.F., Weakley, T.J.R. and Tyler, D.R., Synthesis and characterization of water-soluble 1,2-bis[bis(hydroxyalkylphosphino)ethane ligands and their nickel(II), ruthenium(III), and rhodium(I) complexes, *Inorg. Chem.*, 1996, **35**, 6688; (c) Baxley, G.T., Weakley, T.J.R., Miller, W.K., Lyon, D.K. and Tyler, D.R., Synthesis and catalytic chemistry of two new water-soluble chelating phosphines – comparison of ionic functionalities, *J. Mol. Catal. A*, 1997, **116**, 191.
41. Lavenot, L., Bortoletto, M.H., Roucox, A., Larpent, C. and Patin, H., Synthesis of new hydrophilic phosphines by addition of diphenylphosphine on activated alkenes: characterization of their rhodium complexes, *J. Organomet. Chem.*, 1996, **509**, 9, and references cited therein.
42. Pringle, P.G., Brewin, D., Smith, M.B. and Korboys, K., Metal-catalyzed hydroxyphosphination as a route to water-soluble phosphines, In: Horváth, I. and Joó, F. (Eds.), *Aqueous Organometallic Chemistry and Catalysis*, NATO ASI 3/5, Kluwer, Dordrecht, 1995.
43. Sinou, D. and Amrani, Y., Asymmetric water-soluble diphosphines: synthesis and use in hydrogenation as their rhodium complexes of polyoxadiphosphines, *J. Mol. Catal.*, 1986, **36**, 319.
44. (a) Dieguez, M., Pamies, O. and Claver, C., Ligands derived from carbohydrates for asymmetric catalysis, *Chem. Rev.*, 2004, **104**, 3189; (b) RajanBabu, T.V., Casalnuovo, A.L., Ayers, T.A., Nomura, N., Jin, J., Park, H. and Nandi, M., Ligand tuning as a tool for the discovery of new catalytic asymmetric processes,*Curr. Org. Chem.*, 2003, **7**, 301.
45. (a) Kumar, A., Oehme, G., Roque, J.P., Schwarze, M. and Selke, R., Increase in the enantioselectivity of asymmetric hydrogenation in water influenced by surfactants or polymerized micelles, *Angew. Chem. Int. Ed.*, 1994, **33**, 2197; (b) Selke, R., Carbohydrate phosphinites as chiral ligands for asymmetric synthesis catalyzed by complexes. V. New rhodium(I)-chelates prepared by precipitation from the equilibrium between neutral and cationic complexes and hydrolysis of their bonded ligands, *J. Organomet. Chem.*, 1989, **370**, 241; (c) Grassert, I., Paetzold, E. and Oehme, G., Influence of different types of amphiphiles on the rhodium(I) complex-catalyzed asymmetric hydrogenation of (*Z*)-methyl-α-acetamidocinnamate in aqueous medium,

Tetrahedron, 1993, **49**, 6605; (d) Selke, R., Ohff, M. and Riepe, A., Asymmetric hydrogenation – influence of the structure of carbohydrate derived catalysts on the relative enantioselectivity $Q_{H/Me}$ regarding acid and ester substrates and its inversion. Selectivity increase in water by amphiphiles, *Tetrahedron*, 1996, **52**, 15079; (e) Selke, R., Holz, J., Riepe, A. and Börner, A., Impressive enhancement of the enantioselectivity for a hydroxy-containing rhodium(I) bisphosphine catalyst in aqueous solution by micelle-forming amphiphiles, *Chem. Eur. J.*, 1998, **4**, 769.

46. Grassert, I., Vill, V. and Oehme, G., Investigation of influence of carbohydrate amphiphiles on the complex catalyzed asymmetric hydrogenation of (*Z*)-methyl α-acetamidocinnamate in water, *J. Mol. Catal. A: Chem.*, 1997, **116**, 231.
47. (a) Yonehara, K., Hashizume, T., Mori, K. and Uemura, S., Novel water-soluble bis-phosphinites chiral ligands derived from α,α- and β,β-trehalose. Application to asymmetric hydrogenation of dehydroamino acids and their esters in water or an aqueous/organic biphasic medium, *J. Org. Chem.*, 1999, **64**, 5593; (b) Yonehara, K., Hashizume, T., Mori, K., Ohe, K. and Uemura, S., Highly enantioselective hydrogenation of enamides and itaconic acid in water in the presence of water-soluble rhodium(I) catalyst and sodium dodecyl sulfate, *J. Org. Chem.*, 1999, **64**, 9381.
48. Shin, S. and RajanBabu, T.V., Water-soluble organometallic catalysts from carbohydrates. 1. Diphosphinite-Rh complexes, *Org. Lett.*, 1999, **1**, 1229.
49. (a) Wallace, P.A. and Minnikin, D.E., Synthesis of 2,3-di-*O*-palmitoyl-α,α-trehalose via a novel tri-protected trehalose intermediate, *J. Chem. Soc., Chem. Commun.*, 1993, 1292; (b) Cook, S.J., Khan, R. and Brown, J.M., A simple route to β,β-trehalose via trichloroacetimidates, *J. Carbohydr. Chem.*, 1984, **3**, 343.
50. (a) Burk, M.J., Feaster, J.E. and Harlow, R.L., New chiral phospholanes: synthesis, characterization, and use in asymmetric hydrogenation reactions, *Tetrahedron: Asymmetry*, 1991, **2**, 569; (b) Burk, M.J., Feaster, J.E., Nugent, W.A. and Harlow, R.L., Preparation and use of C2-symmetric bis(phospholanes)-production of α-aminoacid derivatives via highly enantioselective hydrogenation reactions, *J. Am. Chem. Soc.*, 1993, **115**, 10125; (c) Burk, M.J., Harper, T.G.P. and Kalberg, C.S., Highly enantioselective hydrogenation of β-ketoesters under mild conditions, *J. Am. Chem. Soc.*, 1995, **117**, 4423; (d) Burk, M.J., Kalberg, C.S. and Pizzano, A., Rh-DuPHOS-catalyzed enantioselective hydrogenation of enol esters. Application to the synthesis of highly enantioenriched α-hydroxyesters and 1,2-diols, *J. Am. Chem. Soc.*, 1998, **120**, 4345; (e) Burk, M.J., Gross, M.F., Harper, T.G.P., Kalberg, C.S., Lee, J.R. and Martinez, J.P., Asymmetric catalytic routes to chiral building blocks of medicinal interest, *Pure Appl. Chem.*, 1996, **68**, 37, and references therein.
51. (a) Li, W., Zhang, Z., Xiao, D. and Zhang, X., Rhodium-hydroxy bisphospholane catalyzed highly enantioselective hydrogenation of dehydroamino acids and esters, *Tetrahedron Lett.*, 1999, **40**, 6701; (b) Li, W., Zhang, Z., Xiao, D. and Zhang, X., Synthesis of chiral hydroxy phospholanes from D-mannitol and their use in asymmetric catalytic reactions, *J. Org. Chem.*, 2000, **65**, 3489.
52. (a) RajanBabu, T.V., Yan, Y.-Y. and Shin, S., Synthesis, characterization, and applicability of neutral polyhydroxy phospholane derivatives and their rhodium(I) complexes for reactions in organic and aqueous media, *J. Am. Chem. Soc.*, 2001, **123**, 10217; (b) RajanBabu, T.V., Yan, Y.-Y. and Shin, S., Asymmetric catalysis in water: prospects and problems of using hydroxyphosphines and hydroxyphosphinites as ligands, *Curr. Org. Chem.*, 2003, **7**, 1526.
53. Yan, Y.-Y. and RajanBabu, T.V., Highly flexible synthetic routes to functionalized phospholanes from carbohydrates, *J. Org. Chem.*, 2000, **65**, 900.
54. Holz, J., Heller, D., Sturmer, R. and Börner, A., Synthesis of the first water-soluble tetrahydroxydiphosphine Rh(I) catalyst for enantioselective hydrogenation, *Tetrahedron Lett.*, 1999, **40**, 7059.
55. For an especially pertinent discussion, see: Joó, F. and Kathó, A., Hydrogenation, In: Cornils, B. and Herrmann, W.A. (Eds.), *Aqueous-Phase Organometallic Catalysis*, Wiley-VCH, Weinheim, Germany, 1998, Chap. 6.2, pp. 340–372.
56. For use of cobalt salts for diene hydrogenation, see: Lee, J.T. and Alper, H., Regioselective hydrogenation of conjugated dienes catalyzed by hydridopentacyanocobaltate anion using βcyclodextrin as the phase transfer agent and lanthanide halides as promoters, *J. Org. Chem.*, 1990, **55**, 1854. For hydrogenation of terminal olefins using cyclodextrin-containing phosphine:Rh, see: Reetz, M.T. and Waldvogel, S.R., β-Cyclodextrin-modified diphosphanes as ligands for supramolecular rhodium catalysts, *Angew. Chem., Int. Ed. Engl.*, 1997, **36**, 865.
57. (a) Wan, K.T. and Davis, M.E., Ruthenium (II)-sulfonated BINAP: a novel water-soluble asymmetric hydrogenation catalyst, *Tetrahedron: Asymmetry*, 1993, **4**, 2461; (b) Wan, K.T. and Davis, M.E., Asymmetric hydrogenation in water by a rhodium complex of sulfonated 2,2′-bis(diphenylphosphino)-1,1′-binaphthyl (BINAP), *J. Chem. Soc., Chem. Commun.*, 1993, 1262.
58. (a) Amrani, Y., LeComte, L., Sinou, D., Bakos, J., Tóth, I. and Heil, B., Chiral sulfonated phosphines – synthesis and use as ligands in asymmetric hydrogenation using an aqueous organic 2-phase solvent system, *Organometallics*, 1989, **8**, 542. For use of **20**, see: (b) Amrani, Y. and Sinou, D., Synthesis and application in enantioselective hydrogenation of a rhodium complex of an asymmetric water-soluble diphosphine PGE-1-DIOP, *J. Mol. Catal.*, 1984, **24**, 231.

59. (a) Nagel, U. and Kinzel, E., Enantioselective catalytic hydrogenation of α-(acetylamino)cinnamic acid with rhodium phosphane complex in aqueous solution, *Chem. Ber.*, 1986, **119**, 1731.
60. Malmstrom, T. and Andersson, C., Enantioselective hydrogenation in water catalyzed by rhodium phosphine complexes bound to polyacrylic acid, *J. Mol. Catal. A: Chem.*, 1999, **139**, 259.
61. Selke, R., Ohff, M. and Riepe, A., Asymmetric hydrogenation – influence of the structure of carbohydrate derived catalysts on the relative enantioselectivity of $Q_{H/Me}$ regarding acid and ester substrates and its inversion – selectivity increase in water by amphiphiles, *Tetrahedron*, 1996, **52**, 15079.
62. Malmstrom, T., Wendt, O.F. and Andersson, C., pH effects on reaction rates in rhodium catalyzed hydrogenation in water, *J. Chem. Soc., Dalton Trans.*, 1999, 2871.
63. For a review of the synthesis of hydroxyphosphines, see: Holz, J., Quirmbach, M. and Börner, A., Strategies for the synthesis of chiral hydroxyphosphines – a class of versatile ligands and ligand precursors for asymmetric catalysis, *Synthesis*, 1997, 983.
64. (a) Arhancet, J.P., Davis, M.E., Merola, J.S. and Hanson, B.E., Hydroformylation by supported aqueous-phase catalysis: a new class of heterogeneous catalysts, *Nature*, 1989, **339**, 454; (b)Wan, K.T. and Davis, M.E., Design and synthesis of a heterogeneous catalyst, *Nature*, 1994, **370**, 449; (c) Wan, K.T. and Davis, M.E., Asymmetric synthesis of naproxen by supported aqueous-phase catalysis, *J. Catal.*, 1994, **148**, 1; (d) Wan, K.T. and Davis, M.E., Asymmetric synthesis of naproxen by a new heterogeneous catalyst, *J. Catal.*, 1995, **152**, 25.
65. Wolfson, A., Janssens, S., Vankelecom, I., Geresh, S., Gottlieb, M. and Herskowitz, M., Aqueous enantioselective hydrogenation of methyl 2-acetamidoacrylate with Rh-MeDuPhos occluded in PDMS, *Chem. Commun.*, 2002, 388; (b) Wolfson, A., Geresh, S., Gottlieb, M. and Herskowitz, M., Heterogenization of Rh-MeDuPhos by occlusion in polyvinyl alcohol films, *Tetrahedron: Asymmetry*, 2002, **13**, 465.
66. Hoen, R., Leleu, S., Botman, P.N.M., Appelman, V.A.M., Feringa, B.L., Hiemstra, H., Minnaard, A.J. and van Maarseveen, J.H., PegPhos: a monodentate phosphoramidite ligand for enantioselective rhodium-catalyzed hydrogenation in water, *Org. Biomol. Chem.*, 2006, **4**, 613.
67. Refs. 5b and 5c and Allmang, G., Grass, J.M., Grosselin, J.M. and Mercier, C., Catalytic hydrogenation of all-*trans*-retinal to all-*trans*-retinol with ruthenium complexes, *J. Mol. Catal.*, 1991, **66**, L27–L31.
68. Grosselin, J.M. and Mercier, C., Selective hydrogenation of α,β-unsaturated aldehydes in aqueous organic two-phase solvent systems using ruthenium or rhodium complexes of sulfonated phosphines, *Organometallics*, 1991, **10**, 2126, and references cited therein.
69. Joó, F., Kovács, J., Bényei, A.C. and Kathó, A., Solution pH: a selectivity switch in aqueous organometallic catalysis – hydrogenation of unsaturated aldehydes catalyzed by sulfonatophenylphosphane–Ru complexes, *Angew. Chem. Int. Ed.*, 1998, **37**, 969.
70. (a) Joó, F., Csiba, P.J. and Bényei, A.C., Effect of water on the mechanism of hydrogenations catalyzed by rhodium phosphine complexes, *J. Chem. Soc., Chem. Commun.*, 1993, 1602; (b) Avnir, D., Sertchook, H., Blum, J., Joó, F., Kathó, Á., Schumann, H., Weimann, R. and Wernik, S., Sol-gel entrapped lipophilic and hydrophilic ruthenium-, rhodium-, and iridium-phosphine complexes as recyclable isomerization catalysts, *J. Mol. Catal. A*, 1996, **108**, 153; (c) Mebi, C.A. and Frost, B.J., Effect of pH on the biphasic catalytic hydrogenation of benzylidene acetone using $CpRu(PTA)_2H$, *Organometallics*, 2005, **24**, 2339; (d) Mebi, C.A. and Frost, B.J., Aqueous organometallic chemistry: synthesis, structure, and reactivity of the water-soluble metal hydride $CpRu(PTA)_2H$, *Organometallics*, 2004, **23**, 5317.
71. Makihara, N., Ogo, S. and Watanabe, Y., pH-selective hydrogenation of water-soluble carbonyl compounds and alkenes with $[Cp^*Ir(III)(H_2O)_3]^{2+}$ ($Cp^* = \eta^5$-C_5Me_5) as a catalyst precursor in very acidic media, *Organometallics*, 2001, **20**, 497.
72. Ohkuma, T. and Noyori, R., In: Jacobsen, E.N., Pflatz, A. and Yamamoto, H. (Eds.), *Comprehensive Asymmetric Catalysis*, Vol. 1, Springer, Berlin, 1999, Chap. 6.1.
73. Guerreiro, P., Ratovelomanana-Vidal, V., Genét, J.-P. and Dellis, P., Recyclable diguanidinium-BINAP and PEG-BINAP supported catalysts: syntheses and use in Rh(I) and Ru(II) asymmetric hydrogenation reactions, *Tetrahedron Lett.*, 2001, **42**, 3423.
74. Lamouille, T., Saluzzo, C., ter Halle, R., Le Guyader, F. and Lemaire, M., Hydrogenation of ethyl acetoacetate catalyzed by hydrosoluble BINAP derivatives, *Tetrahedron Lett.*, 2001, **42**, 663.
75. Gelpke, A.E.S., Kooijman, H., Spek, A.L. and Hiemstra, H., Synthesis of the dibenzofuran-based diphosphine ligand BIFAP and its water-soluble derivative BIFAPS and their use in ruthenium-catalyzed asymmetric hydrogenation, *Chem. Eur. J.*, 1999, **5**, 2472.
76. (a) King, S.A., Thompson, A.S., King, A.O. and Verhoeven, T.R., An improved procedure for the synthesis and use of $[RuCl_2(BINAP)]_2{\cdot}NEt_3$. Dependence of the ruthenium(II)-BINAP catalyzed asymmetric hydrogenation of β-keto esters on trace amounts of acid, *J. Org. Chem.*, 1992, **57**, 6689; (b) Togni, A., Planar-chiral ferrocenes: synthetic methods and applications, *Angew. Chem. Int. Ed.*, 1996, **35**, 1475.
77. Hashiguchi, S., Fujii, A., Takehara, J., Ikariya, T. and Noyori, R., Asymmetric transfer hydrogenation of aromatic ketones catalyzed by chiral ruthenium(II) complexes, *J. Am. Chem. Soc.*, 1995, **117**, 7562.
78. (a) Rhyoo, H.Y., Yoon, Y.A., Park, H.J. and Chung, Y.K., Use of amino amides derived from proline as chiral ligands in the ruthenium(II)-catalyzed transfer hydrogenation reaction of ketones,*Tetrahedron Lett.*, 2001,

42, 5045; (b) Rhyoo, H.Y., Park, H.J. and Chung, Y.K., The first Ru(II)-catalyzed asymmetric hydrogen transfer reduction of aromatic ketones in aqueous media, *Chem. Commun.*, 2001, 2064; (c) Rhyoo, H.Y., Park, H.J., Suh, W.H. and Chung, Y.K., Use of surfactants in water-soluble ruthenium(II) complex-catalyzed asymmetric hydrogen-transfer reduction of aromatic ketones, *Tetrahedron Lett.*, 2001, **43**, 269.

79. Ma, Y., Liu, H., Chen, L., Cui, X., Zhu, J. and Deng, J., Asymmetric transfer hydrogenation of prochiral ketones in aqueous media with new water-soluble chiral vicinal diamine as ligand, *Org. Lett.*, 2003, **5**, 2103.
80. Li, X., Wu, X., Chen, W., Hancock, F.E., King, F. and Xiao, J., Asymmetric transfer hydrogenation in water with a supported Noyori–Ikariya catalyst, *Org. Lett.*, 2004, **6**, 3321.
81. Liu, P.N., Deng, J.G., Tu, Y.Q. and Wang, S.H., Highly efficient and recyclable heterogeneous asymmetric transfer hydrogenation of ketones in water, *Chem. Commun.*, 2004, 2070.
82. Wang, F., Liu, H., Cun, L.F., Zhu, J., Deng, J.G. and Jiang, Y.Z., Asymmetric transfer hydrogenation of ketones catalyzed by hydrophobic metal-amido complexes in aqueous micelles and vesicles, *J. Org. Chem.*, 2005, **70**, 9424.
83. Wu, X., Li, X., Hems, W., King, F. and Xiao, J., Accelerated asymmetric transfer hydrogenation of aromatic ketones in water, *Org. Biomol. Chem.*, 2004, **2**, 1818.
84. (a) Wu, X., Li, X., King, F. and Xiao, J., Insight into and practical application of pH-controlled asymmetric transfer hydrogenation of aromatic ketones in water, *Angew. Chem. Int. Ed.*, 2005, **44**, 3407. For another related study of Ru complexes, see: (b) Ogo, S., Abura, T. and Watanabe, Y., pH-dependent transfer hydrogenation of ketones with HCOONa as a hydrogen donor promoted by η^6-(C_6Me_6)Ru complexes, *Organometallics*, 2002, **21**, 2964.
85. Wu, X., Vinci, D., Ikariya, T. and Xiao, J.L., A remarkably effective catalyst for the asymmetric transfer hydrogenation of aromatic ketones in water and air, *Chem. Commun.*, 2005, 4447.
86. Canivet, J., Labat, G., Stoeckli-Evans, H. and Süss-Fink, G., Water-soluble arene ruthenium complexes containing a *trans*-1,2-diaminocyclohexane ligand as enantioselective transfer hydrogenation catalysts in aqueous solution, *Eur. J. Inorg. Chem.*, 2005, 4493. For a similar result, see: Canivet, J., Karmazin-Brelot, L. and Suss-Fink, G., Cationic arene ruthenium complexes containing chelating 1,10-phenanthroline ligands, *J. Org. Met. Chem.*, 2005, **690**, 3202.
87. Mao, J., Wan, B., Wu, F. and Lu, S., First example of asymmetric transfer hydrogenation in water induced by chiral amino alcohol hydrochloride, *Tetrahedron Lett.*, 2005, **46**, 7341.
88. (a) Schmitzer, A R., Franceschi, S., Perez, E., Rico-Lattes, I., Lattes, A., Thion, L., Erard, M. and Vidal, C., Reactivity at the interface of chiral amphiphilic dendrimers. High asymmetric reduction by $NaBH_4$ of various prochiral ketones, *J. Am. Chem. Soc.*, 2001, **123**, 5956. For the use of carbohydrate amphiphiles, see: (b) Denis, C , Laignel, B., Plusquellec, D., Le Marouille, J.-Y. and Botrel, A., Highly regio- and stereoselective reductions of carbonyl compounds in aqueous glycosidic media, *Tetrahedron Lett.*, 1996, **37**, 53.
89. (a) Lensink, C. and DeVries, J.G., Improving enantioselectivity by using a mono-sulfonated diphosphine as ligand for homogeneous imine hydrogenation, *Tetrahedron Asymmetry*, 1992, **3**, 235; (b) Lensink, C., Rijnberg, E. and DeVries, J.G., Influence of degree of sulfonation of BDPP upon enantioselectivity in rhodium-BDPP catalyzed hydrogenation reactions in a two phase system, *J. Mol. Catal. A*, 1997, **116**, 199; (c) Bakos, J., Orosz, A., Heil, B., Laghmari, M., Lhoste, P. and Sinou, D., Rhodium(I) sulfonated-BDDP catalyzed asymmetric hydrogenation of imines in aqueous organic 2-phase solvent systems, *J. Chem. Soc., Chem. Commun.*, 1991, 1684. See also Ref. 58a.
90. Zimmermann, B., Herweg, J. and Beller, M., The first efficient hydroaminomethylation with ammonia: with dual metal catalysts and two-phase catalysis to primary amines, *Angew. Chem. Int. Ed.*, 1999, **38**, 2372.
91. For Rh-Skewphos with reverse micelles (up to 87% ee), see: (a) Buriak, J.M. and Osborn, J.A., Studies on catalytic asymmetric imine hydrogenation in the presence of reverse micelles: enhanced enantioselectivity due to surfactant head group coordination, *Organometallics*, 1996, **15**, 3161. For reduction of folic acid derivatives using soluble DIOP derivatives: (b) Brunner, H. and Rosenboen, S., Enantioselective catalyses CXXXV. Stereoselective hydrogenation of folic acid and 2-methylquinoxaline with optically active rhodium(I)–phosphane complexes, *Monatsh. Chem.*, 2000, **131**, 1371; (c) Brunner, H., Bublak, P. and Helget, M., Asymmetric catalysis. 105. Stereoselective hydrogenation of folic acid with immobilized optically active rhodium(I)/diphosphane catalysts, *Chem. Ber.*, 1997, **130**, 55.
92. (a) Margalef-Català, R., Claver, C., Salagre, P. and Fernández, E., Heterogenized iridium complexes for the asymmetric hydrogenation of imines, *Tetrahedron Asymmetry*, 2000, **11**, 1469; (b) Claver, C., Fernández, E., Margalef-Català, R., Medina, F., Salagre, P. and Sueiras, J.E., Studies on the characterization of several iridium- and rhodium-clay catalysts and their activity in imine hydrogenation, *J. Mol. Catal.*, 2001, **201**, 70.
93. Kainz, S., Brinkmann, A., Leitner, W. and Pfaltz, A., Iridium-catalyzed enantioselective hydrogenation of imines in supercritical carbon dioxide, *J. Am. Chem. Soc.*, 1999, **121**, 6421.
94. Bakos, J., Orosz, A., Cserépi, S., Tóth, I. and Sinou, D., Chiral sulfonated phosphines. Rhodium(I)-catalyzed asymmetric hydrogenolysis of epoxides, *J. Mol. Cat. A: Chem.*, 1997, **116**, 85.

95. Bényei., A.C., Lehel, S. and Joó, F.J., Transfer hydrodehalogenation of organic halides catalyzed by water-soluble ruthenium(II) phosphine complexes, *Mol. Catal. A: Chem.*, 1997, **116**, 349.
96. Reviews: (a) Leitner, W., The coordination chemistry of carbon dioxide and its relevance for catalysis: a critical survey, *Coord. Chem. Rev.*, 1996, **153**, 257. Rh/TPPTS: (b) Gassner, F. and Leitner, W., CO_2-activation. 3. Hydrogenation of carbon dioxide to formic acid using water-soluble rhodium catalysts, *J. Chem. Soc., Chem. Commun.*, 1993, 1465. Rh, Ru, TPPMS, PTA: (c) Joó, F., Laurenzy, G., Nadashdi, N. and Elek, J., Homogeneous hydrogenation of aqueous hydrogen carbonate to formate under exceedingly mild conditions – a novel possibility of carbon dioxide activation, *J. Chem. Commun.*, 1999, 971; (d) Laurenzy, G., Joó, F. and Nadashdi, L., Formation and characterization of water-soluble hydrido-ruthenium(II) complexes of 1,3,5-triaza-7-phosphaadamantane and their catalytic activity in hydrogenation of CO_2 and HCO_3^- in aqueous solution, *Inorg. Chem.*, 2000, **39**, 5083. Ru/R_3P in *sc*-CO_2, water: (e) Jessorp, P.G., Hsaio, Y., Ikariya, T. and Noyori, R., Homogeneous catalysis in supercritical fluids: hydrogenation of supercritical carbon dioxide to formic acid, alkyl formates, and formamides, *J. Am. Chem. Soc.*, 1996, **118**, 344.
97. Müller, T. and Bahrmann, H., Separation of catalyst compounds in two-phase and uni-phase systems by membranes, *J. Mol. Catal. A: Chem.*, 1997, **116**, 39.
98. (a) Bergbreiter, D.E., Zhang, L. and Mariagananam, V.M., Smart ligands regulate homogeneously catalyzed-reactions, *J. Am. Chem. Soc.*, 1993, **115**, 9295; (b) Bergbreiter, D.E. and Liu, Y.S., Amphoteric, water-soluble polymer-bound hydrogenation catalysts, *Tetrahedron Lett.*, 1997, **38**, 3703.
99. Martinek, K., Levashov, A.V., Khmelnitsky, Y.L., Klyachko, N.L. and Berezin, I.V., Colloidal solutions of water in organic solvents: a microheterogeneous medium for enzymatic reactions, *Science*, 1982, **218**, 889.
100. Gyarmati, J., Hajdu, C., Dinya, Z., Micskei, K., Zucchi, Z. and Pályi, G., Asymmetric induction by amino acid ligands in chromium(II)-assisted reduction of ketones, *J. Organomet. Chem.*, 1999, **586**, 106.
101. Pámies, O. and Bäckvall, J.-E., Combination of enzymes and metal catalysts. A powerful approach in asymmetric catalysis, *Chem. Rev.*, 2003, **103**, 3247.
102. For a leading reference, see: Willner, I., Maidan, R. and Shapira, M., Thermal and photochemical regeneration of nicotinamide cofactors and a nicotinamide model compound using a water-soluble rhodium phosphine catalyst, *J. Chem. Soc., Perkin Trans. 2*, 1990, 559. (b) For a recent reference, see: Hollmann, F., Kleeb, A., Otto, K. and Schmid, A., Coupled chemoenzymatic transfer hydrogenation catalysis for enantioselective reduction and oxidation reactions, *Tetrahedron: Asymmetry*, 2005, **16**, 3512.
103. For two recent reports and pertinent references, see: (a) Nadasdi, L., Joó, F., Horváth, I. and Vigh, L., Colloidal metal dispersions as catalysts for selective surface hydrogenation of biomembranes. 21. Preparation of nanosize platinum metal catalysts and characterization in hydrogenation of water-soluble olefins and synthetic biomembrane models, *Appl. Catal. A: Gen.*, 1997, **162**, 57; (b) Csabai, P. and Joó, F., Reactivity of the individual lipid classes in homogeneous catalytic hydrogenation of model and biomembranes detected by MALDI-TOF mass spectrometry, *Catal. Commun.*, 2003, **4**, 275.

Chapter 7
Oxidations

Roger A. Sheldon

A wide variety of important catalytic processes – hydrogenation, carbonylation, hydroformylation, olefin metathesis, polymerization and telomerization – can be effectively performed in an aqueous medium. These reactions involve transition metals in low oxidation states coordinated to soft ligands, e.g. phosphines, as the catalytically active species and organometallic compounds as reactive intermediates. Performing such reactions in aqueous/organic biphasic media is simply a matter of replacing the ligands used in organic media with water-soluble equivalents, e.g. sulfonated triarylphosphines. In contrast, catalytic oxidations involve transition metals in high oxidation states as the active species, generally coordinated to relatively simple, hard ligands, e.g. carboxylate. Reactive intermediates tend to be coordination complexes rather than organometallic species.

Furthermore, strong coordination of water to the hard metal center can prevent coordination of a less polar substrate, e.g. a hydrocarbon, resulting in inhibition of the catalyst or deactivation by hydrolysis in aqueous media. On the other hand, coordination of complex nitrogen and/or oxygen-containing ligands can lead to the generation of more active oxidants by promoting the formation of high oxidation states. For example, in heme-dependent oxygenases and peroxidases the formation of active high-valent oxoiron complexes is favored by coordination to a macrocyclic porphyrin ligand.

Because a large variety of oxidation processes can be performed in water, it is necessary to first define our frame of reference for this chapter. A plethora of oxidative enzymatic processes involving dehydrogenases, oxygenases, oxidases and peroxidases as catalysts are generally performed in water.[1] However, enzymatic processes are dealt with in Chapter 10 and will not be covered in any detail here. The discussion will also be largely limited to the use of the green, inexpensive and readily available oxidants – molecular oxygen and hydrogen peroxide – although other oxidants, e.g. hypochlorite, may receive a cursory mention. The aerobic oxidation of water-soluble alcohols, diols and carbohydrates over heterogeneous noble metal catalysts (Pt, Pd, Ru) in water as solvent has been extensively studied and has a long history dating back to the nineteenth century. Indeed, it was the first catalytic process to be studied. However, the subject has been extensively documented elsewhere [2,3] and, hence, shall be largely excluded from our discussion.

If we consider the oxidation of sparingly soluble substrates in an aqueous biphasic medium, we can distinguish two different approaches. In the first category the substrate is dissolved in, or forms itself, an organic phase, while the oxidant, and possibly also the catalyst, is in the aqueous phase. A phase transfer catalyst is employed to transfer the catalyst and/or oxidant to the organic phase where the reaction takes place. Many catalytic oxidations with water-soluble oxidants, such as hydrogen peroxide, hypochlorite, persulfate, etc., are examples of this category. In the second category the catalyst and the oxidant are dissolved in the water phase and the substrate, although it forms a separate organic phase, is sparingly soluble in water and the reaction takes place in the aqueous phase. In short, the fundamental difference between the two categories lies in where the reaction actually takes place, in the organic or the water phase. The substrate and product may be completely (e.g. carbohydrates) or partially dissolved. In the latter case the catalyst is easily recovered and recycled by phase separation.

7.1 Water-soluble ligands

Some examples of water-soluble ligands that have been used in catalytic oxidations with oxygen or hydrogen peroxide are shown in Fig. 7.1. Because much of this research falls into the category of biomimetic oxidations, water-soluble porphyrins and the structurally related phthalocyanines and other tetradentate nitrogen ligands, such as the tetraamido macrocyclic ligand (TAML), have been widely used.

7.2 Oxidations catalyzed by metalloporphyrins and metallophthalocyanines

In nature, oxidoreductases – oxygenases, peroxidases and oxidases – are implicated in a wide variety of degradation processes of biopolymers and xenobiotics. It is not surprising, therefore, that metal complexes of biomimetic ligands such as porphyrins and phthalocyanines

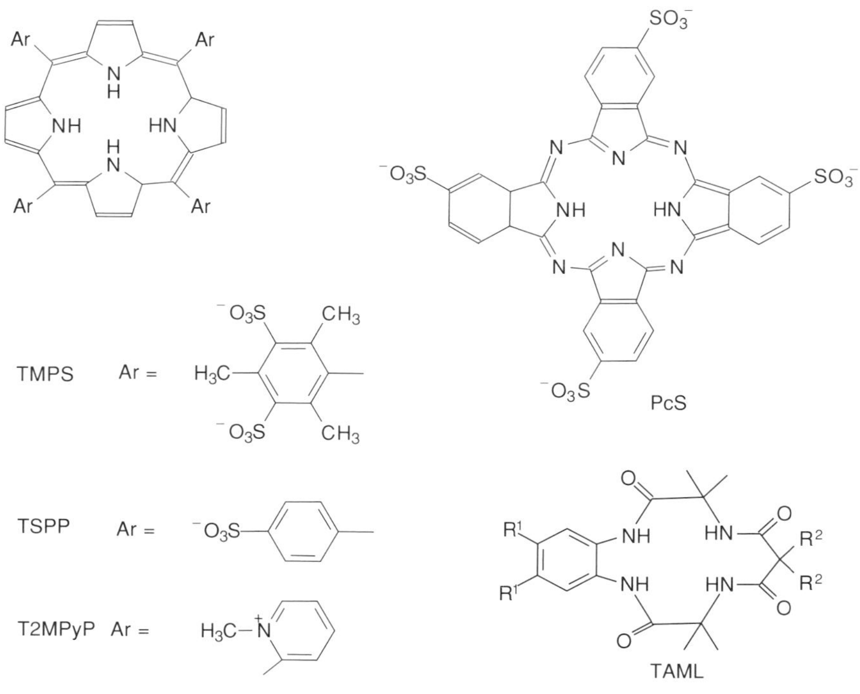

Figure 7.1 Water-soluble ligands.

have been investigated as catalysts for the modification and/or degradation of a variety of biopolymers and organic pollutants. Metal complexes of water-soluble porphyrins and phthalocyanines have been studied as biomimetic catalysts in environment-friendly O_2-based delignification of wood pulp in paper manufacture which involves the oxidative degradation of lignocellulose to produce the desired cellulose. Conventional processes involve the use of Cl_2 or ClO_2 as oxidants and produce effluents containing chlorinated phenols. For example, Wright and coworkers[4] oxidized lignin model compounds such as **1** with O_2 (Eq. 1) in the presence of $Na_3Fe(III)$ (PcTS), $Na_3Co(III)$ (TSPP) and $Na_3Rh(III)$ (TSPP) catalysts. The latter gave the highest rates and selectivities.

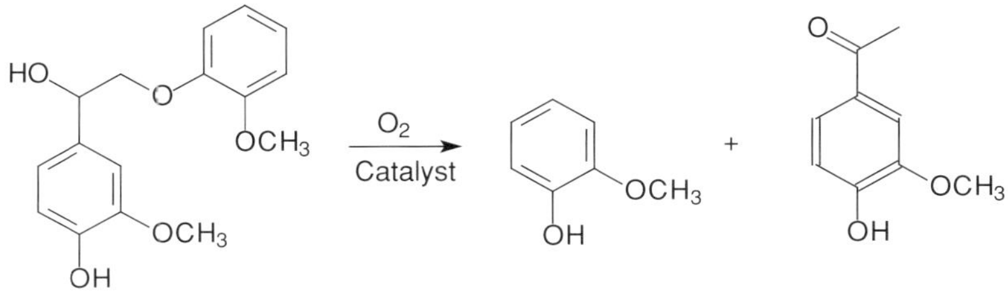

(Eq. 1)

Similarly, Hampton and Ford[5] studied the Fe(PcS)-catalyzed autoxidation of 3,4-dimethoxybenzyl alcohol as a model for delignification. They concluded, however, that the catalyst degrades too fast to be useful for delignification. We also mention, in this context, that water-soluble polyoxometalates such as $PV_2Mo_{10}O_{40}^{5-}$ have also been used as catalysts for delignification with O_2.[6]

Fe(PcS) has also been successfully applied as a catalyst for the oxidative destruction of recalcitrant chlorinated phenol pollutants, such as 2,4,6-trichlorophenol, in wastewater, using H_2O_2 as the primary oxidant.[7,8] Similarly, Fe(TMPS) and Fe(III) and Mn(III) complexes of T2MPyP catalyzed the oxidation of phenols, with $KHSO_5$ as the primary oxidant.[9,10] It was subsequently shown that the oxidative degradation of organic pollutants in wastewater, with hydrogen peroxide and catalytic amounts of Fe(PcS), was markedly accelerated by visible light.[11]

More recently, Sorokin and coworkers showed that Fe(PcS) is an extremely effective catalyst for the waste-free modification of polysaccharides by oxidation with hydrogen peroxide in an aqueous medium.[12,13] Hydrophilic starches, obtained by partial oxidation, are commercially interesting products with many potential applications in the paper and textile industries and as water superabsorbents. Oxidation occurs at the C_6 primary alcohol group or via cleavage of the C_2–C_3 vicinal diol (Fig. 7.2). Current methods for achieving such oxidations involve the use of stoichiometric amounts of oxidants such as hypochlorite, N_2O_4 or periodate, producing copious quantities of inorganic waste.[13] Selective, clean oxidation of native starch by hydrogen peroxide was observed at very low catalyst loadings, as little as 0.003–0.016 mol% FePcS. When $FeSO_4$ was used in the same amount as FePcS no detectable amount of starch oxidation was observed. The oxidation was performed with a suspension of starch in an aqueous solution of catalyst (wet process) or by impregnation of starch powder with catalyst solution (dry process). The combination of a green oxidant (H_2O_2) with very low loadings of a relatively inexpensive iron-based catalyst for the one-step modification of starch has obvious economic and environmental benefits.

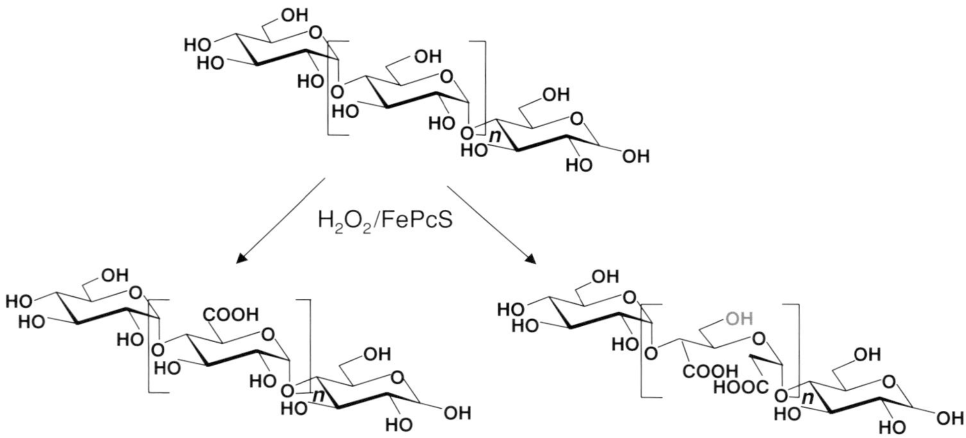

Figure 7.2 Catalytic oxidation of starch with hydrogen peroxide.

A problem associated with the use of porphyrins and, to a lesser extent, phthalocyanines is their susceptibility toward oxidative degradation. Hence, there is a need for analogous macrocyclic ligands which stabilize higher oxidation states of, e.g., iron but are oxidatively stable. Collins and coworkers[14] developed a series of iron(III) complexes of macrocyclic tetradentate nitrogen ligands, so-called TAML oxidant activators, with greatly enhanced stability toward oxidative and hydrolytic degradation (Fig. 7.3).

These ligands were a culmination of work, spanning more than a decade,[15] on the design of macrocyclic ligands which are stable under oxidizing conditions. They are efficient, water-soluble activators of hydrogen peroxide, over a broad pH range, with a wide variety of potential applications, e.g. to replace chlorine bleaching in the pulp and paper industry and for use in water effluent treatment in the textiles industry. Applications in organic synthesis have, as yet, not been explored.

7.3 Epoxidation and dihydroxylation of olefins in aqueous media

The epoxidation of olefins and the related vicinal dihydroxylation are reactions of great industrial importance. The use of commercially available 30% aq. H_2O_2 is obviously

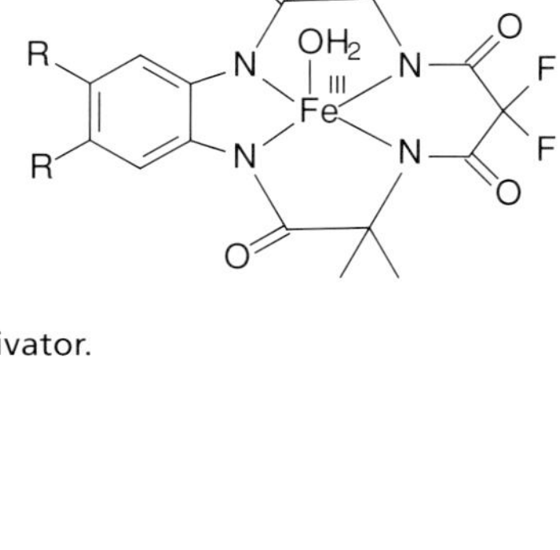

Figure 7.3 TAML oxidant activator.

an environmentally and economically attractive proposition for the epoxidation of most olefins – an alternative to classical methods involving hypochlorite or percarboxylic acids or metal-catalyzed epoxidations with alkyl hydroperoxides.[16] Unfortunately, metal-catalyzed epoxidation with hydrogen peroxide is a pertinent example of a reaction which is generally strongly inhibited by water. Two strategies have been employed to overcome this encumbrance, as noted in the Introduction: using a phase transfer catalyst to transfer the catalyst and the oxidant to the organic phase or the use of ligands which promote the oxidation of the olefin substrate in the aqueous phase.[17]

The use of tungstate, in the presence of phosphate and a tetralkylammonium salt as a phase transfer agent, for the epoxidation of olefins with 30% aq. H_2O_2, in a biphasic dichloroethane/water medium, was first described by Venturello and coworkers in 1983.[18] Ishii and coworkers[19] subsequently reported the use of a heteropolytungstate, $H_3PW_{12}O_{40}$, in an aqueous biphasic system under phase transfer conditions for the epoxidation of olefins with hydrogen peroxide. Later it was shown by Bregault and coworkers[20] that both systems involve the same peroxotungstate complex $(R_4N)_3PO_4[W(O)(O_2)_2]_4$ as the catalytically active species.

Noyori and coworkers[21,22] reported a significant improvement of the original system. An appropriate choice of phase transfer agent, containing a sufficiently lipophilic tetraalkylammonium cation and a bisulfate (HSO_4) anion, in combination with catalytic amounts of $H_2NCH_2PO_3H_2$ and sodium tungstate afforded an effective system for the epoxidation of olefins with hydrogen peroxide in toluene/water or in the absence of an organic solvent (Fig. 7.4).

The same system was used for the direct oxidation of cyclohexene to adipic acid, by oxidation with 4 equiv. of 30% aq. H_2O_2, via the initial formation of cyclohexene oxide (Fig. 7.5).[23] It was further shown that the methodology is applicable to the oxidative cleavage of a range of cyclic olefins.[23]

Reedijk and coworkers[24] recently reported a similar system composed of sodium tungstate, tungstic acid and chloroacetic acid together with methyltrioctylammonium chloride as phase transfer agent, which was effective for the epoxidation of olefins with 50% aq. H_2O_2 without any additional solvent.

Xi and coworkers similarly reported[25] the use of a cetylpyridinium heteropolytungstate, $[C_5H_5NC_{16}H_{33}]_3PO_4[WO_4]_3$, analogous to the Ishii system mentioned above, for the

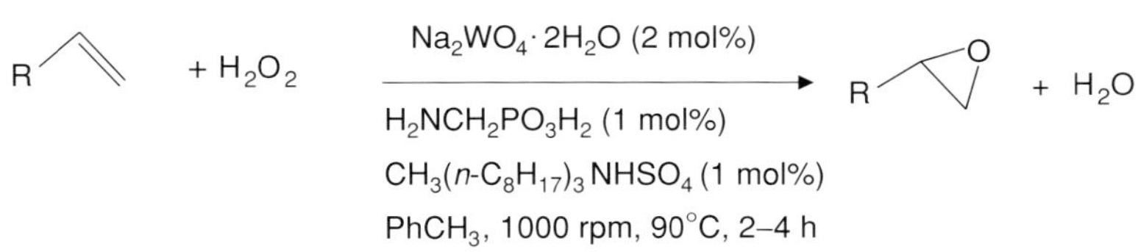

Olefin	Conversion (%)	Yield (%)
1-Octene	96	94
1-Decene	99	99
1-Dodecene	98	97
2-Octene	99	99
Styrene	52	3

Figure 7.4 Solvent- and halide-free epoxidation of olefins with aq. H_2O_2.

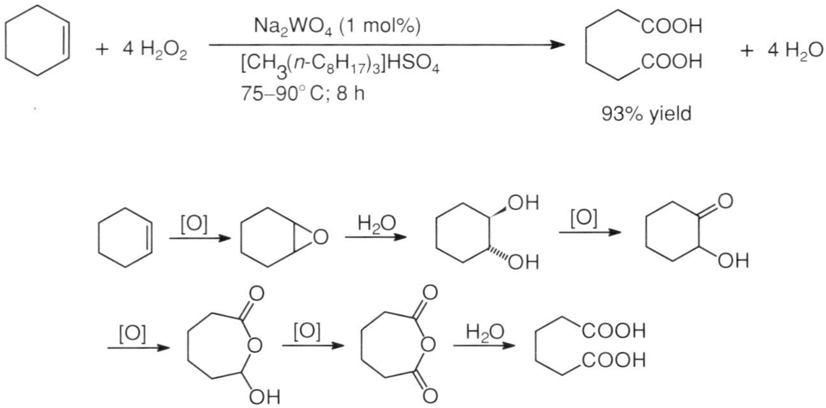

Figure 7.5 A green route to adipic acid by oxidation with aq. H_2O_2.

epoxidation of propylene and other olefins with hydrogen peroxide in an aqueous biphasic system. The catalyst is insoluble in the reaction medium but dissolves on the addition of hydrogen peroxide through the *in situ* formation of a soluble peroxo complex, $[C_5H_5NC_{16}H_{33}]_3PO_4[W(O)_2(O_2)]_4$. When the reaction is complete the original catalyst precipitates from the solution and can be filtered and recycled. Hence, the system was referred to as reaction-controlled phase transfer catalysis.

As noted above, a second strategy to achieve epoxidation with hydrogen peroxide in an aqueous medium involves the deployment of water-soluble metal complexes whereby the reaction presumably takes place in the water phase.[17] For example, water-soluble manganese complexes of 1,4,7-trimethyl-1,4,7-triazacyclononane (**2**) and related ligands, which were originally developed as highly effective catalysts for low-temperature bleaching of stains, were shown to catalyze the selective epoxidation of styrene and 4-vinylbenzoic acid with aq. H_2O_2 in aq. MeOH or water, respectively.[26] However, large amounts of H_2O_2 (10 or more equivalents) were required, indicating that considerable nonproductive decomposition occurs. Subsequently, it was shown that nonproductive decomposition of the hydrogen peroxide could be largely suppressed by the addition of oxalate (Fig. 7.6)[27] or ascorbic acid[28] as cocatalysts, or by anchoring the ligand to a solid support.[29]

More recently, the use of glyoxylic acid methyl ester hemiacetal as a cocatalyst was shown to afford an even more effective epoxidation catalyst, enabling high conversions with only a 30% excess of hydrogen peroxide.[30] Interestingly, the corresponding *cis*-diols were observed as by-products in many cases and a concerted mechanism via a manganese(III)–*cis*-diol complex was proposed to explain their formation.

Burgess and coworkers[31,32] showed that simple manganese(II) salts, such as $MnSO_4$, catalyze the epoxidation of olefins with 30% H_2O_2, in aqueous dimethylformamide or *tert*-butanol, in the presence of sodium bicarbonate. The latter is an essential component of the system because it reacts with the hydrogen peroxide to form percarbonate, HCO_4^- (Fig. 7.7). It was proposed that the percarbonate oxidizes the manganese to a Mn(IV) species which is the active oxidant. Here again, additives such as sodium acetate or salicylic acid had a rate- and selectivity-enhancing effect.[32]

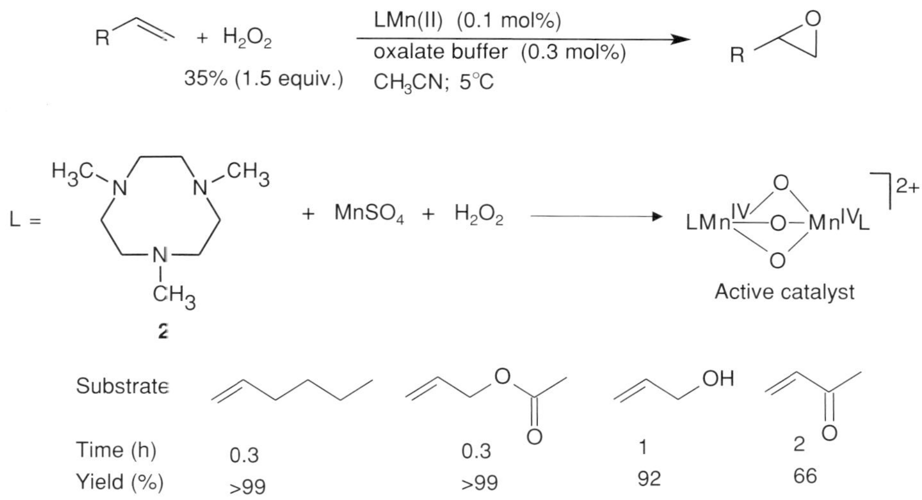

Figure 7.6 Manganese-catalyzed epoxidation with H_2O_2 in water.

An interesting elaboration on this theme is the combination of the $MnSO_4$/bicarbonate system with *in situ* generation of hydrogen peroxide by glucose oxidase-catalyzed oxidation of glucose to afford a chemoenzymatic epoxidation of olefins in an aqueous medium reported by Chan and coworkers.[33] Lipophilic olefins could be epoxidized in a two-phase system by adding the surfactant, sodium dodecyl sulfate. Immobilization of the glucose oxidase by anchoring to silica gel enabled recycling of the enzyme eight times with no significant loss of activity.

Other coordination complexes of manganese and iron, e.g. porphyrin and salen complexes, have been shown[17] to catalyze epoxidations with aqueous hydrogen peroxide but these ligands are prone to rapid deactivation under the oxidizing conditions and have, therefore, limited utility. Ligands containing pyridine and amine coordinating groups fare better under oxidizing conditions. Iron complexes of the pyridyl amine ligands **3** and **4** (Fig. 7.8),

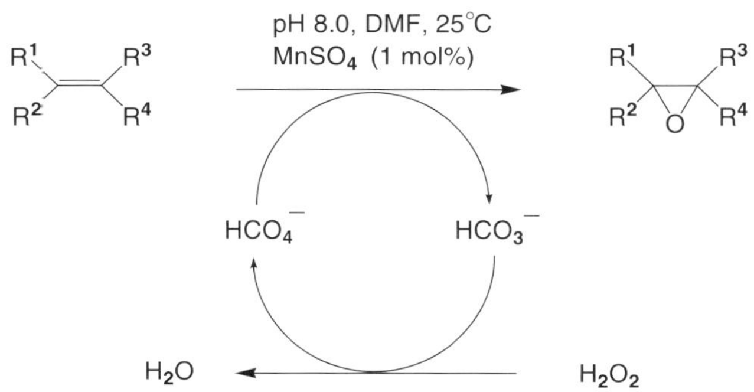

Figure 7.7 $MnSO_4$-catalyzed epoxidation with hydrogen peroxide/bicarbonate.

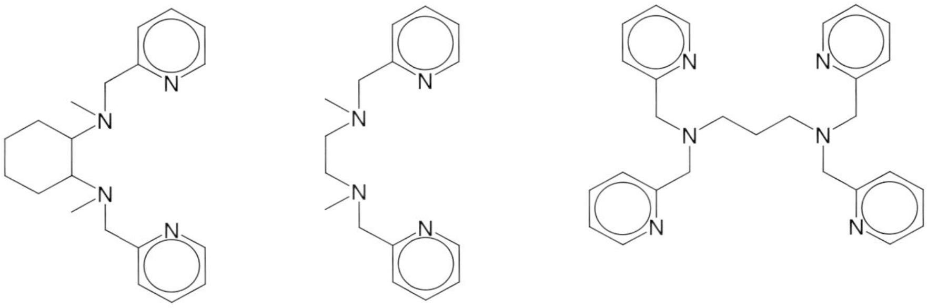

3 4 5

Figure 7.8 Pyridyl amine ligands used in combination with Fe or Mn.

for example, have been investigated by the groups of Que[34] and Jacobsen,[35] respectively, and exhibit moderate olefin epoxidation and/or dihydroxylation activities with aqueous hydrogen peroxide. It should be noted, however, that the reactions are performed in the presence of an organic solvent, generally acetonitrile, and are not strictly speaking examples of catalytic oxidations in water. Activities of these catalysts are also influenced by additives; e.g., the iron complex of **4** gave efficient epoxidation in the presence of acetic acid.[35] High-valent dinuclear iron oxo species are implicated as the active oxidants in these reactions analogous to the putative intermediates in epoxidations mediated by iron-dependent monooxygenases or peroxidases.

Feringa and coworkers[36] similarly showed that manganese complexes of the pyridyl amine ligand **5** were active epoxidation catalysts with 30% H_2O_2, albeit in acetone as solvent. A dinuclear manganese oxo complex was implicated as the active oxidant, analogous to the iron complexes discussed above. We note, however, that none of these catalysts exhibit the high activities observed with the highly active manganese complex of **2** (see above).

More recently, Beller and coworkers[37] have shown that the ruthenium complex **6** (Fig. 7.9) is an effective epoxidation catalyst, for a variety of olefins, with 3 equiv. of 30% H_2O_2 at very low catalyst loadings (0.005 mol%). A tertiary alcohol such as *tert*-amyl alcohol was used as a cosolvent. Based on its high activity and broad scope this system appears to have considerable synthetic potential, which may be adapted to afford effective asymmetric variants in the future. Indeed, a truly effective catalyst, with broad scope, for asymmetric epoxidation with aqueous hydrogen peroxide, preferably in the absence of organic solvents, is still an important and elusive goal in oxidation chemistry.

Some of the systems discussed above (e.g. Fe and Mn) give the *cis*-1,2-diol under certain conditions, via a concerted mechanism, while others, e.g. the tungstate-based systems, can give the *trans*-1,2-diol via acid-catalyzed ring opening of an initially formed epoxide. Recently a simple catalytic, organic solvent- and metal-free system for the oxidation of olefins to the corresponding *trans*-1,2-diols, using 30% H_2O_2 (Fig. 7.10), has been described by Sato and coworkers.[38] The catalyst is a resin-supported sulfonic acid, such as Amberlyst 15 or Nafion^{TM} or the related Nafion–silica composites, and could be recovered by simple filtration and recycled five times without loss of activity.

Similarly, metal-substituted zeolites such as titanium silicalite (TS-1)[39] and titanium beta[40] are recyclable heterogeneous catalysts for the epoxidation and/or dihydroxylation of

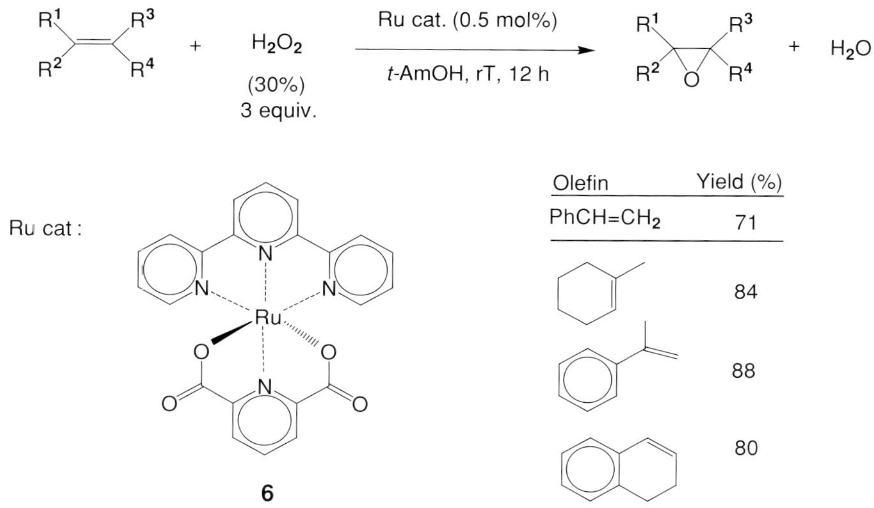

Figure 7.9 Ruthenium-catalyzed epoxidations with 30% H_2O_2.

olefins with aqueous hydrogen peroxide, usually in the presence of an organic cosolvent, such as methanol or *tert*-butyl alcohol, but the reactions can, in principle, be performed in the absence of an added organic solvent. The TS-1-catalyzed epoxidation of propylene with aq. H_2O_2 is currently being scaled up to industrial scale. A serious disadvantage of these molecular sieve catalysts in organic synthesis is that their efficacy is limited to substrates, e.g. in the case of TS-1 linear olefins, which are able to access their micropores.

The osmium-catalyzed vicinal dihydroxylation of olefins with single oxygen donors, typically *tert*-butyl hydroperoxide or *N*-methylmorpholine-*N*-oxide (NMO), has been known for three decades[41] and forms the basis of the Sharpless asymmetric dihydroxylation of olefins. Recently, Sharpless and coworkers reported that particularly electron-deficient olefins are dihydroxylated more efficiently with NMO (Eq. 2) when the pH of

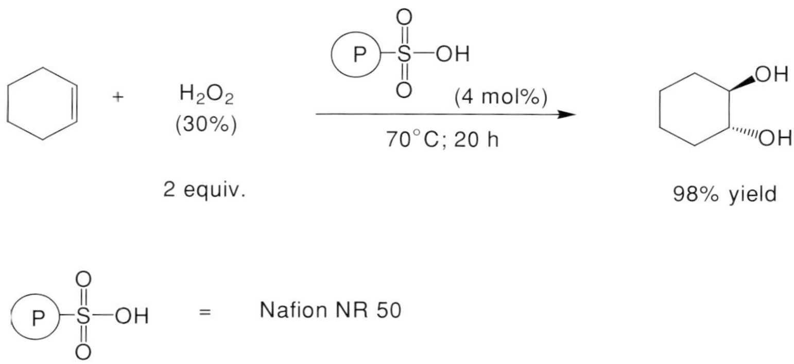

Figure 7.10 Olefin dihydroxylation with 30% H_2O_2 over Nafion resin.

the reaction medium is maintained on the acidic side by adding, e.g., citric acid.[42] The standard conditions involved the use of 1:1 water/*tert*-butyl alcohol mixture as solvent but the authors reported that water without any cosolvent could be used in a number of dihydroxylations.

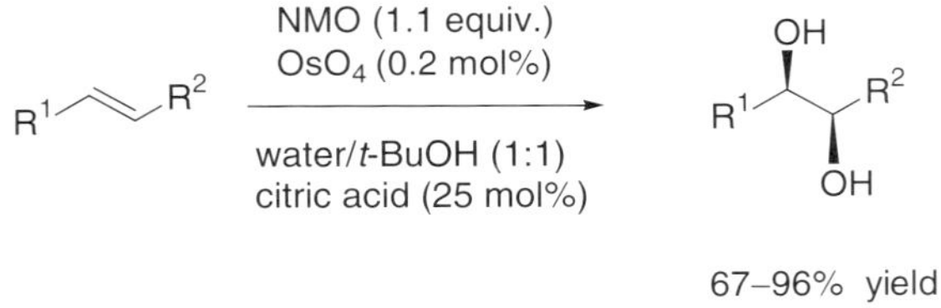

(Eq. 2)

7.4 Alcohol oxidations in aqueous media

Alcohol oxidations to the corresponding carbonyl compounds are pivotal reactions in organic synthesis and catalytic methodologies employing dioxygen (air) or hydrogen peroxide as the primary oxidant are particularly attractive from both an economic and environmental viewpoint. If the reaction can be performed in an aqueous medium, thus avoiding the use of volatile organic solvents, this is an added benefit. As with epoxidation (see above) two strategies have been employed to achieve oxidation in an aqueous medium: using a phase transfer agent to transport the active oxidant to an organic phase or using a water-soluble catalyst whereby the reaction presumably occurs in the water phase, i.e. it is true catalysis in water.

The oxidation of alcohols with aqueous hydrogen peroxide using a tungstate catalyst, in the presence of a tetraalkylammonium salt as a phase transfer agent, in a biphasic system composed of water and 1,2-dichloroethane, was first reported by DiFuria and coworkers in 1986.[43] As with the analogous tungsten-based epoxidation system described above, Noyori and coworkers[22,44] substantially optimized this methodology to afford an extremely effective, chloride- and organic solvent-free system. A combination of 0.002 mol% sodium tungstate and 0.002 mol% of the phase transfer catalyst, methyltrioctylammonium bisulfate, $[CH_3(n\text{-}C_8H_{17})_3N]^+HSO_4^-$, was an effective catalyst for the selective oxidation of alcohols with 1.1 equiv. of 30% H_2O_2 at 90°C in an organic solvent-free medium (Fig. 7.11). As in the analogous epoxidations (see above) the combination of the lipophilic cation with the bisulfate anion is important for activity. Substrate catalyst ratios as high as 400.000 could be used, affording turnover numbers up to 180.000. A wide variety of secondary alcohols afforded the corresponding ketones in high yields. Olefinic alcohols underwent chemoselective oxidation to the corresponding unsaturated ketones. Primary alcohols gave the corresponding carboxylic acid via the intermediate formation of the aldehyde. High yields of carboxylic acids were obtained from a variety of primary alcohols using 2.5 equiv. of H_2O_2. The reactions involve a tetraalkylammonium pertungstate species as the active oxidant.

The palladium(II)-catalyzed aerobic oxidation of alcohols is well documented.[45] A general problem encountered in palladium-catalyzed aerobic oxidations is the sluggish reoxidation of Pd(0) to Pd(II), which results in the agglomeration of the Pd(0) particles to palladium black and accompanying deactivation of the catalyst. In the classical Wacker process for the oxidation of ethylene to acetaldehyde in an aqueous medium, this problem is circumvented

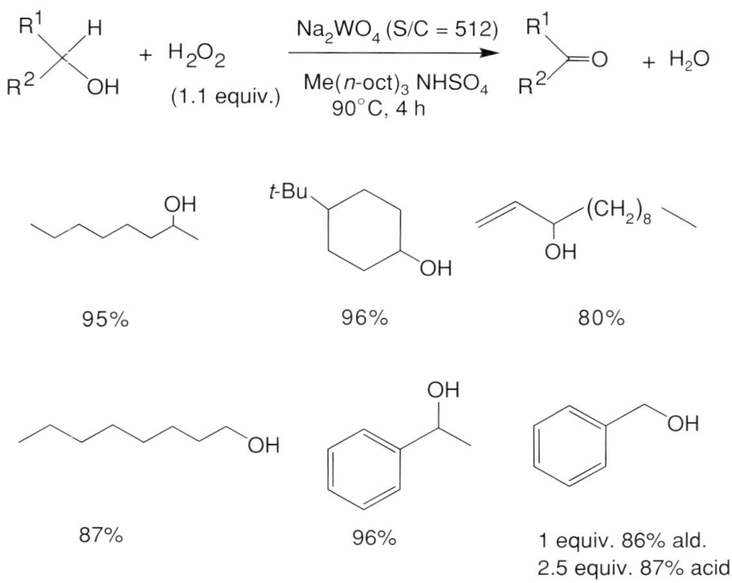

Figure 7.11 Chloride-free and organic solvent-free oxidation of alcohols with aq. H_2O_2.

by adding copper(II).[46] The function of the latter is to reoxidize the Pd(0) to Pd(II) with concomitant formation of Cu(I), which is reoxidized by dioxygen to complete the catalytic cycle (Fig. 7.12).

Substituted olefins are converted to the corresponding ketone and the analogous oxidation of propylene to acetone has been developed to an industrial scale. The Wacker oxidation of higher terminal olefins is fraught with problems: lower rates and complex product mixtures as a result of competing Pd-catalyzed isomerization of the olefin substrate and the formation of chlorinated by-products owing to the high chloride concentrations used. Various approaches have been examined to circumvent these problems, e.g. the use of an organic cosolvent, phase transfer catalysts or a microemulsion system, with varying degrees of success.[46]

Based on our previous work[47] with palladium complexes of sulfonated phosphines as catalysts for carbonylations in aqueous media, we reasoned that analogous water-soluble

$$\mathrm{Pd(II)Cl_4^{2-}} + \mathrm{H_2O} + \mathrm{C_2H_4} \longrightarrow \mathrm{Pd(0)} + 2\,\mathrm{H^+} + 4\,\mathrm{Cl^-} + \mathrm{CH_3CHO}$$

$$\mathrm{Pd(0)} + 2\,\mathrm{Cu(II)Cl_2} + 4\,\mathrm{Cl^-} \longrightarrow \mathrm{Pd(II)Cl_4^{2-}} + 2\,\mathrm{Cu(I)Cl_2^-}$$

$$2\,\mathrm{Cu(I)Cl_2^-} + 2\,\mathrm{H^+} + 0.5\,\mathrm{O_2} \longrightarrow 2\,\mathrm{Cu(II)Cl_2} + \mathrm{H_2O}$$

$$\text{Overall:}\quad \mathrm{C_2H_4} + 0.5\,\mathrm{O_2} \longrightarrow \mathrm{CH_3CHO}$$

Figure 7.12 Wacker oxidation of ethylene in water.

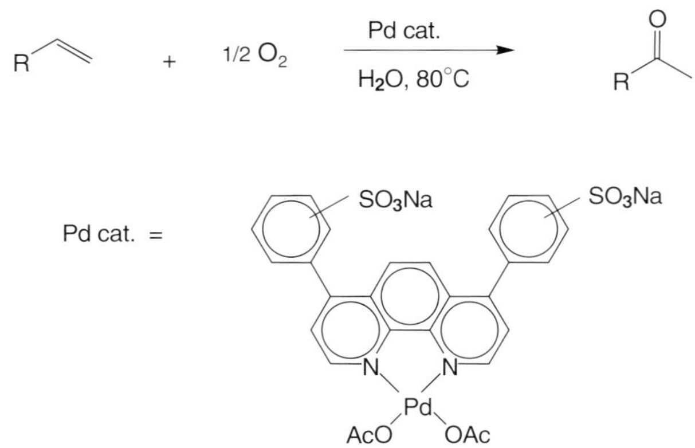

Figure 7.13 Palladium-catalyzed aerobic oxidation of terminal olefins in water.

diamine ligands could possibly stabilize a transient Pd(0) species under oxidizing conditions, and prevent the formation of palladium black. To this end we tested the commercially available sulfonated bathophenanthroline which formed a water-soluble complex when mixed with an aqueous solution of palladium(II) acetate. We showed that this complex was able to catalyze the selective aerobic oxidation of terminal olefins, to the corresponding ketones, in a chloride- and organic solvent-free aqueous medium (Fig. 7.13).[48]

We subsequently found[49–51] that the palladium(II) complex of sulfonated bathophenanthroline and related water-soluble diamine ligands are stable, recyclable catalysts for the aerobic oxidation of alcohols in a two-phase aqueous-organic medium whereby the organic phase consists of the alcohol substrate and the carbonyl product (Fig. 7.14). Reactions were generally complete in 5 h at 100°C/30 bar air with as little as 0.25 mol% catalyst. No organic solvent is required (unless the substrate is a solid) and the product is easily recovered by phase separation. The catalyst is stable and remains in the aqueous phase, facilitating recycling to the next batch.

A wide range of primary and secondary alcohols were oxidized with turnover frequencies ranging from 10 to 100 h^{-1}, depending on the structure and the solubility of the alcohol in water (since the reaction occurs in the water phase the alcohol must be at least sparingly

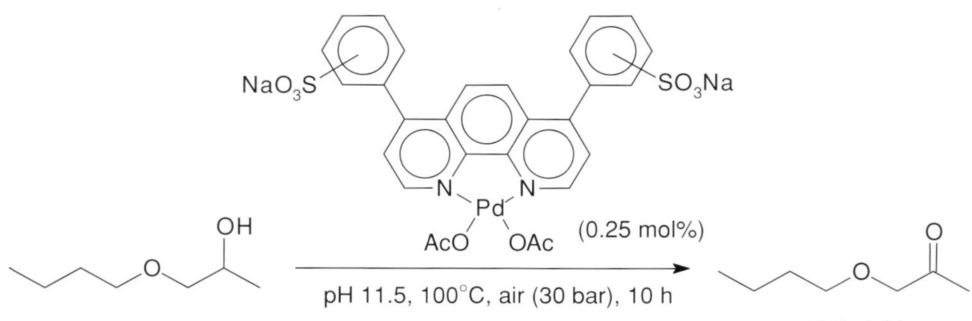

Figure 7.14 Palladium-catalyzed aerobic oxidation of alcohols in water.

soluble in water). Secondary alcohols afforded the corresponding ketones in >99% selectivity in virtually all cases studied. Primary alcohols afforded the corresponding carboxylic acids via further oxidation of the initially formed aldehyde; e.g., 1-hexanol afforded 1-hexanoic acid in 95% yield. It is important to note that this was achieved without the necessity to neutralize the carboxylic acid product with 1 equiv. of base. When the reaction was performed in the presence of 1 mol% of the stable free radical, TEMPO (2,2,6,6-tetramethylpiperidinyloxyl), overoxidation was suppressed and the aldehyde was obtained in high yield; e.g., 1-hexanol afforded hexanal in 97% yield.

A catalytic cycle was proposed[50] for the reaction in which, consistent with the observed half-order in palladium, the active catalyst is formed by initial dissociation of a hydroxyl bridged palladium(II) dimer (Fig. 7.15). This is followed by coordination of the alcohol and β-hydrogen elimination affording the carbonyl product and palladium(0). The latter is reoxidized to palladium(II) by dioxygen.

Compared to most existing systems for the aerobic oxidation of alcohols, the Pd–bathophenanthroline system is an order of magnitude more reactive, requires no organic solvent, involves simple product isolation and catalyst recycling, and has broad scope in organic synthesis. A shortcoming is the requirement that the alcohol substrate should be at least sparingly soluble in water. A second and more general disadvantage from which nearly all catalyst systems seem to suffer is the low tolerance for (coordinating) functional groups in the solvent or the substrate. The bathophenanthroline–$Pd(OAc)_2$ system tolerated only a single ether functionality, and many other functional groups, e.g. containing N or S as heteroatoms, which coordinate more strongly to palladium, were not tolerated. With a view to obtaining superior systems with higher activities and better functional group tolerance and broader substrate scope, we studied electronic[52] and steric[53] effects of substituents in the phenanthroline ligands on the rates and substrate scope of these reactions. Results were in accordance with the proposed mechanism and afforded an optimized

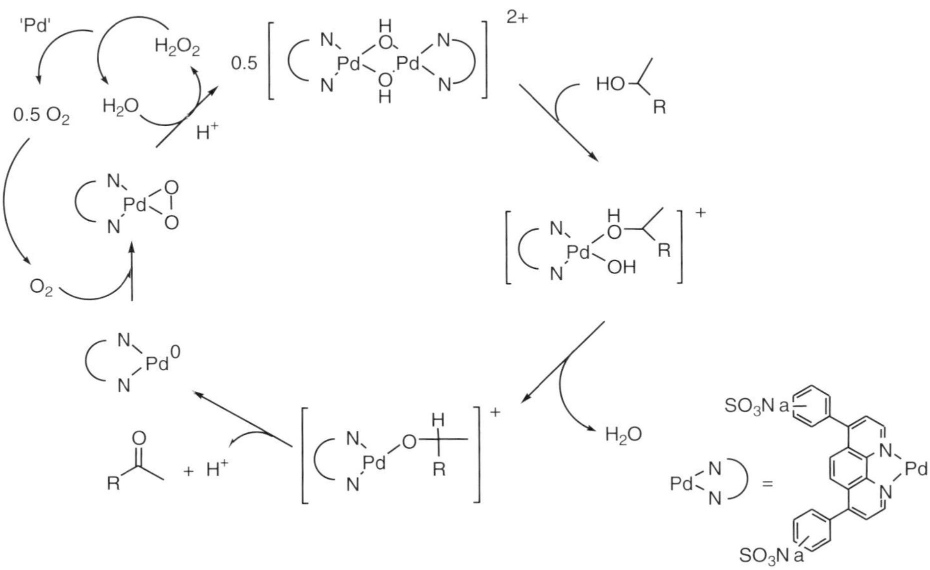

Figure 7.15 Mechanism of Pd-catalyzed aerobic oxidation of alcohols.

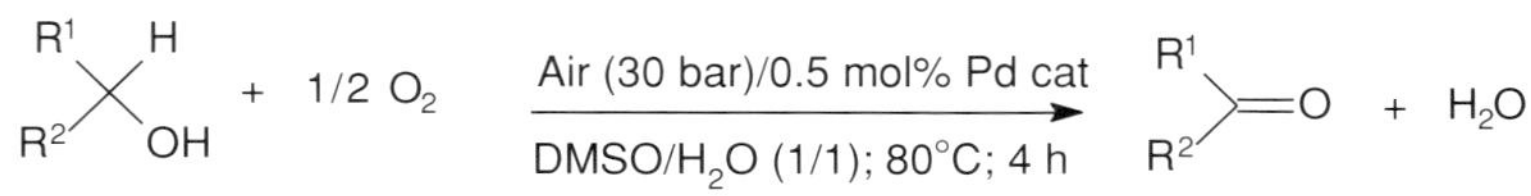

Pd cat :

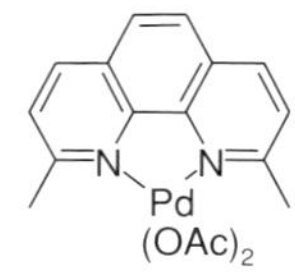

Tolerated functionalities

C=C	C=O	C≡C
OR	SR	SiR_3
(O)S=O	SO_3R	NR_2
CN	$CONH_2$	CO_2R

Figure 7.16 (neocuproin)$Pd(OAc)_2$-catalyzed oxidations of alcohols.

catalyst, (neocuproine)$Pd(OAc)_2$, which was highly active (catalyst loading of 0.1 mol% and turnover frequencies $>1500\ h^{-1}$) in an organic solvent/water mixture, e.g. 1:1 DMSO/water or ethylene carbonate/water, and tolerated a wide variety of functional groups in the alcohol substrate (Fig. 7.16).

Hypervalent iodine compounds, in stoichiometric amounts, are also known to oxidize alcohols and the use of iodosyl benzene or a polymer-supported iodine(III) reagent, in combination with KBr as a cocatalyst, for the oxidation of alcohols in water has been described.[54] More recently a related catalytic system, consisting of $PhIO_2$ (2 mol%), Br_2 (2 mol%) and $NaNO_2$ (1 mol%), for the aerobic oxidation of alcohols in water at 55°C has been described.[55]

The oxidation of alcohols with hypochlorite (household bleach) is widely used in organic synthesis for the oxidation of a broad range of alcohols,[56] including simple carbohydrates[57] and polysaccharides,[58] in the presence of as little as 1 mol% or less of the stable free radical, TEMPO, as catalyst. The stoichiometric oxidation of primary alcohols, to the corresponding aldehydes, by the oxoammonium cation, derived from one-electron oxidation of TEMPO, was first reported by Golubev and coworkers in 1965.[59] The reaction is rendered catalytic in TEMPO by using single oxygen donors such as *m*-chloroperbenzoic acid,[60] persulfate(oxone),[61] periodic acid (H_5IO_6),[62] and sodium hypochlorite[63] to generate the oxoammonium cation *in situ*. In particular, the TEMPO/hypochlorite protocol, using 1 mol% TEMPO in combination with 10 mol% sodium bromide as cocatalyst, in dichloromethane/water at pH 9 and 0°C, has been widely applied in organic synthesis. The method was first described in 1987 by Montanari and coworkers using 4-methoxy TEMPO as the catalyst.[63] The catalytic cycle involves alternating oxidation of the alcohol by the oxoammonium cation and regeneration of the latter by reaction of the TEMPOH with the primary oxidant (hypochlorite). Hence, TEMPO is the catalyst precursor which is presumably oxidized by bromine or chlorine (Fig. 7.17) to the oxoammonium cation which enters the catalytic cycle.

The Montanari protocol, although widely applicable, suffers from several environmental and/or economic drawbacks. It is not waste-free, because at least 1 equiv. of NaCl is produced per molecule of alcohol oxidized and the use of hypochlorite as oxidant can also lead to the formation of chlorinated by-products. Other shortcomings are the use of 10 mol%

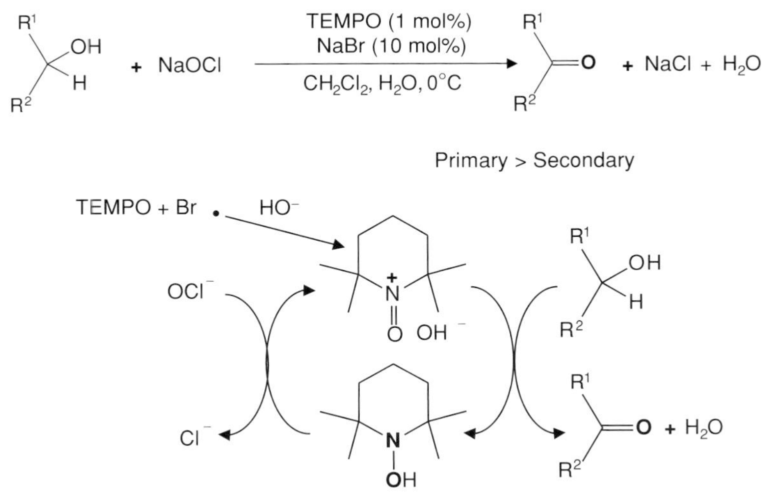

Figure 7.17 TEMPO-catalyzed oxidation of alcohols with NaOCl.

bromide as a cocatalyst and dichloromethane as a solvent. Furthermore, although only 1 mol% is used, TEMPO is rather expensive, which means that efficient recycling is an important issue. Some of these problems were circumvented by replacing the TEMPO with a recyclable oligomeric TEMPO, referred to as PIPO (polymer-immobilized piperidinyloxyl), derived from the commercially available antioxidant and light stabilizer, chimassorb 944, an oligomeric sterically hindered amine (Fig. 7.18). PIPO was shown to be a very effective recyclable catalyst for the oxidation of alcohols, including a wide variety of carbohydrates, with hypochlorite in a bromide- and chlorinated hydrocarbon-free system.[64–67] The reaction is performed with 1 mol% of PIPO and 1.25 equiv. of NaOCl in water as the sole solvent or in a water/*n*-hexane mixture.

Recently, a so-called ion-supported TEMPO was synthesized by building a TEMPO moiety into the side chain of a dialkylimidazolium salt (Fig. 7.19). The resulting material catalyzed the oxidation of alcohols with NaOCl or I_2 in water or an ionic liquid/water mixture.[68]

Although the above recyclable systems offer many economic and environmental benefits, they still suffer from the disadvantage of requiring hypochlorite as a stoichiometric oxidant. Its industrial potential would be significantly enhanced if the latter could be replaced by dioxygen or hydrogen peroxide. Copper salts and complexes in combination with TEMPO have been shown to catalyze the aerobic oxidation of alcohols in acetonitrile or dimethylformamide.[69,70] Furthermore, a copper-dependent oxidase, laccase (EC 1.10.3.2) in combination with TEMPO as a cocatalyst (or so-called mediator), was shown by Galli and coworkers to catalyze the aerobic oxidation of primary benzylic alcohols.[71] The laccase/TEMPO-catalyzed selective aerobic oxidation of the primary alcohol moiety in carbohydrates had been previously reported in two patents.[72,73] It is generally believed that these reactions involve one-electron oxidation of the TEMPO to the oxoammonium cation by the oxidized form of the laccase, followed by reoxidation of the reduced form of laccase by dioxygen.[69]

Laccases are extracellular enzymes that are secreted by white rot fungi and play an important role in the delignification of lignocellulose, the major constituent of wood, by

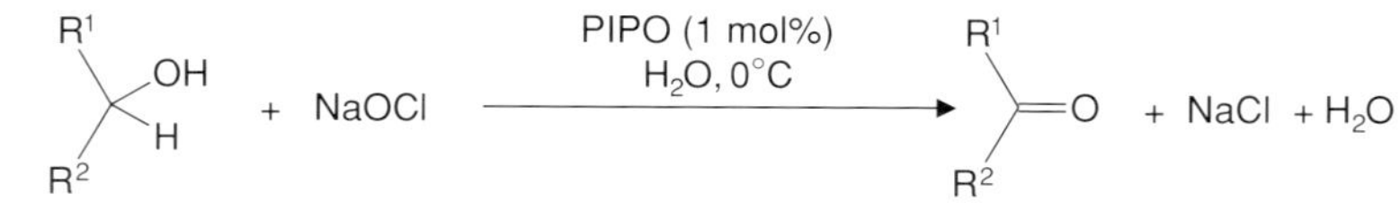

Substrate	Time (min)	Conversion (%)	Selectivity (%)
1-Octanol	45	90	50
	45	80	94[a]
1-Hexanol	45	89	95[a]
2-Octanol	45	99	>99
Cyclooctanol	45	100	>99
Benzyl alcohol	30	100	>99
1-Phenylethanol	30	100	>99

[a]*n*-Hexane as solvent.

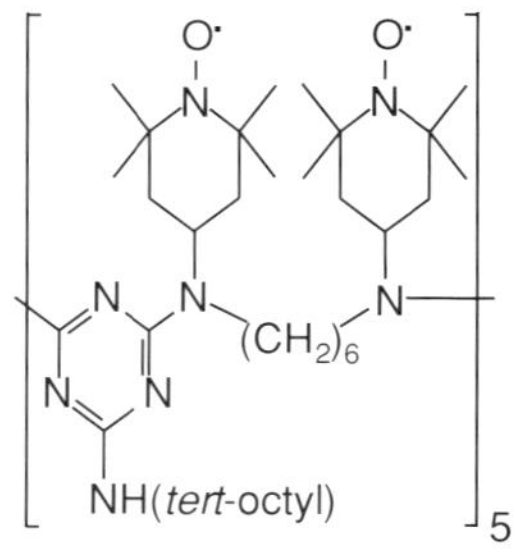

PIPO

Rel. rates:

- $PhCH_2OH/PhCH(OH)CH_3$ 95/4
- 1-octanol/2-octanol 86/1

Figure 7.18 PIPO-catalyzed oxidation of alcohols with NaOCl.

these microorganisms.[74] There is currently considerable commercial interest in laccases for application in pulp bleaching (as a replacement for chlorine) in paper manufacture and remediation of phenol-containing waste streams.[1] The selective oxidation of the primary alcohol groups in starch, to the corresponding carboxyl functionalities, is another reaction of commercial importance (see earlier). The carboxystarch product is a biodegradable alternative for the currently used polyacrylate water superabsorbents, e.g. in baby diapers. It has been shown[72,73] that laccase, in combination with TEMPO or derivatives thereof, is able to catalyze the aerobic oxidation of the primary alcohol moieties in starch but the relatively high enzyme costs form an obstacle to commercialization. Inefficient laccase use is a result of its instability toward the oxidizing reaction conditions. We have recently shown that the stability of the laccase under reaction conditions can be improved by immobilization as a cross-linked enzyme aggregate (CLEA).[75] Indeed, laccase CLEA has broad potential as a catalyst for a variety of applications.

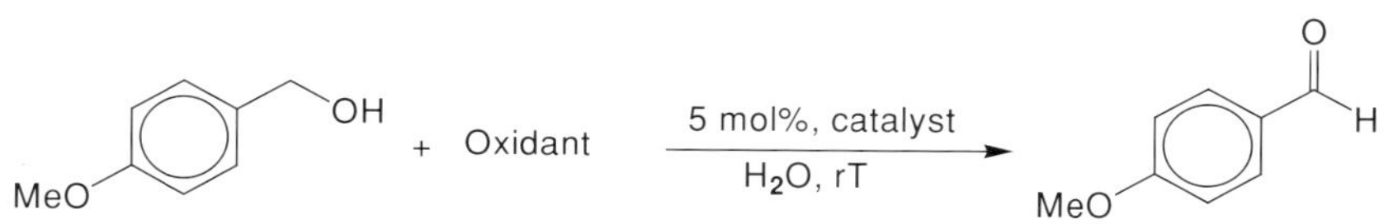

Catalyst :

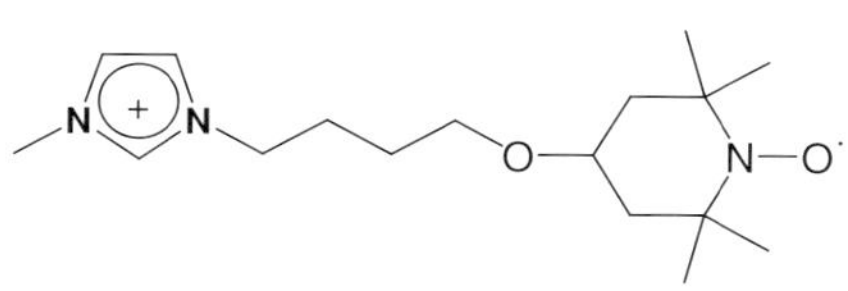

Oxidant	Time (min)	Yield (%)
I_2	40	98
NaOCl	3	96

Figure 7.19 Ion-supported TEMPO as oxidation catalyst.

In other recent developments the combinations of TEMPO with $NaNO_2$ and 1,3-dibromo-5,5-dimethylhydantoin[76] or $FeCl_3$[77] as cocatalysts have been shown to catalyze the aerobic oxidation of alcohols in water at 80°C.

7.5 Aldehyde and ketone oxidations in water

The Baeyer–Villiger oxidation of aldehydes and ketones to the corresponding esters or lactones (Eq. 3) is a widely applied reaction in organic synthesis. Traditionally it is performed with an organic peracid such as peracetic acid. However, the use of a peracid results in the formation of 1 equiv. of the corresponding carboxylic acid salt as waste, which has to be recycled or disposed of. Moreover, organic peracids are expensive and/or hazardous (because of shock sensitivity) which limits their commercial application. The transport and storage of peracetic acid, for example, has been severely restricted, making its use prohibitive. Consequently, increasing attention is focused on the development of procedures deploying aqueous hydrogen peroxide as the primary oxidant, preferably in water as the sole solvent.[78]

(Eq. 3)

R^1, R^2 = alkyl, H

In their studies of the tungstate-catalyzed oxidation of primary alcohols to the corresponding carboxylic acids, via the corresponding aldehydes (see above), Noyori and coworkers discovered that aldehydes could be selectively converted to carboxylic acids by reaction with aqueous hydrogen peroxide in an aqueous/organic biphasic system, in the presence of the acidic, lipophilic phase transfer catalyst, methyltrioctylammonium bisulfate, $[CH_3(n\text{-}C_8H_{17})_3N]^+HSO_4$, without the need for an organic solvent, halide or even a metal catalyst (Fig. 7.20).[22,79] The reactions proceeded via the formation of the aldehyde perhydrate, by addition of hydrogen peroxide to the carbonyl group, which undergoes Brønsted acid-catalyzed rearrangement to the ester (lactone) product and water.

Cyclic ketones were oxidized to the corresponding dicarboxylic acids, via initial Baeyer–Villiger reaction, using hydrogen peroxide in the presence of a catalyst comprising 1 mol%

$$\underset{}{RCHO} + \underset{(30\%)}{H_2O_2} \xrightarrow[90^\circ C, 1\text{–}4\ h]{[CH_3(n\text{-}C_8H_{17})_3N]\ HSO_4\ (0.5\text{–}1\ mol\%)} RCO_2H + H_2O$$

R	H_2O_2 (equiv.)	Yield RCO_2H (%)
n-C_7H_{15}	1.1	85
C_6H_5–CH_2CH_2	1.1	78
C_6H_5	2.5	85
4-$O_2NC_6H_4$	2.5	93
4-$MeOC_6H_4$	2.5	9

Figure 7.20 Catalytic oxidation of aldehydes with aq. H_2O_2.

tungstic acid, H_2WO_4, under organic solvent- and halide-free conditions (Fig. 7.21).[80] For example, cyclohexene afforded adipic acid in 91% isolated yield using 3.3 equiv. of H_2O_2. The acidic nature of the catalyst is crucial because when Na_2WO_4 was used no reaction took place. Since analogous systems can also oxidize alcohols to ketones (see above) reaction of cyclohexanol with 4.4 equiv. of H_2O_2 afforded adipic acid in 87% isolated yield.

Corma and coworkers reported[81] that Sn-beta containing tetrahedrally coordinated Sn(IV) in the zeolite framework is a highly active and selective heterogeneous catalyst for the Baeyer–Villiger oxidation of ketones and aldehydes with aqueous hydrogen peroxide. The Sn-substituted mesoporous silica, Sn-MCM-41, was subsequently shown to exhibit similar activity.[82] The reactions are not strictly speaking oxidations in water as they were performed in a water miscible organic solvent, e.g. dioxane, and it was not clear if they could be successfully carried out in water alone. Evidence was recently presented[83] to support a mechanism involving simultaneous coordination of the carbonyl group of the substrate and the hydrogen peroxide to the catalyst.

7.6 Sulfoxidations in water

Several pharmaceuticals contain a sulfoxide or sulfone moiety and the oxidation of dialkyl sulfides tothe corresponding sulfoxides (or sulfones), including the corresponding enantioselective oxidation of prochiral sulfides, are important reactions in organic synthesis.[84] Noyori and coworkers found that aromatic and aliphatic sulfides are oxidized to the corresponding sulfones, in high yields, using 30% aq. H_2O_2 in an aqueous biphasic system, in the absence of an organic solvent (Fig. 7.22).[22,85] The catalyst consisted of sodium tungstate, phenylphosphonic acid, $PhPO_3H_2$, and a tetraalkylammonium bisulfate as a phase transfer agent. Using a slight excess of H_2O_2, smooth oxidation to the sulfone was observed at 50°C with a substrate/catalyst ratio of 1000–5000 (Fig. 7.22). Olefinic double bonds and primary and secondary alcohol functionalities remained intact under these conditions. Alternatively, the sulfoxide could be obtained, in high yield, by performing the reaction in the absence of the tungstate or at lower temperatures, e.g. 0°C. It was suggested[85] that the function of the phenylphosphonic acid cocatalyst is to increase the reactivity of peroxo ligands by coordination to the W(VI).

Enantioselective sulfoxidation of prochiral sulfides is also of industrial interest, e.g. in the synthesis of the antiulcer drug, esomeprazole, an enantiomerically pure sulfoxide.[84,86]

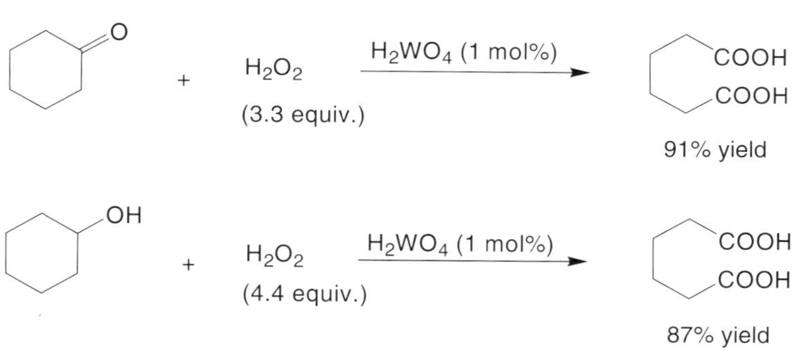

Figure 7.21 Catalytic oxidation of cyclohexanol/cyclohexanone with H_2O_2.

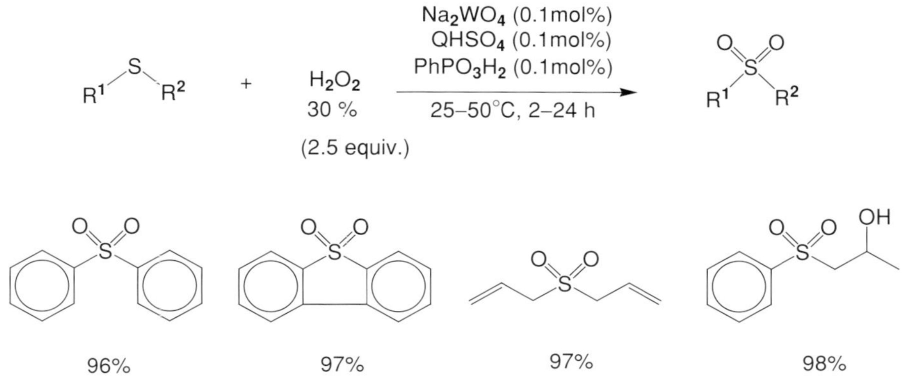

Figure 7.22 Catalytic sulfoxidations with H_2O_2.

The most practical method that is used in the industrial synthesis of esomeprazole involves titanium-catalyzed oxidation with an alkyl hydroperoxide, and a dialkyltartrate as chiral ligand, in an organic solvent such as dichloromethane.[84,86] A variety of oxidoreductases are known to catalyze the enantioselective oxidation of prochiral sulfides, usually as whole-cell biotransformations in aqueous media, but no simple metal complexes have been shown to be effective in water and the development of practical systems employing aqueous hydrogen peroxide as the primary oxidant is still an important challenge. In this context it is worth mentioning the enantioselective sulfoxidation of prochiral sulfoxides catalyzed by the semisynthetic peroxidase, vanadium-phytase,[87] in an aqueous medium.

7.7 Concluding remarks

Important advances have been made in the last few years in the design of effective catalytic systems for oxidations with the green oxidants, dioxygen and hydrogen peroxide, in an aqueous mono- or biphasic system, in the absence of organic solvents. Various practical systems have been reported, based on metal complexes of water-soluble, oxidatively stable ligands or using a phase transfer agent to transport the catalyst and/or active oxidant to the organic phase or employing redox molecular sieves as heterogeneous catalysts. These methodologies constitute green alternatives – clean oxidants, no need for organic solvents, facile product separation and catalyst recycling – for traditional oxidations. In the future we expect that they will be further applied in organic synthesis. In particular, iron complexes offer significant advantages from both an economic and environmental viewpoint but till now few examples of effective iron-based systems have been described.

References

1. For a recent review see: Xu, F. *Ind. Biotechnol.*, 2005, **1**, 38–49.
2. Besson, M. and Gallezot, P., In: Sheldon, R.A. and Bekkum, H. van (Eds.), *Fine Chemicals through Heterogeneous Catalysis*, Wiley-VCH, Weinheim, 2001, pp. 491–506 and 507–518.
3. For oxidations over gold catalysts see: Comotti, M., Della Pina, C., Matarrese, R., Rossi, M. and Siani, A., *Appl. Catal. A Gen.*, 2005, **291**, 204–209; Prati, L. and Porta, F., *Appl. Catal. A Gen.*, 2005, **291**, 199–203.

4. Watson, P.A., Wright, L.J. and Fullerton, T.J., *J. Wood Chem. Technol.*, 1993, **13**, 371, 391, 411.
5. Hampton, K.W. and Ford, W.T., *J. Mol. Catal. A Chem.*, 1996, **113**, 167.
6. Weinstock, I.A., Atilla, R.H., Reiner, R.S., Moen, M.A., Hammel, K.E., Houtman, C.J., Hill, C.L. and Harrup, M.K., *J. Mol. Catal. A Chem.*, 1997, **116**, 59.
7. Sorokin, A., Séris, J.-L. and Meunier, B., *Science*, 1995, **268**, 1163; Sorokin, A., DeSuzzoni-Dezard, S., Poullain, D., Noel, J.P. and Meunier, B., *J. Am. Chem. Soc.*, 1996, **118**, 7410–7411.
8. Meunier, B. and Sorokin, A., *Acc. Chem. Res.*, 1997, **30**, 470–476.
9. Shakla, R.S., Robert, A. and Meunier, B., *J. Mol. Catal. A Chem.*, 1996,**113**, 45.
10. Kamp, N.W.J. and Lindsay Smith, J.R., *J. Mol. Catal. A Chem.*, 1996, **113**, 131.
11. Tao, X., Ma, W., Zhang, T. and Zhao, J., *Angew. Chem. Int. Ed.*, 2001, **40**, 3014–3016.
12. Kachkarova-Sorokina, S.L., Gallezot, P. and Sorokin, A.B., *Chem. Commun.*, 2004, 2844–2845.
13. Sorokin, A.B., Kachkarova-Sorokina, S.L., Donze, C., Pinel, C. and Gallezot, P., *Topics Catal.*, 2004, **27**, 67–76.
14. Collins, T.J., *Acc. Chem. Res.*, 2002, **35**, 782–790, and references cited therein.
15. Collins, T.J., *Acc. Chem. Res.*, 1994, **27**, 279–285.
16. For reviews see: Bregeault, J.M., *Dalton Trans.*, 2003, 3289–3302.
17. For a review of metal-catalyzed epoxidations with hydrogen peroxide see: Lane, B.S. and Burgess, K., *Chem. Rev.*, 2003, **103**, 2437–2473.
18. Venturello, G., Alneri, E. and Ricci, M., *J. Org. Chem.*, 1983, **48**, 3831–3833.
19. Ishii, Y., Yamawaki, K., Ura, T., Yamada, H., Yoshiba, T. and Ogawa, M., *J. Org. Chem.*, 1988, **53**, 3587.
20. Aubry, C., Chottard, G., Platzer, N., Bregault, J.M., Thouvenot, R., Chauveau, F., Huet, C. and Ledon, H., *Inorg. Chem.*, 1991, **30**, 4409.
21. Sato, K., Aoki, M., Ogawa, M., Hashimoto, T. and Noyori, R., *J. Org. Chem.*, 1996, **61**, 8310–8311.
22. Noyori, R., Aoki, M. and Sato, K., *Chem. Commun.*, 2003, 1977–1986.
23. Sato, K., Aoki, M. and Noyori, R., *Science*, 1998, **281**, 1646–1647.
24. Maheswari, P.U., de Hoog, P., Hage, R., Gamez, P. and Reedijk, J., *Adv. Synth. Catal.*, 2005, **347**, 1759–1764.
25. Xi, Z., Zhou, N., Sun, Y. and Li, K., *Science*, 2001, **292**, 1139–1141.
26. Hage, R., Iburg, J.E., Kerschner, J., Koek, J.H., Lempers, E.L.M., Martens, R.J., Racheria, U.S., Russell, S.W., Swarthoff, T., van Vliet, M.R.P., Warnaar, J.B., van der Wolf, L. and Krijnen, B., *Nature*, 1994, **369**, 637; Hage, R., *Recl. Trav. Chim. Pays-Bas* 1996, **115**, 385.
27. De Vos, D.E., Sels, B.F., Reynaers, M., Subba Rao, Y.V. and Jacobs, P.A., *Tetrahedron Lett.*, 1998, **39**, 3221.
28. Berkessel, A. and Sklorz, C.A., *Tetrahedron Lett.*, 1999, **40**, 7965.
29. De Vos, D.E., De Wildeman, S., Sels, B.F., Grobet, P.J. and Jacobs, P.A., *Angew. Chem. Int. Ed.*, 1999, **38**, 980.
30. Brinksma, J., Schmieder, L., van Vliet, G., Boaron, R., Hage, R., De Vos, D.E., Alsters, P.L. and Feringa, B.L., *Tetrahedron Lett.*, 2002, **43**, 2619.
31. Lane, B.S. and Burgess, K., *J. Am. Chem. Soc.*, 2001, **123**, 2933.
32. Lane, B.S., Vogt, M., De Rose, V.J. and Burgess, K., *J. Am. Chem. Soc.*, 2002, **124**, 11947.
33. Tong, K-H., Wong, K-Y. and Chan, T.H., *Tetrahedron*, 2005, **61**, 6009–6014.
34. Chen, K., Costas, M., Kim, J., Tipton, A.K. and Que, L., *J. Am. Chem. Soc.*, 2002, **124**, 3026; Bassan, A., Blomberg, M.R.A., Siegbahn, P.E.M. and Que, L., *J. Am. Chem. Soc.*, 2002, **124**, 11056, and references cited therein.
35. White, M.C., Doyle, A.G. and Jacobsen, E.N., *J. Am. Chem. Soc.*, 2001, **123**, 7194.
36. Brinksma, J., Hage, R., Kerschner, J. and Feringa, B.L., *Chem. Commun.*, 2000, 537.
37. Tse, M.K., Klawonn, M., Bhor, S., Döbler, C., Anilkumar, G., Hugl, H., Maegerlein, W. and Beller, M., *Org. Lett.*, 2005, **7**, 987–990; for an analogous asymmetric epoxidation, see: Anilkumar, G., Bhor, S., Tse, M.K., Klawonn, M., Bitterlich, B., Beller, M. *Tetrahedron Asymm.*, 2005, **16**, 3536–3561.
38. Usui, Y., Sato, K. and Tanaka, M., *Angew. Chem. Int. Ed.*, 2003, **42**, 5623–5625.
39. See Sheldon, R.A. and van Vliet, M.C.A., In: Sheldon, R.A. and Bekkum, H. van (Eds.), *Heterogeneous Catalysis for Fine Chemicals*, Wiley-VCH, Weinheim, 2001, pp. 473–490, and references cited therein.
40. Sato, T., Dakka, J. and Sheldon, R.A., *Stud. Surf. Sci. Catal.*, 1994, **84**, 1853–1860; Corma, A., Navarro, M.T. and Perez-Pariente, J., *Chem. Commun.*, 1994, 147–148.
41. van Rheenen, V., Kelly, R.C. and Cha, P.Y., *Tetrahedron Lett.*, 1976, **23**, 1973.
42. Dupau, P., Epple, R., Thomas, A.A., Fokin, V.V. and Sharpless, K.B., *Adv. Synth. Catal.*, 2002, **344**, 421–433.
43. Bortolini, O., Conte, V., DiFuria, F. and Modena, G., *J. Org. Chem.*, 1986, **51**, 2661.
44. Sato, K., Aoki, M., Takagi, J. and Noyori, R., *J. Am. Chem. Soc.*, 1997, **119**, 12386; Sato, K., Takagi, J., Aoki, M. and Noyori, R., *Tetrahedron Lett.*, 1998, **39**, 7549; Sato, K., Aoki, M., Takagi, J., Zimmermann, K. and Noyori, R., *Bull. Chem. Soc. Jpn.*, 1999, **72**, 2287.
45. For excellent reviews, see: Muzart, J., *Tetrahedron*, 2003, **59**, 5789; Stahl, S.S., *Angew. Chem. Int. Ed.*, 2004, **43**, 3400.
46. For recent reviews of Wacker oxidations, see: Hintermann, L., In: Beller, M. and Bolm, C. (Eds.), *Transition Metals in Organic Synthesis*, Wiley-VCH, Weinheim, 2004, pp. 379–388; Monflier, E. and Mortreux, A., In: Cornils, B. and Herrmann, W. (Eds.), *Aqueous Phase Organometallic Chemistry*, 2nd edn., Wiley-VCH, Weinheim, 2004, pp. 481–487.

47. Papadogianakis, G., Verspui, G., Maat, L. and Sheldon, R.A., *Catal. Lett.*, 1997, **47**, 43.
48. ten Brink, G.-J., Arends, I.W.C.E., Papadogianakis, G. and Sheldon, R.A., *Chem. Commun.*, 1998, 2359–2360; ten Brink, G.-J., Papadogianakis, G., Arends, I.W.C.E. and Sheldon, R.A., *Appl. Catal. A*, 2000, **194–195**, 435.
49. ten Brink, G.-J., Arends, I.W.C.E. and Sheldon, R.A., *Science*, 2000, **287**, 1636.
50. ten Brink, G.-J., Arends, I.W.C.E. and Sheldon, R.A., *Adv. Synth. Catal.*, 2002, **344**, 355.
51. Sheldon, R.A., Arends, I.W.C.E. and ten Brink, G.-J., *Acc. Chem. Res.*, 2002, **35**, 774.
52. ten Brink, G.-J., Arends, I.W.C.E., Hoogenraad, M., Verspui, G. and Sheldon, R.A., *Adv. Synth. Catal.*, 2003, **345**, 497.
53. ten Brink, G.-J., Arends, I.W.C.E., Hoogenraad, M., Verspui, G. and Sheldon, R.A., *Adv. Synth. Catal.*, 2003, **345**, 1341.
54. Tohma, H., Maegawa, T., Takizawa, S. and Kita, Y., *Adv. Synth. Catal.*, 2002, **344**, 328–337.
55. Liu, Z.-L., Yang, L., Mu, R., Liu, Z.Q., Yang, Z. and Liu, Z.G., *Adv. Synth. Catal.*, 2005, **347**, 1333.
56. de Nooy, A.E.J., Besemer, A.C. and van Bekkum, H., *Synthesis*, 1996, 1153–1174.
57. Bragd, P.L., Besemer, A.C. and van Bekkum, H., *J. Mol. Catal. A Chem.*, 2001, **170**, 35–42.
58. Bragd, P.L., van Bekkum, H. and Besemer, A.C., *Top. Catal.*, 2004, **27**, 49–66.
59. Golubev, V.A., Rozantsev, E.G. and Neiman, M.B., *Bull. Acad. Sci. USSR Chem. Ser.*, 1965, **14**, 1898.
60. Cella, J.A., Kelley, J.A. and Kenehan, E.F., *J. Org. Chem.*, 1975, **40**, 1850; Rychovsky, S.D. and Vaidyanathan, R., *J. Org. Chem.*, 1999, **64**, 310.
61. Bolm, C., Magnus, A.S. and Hildebrand, J.P., *Org. Lett.*, 2000, **2**, 1173.
62. Kim, S.S. and Nehru, K., *Synlett*, 2002, 616.
63. Anelli, L., Biffi, C., Montanari, F. and Quici, S., *J. Org. Chem.*, 1987, **522**, 559; see also Anelli, P., Banfi, S., Montanari, F. and Quici, S., *J. Org. Chem.*, 1989, **54**, 2970; Anelli, P.L., Montanari, F. and Quici, S., *Org. Synth.*, 1990, **69**, 212.
64. Dijksman, A., Arends, I.W.C.E. and Sheldon, R.A., *Chem. Commun.*, 2000, 271.
65. Dijksman, A., Arends, I.W.C.E. and Sheldon, R.A., *Synlett*, 2001, 102.
66. Dijksman, A., Arends, I.W.C.E. and Sheldon, R.A., In: Sherrington, D.C. and Kybett, A.P. (Eds.), *Supported Catalysts and Their Applications*, Royal Society of Chemistry, United Kingdom, 2001, pp. 118–124.
67. For other strategies for immobilization of TEMPO see: Bolm, C. and Fey, T., *Chem. Commun.*, 1999, 1795; Fey, T., Fischer, H., Bachmann, S., Albert, K. and Bolm, C., *J. Org. Chem.*, 2001, **66**, 8154; Brunel, D., Lentz, P., Sutra, P., Deroide, B., Fajula, F. and Nagy, J.B., *Stud. Surf. Sci. Catal.*, 1999, **25**, 237; Verhoef, M.J., Peters, J.A. and van Bekkum, H., *Stud. Surf. Sci. Catal.*, 1999, **125**, 465; Brunel, D., Fajula, F., Nagy, J., Deroide, B., Verhoef, M., Veum, L., Peters, J. and van Bekkum, H., *Appl. Catal. A*, 2001, **213**, 73; Gilhespy, M., Lok, M. and Baucherel, X., *Chem. Commun.*, 2005, 2871; Ferreira, P., Hayer, W., Phillips, E., Rippon, D. and Tsang, S.C., *Green Chem.*, 2004, **6**, 310; Ciriminna, R., Blum, J., Avnis, D. and Pagliaro, M., *Chem. Commun.*, 2000, 1441; Ciriminna, R., Bolm, C., Fey, T. and Pagliaro, M., *Adv. Synth. Catal.*, 2002, **344**, 159.
68. Qian, W., Jin, E., Bao, W. and Zhang, Y., *Tetrahedron*, 2006, **62**, 556–562; see also Wu, X., Ma, L., Ding, M. and Gao, L., *Synlett*, 2005, **4**, 607–610.
69. Sheldon, R.A. and Arends, I.W.C.E., *Adv. Synth. Catal.*, 2004, **346**, 1051–1071, and references cited therein.
70. Gamez, P., Arends, I.W.C.E., Reedijk, J. and Sheldon, R.A., *Chem. Commun.*, 2003, 2414; Gamez, P., Arends, I.W.C.E., Sheldon, R.A. and Reedijk, J., *Adv. Synth. Catal.*, 2004, **346**, 805–811.
71. Fabbrini, M., Galli, C., Gentili, P. and Macchitella, D., *Tetrahedron Lett.*, 2001, **42**, 7551.
72. Viikari, L., Niku-Paavola, M.L., Buchert, J., Forssell, P., Teleman, A. and Kruus, K., World Patent WO 992,324,0, 1999.
73. Jetten, J.M., van den Dool, R.T.M., van Hartingsveldt, W. and van Wandelen, M.T.R., World Patent WO 00/50621, 2000.
74. Rochefort, D., Leech, D. and Bourbonnais, R., *Green Chem.*, 2004, 614.
75. For a recent review, see: Sheldon, R.A., Schoevaart, R. and van Langen, L.M., *Biocat. Biotrans.*, 2005, **23**(3/4), 14.
76. Liu, R., Dong, C., Liang, X., Wang, X. and Hu, X., *J. Org. Chem.*, 2005, **70**, 729–731.
77. Wang, N., Liu, R., Chen, J. and Liang, X., *Chem. Commun.*, 2005, 5322–5324.
78. For a review, see: ten Brink, G.-J., Arends, I.W.C.E. and Sheldon, R.A., *Chem. Rev.*, 2004, **104**, 4105–4123.
79. Sato, K., Hyodo, M., Takagi, J., Aoki, M. and Noyori, R., *Tetrahedron Lett.*, 2000, **41**, 1439.
80. Usui, Y. and Sato, K., *Green Chem.*, 2003, **5**, 373–375.
81. Corma, A., Nemeth, L.T., Renz, M. and Valencia, S., *Nature*, 2001, **412**, 423–425.
82. Corma, A., Navarro, M.T., Nemeth, L. and Renz, M., *Chem. Commun.*, 2001, 2190–2191.
83. Boronat, M., Corma, A., Renz, M., Sastre, G. and Viruela, P.M., *Chem. Eur. J.*, 2005, **11**, 6905–6915.
84. Legros, J., Dehli, J.R. and Bolm, C., *Adv. Synth. Catal.*, 2005, **347**, 19–31, and references cited therein.
85. Sato, K., Hyodo, M., Aoki, M., Zheng, X.-Q. and Noyori, R., *Tetrahedron*, 2001, **57**, 2469.
86. Federsel, H.-J. and Larrson, M., In: Blaser, H.-U. and Schmidt, E. (Eds.), *Asymmetric Catalysis on Industrial Scale: Challenges, Approaches and Solutions*, Wiley-VCH, Weinheim, 2004, p. 413.
87. van de Velde, F., Könemann, L., van Rantwijk, F. and Sheldon, R.A., *Chem. Commun.*, 1998, 1891–1892; van de Velde, F., Arends, I.W.C.E. and Sheldon, R.A., *J. Inorg. Biochem.*, 2000, **80**, 81–89.

Chapter 8
Nucleophilic Additions and Substitutions in Water

Denis Sinou

Additions of nucleophiles to unsaturated species such as C═X multiple bonds or substitutions of electrophilic species such as C—X by nucleophiles are among the most common and important reactions in organic chemistry. Most of these reactions are catalyzed by either bases or acids; the use of a base as catalyst generally increases the amount or the strength of the nucleophile, whereas the presence of an acid as catalyst allows the use of neutral nucleophiles by complexation, and so further polarization, of the electrophile.

This chapter will focus on nucleophilic additions to unsaturated bonds catalyzed by bases or acids, excluding the use of organometallic species, which is discussed in Chapter 4, and on nucleophilic substitutions including ring opening of epoxides and allylic substitutions.

8.1 Nucleophilic additions

The addition of nucleophiles such as carbon nucleophiles to electrophiles such as unsaturated carbon–heteroatom bonds is an important reaction in organic chemistry, allowing carbon–carbon bond formation. Although these reactions are usually performed in organic or eventually in organic/aqueous two-phase system, it was recently shown that they could be performed in water alone or in a monophasic water/organic system. This development has also allowed reactions to proceed under less drastic conditions.

8.1.1 The aldol reaction

The aldol or aldol-type reaction is well recognized as one of the most important carbon–carbon bond forming reactions in organic synthesis. As shown in Scheme 8.1, two stereogenic centers could be generated in this aldol reaction. The classical aldol condensation between an aldehyde and a ketone is often catalyzed by a base or an acid. Another approach is the acid-catalyzed cross-aldol reaction of silyl enol ethers with carbonyl compounds, the so-called Mukaiyama reaction.

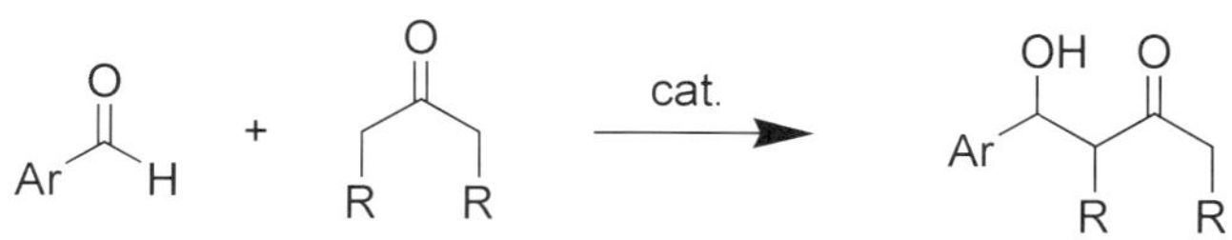

Scheme 8.1

Classical aldol reaction

The classical aldol addition commonly uses a basic catalyst. However, the condensation reaction of various ketones with aryl aldehydes under basic conditions in water gave the expected aldol products along with the dehydration compounds.[1,2] The presence of surfactants led mainly to dehydration products,[3] whereas the use of a water/dioxane two-phase system gave good yields of aldol products with reactive aliphatic aldehydes.[4]

Recent progress in the area of organocatalysis has resulted in the development of direct asymmetric aldol reactions in water. Using primary or secondary amines as catalysts, direct aldol reactions have been executed in good yields,[5,6] and often with high asymmetric induction.[7–11] The influence of anionic micelles[12–14] or of high pressure induced by water freezing,[15] on the direct catalytic asymmetric aldol reaction was also investigated, affording in the latter case the aldol products in enantiomeric excesses (ee's) up to 96% at 200 MPa in the presence of L-proline.

A microporous 1:2 polycondensate obtained by the treatment of anthracenebis (resorcinol) with La(O-*i*-Pr)$_3$ was also successfully applied to this reaction in neat water at neutral pH with efficient recycling of the catalyst.[16]

Various Lewis acids have also been used as catalysts for aldol reactions. Zinc complexes associated with aminoalcohols or aminoesters gave quantitatively the aldol products[17]; when chiral amino acids such as proline, lysine, or arginine were used, ee's up to 56% were obtained (Scheme 8.2).[18]

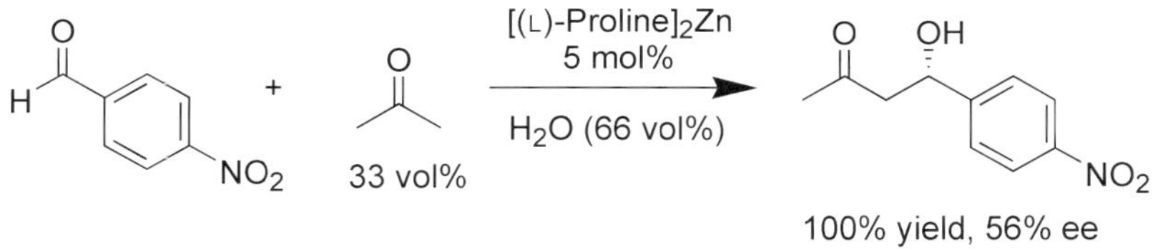

Scheme 8.2

Aldolization of unprotected glycolaldehyde in water in the presence of a Zn-proline catalyst at room temperature and in the absence of strong bases gave the corresponding tetroses and hexoses, but in low ee's.[19]

Mukaiyama aldol reaction

Pioneering work by Lubineau and coworkers showed that the aldol type reaction between a silyl enol ether and an aldehyde (the so-called Mukaiyama aldol reaction) occurred in water at room temperature with high syn stereoselectivity, albeit in low yields.[20,21] However, the development of water-tolerant Lewis acids for this reaction has led to improved rates and chemical yields. Various lanthanides triflates, such as ytterbium triflate [Yb(OTf)$_3$], scandium triflate [Sc(OTf)$_3$], gadolinium triflate [Gd(OTf)$_3$], or lutetium triflate [Lu(OTf)$_3$], have been found to afford the aldol products between various aldehydes and silyl enol ethers in high yields in aqueous media, with good to moderate syn/anti diastereoselectivities (Scheme 8.3, Table 8.1).[22–28]

The catalytic activities of these lanthanide triflates were found to be mainly dependent on two parameters, the hydrolysis constant and the water exchange rate constant.[29] Moreover,

Table 8.1

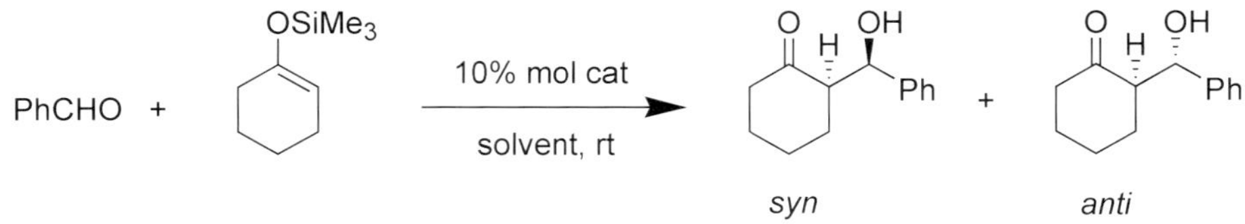

Catalyst	Solvent	Yield (%)	syn/anti
No	H_2O	23	85/15
$La(OTf)_3$	H_2O/THF (1:4)	8	—
$Yb(OTf)_3$	H_2O/THF (1:4)	91	73/27
$Gd(OTf)_3$	H_2O/THF (1:4)	89	77/23
$Lu(OTf)_3$	H_2O/THF (1:4)	88	78/22
$Pr(OTf)_3$	H_2O/THF (1:4)	28	72/28
$Nd(OTf)_3$	H_2O/THF (1:4)	83	68/32
$Eu(OTf)_3$	H_2O/THF (1:4)	34	64/36
$Sm(OTf)_3$	H_2O/THF (1:4)	46	74/26

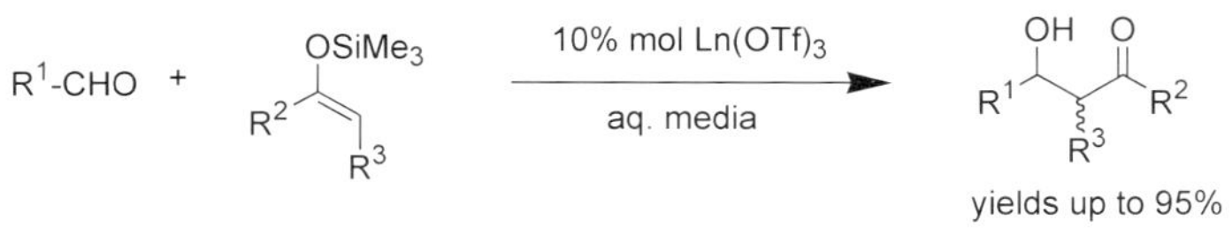

R^1 = H, C_6H_5, C_6H_5-CH_2CH_2, *p*-ClC_6H_4, *p*-$MeOC_6H_4$, *o*-HOC_6H_4, CH_3, CH_2=CH, $ClCH_2$, C_6H_5CO, etc.
Ln = Sc, Y, La, Ce, Pr, Nd, Sm, Eu, Gd, Tb, Dy, Ho, Er, Tm, Yb, Lu

Scheme 8.3

the lanthanide triflates could be quantitatively recovered after the reaction and reused with comparable results. A simple procedure for the easy separation of the products from the catalyst was the use of a water/ethanol/toluene (1:7:4) system.[30]

Recently, Kobayashi noticed that the presence of a small amount of a surfactant such as sodium dodecyl sulfate (SDS) showed a remarkable enhancement of the reactivity in the Mukaiyama-catalyzed aldol reaction in pure water using $Yb(OTf)_3$ or better $Sc(OTf)_3$ as the catalyst; without addition of the surfactant, the reaction was very sluggish (Scheme 8.4).[31] Other surfactants such as calix[6]arene derivatives bearing sulfonate and alkyl groups[32,33] or aromatic and aliphatic anionic surfactants[34] have also been found to be highly effective in the aqueous Mukaiyama aldol reactions in pure water, affording the aldol products in high yields. This was probably due to the formation of micelles which stabilized the labile silyl enol ethers and thus promoted the aldol reaction.

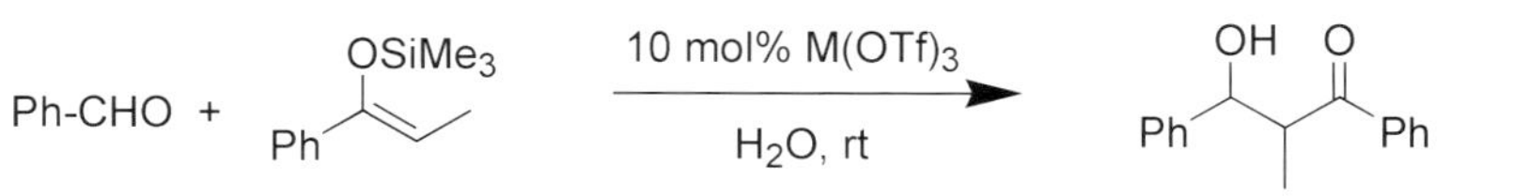

$Yb(OTf)_3$	17% yield (48 h)
$Yb(OTf)_3$ + SDS	50% yield (48 h)
$Sc(OTf)_3$ + SDS	88% yield (4 h)

Scheme 8.4

It was shown that the Lewis acid surfactant-combined catalyst scandium tris(dodecyl sulfate) was an excellent catalyst for the aldol reaction in water without any organic cosolvent, providing yields up to 98%.[27,35,36] The much higher activity observed in water than in organic solvents was probably due to the formation of stable dispersion systems, which included the catalyst and the organic substrates in water. The presence of Brønsted acids, and particularly HCl, drastically accelerated the Mukaiyama aldol reactions mediated by $Sc(O_3SOC_{12}H_{25})_3$ (Scheme 8.5).[37]

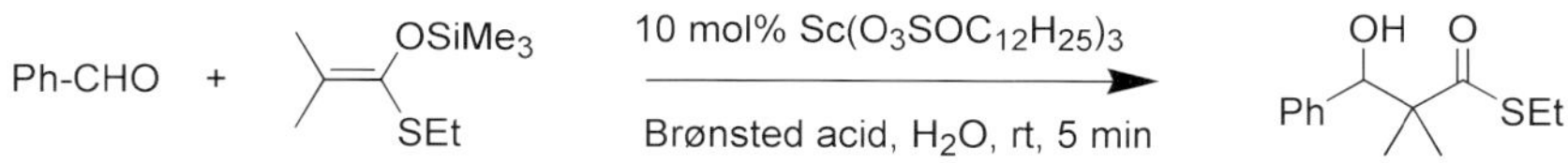

no	16% yield
TsOH (10%)	52% yield
TfOH (10%)	51% yield
HCl (10%)	67% yield
HCl (10%, 15 min)	88% yield

Scheme 8.5

Other metals were also used with more or less success in aqueous aldol reactions. $InCl_3$ was found to be efficient in water only under optimized conditions,[38,39] in the presence of SDS as the surfactant,[40] or in a i-PrOH/H_2O (95:5) mixture.[41]

Boron has been shown to be an efficient catalyst. Various aldehydes and silyl enol ethers afforded the corresponding syn-substituted β-hydroxyketones in high diastereoselectivities (80–94% de) when the reaction was performed in water with 10 mol% Ph_2BOH, surfactant (SDS), and a Brønsted acid (Scheme 8.6). A mechanism involving a boron enolate intermediate generated by a silicon/metal exchange was proposed; the improvement observed in the presence of benzoic acid could be due to an increase of the rate of the Si/B exchange.[42,43]

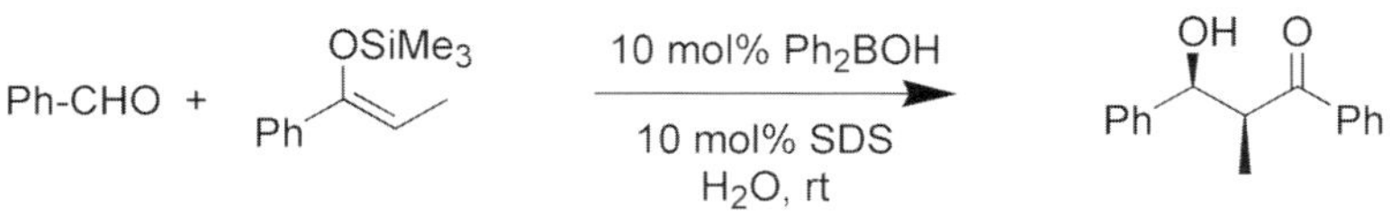

Acid	Time (h)	Yield (%)	syn/anti
No	24	Trace	—
$PhCO_2H$ (0.01 equiv.)	18	90	92/8
CH_3CO_2H	2	93	92/8
HCl	2	76	89/11
TsOH		69	89/11

Scheme 8.6

A supported scandium catalyst prepared from sulfonated polystyrene resin[44] was found to be an effective catalyst for Mukaiyama aldol reactions in water, the use of this solvent being crucial for the reaction. The catalyst was easily recovered by a simple filtration and reused without any loss of catalytic activity (Scheme 8.7). Other similar work includes the use of a nanostructured polymer[45] or a polyoxyethylene/polyoxopropylene resin.[46]

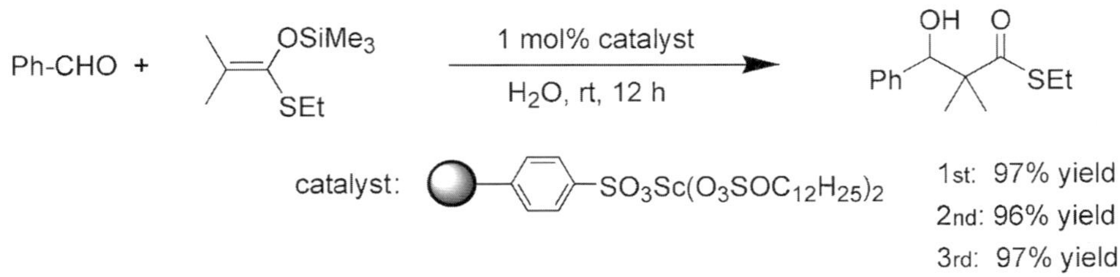

Scheme 8.7

Diastereoselective aldol reactions of various aldehydes with trimethylsilyl enolates have been carried out in neat water using iron(III) chloride and a surfactant with high yields and better diastereoselectivities than with $Sc(OTf)_3$ (Scheme 8.8).[47]

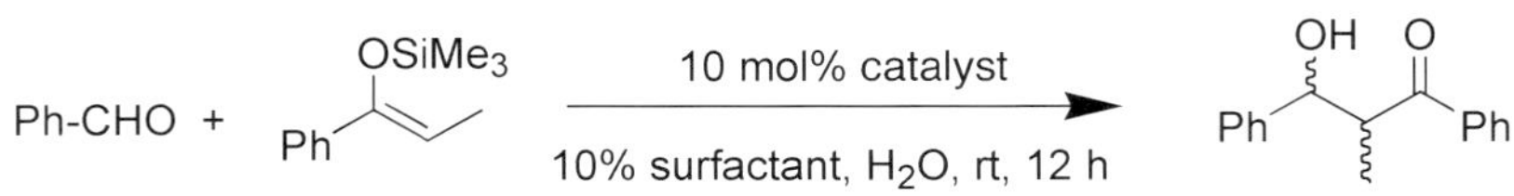

Lewis acid	Surfactant	Yield (%)	syn/anti
$Sc(OTf)_3$	SDS	91	50/50
$FeCl_3$	SDS	83	87/13
$FeCl_3$	$C_8H_{17}C_6H_4SO_3Na$	92	90/10

Scheme 8.8

Montmorillonite K10 also catalyzed the Mukaiyama aldol reactions of aldehydes with silyl enol ethers or ketene silyl acetals in neat water; commercially available montmorillonite K10 can be used without the need for ion-exchange processes and can be reused again after thermal activation.[48,49] Even hydrates of aldehydes such as glyoxylic acid can be used directly in this reaction.

A combination of copper bis(dodecylsulfate) and chiral bis(oxazoline) ligands allowed the reaction to occur in neat water in the presence of a catalytic amount of carboxylic acid with ee's up to 69% for the syn isomer.[50] A chiral bis(oxazoline) supported on a modified poly(ethylene glycol) was also shown to be an effective copper(II) ligand for this aldol reaction in neat water, with 31–63% ee, comparable to those obtained with nonsupported ligands in the same solvent. The solubility of the ligand in water allowed for a highly convenient catalyst recycling procedure.[51]

Recently, Wang and coworkers developed an efficient asymmetric Mukaiyama reaction using $Ga(OTf)_3$, associated with a chiral semi-crown ligand, as the Lewis acid (Scheme 8.9).[52] Enantioselectivities up to 87% were obtained when the reaction was performed in a mixture of water/ethanol (9:1) or even in neat water, with good chemical yields and high diastereoselectivities.

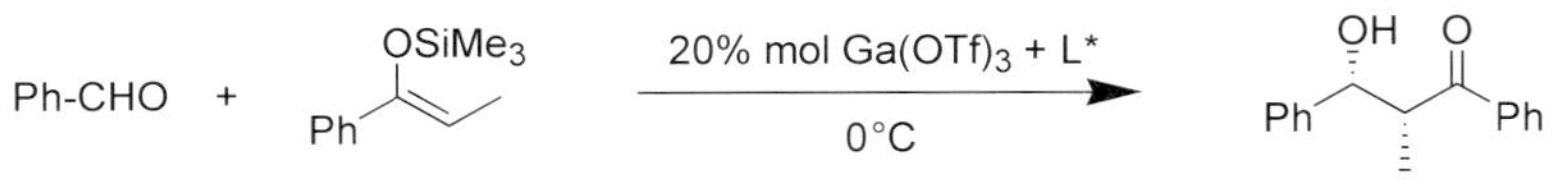

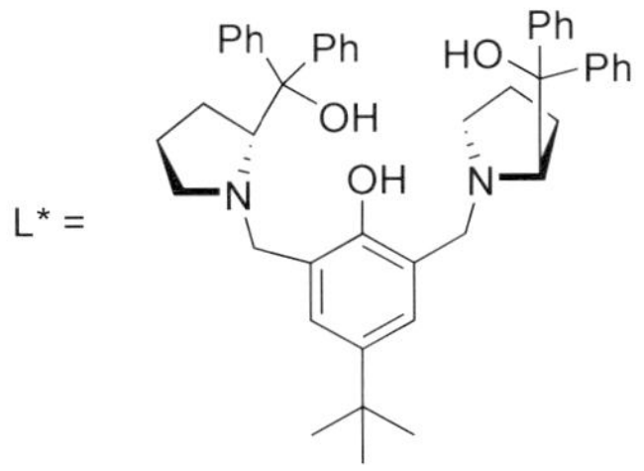

Solvent	Yield (%)	syn/anti	ee syn (%)
EtOH/H_2O (9:1)	89	89/11	87
EtOH/H_2O (1:9)	85	85/15	85
H_2O	41	90/10	84

Scheme 8.9

Catalytic asymmetric hydroxymethylation of silicon enolates with an aqueous formaldehyde solution has been developed by Kobayashi et al. using a bismuth triflate associated with a chiral bipyridine in a DME/H_2O mixture; the reaction proceeded smoothly in the presence of 1 mol% catalyst to afford the hydroxymethylated adducts in high yields and 77–93% ee.[53] Chiral anionic surfactants associated with $Ga(OTf)_3$, $Cu(OTf)_2$, or $Sc(OTf)_3$ catalyzed Mukaiyama aldol reactions in water with moderate to good diastereo- and enantioselectivities.[54]

8.1.2 Michael addition

Nucleophilic addition of carbanions to electrophilic alkenes activated by an electron-withdrawing group (the so-called Michael addition) is one of the most important procedures for the formation of carbon–carbon bonds in organic synthesis. The first examples of Michael addition in water were independently reported by Hajos[55] and Wiechert[56] in the 1970s. 2-Methylcyclopentane-1,3-dione reacted with methyl vinyl ketone to give the corresponding conjugated addition product without the use of any basic catalyst, the product being obtained in higher yield and purity than in methanol in the presence of a base (Scheme 8.10). In this system the Michael addition product cyclized further to give a 5–6 fused-ring system.

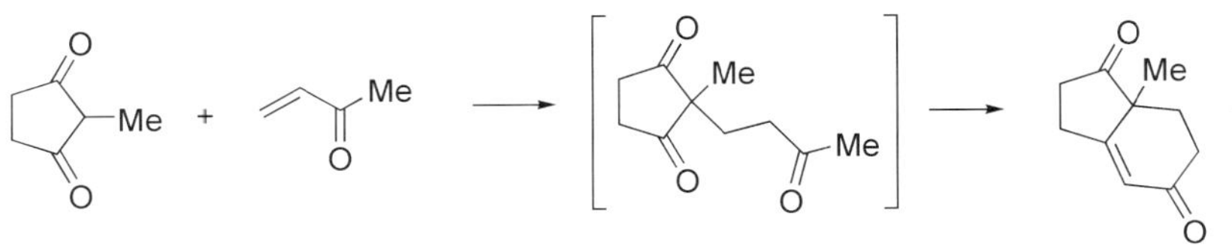

H_2O	87%
MeOH/KOH	54%

Scheme 8.10

The extension of this aqueous Michael reaction to acrolein was performed by Deslongchamps.[57] One application was the total synthesis of 13-α-methyl-14α-hydroxysteroid.

Feringa showed that Michael addition of β-ketoesters and α-nitro esters to enones and α,β-unsaturated aldehydes resulted in a quantitative conversion to the corresponding 1,4-adducts when the reaction was performed in water in the presence of $Yb(OTf)_3$ as the catalyst (Scheme 8.11).[58,59] However the reaction was often sluggish, the conversion being only 10% after 8 h. The use of the Lewis acid surfactant-combined catalyst $Sc(O_3SOC_{12}H_{25})_3$ allowed the reaction to proceed smoothly in water to give the Michael adducts in high yields.[60]

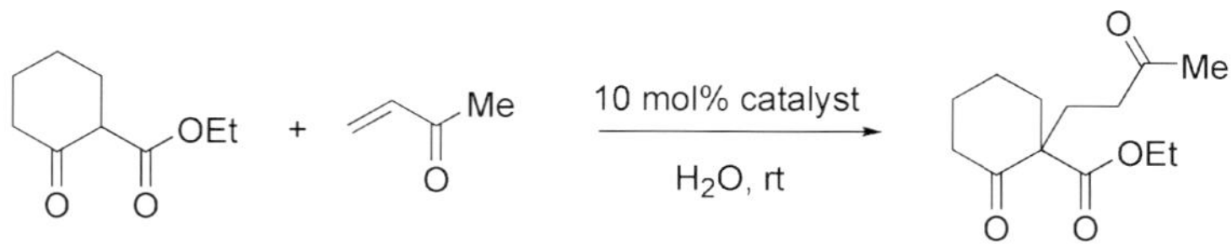

$Yb(OTf)_3$	8 h	10% yield
$Yb(OTf)_3$	5 days	98% yield
$Sc(O_3SOC_{12}H_{25})_3$	8 h	81% yield
$Sc(OTf)_3$ + SDS (30%)	8 h	65% yield

Scheme 8.11

A hydrophobic polymer-supported scandium(III) catalyst was also successfully used in the Michael reaction of unsaturated ketones with silyl enol ethers.[44] Recently, an amphiphilic resin-supported rhodium/phosphine complex was used as catalyst in the 1,4-addition of various boronic acids to enones in water at 25°C. High yields were obtained and the catalyst was easily separated and subjected to a second and third round of reactions with no decrease in activity.[61]

Lubineau and Augé reported the beneficial effect of using water as the solvent in the Michael addition of nitroalkanes to methyl vinyl ketone under neutral conditions.[62] Moreover, addition of glucose or saccharose increased the rate of the reaction. Various nitroalkanes were condensed with different electrophilic alkenes in NaOH (0.025–0.1 M), without any added organic solvent, affording the corresponding adducts with moderate to good yields when a catalytic amount of cetyltrimethylammonium chloride (CTACl) was used (Scheme 8.12).[63]

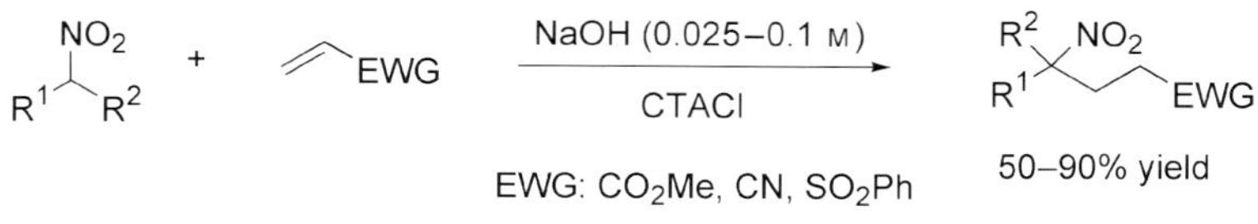

Scheme 8.12

The application of this procedure to the condensation of α-nitroalkanones with α-alkyl-α,β-unsaturated aldehydes afforded, in a one-pot synthesis, functionalized, bridged, and bicyclic lactones containing 10-, 11-, 13-, and 15-membered rings.[64] Recently, Ballini et al. showed that the Michael addition of primary aliphatic nitro compounds to α,β-unsaturated enones performed in aqueous media provided the one-pot synthesis of 1,4-diketones, 1,4-diols, δ-nitroalkanols, or hydroxytetrahydrofurans by appropriate choice of reaction conditions.[65]

The beneficial effect of water-soluble calixarenes as phase-transfer catalysts was noticed in the Michael addition of activated methyl and methylene compounds to α,β-unsaturated ketones and nitriles in aqueous NaOH solution at room temperature.[33]

It is to be noted that Michael additions of dicarbonyl compounds and nitroalkanes to 2-cyclohexen-1-one occurred in water without phase transfer agents by careful control of the pH of the solution.[66]

The Michael addition in aqueous medium involving heteronucleophiles has also been investigated. Amines and thiols added readily to α,β-unsaturated ketones[67] and dehydroalanine amides[68] in neat water without any catalyst. However, addition of a catalytic amount of a surfactant such as SDS generally increased the chemical yields.

Several transition metal-based Lewis acid catalysts such as $FeCl_3$, $CrCl_3$, $SnCl_4$, or $Al(O_3SOC_{12}H_{25})_3$ were shown to be highly effective for aza-type Michael additions of amines such as indoles and pyrroles to α,β-unsaturated compounds in aqueous solution.[69,70]

Addition of thiols to α,β-unsaturated ketones in water was also catalyzed by SDS[67] or β-cyclodextrin.[71] There is also a single report on the stereoselectivity of base-catalyzed Michael additions of thiols to nitro-olefins,[72] where moderate diastereoselectivities were obtained (Scheme 8.13).

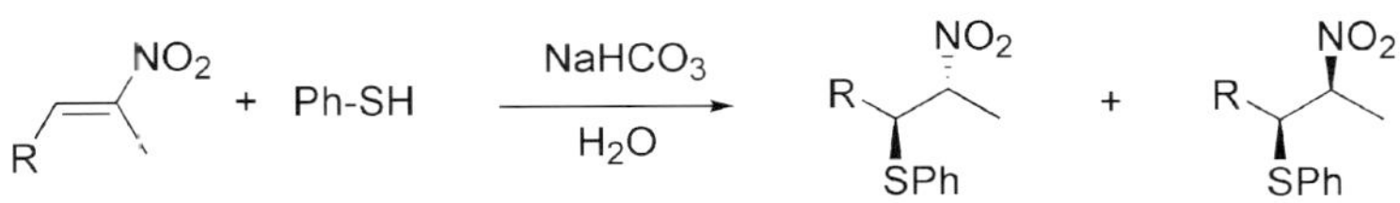

R	Yield (%)	anti/syn
Ph	95	73/27
C_2H_5	70	80/20
i-Pr	58	38/62

Scheme 8.13

The first Lewis acid-catalyzed asymmetric Michael addition in water was developed by Kobayashi et al., who reported ee's up to 83%.[73] Very recent developments show great promise for further improvement of Michael addition reactions in water. In an elegant study, Kaneda and coworkers used montmorillonite-enwrapped metal triflates to execute C—C bond forming Michael additions. When scandium triflate was employed, adducts were obtained in quantitative yield within a 0.5–3 h at or slightly above room temperature. The catalysts were reusable with no appreciable loss in activity.[74] In another recent study, Lindström and coworkers observed a remarkable ligand acceleration effect in aqueous ytterbium triflate-catalyzed Michael additions.[75] A number of 1,2-diamines and 1,2-aminoalcohols were shown to have a positive influence on the rate of the reaction, the most efficient being tetramethylethylenediamine, which induced a nearly 20-fold rate acceleration.

8.1.3 Mannich-type reaction

Mannich-type reactions are among the most useful methodologies for the synthesis of β-aminoketones or esters, which are versatile synthetic intermediates. Lanthanides triflates such as $Sm(OTf)_3$, $Tm(OTf)_3$, and $Sc(OTf)_3$, as well as $Cu(OTf)_2$, have been used as efficient catalysts for Mannich reactions in aqueous micellar media.[76]

Loh and coworkers reported a one-pot Mannich-type reaction between aldehydes, aromatic amines, and silyl enol ethers in water in the presence of 20 mol% indium trichloride, giving the corresponding β-aminoketones in 30–94% yields.[77,78] The use of glyoxylic acid monohydrate directly afforded the β-amino acids. However, reactions using prochiral silyl enol ethers gave very low diastereoselectivities, the syn product being the major isomer. The catalyst could be recovered after completion of the reaction and reused for another reaction without any significant drop in activity.

Kobayashi and coworkers showed that Lewis acid surfactant-combined catalysts such as scandium tris(dodecyl sulfate), $Sc(O_3SC_{12}H_{25})_3$, or copper bis(dodecyl sulfate), $Cu(O_3SC_{12}H_{25})_2$, were efficient catalysts for the three-component Mannich-type reaction, with 73–95% yield being obtained in neat water.[36] Neutral salts such as sodium triflate and sodium iodide catalyzed the condensation reaction in water between preformed imines and silicon enolates, or the three-component Mannich-type reaction using aromatic amines, with 49–93% yields and 0–80% diastereoselectivities.[79] Mechanistic studies indicated that both sodium triflate and the Mannich adduct itself cooperatively promote the reaction.

Under the influence of a catalytic amount of a Brønsted acid, Mannich-type reactions can take place smoothly in aqueous media. For example, it was recently shown that in the presence of HBF_4 and SDS, Mannich-type reactions proceeded in water starting from the preformed aldimine or from the aryl amine and the aldehyde (Scheme 8.14).[80,81] Aliphatic as well as aromatic aldehydes underwent the reaction and afforded the corresponding β-aminocarbonyl compounds in good to high yields. Even chloral and formaldehyde, commercially available as aqueous solutions, could be used as electrophiles. When ketene silyl acetals derived from esters were used, high diastereoselectivities were observed in the condensation with aldimines.[82] Employing ketene silyl acetals derived from an aryl ester gave mainly *anti*-β-amino-α-siloxy esters (92–98% de) in 2-propanol, whereas syn adducts were obtained in a water/SDS mixture starting from a ketene silyl acetal derived from a methyl ester (30–90% de).

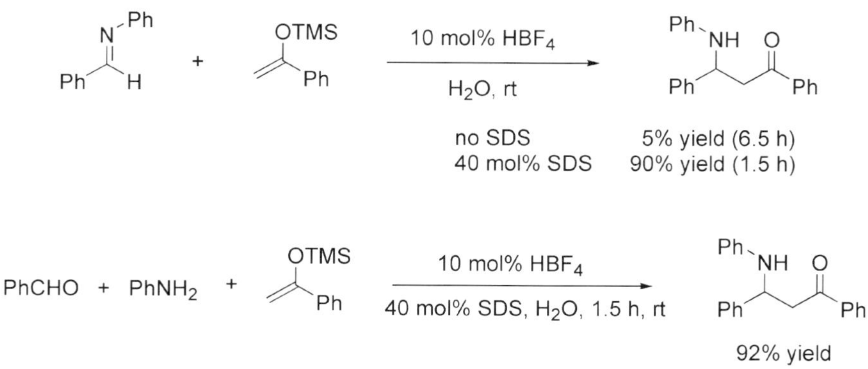

Scheme 8.14

Kobayashi and coworkers have developed a three-component Mannich-type reaction of aldehydes, aromatic amines, and silyl enolates or ketones catalyzed by *p*-dodecylbenzene sulfonic acid (DBSA), a Brønsted acid surfactant-combined catalyst. Various β-aminoketones were obtained in good yields (Scheme 8.15).[83,84] The long alkyl chain of DBSA seems necessary for the formation of a colloidal dispersion, which is assumed to play an essential role in acceleration of the reaction; a combination of *p*-toluene sulfonic acid and SDS afforded the adduct in lower yield (56%). It is noteworthy that this catalyst gave very slow reactions when the condensation was performed in organic solvents, showing the unique property of water to induce hydrophobic interactions between the substrate and the catalyst.

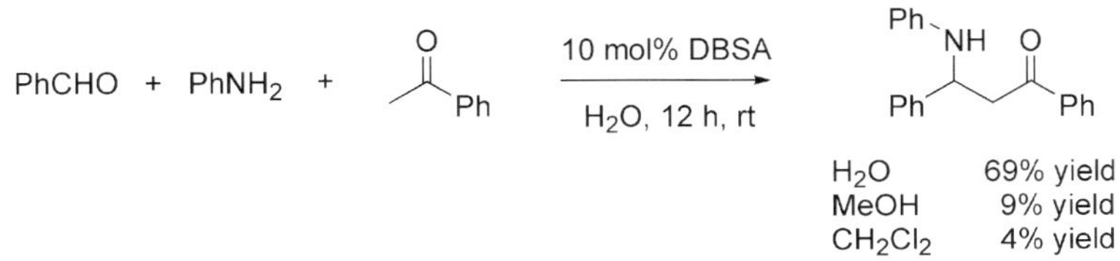

Scheme 8.15

A hydrophobic polystyrene-supported sulfonic acid showed high catalytic activity in water in the three-component Mannich reaction, with 1 mol% of catalyst being enough to catalyze the reaction (Scheme 8.16).[85]

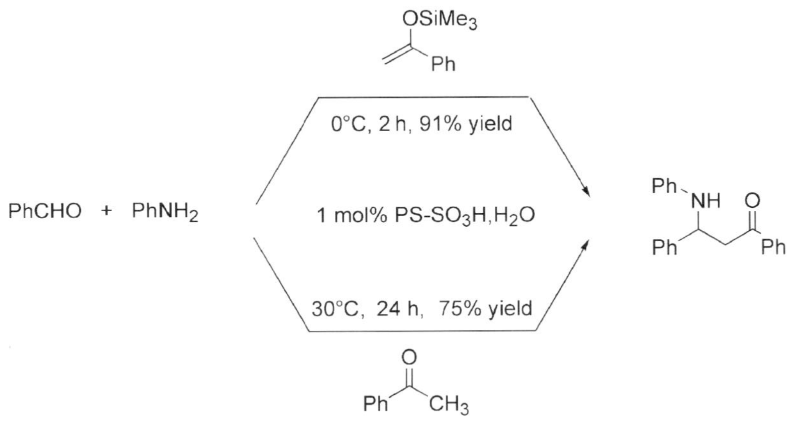

Scheme 8.16

A nanostructured scandium-containing polymer was also successfully used in the condensation of aldehydes, aromatic amines, and silyl enol ethers to give the corresponding β-aminoketones, but the observed diastereoselectivities were moderate.[45] Cai and coworkers reported the use of sulfonated amino acids as efficient Brønsted catalysts in direct diastereo- and regioselective Mannich reactions in water.[86]

The asymmetric version of the Mannich-type reaction has been studied in aqueous conditions. Hydrazono esters reacted with silyl enol ethers in water in the presence of zinc fluoride and a chiral diamine to afford β-aminoketones (Scheme 8.17).[87] In the presence of cetyltrimethylammonium bromide, a cationic surfactant, reactions with silyl enol ethers from α-substituted ketones also gave excellent yields. Syn adducts were obtained starting from the (E)-enolate, and anti adducts from the (Z)-enolate, with diastereoselectivities of 70–80% and ee's up to 97% for the major stereoisomer.

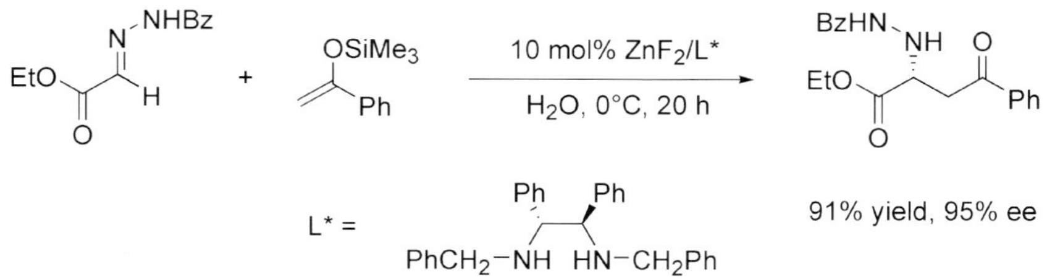

Scheme 8.17

8.2 Nucleophilic substitution

Substitutions of S_N2 type are frequently used for carbon–carbon or carbon–heteroatom bond formation. However, little attention has been devoted to the development of such reactions in water. This is likely due to concerns about competitive hydrolysis of the electrophile in water and S_N2-type reactions being slower in aqueous conditions than in aprotic polar solvents due to the higher cost of desolvation of nucleophiles. We shall discuss the ring opening of epoxides and aziridines, palladium-catalyzed allylic substitutions, as well as acylations and sulfonylations of amines and alcohols.

8.2.1 Ring-opening nucleophilic substitution

Ring opening of oxiranes

Oxirane ring opening with various nucleophiles has been extensively investigated in organic solvents using promoters and catalysts. Performing this reaction in water has received much attention during the past decade.

The ring opening of α,β-epoxycarboxylic acids by bromide and iodine ions has been effectively carried out in water in highly regio- and stereoselective fashion.[88] The $InCl_3$-catalyzed iodolysis of *trans*-β-monoalkylated epoxycarboxylic acids at pH 1.5 gave the corresponding *anti*-β-iodohydrins in 88–95% yields, whereas the iodolysis at pH 4.0 without any added Lewis acid gave the *anti*-β-hydroxy-α-iodocarboxylic acids (Scheme 8.18).

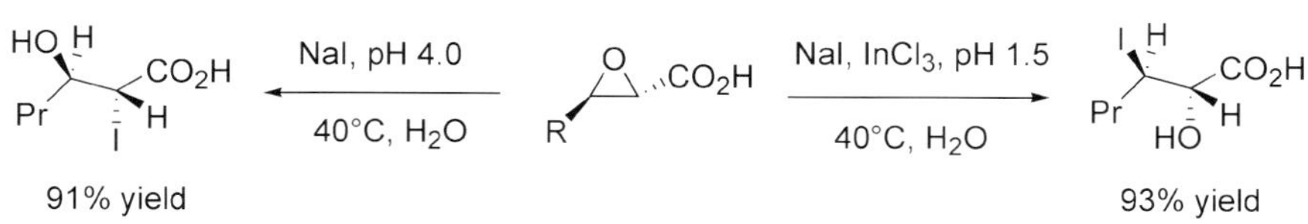

Scheme 8.18

β-Hydroxyethers were obtained with high regioselectivities and in high chemical yields from oxiranes by reaction with phenol in water in the presence of β-cyclodextrin[89] or tributylphosphine.[90]

Thiolysis of alkyl and aryl-1,2-epoxides in water was strongly affected by the pH and the presence of a Lewis acid. In water only, at pH 9, the reaction was sluggish and occurred at the less substituted β-carbon of unsymmetrical epoxides, while at pH 4, in the presence of $InCl_3$, the reaction was very fast and the formation of α-adduct was observed.[91,92] The application of these conditions to α,β-epoxycarboxylic acids gave the corresponding α-carboxy-β-hydroxy- and β-carboxy-α-hydroxysulfides, respectively.[93] The azidolysis of 1,2-epoxides was performed in water alone under heterogeneous conditions[94]; the addition was totally anti-diastereoselective, the reactivity and stereoselectivity being under pH control. The azidolysis of α,β-epoxycyclohexane carboxylic acid was catalyzed by 1 mol% $InCl_3$, or $AlCl_3$ at room temperature in water (pH 4) with the nucleophile reacting mainly at the β-carbon.[95,96]

Cleavage of oxiranes with amines in water has been the subject of several papers recently. For example, the C-3 selective ring opening of aromatic 2,3-epoxy alcohols or epoxides with aromatic amines occurred readily in water at room temperature in the presence of β-cyclodextrin to afford, highly stereoselectively, the corresponding β-aminoalcohols in good yields.[97] Saidi and Azizi reported the aminolysis of epoxides with aliphatic and aromatic amines in pure water, without any catalyst.[98] Reactions with terminal epoxides gave exclusive attack at the less substituted carbon (with the exception of styrene oxide). Yields were high (84–97%) except for aromatic amines, which reacted sluggishly. Similar work was almost simultaneously reported by Wu and Xia.[99]

Kobayashi developed an operationally simple, enantioselective addition of aromatic amines to *meso*-epoxides using a scandium/chiral bipyridine complex in water so that β-aminoalcohols could be obtained in high yields and with ee's up to 96% (Scheme 8.19).[100]

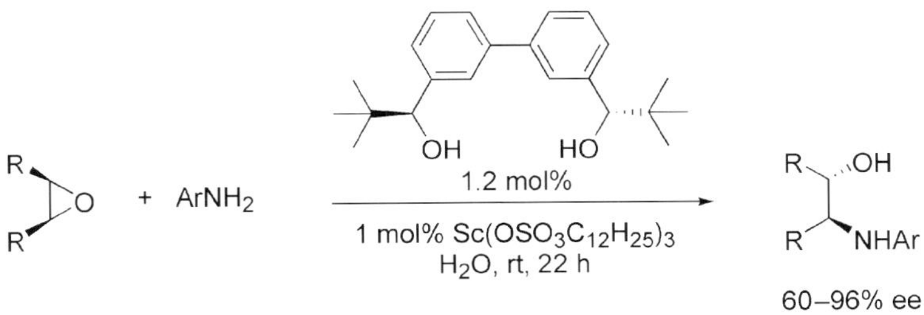

Scheme 8.19

Ring opening of aziridines

Regioselective ring opening of aziridines with different nucleophiles such as tetrabutylammonium halides,[101] lithium halides,[102] amines,[103] azides,[103] and thiophenol[104] occurred in high yields and good regioselectivities in water in the presence of β-cyclodextrin. The reaction of aziridines with NaN_3 or KCN occurred also in neat water in the presence of silica gel 60 in high yields to afford the ring-opening products in a regioselective fashion.[105]

8.2.2 Alkylation reactions

Alkylation reactions of ketones, active methylene compounds, alcohols, and phenols with alkyl halides occurred in quite good yields in aqueous NaOH solution without added organic solvent in the presence of water-soluble calix[*n*]arene ($n = 4$, 6, or 8) containing trimethylammoniomethyl groups acting as a reverse phase-transfer catalyst (Scheme 8.20).[106]

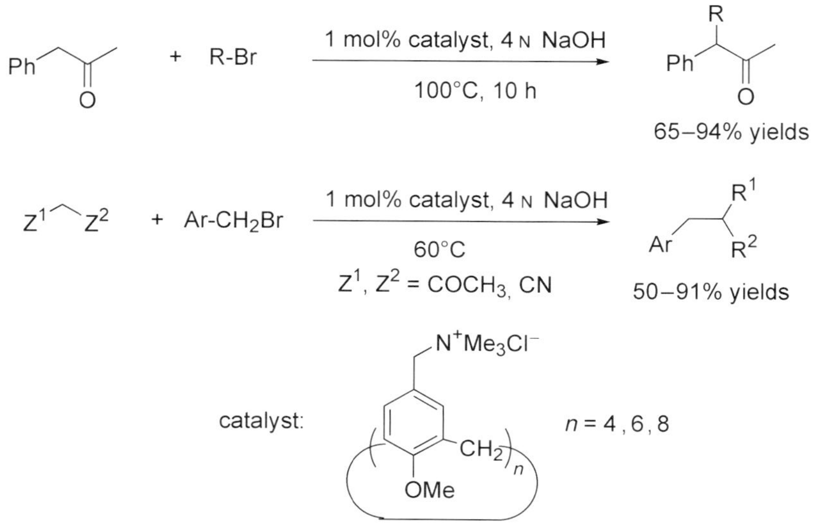

Scheme 8.20

Palladium-catalyzed allylic substitution has been accomplished in water. Sinou and coworkers showed for the first time that a water-soluble palladium(0) catalyst, prepared *in situ* from palladium acetate and $P(C_6H_4\text{-}m\text{-}SO_3Na)_3$, was an efficient catalyst for allylic substitution with various carbon- and heteronucleophiles in an aqueous/organic medium, allowing for easy separation of the product(s) and recycling of the catalyst (Scheme 8.21).[107,108]

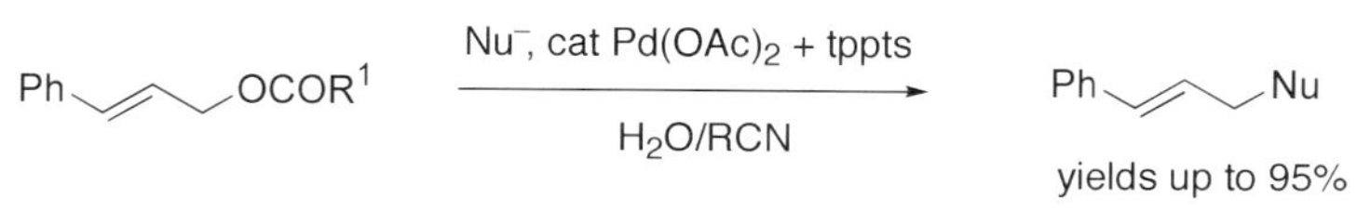

$R^1 = CH_3, OCH_3$; $Nu^- = CH(CO_2Me)_2, CH(COMe)_2, CH(NO_2)CO_2Et, NR_2, N_3, SO_2C_6H_5\text{-}p\text{-}CH_3$

Scheme 8.21

A palladium/tetraphosphine system catalyzed allylic amination in water in good yields, with a very high substrate/catalyst ratio and a turnover number of 980,000.[109]

Pd/C-mediated allylic substitution in water was recently described[110] and applied to various allylic acetates using carbon, nitrogen, sulfur, and oxygen nucleophiles to afford the alkylated products in good yields. The leaching of palladium in water was less than 4 ppm.

A remarkable rate enhancement of palladium-catalyzed allylic alkylation in water using non-water-soluble ligands was observed in the presence of surfactants, and when BINAP was used as the chiral ligand, enantioselectivities up to 94% ee were obtained in the allylic alkylation of 1,3-diphenyl-3-acetoxyprop-1-ene with carbon nucleophiles (Scheme 8.22).[111–113]

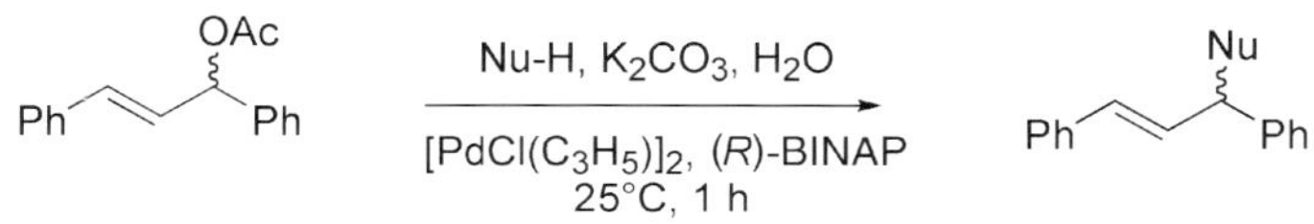

$CH_2(CO_2Me)_2$	no $CTAHSO_4$	45% conv.	85% ee
$CH_2(CO_2Me)_2$	$CTAHSO_4$	100% conv.	91% ee
$AcNHCH(CO_2Et)_2$	no $CTAHSO_4$	9% conv.	—
$AcNHCH(CO_2Et)_2$	$CTAHSO_4$	64% conv.	87% ee

Scheme 8.22

Uemura and coworkers used a carbohydrate-based phosphinite-oxazoline prepared from D-glucosamine as ligand for palladium in the alkylation of 1,3-diphenyl-3-acetoxyprop-1-ene in water or in an aqueous/organic biphasic medium. High yields of alkylation products were obtained with up to 85% ee (Scheme 8.23).[114]

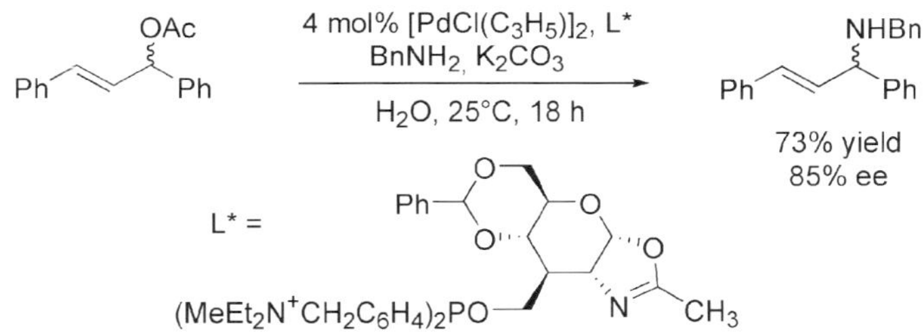

Scheme 8.23

Substitution of allylic alcohols with active methylene compounds as a suspension in water was achieved in excellent yields using $Pd(PPh_3)_4$ in the presence of a carboxylic acid such as 1-adamantanecarboxylic acid.[115] Bergbreiter and Liu used a water-soluble, polymer-bound palladium/phosphine catalyst in water or in mixed aqueous/organic solvents.[116] This polymeric catalyst showed high activity in nucleophilic allylic substitution affording high yields of coupled products, and it could be recycled by solvent or thermal precipitation.

Uozumi and coworkers prepared phosphine/palladium complexes supported on polyethylene glycol–polystyrene graft polymer.[117,118] This amphiphilic resin-supported palladium complex efficiently catalyzed the alkylation of allylic acetates in water with various nucleophiles including 1,3-dicarbonyl compounds, amino acids, sodium azide, sodium sulfinate, phenylboronic acid, and sodium tetraphenylborate to give the corresponding allylic-substituted products in quantitative yields.

Chiral ligand/palladium complexes were found to give high asymmetric induction in these types of reactions.[119–121] Various allylic acetates reacted in water with secondary amines or carbon nucleophiles, in the presence of Li_2CO_3, to give coupling products with up to 98% ee (Scheme 8.24). The recovered resin catalyst was used in two subsequent cycles without any additional charge of palladium and with no loss of catalytic activity or enantioselectivity. It is to be noted that the allylic amination does not take place in organic solvent.

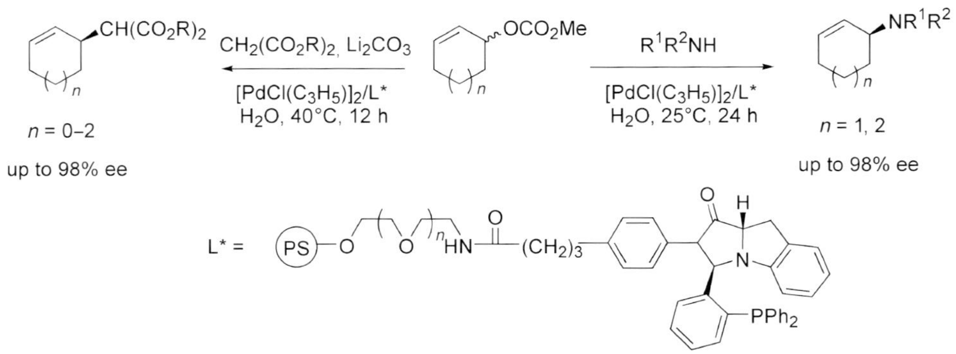

Scheme 8.24

8.2.3 Other types of substitutions

Acylation and sulfonylation of alcohols and amines are useful reactions in organic synthesis, but challenging to perform in aqueous media because of the reactive nature of common electrophilic precursors (e.g. acyl and sulfonyl chlorides, anhydrides) that makes them hydrolyze easily when water is present in large excess. Nevertheless, some recent papers have shown that it is possible to perform these reactions effectively also in water. Patel and coworkers disclosed the acylation of aliphatic and aromatic amines in micellar media.[122] A large number of amides were obtained in high yields by mixing various amines and anhydrides (1.5 equiv.) with SDS in water. In most of the reactions the products precipitated from the reaction mixture within 5 min. An interesting and useful feature was that the reactions were completely chemoselective for amines, leaving thiols and alcohols untouched. Tanabe and coworkers presented an elegant method of overcoming the problem of hydrolysis of sulfonyl chlorides in water by careful control of pH. In fact, efficient tosylations and mesylations of primary alcohols in water were achieved using a combination of KOH and catalytic *N*,*N*-dimethylbenzylamine (Scheme 8.25).[123]

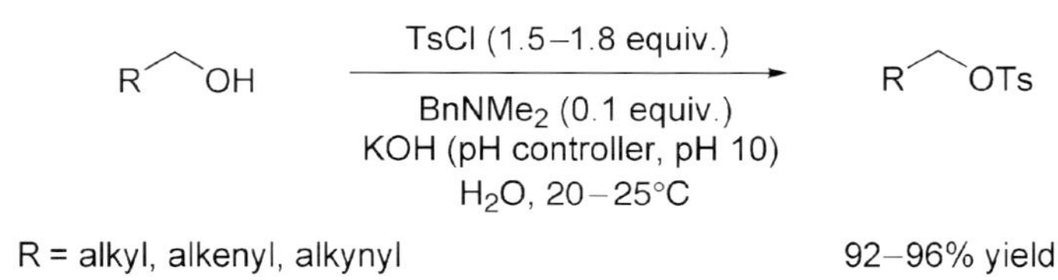

Scheme 8.25

8.3 Conclusion

During the last decade a large amount of nucleophilic addition and substitution reactions have been performed in water alone or in monophasic water/organic solvent systems. The yields obtained, as well as the diastereo- and enantioselectivities, are often as high, and sometimes even better, than those obtained in traditional organic solvents. Sometimes completely new reactivities can be discovered using water as the solvent. Moreover, the substitution of common organic solvents by water should be of real benefit to our health and environment.

References

1. Fringuelli, F., Pani, G., Piermatti, O. and Pizzo, F., Condensation reactions in water of active methylene compounds with arylaldehydes, *Tetrahedron*, 1994, **50**, 11499–11508.
2. Zhang, Z., Dong, Y.-W. and Wang, G.-W., Efficient and clean aldol condensation catalyzed by sodium carbonate in water, *Chem. Lett.*, 2003, **32**, 966–967.
3. Tanemura, K., Suzuki, T., Nishida, Y. and Horaguchi, T., Aldol condensation in water using polyethylene glycol 400, *Chem. Lett.*, 2005, **34**, 576–577.
4. Saimoto, H., Onitsuka, T., Motobe, H., Okabe, S., Takamori, Y., Morimoto, M. and Shigemasa, Y., Reactions of C3 and C4 ketoses with alkenals and alkenones in water, *Tetrahedron Lett.*, 2004, **45**, 8777–8780.
5. Chimni, S.S. and Mahajan, D., Electron deficiency of aldehydes controls the pyrrolidine catalyzed direct cross-aldol reaction of aromatic/heterocyclic aldehydes and ketones in water, *Tetrahedron*, 2005, **61**, 5019–5025.
6. Ji, C., Peng, Y., Huang, C., Wang, N. and Jiang, Y., The influence of acidity on direct aldol reactions catalyzed by pyrrolidine/acid bifunctional organocatalyst, *Synlett*, 2005, 986–990.
7. Sakthivel, K., Notz, W., Bui, T. and Barbas, C.F., III, Amino acid catalyzed direct asymmetric aldol reactions: a bioorganic approach to catalytic asymmetric carbon–carbon bond-forming reactions, *J. Am. Chem. Soc.*, 2001, **123**, 5260–5267.
8. Dickerson, T.J. and Janda, K.D., Aqueous aldol catalysis by a nicotine metabolite, *J. Am. Chem. Soc.*, 2002, **124**, 3220–3221.
9. Pizzarrello, S. and Weber, A.L., Prebiotic aminoacids as asymmetric catalysts, *Science*, 2004, **303**, 1151.
10. Amedjkouh, M., Primary amine catalyzed direct asymmetric aldol reaction assisted by water, *Tetrahedron: Asymmetry*, 2005, **16**, 1411–1414.
11. Mase, N., Nakai, Y., Ohara, N., Yoda, H., Takabe, K., Tanaka, F. and Barbas, C.F., III, Organocatalytic direct asymmetric aldol reactions in water, *J. Am. Chem. Soc.*, 2006, **128**, 734–735.
12. Peng, Y.-Y., Ding, Q.-P., Li, Z., Wang, P.G. and Cheng, J.-P., Proline catalyzed aldol reactions in aqueous micelles: an environmentally friendly reaction system, *Tetrahedron Lett.*, 2003, **44**, 3871–3875.
13. Nyberg, A.I., Usano, A. and Pihko, P.M., Proline-catalyzed ketone–aldehyde aldol reactions are accelerated in water, *Synlett*, 2004, 1891–1896.
14. Torii, H., Nakadai, M., Ishihara, K., Saito, S. and Yamamoto, H., Asymmetric direct aldol reaction assisted by water and a proline-derived tetrazole catalyst, *Angew. Chem. Int. Ed.*, 2004, **43**, 1983–1986.
15. Hayashi, Y., Tsuboi, W., Shoji, M. and Suzuki, N., Application of high pressure, induced by water freezing, to the direct asymmetric aldol reaction, *Tetrahedron Lett.*, 2004, **45**, 4353–4356.
16. Dewa, T., Saiki, T. and Aoyama, Y., Enolization and aldol reactions of ketones with a La^{3+}-immobilized organic solid in water. A microporous enolate mimic, *J. Am. Chem. Soc.*, 2001, **123**, 502–503.
17. Buonora, P.T., Rosauer, K.G. and Dai, L., Control of the aqueous aldol addition under Claisen–Schmidt conditions, *Tetrahedron Lett.*, 1995, **26**, 4009–4012.
18. Darbre, T. and Machuqueiro, M., Zn-proline catalyzed direct aldol reaction in aqueous media, *Chem. Commun.*, 2003, 1090–1091.
19. Kofoed, J., Machuqueiro, M., Reymond, J.-L. and Darbre, T., Zinc-proline catalyzed pathway for the formation of sugars, *Chem. Commun.*, 2004, 1540–1541.
20. Lubineau, A., Water-promoted organic reactions: aldol reaction under neutral conditions, *J. Org. Chem.*, 1986, **51**, 2142–2144.
21. Lubineau, A. and Meyer, E., Water-promoted organic reactions. Aldol reaction of silyl enol ethers with carbonyl compounds under atmospheric pressure and neutral conditions, *Tetrahedron*, 1988, **44**, 6065–6070.

22. Kobayashi, S. and Hachiya, I., The aldol reaction of silyl enol ethers with aldehydes in aqueous media, *Tetrahedron Lett.*, 1992, **33**, 1625–1628.
23. Hachiya, I. and Kobayashi, S., Aqueous reactions with a Lewis acid and an organometallic reagent. The scandium trifluoromethanesulfonate-catalyzed allylation reaction of carbonyl compounds with tetraallyltin, *J. Org. Chem.*, 1993, **58**, 6958–6960.
24. Kobayashi, S. and Hachiya, I., Lanthanide triflates as water-tolerant Lewis acids. Activation of commercial formaldehyde solution and use in the aldol reaction of silyl enol ethers with aldehydes in aqueous media, *J. Org. Chem.*, 1994, **59**, 3590–3596.
25. Kobayashi, S., Sugiura, M., Kitagawa, H. and Lam, W.L., Rare-earth metal triflates in organic synthesis, *Chem. Rev.*, 2002, **102**, 2227–2302.
26. Nagayama, S. and Kobayashi, S., A novel chiral lead(II) catalyst for enantioselective aldol reactions in aqueous media, *J. Am. Chem. Soc.*, 2000, **122**, 11531–11532.
27. Manabe, K., Mori, Y. and Kobayashi, S., Effects of Lewis acid-surfactant-combined catalysts on aldol and Diels–Alder reactions in water, *Tetrahedron*, 1999, **55**, 11203–11208.
28. Kobayashi, S., Nagayama, S. and Busujima, T., Catalytic asymmetric Mukaiyama aldol reactions in aqueous media, *Tetrahedron*, 1999, **55**, 8739–8746.
29. Kobayashi, S. and Manabe, K., Development of novel Lewis acid catalysts for selective organic reactions in aqueous media, *Acc. Chem. Res.*, 2002, **35**, 209–217.
30. Kobayashi, S., Hachiya, I. and Yamanoi, Y., Repeated use of the catalyst in $Ln(OTf)_3$-catalyzed aldol and allylation reactions, *Bull. Chem. Soc. Jpn.*, 1994, **67**, 2342–2344.
31. Kobayashi, S., Wakabayashi, T., Nagayama, S. and Oyamada, H., Lewis acid catalysis in micellar systems. $Sc(OTf)_3$-catalyzed aqueous aldol reactions of silyl enol ethers with aldehydes in the presence of a surfactant, *Tetrahedron Lett.*, 1997, **38**, 4559–4562.
32. Tian, H.-Y., Chen, Y.-J., Wang, D., Zeng, C.-C. and Li, C.-J., Calix[6]arene derivatives bearing sulfonate and alkyl groups as surfactants in $Sc(OTf)_3$-catalyzed Mukaiyama aldol reactions in water, *Tetrahedron Lett.*, 2000, **41**, 2529–2532.
33. Shimizu, S., Shirakawa, S., Suzuki, T. and Sasaki, Y., Water-soluble calixarenes as new inverse phase-transfer catalysts. Their application to aldol-type condensation and Michael addition reactions in water, *Tetrahedron*, 2001, **57**, 6169–6173.
34. Tian, H.-Y., Chen, Y.-J., Wang, D., Bu, Y.-P. and Li, C.-J., The effects of aromatic and aliphatic anionic surfactants on $Sc(OTf)_3$-catalyzed Mukaiyama aldol reactions in water, *Tetrahedron Lett.*, 2001, **42**, 1803–1805.
35. Kobayashi, S. and Wakabayashi, T., Scandium trisdodecylsulfate (STDS). A new type of Lewis acid that forms stable dispersion systems with organic substrates in water and accelerates aldol reactions much faster in water than in organic solvents, *Tetrahedron Lett.*, 1998, **39**, 5389–5392.
36. Manabe, K., Mori, Y., Wakabayashi, T., Nagayama, S. and Kobayashi, S., Organic synthesis inside particles in water: Lewis acid-surfactant-combined catalysts for organic reactions in water using colloidal dispersions as reaction media, *J. Am. Chem. Soc.*, 2000, **122**, 7202–7207.
37. Manabe, K. and Kobayashi, S., Remarkable enhancement of reactivity by Brønsted acids in aldol reactions mediated by Lewis acid-surfactant-combined catalysts in water, *Tetrahedron Lett.*, 1999, **40**, 3773–3776.
38. Loh, T.-P., Pei, J., Koh, K.S.-V., Cao, G.-Q. and Li, X.-R., Indium-trichloride catalyzed Mukaiyama-aldol reaction in water: solubility, aggregation and internal pressure effect, *Tetrahedron Lett.*, 1997, **38**, 3465–3468.
39. Loh, T.-P., Chua, G.-L., Vittal, J.J. and Wong, M.-W., Highly stereoselective indium trichloride-catalysed asymmetric aldol reaction of formaldehyde and a glucose-derived silyl enol ether in water, *Chem. Commun.*, 1998, 861.
40. Kobayashi, S., Busujima, T. and Nagayama, S., On indium(III) chloride-catalyzed aldol reactions of silyl enol ethers with aldehydes in water, *Tetrahedron Lett.*, 1998, **39**, 1579–1582.
41. Muñoz-Muñiz, O., Quintanar-Audelo, M. and Juaristi, E., Reexamination of $CeCl_3$ and $InCl_3$ as activators in the diastereoselective Mukaiyama aldol reaction in aqueous media, *J. Org. Chem.*, 2003, **68**, 1622–1625.
42. Mori, Y., Manabe, K. and Kobayashi, S., Catalytic use of boron source for boron enolate mediated stereoselective aldol reactions in water, *Angew. Chem. Int. Ed.*, 2001, **40**, 2816–2818.
43. Mori, Y., Kobayashi, J., Manabe, K. and Kobayashi, S., Use of boron enolates in water. The first boron enolate-mediated diastereoselective aldol reactions using catalytic boron sources, *Tetrahedron*, 2002, **58**, 8263–8268.
44. Iimura, S., Manabe, K. and Kobayaashi, K., Hydrophobic polymer-supported scandium catalyst for carbon–carbon bond-forming reactions in water, *Tetrahedron*, 2004, **60**, 7673–7678.
45. Gu, W., Zhou, W.-J. and Gin, D.L., A nanostructured, scandium-containing polymer for heterogeneous Lewis acid catalysis in water, *Chem. Mater.*, 2001, **13**, 1949–1951.
46. Graven, A., Grotli, M. and Meldal, M., Towards peptide isostere libraries: aqueous aldol reactions on hydrophilic solid supports, *J. Chem. Soc., Perkin Trans. 1*, 2000, 955–962.

47. Aoyama, N., Manabe, K. and Kobayashi, S., Iron(III) chloride as a water-compatible Lewis acid for diastereoselective aldol reactions in water in the presence of a surfactant, *Chem. Lett.*, 2004, **33**, 312–313.
48. Loh, T.-P. and Li, X.-R., Clay montmorillonite K10 catalyzed aldol-type reaction of aldehydes with silyl enol ethers in water, *Tetrahedron*, 1999, **55**, 10789–10802.
49. Loh, T.-P., Feng, L.-C. and Wei, L.-L., Water-accelerated aldol reaction of ketene silyl acetals with carbonyl compounds, *Tetrahedron*, 2000, **56**, 7309–7312.
50. Kobayashi, S., Mori, Y., Nagayama, S. and Manabe, K., Catalytic asymmetric aldol reactions in water using a chiral Lewis acid-surfactant-combined catalyst, *Green Chem.*, 1999, **1**, 175–177.
51. Benaglia, M., Cinquini, M., Cozzi, F. and Celentano, G., Enantioselective catalysis in water: Mukaiyama-aldol condensation promoted by copper complexes of bisoxazolines supported on poly(ethylene glycol), *Org. Biomol. Chem.*, 2004, **2**, 3401–3407.
52. Li, H.-J., Tian, H.-Y., Chen, Y.-J., Wang, D. and Li, C.-J., Novel chiral gallium Lewis acid catalysts with semi-crown ligands for aqueous asymmetric Mukayama aldol reactions, *Chem. Commun.*, 2002, 2994–2995.
53. Kobayashi, S., Ogino, T., Shimizu, H., Ishikawa, S., Hamada, T. and Manabe, K., Bismuth triflate–chiral bipyridine complexes as water-compatible chiral Lewis acids, *Org. Lett.*, 2005, **7**, 4729–4731.
54. Li, H.-J., Tian, H.-Y., Chen, Y.-J., Wang, D. and Li, C.J., Chiral anionic surfactants for asymmetric Mukaiyama aldol-type reaction in water, *J. Chem. Res. (S)*, 2003, 153–156.
55. Hajos, Z.G. and Parrish, D.R., Synthesis and conversion of 2-methyl-2-(3-oxobutyl)-1,3-cyclopentanedione to the isomeric racemic ketols of the [3.2.1]bicyclooctane and of the perhydroindane series, *J. Org. Chem.*, 1974, **39**, 1612–1615.
56. Eder, U., Sauer, G. and Wiechert, R., Total synthesis of optically active steroids. 6. New type of asymmetric cyclization to optically active steroid CD partial structures, *Angew. Chem., Int. Ed. Engl.*, 1971, **10**, 496–497.
57. Lavallee, J.F. and Deslongchamps, P., One-step construction of a 13α-methyl-14α-hydroxy steroid via a new anionic polycyclization method, *Tetrahedron Lett.*, 1988, **29**, 6033–6036.
58. Keller, E. and Feringa, B.L., Ytterbium triflate catalyzed Michael additions of β-ketoesters in water, *Tetrahedron Lett.*, 1996, **37**, 1879–1882.
59. Keller, E. and Feringa, B.L., Highly efficient ytterbium triflate-catalyzed Michael additions of α-nitro esters in water, *Synlett*, 1997, 842–844.
60. Mori, Y., Kakumoto, K., Manabe, K. and Kobayashi, S., Michael reactions in water using Lewis acid-surfactant-combined catalysts, *Tetrahedron Lett.*, 2000, **41**, 3107–3111.
61. Uozumi, Y. and Nakazono, M., Amphiphilic resin-supported rhodium–phosphine catalysts for C—C bond forming reactions in water, *Adv. Synth. Catal.*, 2002, **344**, 274–277.
62. Lubineau, A. and Augé, J., Water-promoted organic reactions. Michael addition of nitroalkanes to methyvinylketone under neutral conditions, *Tetrahedron Lett.*, 1992, **33**, 8073–8074.
63. Ballini, R. and Bosica, G., Conjugated addition reactions of nitroalkanes with electrophilic alkenes in aqueous media, *Eur. J. Org. Chem.*, 1998, 355–357.
64. Giorgi, G., Miranda, S., Lopez-Alvarado, P., Avendano, C., Rodriguez, J. and Menéndez, J.C., Unique Michael addition-initiated domino reaction for the stereoselective synthesis of functionalized macrolactones from α-nitroketones in water, *Org. Lett.*, 2005, **7**, 2197–2200.
65. Ballini, R., Barboni, L. and Giarlo, G., Nitroalkanes in aqueous medium as an efficient and eco-friendly source for the one-pot synthesis of 1,4-diketones, 1,4-diols, δ-nitroalkanols, and hydroxytetrahydrofurans, *J. Org. Chem.*, 2003, **68**, 9173–9176.
66. Da Silva, F.M., Gomes, A.K. and Jones, J., Jr., Organic reaction in water. Michael addition in water without phase transfer agents, *Can. J. Chem.*, 1999, **77**, 624–627.
67. Firouzabadi, H., Iranpoor, N. and Jafari, A.A., Micellar solution of sodium dodecyl sulfonate (SDS) catalyzes facile Michael addition of amines and thiols to α,β-unsaturated ketones in water under neutral conditions, *Adv. Synth. Catal.*, 2005, **347**, 655–661.
68. Naidu, B.N., Sorenson, M.E., Connolly, T.P. and Ueda, Y., Michael addition of amines and thiols to dehydroalanine amides: a remarkable rate acceleration in water, *J. Org. Chem.*, 2003, **68**, 10098–10102.
69. Firouzabadi, H., Iranpoor, N. and Nowrouzi, F., The facile and efficient Michael addition of indoles and pyrrole to α,β-unsaturated electron-deficient compounds catalyzed by aluminium dodecyl sulfate trihydrate $[Al(Ds)_3]\cdot 3H_2O$ in water, *Chem. Commun.*, 2005, 789–791.
70. Xu, L.-W., Li, L. and Xia, C.-G., Transition-metal-based Lewis acid catalysis of aza-type Michael additions of amines to α,β-unsaturated electrophiles in water, *Helv. Chim. Acta*, 2004, **87**, 1522–1526.
71. Krishnaveni, N.S., Surendra, K. and Rao, K.R., Study of the Michael addition of β-cyclodextrin–thiol complexes to conjugated alkenes in water, *Chem. Commun.*, 2005, 669–671.
72. Da Silva, F.M. and Jones, J., Jr., Organic reaction in water. Part 3: Diastereoselectivity in Michael additions of thiophenol to nitro olefins in aqueous media, *J. Braz. Chem. Soc.*, 2001, **12**, 135–137.
73. Kobayashi, S., Kakumoto, K., Mori, Y. and Kei, M., Chiral Lewis acid-catalyzed enantioselective Michael reactions in water, *Isr. J. Chem.*, 2002, **41**, 247–249.

74. Kawabata, T., Kato, M., Mizugaki, T., Ebitani, K. and Kaneda, K., Monomeric metal aqua complexes in the interlayer space of montmorillonites as strong Lewis acid catalysts for heterogeneous carbon–carbon bond-forming reactions, *Chem. Eur. J.*, 2005, **11**, 288–297.
75. Ding, R., Katebzadeh, K., Roman, L., Bergquist, K.-E. and Lindström, U.M., Expanding the scope of Lewis acid catalysis in water: remarkable ligand acceleration of aqueous ytterbium triflate catalyzed Michael addition reactions, *J. Org. Chem.*, 2006, **71**, 352–355.
76. Kobayashi, S., Busujima, T. and Nagayama, S., $Ln(OTf)_3$- or $Cu(OTf)_2$-catalyzed Mannich-type reactions of aldehydes, amines, and silyl enolates in micellar systems. Facile synthesis of β-amino ketones and esters in water, *Synlett*, 1999, **5**, 545–546.
77. Loh, T.-P. and Wei, L.-L., Novel one-pot Mannich-type reaction in water: indium trichloride-catalyzed condensation of aldehydes, amines and silyl enol ethers for the synthesis of β-amino ketones and esters, *Tetrahedron Lett.*, 1998, **39**, 323–326.
78. Loh, T.-P., Liung, S.B.K.W., Tan, K.-L. and Wei, L.-L., Three component synthesis of β-amino compounds using indium trichloride-catalyzed one-pot Mannich-type reaction in water, *Tetrahedron*, 2000, **56**, 3227–3237.
79. Loncaric, C., Manabe, K. and Kobayashi, S., Mannich-type reactions catalyzed by neutral salts in water, *Adv. Synth. Catal.*, 2003, **345**, 1187–1189.
80. Akiyama, T., Takaya, J. and Kagoshima, H., Brønsted acid-catalyzed Mannich-type reactions in aqueous media, *Adv. Synth. Catal.*, 2002, **344**, 338–347.
81. Akiyama, T., Itoh, J. and Fuchibe, K., Mannich-type reaction catalyzed by HBF_4 in water: effect of the loading of surfactant, *Synlett*, 2002, 1269–1272.
82. Akiayama, T., Takaya, J. and Kagoshima, H., A highly stereo-divergent Mannich-type reaction catalyzed by Brønsted acid in aqueous media, *Tetrahedron Lett.*, 2001, **42**, 4025–4028.
83. Manabe, K. and Kobayashi, S., Mannich-type reactions of aldehydes, amines, and ketones in a colloidal dispersion system created by a Brønsted acid-surfactant-combined catalyst in water, *Org. Lett.*, 1999, **1**, 1965–1967.
84. Manabe, K., Mori, Y. and Kobayashi, S., Three-component carbon–carbon bond-forming reactions catalyzed by a Brønsted acid-surfactant-combined catalyst in water, *Tetrahedron*, 2001, **57**, 2537–2544.
85. Iimura, S., Nobutou, D., Manabe, K. and Kobayashi, S., Mannich-type reactions in water using a hydrophobic polymer-supported sulfonic acid catalyst, *Chem. Commun.*, 2003, 1644–1645.
86. Wu, Y.-S., Cai, J., Hu, Z.-Y. and Lin, G.-X., A new class of metal-free catalysts for direct diastereo- and regioselective Mannich reactions in aqueous media, *Tetrahedron Lett.*, 2004, **45**, 8949–8952.
87. Hamada, T., Manabe, K. and Kobayashi, S., Enantio- and diastereoselective, stereospecific Mannich-type reactions in water, *J. Am. Chem. Soc.*, 2004, **126**, 7768–7769.
88. Amantini, D., Fringuelli, F., Pizzo, F. and Vaccaro, L., Bromolysis and iodolysis of α,β-epoxycarboxylic acids in water catalyzed by indium halides, *J. Org. Chem.*, 2001, **66**, 4463–4467.
89. Surendra, K., Krishnaveni, N.S., Nageswar, Y.V.D. and Rao, K.R., Highly regioselective ring opening of oxiranes with phenoxides in the presence of β-cyclodextrin in water, *J. Org. Chem.*, 2003, **68**, 4994–4995.
90. Fan, R.-H. and Hou, X.-L., Efficient ring-opening reaction of epoxides and aziridines promoted by tributylphosphine in water, *J. Org. Chem.*, 2003, **68**, 726–730.
91. Fringuelli, F., Pizzo, F., Tortoioli, S. and Vaccaro, L., Thiolysis of alkyl- and aryl-1,2-epoxides in water catalyzed by $InCl_3$, *Adv. Synth. Catal.*, 2002, **344**, 379–384.
92. Fringuelli, F., Pizzo, F., Tortoioli, S. and Vaccaro, L., Easy and environmentally friendly uncatalyzed synthesis of β-hydroxy arylsulfides by thiolysis of 1,2-epoxides in water, *Green Chem.*, 2003, **5**, 436–440.
93. Fringuelli, F., Pizzo, F., Tortoioli, S. and Vaccaro, L., $InCl_3$-catalyzed regio- and stereoselective thiolysis of α,β-epoxycarboxylic acids in water, *Org. Lett.*, 2005, **7**, 4411–4414.
94. Fringuelli, F., Piermatti, O., Pizzo, F. and Vaccaro, L., Ring opening of epoxides with sodium azide in water. A regioselective pH-controlled reaction, *J. Org. Chem.*, 1999, **64**, 6094–6096.
95. Fringuelli, F., Pizzo, F. and Vaccaro, L., Lewis-acid catalyzed organic reactions in water. The case of $AlCl_3$, $TiCl_4$, and $SnCl_4$ believed to be unusable in aqueous medium, *J. Org. Chem.*, 2001, **66**, 4719–4722.
96. Fringuelli, F., Pizzo, F. and Vaccaro, L., Azidolysis of α,β-epoxycarboxylic acids. A water-promoted process efficiently catalyzed by indium trichloride at pH 4.0, *J. Org. Chem.*, 2001, **66**, 3554–3558.
97. Surendra, K., Krishnaveni, N.S. and Rao, K.R., The selective C-3 opening of aromatic 2,3-epoxy alcohols/epoxides with aromatic amines catalyzed by β-cyclodextrin in water, *Synlett*, 2005, 5806–5810.
98. Azizi, N. and Saidi, M.R., Highly chemoselective addition of amines to epoxides in water, *Org. Lett.*, 2005, **7**, 3649.
99. Wu, J. and Xia, H.-G., Tertiary amines as highly efficient catalysts in the ring-opening reactions of epoxides with amines or thiols in H_2O: expeditious approach to β-aminoalcohols and β-aminothioethers, *Green Chem.*, 2005, **7**, 708–710.
100. Azoulay, S., Manabe, K. and Kobayashi, S., Catalytic asymmetric ring opening of meso-epoxides with aromatic amines in water, *Org. Lett.*, 2005, **7**, 4593–4595.

101. Narender, M., Surendra, K., Krishnaveni, N.S., Reddy, M.S. and Rao, K.R., A facile regioselective ring opening of aziridines to haloamines using tetrabutyl ammonium halides in the presence of β-tyclodextrin in water, *Tetrahedron Lett.*, 2004, **45**, 7995–7997.
102. Krishnaveni, N.S., Surendra, K., Narender, M., Nageswar, Y.V.D. and Rao, K.R., Highly efficient regioselective ring opening of aziridines to β-halo amines in the presence of β-cyclodextrin in water, *Synthesis*, 2004, 501–502.
103. Reddy, M.A., Reddy, L.R., Bhanumathi, N. and Rao, K.R., An efficient biomimetic cleavage of aziridines with nucleophiles catalyzed by β-cyclodextrin in water, *Chem. Lett.*, 2001, 246–247.
104. Reddy, M.S., Narender, M. and Rao, K.R., Regioselective ring opening of aziridines with thiophenol in the presence of β-cyclodextrin in water, *Synlett*, 2005, 489–490.
105. Minakata, S., Kano, D., Oderaotoshi, Y. and Komatsu, M., Silica–water reaction media: its application to the formation and ring opening of aziridines, *Angew. Chem. Int. Ed.*, 2004, **43**, 79–81.
106. Shimizu, S., Suzuki, T., Shirakawa, S., Sasaki, Y. and Hirai, C., Water-soluble calixarenes as new inverse phase-transfer catalysts. Their scope in aqueous biphasic alkylations and mechanistic implications, *Adv. Synth. Catal.*, 2002, **344**, 370–378.
107. Safi, M. and Sinou, D., Palladium(0)-catalyzed substitution of allylic substrates in a two-phase aqueous–organic medium, *Tetrahedron Lett.*, 1991, **32**, 2025–2028.
108. Blard, E., Genêt, J.P., Safi, M., Savignac, M. and Sinou, D., Palladium-catalyzed substitution of allylic substrates in an aqueous–organic medium, *Tetrahedron*, 1994, **50**, 505–514.
109. Feuerstein, M., Laurenti, D., Doucet, H. and Santelli, M., Palladium-tetraphosphine catalyzed allylic substitution in water, *Tetrahedron Lett.*, 2001, **42**, 2313–2315.
110. Felpin, F.-X. and Landais, Y., Practical Pd/C-mediated allylic substitution in water, *J. Org. Chem.*, 2005, **70**, 6441–6446.
111. Kobayashi, S., Lam, W.W.-L. and Manabe, K., Remarkable rate enhancement of palladium-catalyzed allylic alkylation in water using a colloidal dispersion system, *Tetrahedron Lett.*, 2000, **41**, 6115–6119.
112. Sinou, D., Rabeyrin, C. and Nguefack, C., Catalytic asymmetric alkylation in water in the presence of surfactants, *Adv. Synth. Catal.*, 2003, **345**, 357–363.
113. Rabeyrin, C. and Sinou, D., Catalytic asymmetric alkylation in water in the presence of surfactants: influence of the nature of the nucleophile and the allylic acetate on the activity and enantioselectivity, *Tetrahedron: Asymmetry*, 2003, **14**, 3891–3897.
114. Tomohiro, H., Koji, Y., Kouichi, O. and Uemura, S., A novel amphiphilic chiral ligand derived from D-glucosamine. Application to palladium-catalyzed asymmetric allylic substitution reaction in an aqueous or an organic medium, allowing for catalyst recycling, *J. Org. Chem.*, 2000, **65**, 5197–5201.
115. Manabe, K. and Kobayashi, S., Palladium-catalyzed, carboxylic acid-assisted allylic substitution of carbon nucleophiles with allyl alcohols as allylating agents in water, *Org. Lett.*, 2003, **5**, 3241–3244.
116. Bergbreiter, D.E. and Liu, Y.-S., Water-soluble polymer-bound, recoverable palladium(0)–phosphine catalysts, *Tetrahedron Lett.*, 1997, **38**, 7843–7846.
117. Danjo, H., Tanaka, D., Hayashi, T. and Uozumi, Y., Allylic substitution in water catalyzed by amphiphilic resin-supported palladium–phosphine complexes, *Tetrahedron*, 1999, **55**, 14341–14352.
118. Uozumi, Y., Danjo, H. and Hayashi, T., Cross-coupling of aryl halides and allyl acetates with arylboron reagents in water using an amphiphilic resin-supported palladium-catalyst, *J. Org. Chem.*, 1999, **64**, 3384–3388.
119. Uozumi, Y., Danjo, H. and Hayashi, T., Palladium-catalyzed asymmetric allylic substitution in aqueous media using amphiphilic resin-supported MOP ligands, *Tetrahedron Lett.*, 1998, **39**, 8303–8306.
120. Uozumi, Y. and Shibatomi, K., Catalytic asymmetric allylic alkylation in water with a recyclable resin-supported P,N-chelating palladium complex, *J. Am. Chem. Soc.*, 2001, **123**, 2919–2920.
121. Uozumi, Y., Tanaka, H. and Shibatomi, K., Asymmetric allylic amination in water catalyzed by an amphiphilic resin-supported chiral palladium complex, *Org. Lett.*, 2004, **6**, 281–283.
122. Naik, S., Bhattacharjya, G., Talukdar, B. and Patel, B.K., Chemoselective acylation of amines in aqueous media, *Eur. J. Org. Chem.*, 2004, 1254–1260.
123. Morita, J., Nakatsuji, H., Misaki, T. and Tanabe, Y., Water-solvent method for tosylation and mesylation of primary alcohols promoted by KOH and catalytic amines, *Green Chem.*, 2005, **7**, 711–715.

Chapter 9

Reactions in Nearcritical Water

C.L. Liotta, J.P. Hallett, P. Pollet, and C.A. Eckert

Nearcritical water (NCW) offers exciting possibilities as a benign solvent for the synthesis of organic chemicals, followed by opportunities for facile separations. As such, it constitutes an important solvent for the implementation of sustainable processing in the chemical industry. Because water is an abundant as well as a nontoxic solvent, it is attractive from both a green chemistry and an economic processing point of view. Water is one of only two solvents (along with carbon dioxide) that are not regulated for incorporation into food and drug products. Also, solvent processing losses from the use of water do not represent net emissions of pollutants to the environment. Additionally, water is a low-cost solvent and processing techniques for using water as a solvent have been established for centuries. Since separations typically comprise 60–80% of the cost of a chemical process, the facile separations offered by NCW represent significant processing advantages over traditional organic solvents. Because of this unique combination of environmental and economic benefits, NCW is a solvent at the forefront of sustainable technology.

NCW is defined as water that has been heated to a temperature range of 200–300°C where its properties have begun to differ significantly from those of ambient water. NCW has been called by a variety of names, including 'high-temperature water', 'hot water', 'subcritical water', and 'near-subcritical water'. In this chapter NCW will be used for consistency. All liquids exhibit dilation (a rapid decrease in density) as they approach their critical points. In the NCW region, water is not yet a supercritical fluid, but a significant change in properties such as the density decline has already begun to occur. Thus the properties of NCW are intermediate between ambient and supercritical water. Supercritical water (SCW) has been the subject of much investigation, primarily for waste remediation by SCW oxidation. This oxidation process typically operates at 400–500°C and pressures of 20–50 MPa, where it provides an excellent medium for the destruction of chemicals, but is much less useful for synthesis. In contrast, NCW is greatly superior for making chemicals, and it has the advantage of doing this at lower temperatures and pressures than SCW.

This chapter provides insights into potential chemical processes available in NCW and the underlying physical properties that make these processes possible. Its purpose is to provide the reader with an overview of the exciting opportunities provided by this novel reaction medium and to show how these opportunities can be exploited for chemical processing, including both reactions and separations. Relevant and illustrative examples from the recent chemical literature will be presented. Additionally, discussions are presented concerning how the properties of water in the nearcritical temperature range differ from water at ambient temperatures and how changes in temperature and pressure can be used to alter the physical and chemical properties of a reaction system.

A wide range of chemical transformations have been performed in NCW. By far the most common type of reactions studied to date has been hydrolyses. However, more recent studies have focused on bond-making reactions such as alkylations, acylations, skeletal rearrangements, and condensation reactions. While this chapter discusses several of these reaction classes in detail, it is intended to be a selective, and not an exhaustive, review of the chemical literature, as several excellent reviews are available.[1–7]

9.1 Characterization of NCW

9.1.1 Physical and thermodynamic properties of NCW

There are several important differences in the physiochemical properties of NCW when compared with ambient water. The most important are the changes in the structure of water due to the reduction in hydrogen bonding (exothermic) with increased temperature. Many experimental and theoretical studies have attempted to quantify the structure of water at elevated (mostly supercritical) temperatures.[1] These studies show that water loses approximately 55–60% of its hydrogen-bonding network as the temperature is increased from 25 to 300°C, with a corresponding reduction in molecular ordering.[8] The ordered hydrogen-bonding network gives rise to the fundamental characteristics of ambient water, and its breakdown contributes to important changes in the properties of NCW, such as the density, dielectric constant, and dissociation constant.

This breakdown of hydrogen bonding in NCW effects a variety of changes in the physical properties of water; the most apparent physical property change is that of density. Following the saturation vapor pressure of water in Fig. 9.1, the density decreases from 1 g/cm^3 at 25°C to about 0.75 g/cm^3 at 300°C.[9] In contrast, the density of SCW just above the critical point is around 0.1 g/cm^3. Naturally, the density of water is easily tuned by adjusting the temperature and pressure of the system, and since the compressibility goes to infinity at the critical point, in this region the tuning becomes more sensitive. For example, increasing

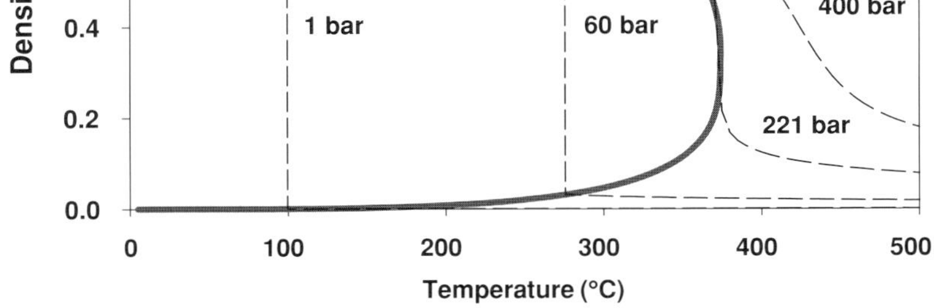

Figure 9.1 Density of water as a function of temperature from 0 to 500°C at various pressures. Correlation from NIST steam tables.[9]

the system pressure from saturation to 1 kbar at 200°C increases the density of water by only 10% ($\rho = 0.85$ g/cm^3 at 40 bar, $\rho = 0.95$ g/cm^3 at 1 kbar). However, at 300°C, an increase of nearly 25% occurs ($\rho = 0.7$ g/cm^3 at 70 bar, $\rho = 0.9$ g/cm^3 at 1 kbar) and nearly 50% at 350°C ($\rho = 0.55$ g/cm^3 at 150 bar, $\rho = 0.8$ g/cm^3 at 1 kbar).[9] This decrease in density with temperature is also a contributing factor to other property changes of NCW, such as the dielectric constant and dissociation constant. It is also a major factor in improved transport in NCW, as a reduction in density corresponds to an increase in diffusion.[10] The higher density of NCW as compared with SCW also contributes to the predominance of ionic as opposed to free radical mechanisms in NCW.[11] Density control by pressure-tuning is an important feature of SCW operations; however, for NCW the pressures required for significant density effects are too large to be of practical interest. But, such density changes can be used in research to give insights into the nature of molecular events.

In addition, the dielectric constant of NCW is vastly reduced when compared to water at ambient conditions, as shown in Fig. 9.2. Uematsu and Franck[12] correlated the dielectric constant of water as a function of both temperature and density. For example, at 300°C and saturation pressure, the dielectric constant of water is approximately 20, a nearly 75% reduction from the value of 78 at ambient conditions. This dielectric constant most closely corresponds to a moderately polar solvent, such as acetone (dielectric constant = 21.4 at 25°C). This reduction in dielectric constant, resulting from a major decrease of the hydrogen bonding relative to ambient water, enables a greatly enhanced solubility of nonpolar organic species in NCW. Increasing the density can increase the dielectric properties of NCW, but kilobar pressures are required. For example, increasing the density at 300°C from saturation conditions ($\rho \sim 0.75$ g/cm^3) to the density of ambient water ($\rho = 1$ g/cm^3) nearly doubles the dielectric constant to approximately 35.

The reduced dielectric constant of NCW, while increasing the solubility of nonpolar organics, reduces the solubility of inorganic salts. Therefore, for some applications of NCW as a reaction medium, a careful balance must be struck for the desired solubility of ionic and nonionic species to be achieved. Although pressure is a possible variable, normally this will

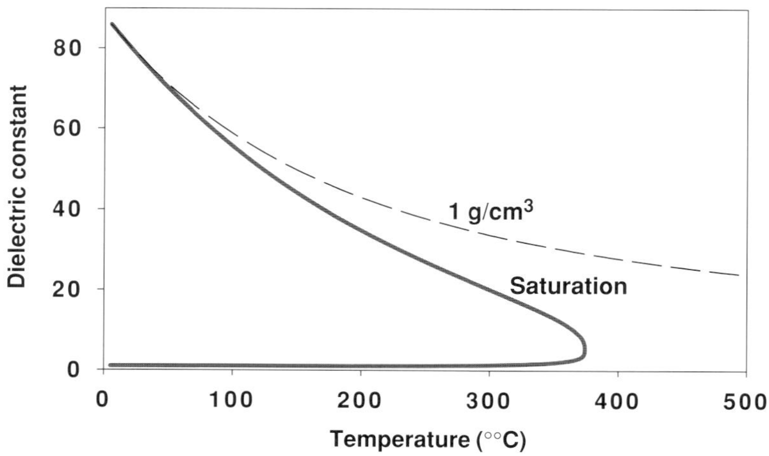

Figure 9.2 Dielectric constant of water as a function of temperature from 0 to 500°C. Correlation from Uematsu and Franck.[12]

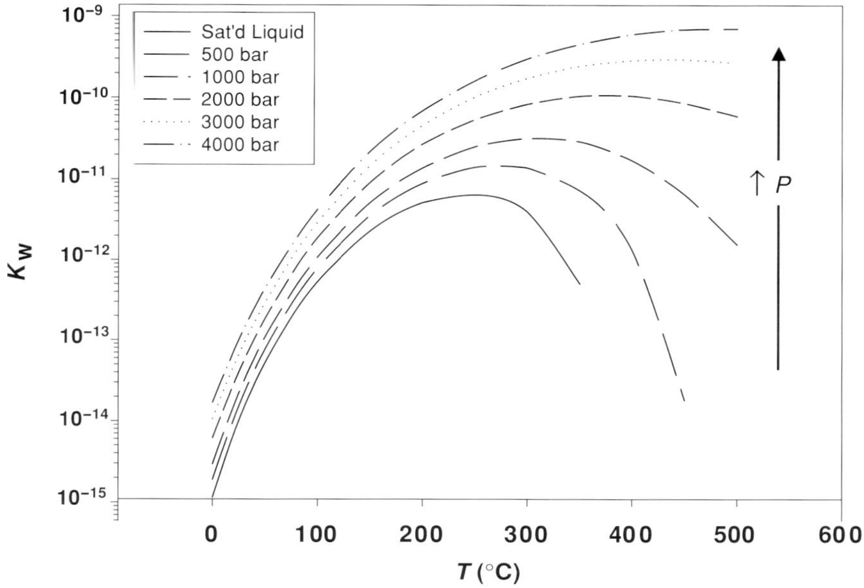

Figure 9.3 Dissociation constant of water as a function of temperature from 0 to 600°C at various pressures. Correlation from Marshall and Franck.[13]

be done by adjustments in temperature, and this ability to tune properties offers important advantages for NCW in reaction processes.

A major advantage of using NCW as a reaction solvent has grown from opportunities related to the increase in dissociation constant. The K_W of water increases by as much as 3 orders of magnitude, from 10^{-14} at 25°C to nearly 10^{-11} at around 250°C (at saturation density),[13] where a maximum occurs, as displayed in Fig. 9.3. This maximum can be increased by about another order of magnitude (and shifted to higher temperature) by increasing the density of the solvent – but again this requires very high pressure and is predominantly of academic interest. The presence of the maximum constitutes additional evidence consistent with our interpretation of other effects discussed previously. Increased temperature renders dissociation more favorable, but the decrease in dielectric constant reduces the ability of the medium to solvate the resulting ions. This is also why the maximum can be increased by increasing density, as the dielectric constant also increases.

The reason for interest in the high dissociation constant of NCW is that it provides a means of performing both acid and base catalysis in NCW without adding mineral acids or bases. By simply heating the solution, the dissociation of water increases, providing an enormous increase in the concentration of hydronium and hydroxide ions, resulting in a catalytic medium for the chemical transformation. Cooling the mixture restores the ambient ion concentrations, representing a self-neutralizing catalytic medium. This is important for ease of separations (no catalyst to be recovered), elimination of processing steps (no neutralization required) and waste reduction (no salt disposal).

To perform chemical transformations on organic substrates in NCW, the medium must be able to dissolve those substrates in sufficient quantities for homogeneous reaction to take place. Solubility is achieved for even nonpolar organic substrates through heating of the mixture. At the heart of this solubility change is a dramatic reversal of solubility with temperature, providing not only the opportunity to dissolve both nonpolar and ionic substrates at high temperature, but also the ability to phase separate easily organic products by simply cooling the postreaction mixture. This enables a simple decantation of products to replace more difficult forms of product recovery.

The solubility of organic species in NCW varies greatly with the functionality of the organic molecule. The upper critical solution temperature (UCST) is a useful parameter for examining these differences. The UCST is the temperature above which the organic substrate and water are miscible in all proportions (at the bubble pressure of the mixture). Polar organics, such as acetonitrile, have very low UCSTs (−1°C for acetonitrile).[14] Functionalized nonpolar organics have much higher UCTSs, e.g. acetophenone (228°C), 1-octanol (278°C), anisole (291°C).[15] Nonpolar organics can be much higher, as evidenced by benzene (305°C) toluene (310°C), and *n*-hexane (355°C).[16] The UCST is an important measure, but it does not represent the entire solubility phenomenon. Benzene is completely miscible at 305°C, despite having an ambient solubility in water of 500 ppm.[17] *n*-Hexane solubility increases by almost 5 orders of magnitude from ambient water to NCW.[16] The structural effects are much less significant, as the UCST for 1-hexanol (221°C) and 2-hexanol (230°C) are quite similar.[18] Most phase equilibrium models for the solubility of organics in water at elevated temperatures are extremely poor[19]; thus data acquisition at elevated temperature are quite important.

Some example solubility curves for organics in water as a function of temperature are provided in Figs. 9.4–9.6. We choose to discuss four examples of solubility behavior for low

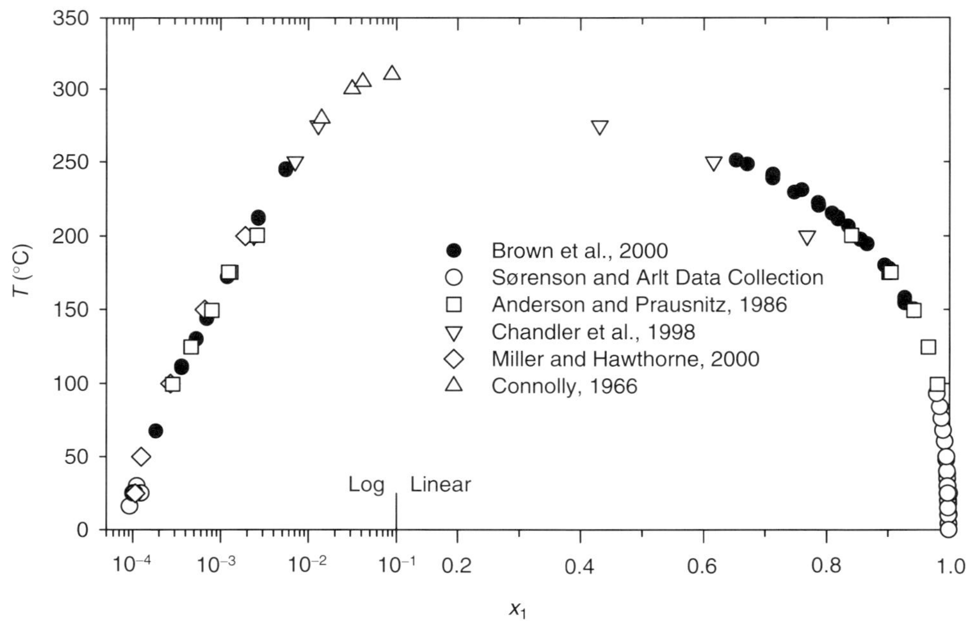

Figure 9.4 Liquid–liquid equilibria for toluene (component 1)/water (component 2). The data were taken at a variety of pressures.[15,20–23]

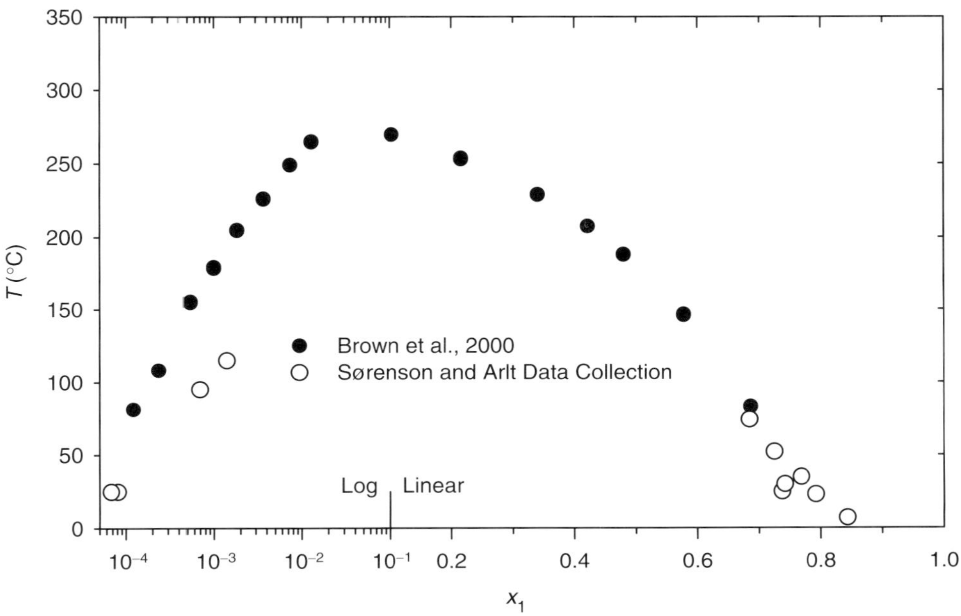

Figure 9.5 Liquid–liquid equilibria for 1-octanol (component 1)/water (component 2). The data were taken at a variety of pressures.[15,23]

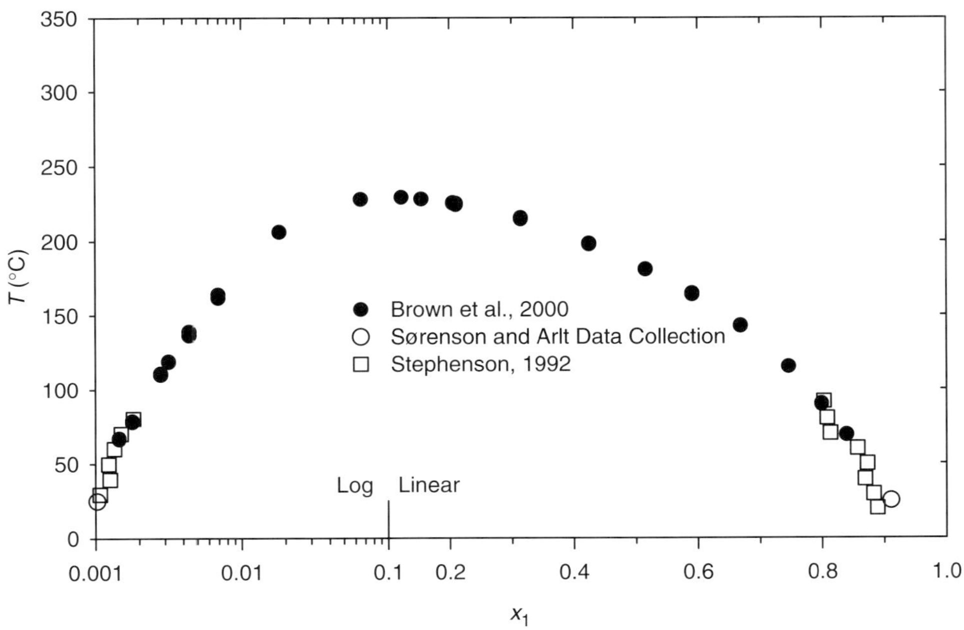

Figure 9.6 Liquid–liquid equilibria for acetophenone (component 1)/water (component 2). The data were taken at a variety of pressures.[15,24]

polarity organic species with different functionalities: toluene, 1-octanol, acetophenone, and anisole. By comparing the phase behavior of different organic functionalities with NCW, we can learn a great deal about the ease of separating the corresponding reaction mixtures. In Fig. 9.4, the mutual solubilities of toluene and water are represented.[15,20–23] Toluene is a nonpolar aromatic compound; as such its solubility in ambient water is quite limited, only $x = 0.0001$ at 25°C. The solubility then increases fairly exponentially with temperature up to $x = 0.014$ at 280°C and even more rapidly up to the UCST at ~310°C. This log-linear solubility relationship is typical of nonpolar compounds in water, and the rapid increase near the UCST is typical for the aqueous phase solubility. It is important to note that the solubility of water in the *organic*-rich side of the phase diagram is still low: $x = 0.0025$; however, the solubility increase with temperature is much more smooth and gradual. This is generally the case for most aqueous/nonpolar organic pairs. The very low mutual miscibility at room temperature leads to excellent separations potential for an NCW system, and the extreme change in solubility behavior between 280 and 310°C suggests that operating conditions must be properly chosen to ensure miscibility throughout any reaction process. The rapid drop in solubility with just a 30°C change in temperature further indicates that for an economic phase separation, full cooling to ambient temperature may be unnecessary. This behavior for toluene is comparable to that of Fig. 9.5 for a long-chain alkyl alcohol, 1-octanol.[15,23] Despite the hydroxyl functionality, the 25°C solubility of 1-octanol in water is slightly lower than toluene, $x = 0.000075$, because the long alkyl chain effect is dominant. However, on the organic-rich side the solubility is much higher, $x = 0.27$ at 25°C, because of hydrogen bonding, leading to a less drastic phase split at low temperature. The changes in 1-octanol aqueous solubility are similarly rapid for this system as for toluene/water, with $x = 0.013$ at 265°C and full miscibility at 278°C. As another example, ketones, such as acetophenone (Fig. 9.6),[15,23,24] offer a higher degree of hydrogen bonding potential and slightly higher polarity, the solubility in water is higher ($x = 0.001$ at 25°C) and the solubility increase much steeper than in the other systems. As is so common with the majority of liquid–liquid equilibria, the temperature behavior is very flat at the top of the curve (UCST) ($x = 0.02$ at 225°C, UCST = 230°C), providing an excellent opportunity for a highly efficient phase separation.

A more interesting case is demonstrated for anisole/water (Fig. 9.7).[15,24] Anisole is the simplest aromatic ether, and as such is somewhat similar to toluene in terms of aqueous solution behavior. The solubility of anisole in water at 25°C is slightly higher ($x = 0.00032$) than for toluene, as is that of water in anisole ($x = 0.012$). However, the temperature effect on solubility is different. Because of the basicity of the ether oxygen, the solubility of anisole in water goes through a minimum at around 40°C, indicative of a closed-loop system. We speculate that this behavior is caused by the formation of anisole/water unlike-pair hydrogen bonds at the lower temperatures, thus tending to solvate better the organic. However, for this pair, the lower critical solution temperature (below which the pair is again miscible in all proportions) is below the mixture freezing point, and thus experimentally inaccessible. However, the intermolecular forces leading to this behavior could have an enormous effect on the separation, as the existence of an optimum phase separation temperature above ambient impacts the separation procedure. After heating past this minimum in solubility, the solubility increases as normal (log-linear) until $x = 0.013$ at 235°C, and then proceeds to a UCST at 291°C, behavior somewhat between 1-octanol and toluene. Because of the closed-loop phase diagram, the organic-rich solubility curve is much steeper than usual, with $x = 0.01$ at 25°C at $x = 0.028$ at 90°C, before a rapid rise at higher temperature.

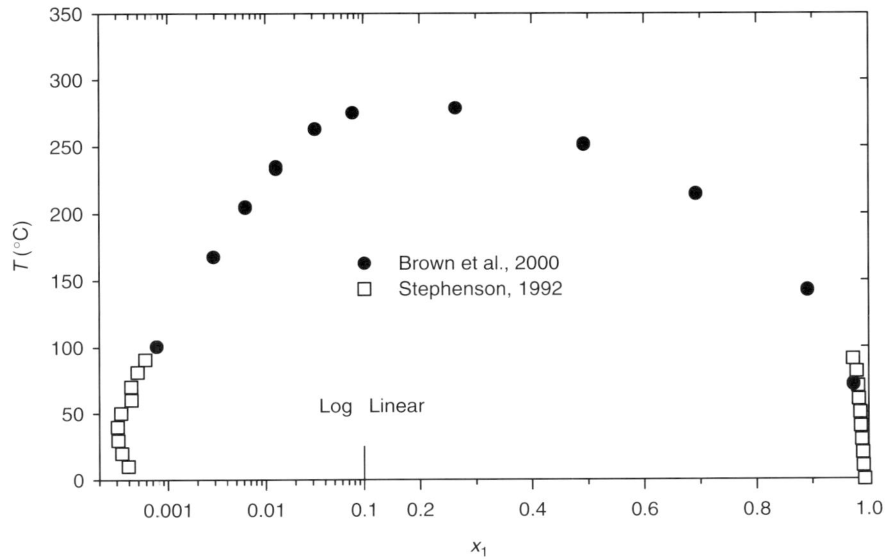

Figure 9.7 Liquid–liquid equilibria for anisole (component 1)/water (component 2). The data were taken at a variety of pressures.[15,24]

This closed-loop behavior is typical of hydrogen-bond acceptor molecules (nicotine is an example), as the balance between effects of enthalpy and entropy leads to higher solubility at lower temperatures, then a drop as hydrogen bonding begins to break down, then a second increase as the entropic forces take control. The use of this type of behavior makes certain functional groups (ethers, esters) most useful for separations in NCW systems.

9.1.2 Solvatochromic characterization of NCW

The Kamlet–Taft method is a powerful linear free-energy relationship for understanding the role of various solvent/solute interactions. It defines a set of parameters associated with a particular solvent by using solvatochromism. This is a process of assessing the local environment around a probe molecule, by measuring the solvent-induced spectral shift of certain spectroscopic indicators,[8–10,25,26] which reflect the chromophore's molecular environment, characterizing the interaction of solvent molecules with an electronic transition of an indicator.

The three parameters that are central to the method are π^*, α, and β. The π^* parameter provides a comprehensive measure of the ability of a solvent to stabilize a solute molecule based on the dielectric effects. It is a quantitative index of solvent dipolarity and polarizability. The acidity α for a solvent is a measure of its strength as a hydrogen-bond donor HBD, its ability to donate a proton in a solvent-to-solute hydrogen bond. The estimation of α is based on the experimental determination of π^* and $E_T(30)$. The $E_T(30)$ scale, developed by Reichardt et al.,[26] indicates a solvent strength by combining polarity and HBD acidity, which itself serves as a useful solvent parameter for physicochemical correlations in a wide range of solvents.[28–30] The basicity parameter β denotes the solvent's hydrogen-bond acidity HBA or an index of the solvent's ability to accept a proton in a solute-to-solvent hydrogen bond. Specialized indicators have been developed for the determination of the π^*, $E_T(30)$, α, and β solvent parameters. The Kamlet–Taft parameters have been used to correlate a variety of configurational properties in solution, including solubilities, partition coefficients, and reaction rates, among others. One form of the generalized equation is shown as follows:

$$XYZ = (XYZ)_0 + s(\pi^* + d\delta) + a\alpha + b\beta \quad \text{(Eq. 9.1)}$$

where XYZ and $(XYZ)_0$ denote the values of the solvent-dependent physicochemical property in a given solvent and in a reference solvent (gas or inert solvent), respectively. δ is a polarity correction parameter for aromatic ($\delta = 1.0$), polyhalogenated aliphatic ($\delta = 0.5$), and all other solvents ($\delta = 0$), s, d, a, and b are solvent-independent coefficients indicating the susceptibility of the property to the applied parameters. This multiparameter approach assumes additivity of three aspects of solvation: nonspecific van der Waals interactions, hydrogen-bond donation, and hydrogen-bond acceptance. Lu et al.[25] have reported the Kamlet–Taft parameters for water at temperatures ranging from 25 to 275°C. The π^* values and density of saturated liquid water are shown in Fig. 9.8.[25,26] The dipolarity/polarizability of liquid water continuously decreases as the temperature increases, as indicated by the π^* value of 0.69 at 275°C in comparison with 1.08 at 25°C.

Nearcritical water at 275°C has a dipolarity/polarizability comparable to that of ambient acetic acid, and the observed decline in π^* is consistent with the change in density. The $E_T(30)$ values shown in Fig. 9.9 decrease in a linear fashion with increasing temperature. Figure 9.10 shows that the α values for liquid water decrease from 1.16 at 25°C to 0.84 at 275°C. This trend can be attributed to a decrease in hydrogen bonding between the solvent and solute molecules as the temperature is elevated. Bennett and Johnston[31] reported that at temperatures as high as 380°C, hydrogen bonding between water and acetone is relatively still intact at $0.5\rho_c$ and changes little with increased density. The results reported in Fig. 9.10 indicate that NCW exhibits considerable HBD acidity, comparable to that of ambient ethanol. The β values of water are shown in Fig. 9.11. In contrast to π^* and α, β increases only slightly from 25 to 275°C.

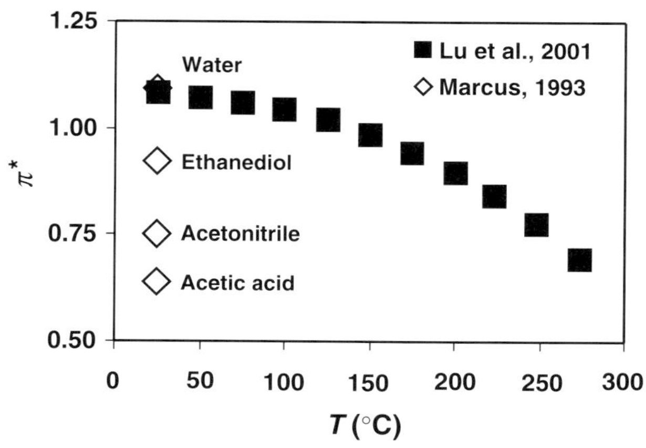

Figure 9.8 Dipolarity/polarizability (π^*) and density of saturated liquid water as a function of temperature.[25,26]

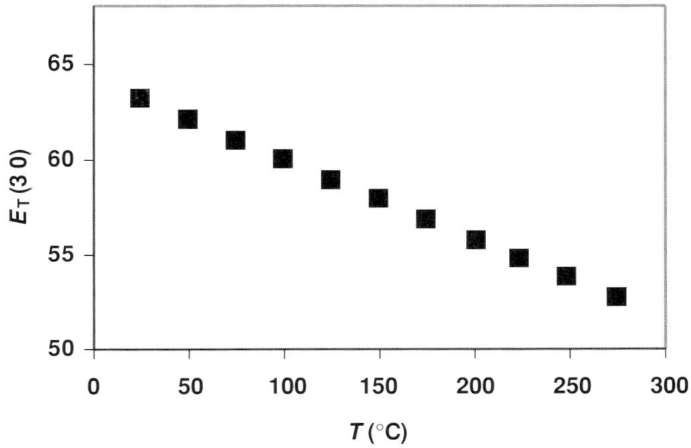

Figure 9.9 $E_T(30)$ parameters of saturated liquid water at different temperatures.[25]

It has been reported that 4-nitroaniline and *N,N*-dimethyl-4-nitroaniline react in NCW to produce aniline and *N,N*-dimethylaniline, respectively (Fig. 9.12).[32] The rates associated with these reactions have been studied at temperatures ranging from 200 to 275°C. The reactions are acid catalyzed, and since nitric acid is a product in both cases, the reaction kinetics follow an autocatalytic pathway. For the autocatalytic reaction.

$$A + H^+ \rightarrow B + 2H^+$$

an autocatalysis function $f(x)$ is defined as

$$f(x) = \ln\,[M + xA]\,/\,[M(1 - xA)] = Ca_0\,(M + 1)\;kt \qquad \text{(Eq. 9.2)}$$

where $M = CH_0^+/Ca_0$, x is the mole fraction, and C_0 is the initial concentration. As shown in Fig. 9.13, the autocatalysis function $f(x)$ vs time for 4-nitroaniline and

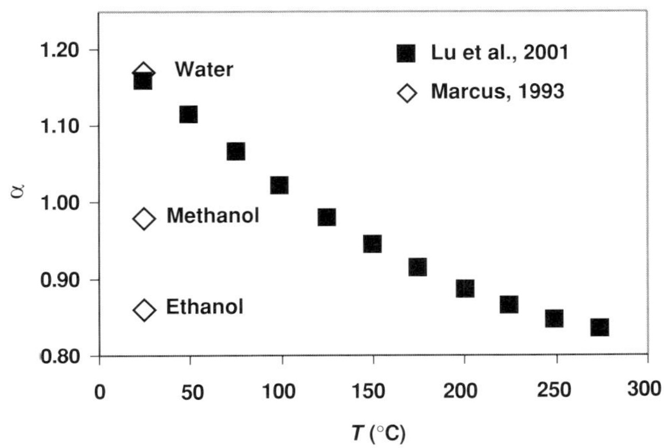

Figure 9.10 HBD acidity α of saturated liquid water as function of temperature.[25,26]

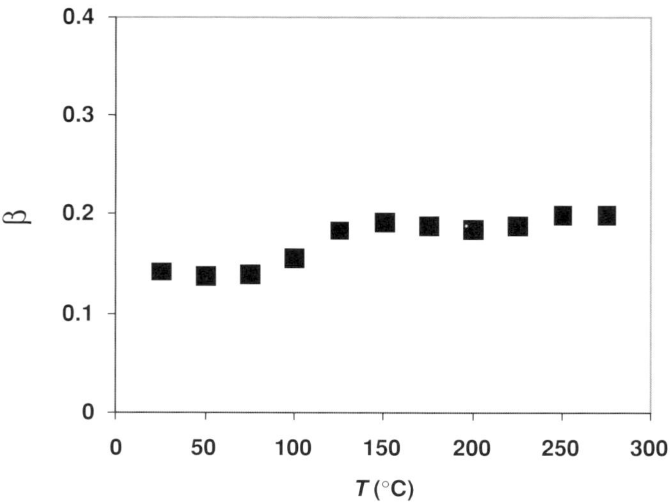

Figure 9.11 HBA basicity β of saturated liquid water as a function of temperature.[25]

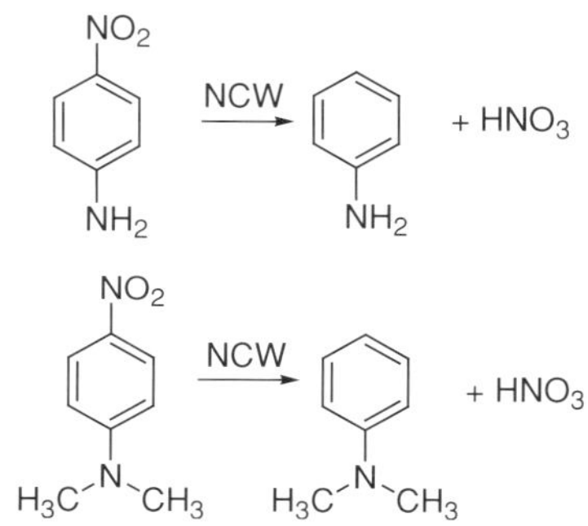

Figure 9.12 Reactions of 4-nitroaniline and *N,N*-dimethyl-4-nitroaniline in near critical water.

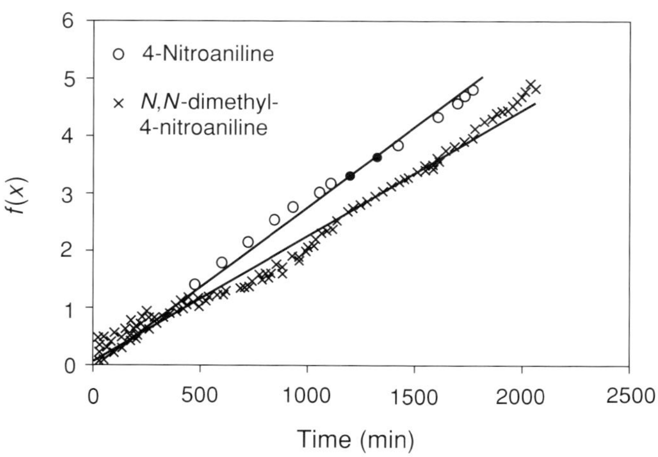

Figure 9.13 Autocatalysis function vs time for 4-nitroaniline and *N,N*-dimethyl-4-nitroaniline at 200°C.[32] The autocatalysis function is $f(x) = \ln[M + xA]/[M(1 - xA)] = Ca_0\,(M + 1)\,kt$.

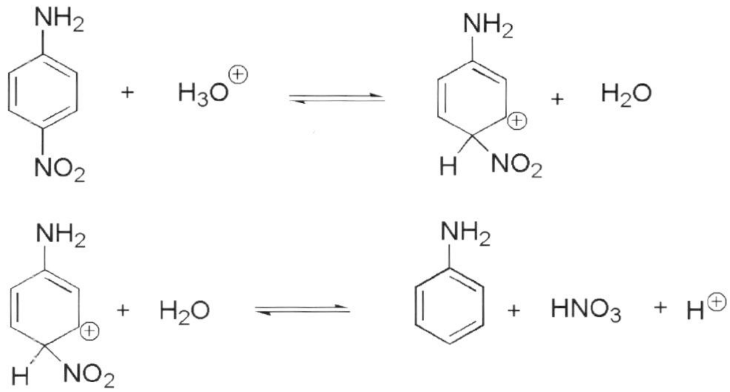

Figure 9.14 Acid-catalyzed mechanism for the hydrolysis of 4-nitroaniline in NCW.

N,*N*-dimethyl-4-nitroaniline at 200°C is linear with a slope of $\mathrm{Ca}_0(M+1)k$. The mechanism is shown in Fig. 9.14. The Arrhenius plots and the accompanying E_{act} values are shown in Fig. 9.15.

A linear solvation energy relationship (LSER) can be applied for kinetic correlations and written as

$$\ln k = (\ln k)_0 + s\pi^* + a\,\alpha + b\,\beta \qquad \text{(Eq. 9.3)}$$

in which k is the rate constant in a given solvent, and k_0 is in a reference solvent. When the concentration of hydronium ion ($[\mathrm{H}^+]$) remains constant, the reaction rate is described by

$$v = k[\mathrm{H}^+][\mathrm{R}] = k'[\mathrm{R}] \qquad \text{(Eq. 9.4)}$$

in which the pseudo-first-order rate constant $k' = k[\mathrm{H}^+]$. k is a function of temperature only, while k' is dependent on temperature and $[\mathrm{H}^+]$. For the hydrolysis reaction, we define cyclohexane at ambient temperature as the reference solvent, which leads to a zero value of $(\ln k')_0$. The HBA term β is negligible because water as a hydrogen-bond donor has a

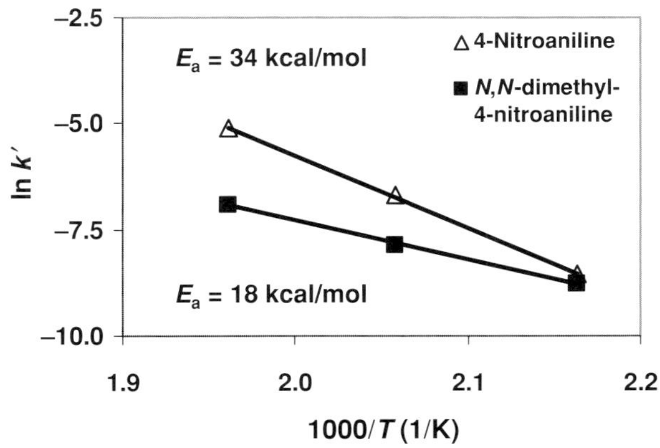

Figure 9.15 Arrhenius plots of the hydrolyses of 4-nitroaniline and *N*,*N*-dimethyl-4-nitroaniline in NCW.[32]

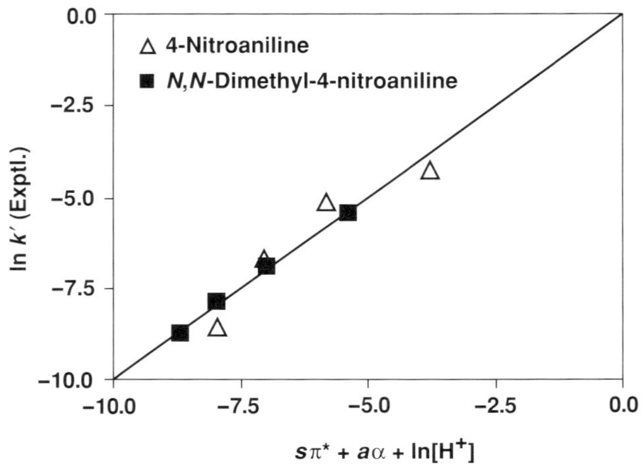

Figure 9.16 LSER correlation of initial rate constant with Kamlet–Taft solvent parameters.[32]

comparable interaction with the reactant and with the intermediate. Therefore, it contributes little to the kinetics observed. Thus, an applicable LSER for the hydrolysis reaction can be expressed as

$$\ln k' = \ln[\mathrm{H}^+] + s\pi^* + a\alpha + b\beta \qquad \text{(Eq. 9.5)}$$

in which $b = 0$. $[\mathrm{H}^+]$ was calculated from the ionization constants of liquid water at the corresponding temperature.[13] Values of the susceptibility coefficients in Eq. 9.5 were obtained by multiple regression of k' and Kamlet–Taft parameters ($s = -29.1$, $a = 35.0$ for 4-nitroaniline; $s = -23.1$, $a = 28.1$ for *N,N*-dimethyl-4-nitroaniline). Figure 9.16 shows the consistency between experimental and correlated k' (x axis). The correlation for *N,N*-dimethyl-4-nitroaniline (non-HBD) appears to be better than that for 4-nitroaniline. The result is shown as Fig. 9.17 and the coefficients m (–1.37, –1.03) and n (81.1, 61.7) were

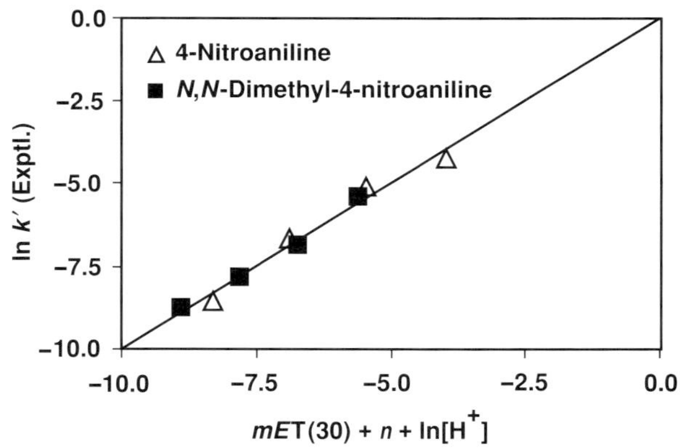

Figure 9.17 Correlation of initial rate constant of the hydrolysis reactions in NCW with $E_T(30)$ solvent parameter.[32]

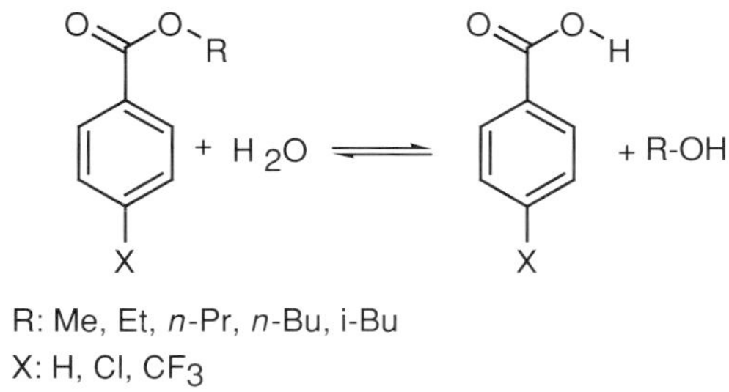

Figure 9.18 Hydrolysis of substituted benzoate esters.

regressed for 4-nitroaniline and N,N-dimethyl-4-nitroaniline, respectively. It appears to give reasonably consistent results. Thus, $E_T(30)$ is also a useful solvent parameter for NCW.

Similarly, k' may be correlated in terms of $E_T(30)$.

$$\ln k' = \ln[H^+] + mE_T(30) + n \qquad \text{(Eq. 9.6)}$$

9.2 Reactions in NCW

9.2.1 Hydrolysis of ester and ether

The main products formed from the hydrolysis of di-n-butyl phthalate[33] and ethyl acetate[34] in NCW are the corresponding acids and alcohols. The formation of thermolysis products, such as alkenes and oxides of carbon, is considerably slower in NCW than when performed neat.

Lesutis et al.[35] have reported the kinetics associated with the hydrolysis of a series of benzoate esters in NCW (Fig. 9.18). Conversion vs time yielded S-shaped curves (Fig. 9.19), which suggested an autocatalytic mechanism. The $A_{AC}2$ mechanism for acid-catalyzed

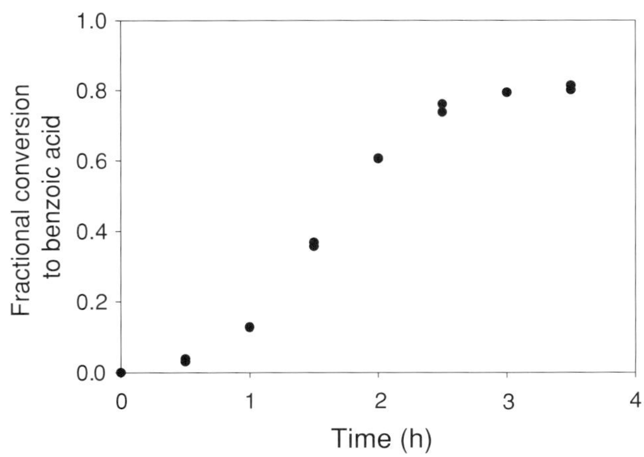

Figure 9.19 Fractional conversion of n-propyl benzoate vs time.[35]

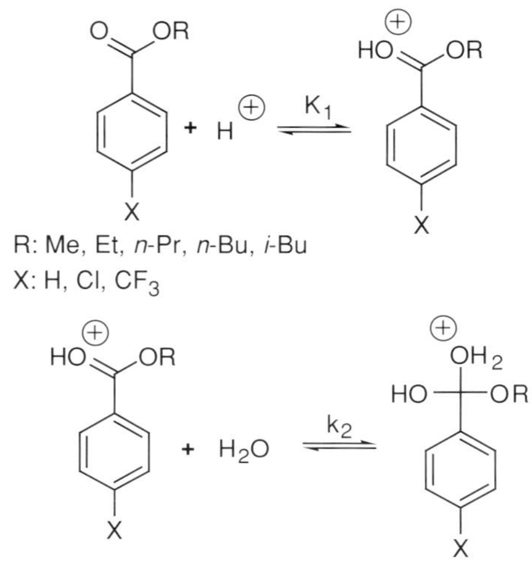

Figure 9.20 $A_{AC}2$ mechanism for acid-catalyzed ester hydrolysis.

hydrolysis of esters conforms to the following rate expression:

$$\mathrm{d}x/\mathrm{d}t = k[\text{ester}][\mathrm{H_2O}][\mathrm{H^+}] \quad \text{(Eq. 9.7)}$$

where $k = k_2 K_1$[36], k_2 is the rate constant for the addition of water to the protonated ester, and K_1 is the equilibrium constant for the protonation of the ester (Fig. 9.20). The second step is postulated to be rate controlling and its product equilibrates rapidly to the benzoic acid and the alcohol. The concentration of protons is determined from the dissociation constants of water and the particular benzoic acid at 250°C. The dissociation constant of benzoic acid at 250°C is 3.7×10^{-6} M.[37] The Hammett ρ value for this dissociation increases by only 2% from 25 to 250°C[38] and the dissociation constants for substituted benzoic acids are calculated from the Hammet relationship using a ρ value of 1.02. Table 9.1 summarizes these rate constants as products of k_2 and K_1^*, where K_1^* is the equilibrium constant for the protonation of the nonsubstituted ester. It is assumed that the effect of substituents on the equilibrium constant for ester protonation is the same as the effect for acid dissociation. Rates of hydrolysis of isobutyl benzoate are also reported at 260 and 300°C. Using the dissociation constants of benzoic acid at these temperatures,[37,38] the activation energy is

Table 9.1 Rate constants at 250°C for the hydrolysis of benzoate esters, where K_1^* is the equilibrium constant for the protonation of nonsubstituted ester[35]

Substituent	σ	$k_1 \times 10^4$ (L/(mol h))
4-Methyl	−0.14	0.62 ± 0.03
3,5-Dimethyl	−0.12	0.61 ± 0.02
H	0	1.31 ± 0.10
4-Phenyl	+0.05	2.04 ± 0.15
3-Hydroxyl	+0.13	2.11 ± 0.32

Figure 9.21 Hydrolysis of substituted anisoles.

calculated to be 24 ± 3 kcal/mol, which is consistent with the activation energies obtained for the acid-catalyzed hydrolyses of substituted benzoate esters in low-temperature aqueous solvents with a stoichiometric addition of acid.[39] As shown in Table 9.1, the rates of hydrolysis decrease as the length of the alcohol portion of the ester increases. Branching of the alcohol slows the rate even more. Moreover, the rate of hydrolysis of substituted isobutyl benzoates is independent of the substituent; the ρ value is close to zero. This negligible substituent effect is a well-documented characteristic of acid-catalyzed ester hydrolyses in aqueous solvents at lower temperatures. The ρ value for acid-catalyzed ester hydrolyses is 0, but the ρ value for base-catalyzed ester hydrolyses is 2.4. Thus, the acidity of NCW is sufficient to initiate hydrolysis of the benzoate esters.

The rates of hydrolysis of a series of substituted anisoles have also been reported.[40] The reactions are shown in Fig. 9.21, and the rate constants, along with the σ values, are summarized in Table 9.2. A Hammett plot produces a ρ value of 2.3 (Fig. 9.22), showing that the mechanism of reaction involves a simple S_N2 displacement by water on the methyl group of the substituted anisole (Fig. 9.23). Finally, the rates of hydrolysis of anisole are found to be substantially faster than that of the more sterically hindered phenetole (Fig. 9.24), consistent with the postulated S_N2 mechanism.

Hydrolysis reactions have been widely reported in NCW for low-molecular-weight molecules as well as for polymeric materials. Mandoki[41] reported a process for depolymerizing condensation polymers using NCW without addition of bases or acids. More particularly, polyethylene terephthalate, polybutene terephthalate, nylon 6, and nylon 66 were hydrolytically depolymerized (Fig. 9.25).

Siskin et al.[42] similarly hydrolyzed polyacrylonitrile to low-molecular-weight oligomeric materials, with the generation of ammonia instead of the toxic hydrogen cyanide formed by conventional thermolysis processes. Holliday et al.[43] reported that triglyceride-based vegetable oils can be hydrolyzed into their fatty acids constituents. The authors studied NCW and SCW. Although the conversion yields are comparable in both medium, NCW

Table 9.2 Rate constants for S_N2 nucleophilic attack by water of substituted anisoles[40]

R	X	$k_2 K_1^*$ ($L^2/(mol^2\ h)$)
Methyl	H	26.9 ± 2.5
Ethyl	H	25.7 ± 0.9
n-Propyl	H	10.4 ± 0.5
n-Butyl	H	17.1 ± 0.6
Isobutyl	H	6.7 ± 0.3
Isobutyl	Cl	7.4 ± 0.6
Isobutyl	CF_3	7.0 ± 0.5

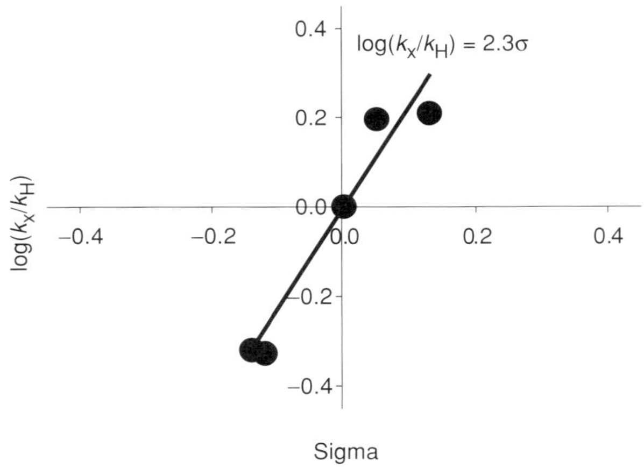

Figure 9.22 Hammett plot for anisole hydrolysis.[40]

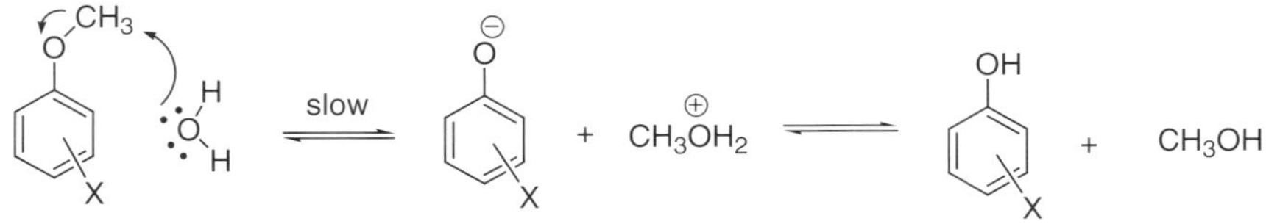

Figure 9.23 Mechanism for the hydrolysis of anisole.

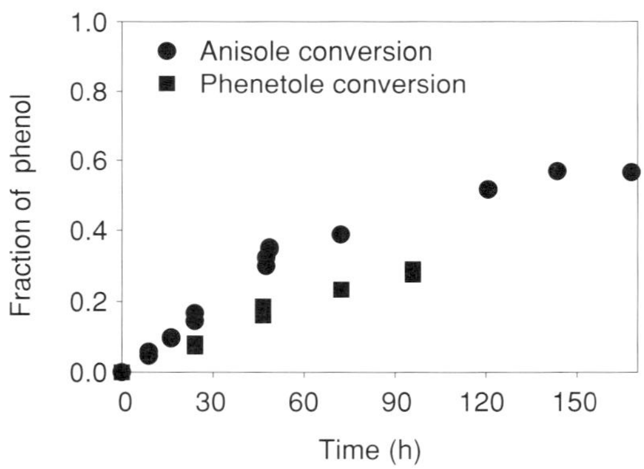

Figure 9.24 Comparison of anisole and phenetole hydrolyses at 300°C.[40]

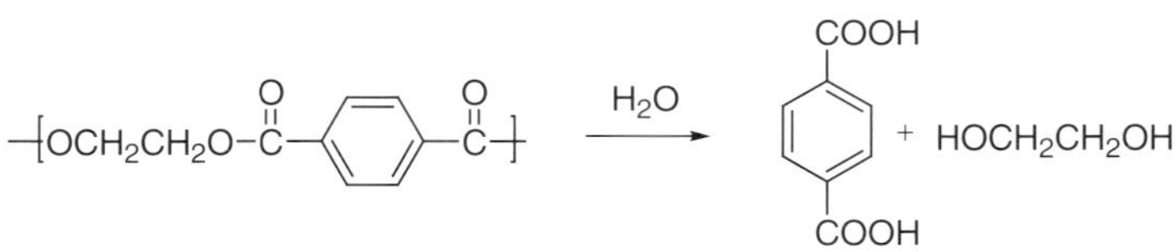

Figure 9.25 Hydrolysis of polyethylene terephthalate.

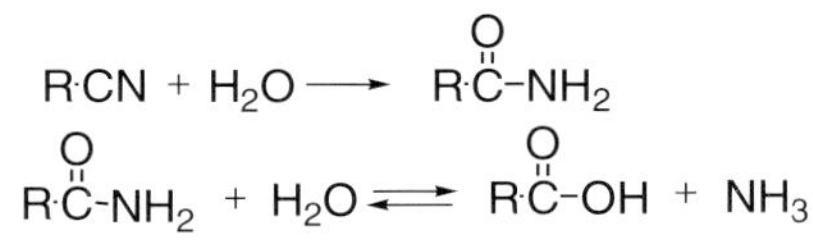

Figure 9.26 Reaction of nitriles in NCW.

induces significantly less degradation of the fatty acid products. Additionally, finding new ways to transform biomass to reusable chemicals has become critical to most countries. In that context, Minowa et al.[44] have shown that the hydrolysis of cellulose to glucose in NCW can be achieved without addition of catalyst.

9.2.2 Hydrolysis of nitriles

Alkyl and aryl nitriles readily hydrolyze when submitted to NCW conditions. The hydrolysis is a multistep sequence as shown in Fig. 9.26. For instance, Katritzky et al.[45] have reported that benzonitrile is converted to benzamide and benzoic acid at 250°C over a period of 5 days, and they conclude that the amide and the acid were in equilibrium. Under these conditions some decarboxylation can also occur. An et al.[2] have reported the product distribution for the hydrolysis of benzonitrile as a function of time and temperature. Specifically, the ratio of benzamide to benzoic acid varied as follows: after 1 h at 250°C, the distribution was 5:4. However, at 280°C after 1 h, the ratio was 1:1, and became 1:25 when the reaction time was extended to 6 h. Alkylnitriles exhibit similar behaviors; Siskin et al. reported that at 250°C for 2.5 days decanonitrile quantitatively yields two major products, decanoic acid and decanoamide.[46] When octanenitrile was hydrolyzed to octanoic acid amide and octanoic acid, the reaction was slightly slower than that of benzonitrile. Only 29% conversion took place in 1 h at 290°C.[2] The limited solubility of octanenitrile in water, even in NCW conditions, was suggested as a possible factor for the slow reaction. Again the product distribution was dependent on the residence time and the temperature.

Iyer and Klein[47] reported the reaction of benzonitrile in NCW at 330°C at a variety of pressures, yielding as products butyramide, butyric acid, and ammonia (Fig. 9.26). Figure 9.27 summarizes the kinetics of the disappearance of butyronitrile as a function of pressure. The rate of hydrolysis increases with increasing pressure. Figure 9.27 also shows that the kinetic profile has an induction period attributed to autocatalysis by the butyric acid formed during the hydrolysis. The authors clearly show that in the closed system an equilibrium exists between the butyric acid/ammonia and the butyramide. The rate constants for the various steps in the hydrolysis are summarized in Table 9.3.

They report activation volumes for k_1 and k_2 at a pressure of 128.5 bar of –362 and –231 cm^3/mol, respectively. The activation volume is a useful mechanistic probe for reactions in liquids; however, in compressible fluids the mechanistic contribution is far overshadowed by that of solvation. The values of NCW reported are intermediate between the corresponding values in normal liquid solvents and about an order of magnitude lower than those observed in supercritical solvents. At higher pressures, where water is less compressible, the activation volumes varied between –19.3 and –12 cm^3/mol, values comparable in magnitude to those in liquid solvents.

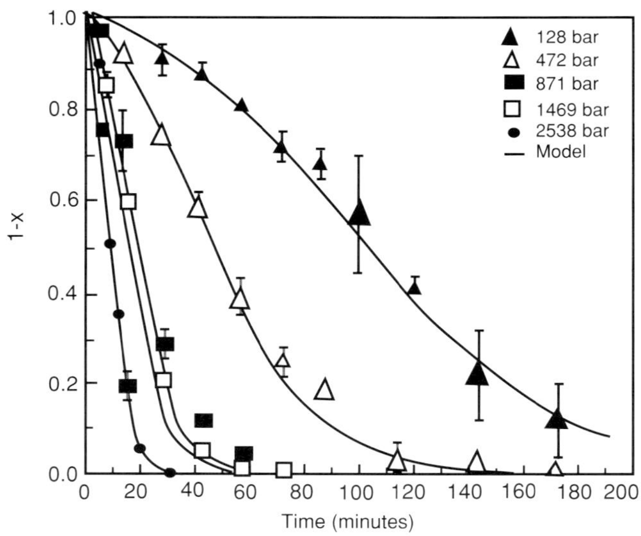

Figure 9.27 Pressure dependence of the disappearance kinetics of butyronitrile. Reprinted from Ref. 47. Copyright 1997, with permission from Elsevier.

Katritzky et al.[45] and Hallett[18] reported the hydrolysis of 3-cyanopyridine in NCW at 200 and 250°C. In addition to the corresponding amide and carboxylic acid products, decarboxylation to pyridine was also observed (Fig. 9.28). Quantitative conversion of the 3-cyanopyridine was typically achieved in 3 h at 250°C (Fig. 9.29). Kinetic results indicate a temperature-dependent maximum in the production of both nicotinamide (before hydrolysis to nicotinic acid initiates) and at a later time of nicotinic acid (before decarboxylation to pyridine takes place). Figure 9.29 presents the results of the conversion of 3-cyanopyridine at 250°C. At this temperature, a maximum nicotinamide yield of 18% was found to occur after 1 h at which point the conversion of 3-cyanopyridine was 64%. It is clear that the amide hydrolyzes at a faster rate than the nitrile. A nicotinic acid yield of 96% was detected after 3 h, at which point the amide yield had dropped to 2.0%. The pyridine yield grew steadily to 1.8% after 3 h. Within 1 week, the yield of pyridine was quantitative. The experimental evidence suggests reaction times and temperatures could be optimized for production of either nicotinamide (lower reaction temperatures and shorter reaction times) or nicotinic acid (longer times and higher temperatures). Also, decarboxylation of the acid can be effectively suppressed by lowering the temperature; no detectable pyridine existed after 6 h at

Table 9.3 Rate constants for the four-step reaction of butyronitrile hydrolysis[47]

P (bar)	k_1 ($M^{-1}\ s^{-1}$)	k_2 ($M^{-1}\ s^{-1}$)	k_3 ($M^{-1}\ s^{-1}$)	k_{-2} ($M^{-1}\ s^{-1}$)
128.5	0.01	0.3934	40.94	11.88
471.8	0.025	0.6668	63.39	24.39
870.3	0.055	1.2936	148.39	44.64
1468.7	0.0834	1.0986	139.96	38.81
2538.8	0.1234	1.902	280.06	55.32

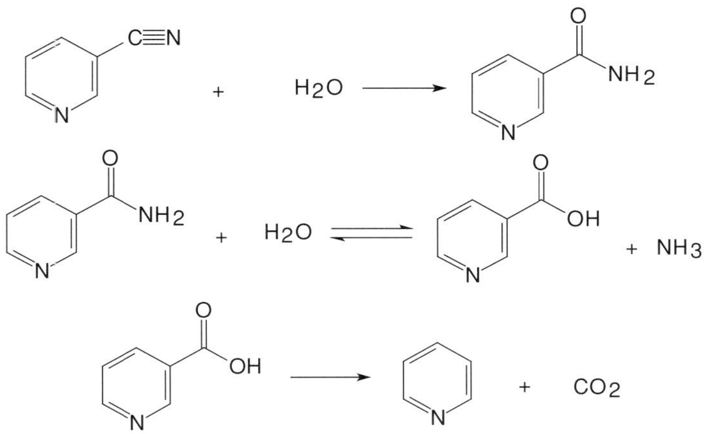

Figure 9.28 Hydrolysis of 3-cyanopyridine.

200°C despite 98% conversion of starting material. Additionally, separation is easy as the products are insoluble in water at ambient temperature. The separation of nicotinic acid from nicotinamide will also be facile by fractional crystallization, as the melting points differ by 100°C.

9.2.3 Hydration of β-pinene

The reaction of pinene to form α-terpineol has been known since the late nineteenth century. Pinene is relatively abundant in nature, with the major source being pine trees. α- and β-pinene (**1**) will hydrolyze in aqueous acid to form a tertiary carbocation, as shown in Fig. 9.30, followed by a series of carbonium ion rearrangements, which results in a multitude of products. The major product of the reaction is α-terpineol (**2**); however, as a consequence of carbonium ion rearrangements, terpinen-4-ol (**3**) and γ-terpineol (**4, 5**) are also produced. These monocyclic alcohols are collectively known as 'terpineol.' Smaller amounts of the bicyclic alcohols fenchol (**6**) and borneol (**7**) are also formed. Undesirable side reactions

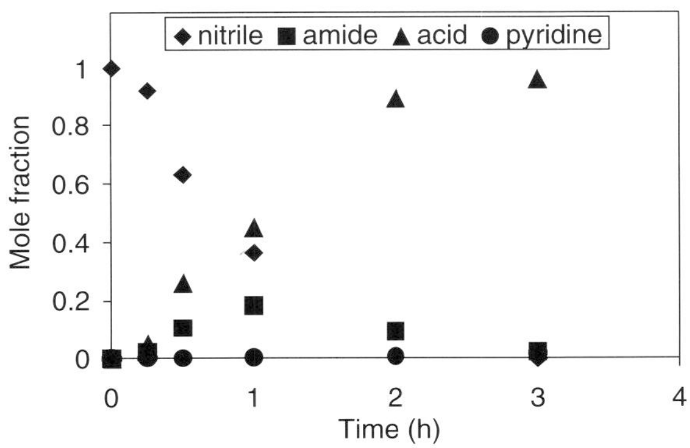

Figure 9.29 Results of hydrolysis of 3-cyanopyridine in NCW at 250°C.[18]

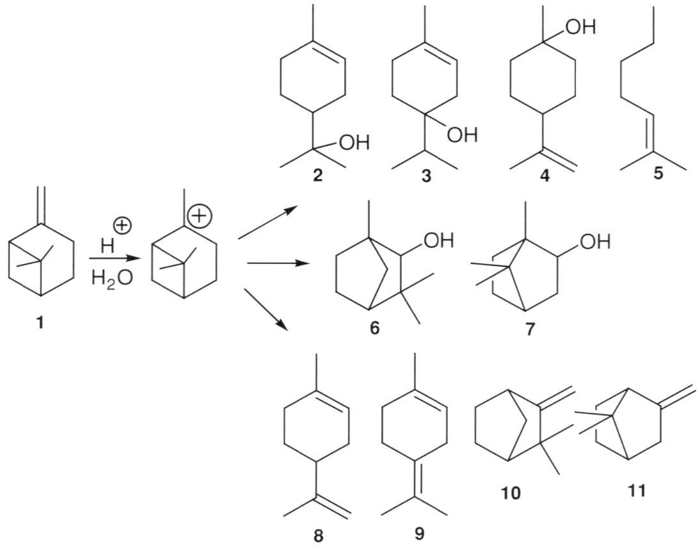

Figure 9.30 β-Pinene acid-catalyzed hydrolysis.

producing hydrocarbons are formed and are difficult to eliminate.[48] Terpineol and the other alcohols are used in the flavor and fragrance industry in perfumes, toiletries, cleaning solutions, and beverages.

Chamblee et al.[49] have reported the reaction of β-pinene with water at 200 and 250°C. The distribution of products obtained was identical to those obtained under aqueous acid-catalyzed conditions (Fig. 9.30). As a consequence, it is assumed that the reaction in hot water proceeded via carbonium ion intermediates. The reaction with β-pinene in water at 200°C is relatively fast with 90% conversion in 20 min. However, the yield of terpineol is small (10%) vs the formation of hydrocarbons, which are the major products formed under these conditions. Reactions run at 250°C show even greater hydrocarbon formation.

9.2.4 Elimination reactions

Although an aqueous environment may not seem to be an appropriate medium for the dehydration of alcohols, such transformation can proceed surprisingly well in NCW. For instance, Khulmann et al.[50] reported that cyclohexanol undergoes complete dehydration at 250–300°C and that the acid-catalyzed conversion is enhanced by the addition of traces of acid (0.02 wt%, 5 mM at rt). Xu and coworkers[51–55] reported a mechanistic studying regarding the dehydration of *tert*-butanol. In NCW at 250°C, *tert*-butanol reacts rapidly to form an equilibrium mixture of *tert*-butanol and isobutene. The rate of reaction can be enhanced by the addition of trace amounts of sulfuric acid. The authors used detailed mathematical modeling to determine the mechanism of dehydration and whether water ($K_w = 10^{-11}$) or *tert*-butanol ($pK_a = 9$ at 250°C) acts as the dominant source of protons. The authors fit the calculated values to the experimental data and concluded that water was not playing a significant role in the acid-catalyzed dehydration. The authors' proposed mechanism for the acid-catalyzed dehydration of *tert*-butanol is provided as Fig. 9.31. An

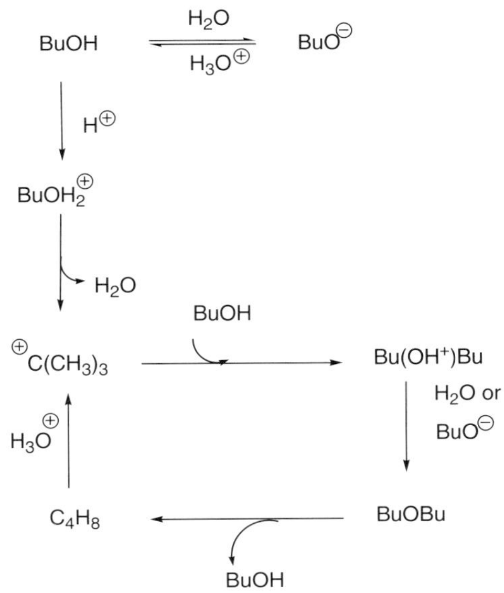

Figure 9.31 Acid-catalyzed reaction of *tert*-butanol (BuOH) in NCW at 250°C.

important detail that emerges from these reports is that the entire reaction time at 250°C was observed to be only 30 s. This has major implications for alkylation reactions in NCW, which have typically utilized alcohols as alkylation reagents and take place over as much as 20 h.[56,57]

Dehydration reactions are important for the synthesis of commodity ethers, such as tetrahydrofuran, which was synthesized from 1,4-butanediol at low yield in NCW.[58] It is also important to note that the reverse reaction can be performed in NCW, although at a greatly reduced yield. For example, the conversion of alkenes to alcohols in NCW has been reported to proceed to $<10\%$ equilibrium conversion.[2]

Decarboxylation reactions are also a common type of elimination in NCW. Both aliphatic and aromatic carboxylic acids will undergo elimination of CO_2 in NCW. Since decarboxylations produce CO_2, the resulting carbonic acid can have an accelerating effect on numerous acid-catalyzed processes. Carlsson et al.[59] studied the conversion of citric and itaconic acids to methacrylic acid, suggesting decarboxylation of acotinic acid yielding methacrylic acid (Fig. 9.32). The authors reported NMR evidence that supports

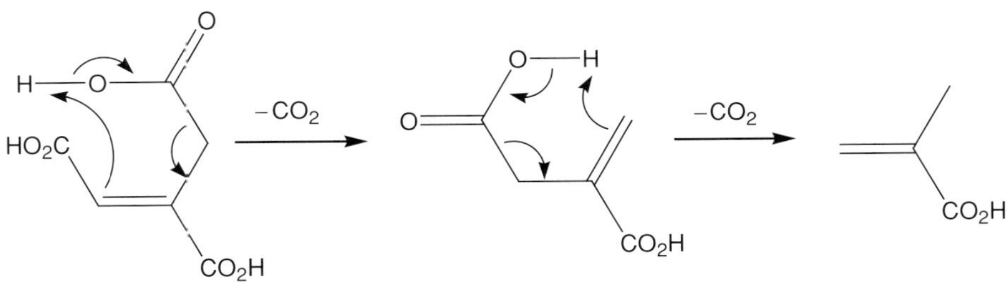

Figure 9.32 Proposed mechanism of the methacrylic acid formation from acotinic acid.

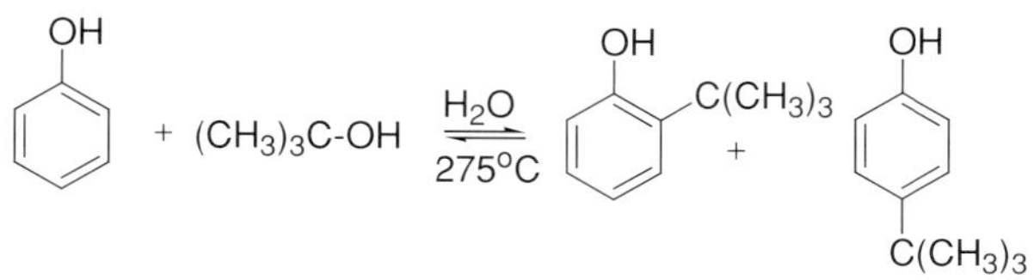

Figure 9.33 Reaction of phenol with *tert*-butyl alcohol in NCW.

the proposed mechanism. Specifically, a carbon-labeled itaconic acid was used to study the decarboxylation pathways and the formation of methacrylic acid. In addition to decarboxylations, which are typically slow, dehydrohalogenations have been reported to occur in SCW and may be possible in NCW.[3]

9.2.5 Friedel–Crafts alkylation reactions

Friedel–Crafts alkylation reactions are useful for attaching carbon functionalities to aromatic ring systems. They are usually conducted in the presence of a Lewis acid such as $AlCl_3$ and BF_3 or protic acids such as H_2SO_4, HF, and H_3PO_4, which must be subsequently neutralized and separated from the product during the isolation procedure. By employing NCW to replace the required acid catalyst, the need for expensive base neutralization, catalyst regeneration, and disposal of salt by-products is eliminated.

The reaction of phenol with *tert*-butanol in water at 250, 275, and 300°C to produce 2-*tert*-butylphenol and 4-*tert*-butylphenol has been reported by Chandler et al. (Fig. 9.33).[56,57] The mole fraction of product yields are shown as a function of time at each temperature in Figs. 9.34–9.36. When the products were subjected to NCW at 275°C, phenol was produced, indicating that the reaction is reversible. Small quantities (5%) of 2,4-di-*tert*-butylphenol are also formed. The reaction kinetics were described using a simple reaction network involving two reversible, first-order reactions (Fig. 9.37).

Xu et al. [55] reported that under neutral conditions in water in the temperature range of 225–320°C, *tert*-butanol undergoes rapid dehydration to form isobutylene. The alkylation

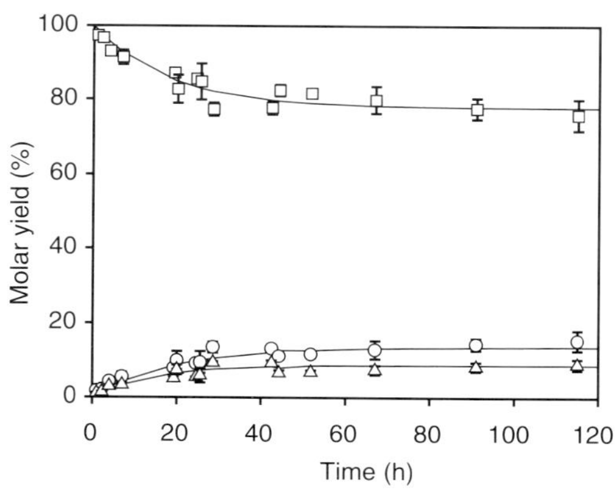

Figure 9.34 Mole fraction product yields as a function of time for the reaction of phenol with *tert*-butyl alcohol in water at 250°C and 172 bar: (□) phenol; (○) 2-*tert*-butylphenol; (△) 4-*tert*-butylphenol.[57]

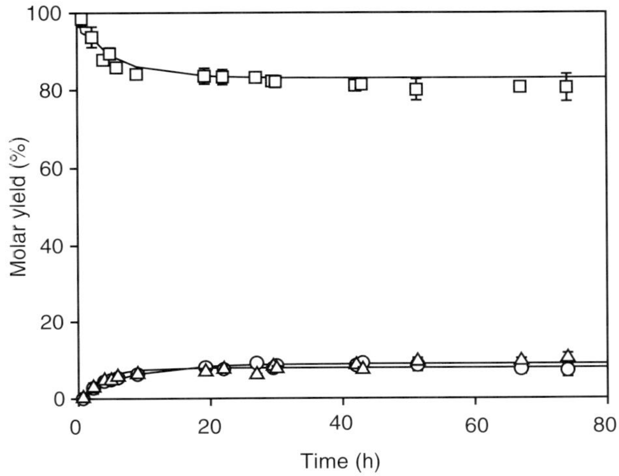

Figure 9.35 Mole fraction product yields as a function of time for the reaction of phenol with *tert*-butyl alcohol in water at 275°C and 172 bar: (□) phenol; (○) 2-*tert*-butylphenol; (△) 4-*tert*-butylphenol.[57]

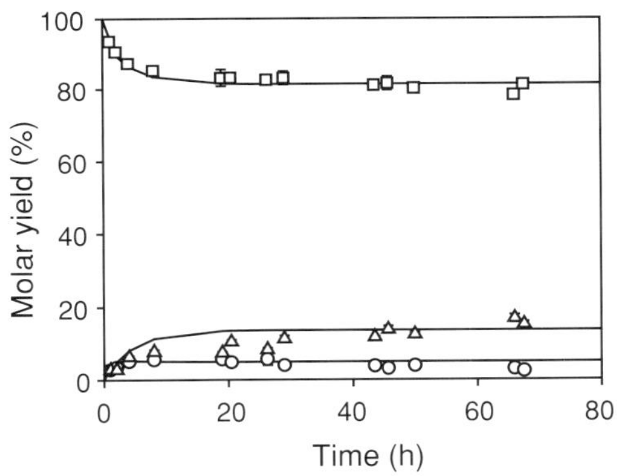

Figure 9.36 Mole fraction product yields as a function of time for the reaction of phenol with *tert*-butyl alcohol in water at 300°C and 172 bar: (□) phenol; (○) 2-*tert*-butylphenol; (△) 4-*tert*-butylphenol.[57]

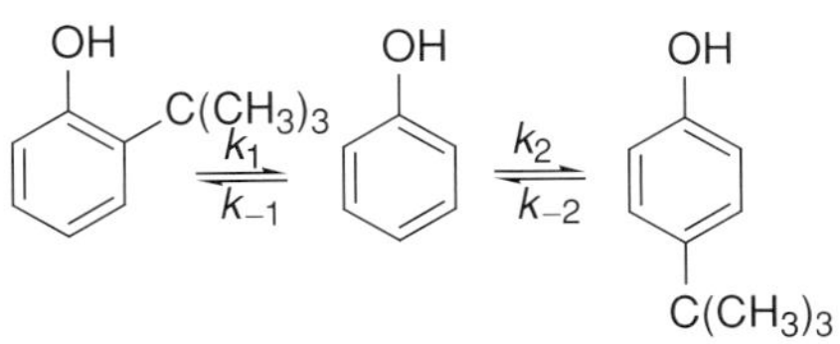

Figure 9.37 Model reaction network for the reversible alkylation reaction of phenol to form 2-*tert*-butylphenol and 4-*tert*-butylphenol in NCW.

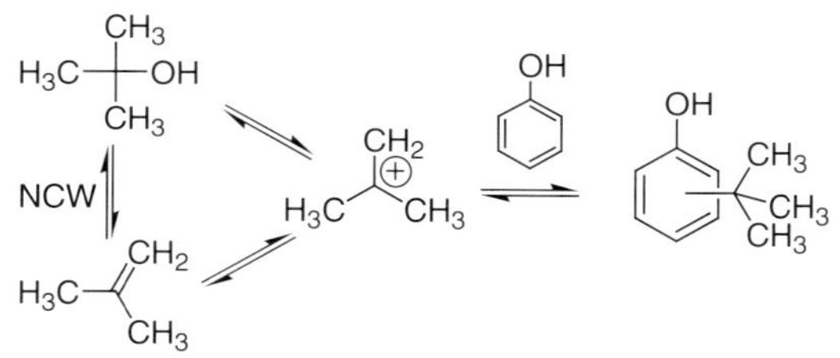

Figure 9.38 Reaction sequence of the Friedel–Crafts alkylation of phenol from *tert*-butanol or isobutylene.

can occur through the incipient tertiary carbonium ion formed from either the *tert*-butanol or the isobutylene (Fig. 9.38). However, because the *tert*-butanol was present in large excess, the reaction of phenol to form both 2-*tert*-butylphenol and 4-*tert*-butylphenol was assumed to be pseudo-first order in phenol. Additionally, 2,4-di-*tert*-butylphenol was produced in very low concentrations and was not included in the proposed reaction network.

Table 9.4 summarizes the pseudo-first-order rate constants for each of the temperatures studied. The activation energies were calculated from Arrhenius plots with values reported in Figs. 9.39 and 9.40. Based on the difference in activation energies, the heat of reaction of the forward reaction of phenol to form 2-*tert*-butylphenol was calculated to be −13 kcal/mol and heat of the forward reaction of phenol to form 4-*tert*-butylphenol was calculated to be +4 kcal/mol. The difference in the heats of reaction between the formation of the ortho-substituted product and the para-substituted product is large, and for comparison, the heats of reaction were calculated from heats of formation of the reactants and products. The calculated heat of reaction for the formation of the ortho-isomer was reported to be −5 kcal/mol. The corresponding heat of formation of the para-isomer was estimated to be approximately the same value; thus, the calculated heat of reaction was not consistent with those determined experimentally. It was emphasized, however, that the calculated values were for a gas-phase reaction at 25°C. The reported reactions, in contrast, were in highly nonideal aqueous solutions at high temperatures. Table 9.4 shows that the rate of formation of 2-*tert*-butylphenol was faster than the rate of formation of 4-*tert*-butylphenol at all reaction temperatures. In addition, the activation energies and heats of reaction revealed that the formation of 2-*tert*-butylphenol was exothermic, and thus, the equilibrium concentration of the 2-isomer actually decreased with increasing temperature. Conversely, the formation of 4-*tert*-butylphenol was found to be endothermic and the equilibrium concentration increased with increasing temperature.

The reaction of *p*-cresol with *tert*-butanol to form 2-*tert*-butyl-4-methylphenol has also been reported[56] (Fig. 9.41) at 250, 275, and 300°C, and the mole fraction product yields are shown as a function of time in Figs. 9.42, 9.43, and 9.44. The product produced *p*-cresol

Table 9.4 Pseudo-first-order rate constants for the alkylation reaction of phenol with *tert*-butyl alcohol in NCW[57]

Temperature (°C)	$k_1 \times 10^5$ (s^{-1})	$k_{-1} \times 10^5$ (s^{-1})	$k_2 \times 10^5$ (s^{-1})	$k_{-2} \times 10^5$ (s^{-1})
250	0.19	1.1	0.17	1.4
275	0.61	6.4	0.36	3.3
300	2.1	35	0.92	5.4

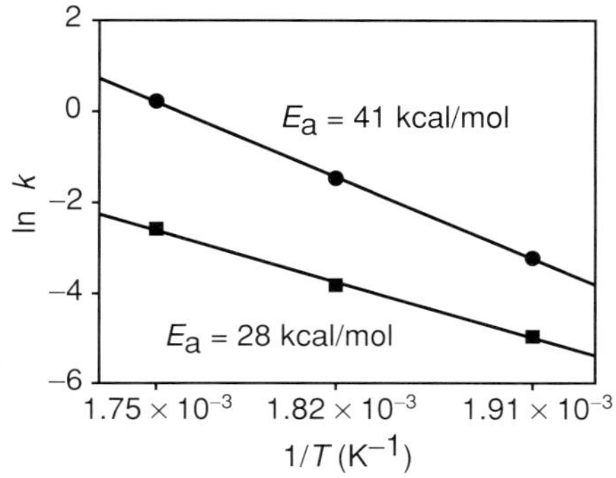

Figure 9.39 Arrhenius plot for the forward and reverse rate constants for the alkylation of phenol to form 2-*tert*-butylphenol in NCW: (■) k_1; (•) k_{-1}^{57}.

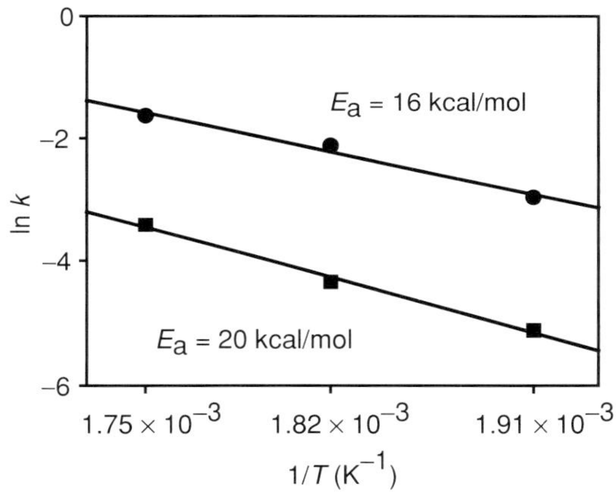

Figure 9.40 Arrhenius plot for the forward and reverse rate constants for the alkylation of phenol to form 4-*tert*-butylphenol in NCW: (■)k_2; (•) k_{-2}^{57}.

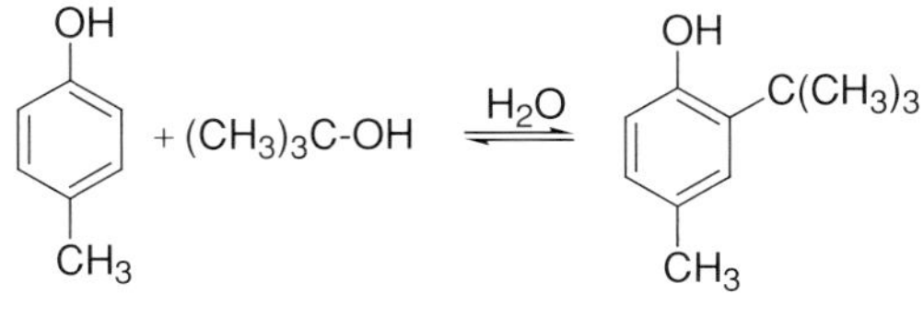

Figure 9.41 Reaction of *p*-cresol with *tert*-butanol in NCW.

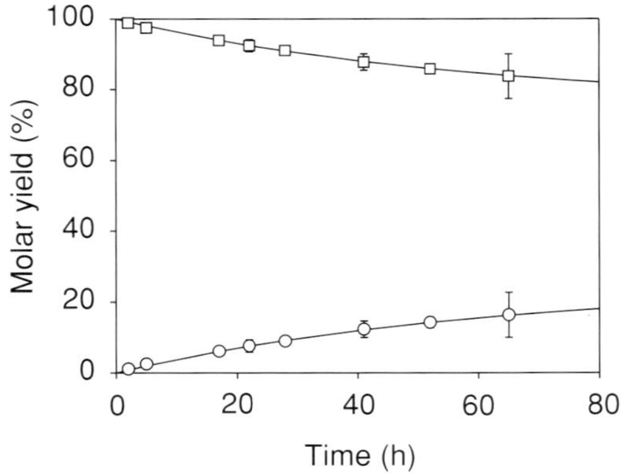

Figure 9.42 Mole fraction product yields as a function of time for the reaction of *p*-cresol with *tert*-butyl alcohol in water at 250°C and 172 bar: (□) *p*-cresol; (○) 2-*tert*-butyl-4-methylphenol.[57]

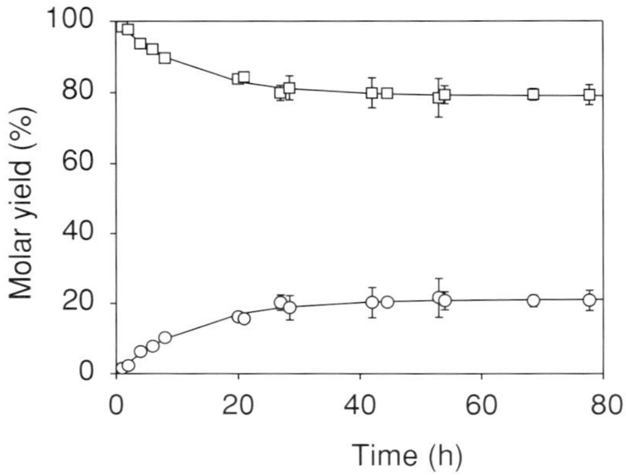

Figure 9.43 Mole fraction product yields as a function of time for the reaction of *p*-cresol with *tert*-butyl alcohol in water at 275°C and 172 bar: (□) *p*-cresol; (○) 2-*tert*-butyl-4-methylphenol.[57]

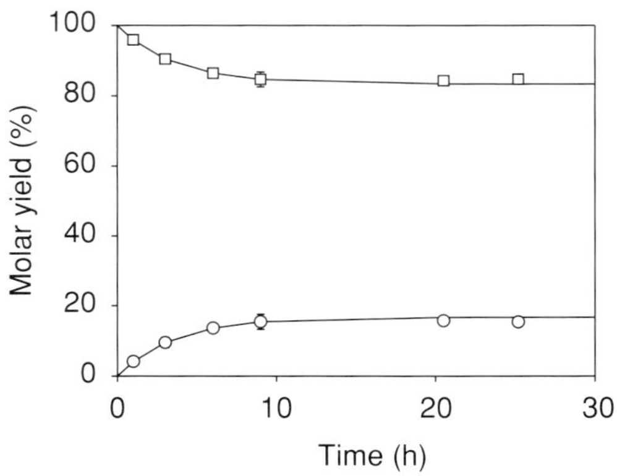

Figure 9.44 Mole fraction product yields as a function of time for the reaction of *p*-cresol with *tert*-butyl alcohol in water at 300°C and 172 bar: (□) *p*-cresol; (○) 2-*tert*-butyl-4-methylphenol.[57]

when subjected to NCW at 275°C, indicating that the reaction was reversible. The pseudo-first-order rate constants, summarized in Table 9.5, increased with increasing temperature and the effect of temperature on the forward and reverse rate constants along with the activation energies are shown in Fig. 9.45. For the forward reaction, the heat of reaction was calculated to be –6 kcal/mol. The corresponding value calculated from heats of formation was –3 kcal/mol. As in the previous case, this latter value is for a gas-phase reaction at 25°C. Finally, the reaction of *tert*-butanol with *p*-cresol reached equilibrium at approximately 1 h, which is much faster than the corresponding reaction with phenol.

The reactions of phenol with isopropyl alcohol and *n*-propyl alcohol occurred slowly,[56] and the major products with isopropyl alcohol are 2-isopropylphenol and 2,6-diisopropylphenol. The mole fraction yield as a function of time is shown in Fig. 9.46.

These results may be contrasted to the reaction of phenol with *tert*-butyl alcohol where the second *tert*-butyl group went to the 4-position of the phenol, probably due to steric effects. In the case of isopropyl alcohol, steric effects are less severe and the second isopropyl group went to the 6-position. Reaction of phenol with 1-propanol proceeded even more slowly than the isopropyl alcohol. The yield of products was less than 5% yield over a time period of 144 h, with the major product 2-isopropylphenol plus very small amounts of 2,6-diisopropylphenol and 2-*n*-propylphenol. Clearly, the incipient primary carbonium ion rearranged to the more stable secondary carbonium ion.

In somewhat related work, Siskin et al.[60] reported that 4-hydroxyphenyl benzyl ether undergoes complete conversion in decalin and water at 343°C (Fig. 9.47) during 2 days to give toluene (63 and 44%), hydroquinone (32 and 35%), and 2-benzylhydroquinone (4 and 12%). It was postulated that the intermediate benzyl alcohol was formed, which subsequently reacted with hydroquinone to produce the 2-benzylhydroquinone. The benzyl carbonium ion was proposed as the reactive species in the aromatic electrophilic substitution reaction.

9.2.6 Friedel–Crafts acylation reactions

Like Friedel–Crafts alkylation reactions, the corresponding acylation reactions require the presence of Lewis or Brønsted acids. Unlike the alkylation process, acylations usually require more than stoichiometric quantities of the acid and the electrophilic reaction partner is usually an expensive and hydrolytically unstable acid chloride.[61] As a consequence, these Lewis and Brønsted acids require neutralization and subsequent expensive disposal. For instance, the conventional acylation of phenol with acetyl chloride consumes more than stoichiometric quantities of $AlCl_3$. Neutralizing this $AlCl_3$ requires land-filling several pounds of

Table 9.5 Pseudo-first-order rate constants for the alkylation reaction of *p*-cresol with *tert*-butyl alcohol in NCW[57]

Temperature (°C)	$k_1 \times 10^5$ (s^{-1})	$k_{-1} \times 10^5$ (s^{-1})
250	0.11	0.35
275	0.48	1.8
300	1.3	6.3

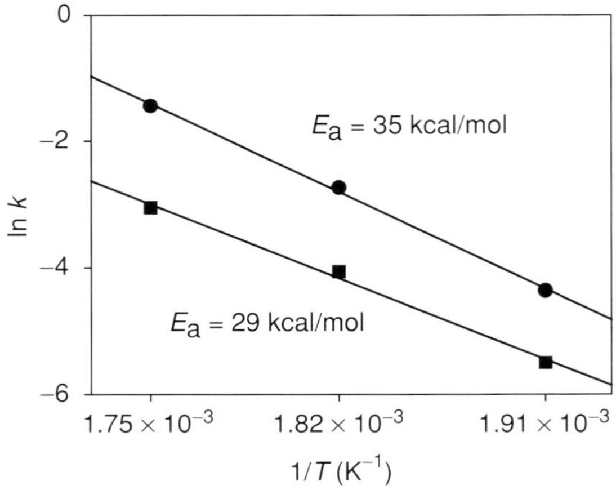

Figure 9.45 Arrhenius plot for the forward and reverse rate constants for the alkylation of *p*-cresol to form 2-*tert*-butyl-4-methylphenol in NCW: (■) k_1; (•) k_{-1}^{57}.

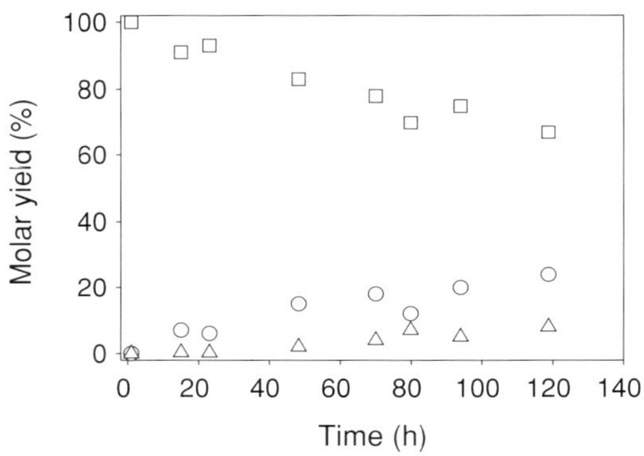

Figure 9.46 Mole fraction product yields as a function of time for the reaction of phenol with isopropyl alcohol in water at 275° C.•: phenol. ■: 2-isopropylphenol. Δ: 2,6-diisopropylphenol.[56]

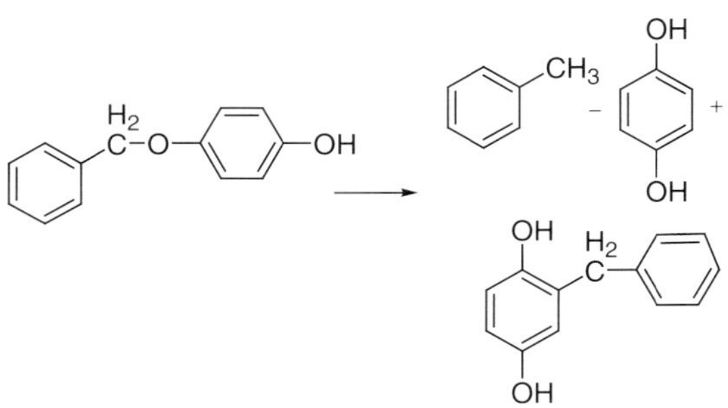

Figure 9.47 Conversion of 4-hydroxyphenyl benzyl ether in decalin/water at 343°C.

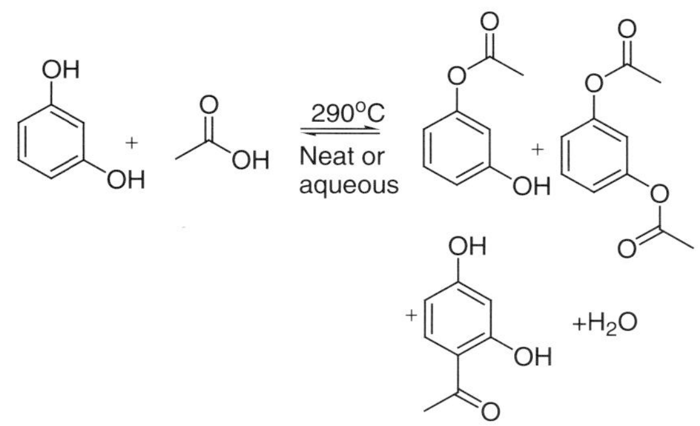

Figure 9.48 Acetylation of phenol and resorcinol in NCW and neat acetic acid.

$Al(OH)_3$ for every pound of product produced.[62] In addition, many of these acylation reactions require polar organic solvents, such as methylene chloride or nitrobenzene, that can dissolve simultaneously the reactant as well as the reactant/catalyst complex.[61,63]

Brown et al.[64] have reported that phenol and resorcinol have been successfully acetylated using acetic acid in NCW and in neat acetic acid in the temperature range of 250–300°C without added catalyst (Fig. 9.48). As is the case for the alkylation reaction discussed previously, the acetylations of phenol and resorcinol are also reversible at high temperatures in the presence of water, but with less favorable equilibrium yields of desired acetylated product. In aqueous acetic acid at 290°C, phenol was primarily converted to 2-hydroxyacetophenone, 4-hydroxyacetophenone, and phenyl acetate in approximately equal amounts, with a combined equilibrium yield of less than 1%. Under the same conditions, resorcinol was converted to primarily 2,4-dihydroxyacetophenone with an equilibrium yield of 4% (Fig. 9.49). Acetylations are generally irreversible with traditional Friedel–Crafts catalysts due to the complexation of the acid catalyst with the carbonyl oxygen of the product.

When NCW is used as the medium for the reaction, however, the products are not stabilized by complexation with catalyst and are free to revert to the starting material. To determine the effect of water on the equilibrium limitation of these reactions, the stability

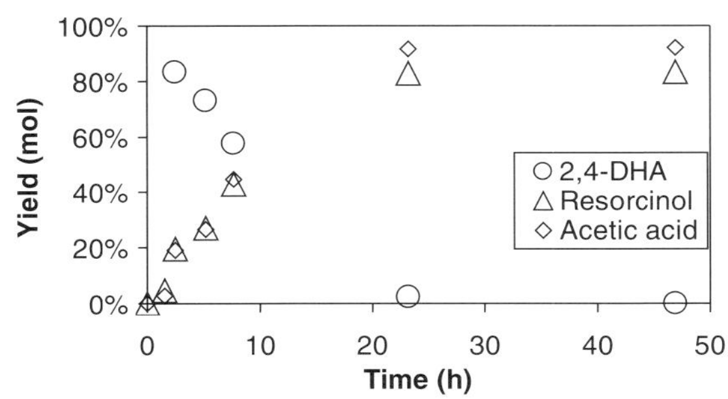

Figure 9.49 Back reaction of 2,4-dihydroxyacetophenone (2,4-DHA,○) to resorcinol (△), and acetic acid (◇) using water as a solvent at 250°C.[64]

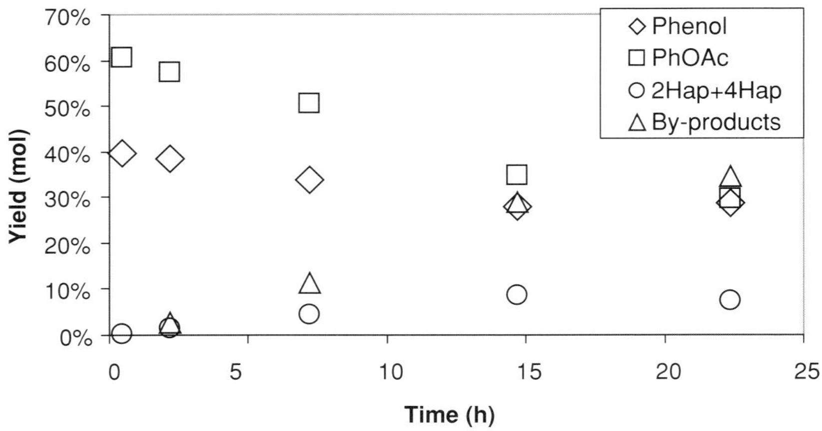

Figure 9.50 Reaction of phenol (◇) and acetic acid to phenyl acetate (PhOAc, □), 2′-hydroxyacetophenone + 4′-hydroxyacetophenone (2Hap + 4Hap, ○), and by-products (△) at 290°C.[64]

of the products was checked in water at nearcritical conditions. The product of the forward reaction of resorcinol and acetic acid, 2,4-dihydroxyacetophenone, was placed in liquid water at 250°C. The concentration vs time data of the almost complete conversion of 2,4-dihydroxyacetophenone back to the corresponding amounts of resorcinol and acetic acid are shown in Fig. 9.49.

The products of phenol acetylation, 2-hydroxyacetophenone and 4-hydroxyacetophenone, were also converted to phenol and acetic acid when high-temperature water was used as the solvent.

The tunable solvent properties of NCW allow for facile separation of products upon cooling, but excess water imposes a too severe equilibrium limitation on this reaction. To address the equilibrium limitation in the presence of excess water, the acetylations of phenol and resorcinol were run in neat acetic acid. At 290°C, phenol was converted to 2-hydroxyacetophenone, 4-hydroxyacetophenone, and phenyl acetate with a combined equilibrium yield of 8 mol% (Fig. 9.50). By-products included 2-methylchromone and 4-methylcoumarine. Under the same conditions, resorcinol was successfully converted to primarily 2,4-dihydroxyacetophenone, with an equilibrium yield of more than 50% in less than 12 h. The yield of 2,4-dihydroxyacetophenone vs time is shown in Fig. 9.51.

Attempts at the intramolecular cyclization of 2-benzoylbenzoic acid in water at 275°C proved to be unsuccessful. This result should not be surprising since the electrophile must successfully attack an electron-deficient ring.

9.2.7 Condensation reactions

A variety of condensation reactions have been explored in NCW in order to elucidate the catalytic behavior of this medium. Nolen et al.[65] have reported the results of the Claisen–Schmidt condensation between benzaldehyde and butanone to form α,β-unsaturated ketones. Due to the asymmetry of the butanone with respect to the carbonyl group, two products are formed (Fig. 9.52). It has been reported that under traditional acidic

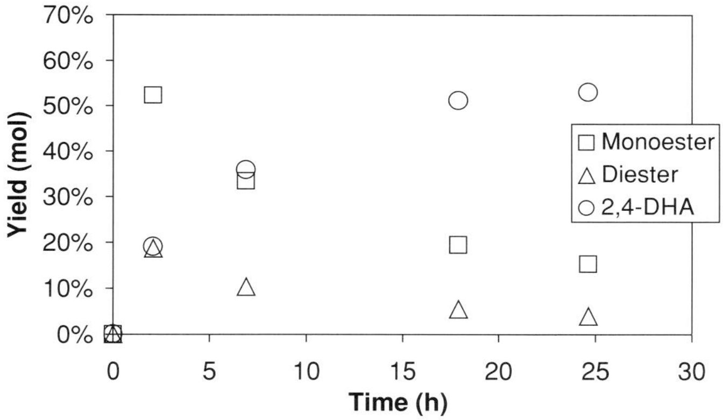

Figure 9.51 Reaction of resorcinol and acetic acid to 2,4-dihydroxyacetophenone (2,4-DHA,○), resorcinol monoacetate (□), and resorcinol diacetate (△) at 290°C.[64]

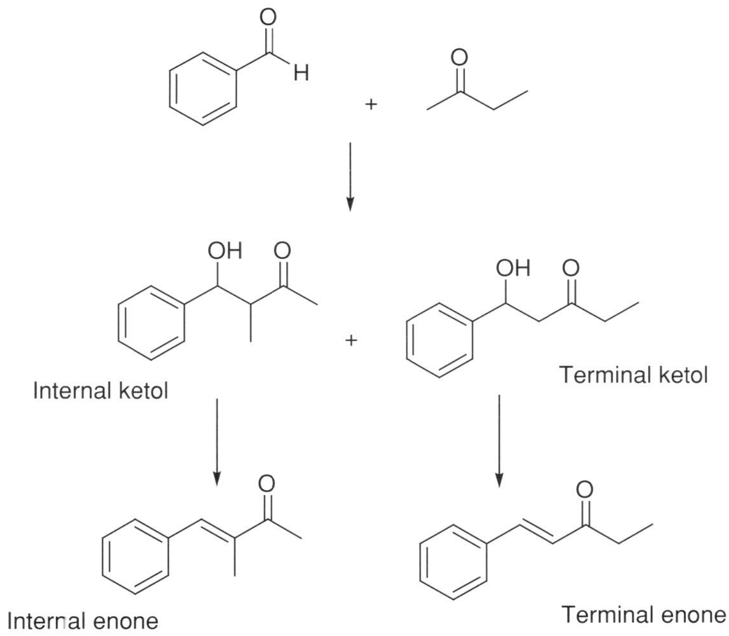

Figure 9.52 Claisen–Schmidt reaction of benzaldehyde with 2-butanone showing the two possible condensation products 4-phenyl-3-methyl-3-buten-2-one (internal enone) and 1-phenyl-1-penten-3-one (terminal enone), as well as their precursor intermediate ketols.

Figure 9.53 Internal and terminal enols of 2-butanone.

conditions the terminal enone condensation product is selectively formed, while under basic conditions the internal enone condensation product is selectively formed.[66] Although it is possible for both products to be formed via acid or base catalysis, a single dominant species is favored for each set of conditions.[67] Stiles et al.[68] verified this through a series of investigations conducted at ambient temperatures in which the ketol intermediates were subjected to conventional acidic and basic conditions to determine if the ketols dehydrate to form their corresponding unsaturated products or are cleaved to the starting materials. Under basic conditions, the internal ketol was unstable and decomposed to form the starting material, while the terminal ketol was dehydrated to form a single product, the terminal enone product (Fig. 9.53). Under acidic conditions, both ketols resulted in the formation of their corresponding α,β-unsaturated carbonyl products; however, with an acidic environment, the internal enol intermediate of butanone was more stable than the terminal intermediate, resulting in the preferential formation of the internal ketol and therefore internal enone product. The only deviation from this trend was observed under very strongly acidic or basic conditions where both α,β-unsaturated carbonyl products were formed. Even under these extreme conditions, the product derived from the more stable enol for a given set of conditions was favored by as much as 9 to 1 over the product derived from the less stable enol.[69,70]

The Claisen–Schmidt reaction between benzaldehyde and butanone was conducted in NCW over a temperature range of 250–300°C without the addition of added acid or base.[65] A 10-fold molar excess of 2-butanone to benzaldehyde was used in order to minimize the formation of higher adducts.[67] The results at temperature of 250, 275, and 300°C are shown in Figs. 9.54, 9.55, and 9.56, respectively.

Both 4-phenyl-3-methyl-3-buten-2-one and 1-phenyl-1-penten-3-one were formed. Secondary adducts resulting from the addition of another benzaldehyde or butanone molecule to the primary adducts were also detected. The formation of 1-phenyl-1-penten-3-one was dominant over the formation of 4-phenyl-3-methyl-3-buten-2-one. A comparison between the selectivities of the internal and terminal enone products is shown in Fig. 9.57 for the three temperatures. The selectivity of the terminal enone product decreases with increasing temperature, while the selectivity of the internal enone product remains relatively constant. In order to determine if equilibrium limitations exist for the Claisen–Schmidt condensation, the stability of 4-phenyl-3-methyl-3-buten-2-one was investigated in NCW at 250°C over a period of 30 h. The results indicated that small quantities of benzaldehyde and butanone were formed along with a large concentration of other unidentified products resulting from the self-condensation of the 4-phenyl-3-methyl-3-buten-3-one. Of significance is the fact that no 1-phenyl-1-penten-3-one was detected during the investigation of the back reaction, indicating that isomerization did not occur between internal and terminal enone products. It is interesting to note that when the condensation was carried out at 250°C in the presence of HCl the yield of 4-phenyl-3-methyl-3-buten-2-one was 30% after 15 min in comparison to 0% in the absence of HCl during the same period of time. There was also a pronounced

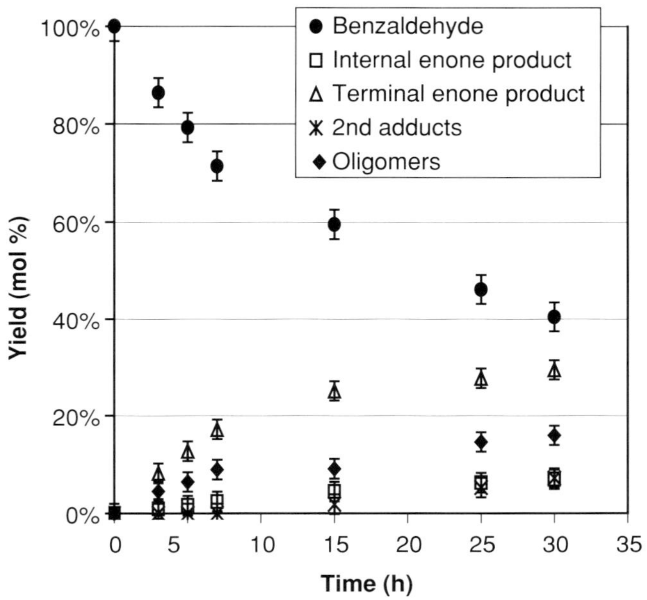

Figure 9.54 Production of 1-phenyl-1-penten-3-one (terminal enone product) and 4-phenyl-3-methyl-3-buten-2-one (internal enone product) from the Claisen–Schmidt condensation of benzaldehyde with 2-butanone as a function of time at 250°C. Formation of secondary condensation products and higher addition products (oligomers) is also shown. The reactant molar ratio (benzaldehyde/2-butanone/water) was 1:10:55.[65]

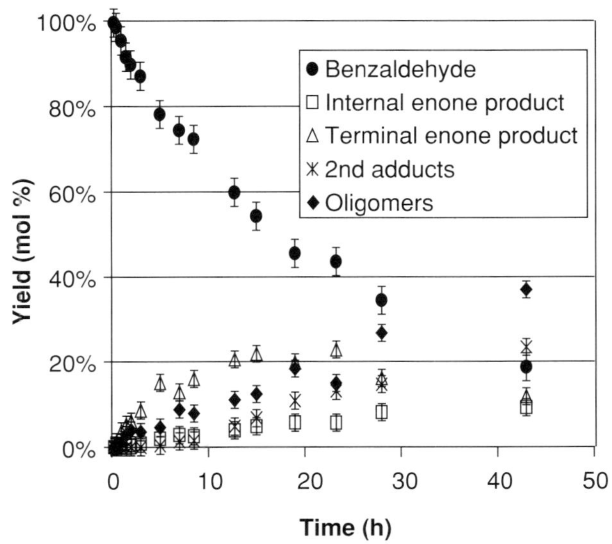

Figure 9.55 Production of 1-phenyl-1-penten-3-one (terminal enone product) and 4-phenyl-3-methyl-3-buten-2-one (internal enone product) from the Claisen–Schmidt condensation of benzaldehyde with 2-butanone as a function of time at 275°C. Formation of secondary condensation products and higher addition products (oligomers) is also shown. The reactant molar ratio (benzaldehyde/2-butanone/water) was 1:10:55.[65]

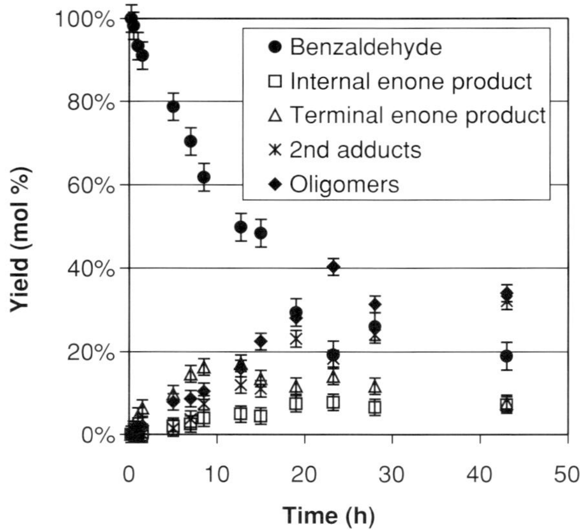

Figure 9.56 Production of 1-phenyl-1-penten-3-one (terminal enone product) and 4-phenyl-3-methyl-3-buten-2-one (internal enone product) from the Claisen–Schmidt condensation of benzaldehyde with 2-butanone as a function of time at 300°C. Formation of secondary condensation products and higher addition products (oligomers) is also shown. The reactant molar ratio (benzaldehyde/2-butanone/water) was 1:10:55.[65]

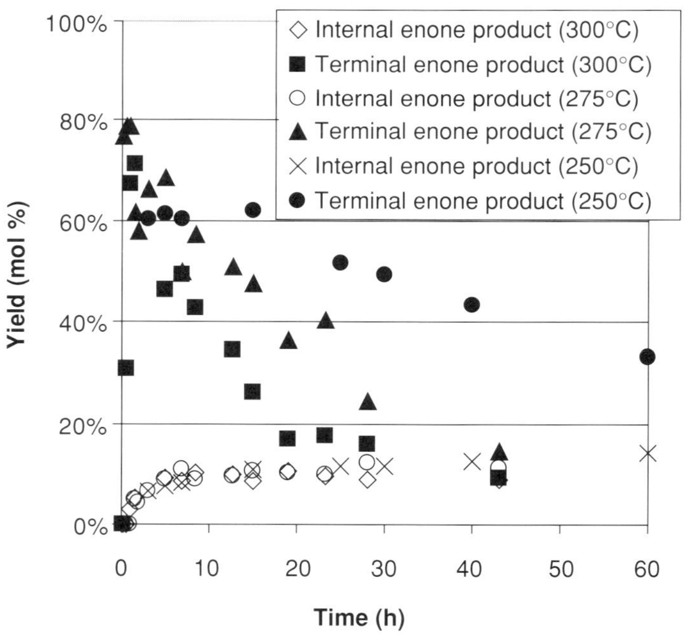

Figure 9.57 Selectivity comparison between the internal and terminal enone products at 250, 275, and 300°C. The reactant molar ratio (benzaldehyde/2-butanone/water) was 1:10:55.[65]

Table 9.6 Rate constants for the formation of 1-phenyl-1-penten-3-one (terminal enone product, k_1) and 4-phenyl-3-methyl-3-buten-2-one (internal enone product, k_2) from the Claisen–Schmidt condensation of benzaldehyde with 2-butanone[65]

T (°C)	k_1' (s^{-1})	k_1' (L/(mol s))	k_2' (s^{-1})	k_1 (L/(mol s))
250	190.15	38.03	21.26	4.25
275	219.29	43.86	28.36	5.67
300	299.94	59.99	34.72	6.94

increase in the concentration of 1-phenyl-1-penten-3-one from 1 to 10% in the absence and in the presence of HCl, respectively. The reaction kinetics for the condensation of benzaldehyde with butanone are described using first-order rate models for the formation of each of the primary condensation products. k_1' is the pseudo-first-order rate constant for the formation of 1-phenyl-1-penten-3-one, k_2' is the pseudo-first-order rate constant for the formation of 4-phenyl-3-methyl-3-buten-2-one and k_1 and k_2 are the corresponding second-order rate constants. The results are summarized in Table 9.6. The activation energies were calculated to be 5.40 kcal/mol for the formation of 1-phenol-1-penten-3-one and 5.9 kcal/mol for the formation of 4-phenyl-3-methyl-3-buten-2-one. The activation energy for the formation of 1-phenyl-1-penten-3-one is similar to the value reported by Gettler and Hammett ($E_a = 4.8$ kcal/mol) for the reaction conducted in basic aqueous dioxane over a temperature range of 25–50°C.[67]

Nolen et al.[65] also reported the self-condensation reaction of butyraldehyde and the cross-aldol condensation of benzaldehyde with acetone (Figs. 9.58 and 9.59) at 250°C. The butyraldehyde self-condensation produced a number of products, including 2-ethyl-2-hexenal, 2-butyl-2-butenal, and 2-ethylhexanal. The results from the condensation of butyraldehyde indicate that a 40% yield of 2-ethyl-2-hexenal is achieved before the formation of by-products becomes dominant. In addition, investigations of the back reaction show that a substantial quantity of butyraldehyde is formed when 2-ethyl-2-hexenal is subjected to water at 250°C. The condensation reaction of benzaldehyde with acetone produced a 15% yield of *trans*-4-phenyl-3-buten-2-one in 5 h and very small quantities of *trans,trans*-dibenzylidene acetone during this same period of time. The authors suggest that the low yield could be a result of equilibrium limitations.

9.2.8 Rearrangements

The rearrangement of pinacol to pinacolone has been reported in NCW by Khulmann et al.[71] (Fig. 9.60). The rearrangement took place in 60 min at 275°C with negligible alkene formation. In contrast to the use of NCW, classical methods used to promote the rearrangement required boiling 25% sulfuric acid for 3 h.[72]

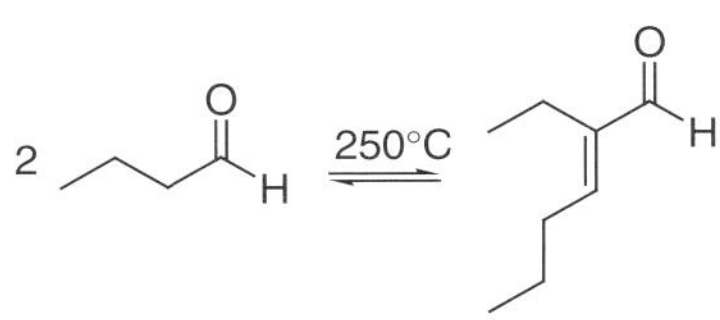

Figure 9.58 Self-condensation of butyraldehyde.

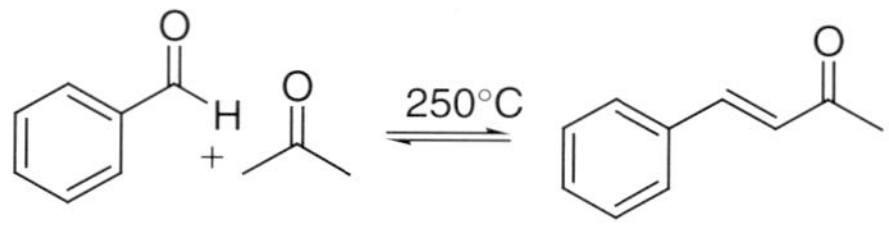

Figure 9.59 Cross-aldol condensation of benzaldehyde with acetone.

The Claisen rearrangement of allyl phenyl ether in NCW was first reported by An et al.[73] (Fig. 9.61). At 200 and 240°C for a period of 10 min, the conversion to 2-allylphenol increased significantly, 10 and 84% respectively. At higher temperatures (245 and 250°C) and longer reaction time (60 min), an array of products appeared. These products included phenol, 2-(2-hydroxyprop-1-yl)-phenol, and 2-methyl-2,3-dihydrobenzofuran. The results are summarized in Table 9.7. 2-Methyl-2,3-dihydrobenzofuran was identified as the thermodynamic product. Figure 9.62 summarizes the proposed mechanistic pathways to each of the products.[74]

9.2.9 Hydrogen/deuterium exchange

Kuhlmann et al.[50] reported the reaction of 2,5-dimethylfuran in deuterium oxide, including extensive deuteration of the methyl and methylene groups of the product, 2,5-hexanedione (Fig. 9.63). When 2,5-hexanedione was subjected to the same conditions, extensive deuteration again resulted. Therefore, it was concluded that deuterium exchange took place after the formation of the dione. Table 9.8 lists the sites and degree of deuteration observed for a variety of ketones at temperature from 200 to 300°C for 30–60 min. The ionic mechanism for the deuterium exchange of a specific example, pinacolone, is illustrated in Fig. 9.64.[50] Deuterium exchange takes place readily and in nearly quantitative fashion in the α and α' (where applicable) positions in ketones. The authors highlighted that the operating enol-keto tautomerism led exclusively to hydrogen exchange with no aldol products. A detailed study of the deuterium exchange of pinacolone showed that the reaction is pseudo-first order (Fig. 9.65). At 225°C, the reported rate constant was $(8.9 \pm 0.5) \times 10^{-3}\ \text{min}^{-1}$.

In contrast to ketones, deuterium oxide treatment of methyl (300°C, 93 h), isopropyl (200°C, 30 min), and neopentyl alcohol (300°C, 60 min) did not result in deuterium exchange of C—H hydrogens. The same negative results were reported with ethylene glycol (300°C, 60 min) and pentaerythrol (250°C, 60 min).[50] Deuterium exchange of hydrocarbons such as toluene and triphenylmethane under similar reaction conditions was not observed. Interestingly, Streitwieser et al.[75] noted that the limit for deuterium exchange in water at temperature of 200–300°C and reaction time less than 2 h appears to be for substrates whose pK_a is below 40.

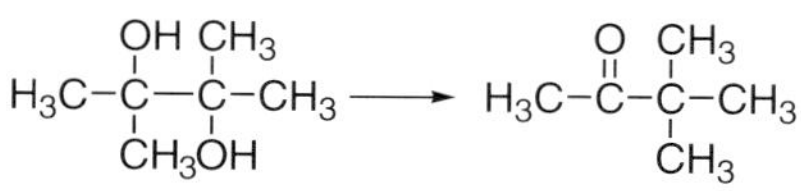

Figure 9.60 Pinacol rearrangement to form pinacolone.

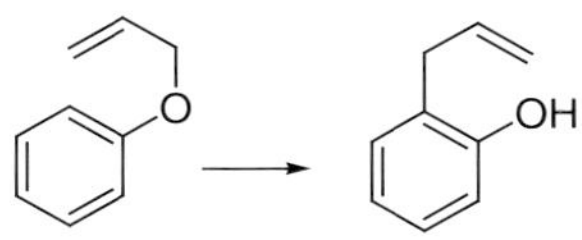

Figure 9.61 Claisen rearrangement of the allyl phenyl ether to the 2-allylphenol.

Table 9.7 Claisen rearrangement of allylphenyl ether[73]

Starting material	Temperature (°C)	Time (min)	Major products (ratio)	Yield (%)
Allyl phenyl ether	240	10	2-allylphenol	84
Allyl phenyl ether	245	60	2-allylphenol/ 2-(2-hydroxyprop-1-yl)-phenol/ 2-methyl-2,3-dihydrobenzofuran (3:2:1.5)	72
Allyl phenyl ether	250	60	2-methyl-2,3-dihydrobenzofuran	77

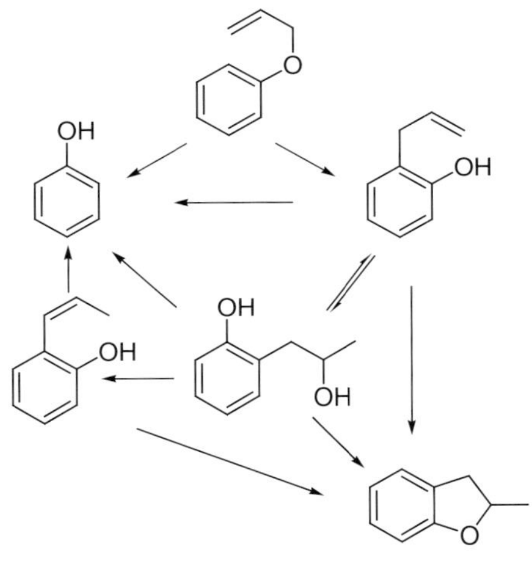

Figure 9.62 Reaction pathways of allyl phenyl ether in NCW.

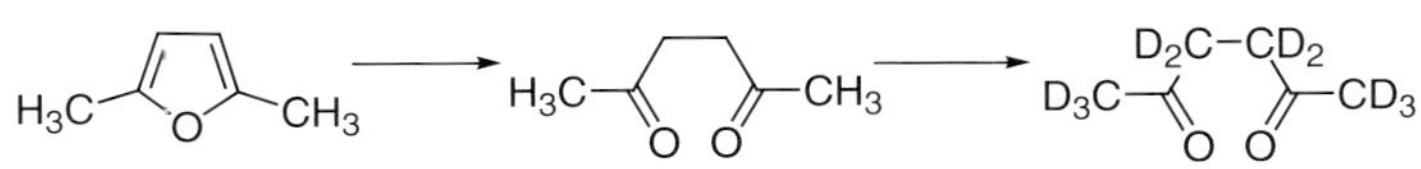

Figure 9.63 Ring cleavage followed by H/D exchange of 2,5-dimethylfuran.

Table 9.8 Hydrogen/deuterium exchange in ketones[50]

Compound	% D (position)	Temperature (°C)	Time (min)
Pinacolone[a]	100 (α, CH_3)	277	60
Acetone	97 (α, α', CH_3)	200	60
Cyclopentanone[b]	100 (α, α', CH_2)	225	30
1,4-Cyclohexanedione	100 (α, α', CH_2)	225	30
Acetophenone	>88 (α, CH_3)	250	60
Deoxybenzoin	99 (α, CH_2)	250	30

[a] Exchange observed in the rearrangement product of pinacol.
[b] Exchange observed in the hydrolysis product of the corresponding ethylene ketal.

9.2.10 General acid/base reactions

In order to understand the catalytic capabilities of NCW, it is important to know the underlying mechanisms relevant to these capabilities. The increased dissociation constant of liquid water opens up a plethora of opportunities for acid and base catalysis. Recent efforts taken by Savage[76] have attempted to uncover whether this is general or specific acid and base catalysis through a variety of model reaction studies.[77] This was done by exploring the effect of changing pH on reactions in NCW. In the cleavage of bisphenol A to form p-isopropenylphenol in NCW (Fig. 9.66), the pH was varied by adding small amounts of H_2SO_4, HCl or NaOH to adjust the pH from the natural value in water at 250°C, pH 5.7. By using a literature model for the effects of temperature on the dissociation of acids and bases,[78,79] the effective pH at 250°C could be estimated. Figure 9.67 demonstrates the existence of a minimum in the pseudo-first-order rate constant as a function of system pH, with a broad, flat region in the range of $3.5 < \text{pH} < 7.5$. This result indicates that the bisphenol A cleavage is due to specific acid catalysis at low pH and specific base catalysis at high pH in NCW – results confirmed by modeling efforts in the same paper. This leaves the flat region of pH-independent behavior (Fig. 9.67) to be determined. To explain the behavior in this region, the authors modeled the pseudo-first-order rate constant using

$$k' = k_A\left[H_3O^+\right] + k_B\left[OH^-\right] + k_W$$

where k_A and k_B are the second-order-rate constants for specific acid and specific base catalysis, respectively, and k_W is the pseudo-first-order rate constant resulting from general acid or general base catalysis due to water. The values giving the best fit to the data were $k_A = 0.094 \pm 0.053$ L/(mol s), $k_B = 0.19 \pm 0.12$ L/(mol s), and $k_W = 2.0 \times 10^{-4} \pm 6.8 \times 10^{-5}\ s^{-1}$. The high value of k_W indicates that neat (neutral pH) NCW mainly performs general acid or general base catalysis due to water. The relatively higher value of k_B compared to k_A was concluded to reveal general base catalysis as dominant under neutral pH conditions.

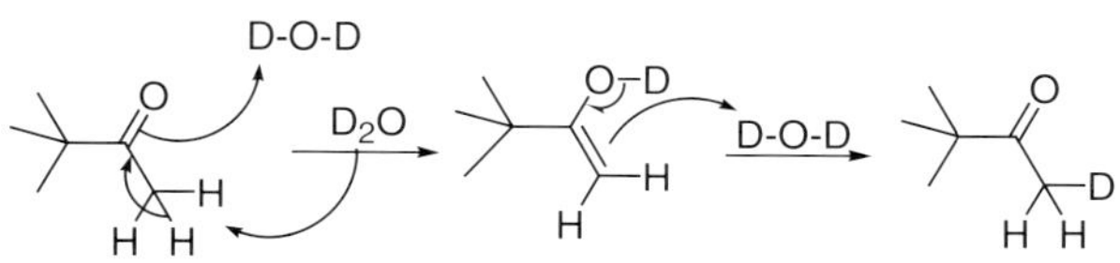

Figure 9.64 Mechanism for deuterium exchange for pinacolone.

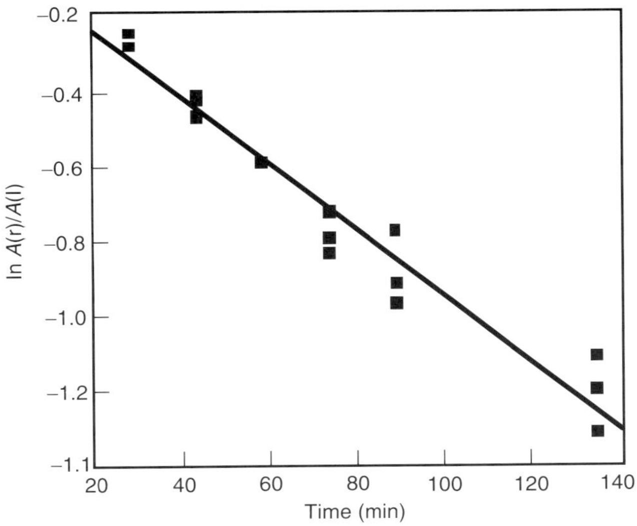

Figure 9.65 Pinacolone deuterium exchange at 0.5 M in NCW (225°C). Reprinted with permission from Ref. 50. Copyright 1994 American Chemical Society.

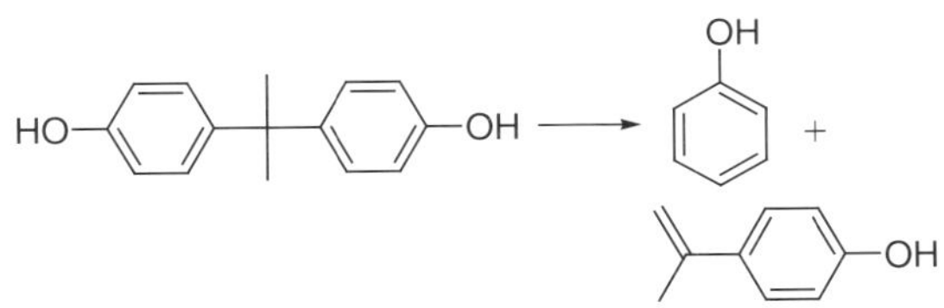

Figure 9.66 Reaction of bisphenol A in NCW to form *p*-isopropenylphenol and phenol.

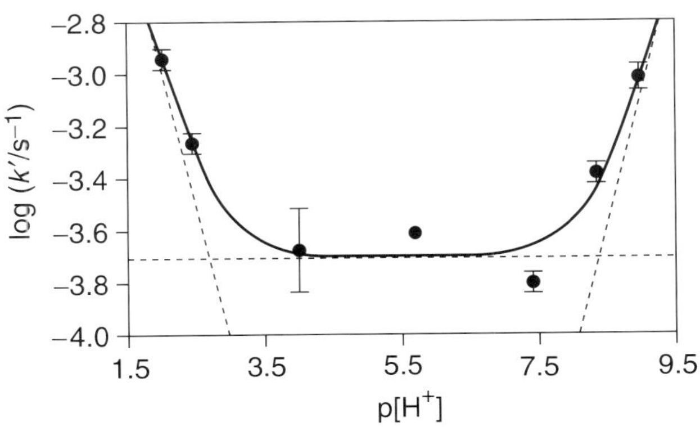

Figure 9.67 Effect of added acid and base on pseudo-first-order rate constant for bisphenol A disappearance in NCW at 250°C.[77] Reprinted with permission from Ref. 77. Copyright 2004 American Chemical Society.

Supporting evidence for this conclusion was obtained by comparing the cleavage of bisphenol A with that of *p*-cumylphenol in neutral NCW. *p*-Cumylphenol reacts in the presence of acid to give α-methylstyrene, the analogue of *p*-isopropenylphenol, as well as phenol. In the presence of base, however, *p*-cumylphenol is stable up to 300°C. The lack of *p*-cumylphenol reactivity in neutral NCW supports the conclusion that general base catalysis is the dominant mechanism for this reaction.

In other reactions, such as aldol condensations,[80] Savage concluded that specific acid or base catalysis was dominant at near-neutral pH. The variance of these results for different reactions illuminates the versatility of NCW as a reaction medium.

9.3 Reactions in high-temperature water enriched with CO_2

It has already been mentioned that the K_w in NCW is several orders of magnitude greater than in water at room temperature. Thus, as shown previously, acid and base catalysis can be facilitated without the use of additional acid. Certainly CO_2 reacts with water to form carbonic acid and, as a consequence, the concentration of hydronium ion in NCW can be increased by enriching the medium with CO_2. From an environmental point of view this procedure will not only facilitate specific acid-catalyzed reactions but will not require neutralization of the acid after the reaction is complete. A simple cooling and depressurization will eliminate the CO_2 and phase separates the product(s) of reaction. Thus, Aleman et al.[81] have reported that the conversion of mesitoic acid to mesitylene over a period of 120 min at 250°C increased from 50 to 80% in the presence of 10 bar (rt) of CO_2. Hunter and Savage[82] reported the dehydration of cyclohexanol in water at 250 and 275°C and the reaction of *p*-cresol with *tert*-butanol in water at 275°C in the absence and presence of CO_2. Their results indicated that in the presence of CO_2 the rate of dehydration of the cyclohexanol increased by more than a factor of 2 and the rate of formation of 2-*tert*-butyl-4-methylphenol increased 40–120%. Modest increases in rate were reported for the hydration of cyclohexene to cyclohexanol.

9.4 Limitations and safety

Although NCW provides a number of benefits over traditional chemical processes, there are limitations. Many reactions produce water as a by-product, such as the Friedel–Crafts reactions, and may be equilibrium-limited in an NCW system due to the extreme amount of water present. It may be possible to use a temperature between the solubility of the reactants and products so that the products fall out of solution as they are formed, which would help drive the reaction to completion.

The processing conditions for an NCW operation are also different from most traditional syntheses, resulting in the need for more robust equipment. This equipment needs to be capable of resisting corrosion (due to water dissociation) and temperatures and pressures up to 350°C and 10 MPa. Fortunately, equipment for the generation and handling of steam is already widely available, but corrosion is basically an electrochemical process which will proceed far better in even dilated liquid water than in a compressed gas. The presence of O_2 can augment corrosion, and the presence of halides, especially chloride, gives an even greater enhancement of corrosion. Thus the cost for NCW equipment will be higher than the equipment designed for a process operated at ambient conditions but is balanced by the reduced separation cost, fewer steps, and lesser waste stream.

A related example of these offsetting effects is the successful implementation of SCW oxidation technology, utilizing titanium-clad reactors in a cost-effective manner. For SCW oxidation, the added benefit of using the exothermic destruction reactions to provide heat to the system also provides a major reduction in processing costs.

Safety is always a major concern in high-pressure systems because of the enormous energy storage. While this is extremely large for any compressed gas or supercritical fluid, NCW also is somewhat safer as the built-up energy of the medium is much lower than in a supercritical fluid, reducing the danger from a catastrophic failure. Based upon the physical and chemical properties of NCW and the ability to manipulate these properties by changes in temperature or pressure, NCW is a very promising technology for the replacement of traditional organic solvents.

9.5 Conclusion

Increasing environmental awareness as well as the politicization of environmental and public health issues is pushing the chemical and pharmaceutical industries toward greener and more sustainable processes. Not only are these industries required to reduce waste and remove even trace contaminants from final products, but they must also provide incontrovertible evidence of having done so. Nearcritical water offers exciting possibilities for chemical processing. The potential to replace environmentally undesirable solvents and eliminate many hundreds of millions of tons a year of waste makes NCW an attractive solvent for environmentally benign synthesis. In addition, NCW also offers exciting possibilities in catalysis. The ability to perform acid- and base-catalyzed processes without the addition of acids or bases enables a simplification of the traditional techniques involved. Additionally, the tunability of NCW allows for a unique control of reaction rates and selectivities not available through traditional processing techniques. The potential advantages of reactions run in NCW include replacing environmentally undesirable catalysts, eliminating unwanted by-products, recycling, improved selectivity, and elimination of mass transfer limitations by changing from heterogeneous to homogeneous systems. Considering the large potential impact of environmentally benign chemical processing, only a preliminary study of chemical synthesis in high-temperature water has been done to date. Candidate processes need to be identified that can either avoid or take advantage of the thermal degradation and hydrolysis reactions that take place in NCW. Along with the proper scouting of potential reactions and processes for NCW, the elements necessary for industrial scale-up such as the phase equilibria, reaction kinetics, reaction equilibria, and analytical process monitoring techniques need to be determined and compiled.

References

1. Akiya, N. and Savage, P.E., Roles of water for chemical reactions in high-temperature water, *Chem. Rev.*, 2002, **102**(8), 2725–2750.
2. An, J., Bagnell, L., Cablewski, T., Strauss, C.R. and Trainor, R.W., Application of high-temperature aqueous media for synthetic organic reactions, *J. Org. Chem.*, 1997, **62**, 2505–2511.

3. Katritzky, A.R. and Allin, S.M., Aquathermolysis: reactions of organic compounds with superheated water, *Acc. Chem. Res.*, 1996, **29**, 399–406.
4. Katritzky, A.R., Nichols, D.A., Siskin, M., Murugan, R. and Balasubramanian, M., Reactions in high-temperature aqueous media, *Chem. Rev.*, 2001, **101**(4), 837–892.
5. Savage, P.E., Organic chemical reactions in supercritical water, *Chem. Rev.*, 1999, **99**, 603–621.
6. Savage, P.E., Gopalan, S., Mizan, T.I., Martino, C.J. and Brock, E.E., Reactions at supercritical conditions: applications and fundamentals, *AIChE J.*, 1995, **41**(7), 1723–1778.
7. Siskin, M. and Katrizky, A.R., Reactivity of organic compounds in superheated water: general background, *Chem. Rev.*, 2001, **101**(4), 825–835.
8. Hoffmann, M. and Conradi, M.S., Are there hydrogen bonds in supercritical water? *J. Am. Chem. Soc.*, 1997, **119**(16), 3811–3817.
9. Harvey, A.H. and Klein, S.A., *NIST/ASME Steam Properties*, 2.01 edn., NIST, Washington, DC, 1996.
10. Lamb, W.J., Hoffman, G.A. and Jonas, J., Self-diffusion in compressed supercritical water, *J. Chem. Phys.*, 1981, **74**(12), 6875–6880.
11. Antal, M.J., Jr., Brittain, A., DeAlmeida, C., Ramayya, S. and Roy, J.C., Heterolysis and homolysis in supercritical water, In: *Supercritical Fluids: Chemical and Engineering Principles and Applications*, ACS Symposium Series 329, American Chemical Society, Washington, DC, 1987, pp. 77–86.
12. Uematsu, M. and Franck, E.U., Static dielectric constant of water and steam, *J. Phys. Chem. Ref. Data*, 1980, **9**, 1291–1306.
13. Marshall, W.L. and Franck, E.U., Ion product of water substance, *J. Phys. Chem. Ref. Data*, 1981, **10**(2), 295–304.
14. Szydlowski, J. and Szykula, M., Isotope effect on miscibility of acetonitrile and water, *Fluid Phase Equilib.*, 1999, **154**(1), 79–87.
15. Brown, J.S., Hallett, J.P., Bush, D. and Eckert, C.A., Liquid–liquid equilibria for binary mixtures of water + acetophenone, +1-octanol, + anisole, and + toluene from 370 K to 550 K, *J. Chem. Eng. Data*, 2000, **45**, 846–850.
16. Connolly, J.F., Solubility of hydrocarbons in water near the critical solution temperature, *J. Chem. Eng. Data*, 1966, **11**, 13–16.
17. Sherman, S.R., Trampe, D.B., Bush, D.M., Schiller, M., Eckert, C.A., Dallas, A.J., Li, J. and Carr, P.W., Compilation and correlation of limiting activity coefficients of nonelectrolytes in water, *Ind. Eng. Chem. Res.*, 1996, **35**(4), 1044–1058.
18. Hallett, J.P., *Enhanced Recovery of Homogeneous Catalysts through Manipulation of Phase Behavior*, Georgia Institute of Technology, Atlanta, GA, 2002.
19. Hooper, H.H., Michel, S. and Prausnitz, J.M., Correlation of liquid–liquid equilibria for some water–organic liquid systems in the region 20–250°C, *Ind. Eng. Chem. Res.*, 1988, **27**(11), 2182–2187.
20. Anderson, F.E. and Prausnitz, J.M., Mutual solubilities and vapor pressures for binary and ternary aqueous systems containing benzene, toluene, *m*-xylene, thiophene, and pyridine in the region 100–200°C, *Fluid Phase Equilib.*, 1986, **32**, 63–76.
21. Chandler, K., Eason, B., Liotta, C.L. and Eckert, C.A., Phase equilibria for binary aqueous systems from a near-critical water reaction apparatus, *Ind. Eng. Chem. Res.*, 1998, **37**, 3515–3518.
22. Miller, D.J. and Hawthorne, S.B., Solubility of liquid organics of environmental interest in subcritical (hot/liquid) water from 298 K to 473 K, *J. Chem. Eng. Data*, 2000, **45**, 78–81.
23. Sørenson, J.M. and Arlt, W. (Eds.), *Liquid–Liquid Equilibrium Data Collection*, Deutsche Gesellschaft fur Shemisches Apparatewesen, Frankfurt, Germany, 1979.
24. Stephenson, R.M., Mutual solubilities: water–ketones, water–ethers, and water–gasoline–alcohols, *J. Chem. Eng. Data*, 1992, **37**, 80–95.
25. Lu, J., Brown, J.S., Liotta, C.L. and Eckert, C.A., Polarity and hydrogen-bonding of ambient to near-critical water: Kamlet–Taft solvent parameters, *Chem. Commun.*, 2001, (7), 665–666.
26. Marcus, Y., The properties of organic liquids that are relevant to their use as solvating solvents, *Chem. Soc. Rev.*, 1993, **22**, 409–416.
27. Reichardt, C., *Solvents and Solvent Effects in Organic Chemistry*, 2nd edn., Weinheim-VCH, Germany, 1988.
28. Kamlet, M.J. and Taft, R.W., The solvatochromic comparison method. I. The β-scale of solvent hydrogen-bond acceptor (HBA) basicities, *J. Am. Chem. Soc.*, 1976, **98**, 377.
29. Kamlet, M.J. and Taft, R.W., The solvatochromic comparison method. II. The α-scale of solvent hydrogen-bond donor (HBD) acidities, *J. Am. Chem. Soc.*, 1976, **98**, 2886.
30. Reichardt, C., Solvatochromic dyes as solvent polarity indicators, *Chem. Rev.*, 1994, **94**, 2319.
31. Bennett, G.E. and Johnston, K.P., UV–visible absorbance spectroscopy of organic probes in supercritical water, *J. Phys. Chem.*, 1994, **98**(2), 441–447.
32. Lu, J., Brown, J.S., Boughner, E.C., Liotta, C.L. and Eckert, C.A., Solvatochromic characterization of near-critical water as a benign reaction medium, *Ind. Eng. Chem. Res.*, 2002, **41**(12), 2835–2841.
33. Penninger, J.M.L., Reactions of di-*n*-butyl phthalate in water at near-critical temperature and pressure, *Fuel*, 1988, **67**(4), 490–496.

34. Bröll, D., Kaul, C., Kramer, A., Krammer, P., Richter, T., Jung, M., Vogel, H. and Zehner, P., Chemistry in supercritical water, *Angew. Chem. Int. Ed.*, 1999, **38**, 2998–3014.
35. Lesutis, H.P., Gläser, R., Liotta, C.L. and Eckert, C.A., Acid/base-catalyzed ester hydrolysis in near-critical water, *Chem. Commun.*, 1999, 2063–2064.
36. Lowry, T.H. and Richardson, K.S., *Mechanism and Theory in Organic Chemistry*, 3rd edn., Harper-Collins, New York, 1987, pp. 717–723.
37. Read, A.J., Ionization constants of benzoic acid from 25 to 250°C and to 2000 bar, *J. Solut. Chem.*, 1981, **10**(7), 437–450.
38. Shock, E.L., Organic acids in hydrothermal solutions: standard molal thermodynamic properties of carboxylic acids and estimates of dissociation constants at high temperatures and pressures, *Am. J. Sci.*, 1995, **295**(5), 496–580.
39. Timm, E.W. and Hinshelwood, C.N., The activation energy of organic reactions. III. The kinetics of acid hydrolysis of esters, *J. Am. Chem. Soc. Abs.*, 1938, 862–869.
40. Patrick, H.R., Gläser, R., Griffith, K., Liotta, C.L. and Eckert, C.A., Near-critical water: a benign medium for catalytic reactions, *Ind. Eng. Chem. Res.*, 2001, **40**, 6063–6067.
41. Mandoki, J.W., Depolymerization of condensation polymers, US Patent 83-563812, 1986 (Celanese Mexicana S.A., Mex).
42. Siskin, M., Saleh, R.Y. and Knudsen, G.A., Process for depolymerizing polyacrylonitrile without forming hydrogen cyanide. EP Patent 92-311505, 1993 (Exxon Research and Engineering Co., USA).
43. Holliday, R.L., King, J.W. and List, G.R., Hydrolysis of vegetable oils in sub- and supercritical water, *Ind. Eng. Chem. Res.*, 1997, **36**(3), 932–935.
44. Minowa, T., Zhen, F. and Ogi, T., Cellulose decomposition in hot-compressed water with alkali or nickel catalyst, *J. Supercrit. Fluid*, 1998, **13**(1–3), 253–259.
45. Katritzky, A.R., Lapucha, A.R. and Siskin, M., Aqueous high-temperature chemistry of carbo- and heterocycles. 12. Benzonitriles and pyridinecarbonitriles, benzamides and pyridinecarboxamides, and benzylamines and pyridylamines, *Energy Fuel*, 1990, **4**(5), 555–561.
46. Siskin, M., Brons, G., Katritzky, A.R. and Balasubramanian, M., Aqueous organic chemistry. 1. Aquathermolysis: comparison with thermolysis in the reactivity of aliphatic compounds, *Energy Fuel*, 1990, **4**(5), 475–482.
47. Iyer, S.D. and Klein, M.T., Effect of pressure on the rate of butyronitrile hydrolysis in high-temperature water, *J. Supercrit. Fluid*, 1997, **10**(3), 191–200.
48. Gladden, G.W. and Watson, G., Terpineol production – the chemistry and the reactions involved, *Perfumery Essent. Oil Rec.*, 1964, **55**(11), 793–804.
49. Chamblee, T.S., Weikel, R.R., Nolen, S.A., Liotta, C.L. and Eckert, C.A., Reversible in situ acid formation for β-pinene hydrolysis using CO_2 expanded liquid and hot water, *Green Chem.*, 2004, **6**(8), 382–386.
50. Kuhlmann, B., Arnett, E. and Sisken, M., Classical organic reactions in pure superheated water, *J. Org. Chem.*, 1994, **59**, 3098–3101.
51. Antal, M.J., Jr., Carlsson, M., Xu, X. and Anderson, D.G.M., Mechanism and kinetics of the acid-catalyzed dehydration of 1- and 2-propanol in hot compressed liquid water, *Ind. Eng. Chem. Res.*, 1998, **37**(10), 3820–3829.
52. Habenicht, C., Kam, L.C., Wilschut, M.J. and Antal, M.J., Jr., Homogeneous catalysis of ethyl *tert*-butyl ether formation from *tert*-butyl alcohol in hot, compressed liquid ethanol, *Ind. Eng. Chem. Res.*, 1995, **34**(11), 3784–3792.
53. Narayan, R. and Antal, M.J., Jr., Influence of pressure on the acid-catalyzed rate constant for 1-propanol dehydration in supercritical water, *J. Am. Chem. Soc.*, 1990, **112**(5), 1927–1931.
54. Xu, X. and Antal, M.J., Jr., Kinetics and mechanism of isobutene formation from *t*-butanol in hot liquid water, *AIChE J.*, 1994, **40**(9), 1524–1532.
55. Xu, X., Antal, M.J., Jr. and Anderson, D.G.M., Mechanism and temperature-dependent kinetics of the dehydration of *tert*-butyl alcohol in hot compressed liquid water, *Ind. Eng. Chem. Res.*, 1997, **36**(1), 23–41.
56. Chandler, K., Deng, F., Dillow, A.K., Liotta, C.L. and Eckert, C.A., Alkylation reactions in near-critical water in the absence of acid catalysts, *Ind. Eng. Chem. Res.*, 1997, **36**(12), 5175–5179.
57. Chandler, K., Liotta, C.L. and Eckert, C.A., Tuning alkylation reactions with temperature in near-critical water, *AIChE J.*, 1998, **44**(9), 2080–2087.
58. Richter, T. and Vogel, H., The dehydration of 1,4-butanediol to tetrahydrofuran in supercritical water, *Chem. Eng. Technol.*, 2001, **24**(4), 340–343.
59. Carlsson, M., Habenicht, C., Kam, L.C., Antal, M.J., Jr., Bian, N., Cunningham, R.J. and Maitland, J., Jr., Study of the sequential conversion of citric to itaconic to methacrylic acid in near-critical and supercritical water, *Ind. Eng. Chem. Res.*, 1994, **33**(8), 1989–1996.
60. Siskin, M., Brons, G., Vaughn, S.N., Katritzky, A.R. and Balasubramanian, M., Aqueous organic chemistry. 3. Aquathermolysis: reactivity of ethers and esters, *Energ. Fuel*, 1990, **4**(5), 488–492.
61. Olah, G.A., *Friedel–Crafts and Related Reactions*, Interscience Publishers, New York, 1964.

62. Harrington, P.J. and Lodewijk, E., Twenty years of naproxen technology, *Org. Process Res. Dev.*, 1997, **1**(1), 72–76.
63. Kabbe, H.J. and Widdig, A., Syntheses and reactions of 4-chromanones, *Angew. Chem. Int. Ed.*, 1982, **21**, 247.
64. Brown, J.S., Gläser, R., Liotta, C.L. and Eckert, C.A., Acylation of activated aromatics without added acid catalyst, *Chem. Commun.*, 2000, 1295–1296.
65. Nolen, S.A., Liotta, C.L., Eckert, C.A. and Glaeser, R., The catalytic opportunities of near-critical water: a benign medium for conventionally acid and base catalyzed condensations for organic synthesis, *Green Chem.*, 2003, **5**(5), 663–669.
66. Carey, F.A. and Sundberg, R.J., *Advanced Organic Chemistry, Part B: Reactions and Synthesis*, 3rd edn., Plenum Press, New York, 1990.
67. Gettler, J.D. and Hammett, L.P., Rates and temperature coefficients of the hydroxyl ion catalyzed aldol condensation of benzaldehyde with methyl ethyl ketone and acetone, *J. Am. Chem. Soc.*, 1943, **65**, 1824–1829.
68. Stiles, M., Wolf, D. and Hudson, G.V., Catalyst selectivity in the reactions of unsymmetrical ketones: reaction of butanone with benzaldehyde and *p*-nitrobenzaldehyde, *J. Am. Chem. Soc.*, 1959, **81**, 628–632.
69. Fine, S.A. and Pulaski, P.D., Reexamination of the Claisen–Schmidt condensation of phenylacetone with aromatic aldehydes, *J. Org. Chem.*, 1973, **38**(9), 1747–1749.
70. Noyce, D.S. and Snyder, L.R., Carbonyl reactions. IV. The kinetics of the acid-catalyzed reaction of anisaldehyde with methyl ethyl ketone, *J. Am. Chem. Soc.*, 1958, **80**, 4033–4037.
71. Kuhlmann, B., Arnett, E.M. and Siskin, M., H–D exchange in pinacolone by deuterium oxide at high temperature and pressure, *J. Org. Chem.*, 1994, **59**(18), 5377–5380.
72. Boeseken, J. and van Tonningen, W.R., The increase of the conductivity of α-ketonic acids by boric acid as a consequence of the formation of α-hydroxylated acids by hydration, $RC(OH)_2CO_2H$, *Recueil des Travaux Chimiques des Pays-Bas et de la Belgique*, 1920, **39**, 187–190.
73. An, J., Bagnell, L., Cablewski, T., Strauss, C.R. and Trainor, R.W. Applications of high-temperature aqueous media for synthetic organic reactions. *J. Org. Chem.*, 1997, **62**(8), 2505–2511.
74. Bagnell, L., Cablewski, T., Strauss, C.R. and Trainor, R.W., Reactions of allyl phenyl ether in high-temperature water with conventional and microwave heating, *J. Org. Chem.*, 1996, **61**, 7355–7359.
75. Streitwieser, A., Jr., Granger, M.R., Mares, F. and Wolf, R.A., Acidity of hydrocarbons. XLVIII. Kinetic acidities of mono-, di-, and triarylmethanes toward lithium cyclohexylamide, *J. Am. Chem. Soc.*, 1973, **95**(13), 4257–4261.
76. Hunter, S.E. and Savage, P.E., Recent advances in acid- and base-catalyzed organic synthesis in high-temperature liquid water, *Chem. Eng. Sci.*, 2004, **59**(22/23), 4903–4909.
77. Hunter, S.E. and Savage, P.E., Kinetics and mechanism of *p*-isopropenylphenol synthesis via hydrothermal cleavage of bisphenol A, *J. Org. Chem.*, 2004, **69**(14), 4724–4731.
78. Patterson, C.S., Busey, R.H. and Mesmer, R.E., Second ionization of carbonic acid in sodium chloride media to 250°C, *J. Solut. Chem.*, 1984, **13**(9), 647–661.
79. Patterson, C.S., Slocum, G.H., Busey, R.H. and Mesmer, R.E., Carbonate equilibriums in hydrothermal systems: first ionization of carbonic acid in sodium chloride media to 300°C, *Geochim. Cosmochim. Ac.*, 1982, **46**(9), 1653–1663.
80. Comisar, C.M. and Savage, P.E., Kinetics of crossed aldol condensations in high-temperature water, *Green Chem.*, 2004, **6**(4), 227–231.
81. Aleman, P. A., Boix, C., Poliakoff, M., Hydrolysis and saponification of methy1 benzoates, *Green Chem.*, 1999, **1**(2), 65–68.
82. Hunter, S. E. Savage, P. E. Acid-Catalyzed Reactions in Carbon Dioxide-Enriched High-Temperature Liquid Water. *Ind. & Eng. Chem. Res.*, 2003, **42**(2), 290–294.

Chapter 10
Biocatalysis in Water

Kaoru Nakamura and Tomoko Matsuda

Biocatalysts have been widely used in synthesis to make useful organic compounds effectively, and are now considered efficient complements to metal catalysts.[1] Due to the development of methods in biotechnology that produce stable and easy-to-use enzymes in large quantities, many biocatalysts are commercially available as reagents for organic synthesis, while others have to be obtained by cultivating the microorganism of which the stock culture is commercially available. Moreover, various kinds of reactions can be catalyzed by enzymes because biotechnology enables the production of various types of enzymes. When an enzyme does not have satisfactory activity, substrate specificity or selectivity, genetic engineering can modify the enzyme by introducing a mutation that is suitable for the desired reaction, just as the organic chemist can modify ligands to improve the function of metal catalysts. Now, because molecular biology techniques have become popular, even organic chemists can overexpress and mutate enzymes to improve the reactivity and selectivity.

Both isolated enzymes, either immobilized or not, and whole cells have been used for organic synthesis. As a solvent for biocatalytic reactions, water has been most widely used, of course because the nature developed and evolved the biocatalysts to be used in water however, some reactions have been conducted in organic solvents. In this chapter, characteristics and examples of biocatalysis in water are introduced.

10.1 Basic aspects of biocatalysis

10.1.1 Reaction classification

Various types of organic reactions can be conducted using biocatalysts. Typical enzymatic reactions used for organic synthesis are shown in Fig. 10.1. In particular, hydrolytic enzymes for kinetic resolutions of racemates have been utilized widely because of their high level of stability, wide substrate specificities, lack of cofactor requirements and high level of availability.

10.1.2 Kinetics of enzymatic reactions

Enzymatic reactions usually proceed as shown in Fig. 10.2. Initially, an enzyme–substrate complex is formed, which is then converted into an enzyme–product complex, followed by product release from the enzyme. Figure 10.2 shows the relationship between the reaction rate and the substrate concentration, following the Michaelis–Menten equation. Although one typical problem for chemical reactions in water may be that the solubility of hydrophobic

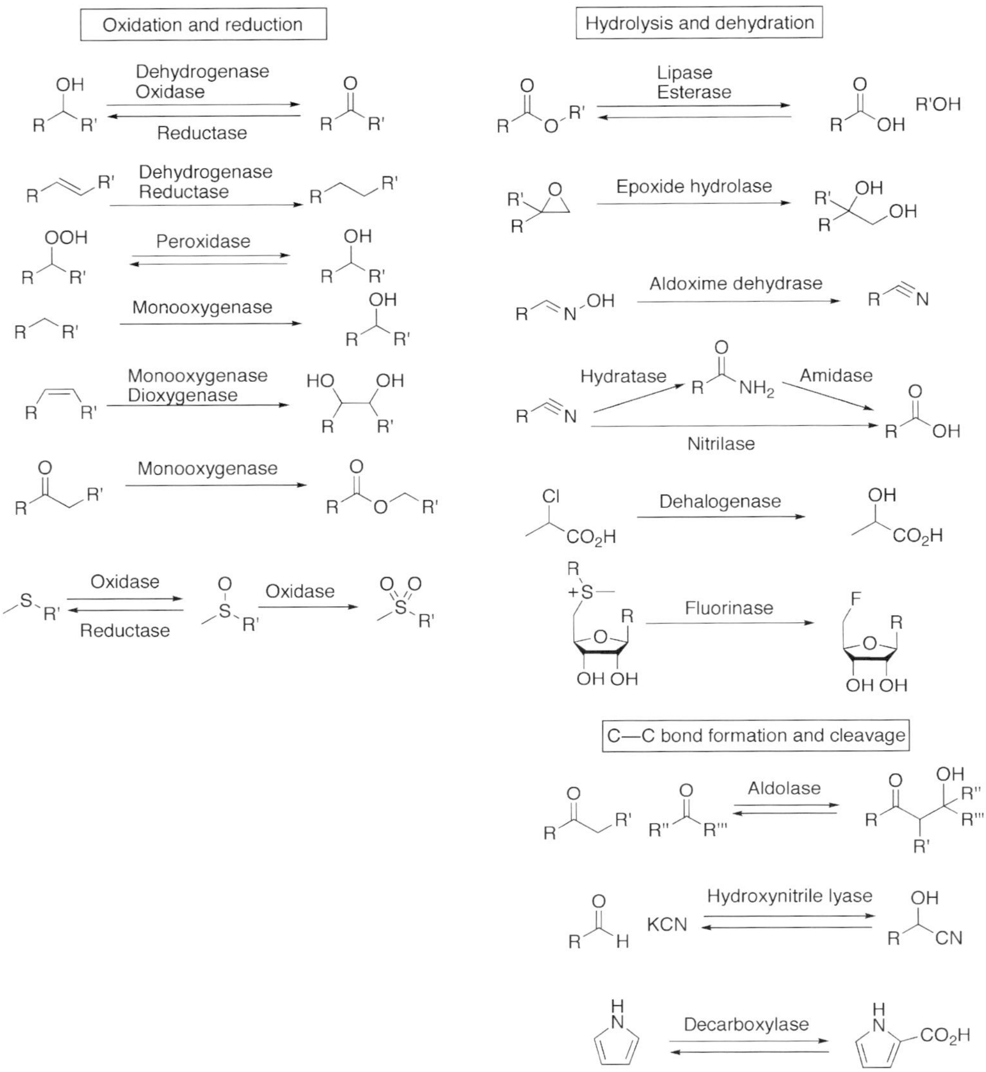

Figure 10.1 Biocatalytic reactions.

substrates in water is poor and so the effective substrate concentration is low, this is much less problematic for enzymatic reactions. The reaction rate of an enzymatic reaction is high even at a low substrate concentration, as shown in Fig. 10.2. The problem associated with enzymatic reactions is that substrate concentration sometimes has to be decreased because substrate or product might cause inhibition of the enzyme since enzymes display feedback inhibition properties. When enzyme inhibition becomes a problem, enzymes must be developed using biotechnology methods that tolerate a higher substrate concentration or, alternatively, better enzymes have to be found from nature.

$$\mathrm{E + S} \underset{k_{-1}}{\overset{k_1}{\rightleftharpoons}} \mathrm{ES} \overset{k_2}{\rightleftharpoons} \mathrm{EP} \rightleftharpoons \mathrm{E + P}$$

E: enzyme
S: substrate
P: product

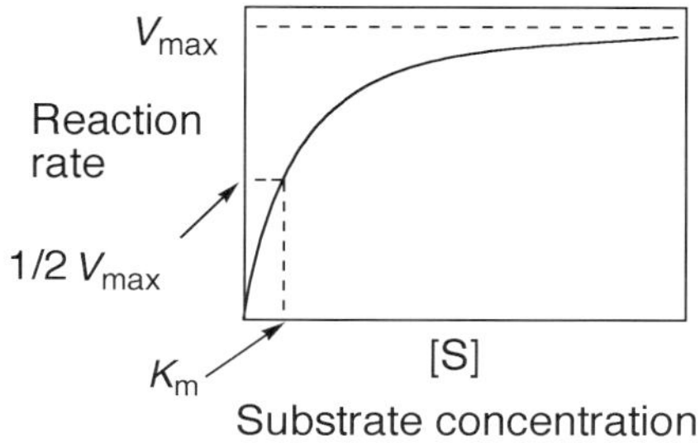

$$\text{Rate} = \frac{k_2\mathrm{ET[S]}}{\frac{k_{-1} + k_2}{k_1} + [\mathrm{S}]}$$

ET = total enzyme added

$$K_m = \frac{k_{-1} + k_2}{k_1}$$

Figure 10.2 Kinetics of the enzymatic reactions.

10.1.3 Reaction mechanism

The reaction mechanisms of some enzymatic reactions are known in detail due to the development of powerful structure elucidation tools, such as X-ray and NMR; enzymatic reaction mechanisms are no longer in a black box. One of the most studied hydrolytic enzymes is chymotrypsin, which represents a group of serine proteases. It catalyzes the hydrolysis of peptides to amino acids, and the reaction mechanism is shown in Fig. 10.3. Two amino acid residues of the enzyme, Asp and His, locate together to facilitate nucleophilic attack of Ser on the carbonyl carbon of the substrate. The reaction proceeds through a tetrahedral transition state, cleavage of the peptide bond and rapid diffusion of the amine moiety to leave the acyl-enzyme intermediate, followed by hydrolysis to give a carboxylic acid.

10.1.4 Selectivities

In organic synthesis, the achievement of high selectivity is still a major challenge. For this purpose, enzymes can play an important role in the development of synthetic organic chemistry because the most eminent characteristic of enzymatic reactions is the high chemo-, regio- and enantioselectivity. An enzyme recognizes the structure of a substrate in two stages: incorporation of the substrate and reaction of the substrate. Why, in a reaction with racemic compounds, one of the enantiomers is expected to react faster than the other enantiomers is shown in Fig. 10.4. For the reaction of chymotrypsin (see Fig. 10.4(a)), only L-peptides can form a stable enzyme–substrate complex, and thus very high enantioselectivity is observed. Oxidoreductases also form the enzyme–substrate complex of only one enantiomer, and so enantioselectivities are high when isolated enzymes are used for reactions instead of whole cells containing both *R*- and *S*-specific enzymes, which leads to overall low enantioselectivities.

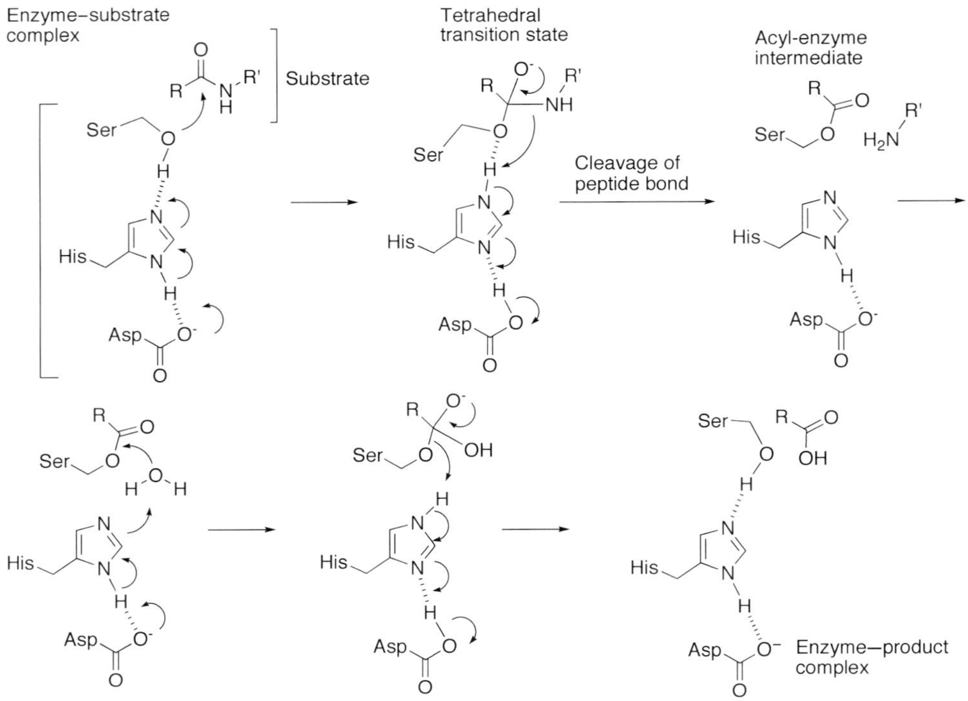

Figure 10.3 Reaction mechanism of hydrolysis of a peptide by chymotrypsin.

In the case of lipases and esterases, chiral recognition is not so strict, and both enantiomers are incorporated to form the enzyme–substrate complex. However, one of the enantiomers reacts slower because it lacks crucial hydrogen-bond interaction, for example in the hydrolysis of menthol acetate, between the substrate menthol and the enzyme histidine group for the reaction to proceed further (Fig. 10.4(b)).[2a,b] This explanation was also supported by observations in the esterification reaction of 1-phenylethanol by lipases.[2c,d] K_m values, showing the easiness of the enzyme–substrate formation, of the slow- and fast-reacting enantiomers were almost the same (K_m(slow):K_m(fast) = 1:0.3–1). However, V_{max} values were significantly different (V_{max}(slow):V_{max}(fast) = 1:150–500).

One of the most clearly distinguishable features of biocatalytic reactions compared with those using chemical catalysts is that biocatalysts can recognize a remote stereogenic center apart from the reaction center of a substrate. As shown in Fig. 10.5, biocatalysts can discriminate between enantiomers where the stereogenic carbon is six bonds away from the reaction center.

10.1.5 Experimental conditions

The conditions for enzymatic reactions are usually very mild. The solvents employed when isolated enzymes are used are typically aqueous buffer around pH of 7 or hydrophobic organic solvent such as hexane or isopropyl ether. In whole-cell reactions distilled water or tap water is normally used.

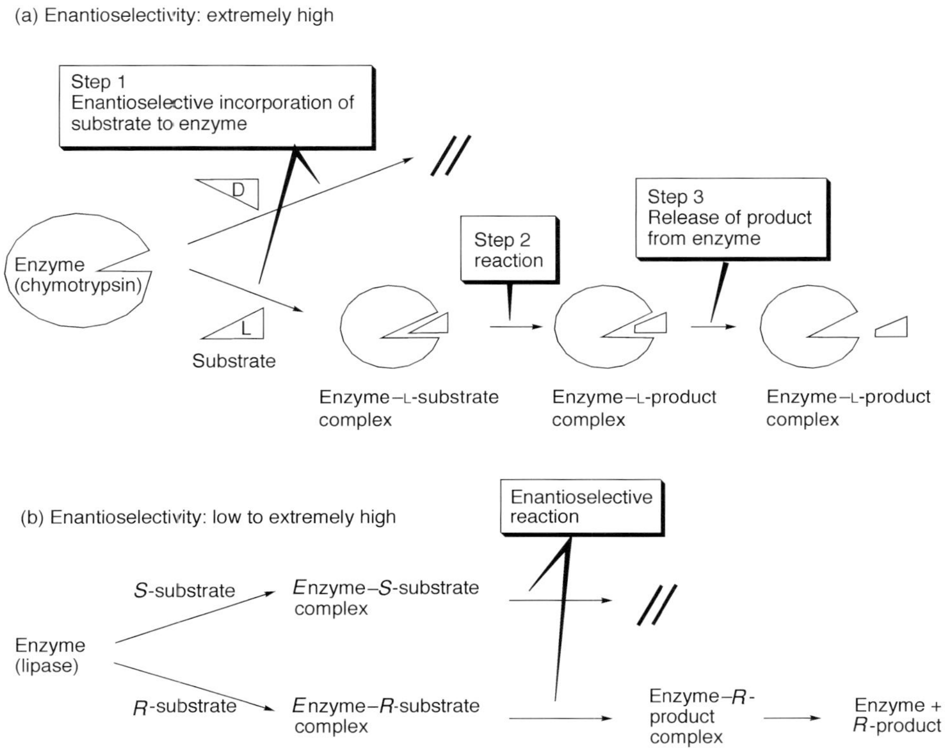

Figure 10.4 Two key steps for recognition of enantiomers by enzymes.

The reaction temperature is around room temperature, although some enzymes from extremophiles, microorganisms from extreme environmental conditions, can react very effectively at temperatures up to 120°C and down to –50°C. Sometimes, enantioselectivity can be improved by a decrease or increase in the reaction temperature.

To improve the efficiency of an enzymatic reaction various additives are often used. In reductions, for example, sugar is used as a hydrogen source. When an isolated enzyme is used, the addition of one or several cofactors is sometimes necessary. In whole-cell reductions, however, this is not needed as the cell already contains all the required cofactors.

10.2 Reduction

10.2.1 Stereochemistry of hydride transfer

Dehydrogenases and reductases are enzymes that catalyze the reduction of carbonyl groups. The natural substrates of the enzymes are alcohols, such as ethanol, lactate, glycerol, and the corresponding carbonyl compounds; however, unnatural ketones can also be reduced enantioselectively. To exhibit catalytic activities, the enzymes require a coenzyme such as NADH or NADPH from which a hydride is transferred to the substrate carbonyl carbon.

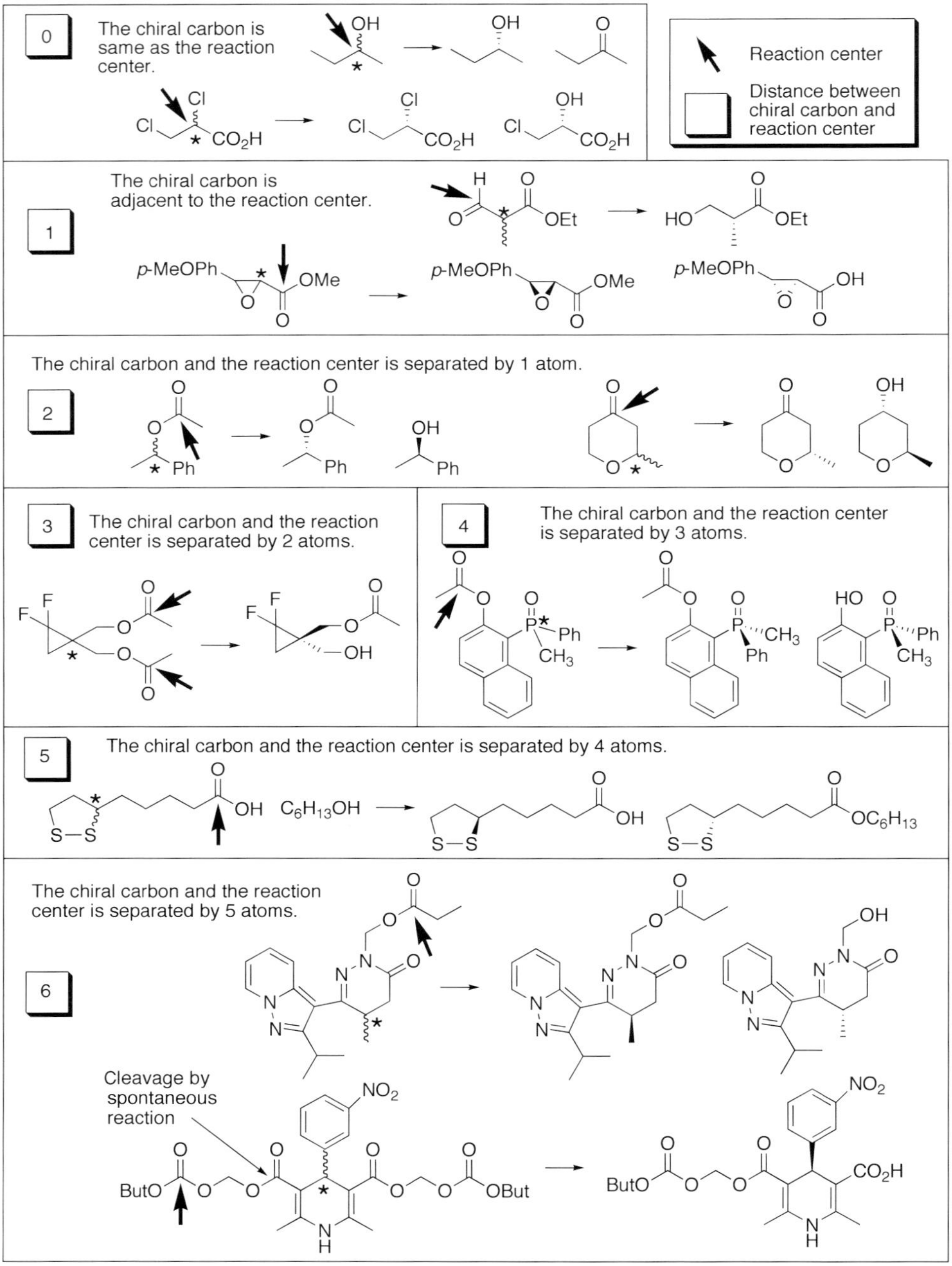

Figure 10.5 Recognition of stereocenters at various distances from reaction centers.

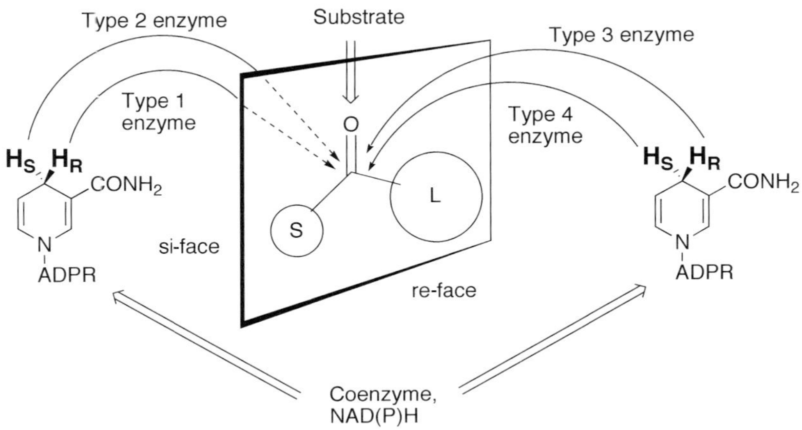

Figure 10.6 Stereochemistry of the hydride transfer from NAD(P)H to the carbonyl carbon on the substrate (S is a small group and L is a large group).

There are four stereochemical patterns that enable the transfer of the hydride from the coenzyme, NAD(P)H, to the substrate, as shown in Fig. 10.6.[3] With type 1 enzyme[3a,b] and type 2 enzyme,[3c–f] the hydride attacks the si-face of the carbonyl group, whereas with type 3 enzyme[3g–l] and type 4 enzyme, the hydride attacks the re-face, which results in the formation of (*R*)- and (*S*)-alcohols, respectively. On the other hand, type 1 and type 3 enzymes transfer the pro-(*R*)-hydride of the coenzyme, and type 2 and type 4 enzymes use the pro-(*S*)-hydride. Examples of type 1 to type 3 enzymes are as follows. Type 4 enzyme is not popular.

Type 1 enzyme:	*Pseudomonas* sp. alcohol dehydrogenase[3a]
	Lactobacillus kefir alcohol dehydrogenase[3b]
Type 2 enzyme:	*Geotrichum candidum* glycerol dehydrogenase[3c–e]
	Mucor javanicus dihydroxyacetone reductase[3f]
Type 3 enzyme:	Yeast alcohol dehydrogenase[3g]
	Horse liver alcohol dehydrogenase[3h–k]
	Moraxella sp. alcohol dehydrogenase[3l]

10.2.2 Baker's yeast-catalyzed reaction

One of the most useful biocatalysts is the baker's yeast. It has been widely used to synthesize chiral intermediates.[4] For example, baker's yeast reduction of diketones proceeds highly regio- and enantioselectively with aliphatic diketones, 2,2-disubstituted cycloalkanediones and spiro diones, producing enantiomerically enriched hydroxyketones (Fig. 10.7). Starting from a chiral hydroxyketone, many terpenes have been enantioselectively synthesized by Mori et al., as shown in Fig. 10.7(a).[4a,b]

The enantioselectivity of the baker's yeast reduction is excellent for the examples in Fig. 10.7. However, it is not always high, because there are many enzymes existing in the whole cell and some of them are *S*-selective and others are *R*-selective. For example, in the reduction of β-keto esters, the enantioselectivity is low and (*S*)-alcohols are produced when the ester

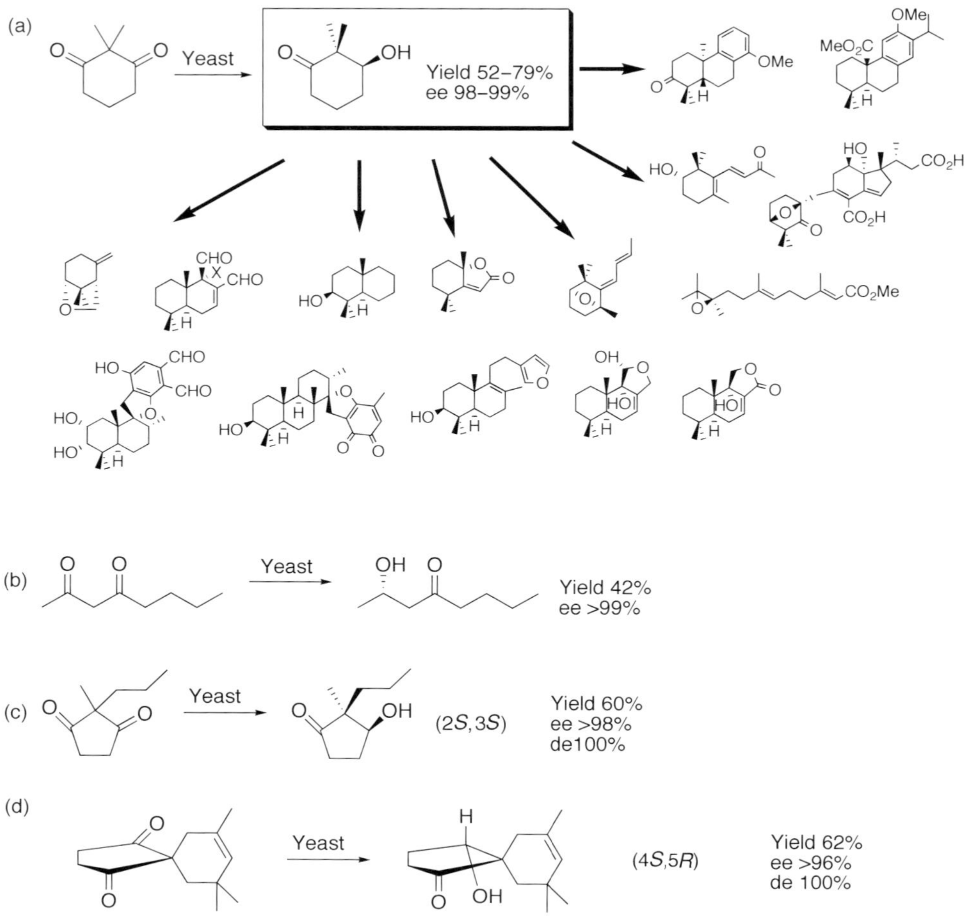

Figure 10.7 Baker's yeast-catalyzed asymmetric reduction.[4a–e]

moiety is small, but it is excellent and (*R*)-alcohols are produced when the ester moiety is large. The enantioselectivity is dependent on the size of the ester moiety, and can thus be controlled by modifying the substrate structures (Fig. 10.8).

10.2.3 Overexpression of key reductases from baker's yeast in *Escherichia coli*

The problem of low enantioselectivity of whole-cell reduction possessing more than two enzymes with opposite enantioselectivities has been solved by overexpressing the individual enzymes in *E. coli*. Now, 18 key reductases from baker's yeast have been overproduced in *E. coli* as glutathione S-transferase (GST) fusion proteins. Therefore, rapid identification of synthetically useful biocatalysts is possible.[5] A set of fusion proteins consisting of GST linked to the N-terminus of putative dehydrogenases produced by baker's yeast was screened for the reduction of various substrates. For example, ethyl 2-oxo-4-phenylbutyrate was reduced

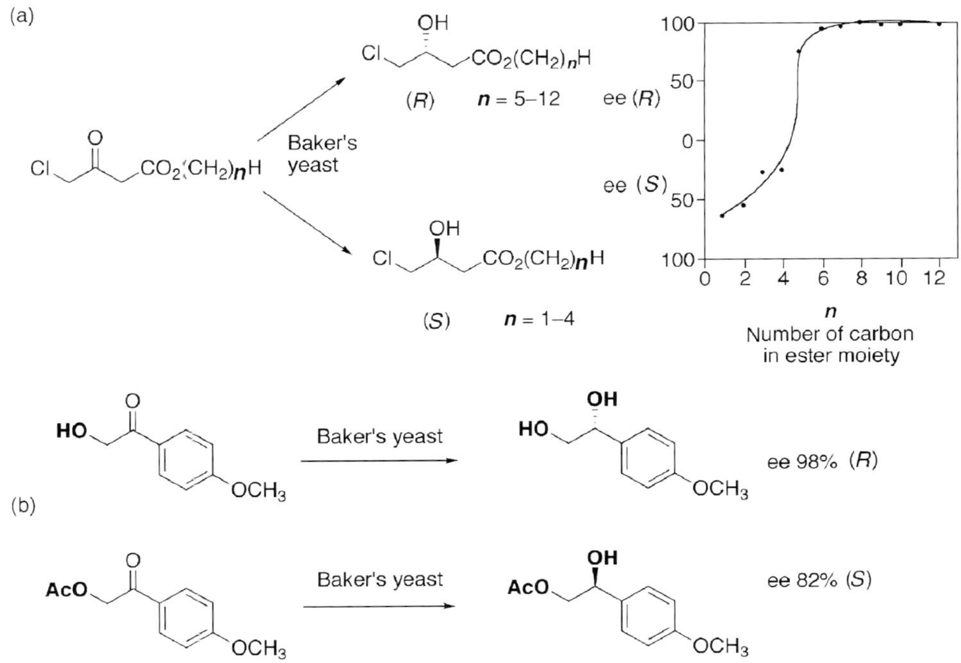

Figure 10.8 Control of enantioselectivity of baker's yeast (whole cell) catalyzed reduction.[4f–i]

rapidly in the presence of NADH and NADPH by two dehydrogenases, Ypr1p and Gre2p, providing the (*R*)- and (*S*)-alcohol, respectively, with high stereoselectivities (Fig. 10.9(a)).[5a] The same enzymes were overexpressed in their native forms in *E. coli* and growing cells of the engineered strains could also be used to carry out the reductions without the need for exogenous cofactor.

A representative set of α- and β-keto esters was also tested as substrates (11 total) for each purified fusion protein (Fig. 10.9(b) and (c)).[5b] The stereoselectivities of β-keto ester reductions depended on both the identity of the enzyme and the substrate structure, and some reductases yielded both L- and D-alcohols with high stereoselectivities. While α-keto esters were generally reduced with lower enantioselectivities, it was possible to identify pairs of yeast reductases that delivered both enantiomers in optically pure form.

Furthermore, two enantiocomplementary baker's yeast enzymes were also used for the reduction of an α-chloro-β-keto ester to yield precursors for both enantiomers of the *N*-benzoyl phenylisoserine Taxol side chain (Fig. 10.10). These results demonstrate the power of genomic fusion protein libraries to identify appropriate biocatalysts rapidly and expedite process development.[5f]

10.2.4 Asymmetric reduction by *Geotrichum candidum*

Other microorganisms have also been used for the asymmetric reduction of carbonyl compounds. Simple aliphatic ketones as well as aromatic ketones can be reduced with very high enantioselectivity by using biocatalysts.[6] For example, aliphatic ketones such as

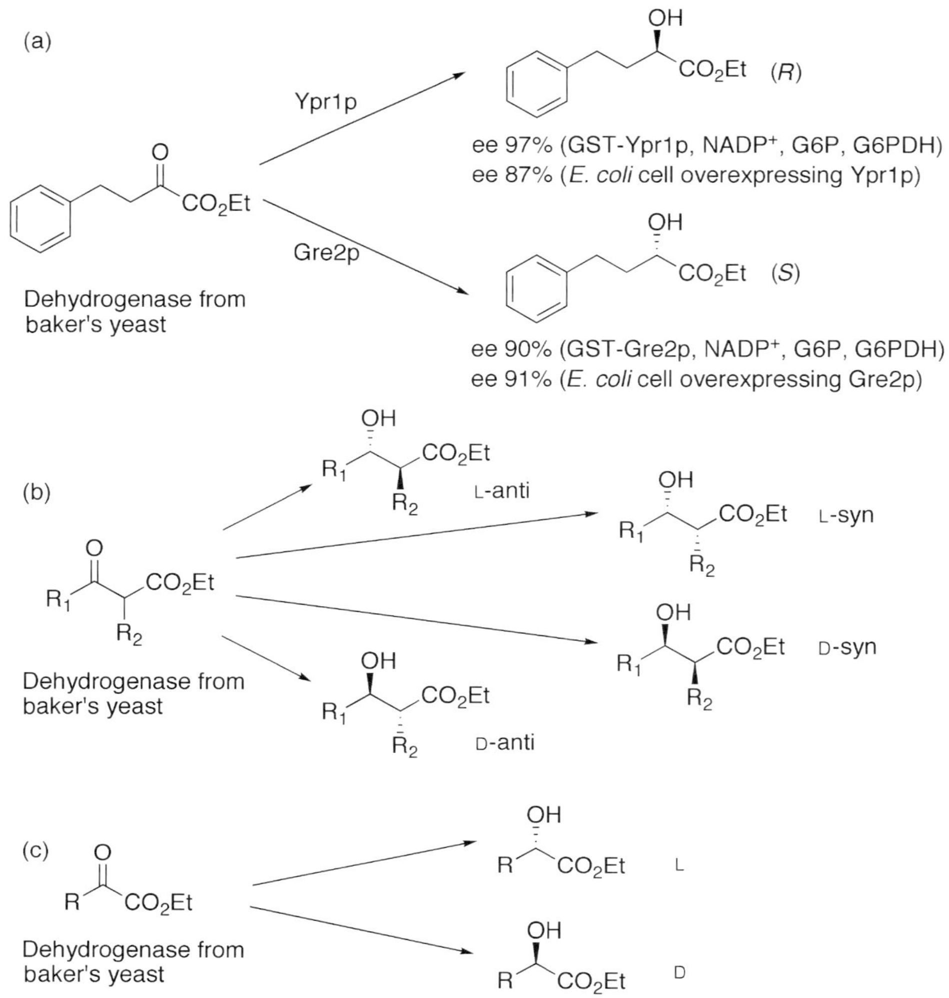

Figure 10.9 Identification of appropriate biocatalysts from fusion protein libraries from baker's yeast.[5]

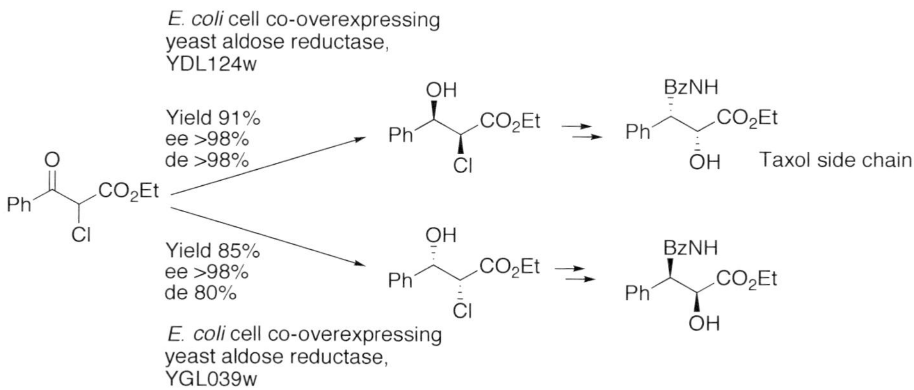

Figure 10.10 Use of engineered *E. coli* cell to synthesize both Taxol side-chain antipodes.[5f]

2-pentanone, 2-butanone and 3-hexanone were reduced with excellent enantioselectivity to the corresponding (S)-alcohols by using the dried cells of *G. candidum* (Table 10.1). The dried-cell *G. candidum* system can distinguish between two alkyl groups with a difference of a single methylene unit in the reduction of 3-hexanone (Fig. 10.11).[6c] It has been applied to the reduction of aromatic ketones, and the substrate specificity of the system is shown in Table 10.2. For the reduction of trifluoromethyl ketone, the opposite enantiomers were obtained because different enzyme in the dried cell catalyzes the reaction.

Table 10.1 Reduction of aliphatic ketones by the dried cell of *G. candidum*, NAD^+ and secondary alcohol[6]

Product	Yield (%)	ee (%)
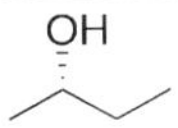	73	94
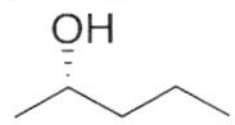	97	>99
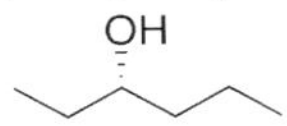	35	98
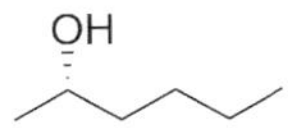	89	>99
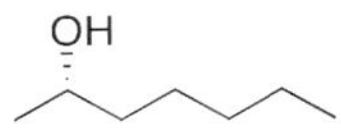	89	>99
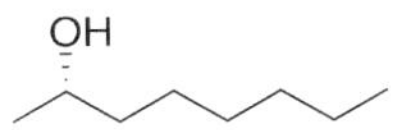	87	>99
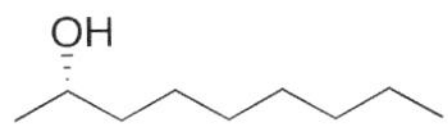	87	>99
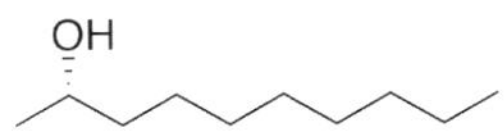	85	>99
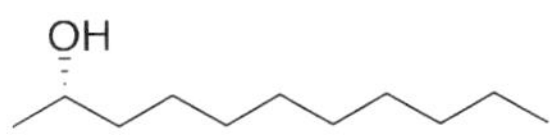	60	>99
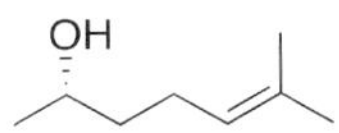	90	99
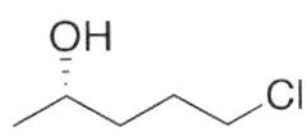	92	99
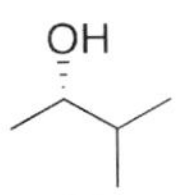	96	98
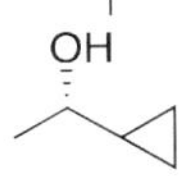	34	>99

Table 10.2 Reduction of ketones by the dried cell of *G. candidum*, NAD(P)$^+$ and secondary alcohol[6]

Product		Yield (%)	ee (%)
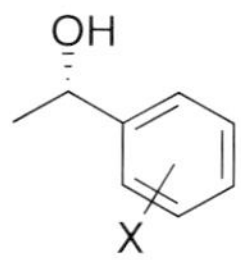	X = H	89	>99 (*S*)
	o-F	>99	>99 (*S*)
	m-F	95	>99 (*S*)
	p-F	74	>99 (*S*)
	o-Cl	>99	>99 (*S*)
	m-Cl	95	99 (*S*)
	p-Cl	62	>99 (*S*)
	o-Br	97	>99 (*S*)
	m-Br	92	>99 (*S*)
	p-Br	95	>99 (*S*)
	o-Me	96	>99 (*S*)
	m-Me	86	>99 (*S*)
	p-Me	78	>99 (*S*)
	o-MeO	84	>99 (*S*)
	m-MeO	90	>99 (*S*)
	p-MeO	29	>99 (*S*)
	o-CF_3	6	97 (*S*)
	m-CF_3	96	>99 (*S*)
	p-CF_3	73	>99 (*S*)
	1',2',3',4',5'-F_5	62	>99 (*S*)
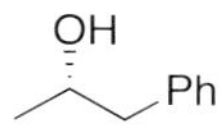		96	>99 (*S*)
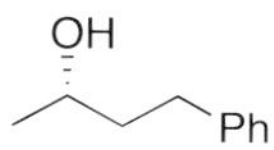		93	>99 (*S*)
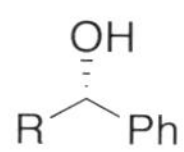	*R* = Et	41	>99 (*S*)
	Pro	0	—
	i-Pro	12	99 (*S*)
	t-Bu	1	—
	$MeOCH_2$	8	>99 (*R*)
	$ClCH_2$	80	98 (*R*)
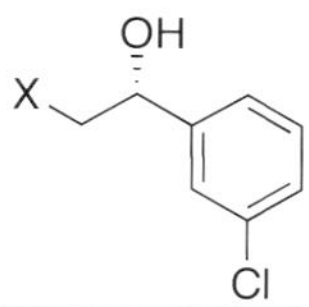	X = Cl	94	98 (*R*)
	Br	95	93

10.2.5 Hydrogen sources

One of the advantages of using biocatalysts for reductions is the use of natural sources for the reduction power. For the baker's yeast reduction, glucose has been used. Light energy can also be used to provide hydrogen for the reduction. Photochemical methods[7] that employ light energy to regenerate NAD(P)H, for example, by the use of a cyanobacterium, a photosynthetic biocatalyst, have been developed to provide an environment-friendly system.

Table 10.2 Reduction of ketones by the dried cell of *G. candidum*, NAD(p)$^+$ and secondary alcohol[6]

Product		Yield (%)	ee (%)
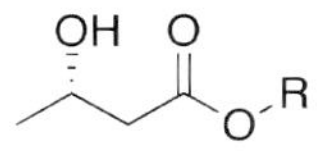	*R* = Me	>99	>99 (*S*)
	Et	>99	>99 (*S*)
	t-Bu	>99	>99 (*S*)
	neo-Pentyl	>99	>99 (*S*)
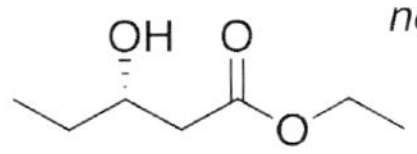		72	>99 (*S*)
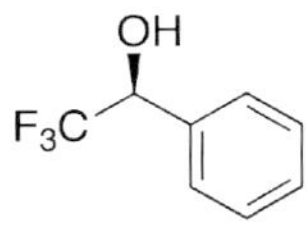	*R* = H	>99	98 (*S*)
	Cl	91	>99 (*S*)
	Br	82	>99 (*S*)
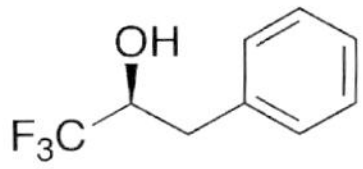		55	97 (*S*)
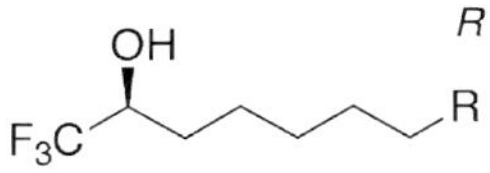	*R* = Me	>99	96 (*S*)
	Et	>99	96 (*S*)
	Pr	>99	>99 (*S*)
	Bu	>99	98 (*S*)
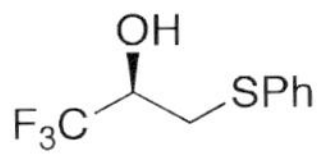		97	98 (*R*)
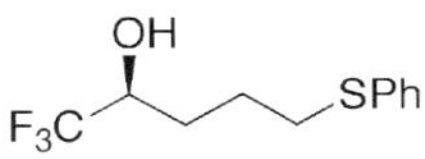		98	>99 (*S*)
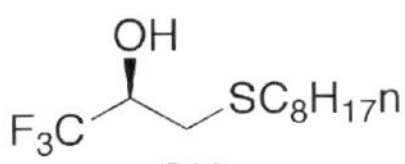		60	96 (*R*)
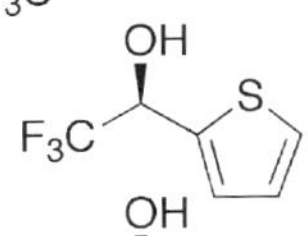		99	>99 (*R*)
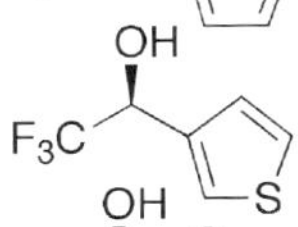		99	>99
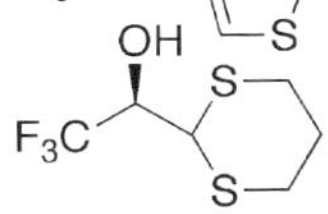		67	>99

In this way, the reduction of acetophenone derivatives occurred more effectively under illumination than in the dark (Fig. 10.12). The light energy harvested by the cyanobacterium is converted into chemical energy in the form of NADPH through an electron transfer system, and, consequently, the chemical energy (NADPH) is used to reduce the substrate to chiral alcohol (96 to >99% ee). The light energy, which is usually employed to reduce CO_2 to synthesize organic compounds in the natural environment, was used to reduce the substrate in this case.

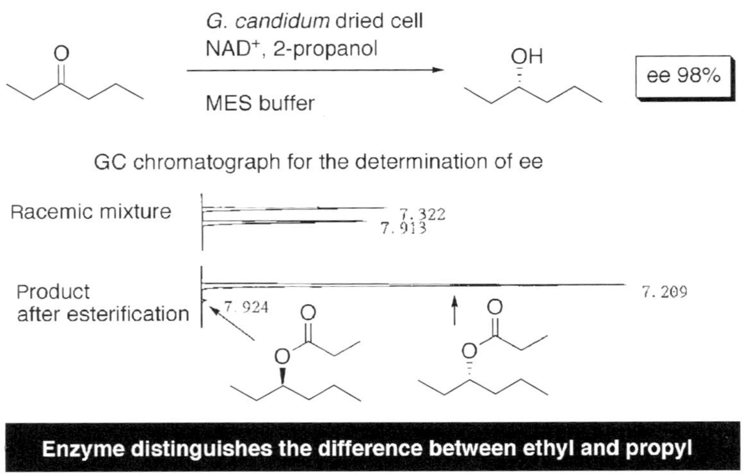

Figure 10.11 Reduction of 3-hexanone by the dried cell of *G. candidum*, NAD^+ and secondary alcohol.[6c]

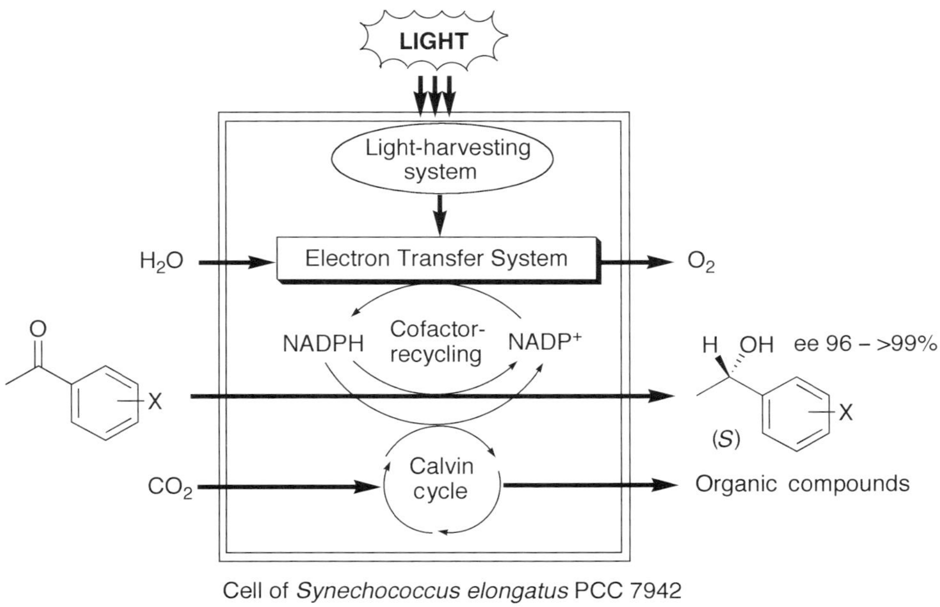

Figure 10.12 Reduction of ketone with photosynthetic biocatalyst using light energy.[7]

10.2.6 Reduction of carbon–carbon double bonds

The reduction of carbon–carbon double bonds to single bonds has been studied with various substrates.[8] For example, Ohta et al. demonstrated that the reduction of a number of 1-nitro-1-alkenes by fermenting baker's yeast was enantioselective, resulting in the formation of optically active 1-nitroalkanes as shown in Fig. 10.13(a).[8f] On the other hand, Fuganti et al. reduced α,β-unsaturated γ-lactones to produce enantiomerically pure (+)-(*R*)-goniothalamin (Fig. 10.13(b)), which shows central nervous system activity. They also performed the kinetic resolution of the corresponding embryotoxic epoxide with yeast.[8g]

One of the most studied enzymes for the reduction of carbon–carbon double bonds is the 'old yellow enzyme' from yeast,[8c–e] which has been shown to efficiently catalyze the NADPH-linked reduction of nitro-olefins in a stepwise fashion (Fig. 10.13(c)). The first step involves hydride transfer from the enzyme-reduced flavin to the β-carbon of the nitro-olefin, which forms a nitronate intermediate that is freely dissociable from the enzyme. The second step, protonation of the nitronate at the α-carbon to form the final nitroalkane product, is also catalyzed by the enzyme.

Photosynthetic microorganisms and plant cell cultures are very important sources of enzymes for the reduction of olefins.[8a,i] For example, Hirata et al. has found that reduction of cyclohexenones with *Nicotiana tabacum* p90 reductase and *Nicotiana tabacum* p44 reductase affords (*S*)- and (*R*)-alkylcyclohexanones, respectively, with excellent enantioselectivities as shown in Fig. 10.13(d). They also found two enone reductases from *Astasia longa*, a nonchlorophyllous cell line classified in Euglenales, and studied the mechanism. Both catalyzed enantiospecific trans-addition of hydrogen atoms to carvone from the si-face at the α-position and from the re-face at the β-position.

10.2.7 Reduction of hydroperoxides

Peroxidases have been used for the kinetic resolution of racemic peroxides to give chiral peroxides.[9] For example, a horseradish peroxidase selectively recognized sterically uncumbered (*R*)-alkyl aryl hydrogen peroxides, which allowed kinetic resolution to provide (*R*)-alcohol and (*S*)-peroxide (Fig. 10.14(a)).[9a] However, poor enzyme recognition was observed with hydroperoxides possessing larger R_2 groups such as propyl or butyl moiety. This reaction was conveniently performed on a preparative scale to provide optically pure hydroperoxides. Aliphatic hydroperoxides were also resolved using the horseradish peroxidase (Fig. 10.14(b)–(d)).[9b,c,e] R-Enantiomers were recognized enantioselectively to give the corresponding (*R*)-alcohols, leaving (*S*)-hydroperoxides untouched.

10.2.8 Reduction of sulfoxides

Preparation of chiral sulfoxides can be achieved by the biocatalytic reduction of sulfoxides.[10] One example is the reduction of alkyl aryl sulfoxides by intact cells of *Rhodobacter sphaeroides* f. sp. *denitrificans* (Fig. 10.15).[10a] In the reduction of methyl para-substituted phenyl sulfoxides, *S*-enantiomers were exclusively deoxygenated while enantiomerically pure *R*-isomers were recovered in good yield. For poor substrates such as ethyl phenyl sulfoxide, the repetition

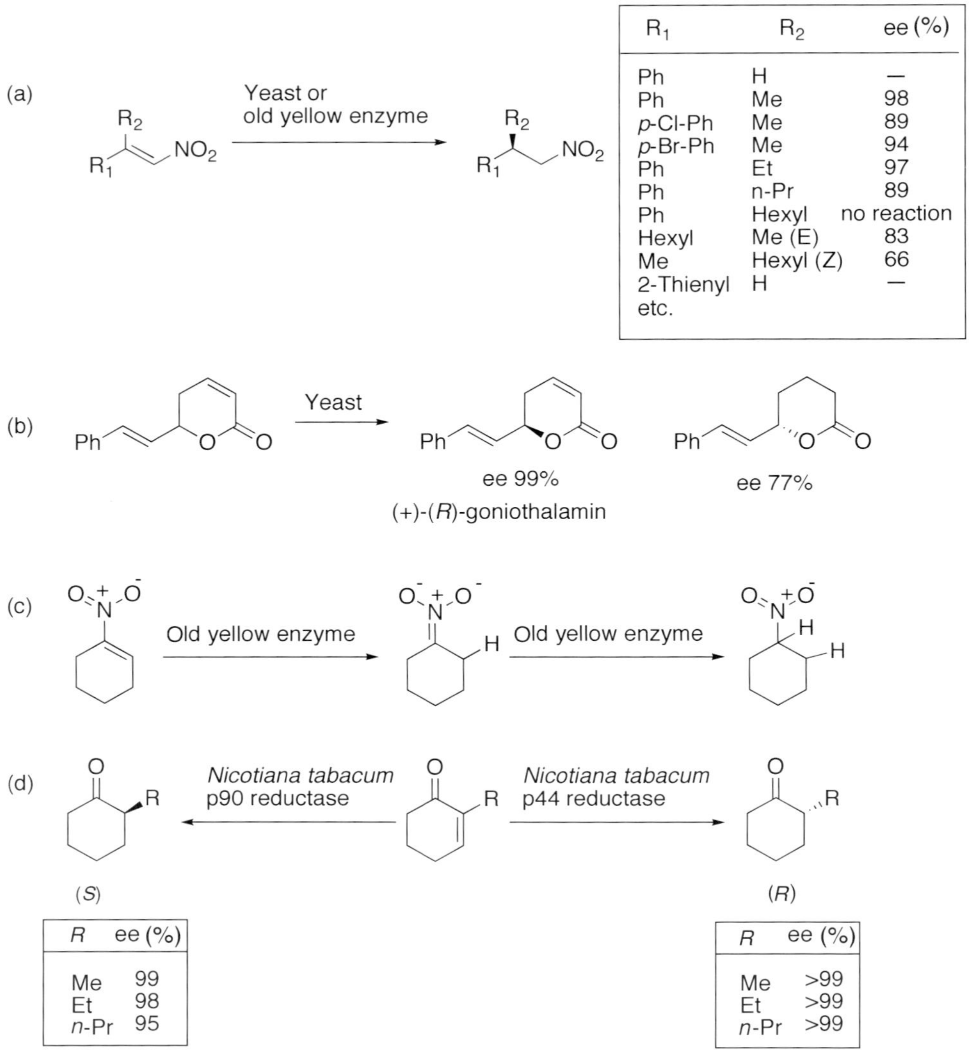

R_1	R_2	ee (%)
Ph	H	—
Ph	Me	98
p-Cl-Ph	Me	89
p-Br-Ph	Me	94
Ph	Et	97
Ph	n-Pr	89
Ph	Hexyl	no reaction
Hexyl	Me (E)	83
Me	Hexyl (Z)	66
2-Thienyl	H	—
etc.		

R	ee (%)
Me	99
Et	98
n-Pr	95

R	ee (%)
Me	>99
Et	>99
n-Pr	>99

Figure 10.13 Reduction of olefins.[8c–g,i]

of the incubation after removing the toxic product was effective in enhancing the enantiomeric excess (ee) of the recovered *R*-enantiomers to 100%.

10.3 Oxidation

10.3.1 Oxidation of alcohols

Various microorganisms and isolated enzymes have also been used for the oxidation of alcohols (Fig. 10.16).[11] For example, horse liver alcohol dehydrogenase was used for the

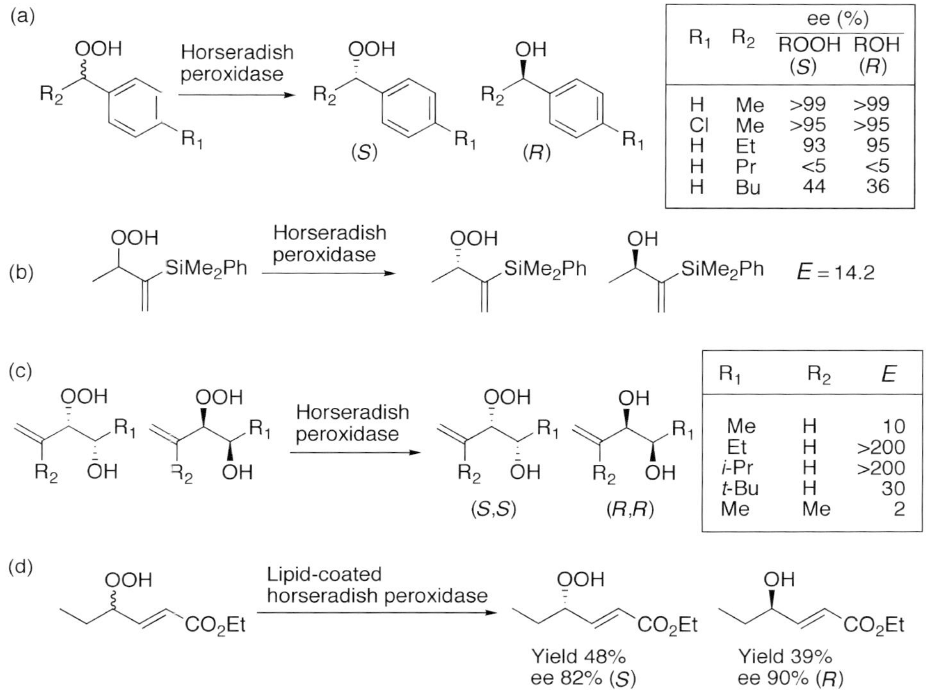

(a)

R_1	R_2	ee (%) ROOH (S)	ee (%) ROH (R)
H	Me	>99	>99
Cl	Me	>95	>95
H	Et	93	95
H	Pr	<5	<5
H	Bu	44	36

(c)

R_1	R_2	E
Me	H	10
Et	H	>200
i-Pr	H	>200
t-Bu	H	30
Me	Me	2

Figure 10.14 Reductions of hydroperoxides to alcohols for the synthesis of chiral hydroperoxides.[9]

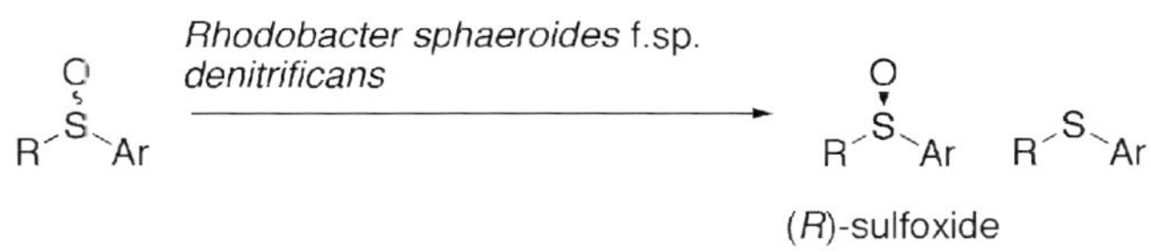

R	Ar	(R)-sulfoxide Yield (%)	ee (%)
Me	Ph	46	100
Me	*p*-Me-C_6H_4	40	100
Me	*p*-Br-C_6H_4	43	100
Me	*p*-MeO-C_6H_4	47	>99
Me	$PhCH_2$	41	90
Et	Ph	41	100
n-Pr	Ph	54	21

Figure 10.15 Reduction of sulfur compounds for the synthesis of *S*-chiral compounds.[10a]

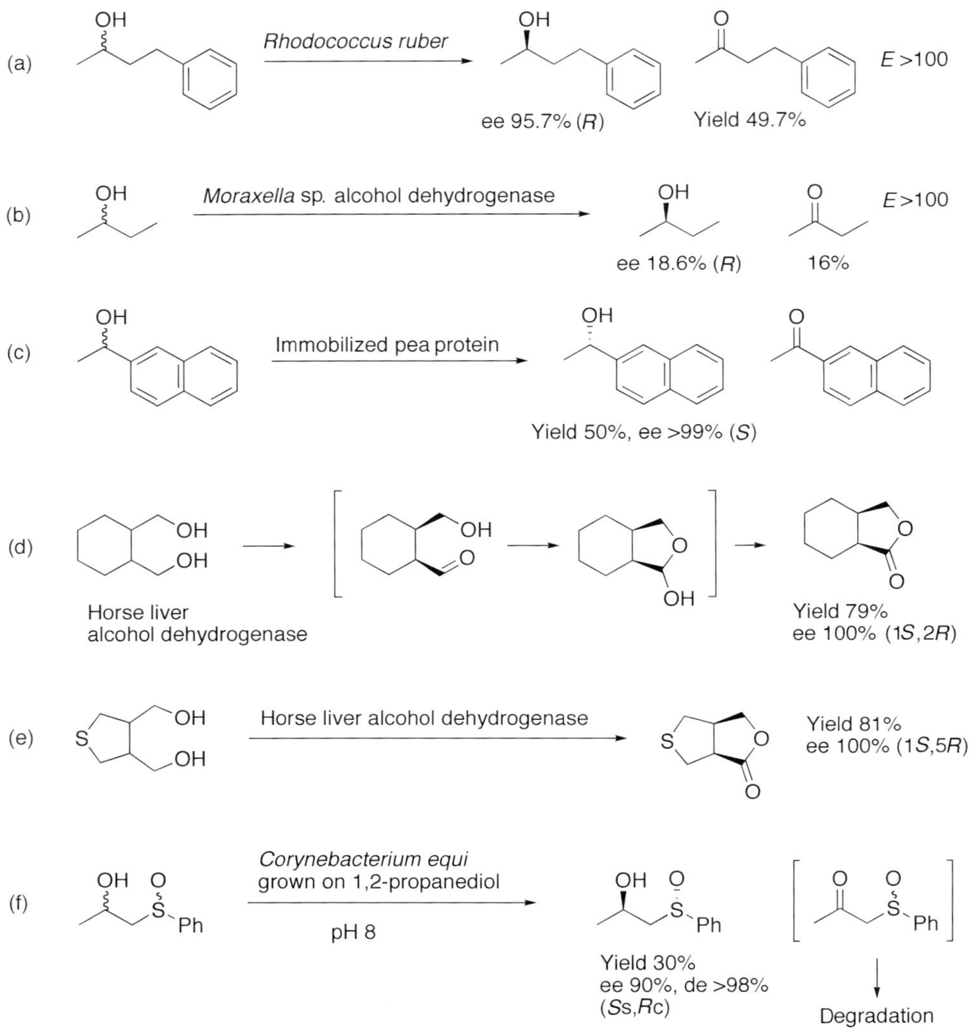

Figure 10.16 Enantioselective oxidations of alcohols.[11]

oxidation of meso diols (Fig. 10.16(d) and (e)).[11e,f] When one of the hydroxyl groups was oxidized, cyclization proceeded spontaneously, followed by the enzyme-catalyzed oxidation, giving chiral lactones.

10.3.2 Hydroxylation

Monooxygenases, enzymes that transfer one oxygen atom to the substrate from molecular oxygen, and dioxygenases, enzymes that transfer both oxygen atoms from molecular oxygen, have been used for oxidations. Examples of such reactions are shown in Fig. 10.17.[12]

Regioselectivities and enantioselectivities are excellent. As shown in Fig. 10.17(g) and (h), oxidation of thioanisole by recombinant *E. coli* JM 109, an organism that overexpresses toluene dioxygenase, afforded (*R*)-sulfoxide in 92% ee, while the reaction with *p*-bromothioanisole by the same microbe provided the corresponding (2*R*,3*S*)-diene diol in good yield.[12l] Interestingly, although the same biocatalyst was used, the reaction course was different. For the hydroxylation shown in Fig. 10.17(i), the enantioselectivities can be controlled by pH; the S-enantiomer was formed at pH 2 and the *R*-enantiomer at pH 6. In this reaction, the initial step is epoxide formation followed by fast chemical hydrolysis at pH 2 yielding the S-enantiomer, whereas at pH 6 enzymatic hydrolysis is faster than chemical hydrolysis and so the R-enantiomer is obtained.[12n]

10.3.3 Baeyer–Villiger oxidations

Monooxygenases have been used in Baeyer–Villiger oxidations to obtain optically active lactones (Fig. 10.18).[13] A cyclohexanone monooxygenase from *Acinetobacter calcoaceticus* has been widely used. In the oxidation of 2-oxabicyclo[3.2.0]-heptan-7-one by *A. calcoaceticus* (Fig. 10.18(a), $n = 0$, $m = 2$), the S,S-substrate led to the normal lactone, whereas the R,R-substrate gave the abnormal lactone, with ee's of 90 and >98%, respectively. Other substrates proceeded with a similar stereochemical outcome.

Recently, the enzyme was expressed in baker's yeast and used for the asymmetric oxidation of alkyl cyclohexanones.[13c] The use of designer baker's yeast combined the advantages of using purified enzymes (single catalytic species, no overmetabolism) with the benefits of whole-cell reactions (experimentally simple, no cofactor regeneration necessary). The enantioselectivities observed with the recombinant enzyme were the same as those with the original, isolated enzyme.[13g]

10.3.4 Oxidation of sulfur compounds

Chiral compounds have been synthesized through oxidation of sulfur compounds.[10b,11g,14] For example, a highly enantioselective oxidation of benzhydrylsulfanyl acetic acid to the corresponding (*S*)-sulfinyl carboxylic acid was achieved in 89% yield and 99.1% ee employing the fungus *Beauveria bassiana*. This product was amidated using the bacteria *Bacillus subtilis* to afford (*S*)-modafinil in 68% yield and 100% ee (Fig. 10.19(a)).[14a] In another example, *Rhodococcus equi* was used for oxidation of sulfinic acid (Fig. 10.19(b)).[10b] Incubation of racemic arenesulfinic acid esters with the bacterial strain gave the corresponding sulfonates, leaving optically pure R-substrate (>99% ee). For the reaction of phenylsulfinyl propanone with *R. equi*, both oxidation of sulfoxide and reduction of carbonyl group proceeded (Fig. 10.19(c)).[11g] The carbonyl group of (*R*)-sulfoxide was reduced to the corresponding (*R*s,*S*c)-alcohols without the reaction of sulfoxide group, while (*S*)-sulfoxide was oxidized first to the corresponding sulfone and then the carbonyl group was reduced to the corresponding (*S*)-alcohol.

10.3.5 Oxidative polymerization

Various enzymes, such as laccase and peroxidase, have been used for the synthesis of polymers. For example, peroxidase catalyzes oxidative polymerization as shown in Fig. 10.20.[15]

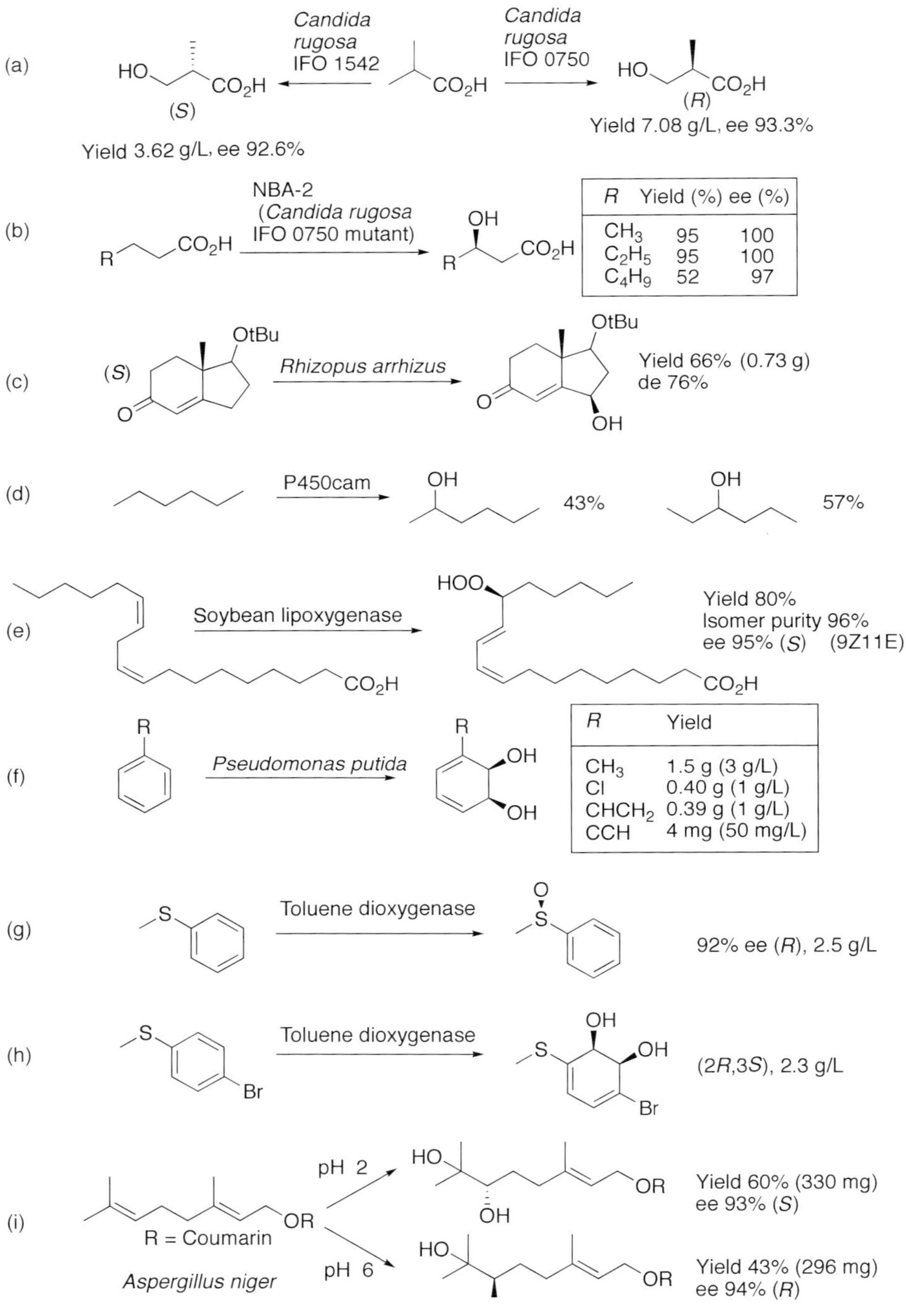

R	Yield (%)	ee (%)
CH_3	95	100
C_2H_5	95	100
C_4H_9	52	97

R	Yield
CH_3	1.5 g (3 g/L)
Cl	0.40 g (1 g/L)
$CHCH_2$	0.39 g (1 g/L)
CCH	4 mg (50 mg/L)

Figure 10.17 Hydroxylation.

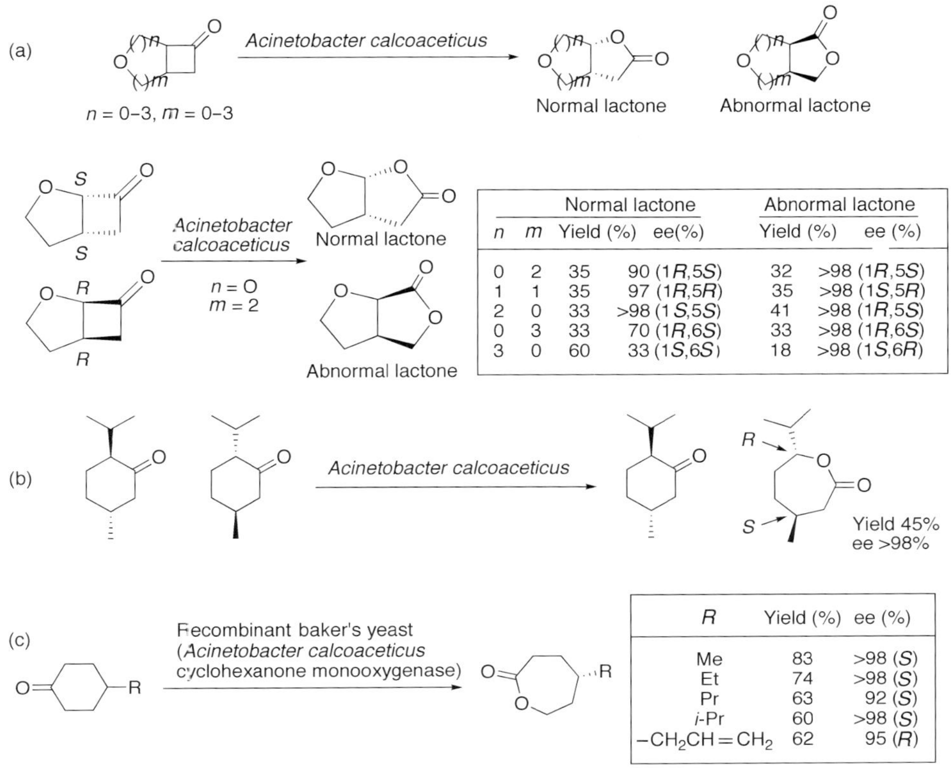

		Normal lactone		Abnormal lactone	
n	*m*	Yield (%)	ee(%)	Yield (%)	ee (%)
0	2	35	90 (1*R*,5*S*)	32	>98 (1*R*,5*S*)
1	1	35	97 (1*R*,5*R*)	35	>98 (1*S*,5*R*)
2	0	33	>98 (1*S*,5*S*)	41	>98 (1*R*,5*S*)
0	3	33	70 (1*R*,6*S*)	33	>98 (1*R*,6*S*)
3	0	60	33 (1*S*,6*S*)	18	>98 (1*S*,6*R*)

R	Yield (%)	ee (%)
Me	83	>98 (*S*)
Et	74	>98 (*S*)
Pr	63	92 (*S*)
i-Pr	60	>98 (*S*)
$-CH_2CH=CH_2$	62	95 (*R*)

Figure 10.18 Baeyer–Villiger oxidations.

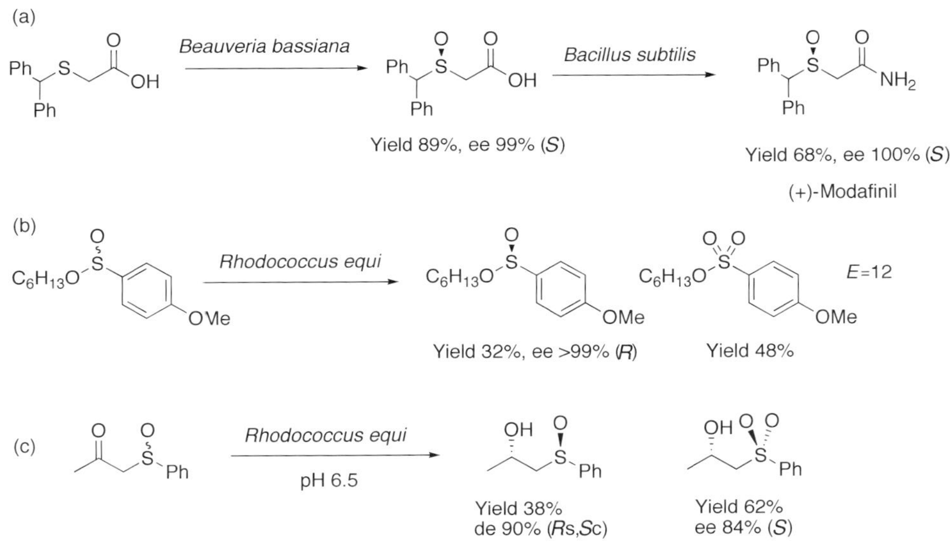

Figure 10.19 Asymmetric oxidation of sulfur compounds.

The regioselectivity of the polymerization could be controlled when the diphenolic moiety of caffeic acid, immobilized on 4-aminothiophenol-functionalized gold surfaces, was polymerized by horseradish peroxidase and hydrogen peroxide (Fig. 10.20(c)).[15b]

10.4 Hydrolysis of esters

Various lipases and esterases have been used for the enantioselective hydrolysis of esters. For example, *Burkholderia cepacia* lipases (PS, Amano Enzyme Inc.) and *Candida antarctica* lipases (CAL, Novozymes) have been widely used for their wide substrate specificities, high

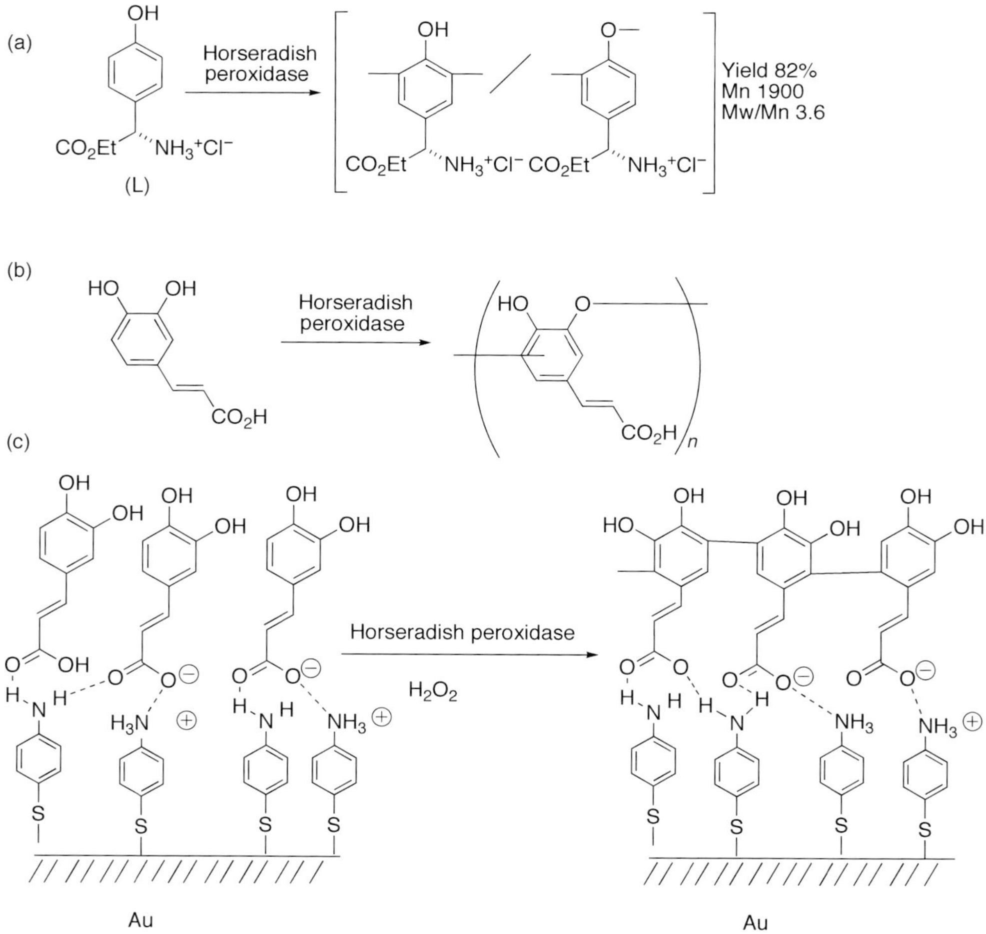

Figure 10.20 Peroxidase-catalyzed oxidative coupling for polymer synthesis.

activities and chemo-, regio- and enantioselectivities. The origins and abbreviations of lipases introduced here are as follows:

PS = *Burkholderia cepacia* (*Pseudomonas cepacia*)
AK = *Pseudomonas* sp.
LIP = *Pseudomonas aeruginosa*
AH = *Pseudomonas* sp.
CRL = *Candida rugosa*
CAL = *Candida antarctica*
QL = *Alcaligenes* sp.

10.4.1 *E*-value

The enantioselectivity of a reaction is usually expressed in terms of ee of the product. However, when a racemic substrate reacts in a selective enzymatic reaction, the ee of the product as well as that of the substrate will depend on total conversion. A better measure of enantioselectivity in this type of reaction is the *E*-value, which is the ratio of the specificity constants of the two enantiomers.[16] When *E* is more than 50, the reaction is often selective enough for practical use.

When *S*-substrate reacts faster than *R*-substrate,

$$E = (V_{max}/K_m)_S \, / \, (V_{max}/K_m)_R$$

where V_{max} is the maximal velocity and K_m is the Michaelis constant.

For the reaction of racemates

E = initial reaction rate for *S*/initial reaction rate for *R*

For example, if $E = 50$ and conversion is at 55%, then ee(substrate) = 99% and ee(product) = 82%. When $E = 20$, and conversion is at 62%, then ee(substrate) = 99% and ee(product) = 60%.

10.4.2 Synthesis of chiral compounds by enzymatic hydrolysis of esters

Various optically active compounds have been synthesized by enzymatic hydrolysis of esters.[17–24] Some examples are shown in Fig. 10.21. The kinetic resolution of racemic esters afforded an intermediate in the synthesis of a β-blocker (Fig. 10.21(a))[17a] and phosphodiesterase inhibitor (Fig. 10.21(b)).[17b] Useful chiral acids can be obtained through the resolution of meso esters by pig liver esterase-catalyzed hydrolysis (Fig. 10.21(c)).[17d,e]

10.4.3 Hydrolysis of sterically hindered esters

Several enzymes can react with sterically hindered compounds. For example, the acetate of a tertiary alcohol can be hydrolyzed by papain in buffer–ethyl acetate (pH 6.5) with excellent *E*-value (>400), affording the (*R*)-alcohol in 49% yield and 98% ee and the (*S*)-acetate in 50% yield and 99% ee.[17f] The remaining (*S*)-acetate was converted to 20(*S*)-camptothecin, which is a pentacyclic alkaloid with potent antitumor activity isolated from *Camptotheca acuminata*. The product (*R*)-acetate can be converted to the starting racemic acetate chemically in 70% yield (Fig. 10.22).

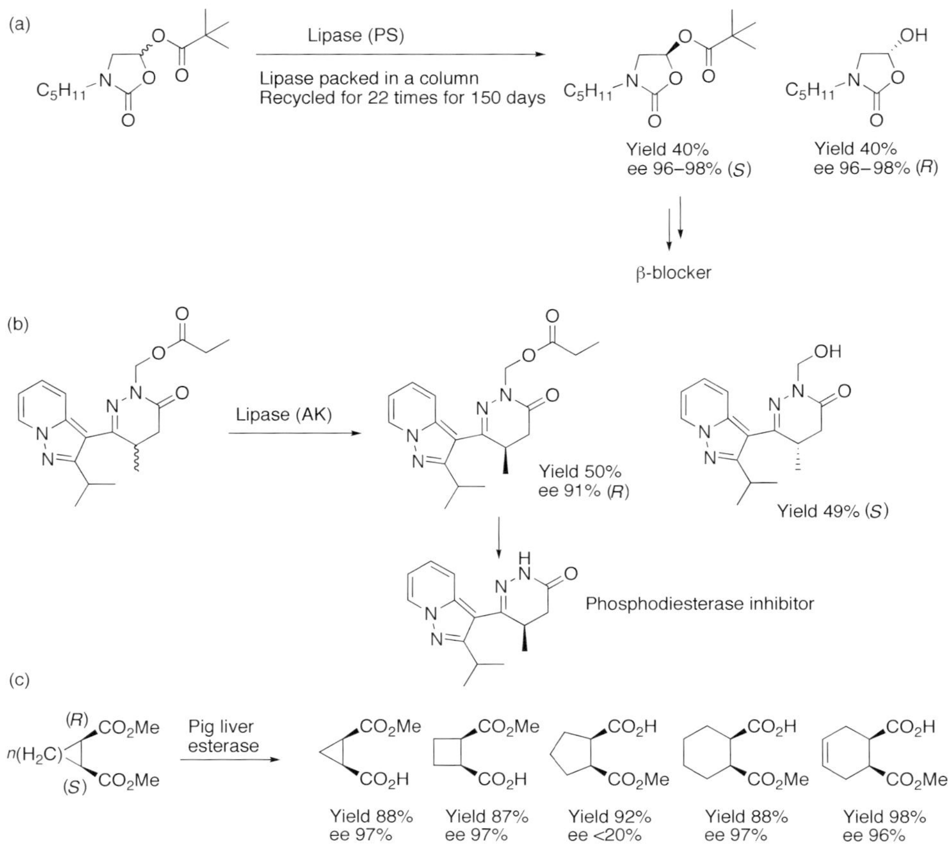

Figure 10.21 Synthesis of optically active compounds through hydrolysis of esters.

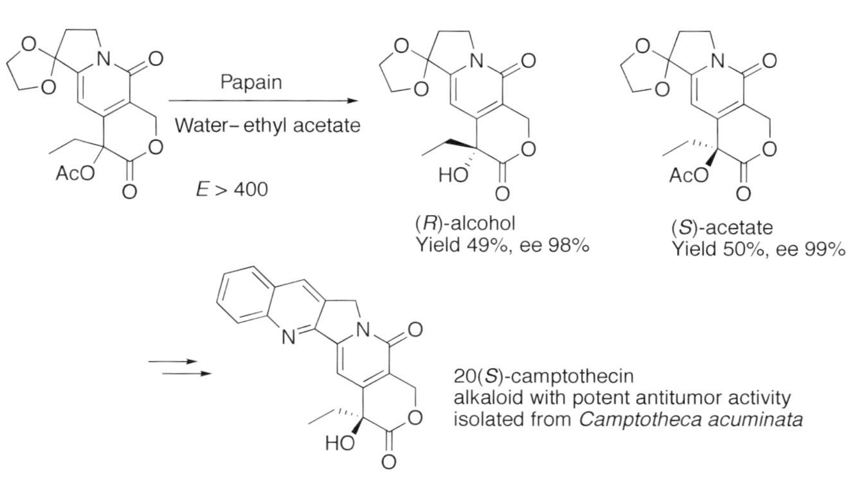

Figure 10.22 Hydrolysis of sterically hindered ester.

10.4.4 Hydrolysis of esters with fluorine functionalities

Chiral fluorinated compounds have been of increasing interest because of the unique influence of fluorine atoms on physical and biological properties. Toward this end, various racemic fluorinated esters have been optically resolved (Fig. 10.23).[18] Kitazume and coworkers prepared trifluoromethyl carbinols through enzymatic resolution of the corresponding acetates by *Candida rugosa* lipase.[18a] Optically active trifluoromethyl alcohol carrying a sulfur group was prepared through resolution, with excellent enantioselectivity, by using *B. cepacia* lipase (lipase PS).[18b] The product was converted to useful chiral compounds including materials for liquid crystals.

10.4.5 Methods of controlling reactivity and enantioselectivity

There are several methods to improve reactivities and selectivities of enzymatic reactions in case they are not sufficient for practical use. Modification of substrate structure and reaction conditions as well as screening and genetic modification of enzymes can improve reactions. For example, although hydrolysis of dimethyl β-aminoglutarate by pig liver esterase afforded the (R)-monoester in 40% ee, modification of the amino moiety with a benzyloxycarbonyl group reversed the stereoselectivity and afforded the (S)-monoester in >96% ee and 93% chemical yield (Fig. 10.24(a)).[19a] In the lipase-catalyzed hydrolysis of cyano acetate, a method to improve the enantioselectivity by using an additive was developed. The addition of a

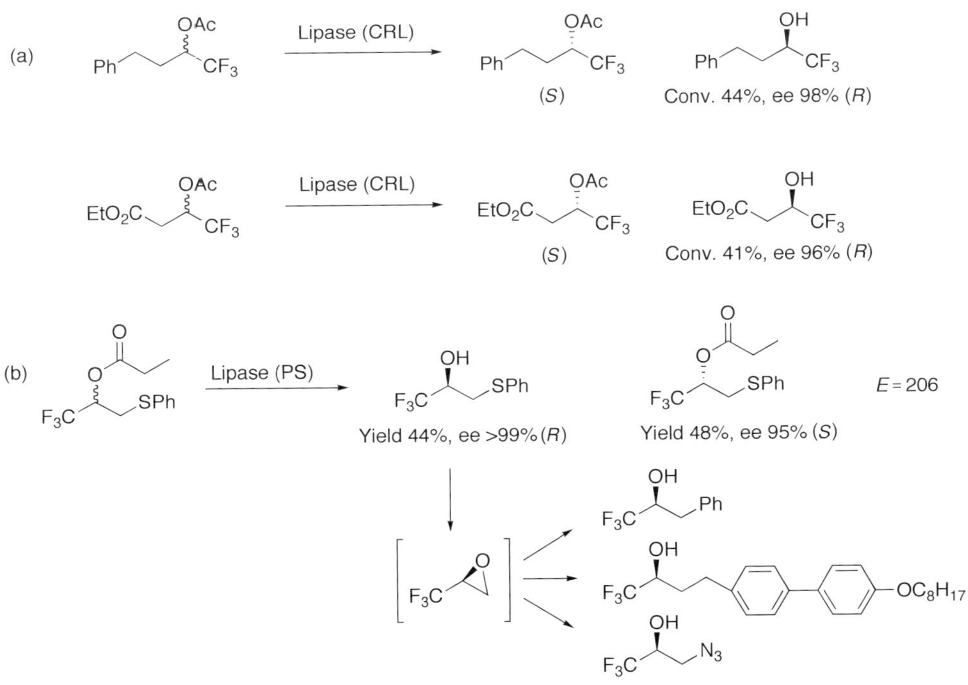

Figure 10.23 Examples of chiral resolution through hydrolysis of esters containing fluorine atoms.[18]

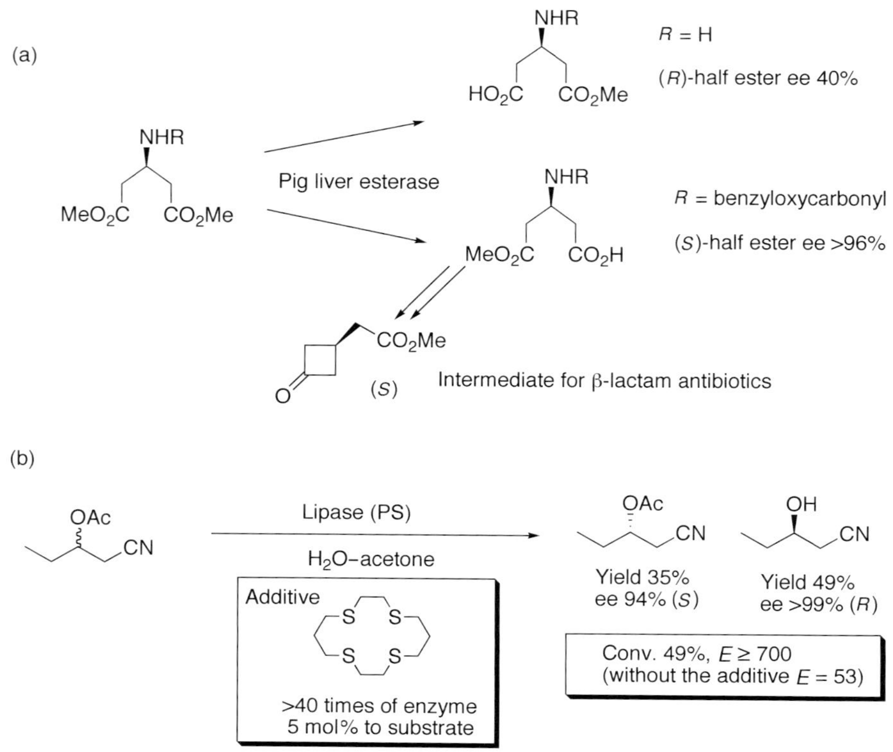

Figure 10.24 Improvement in reactivity and enantioselectivity by changing substrate structure or using an additive.

thiacrown ether dramatically improved the enantioselectivity from $E = 53$ to $E \geq 700$ (Fig. 10.24(b)).[19b,c]

10.4.6 Control of reactivity and enantioselectivity by genetic engineering

Genetic engineering of hydrolytic enzymes has been widely conducted to improve activities, stabilities and enantioselectivities. Examples of point mutation and directed evolution are shown in Fig. 10.25.[20] A mutant of lipase (PS) that had three amino acids exchanged (F221L, V226L, L287I) by site-specific mutagenesis showed inversion of enantioselectivity in the hydrolysis of 1,4-dihydropyridine (Fig. 10.25(a)).[20a] The positions of two of the mutations, V226L and L287I, were close to the reaction center (catalytic triads) of the lipase, but F221L was located at the end of a remote β-sheet. Nevertheless, the double mutant (V226L, L287I) showed no change in enantioselectivity.

The improvement in enantioselectivity by directed evolution of *Pseudomonas aeruginosa* lipase is shown in Fig. 10.25(b).[20b] The combination of different mutagenesis methods (error-prone PCR and site-specific saturation mutagenesis) improved the enantioselectivity from $E = 1.1$ to $E = 25.8$.

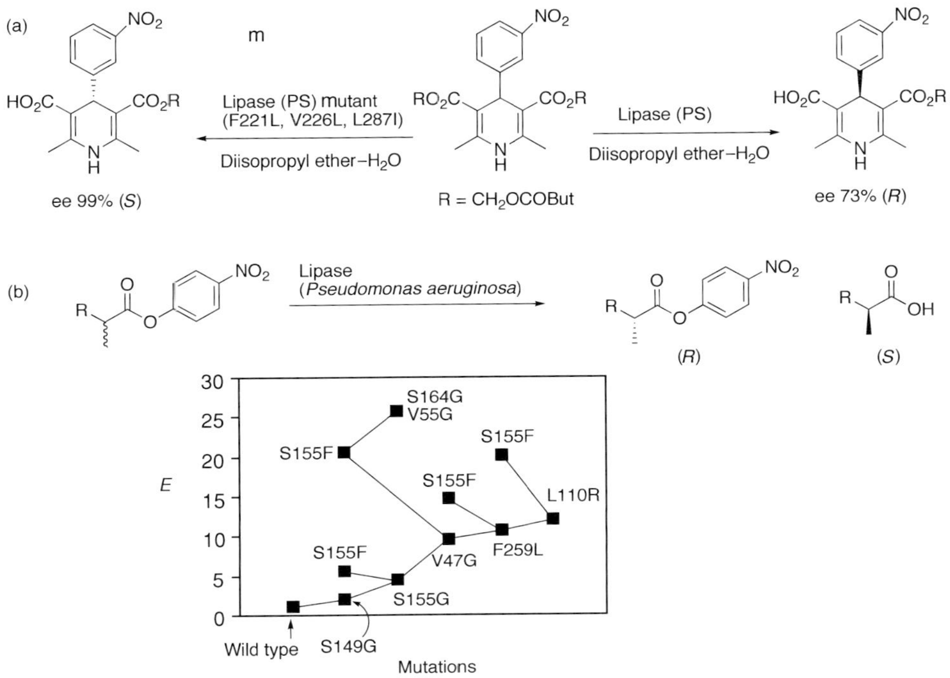

Figure 10.25 Genetic engineering of enzymes for hydrolysis of esters.

10.4.7 Hollow-fiber membrane reactor for lipase-catalyzed hydrolysis: synthesis of diltiazem

A key intermediate in the synthesis of a useful coronary vasodilator, diltiazem [(−)-3-phenylglycidic acid ester, (−)-MPGM], was synthesized through a lipase-catalyzed optical resolution (Fig. 10.26).[21] (+)-MPGM was enantioselectively hydrolyzed to give the corresponding (+)-(2*S*,3*R*)-epoxyacid and unreacted (−)-MPGM. The undesired acid was removed efficiently by spontaneous decarboxylation followed by the formation of water-soluble complex with $NaHSO_3$. A hollow-fiber membrane reactor was developed for this purpose.[21e] The reactor had a toluene layer, an enzyme-containing sponge layer and an aqueous layer (Fig. 10.26(b)). The pressure of the toluene layer was kept slightly higher than that of the aqueous layer, so that the substrate moved to the sponge layer but the water did not move to the toluene layer, preventing the decomposition of MPGM.

10.4.8 Lipase-catalyzed optical resolution coupled with *in situ* inversion: synthesis of prallethrin (pyrethroid), etc.

Normally, optical resolution of a racemate gives 50% of the desired product at 100% ee. However, when resolutions are coupled with reactions that proceed with retention of the substrate configuration but inversion of product, or vice versa, then 100% yield and 100% ee can, in theory, be achieved.[22] This strategy was used for the synthesis of pyrethroids using microbial

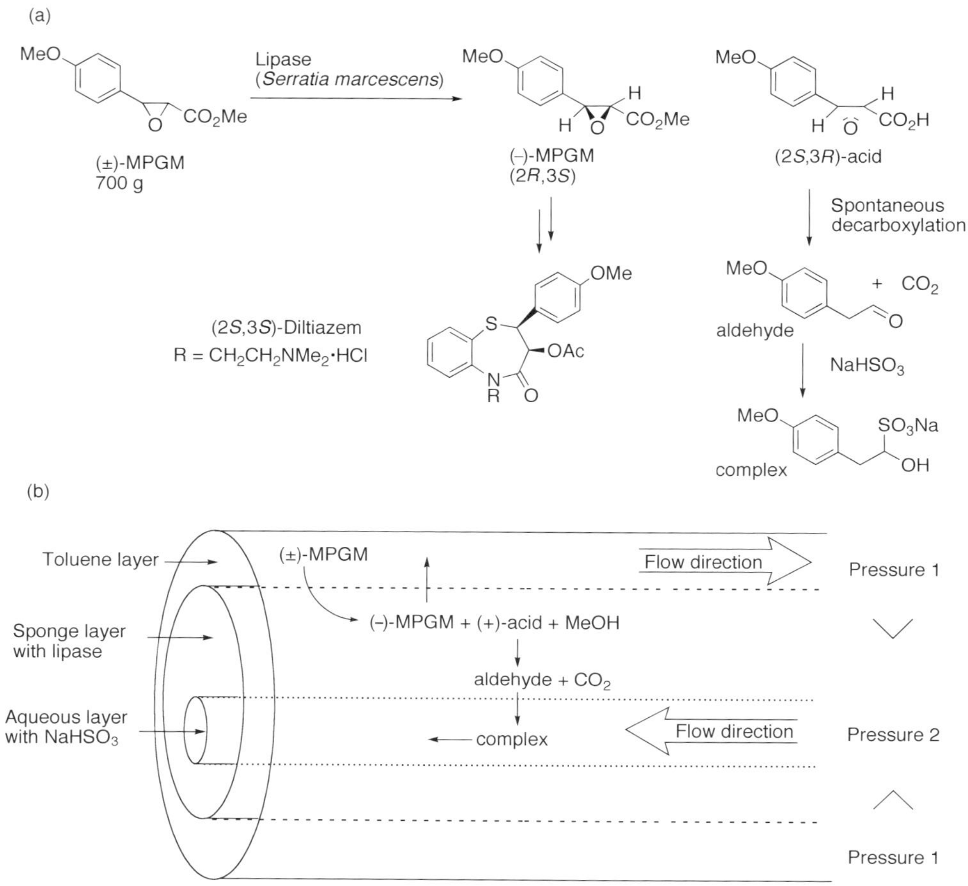

Figure 10.26 Optical resolution of 3-phenylglycidic acid methyl ester (MPGM) by lipase and separation of the optically active product, (–)-MPGM, from the undesired acid using a hollow-fiber membrane reactor.[21]

lipases (Fig. 10.27(a)).[22a,b] Hydrolysis of 4-acetoxy-3-methyl-2-(2-propynyl)-cyclopent-2-enone was catalyzed by *Arthrobacter* lipase in a biphasic liquid reaction system of water and the insoluble substrate. The hydrolysis proceeded even at a substrate concentration of 80 w/v%. After the enzymatic hydrolysis, the resulting mixture was mesylated and hydrolyzed with aqueous $CaCO_3$, leading to inversion of (*R*)-mesylate and retention of (*S*)-acetate, affording the corresponding (*S*)-alcohol, an important intermediate for the preparation of prallethrin, in 83% yield and 91.4% ee.

10.4.9 Recognition of fluorinated functionalities from unfluorinated group: H vs F

Enzymes can discriminate between fluorine and hydrogen atoms on molecules despite of the fact that fluorine is only slightly larger than hydrogen. For example, hydrolysis of

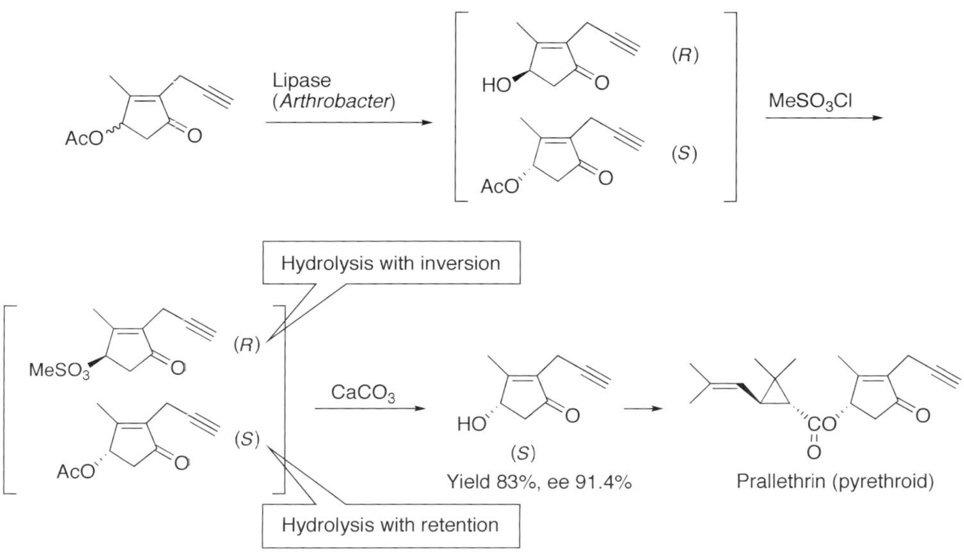

Figure 10.27 Lipase-catalyzed resolution coupled with a chemical inversion.[22a,b]

cis-ethyl-2-fluorocyclopropanecarboxylate by *Micrococcus* sp. (DSC 4011) afforded the corresponding (1*R*,2*R*)-acid in 76% ee remaining the (1*S*,2*S*)-acetate in 98% ee ($E = 36$). The latter was converted to the important key intermediate for a new-generation antibacterial quinolone drug (Fig. 10.28(a)).[23a] The reactions of fluorinated meso compounds were conducted to obtain fluorinated amino acid (Fig. 10.28(b)). Esterification of meso alcohol gave the corresponding (*R*)-amino acid, whereas the hydrolysis gave the corresponding *S*-product.[23b]

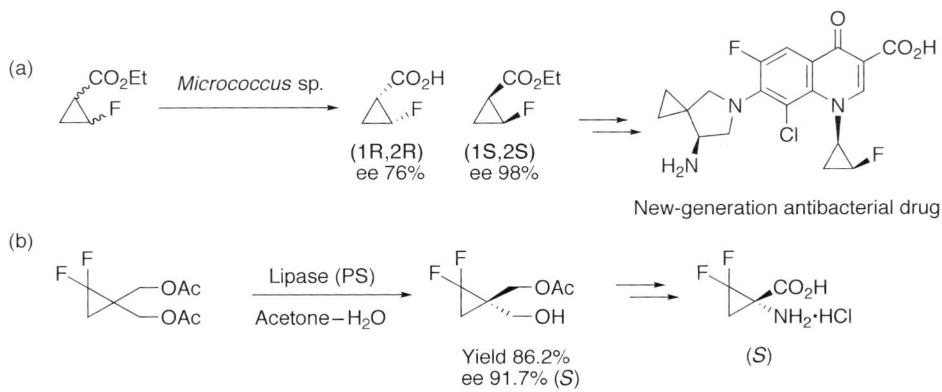

Figure 10.28 Recognition of the difference between H and F separated from the reaction center in the hydrolysis.

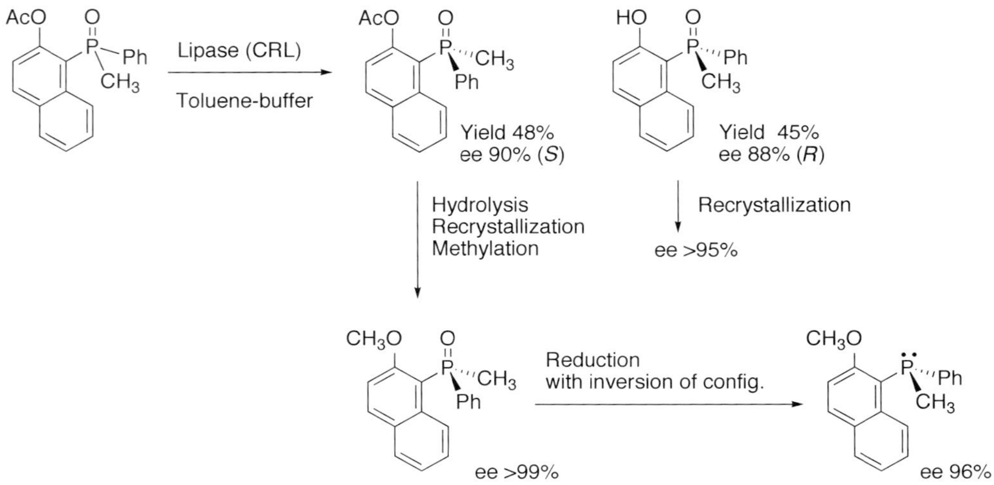

Figure 10.29 Optical resolution of P-chiral compounds with hydrolytic enzymes.[24a]

10.4.10 P-chiral and S-chiral compounds

P-chiral phosphines, which are potential ligands for transition metal-catalyzed reactions, were synthesized through lipase-catalyzed optical resolution of the corresponding racemic phosphine oxide compounds (Fig. 10.29).[24a] For example, lipase from *C. rugosa* (CRL) was used for the enantioselective hydrolysis of acetoxynaphthyl phosphine oxide (Fig. 10.29(a)). The *R*-enantiomer was hydrolyzed selectively, leaving the (*S*)-acetoxy compound, which was further subjected to chemical hydrolysis. Both enantiomeric phosphine oxides were obtained in >95% after recrystallization. Methylation followed by reduction with triethyl amine/trichlorosilane, with inversion of configuration, yielded the desired chiral phosphine.

Chiral sulfoxides have also been efficiently prepared through lipase-catalyzed enantioselective hydrolysis of a racemate (Fig. 10.30).[24b]

10.5 Other types of hydrolysis, dehydration and halogenation

Various enzymes have been used for the hydrolysis of epoxides, amides and nitriles. Some examples are given here. Enzymatic dehydration reaction in water, dehalogenation, as well as halogenation are also discussed.

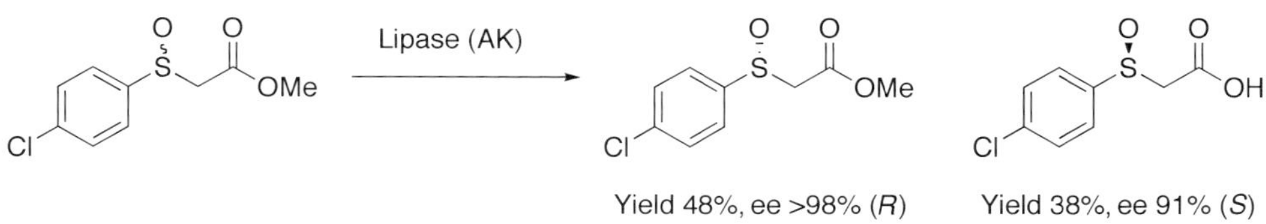

Figure 10.30 Optical resolution of a chiral sulfoxide by enzymatic hydrolysis.[24b]

10.5.1 Hydrolysis of epoxides

Epoxides are important intermediates in many syntheses of bioactive compounds; thus, the demand for chiral epoxides is increasing. An epoxide hydrolase can hydrolyze epoxides enantioselectively (Fig. 10.31).[25] For example, *Aspergillus niger* was used for the hydrolysis of carvone epoxide (Fig. 10.31(a)).[25a] In the reaction of styrene oxide, the nucleophilic reaction with *Beauveria sulfurescents* occurred on the chiral carbon through inversion and that with *A. niger* occurred on the adjacent carbon with retention of configuration at the chiral carbon. Therefore, the reaction of racemic styrene oxide with a combination of these two microorganisms gave the corresponding (*R*)-diol in 92% yield and 89% ee (Fig. 10.31(b)).[25b]

10.5.2 Hydrolysis of amide and nitrile

The enantioselective enzymatic hydrolysis of amides is widely studied. These reactions are catalyzed by acylases, amidases and lipases. Some examples are shown in Fig. 10.32.[26] Aspartame, an artificial sweetener, is synthesized by thermolysin, a protease (Fig. 10.32(a)).[26a] In this reaction, the L-enantiomer of racemic phenylalanine methyl ester reacts selectively with the α-carboxyl group of N-protected L-aspartate. The synthesis is greatly simplified by the fact that the racemic phenylalanine methyl ester can be employed in the reaction, thus avoiding the process of separation. In addition, the high selectivity for condensation at the α-carboxyl group made protection of the γ-carboxyl group unnecessary. Enzyme-catalyzed hydrolysis of amide was applied to the synthesis of levofloxacin, an antibacterial agent with potent activity against both Gram-positive and Gram-negative bacteria. Hydrolysis with *Bacillus* sp. (DSC 1012) afforded (*S*)-amine in 99% ee and unreacted (*R*)-amide in 92% ee ($E = 154$). The (*S*)-amine was converted to levofloxacin.[26c]

Enantioselective hydrolysis of nitriles into amides or acids has primarily been catalyzed by various *Rhodococus* and *Pseudomonas* species (Fig. 10.33).[27] Prochiral compounds were also hydrolyzed to give the corresponding acid in high yield and ee (Fig. 10.33(d), (e) and (g)).[27b,f–h] In the reaction of substituted malononitrile with *Rhodococcus rhodochrous*, the first hydrolysis step leading to diamide proceeded without enantiodiscrimination, but further hydrolysis of the diamide proceeded with high enantioselectivity, affording the (*R*)-acid in 92% yield and 96% ee (Fig. 10.33(g)).[27i]

10.5.3 Dehydration in water for the synthesis of nitriles

In an aqueous medium, hydrolysis proceeds faster than dehydration because water is in large excess. However, in some cases, dehydration is preferred over hydrolysis. Remarkable results were reported by Kato et al. (Fig. 10.34), who demonstrated that aldoxime dehydratase from *Bacillus* sp. efficiently converted aldoximes into the corresponding nitriles in water.[28]

10.5.4 Desulfonation

Regioselective sulfonation of sugars is an important method in organic synthesis. By chemical (organotin) methods, sulfonation of *p*-nitrophenyl β-D-galactopyranoside affords the corresponding 3,6-disulfonated derivative and selective synthesis of the 6-sulfonate is difficult. However, sulfatase could be used to regioselectively hydrolyze the 3-sulfonate group

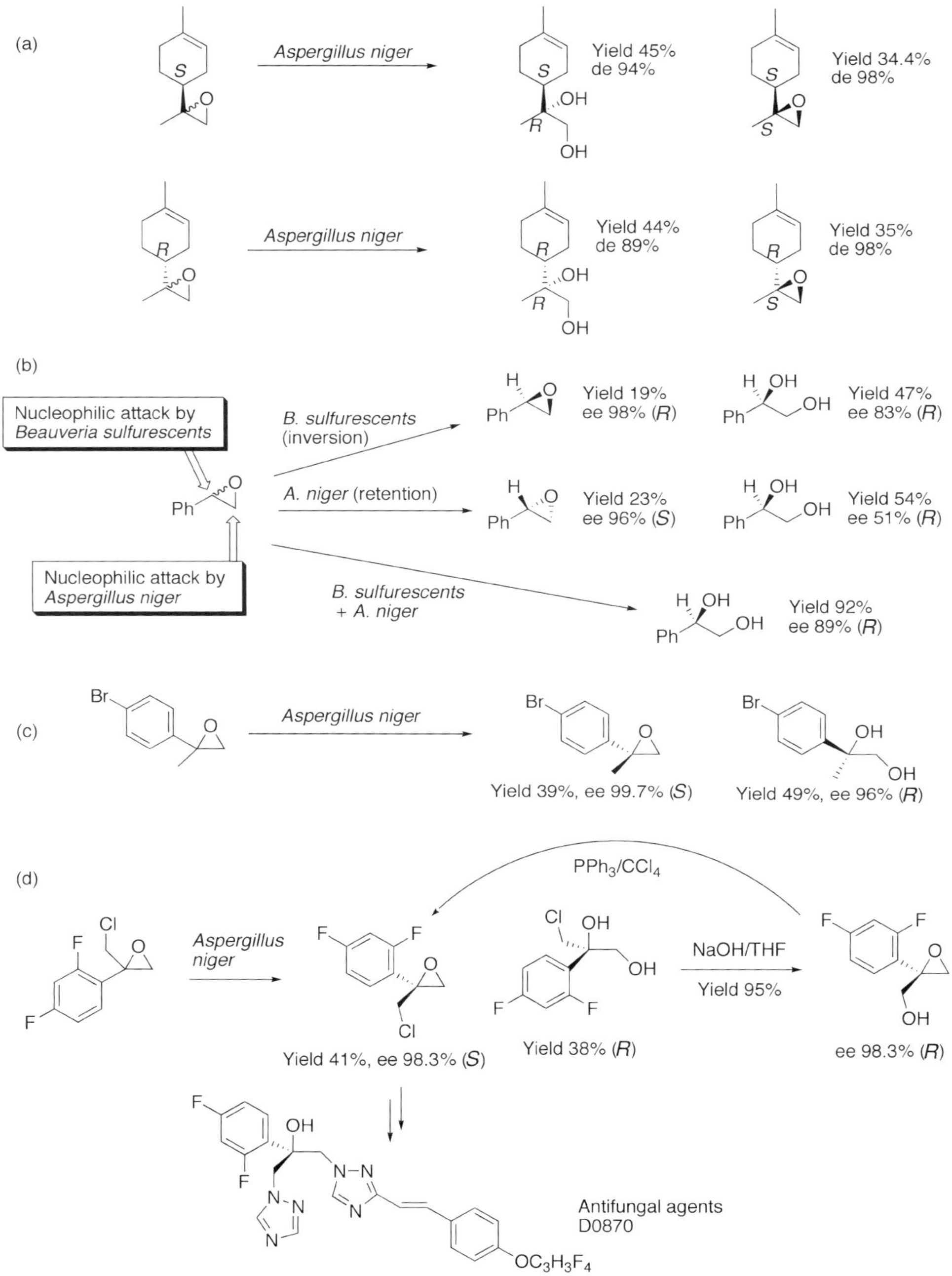

Figure 10.31 Epoxide hydrolase-catalyzed resolutions.

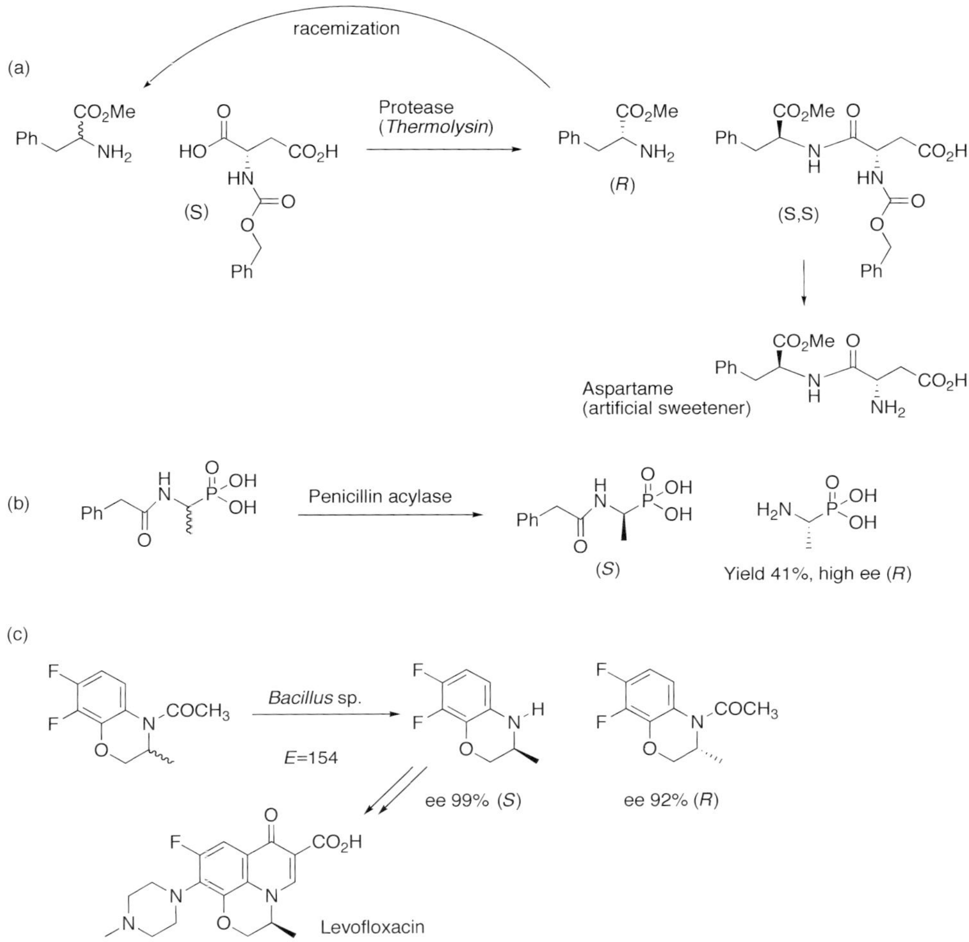

Figure 10.32 Enantioselective hydrolysis of amides.

of the 3,6-disulfonated compound to give the 6-sulfonated sugar in 94% yield. The corresponding lactoside was also prepared with the same method (Fig. 10.35).[29]

10.5.5 Direct glycosylation

Direct glycosylation is difficult to perform by chemical methods unless the sugar is protected. However, enzymatic systems can enable direct glycosylation reactions.[30] The reaction of 1,8-octanediol and glucose using β-glucosidase (EC 3.2.1.21) from almonds gave the glucoside in 58% yield. The product was further transformed into rhodiooctanoside, a natural product known from traditional Chinese medicine (Fig. 10.36).

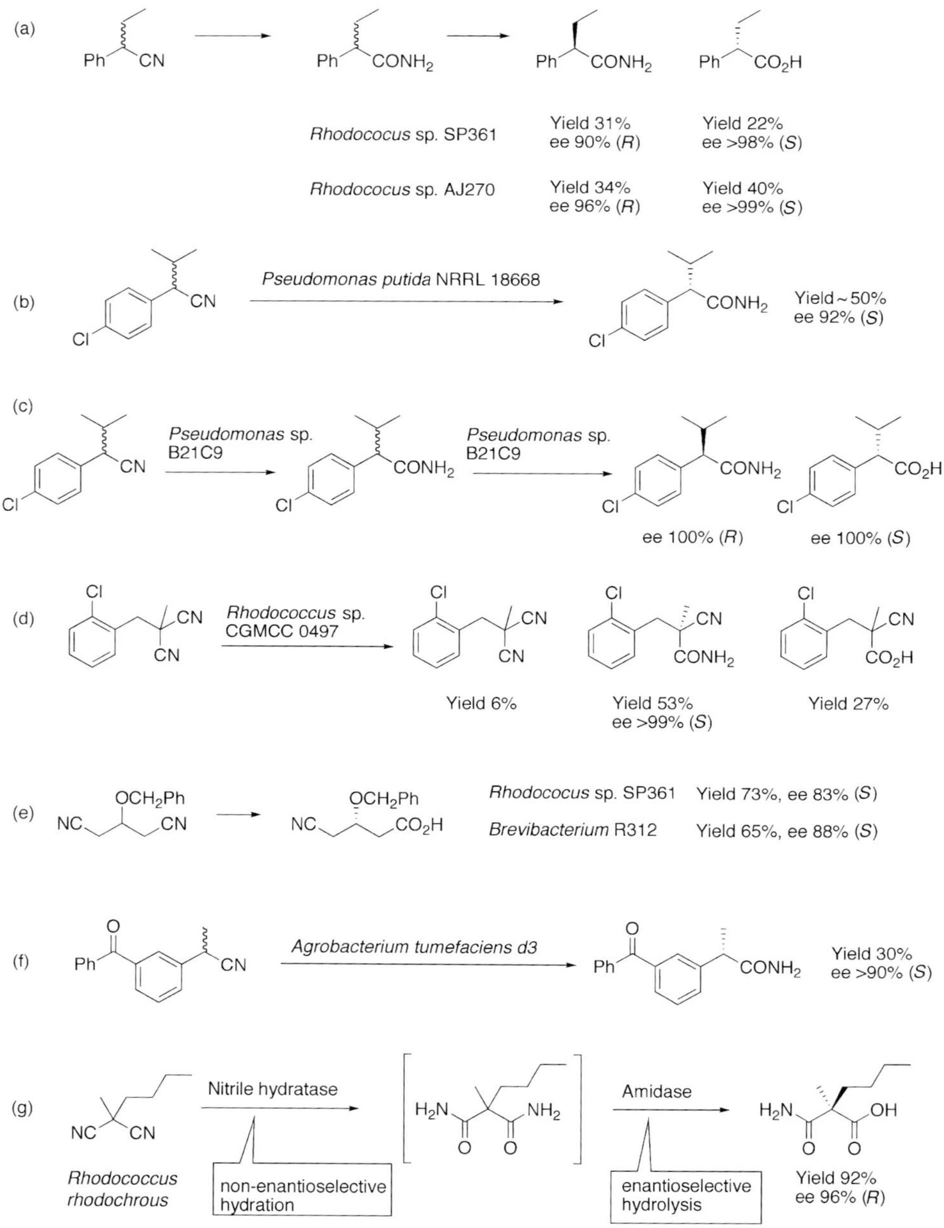

Figure 10.33 Enantioselective hydrolysis of nitriles.

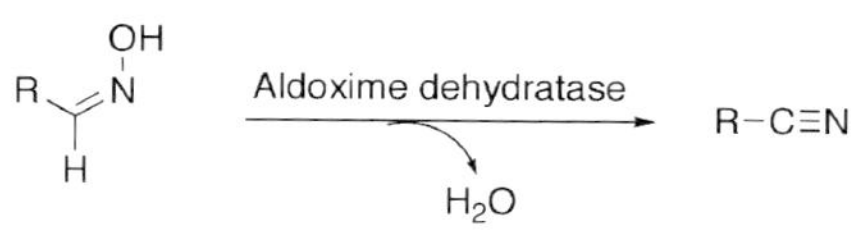

Figure 10.34 Dehydration in water for the synthesis of nitriles from aldoximes.

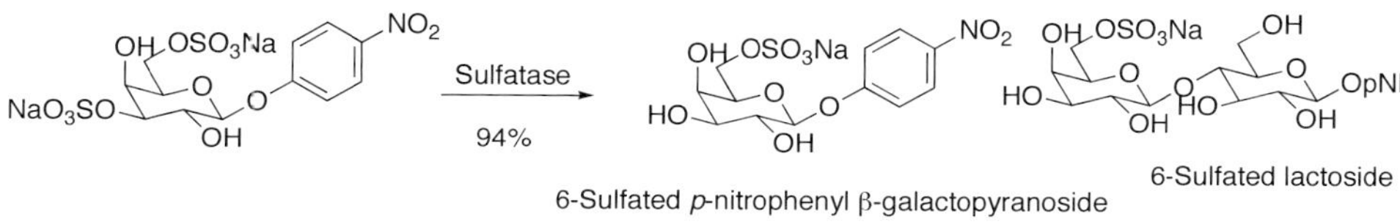

Figure 10.35 Regioselective desulfation.[29]

10.5.6 Dehalogenation

A novel entry to enantiomerically pure alcohols from racemic compounds is the use of dehalogenases.[31] For example, the L-2-halo acid dehalogenase *Pseudomonas putida* was used for the synthesis of D-3-chlorolactic acid from racemic 2,3-dichloropropionic acid (Fig. 10.37(a)).[31a–d] The enzyme catalyzes hydrolytic release of halogen from 2-halocarboxylic acids and produces 2-hydroxy acids with inversion of configuration. L-2-Halo acid dehalogenase acts on the L-isomer of 2-halo acids and produces D-2-hydroxy acid in excellent ee.

(*R*)- and (*S*)-2,3-Dichloro-1-propanol have been prepared by the enantioselective degradation of the S- or R-enantiomers by *Alcaligenes* sp. and *Pseudomonas* sp., respectively (Fig. 10.37(b)).[31e,f] The selectivities were excellent but the yield did not exceed 50% because half of the starting material was degraded.

In the reaction using halohydrin dehalogenase from *Agrobacterium radiobacter* (Fig. 10.37(c)), the R-enantiomer was converted to the corresponding epoxide, which was further converted to (*S*)-diol (ee 91%) by epoxide hydrolase from the same organism to prevent attack of chloride at the β-position.[31g] The unreacted (*S*)-dichloropropanol was obtained in enantiomerically pure form (ee >99%).

10.5.7 Fluorination

Fluorination of organic compounds using fluoride ion can be catalyzed by enzymes (Fig. 10.38).[32] For example, fluorinase from *Streptomyces cattleya* catalyzes C—F bond formation in the reaction of S-adenosyl-L-methionine with fluoride to yield 5′-fluoro-5′-deoxyadenosine.[32a–c] Although their substrate scope is limited at present, these kinds of enzymes may come to play an important role for fluorination in the future due to the ease, mildness and safety of the reaction conditions and the use of fluoride as fluorine source.

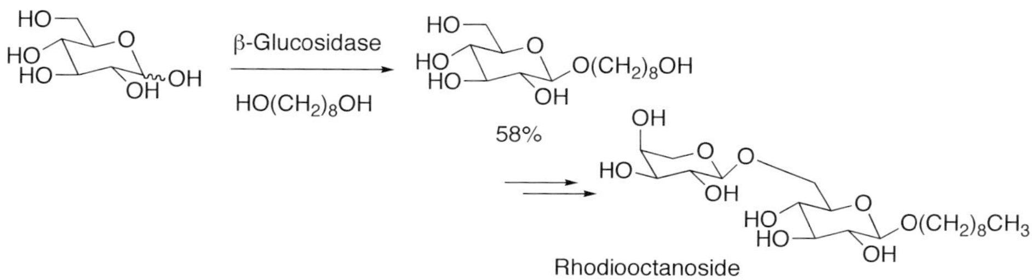

Figure 10.36 A β-glucosidase-catalyzed glycosylation reaction.

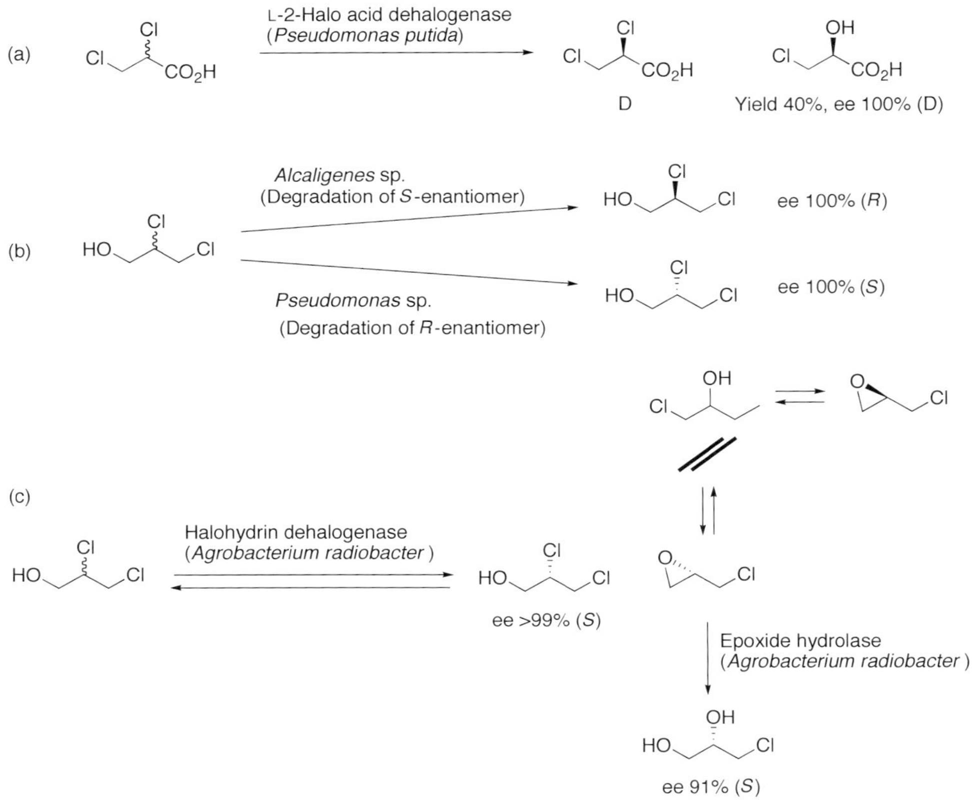

Figure 10.37 Optical resolution through dehalogenation.

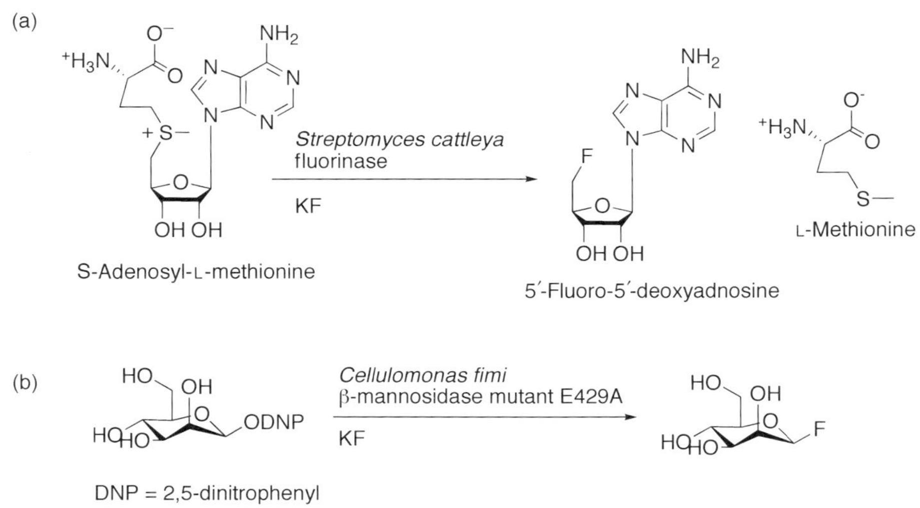

Figure 10.38 Fluorination.

10.6 C—C bond formations

Carbon–carbon bond forming reactions can be catalyzed by enzymes such as aldolases for aldol reactions, hydroxynitrile lyase for cyanohydrin synthesis and decarboxylases for carboxylations. Some examples are shown in this section.[33–35]

10.6.1 Aldol reactions

Aldol reactions have been catalyzed by aldolases as well as by catalytic antibodies.[33] For example, L-threonine aldolase was applied to C—C bond formation of an aldehyde with glycine. The resulting adduct could be further converted to a precursor of *N*-acetyl-4-deoxy-D-mannosamine, a potent inhibitor of *N*-acetylneuraminic acid synthetase (Fig. 10.39(a)).[33a]

Aldol reaction using monoclonal aldolase antibodies, generated against a ketosulfone hapten by reactive immunization, was used to catalyze rapid and highly enantioselective retro-aldol reaction of a thiazole aldol, providing optically pure aldol by kinetic resolution. The product was used for the synthesis of epothilone E (Fig. 10.39(b) and (c)).[33b]

10.6.2 Cyanohydrin synthesis

With the use of biocatalysts, the preparation of chiral cyanohydrins is possible. (*R*)- as well as (*S*)-cyanohydrins are now easily available as a result of the excellent accessibility, the relatively high level of stability and the easy handling of hydroxynitrile lyases (HNLs).[34] An example of the synthesis of (*S*)-cyanohydrins is shown in Fig. 10.40. The optimization of reaction conditions (solvent, temperature and site-directed mutagenesis) has enabled HNL-catalyzed preparation of optically active cyanohydrins on an industrial scale.

10.6.3 Carboxylations

Carboxylation of organic molecules using CO_2 has received attention as an environmentally benign synthetic method.[35] A decarboxylase, an enzyme from *Bacillus megaterium* that catalyzes the elimination of CO_2 from organic molecules, has been found to also catalyze the reverse CO_2 fixation reactions.[35a–d] The substrate scope of this enzyme, however, is limited to pyrrole, which is carboxylated to form pyrrole-2-carboxylate (Fig. 10.41(a)). For the carboxylations of phenol and catechol, decarboxylases from *Clostridium hydroxybenzoicum* have been isolated and employed (Fig. 41(b) and(c)).[36e,f]

10.7 Dynamic kinetic resolution

Dynamic kinetic resolution of racemates to obtain 100% yield of products with 100% ee is theoretically possible when the substrate racemizes but the product does not. Some examples are shown in this section. Deracemization reactions where racemic compounds are converted to enantiomerically pure form without changing the chemical structure will also be discussed.

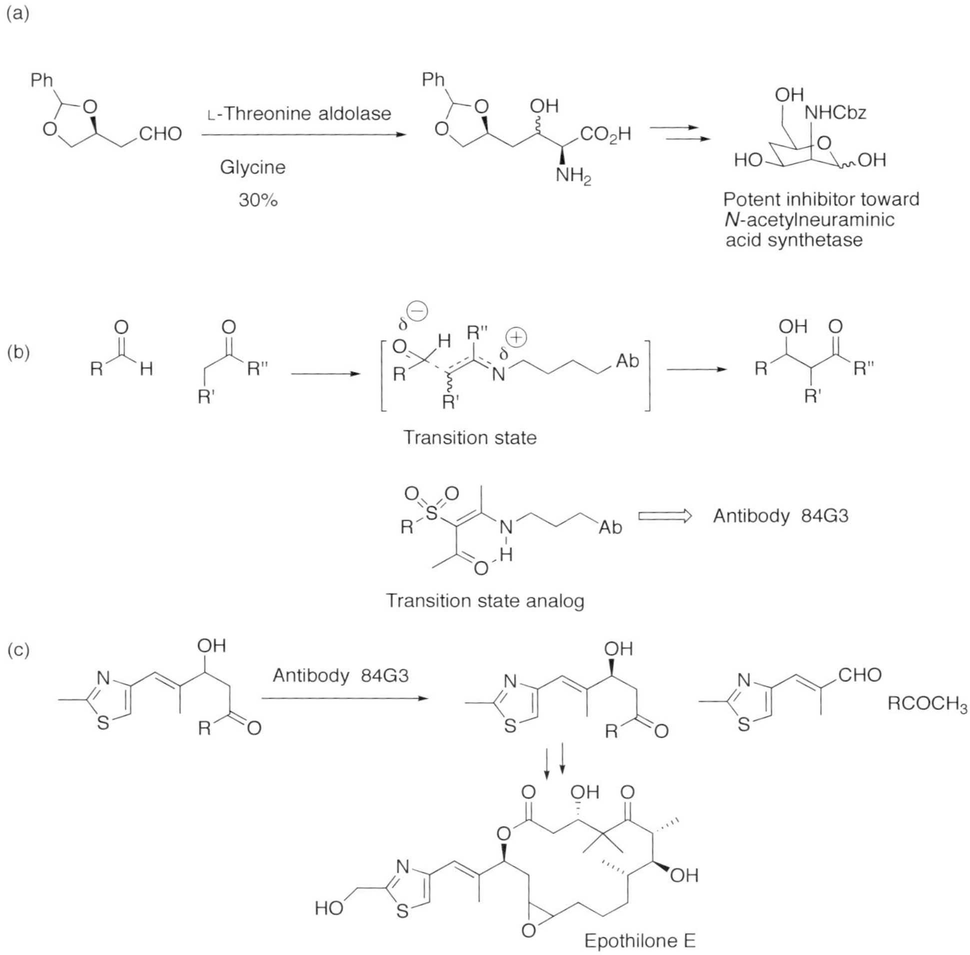

Figure 10.39 Aldolase-catalyzed C—C bond formation and cleavage.

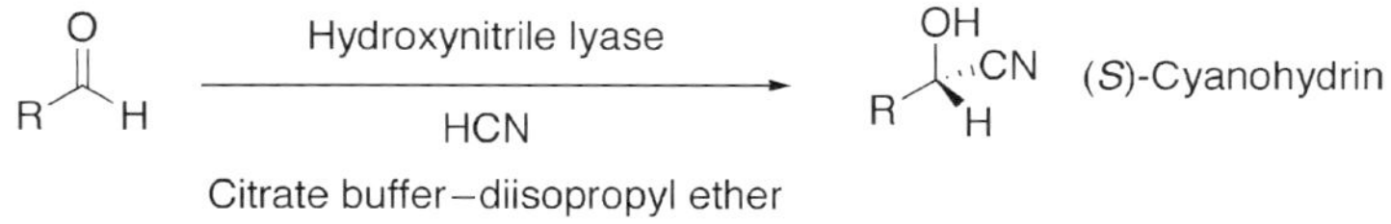

R	Yield (%)	ee (%)	*R*	Yield (%)	ee (%)
C_2H_5	86	91	C_6H_5	100	98
C_4H_9	100	91	4-$CH_3OC_6H_4$	82	98
E-$H_7C_3CH{=}C$	82	97	3-Thienyl	98	98

Figure 10.40 Synthesis of (*S*)-cyanohydrin by hydroxynitrile lyase.[34]

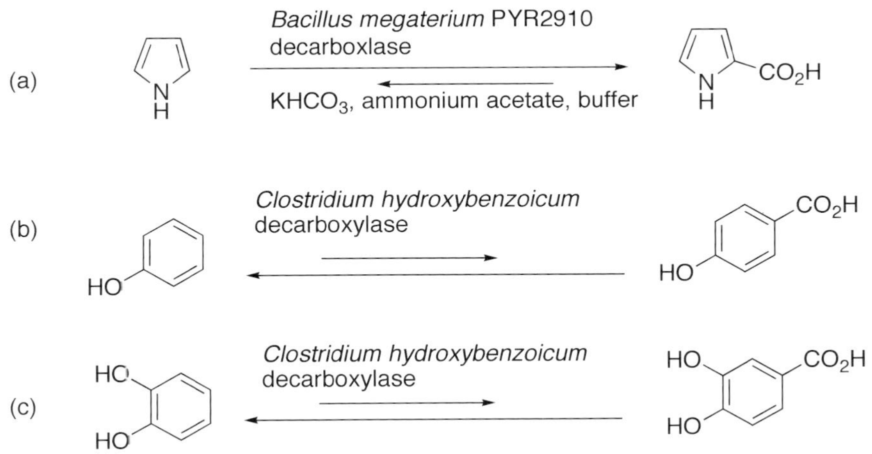

Figure 10.41 Carboxylation by decarboxylase using CO_2 as a carbon source.

10.7.1 Dynamic kinetic resolution of racemic ketones through asymmetric reduction

Dynamic kinetic resolution of racemic ketones can be achieved through asymmetric reduction.[36] For example, baker's yeast reduction of (*R*/*S*)-2-(4-methoxyphenyl)-1,5-benzothiazepin-3,4(2*H*,5*H*)-dione gave only one out of four possible isomers as shown in Fig. 10.42(a).[36a] Only (*S*)-ketone was recognized by the enzyme as a substrate and reduction of the ketone proceeded diastereoselectively to give the enantiomerically pure (2*S*,3*S*)-alcohol. The resulting product was used for the synthesis of (2*S*,3*S*)-diltiazem, a coronary vasodilator.

Dynamic kinetic resolution of α-alkyl-β-keto esters was conducted successfully using biocatalysts (Fig. 10.42(b)).[36c] For the reduction of ethyl 2-methyl-3-oxobutanoate, a method of extensive screening was used to find the suitable microorganism. As a result, *Klebsiella pneumoniae* IFO 3319, out of 450 bacterial strains, was found to give the corresponding (2*R*,3*S*)-hydroxy ester with 99% de and >99% ee in kilogram scale quantitatively.

The dynamic resolution of an aldehyde is shown in Fig. 10.42(c).[36d–g] The racemization of starting aldehyde and enantioselective reduction of the carbonyl group by baker's yeast resulted in the formation of a chiral carbon. The ee of the product was improved from 19 to 90% by changing the ester moiety from the isopropyl group to the neopentyl group.[36e] Other biocatalysts were also used to perform the dynamic kinetic resolution through reduction (Fig. 10.42(c)). For example, *Thermoanaerobium brockii* reduced the aldehyde with moderate enantioselectivity,[36f] while *Candida humicola*, after screening 107 microorganisms, was found to give the (*R*)-alcohol in 98.2% ee with the methyl ester.[36g]

10.7.2 Dynamic kinetic resolution using hydrolytic enzymes

There are various biocatalytic hydrolysis and esterification reactions that result in dynamic kinetic resolution (Fig. 10.43).[37] For example, a protease-catalyzed ester hydrolysis

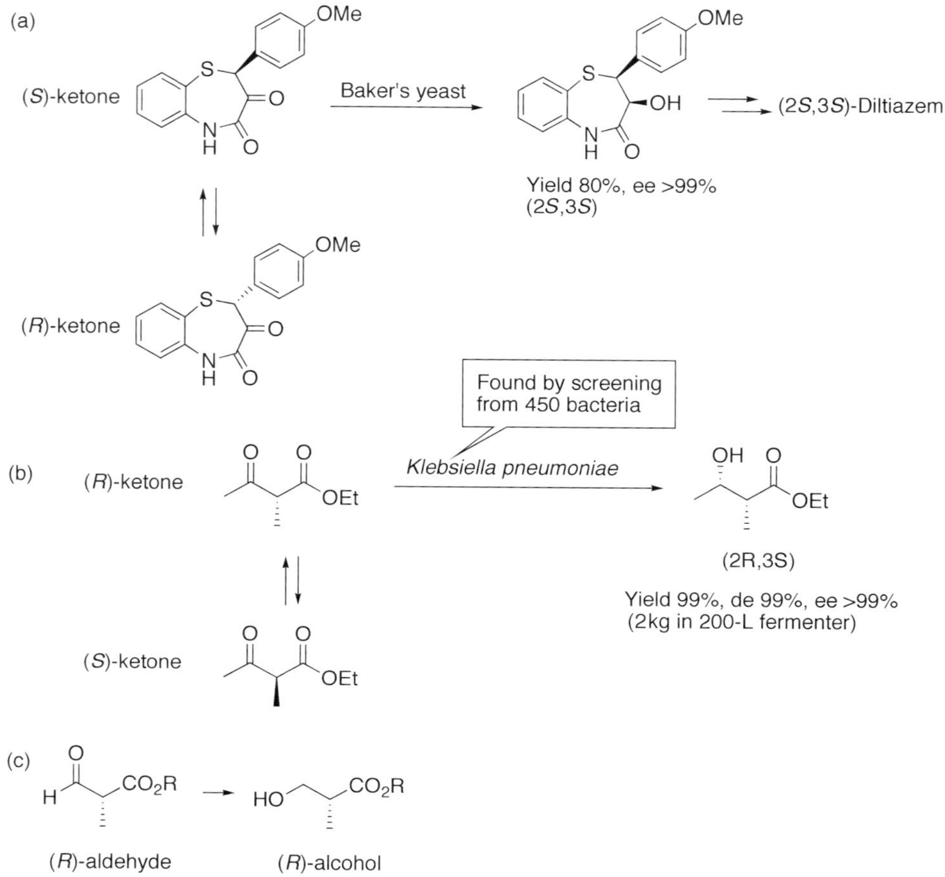

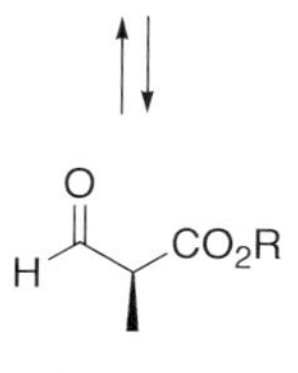

Biocatalysts	*R*	Yield (%)	ee (%)
Baker's yeast	$—CH_2CH_3$	70-80	60–65
Baker's yeast	$—CH(CH_3)_2$	49	19
Baker's yeast	$—CH_2CH(CH_3)_2$	84	64
Baker's yeast	$—CH_2C(CH_3)_3$	78	90
Thermoanaerobium brockii	$—CH_2CH_3$	50–80	72
Candida humicola	$—CH_3$	—	98.2
Candida humicola	$—CH_2CH_3$	—	73.6

Figure 10.42 Dynamic kinetic resolutions by reduction of ketones and aldehydes.

was applied in the synthesis of (*S*)-ketorolac, an anti-inflammatory and analgesic drug (Fig. 10.43(a)). The hydrogen at the α-position of ketorolac ethyl ester is more acidic than that of the product acid.[37a] Therefore, at pH 9.7, the substrate racemized but the product did not. By taking advantage of this acidity difference, (*S*)-ketorolac could be obtained in 92% yield and 85% ee. In the example shown in Fig. 10.43(b), bromide was employed to

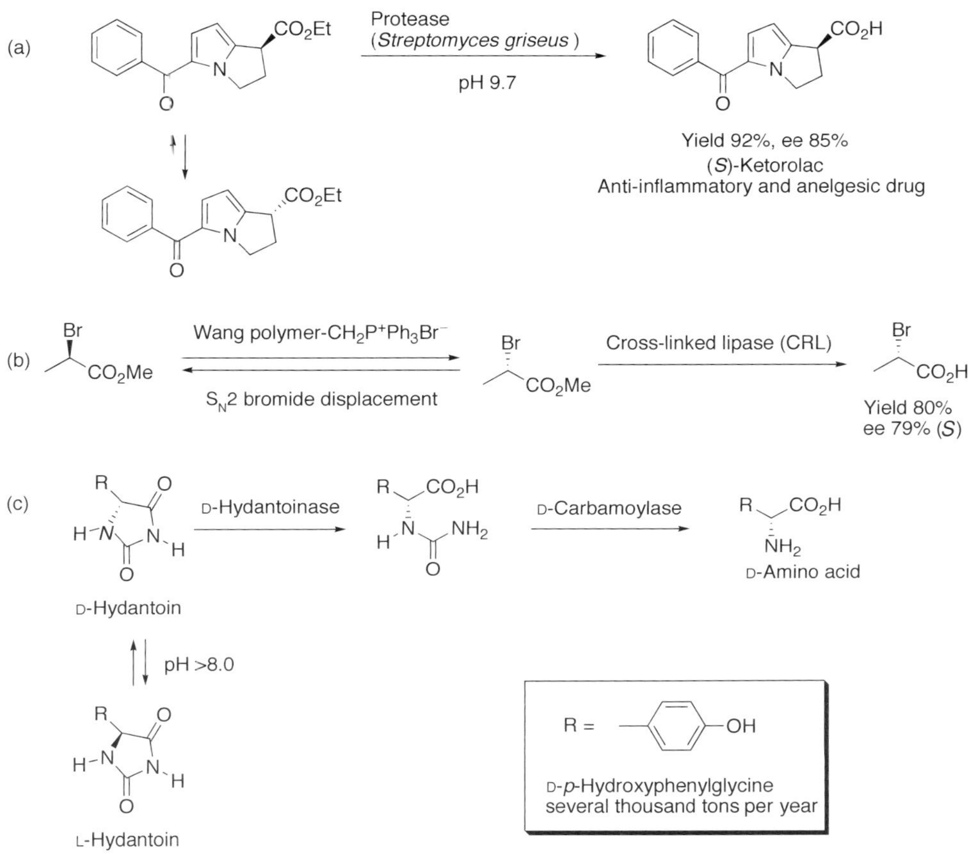

Figure 10.43 Dynamic kinetic resolutions by hydrolysis.

racemize an α-bromo ester more rapidly than the corresponding acid (carboxylate), and this rate difference allowed for dynamic kinetic resolution when the reaction was combined with lipase-catalyzed hydrolysis.[37b] This method to obtain chiral compounds was also used for the synthesis of D-amino acids (Fig. 10.43(c)).[37c] Since hydantoins (imidazolidine-2,4-diones) are easily racemized at conditions of pH >8.0, the use of D-specific hydantoinase and carbamoylase affords D-amino acids. D-Phenylglycine and D-*p*-hydroxyphenylglycine are important in the synthesis of semisynthetic penicillins and are produced in several thousand tons per year using the hydantoinase process.

10.7.3 Deracemization

Deracemization reactions, which convert racemic compounds into chiral form in one step without changing the chemical structure, can be performed using microorganisms that contain several different stereospecific enzymes (Fig. 10.44).[38] For example, for deracemization of 1,2-pentandiol (Fig. 10.44(a)), the R-specific NADH enzyme in *Candida parapsilosis* was

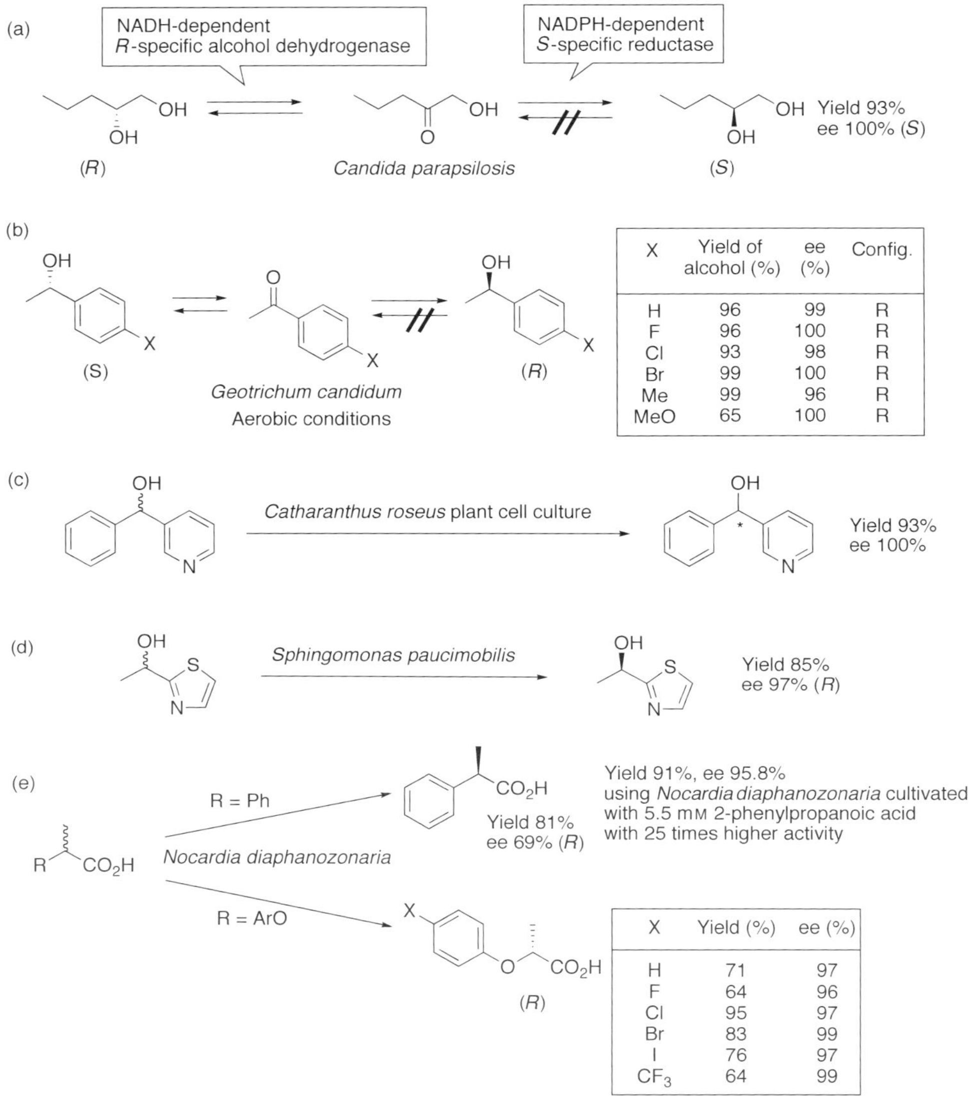

Figure 10.44 Deracemization reactions.

reversible and *S*-specific NADPH enzyme in the same microorganism was irreversible.[38a] Therefore, whole-cell reaction of racemic 1,2-pentandiol gave the (*S*)-diol in 93% yield and 100% ee. For the deracemization of phenylethanol derivatives using *Geotrichum candidum* under aerobic conditions (Fig. 10.44(b)), the *S*-specific enzyme was reversible and the *R*-enzyme irreversible, so (*R*)-alcohols accumulated when the cells and racemic alcohols were mixed in water.[38b,c] Para-substituted phenylethanol derivatives gave better results than meta-substituted derivatives. Racemic acids have also been deracemized through deprotonation and protonation using *Nocardia diaphanozonaria* (Fig. 10.44(e)).[38f–i] In this reaction, subtle differences in the structure of the substrate determined the enantioselectivity.[38f]

10.8 Conclusion

Examples of the use of biocatalysts in water for organic synthesis have been described in this chapter. As shown, biocatalysts can be very suitable for organic synthesis in water, of course because they are by nature designed to work in water. The most important characteristics of biocatalysis include the high chemo-, regio- and enantioselectivities as well as mildness of reaction conditions. The most suitable reaction temperature, for example, is around room temperature. Since biocatalysts are from renewable natural resources and are biologically degradable, they are suitable catalysts from the viewpoint of a sustainable society. Man-made catalysts and biocatalysts will complement each other to provide efficient processes for environment-friendly organic synthesis of the future.

References

1. (a) Fessner, W.-D. (Ed.), *Biocatalysis from Discovery to Application*, Springer, Berlin, 2000; (b) Drauz, K. and Waldmann, H. (Eds.), *Enzyme Catalysis in Organic Synthesis: A Comprehensive Handbook*, Wiley-VCH Verlag GmbH, Weinheim, Germany, 2002.
2. (a) Kazlauskas, R.J., Weissfloch, A.N.E., Rappaport, A.T. and Cuccia, L.A., A rule to predict which enantiomer of a secondary alcohol reacts faster in reactions catalyzed by cholesterol esterase, lipase from *Pseudomonas cepacia*, and lipase from *Candida rugosa*, *J. Org. Chem.*, 1991, **56**, 2656–2665; (b) Cygler, M., Grochulski, P., Kazlauskas, R.J., Schrag, J.D., Bouthillier, F., Rubin, B., Serreqi, A.N. and Gupta, A.K., A structural basis for the chiral preferences of lipases, *J. Am. Chem. Soc.*, 1994, **116**, 3180–3186; (c) Nakamura, K., Kawasaki, M. and Ohno, A., Lipase-catalyzed transesterification of aryl-substituted alkanols in an organic solvent, *Bull. Chem. Soc. Jpn.*, 1996, **69**, 1079–1085; (d) Ema, T., Kobayashi, J., Maeno, S., Sakai, T. and Utaka, M., Origin of the enantioselectivity of lipases explained by a stereo-sensing mechanism operative at the transition state, *Bull. Chem. Soc. Jpn.*, 1998, **71**, 443–453.
3. (a) Bradshaw, C.W., Fu, H., Shen, G.-J. and Wong, C.-H., A *Pseudomonas* sp. alcohol dehydrogenase with broad substrate specificity and unusual stereospecificity for organic synthesis, *J. Org. Chem.*, 1992, **57**, 1526–1532; (b) Bradshaw, C.W., Hummel, W. and Wong, C.-H., *Lactobacillus kefir* alcohol dehydrogenase: a useful catalyst for synthesis, *J. Org. Chem.*, 1992, **57**, 1532–1536; (c) Nakamura, K., Shiraga, T., Miyai, T. and Ohno, A., Stereochemistry of NAD(P)-coenzyme in the reaction catalyzed by glycerol dehydrogenase, *Bull. Chem. Soc. Jpn.*, 1990, **63**, 1735–1737; (d) Nakamura, K., Takano, S., Terada, K. and Ohno, A., Asymmetric reduction of butyl pyruvate catalyzed by immobilized glycerol dehydrogenase in organic–aqueous biphasic media, *Chem. Lett.*, 1992, 951–954; (e) Nakamura, K., Yoneda, T., Miyai, T., Ushio, K., Oka, S. and Ohno, A., Asymmetric reduction of ketones by glycerol dehydrogenase from *Geotrichum*, *Tetrahedron Lett.*, 1988, **29**, 2453–2454; (f) Dutler, H., Van Der Baan, J.L., Hochuli, E., Kis, Z., Taylor, K.E. and Prelog, V., Dihydroxyacetone reductase from *Mucor javanicus*. 1. Isolation and properties, *Eur. J. Biochem.*, 1977, **75**, 423–432; (g) Weinhold E.G., Glasfeld, A., Ellington, A.D. and Benner, S.A., Structural determinants of stereospecificity in yeast alcohol dehydrogenase, *Proc. Natl. Acad. Sci. U.S.A.*, 1991, **88**, 8420–8424; (h) Jones, J.B., Enzymes in organic synthesis, *Tetrahedron*, 1986, **42**, 3351–3403; (i) Jones, J.B. and Beck, J.F., Asymmetric synthesis and resolutions using enzymes, In: Jones, J.B., Sih, C.J. and Perlman, D. (Eds.), *Applications of Biochemical Systems in Organic Chemistry*, John Wiley and Sons, New York, 1976, pp. 107–401; (j) Davies, J. and Jones, J.B., Enzymes in organic synthesis. 16. Heterocyclic ketones as substrates of horse liver alcohol dehydrogenase – stereospecific reductions of 2-substituted tetrahydrothiopyran-4-ones, *J. Am. Chem. Soc.*, 1979, **101**, 5405–5410; (k) Lam, L.K.P., Gair, I.A. and Jones, J.B., Enzymes in organic synthesis. 41. Stereoselective horse liver alcohol dehydrogenase catalyzed reduction of heterocyclic bicyclic ketones, *J. Org. Chem.*, 1988, **53**, 1611–1615; (l) Velonia, K., Tsigos, I., Bouriotis, V. and Smonou, I., Stereospecificity of hydrogen transfer by the NAD(+)-linked alcohol dehydrogenase from the antarctic psychrophile *Moraxella* sp. TAE123, *Bioorg. Med. Chem. Lett.*, 1999, **9**, 65–68.

4. (a) Mori, K. and Mori, H., Yeast reduction of 2,2-dimethylcyclohexane-1,3-dione: (*S*)-(+)-3-hydroxy-2,2-dimethylcyclohexnone, *Org. Synth.*, 1993, **8**, 312–315; (b) Mori, K., Takayama, S. and Yoshimura, S., Preparative bioorganic chemistry. 17. Reduction of a prochiral diketone, 9-methey-*trans*-decalin-1,8-dione, with bakers' yeast, *Liebigs Ann. Chem.*, 1993, 91–95; (c) Brooks, D.W., Mazdiyasni, H. and Grothaus, P.G., Asymmetric microbial reduction of prochiral 2,2-disubstituted cycloalkanediones, *J. Org. Chem.*, 1987, **52**, 3223–3232; (d) Ohta, H., Ozaki, K. and Tuchihashi, G., Regio- and enantioselective reduction of α,γ-diketones by fermenting bakers' yeast, *Agric. Biol. Chem.*, 1986, **50**, 2499–2502; (e) Zhu, Y.-Y. and Burnell, D.J., Enantio- and diastereoselectivity in the reduction of spiro[4.5]decane-1,4-dione derivatives with baker's yeast, *Tetrahedron: Asymmetry*, 1996, **7**, 3295–3304; (f) Zhou, B.-N., Gopalan, A.S., VanMiddlesworth, F., Shieh, W.-R. and Sih, C.J., Stereochemical control of yeast reductions. 1. Asymmetric synthesis of L-carnitine, *J. Am. Chem. Soc.*, 1983, **105**, 5925–5926; (g) Chen, C.-S., Zhou, B.-N., Girdaukas, G., Shieh, W.-R., VanMiddlesworth, F., Gopalan, A.S. and Sih, C.J., Stereochemical control of yeast reductions. 2. Quantitative treatment of the kinetics of competing enzyme systems for a single substrate, *Bioorg. Chem.*, 1984, **12**, 98–117; (h) Shieh, W.-R., Gopalan, A.S. and Sih, C.J., Stereochemical control of yeast reductions. 5. Characterization of the oxidoreductases involved in the reduction of β-keto esters, *J. Am. Chem. Soc.*, 1985, **107**, 2993–2994; (i) Ferraboschi, P., Grisenti, P., Manzocchi, A. and Santaniello, E., Baker's yeast-mediated preparation of optically active aryl alcohols and diols for the synthesis of chiral hydroxy acids, *J. Chem. Soc., Perkin Trans. 1*, 1990, 2469–2474.
5. (a) Kaluzna, I., Andrew, A.A., Bonilla, M., Martzen, M.R. and Stewart, J.D., Enantioselective reductions of ethyl 2-oxo-4-phenylbutyrate by *Saccharomyces cerevisiae* dehydrogenases, *J. Mol. Catal., B: Enzymatic*, 2002, **17**, 101–105; (b) Kaluzna, I.A., Matsuda, T., Sewell, A.K. and Stewart, J.D., Systematic investigation of *Saccharomyces cerevisiae* enzymes catalyzing carbonyl reductions, *J. Am. Chem. Soc.*, 2004, **126**, 12827–12832; (c) Homann, M.J., Vail, R.B., Previte, E., Tamarez, M., Morgan, B., Dodds, D.R. and Zaks, A., Rapid identification of enantioselective ketone reductions using targeted microbial libraries, *Tetrahedron*, 2004, **60**, 789–797; (d) Carballeira, J.D., Álvarez, E., Campillo, M., Pardo, L. and Sinisterra, J.V., *Diplogelasinospora grovesii* IMI 171018, a new whole cell biocatalyst for the stereoselective reduction of ketones, *Tetrahedron: Asymmetry*, 2004, **15**, 951–962; (e) Kambourakis, S. and Rozzell, J.D., Ketoreductases in the synthesis of valuable chiral intermediates: application in the synthesis of α-hydroxy β-amino and β-hydroxy γ-amino acids, *Tetrahedron*, 2004, **60**, 663–669; (f) Feske, B.D., Kaluzna, I.A. and Stewart, J.D., Enantiodivergent, biocatalytic routes to both Taxol side chain antipodes, *J. Org. Chem.*, 2005, **70**, 9654–9657.
6. (a) Nakamura, K., Matsuda, T. and Harada, T., Chiral synthesis of secondary alcohols using *Geotrichum candidum*, *Chirality*, 2002, **14**, 703–708; (b) Nakamura, K., Kitano, K. Matsuda, T. and Ohno, A., Asymmetric reduction of ketones by the acetone powder of *Geotrichum candidum*, *Tetrahedron Lett.*, 1996, **37**, 1629–1632; (c) Matsuda, T., Nakajima, Y., Harada, T. and Nakamura, K., Asymmetric reduction of simple aliphatic ketones with dried cells of *Geotrichum candidum*, *Tetrahedron: Asymmetry*, 2002, **13**, 971–974; (d) Nakamura, K., Matsuda, T., Itoh, T. and Ohno, A., Different stereochemistry for the reduction of trifluoromethyl ketones and methyl ketones catalyzed by alcohol dehydrogenase from *Geotrichum*, *Tetrahedron Lett.*, 1996, **37**, 5727–5730; (e) Matsuda, T., Harada, T., Nakajima, N., Itoh, T. and Nakamura, K., Two classes of enzymes of opposite stereochemistry in an organism: one for fluorinated and another for nonfluorinated substrates, *J. Org. Chem.*, 2000, **65**, 157–163; (f) Nakamura, K., Matsuda, T., Shimizu, M. and Fujisawa, T., Asymmetric reduction of trifluoromethyl ketones containing a sulfur functionality by the alcohol dehydrogenase from *Geotrichum*, *Tetrahedron*, 1998, **54**, 8393–8402; (g) Nakamura, K., Takenaka, K., Fuji, M. and Ida, Y., Asymmetric synthesis of both enantiomers of secondary alcohols by reduction with a single microbe, *Tetrahedron Lett.*, 2002, **4**, 3629–3631.
7. (a) Nakamura, K., Yamanaka, R., Tohi, K. and Hamada, H., Cyanobacterium-catalyzed asymmetric reduction of ketones, *Tetrahedron Lett.*, 2000, **41**, 6799–6802; (b) Nakamura, K. and Yamanaka, R., Light mediated cofactor recycling system in biocatalytic asymmetric reduction of ketone, *Chem. Commun.*, 2002, 1782–1783; (c) Nakamura, K. and Yamanaka, R., Light-mediated regulation of asymmetric reduction of ketones by a cyanobacterium, *Tetrahedron: Asymmetry*, 2002, **13**, 2529–2533.
8. (a) Shimoda, K. and Hirata, T., Biotransformation of enones with biocatalysts – two enone reductases from *Astasia longa*, *J. Mol. Catal., B: Enzymatic*, 2000, **8**, 255–264; (b) D'Arrigo, P., Lattanzio, M., Fantoni, G.P. and Servi, S., Chemo-enzymatic synthesis of the active enantiomer of the anorressant 2-benzylmorpholine, *Tetrahedron: Asymmetry*, 1998, **9**, 4021–4026; (c) Stott, K., Saito, K., Thiele, D.J. and Massey, V., Old yellow enzyme – the discovery of multiple isozymes and a family of related proteins, *J. Biol. Chem.*, 1993, **268**, 6097–6106; (d) Vaz, A.D.N., Chakraborty, S. and Massey, V., Old yellow enzyme – aromatization of cyclic enones and the mechanism of a novel dismutation reaction, *Biochemistry*, 1995, **34**, 4246–4256; (e) Meah, Y. and Massey, V., Old yellow enzyme: stepwise reduction of nitro-olefins and catalysis of aci-nitro tautomerization, *Proc. Natl. Acad. Sci. U.S.A.*, 2000, **97**, 10733–10738; (f) Ohta, H., Kobayashi, N. and Ozaki, K., Asymmetric reduction of nitro olefins by fermenting bakers' yeast, *J. Org. Chem.*, 1989, **54**, 1802–1804; (g) Fuganti, C., Pedrocchi-Fantoni, G., Sarra, A. and Servi, S., Stereochemistry of bakers' yeast mediated reduction of α,β-unsaturated δ-lactones in the goniothalamin series, *Tetrahedron: Asymmetry*, 1994, **5**, 1135–1138; (h) van Dyk, M.S., van Rensburg, E., Rensburg, I.P.B. and Moleleki, N., Biotransformation of monoterpenoid ketones by yeasts and yeast-like fungi, *J. Mol. Catal., B: Enzymatic*, 1998, **5**, 149–154; (i) Hirata, T., Shimoda,

K. and Gondai, T., Asymmetric hydrogenation of the C—C double bond of enones with the reductases from *Nicotiana tabacum*, *Chem. Lett.*, 2000, 850–851; (j) Fronza, G., Fuganti, C., Mendozza, M., Rallo, R.S., Ottolina, G. and Joulain, D., Stereochemistry of the double bond saturation in the formation in baker's yeast of 4-(4-hydroxyphenyl)-2-butanone (raspberry ketone), *Tetrahedron*, 1996, **52**, 4041–4052; (k) Takabe, K., Tanaka, M., Sugimoto, M., Yamada, T. and Yoda, H., Asymmetric reduction of 3-hydroxymethylbutanolide derivatives by bakers' yeast – a new approach to the synthesis of factor-1, *Tetrahedron: Asymmetry*, 1992, **3**, 1385–1386; (l) Fronza, G., Fogliato, G., Fuganti, C., Lanati, S., Rallo, R. and Servi, S., On the mode of bakers' yeast reduction of benzylidencyclohexanone, *Tetrahedron Lett.*, 1995, **36**, 123–124; (m) Hogberg, H.E., Hedenstrom, E., Fagerhag, J. and Servi, S., Enantioselective synthesis of (*S*)-2-methyl-1-alkanols via bakers' yeast mediated reduction of α-methyl-2-thiophenepropenals, *J. Org. Chem.*, 1992, **57**, 2052–2059; (n) Sakai, T., Matsumoto, S., Hidaka, S., Imajo, N., Tsuboi, S. and Utaka, M., Asymmetric reduction of carbon–carbon double bonds of conjugated enones with fermenting bakers' yeast, *Bull. Chem. Soc. Jpn.*, 1991, **64**, 3473–3475; (o) Varie, D.L., Brennan, J., Briggs, B., Cronin, J.S., Hay, D.A., Rieck, J.A., III and Zmijewski, M.J., Bioreduction of (*R*)-carvone and regioselective Baeyer–Villiger oxidations: application to the asymmetric synthesis of cryptophycin fragment A, *Tetrahedron Lett.*, 1998, **39**, 8405–8408; (p) Das, B., Kashinatham, A. and Madhusadhan, P., Baker's yeast treatment of some naturally occurring amides, *Tetrahedron Lett.*, 1997, **38**, 7457–7458; (q) Matsumoto, K., Kawabata, Y., Takahashi, J., Fujita, Y. and Hatanaka, M., Asymmetric reduction of α,β-unsaturated cyclic ketones by a yeast, *Chem. Lett.*, 1998, 283–284.

9. (a) Adam, W., Hoch, U., Lazarus, M., Saha-Möller, C.R. and Schreier, P., Enzyme-catalyzed asymmetric synthesis – kinetic resolution of racemic hydroperoxides by enantioselective reduction to alcohols with horseradish peroxidase, *J. Am. Chem. Soc.*, 1995, **117**, 11898–11901; (b) Adam, W., Mock-Knoblauch, C. and Saha-Möller, C.R., Biocatalytic kinetic resolution of hydroperoxy vinylsilanes by horseradish peroxidase (HRP) and lipases: a comparative study, *Tetrahedron: Asymmetry*, 1997, **8**, 1947–1950; (c) Adam, W., Lazarus, M., Hoch, U., Korb, M.N., Saha-Möller, C.R. and Schreier, P., Horseradish peroxidase-catalyzed enantioselective reduction of racemic hydroperoxy homoallylic alcohols: a novel enzymatic method for the preparation of optically active, unsaturated diols and hydroperoxy alcohols, *J. Org. Chem.*, 1998, **63**, 6123–6127; (d) Adam, W., Boss, B., Harmsen, D., Lukacs, Z., Saha-Möller, C.R. and Schreier, P., Kinetic resolution of chiral hydroperoxides: hydrogen-peroxide-mediated screening of peroxidase-active soil bacteria, *J. Org. Chem.*, 1998, **63**, 7598–7599; (e) Nagatomo, H., Matsushita, Y., Sugamoto, K. and Matsui, T., Enantioselective reduction of γ-hydroperoxy-α,β-unsaturated carbonyl compounds catalyzed by lipid-coated peroxidase in organic solvents, *Tetrahedron: Asymmetry*, 2003, **14**, 2339–2350.
10. (a) Abo, M., Okubo, A. and Yamazaki, S., Preparative asymmetric deoxygenation of alkyl aryl sulfoxides by *Rhodobacter sphaeroides* f. sp. *denitrificans*, *Tetrahedron: Asymmetry*, 1997, **8**, 345–348; (b) Kawasaki, T., Watanabe, N., Sugai, T. and Ohta, H., Enantioselective oxidation of sulfinic acid esters with the aid of a microorganism, *Chem. Lett.*, 1992, 1611–1614.
11. (a) Stampfer, W., Kosjek, B., Moitzi, C., Kroutil, W. and Faber, K., Biocatalytic asymmetric hydrogen transfer, *Angew. Chem. Int. Ed.*, 2002, **41**, 1014–1017; (b) Stampfer, W., Kosjek, B., Faber, K. and Kroutil, W., Biocatalytic asymmetric hydrogen transfer employing *Rhodococcus* ruber DSM 44541, *J. Org. Chem.*, 2003, **68**, 402–406; (c) Velonia, K., Tsigos, I., Bouriotis, V. and Smonou, I., Stereospecificity of hydrogen transfer by the NAD(+)-linked alcohol dehydrogenase from the antarctic psychrophile *Moraxella* sp. TAE123, *Bioorg. Med. Chem. Lett.*, 1999, **9**, 65–68; (d) Nagaoka, H., Chiral resolution function with immobilized food proteins, *Biotechnol. Prog.*, 2003, **19**, 1149–1155; (e) Jakovac, I.J., Goodbrand, H.B., Lok, K.P. and Jones, J.B., Enzymes in organic synthesis. 24. Preparation of enantiomerically pure chiral lactones via stereospecific horse liver alcohol dehydrogenase catalyzed oxidations of monocyclic meso diols, *J. Am. Chem. Soc.*, 1982, **104**, 4659–4665; (f) Ng, G.S.Y., Yuan, L.-C., Jakovac, I.J. and Jones, J.B., Enzymes in organic synthesis. 29. Preparation of enantiomerically pure *cis*-2,3-dimethyl and 2,4-dimethyl lactones via horse liver alcohol dehydrogenase catalyzed oxidations, *Tetrahedron*, 1984, **40**, 1235–1243; (g) Ohta, H., Kato, Y. and Tsuchihashi, G., Studies on enzymic oxidation of sulfides. V. Switching of the direction of enzyme-mediated oxidation and reduction of sulfur-substituted 2-propanols and 2-propanones, *J. Org. Chem.*, 1987, **52**, 2735–2739.
12. (a) Hasegawa, J., Ogura, M., Hamaguchi, S., Shimazaki, M., Kawaharada, H. and Watanabe, K., Studies on β-hydroxycarboxylic acids. 1. Stereoselective conversion of isobutyric acid to β-hydroxyisobutyric acid by microorganisms, *J. Ferment. Technol.*, 1981, **59**, 203–208; (b) Hasegawa, J., Ogura, M., Kanema, H., Kawaharada, H. and Watanabe, K., Studies of β-hydroxycarboxylic acids. 5. Production of D-β-hydroxycarboxylic acids from the corresponding carboxylic acids by a mutant of *Candida rugosa*, *J. Ferment. Technol.*, 1983, **61**, 37–42; (c) Arseniyadis, S., Ouazzani, J., Rodriguez, R., Rumbero, A. and Ourisson, G., Biohydroxylations of hydrindenone derivatives, *Tetrahedron Lett.*, 1991, **32**, 3573–3576; (d) Stevenson, J.-A., Westlake, A.C.G., Whittock, C. and Wong, L.-L., The catalytic oxidation of linear and branched alkanes by cytochrome P450(cam), *J. Am. Chem. Soc.*, 1996, **118**, 12846–12847; (e) Bell, S.G., Sowden, R.J. and Wong, L.-L., Engineering the haem monooxygenase cytochrome P450(cam) for monoterpene oxidation, *Chem. Commun.*, 2001, 635–636; (f) Jones, J.P., O'Hare, E.J. and Wong, L.L., The oxidation of polychlorinated benzenes by genetically engineered cytochrome P450(cam): potential applications in bioremediation, *Chem. Commun.*, 2000, 247–248; (g) Li, Z., Feiten, H.J., Chang, D., Duetz, W.A., van Beilen, J.B. and Witholt, B., Preparation

of (R)- and (S)-N-protected 3-hydroxypyrrolidines by hydroxylation with *Sphingomonas* sp. HXN-200, a highly active, regio- and stereoselective, and easy to handle biocatalyst, *J. Org. Chem.*, 2001, **66**, 8424–8430; (h) Iacazio, G., Langrand, G., Baratti, J., Buono, G. and Triantaphylides, C., Preparative enzymatic synthesis of linoleic acid (13*S*)-hydroperoxide using soybean lipoxygenase-1, *J. Org. Chem.*, 1990, **55**, 1690–1691; (i) Martini, D., Buono, G. and Iacazio, G., Regiocontrol of soybean lipoxygenase oxygenation. Application to the chemoenzymatic synthesis of methyl 15(*S*)-HETE and methyl 5(*S*),15(*S*)-diHETE, *J. Org. Chem.*, 1996, **61**, 9062–9064; (j) Hudlicky, T., Luna, H., Barbieri, G. and Kwart, L.D., Enantioselective synthesis through microbial oxidation of arenes. 1. Efficient preparation of terpene and prostanoid synthons, *J. Am. Chem. Soc.*, 1988, **110**, 4735–4741; (k) Gibson, D.T., Gschwendt, B., Yeh, W.K. and Kobal, V.M., Initial reactions in oxidation of ethylbenzene by *Psuedomonas putida*, *Biochemistry*, 1973, **12**, 1520–1528; (l) Finn, K.J., Cankar, P., Jones, T.R.B. and Hudlicky, T., Enzymatic oxidation of thioanisoles: isolation and absolute configuration of metabolites, *Tetrahedron: Asymmetry*, 2004, **15**, 2833–2836; (m) Finn, K.J., Rochon, L. and Hudlicky, T., Processing of cyclopropylarenes by toluene dioxygenase: isolation and absolute configuration of metabolites, *Tetrahedron: Asymmetry*, 2005, **16**, 3606–3613; (n) Zhang, X., Archelas, A., Meou, A. and Furstoss, R., Microbiological transformations. 21. An expedient route to both enantiomers of marmin and epoxyauraptens via microbiological dihydroxylation of 7-geranyloxycoumarin, *Tetrahedron: Asymmetry*, 1991, **2**, 247–250.

13. (a) Petit, F. and Furstoss, R., Microbiological transformations. 25. Enantioselective Baeyer–Villiger oxidation as a tool for the synthesis of enantiopure bicyclic furofuran and pyrofuran chirons, *Tetrahedron: Asymmetry*, 1993, **4**, 1341–1352; (b) Alphand, V. and Furstoss, R., Microbiological transformations. 23. A surprising regioselectivity of microbiological Baeyer–Villiger oxidations of menthone and dihydrocarvone, *Tetrahedron: Asymmetry*, 1992, **3**, 379–382; (c) Stewart, J.D., Reed, K.W., Martinez, C.A., Zhu, J., Chen, G. and Kayser, M.M., Recombinant baker's yeast as a whole-cell catalyst for asymmetric Baeyer–Villiger oxidations, *J. Am. Chem. Soc*, 1998, **120**, 3541–3548; (d) Stewart, J.D., Reed, K.W., Zhu, J., Chen, G. and Kayser, M.M., A 'designer yeast' that catalyzes the kinetic resolutions of 2-alkyl-substituted cyclohexanones by enantioselective Baeyer–Villiger oxidations, *J. Org. Chem.*, 1996, **61**, 7652–7653; (e) Kayser, M.M., Chen, G. and Stewart, J.D., Enantio- and regioselective Baeyer–Villiger oxidations of 2- and 3-substituted cyclopentanones using engineered bakers' yeast, *J. Org. Chem.*, 1998, **63**, 7103–7106; (f) Mihovilovic, M.D., Chen, G., Wang, S., Kyte, B., Rochon, F., Kayser, M.M. and Stewart, J.D., Asymmetric Baeyer–Villiger oxidations of 4-mono- and 4,4-disubstituted cyclohexanones by whole cells of engineered *Escherichia coli*, *J. Org. Chem.*, 2001, **66**, 733–738; (g) Taschner, M.J., Black, D.J. and Chen, Q.-Z., The enzymatic Baeyer–Villiger oxidation – a study of 4-substituted cyclohexanones, *Tetrahedron: Asymmetry*, 1993, **4**, 1387–1390; (h) Berezina, N., Alphand, V. and Furstoss, R., Microbiological transformations. Part 51. The first example of a dynamic kinetic resolution process applied to a microbiological Baeyer–Villiger oxidation, *Tetrahedron: Asymmetry*, 2002, **13**, 1953–1955.
14. Olivo, H.F., Osorio-Lozada, A. and Peeples, T.L., Microbial oxidation/amidation of benzhydrylsulfanyl acetic acid. Synthesis of (+)-modafinil, *Tetrahedron: Asymmetry*, 2005, **16**, 3507–3511.
15. (a) Fukuoka, T., Tachibana, Y., Tonami, H., Uyama, H. and Kobayashi, S., Enzymatic polymerization of tyrosine derivatives. Peroxidase- and protease-catalyzed synthesis of poly(tyrosine)s with different structures, *Biomacromolecules*, 2002, **3**, 768–774; (b) Xu, P., Uyanma, H., James, E.W., Kobayashi, S. and David, L.K., Peroxidase-catalyzed in situ polymerization of surface orientated caffeic acid, *J. Am. Chem. Soc.*, 2005, **127**, 11745–11753.
16. Chen, C.-S., Fujimoto, Y., Girdaukas, G. and Sih, C.J., Quantitative analysis of biochemical kineticresolutions of enantiomers, *J. Am. Chem. Soc.*, 1982, **104**, 7294–7299.
17. (a) Hamaguchi, S., Asada, M., Hasegawa, J. and Watanabe, K., Stereospecific hydrolysis of 2-oxazolidinone esters and separation of products with an immobilized column, *Agric. Biol. Chem.*, 1985, **49**, 1661–1667; (b) Yoshida, N., Aono, M., Tsubuki, T., Awano, K. and Kobayashi, T., Enantioselective synthesis of a chiral pyridazinone derivative by lipase-catalyzed hydrolysis, *Tetrahedron: Asymmetry*, 2003, **14**, 529–535; (c) Iriuchijima, S. and Kojima, N., Asymmetric hydrolysis of esters by biochemical methods. 3. Asymmetric hydrolysis of (±)-1,2-diacetoxy-3-chloropropane and its related compounds with lipase – synthesis of optically pure (*S*)-propanolol, *Agric. Biol. Chem.*, 1982, **46**(5), 1153–1157; (d) Sabbioni, G. and Jones, J.B., Enzymes in organic synthesis. 39. Preparations of chiral cyclic acid esters and bicyclic lactones via stereoselective pig-liver esterase catalyzed hydrolysis of cyclic meso diesters, *J. Org. Chem.*, 1987, **52**, 4565–4570; (e) Kobayashi, S., Kamiyama, K., Iimori, T. and Ohno, M., Creation of novel chiral synthons with enzymes and applications to natural product synthesis. 15. Efficient introduction of chiral centers into cyclohexane ring, *Tetrahedron Lett.*, 1984, **25**, 2557–2560; (f) Imura, A., Itoh, M. and Miyadera, A., Enantioselective synthesis of 20(*S*)-camptothecin using an enzyme-catalyzed resolution, *Tetrahedron: Asymmetry*, 1998, **9**, 2285–2291.
18. (a) Lin, J.T., Yamazaki, T. and Kitazume, T., A microbially based approach for the preparation of chiral molecules possessing the trifluoromethyl, *J. Org. Chem.*, 1987, **52**, 3211–3217; (b) Shimizu, M., Sugiyama, K. and Fujisawa, T., Highly stereocontrolled access to 1,1,1-trifluoro-2,3-epoxypropane via lipase-mediated netic resolution of 1,1,1-trifluoro-3-(phenylthio)propan-2-ol and its application, *Bull. Chem. Soc. Jpn.*, 1996, **69**, 2655–2659.
19. (a) Ohno, M., Kobayashi, S., Iimori, T., Wang, Y.-F. and Izawa, T., Synthesis of (*S*)- and (*R*)-4-[(methoxycarbonyl)methyl]-2-azetidinone by chemicoenzymatic approach, *J. Am. Chem. Soc.*, 1981, **103**,

2405–2406; (b) Itoh, T., Takagi, Y., Murakami, T., Hiyama, Y. and Tsukube, H., Crown ethers as regulators ofenzymatic reactions: enhanced reaction rate and enantioselectivity in lipase-catalyzed hydrolysis of 2-cyano-1-methyl ethyl acetate, *J. Org. Chem.*, 1996, **61**, 2158–2163; (c) Itoh, T., Mitsukura, K., Kanphai, W., Takagi, Y., Kihara, H. and Tsukube, H., Thiacrown ether technology in lipase-catalyzed reaction: scope and limitation for preparing optically active 3-hydroxyalkanenitriles and application to insect pheromone synthesis, *J. Org. Chem.*, 1997, **62**, 9165–9172.

20. (a) Hirose, Y., Kariya, K., Nakanishi, Y., Kurono, Y. and Achiwa, K., Inversion of enantioselectivity in hydrolysis of 1,4-dihydropyridines by point tation of lipase PS, *Tetrahedron Lett.*, 1995, **36**, 1063–1066; (b) Liebeton, K., Zonta, A., Schimossek, K., Nardini, M., Lang, D., Dijkstra, B.W., Reetz, M.T. and Jaeger, K.E., Directed evolution of an enantioselective lipase, *Chem. Biol.*, 2000. **7**, 709–718.
21. (a) Shibatani, T., Omori, K., Akatsuka, H., Kawai, E. and Matsumae, H., Enzymatic resolution of diltiazem intermediate by *Serratia marcescens* lipase: molecular mechanism of lipase secretion and its industrial application, *J. Mol. Catal., B: Enzymatic*, 2000, **10**, 141–149; (b) Furutani, T., Furui, M., Mori, T. and Shibatani, T., Production of (2*R*,3*S*)-3-(4-methoxyphenyl) glycidic acid methyl ester using an emulsion bioreactor containing lipase from *Serratia marcescens*, *Appl. Biochem. Biotechnol.*, 1996, **59**, 319–328; (c) Furui, M., Furutani, T., Shibatani, T., Nakamoto, Y. and Mori, T., A membrane bioreactor combined with crystallizer for production of optically active (2*R*,3*S*)-3-(4-methoxyphenyl)-glycidic acid methyl ester, *J. Ferment. Bioeng.*, 1996, **81**, 21–25; (d) Matsumae, H. and Shibatani, T., Purification and characterization of the lipase from *Serratia marcescens* SR41-8000 resposible for asymmetric hydrolysis of 3-phenylglycidic acid esters, *J. Ferment. Bioeng.*, 1994, **77**, 152–158; (e) Matsumae, H., Furui, M., Shibatani, T. and Tosa, T., Production of optically active 3-phenylglycidic acid ester by the lipase from *Serratia marcescens* on a hollow-fiber membrane reactor, *J. Ferment. Bioeng.*, 1994, **78**, 59–63; (f) Matsumae, H., Furui, M. and Shibatani, T., Lipase-catalyzed asymmetric hydrolysis of 3-phenylglycidic acid ester, the key intermediate in the synthesis of diltiazem hydrochloride, *J. Ferment. Bioeng.*, 1993, **75**, 93–98; (g) Desai, S.B., Argade, N.P. and Ganesh, K.N., Remarkable chemo-, regio-, and enantioselectivity in lipase-catalyzed hydrolysis: efficient resolution of (±)-*threo*-ethyl 3-(4-methoxyphenyl)-2,3-diacetoxypropionate leading to chiral intermediates of (+)-diltiazem, *J. Org. Chem.*, 1996, **61**, 6730–6732.
22. (a) Mitsuda, S., Umemura, T. and Hirohara, H., Preparation of an optically pure secondary alcohol of synthetic pyrethroids using microbial lipases, *Appl. Microbiol. Biotechnol.*, 1988, **2**, 310–315; (b) Mitsuda, S. and Nabeshima, S., Enzymatic optical resolution of a synythetic pyrethroid alcohol. Enantioselective transesterifiction by lipase in organic solvent, *Recl. Trav. Chim. Pays-Bas*, 1991, **110**(5), 151–154; (c) Wallner, A., Mang, H., Glueck, S.M., Steinreiber, A., Mayer, S.F. and Faber, K., Chemo-enzymatic enantio-convergent asymmetric total synthesis of (*S*)-(+)-dictyoprolene using a kinetic resolution-stereoinversion protocol, *Tetrahedron: Asymmetry*, 2003, **14**, 2427–2432.
23. (a) Imura, A., Itoh, M. and Miyadera, A., Resolution of *cis*-2-fluorocyclopropanecarboxylic acid by a microbial enantioselective hydrolysis, *Tetrahedron: Asymmetry*, 1998, **9**, 3047–3052; (b) Kirihara, M., Kawasaki, M., Takuwa, T., Kakuda, H., Wakikawa, T., Takeuchi, Y. and Kirk, K.L., Efficient synthesis of (*R*)- and (*S*)-1-amino-2,2-difluorocyclopropanecarboxylic acid via lipase-catalyzed desymmetrization of prochiral precursors, *Tetrahedron: Asymmetry*, 2003, **14**(12), 1753–1761.
24. (a) Serreqi, A.N. and Kazlauskas, R.J., Kinetic resolution of phosphines and phosphine oxides with phosphorus stereocenters by hydrolases, *J. Org. Chem.*, 1994, **59**, 7609–7615; (b) Burgess, K. and Henderson, I., A facile route to homochiral sulfoxides, *Tetrahedron Lett.*, 1989, **30**, 3633–3636.
25. (a) Chen, X.-J., Archelas, A. and Furstoss, R., Microbiological transformations. 27. The first examples for preparative-scale enantioselective or diastereoselective epoxide hydrolyses using microorganisms – an unequivocal access to all 4 bisabolol, *J. Org. Chem.*, 1993, **58**, 5528–5532; (b) Pedragosa-Moreau, S., Archelas, A. and Furstoss, R., Microbiological transformations. 28. Enantiocomplementary epoxide hydrolyzes as a preparative access to both enantiomers of styrene oxide, *J. Org. Chem.*, 1993, **53**, 5533–5536; (c) Cleij, M., Archelas, A. and Furstoss, R., Microbiological transformations. Part 42. A two-liquid-phase preparative scale process for an epoxide hydrolase catalysed resolution of *para*-bromo-α-methyl styrene oxide. Occurrence of a surprising enantioselectivity enhancement, *Tetrahedron: Asymmetry*, 1998, **9**, 1839–1842; (d) Monfort, N., Archelas, A. and Furstoss, R., Enzymatic transformations. Part 53. Epoxide hydrolase-catalysed resolution of key synthons for azole antifungal agents, *Tetrahedron: Asymmetry*, 2002, **13**, 2399–2401; (e) Karboune, S., Archelas, A., Furstoss, R. and Baratti, J., Immobilization of epoxide hydrolase from *Aspergillus niger* onto DEAE-cellulose: enzymatic properties and application for the enantioselective resolution of a racemic epoxide, *J. Mol. Catal., B: Enzymatic*, 2005, **32**, 175–183.
26. (a) Roberts, S.M., Turner, N.J., Willetts, A.J. and Turner, M.K., *Introduction to Biocatalysis Using Enzymes and Microorganisms*, Cambridge University Press, Cambridge, UK, 1995, pp. 175–178; (b) Solodenko, V.A., Kasheva, T.N., Kukhar, V.P., Kozlova, E.V., Mironenko, D.A. and Svedas, V.K., Preparation of optically active 1-aminoalkylphosphonic acids by stereoselective enzymatic hydrolysis of racemic N-acylated 1-aminoalkylphosphonic acids, *Tetrahedron*, 1991, **47**, 3989–3998; (c) Miyadera, A. and Imura, A., Enantioselective synthesis of a key intermediate of levofloxacin using microbial resolution, *Tetrahedron: Asymmetry*, 1999, **10**, 119–123.

27. (a) Yamada, H., Shimizu, S. and Kobayashi, M., Hydratases involved in nitrile conversion: screening, characterization and application, *Chem. Rec.*, 2001, **1**, 152–161; (b) Beard, T., Cohen, M.A., Parratt, J.S. and Turner, N.J., Stereoselective hydrolysis of nitriles and amides under mild conditions using a whole-cell catalyst, *Tetrahedron: Asymmetry*, 1993, **4**, 1085–1104; (c) Wang, M.X., Lu, G., Ji, G.J., Huang, Z.T., Meth-Cohn, O. and Colby, J., Enantioselective biotransformations of racemic α-substituted phenylacetonitriles and phenylacetamides using *Rhodococcus* sp. AJ270, *Tetrahedron: Asymmetry*, 2000, **11**, 1123–1135; (d) Fallon, R.D., Stieglitz, B. and Turner, I., Jr., A *Pseudomonas putida* capable of stereoselective hydrolysis of nitriles, *Appl. Microbiol. Biotechnol.*, 1997, **47**, 156–161; (e) Masutomo, S., Inoue, A., Kumagai, K., Murai, R. and Mitsuda, S., Enantioselective hydrolysis of (*RS*)-2-isopropyl-4′-chlorophenylacetonitrile by *Pseudomonas* sp. B21C9, *Biosci. Biotechnol. Biochem.*, 1995, **59**, 720–722; (f) Wu, Z.L. and Li, Z.Y., Enantioselective biotransformation of α,α-disubstituted dinitriles to the corresponding 2-cyanoacetamides using *Rhodococcus* sp. CGMCC 0497, *Tetrahedron: Asymmetry*, 2003, **14**, 2133–2142; (g) Maddrell, S.J., Turner, N.J., Kerridge, A., Willetts, A.J. and Crosby, J., Nitrile hydratase enzymes in organic synthesis: enantioselective synthesis of the lactone moiety of the mevinic acids, *Tetrahedron Lett.*, 1996, **37**(33), 6001–6004; (h) Bauer, R., Knackmuss, H.J. and Stolz, A., Enantioselective hydration of 2-arylpropionitriles by a nitrile hydratase from *Agrobacterium tumefaciens* strain d3, *Appl. Microbiol. Biotechnol.*, 1998, **49**, 89–95; (i) Yokoyama, M., Sugai, T. and Ohta, H., Asymmetric hydrolysis of a disubstituted malononitrile by the aid of a microorganism, *Tetrahedron: Asymmetry*, 1993, **4**, 1081–1084.
28. (a) Kato, Y., Nakamura, K., Sakiyama, H., Mayhew, S.G. and Asano, Y., Novel heme-containing lyase, phenylacetaldoxime dehydratase from *Bacillus* sp. strain OxB-1: purification, characterization, and molecular cloning of the gene, *Biochemistry*, 2000, **39**(4), 800–809; (b) Kobayashi, K., Yoshioka, S., Kato, Y., Asano, Y. and Aono, S., Regulation of aldoxime dehydratase activity by redox-dependent change in the coordination structure of the aldoxime–heme complex, *J. Biol. Chem.*, 2005, **280**, 5486–5490.
29. Uzawa, H., Minoura, N., Toba, T., Nishida, Y., Kobayashi, K. and Hiratani, K., Convenient synthetic approach towards regioselectively sulfated sugars using limpet and abalone sulfatase-catalyzed desulfation, *Chem. Commun.*, 1998, (21), 2311–2312.
30. Akita, H., Kawahara, E. and Kato, K., Chemoenzymatic synthesis of rhodiooctanoside isolated from Chinese medicines, rhodiolae radix, *Tetrahedron: Asymmetry*, 2004, **15**, 1623–1629.
31. (a) Onda, M., Motosugi, K. and Nakajima, H., A new approach for enzymatic synthesis of D-3-chlorolactic acid from racemic 2,3-dichloropropionic acid by halo acid dehalogenase, *Agric. Biol. Chem.*, 1990, **54**, 3031–3033; (b) Motosugi, K., Esaki, N. and Soda, K., Enzymatic preparation of D-lactic and L-lactic acid from racemic 2-chloropropionic acid, *Biotechnol. Bioeng.*, 1984, **26**, 805–806; (c) Motosugi, K., Esaki, N. and Soda, K., Purification and properties of 2-halo acid dehalogenase from *Pseudomonas putida*, *Agric. Biol. Chem.*, 1982, **46**, 837–838; (d) Motosugi, K., Esaki, N. and Soda, K., Purification and properties of a new enzyme, DL-2-halo acid dehalogenase from *Pseudomonas* sp., *J. Bacteriol.*, 1982, **150**, 522–527; (e) Kasai, N., Suzuki, T. and Furukawa, Y., Optically active chlorohydrins as chiral C3 and C4 building units: microbial resolution and synthetic applications, *Chirality*, 1998, **10**, 682–692; (f) Kasai, N. and Suzuki, T., Industrialization of the microbial resolution of chiral C-3 and C-4 synthetic units: from a small beginning to a major operation, a personal account, *Adv. Synth. Catal.*, 2003, **345**, 437–455; (g) Spelberg, J.H.L., Vlieg, J.E.T.V., Bosma, T., Kellogg, R.M. and Janssen, D.B., A tandem enzyme reaction to produce optically active halohydrins, epoxides and diols, *Tetrahedron: Asymmetry*, 1999, **10**, 2863–2870.
32. (a) O'Hagan, D., Schaffrath, C., Cobb, S.L., Hamiltont, J.T.G. and Murphy, C.D., Biosynthesis of an organofluorine molecule, *Nature*, 2002, **416**, 279; (b) Dong, C., Huang, F., Deng, H., Schaffrath, C., Spencer, J.B., O'Hagan, D. and Naismith, J.H., Crystal structure and mechanism of a bacterial fluorinating enzyme, *Nature*, 2004, **427**, 561–565; (c) Schaffrath, C., Deng, H. and O'Hagan, D., Isolation and characterisaton of 5′-fluorodeoxyadenosine synthase, a fluorination enzyme from *Streptomyces cattleya*, *FEBS Lett.*, 2003, **547**, 111–114; (d) Zechel, D.L., Reid, S.P., Nashiru, O., Mayer, C., Stoll, D., Jakeman, D.L., Warren, R.A.J. and Withers, S.G., Enzymatic synthesis of carbon–fluorine bonds, *J. Am. Chem. Soc.*, 2001, **123**, 4350–4351.
33. (a) Miura, T. and Kajimoto, T., Application of L-threonine aldolase-catalyzed reaction to the preparation of protected 3*R*,5*R*-dihydroxy-L-homoproline as a mimetic of idulonic acid, *Chirality*, 2001, **13**, 577–580; (b) Sinha, S.C., Sun, J., Miller, G., Barbas, C.F., III and Lerner, R.A., Sets of aldolase antibodies with antipodal reactivities, formal synthesis of epothilone E by large-scale antibody-catalyzed resolution of thiazole aldol, *Organic Lett.*, 1999, **1**, 1623–1626; (c) Zhong, G.F., Shabat, D., List, B., Anderson, J., Sinha, S.C., Lerner, R.A. and Barbas, C.F., III, Catalytic enantioselective retro-aldol reactions: kinetic resolution of β-hydroxyketones with aldolase antibodies, *Angew. Chem. Int. Ed.*, 1998, **37**, 2481–2484.
34. Effenberger, F., Förster, S. and Wajant, H., Hydroxynitrile lyases in stereoselective catalysis, *Curr. Opin. Biotechnol.*, 2000, **11**, 532–539.
35. (a) Wieser, M., Yoshida, T. and Nagasawa, T., Microbial synthesis of pyrrole-2-carboxylate by *Bacillus megaterium* PYR2910, *Tetrahedron Lett.*, 1998, **39**, 4309–4310; (b) Wieser, M., Fujii, N., Yoshida, T. and Nagasawa, T., Carbon dioxide fixation by reversible pyrrole-2-varboxylate decarboxylase from *Bacillus megaterium* PYR2910, *Eur. J. Biochem.*, 1998, **257**, 495–499; (c) Yoshida, T. and Nagasawa, T., Enzymatic functionalization of aromatic N-heterocycles: hydroxylation and carboxylation, *J. Biosci. Bioeng.*, 2000, **89**, 111–118;

(d) Wieser, M., Yoshida, T. and Nagasawa, T., Carbon dioxide fixation by reversible pyrrole-2-carboxylate decarboxylase and its application, *J. Mol. Catal., B: Enzymatic*, 2001, **11**, 179–184; (e) He, Z. and Wiegel, J., Purification and characterization of an oxygen-sensitive, reversible 3,4-dihydoxybenzoate decarboxylase from *Clostridium hydroxybenzoicum*, *Eur. J. Biochem.*, 1995, **229**, 77–82; (f) He, Z. and Wiegel, J., Purification and characterization of an oxygen-sensitive reversible 4-hydroxybenzoate decarboxylase from *Clostridium hydroxybenzoicum*, *J. Bacteriol.*, 1996, **178**(12), 3539–3543.

36. (a) Kometani, T., Sakai, Y., Matsumae, H., Shibatani, T. and Matsuno, R., Production of (2*S*,3*S*)-2,3-dihydro-3-hydroxy-2-(4-methoxyphenyl)-1,5-benzothiazepin-4(5*H*)-one, a key intermediate for diltiazem synthesis, by bakers' yeast-mediated reduction, *J. Ferment. Bioeng.*, 1997, **84**, 195–199; (b) Matsumae, H., Douno, H., Yamada, S., Nishida, T., Ozaki, Y., Shibatani, T. and Tosa, T., Microbial asymmetric reduction of (*RS*)-2-(4-methoxyphenyl)-1,5-benzothiazepin-3,4-dione, the key intermediate in the synthesis of diltiazem hydrochloride, *J. Ferment. Bioeng.*, 1995, **79**, 28–32; (c) Miya, H., Kawada, M. and Sugiyama, Y., Stereoselective reduction of ethyl 2-methyl-3-oxobutanoate by bacteria, *Biosci. Biotech. Biochem.*, 1996, **60**, 95–98; (d) Züger, M.F., Giovannini, F. and Seebach, D., Preparation of ethyl (*R*)-(−)-3-hydroxy-2-methylpropanoate by yeast reduction of ethyl α-formylpropanoate, *Angew. Chem., Int. Ed. Engl.*, 1983, **22**, 1012–1012; (e) Nakamura, K., Miyai, T., Ushio, K., Oka, S. and Ohno, A., Enantioselective reduction of 2-methyl-3-oxopropionate by bakers' yeast, *Bull. Chem. Soc. Jpn.*, 1988, **61**, 2089–2093; (f) Seebach, D., Züger, M.F., Giovannini, F., Sonnleitner, B. and Fiechter, A., Preparative microbial reduction of β-oxoesters with *Thermoanerobium brockii*, *Angew. Chem., Int. Ed. Engl.*, 1984, **23**, 151–151; (g) Matzinger, P.K. and Leuenberger, H.G.W., Screening for microorgnisms performing the stereoselecitve reduction of α-formayl-esters, *Appl. Microbiol. Biotechnol.*, 1985, **22**, 208–210.

37. (a) Fülling, G. and Sih, C.J., Enzymatic second order asymmetric hydrolysis of ketorolac esters in situ racemization, *J. Am. Chem. Soc.*, 1987, **109**, 284–2846; (b) Jones, M.M. and Williams, J.M.J., Dynamic kinetic resolution in the hydrolysis of an α-bromo ester, *Chem. Commun.*, 1998, 2519–2520; (c) Pietzsch, M. and Syldatk, C., Hydrolysis and formation of hydantoins, In: Drauz, K. and Waldmann, H. (Eds.), *Enzyme Catalysis in Organic Synthesis: A Comprehensive Handbook*, Wiley-VCH Verlag GmbH, Weinheim, Germany, 2002, pp. 761–799.

38. (a) Hasegawa, J., Ogura, M., Tsuda, S., Maemoto, S., Kutsuki, H. and Ohashi, T., High-yield produciton of optically-active 1,2-diols from the corresponding racemates by microbila stereoinversion, *Agric. Biol. Chem.*, 1990, **54**, 1819–1827; (b) Nakamura, K., Inoue, Y., Matsuda, T. and Ohno, A., Microbial deracemization of 1-arylethanol, *Tetrahedron Lett.*, 1995, **36**, 6263–6266; (c) Nakamura, K., Fujii, M. and Ida, Y., Stereoinversion of arylethanols by *Geotrichum candidum*, *Tetrahedron: Asymmetry*, 2001, **12**, 3147–3153; (d) Takemoto, M. and Achiwa, K., Deracemization of racemic 4-pyridyl-1-ethanol by *Catharanthus roseus* cell cultures, *Phytochemistry*, 1998, **49**, 1627–1629; (e) Allan, G.R. and Carnell, A.J., Microbial deracemization of 1-aryl and 1-heteroaryl secondary alcohols, *J. Org. Chem.*, 2001, **66**, 6495–6497; (f) Kato, D., Mitsuda, S. and Ohta, H., Microbial deracemization of α-substituted carboxylic acids, *Org. Lett.*, 2002, **4**, 371–373; (g) Kato, D., Miyamoto, K. and Ohta, H., Microbial deracemization of α-substituted carboxylic acids: control of the reaction path, *Tetrahedron: Asymmetry*, 2004, **15**, 2965–2973; (h) Kato, D., Mitsuda, S. and Ohta, H., Microbial deracemization of α-substituted carboxylic acids: substrate specificity and mechanistic investigation, *J. Org. Chem.*, 2003, **68**, 7234–7242; (i) Mitsukura, K., Yoshida, T. and Nagasawa, T., Synthesis of (*R*)-2-phenylpropanoic acid from its racemate through an isomerase-involving reaction by *Nocardia diaphanozonaria*, *Biotechnol. Lett.*, 2002, **24**, 1615–1621.

Chapter 11

Chemistry 'On Water' – Organic Synthesis in Aqueous Suspension

Sridhar Narayan, Valery V. Fokin, and K. Barry Sharpless

Water possesses many of the ideal characteristics for a reaction medium – it is readily available at negligible cost, is safe and convenient to use, and has low environmental impact.[1–3] In addition, by virtue of its unique physical and chemical properties, water is known to enhance the rates and efficiencies of a wide variety of organic reactions.[4,5] In spite of these potential advantages, water is not commonly used as a solvent for organic reactions, mostly because most organic compounds do not dissolve in water to a significant extent, and solubility is generally considered a prerequisite for reactivity. Consequently, a number of strategies have been developed in order to carry out preparative reactions in water, many of which are described in other chapters of this book. Most commonly, organic cosolvents are used to promote the water solubility of hydrophobic substrates.[5,6] In other cases, substrate modification is undertaken so as to render the resulting compound at least partially water soluble.[7] However, such machinations usually tend to diminish the advantages in cost, simplicity of reaction conditions, and ease of workup and product isolation that water has over traditional solvents.

In this chapter, we describe a completely different approach for the use of water as a medium for organic reactions. We have found that a wide variety of organic reactions can be conveniently carried out simply by stirring the neat reactants in an aqueous suspension.[8] We have termed such reactions 'on water', and define the 'on water' reactions as those fulfilling the following requirements:

1. Water is the only reaction medium – i.e., no organic cosolvents are employed.
2. Reactions involve insoluble substrates; hence, a suspension results when the reactants are stirred with water.
3. The term 'on water' refers to preparative reactions; i.e., the substrates are present in concentrations of approximately 0.1 M or above.

In many cases, we have observed considerable rate acceleration in reactions carried out under these conditions over those in organic solvents.[8] Moreover, significant rate increase is observable 'on water', over reactions carried out in the absence of any solvent, indicating that rate acceleration is not merely a consequence of increased concentration. The degree of 'on water' acceleration varies between different reaction classes, although in the examples we have studied to date, 'on water' reactions are at least as fast as in other solvents. In particular, the reactions of azodicarboxylates with olefins, dienes, and other unsaturated hydrocarbons represent dramatic examples of the 'on water' phenomenon. Consequently, we have studied these reactions in some detail (*vide infra*).

Even when rate acceleration is modest, there are two major advantages to carrying out reactions in this manner. First, water is an excellent heat sink due to its large heat capacity, which is extremely helpful for exothermic processes, especially on large scale. Second, reactions of insoluble substrates usually lead to the formation of water-insoluble products. In such cases, product isolation simply involves filtration of solid products or phase separation, in case of liquids. We hope that our studies will encourage other researchers to utilize the 'on water' method to their advantage.

11.1 Background

Our recent interest in the use of water as a medium for organic reactions stems from our click chemistry endeavors.[9] During the past few years, we have been interested in the modular synthesis of functional molecules using a small collection of nearly perfect reactions. We have termed such processes click reactions, and have found that many of the reactions that meet click standards often proceed better in water than in an organic solvent. We have attributed these observations to one or more of the following factors:

1. Reactions between organic species in aqueous solution can have higher apparent rate constants than the same processes in organic media. Among the many explanations offered for such phenomena,[10] we call particular attention to the notion that the free energies of organic molecules are substantially greater when poorly solvated in water, often imparting increased reactivity, which compensates for the low concentration of the participants.
2. Nucleophilic additions to epoxide and aziridine electrophiles (as well as aziridinium and episulfonium ions) are favored by solvents able to respond almost continuously to the rapidly changing hydrogen bonding networks along an optimal reaction coordinate.[11] In this respect, water is unique.
3. Two important subsets of olefin and acetylene click reactions are oxidations by electrophilic reagents and cycloaddition reactions. These processes are either concerted, or involve polarizable nucleophiles/electrophiles, so that water is rarely an interfering medium.[12–14] In fact, the use of water as 'solvent' may offer the greatest leverage for differentiating the reactivities of competing 'hard' (nonpolarizable) and 'soft' (polarizable) species.
4. A highly favorable reaction of two solutes (say at 0.1 M concentration) is usually much faster than a low-driving-force side reaction of one of the solutes with solvent water (55 M). The Schotten–Baumann method for making amides from acyl or sulfonyl halides in water is a well-known example.[15]
5. The greatest benefit of the use of water from the standpoint of chemical reactivity is that most hydroxyl and amide N—H groups are rendered 'invisible' and do not interfere with reactions performed in water. As a consequence, the installation and removal of protecting groups is avoided.

In this section, we discuss a few early examples from our laboratory, where water proved to be the optimal medium in a variety of processes. Nucleophilic opening of three-membered rings such as epoxides, aziridines, and episulfides are excellent click reactions because competing elimination processes are stereoelectronically disfavored.[9] As a result, high yields of ring-opened products are obtained. We have used the nucleophilic opening of epoxides

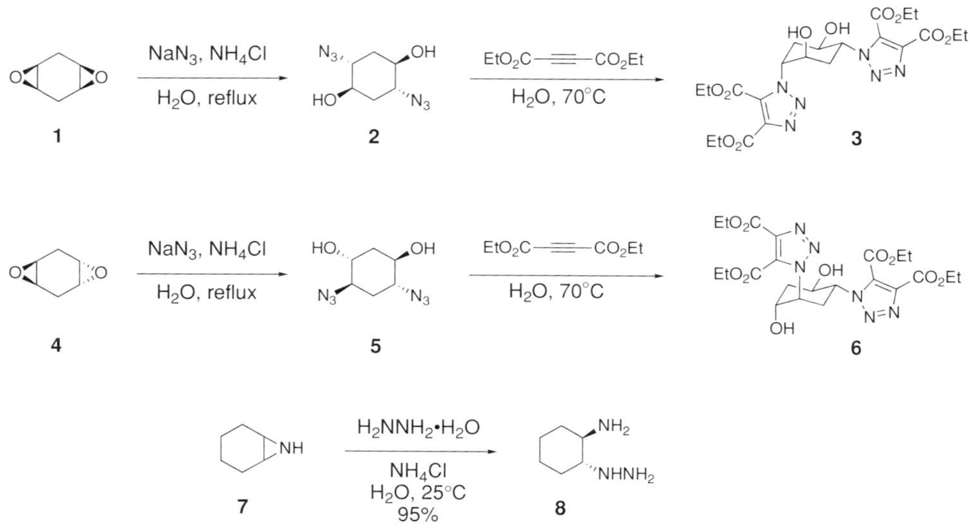

Figure 11.1 Epoxide and aziridine opening reactions in water.

with azide anion to generate a variety of useful azide-containing products, which readily participate in subsequent transformations. As shown in Fig. 11.1, the isomeric diepoxides **1** and **4** can be opened with NaN_3 in water containing NH_4Cl (buffered conditions) to obtain **2** and **5** as single regioisomers. The cycloaddition of aliphatic azides with alkynes, like many other Huisgen 1,3-dipolar cycloadditions, is among the most reliable bond-forming processes known and can be carried out in virtually any solvent.[16,17] Nevertheless, pure water serves as an ideal medium for carrying out these highly exothermic processes. The product bistriazoles **3** and **6**, which are crystalline solids, are isolated simply by filtration. Even the opening of unactivated aziridines, such as cyclohexane aziridine **7** with hydrazine proceeds readily in water at ambient temperature, using a catalytic amount of NH_4Cl.

The nucleophilic opening of the aziridinium ion derived from **9** (Fig. 11.2) is another striking example where the beneficial effects of water are readily apparent.[18–20] A significant improvement in both yield and regioselectivity was observed upon increasing the water content of the reaction mixture, with the best results obtained using pure water. Aziridinum ion reactions are particularly well suited to the aqueous phase, because the three-membered ring intermediates bear a positive charge, balanced by the anionic counterion that served as the leaving group. When this leaving group is chloride or sulfonate, the favorable effects of water are most apparent, probably a result of the strong protic solvation of the harder anions in water.[21] By contrast, little improvement is noted upon changing the solvent from CH_3CN to water in displacement reactions where the leaving group is bromide or iodide.

We have previously reported on the excellent properties of the mustard-type sulfur compound **12** (Fig. 11.3) as a scaffold for the synthesis of functional molecules.[22,23] The nucleophilic displacement of the chlorides in **12** occurs in high efficiency with a variety of nucleophiles using 'on water' conditions. These reactions occur via episulfonium ion intermediates, and are therefore uniquely assisted by the aqueous medium.[22]

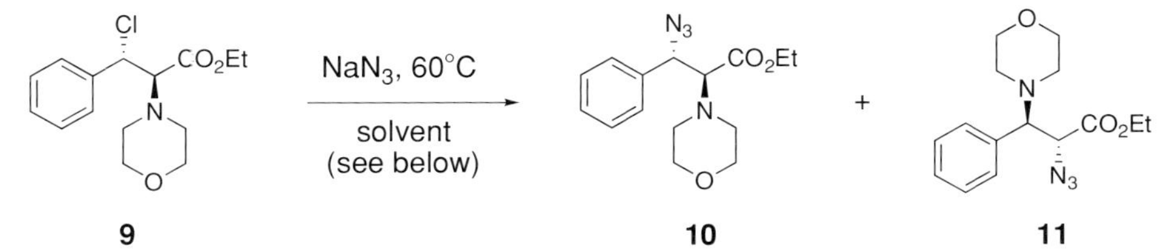

Solvent	Time	Yield	**10:11**
CH_3CN	12 h	85%	60:40
$EtOH{\bullet}H_2O$ (4:1)	12 h	97%	88:12
$EtOH{\bullet}H_2O$ (1:4)	6 h	92%	93:7
H_2O	6 h	>99%	95:5

Figure 11.2 Influence of water on the nucleophilic opening of aziridinium ions with azide.

The direct addition of azide to aliphatic and aromatic nitriles allows convenient access to a variety of tetrazoles, which are useful compounds for a variety of applications. We have found that this transformation can be carried out simply by heating the nitrile substrates (e.g. **17** and **19**; Fig. 11.4) with sodium azide and a zinc salt in water.[24] Under these conditions, the zinc salts of the tetrazole products are formed, from which the 1*H*-tetrazole products (**18** and **20**) can be generated by acidification of the solution.

Although we began this section with examples from our laboratory, it must be recognized that the use of water in organic synthesis has a long and interesting history. Indeed, the special nature of water as a solvent for certain reactions has been documented, and a number of excellent procedures have been reported. In the remainder of this section, we

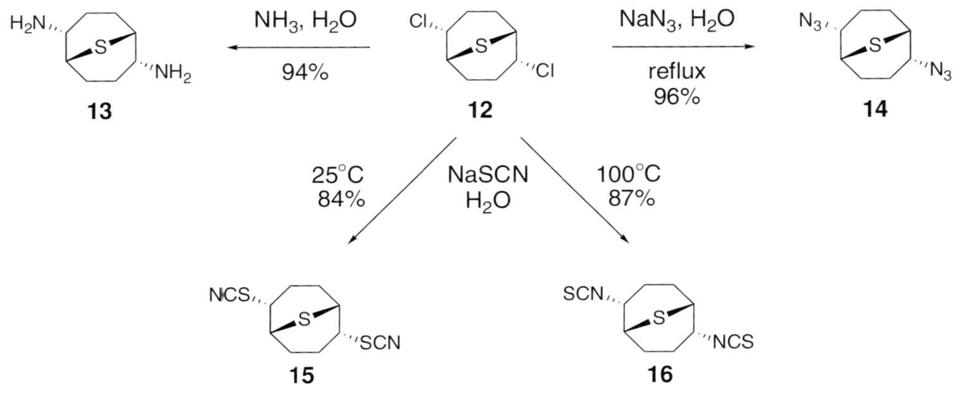

Figure 11.3 Nucleophilic displacements on dichloride **12** in aqueous media.

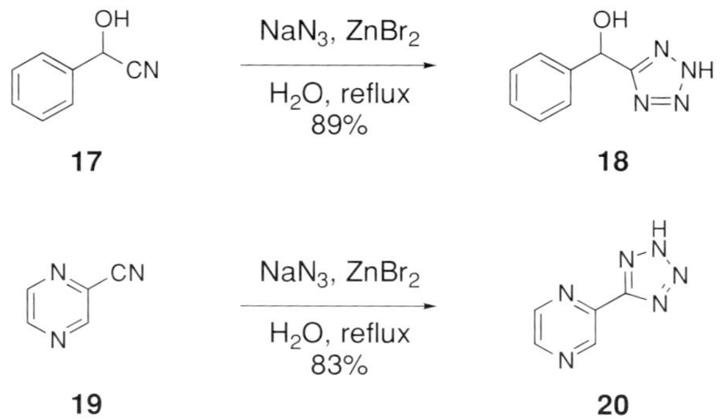

Figure 11.4 Synthesis of tetrazoles from nitriles and NaN_3 in water promoted by zinc.

highlight a number of reports from other laboratories on the use of experimental conditions essentially like the 'on water' protocol. This account is not intended to be comprehensive; rather it includes some representative examples, and serves to illustrate the effectiveness and convenience of this method for carrying out synthetically useful transformations.

The Schotten–Baumann reaction of acyl and sulfonyl halides with amines and alcohols is a classic example of the beneficial use of an aqueous medium.[15] The original conditions of Schotten and Baumann (first reported in the 1880s) used only aqueous base for carrying out the acylations of amines and alcohols (Fig. 11.5).[25,26] However, a modified version that involves the use of an organic solvent along with water in a biphasic solvent system is now commonly used.[27]

One of the most remarkable 'on water' type processes was reported by Guss and Rosenthal in 1955.[28] They showed that bromohydrins could be prepared simply by vigorously stirring the olefin substrates with *N*-bromosuccinimide (NBS) in water (Fig. 11.6). The product bromohydrins cleanly separate from the water phase, while the succinimide by-product remains in the aqueous layer. A number of olefins were efficiently oxidized in this manner. These authors also reported that NBS could be precipitated from the crude aqueous layer (50% recovery) simply by adding bromine to it.

Guss and Rosenthal also showed that the corresponding epoxides could be accessed by heating the product bromohydrins in aqueous NaOH solution.[28] This is an efficient method for the synthesis of racemic epoxides, and is especially valuable in case of acid-sensitive

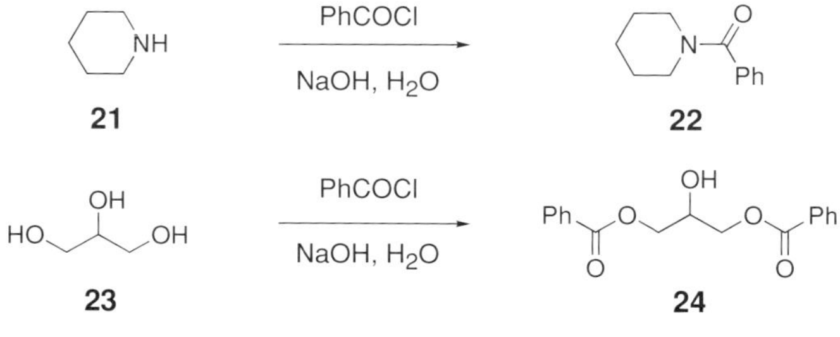

Figure 11.5 The Schotten–Baumann acylation of amines and alcohols.

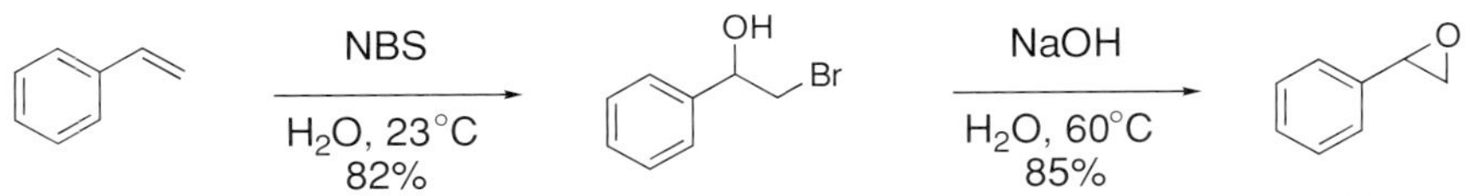

24 **25** **26**

Figure 11.6 Formation of bromohydrins and epoxides from olefins in water.

epoxides. In our own laboratory, we have found that the formation of epoxides in this manner can be carried out in a one-pot two-step process, which involves addition of base to the reaction mixture and stirring at ambient temperature, once the conversion of the olefin to the bromohydrin is complete.

The Diels–Alder cycloaddition is an extremely useful synthetic transformation, and the effect of aqueous solvent on this class of reactions has been studied extensively.[29] As early as 1939, Hopff and Rautenstrauch disclosed that Diels–Alder reactions could be carried out efficiently in an 'aqueous dispersion'.[30] A number of examples were described in their patent, and reactions were typically carried out in the presence of dispersing or emulsifying agents. The first quantitative studies on the effect of water on the rates of organic reactions were reported by Breslow, who showed that certain Diels–Alder reactions were significantly accelerated in water over reactions in organic solvent.[31–33] These studies were carried out in dilute homogeneous solution to ensure the solubility of the reaction substrates.[33] In the ensuing years, a number of other researchers have used water as a solvent for preparative Diels–Alder reactions, most commonly by using organic cosolvents. These studies are discussed elsewhere in this book. However, we call attention to a noteworthy study by Grieco et al. on the Diels–Alder reaction en route to the quassinoid family of natural products.[34] A dramatic solvent effect was observed on the rate and *endo* selectivity of the cycloaddition of enal **27** with diene **28**, and the best results were obtained when the dienophile and the sodium salt of the diene carboxylic acid (**28**, R = Na) were combined in aqueous suspension at room temperature (Fig. 11.7).

1,3-Dipolar cycloadditions are extremely useful transformations for the synthesis of heterocycles,[16,17] and it is not surprising that water can play a beneficial role in these processes.[12–14] Chemists at Novartis have reported an elegant method for the synthesis of cyanotriazoles from azides and 2-chloroacrylonitrile (**32**; Fig. 11.8) – a cyanoacetylene equivalent.[35] The major challenge in carrying out these cycloadditions is that **32** can polymerize under both acidic and basic conditions. Upon cycloaddition of **32** with azides such as **31**, the intermediate triazoline loses HCl, which increases the acidity of the reaction medium favoring the polymerization process. When an organic solvent is used, HCl remains in the reaction mixture leading to polymerization of **32** and decreasing the yield. Moreover, the reaction proceeds more slowly partly due to dilution.

Instead, when the reactants are heated together with water, a two-phase system results where the neat reactants **31** and **32** form the organic phase, while water constitutes the other phase. During the course of the reaction, the generated HCl is continuously extracted into the aqueous phase, thus reducing the propensity of **32** to polymerize. In addition, the reaction rate is highest in the two-phase system.[35]

Continuing interest in the parallel synthesis of compound libraries has focused attention on multicomponent reactions such as the Passerini and Ugi reactions. Pirrung and Sarma have found that a number of such reactions can be carried out in water, even when one or

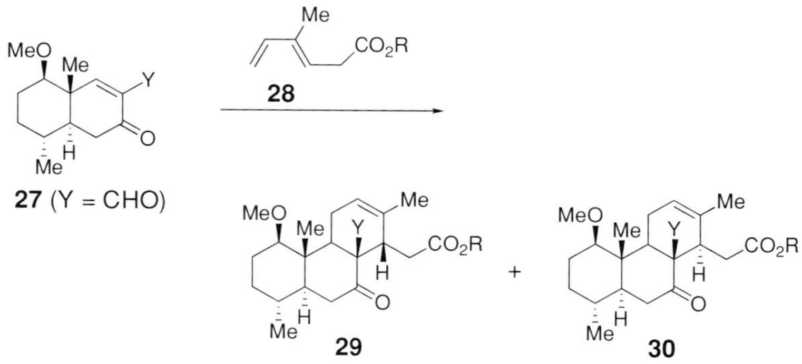

R	Solvent	Diene conc.	Time	Yield	**29**:**30**
Et	Benzene	1 M	288 h	52%	0.85
Et	Neat	–	144 h	69%	1.3
Et	H_2O	1 M	168 h	82%	1.3
Na	H_2O	1 M	8 h	83%	2.0
Na	H_2O	2 M	5 h	100%	3.0

Figure 11.7 Grieco's study on the aqueous Diels–Alder reaction en route to the quassinoid natural products.

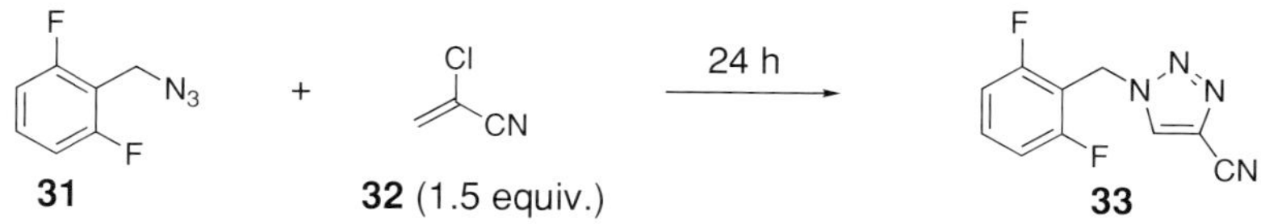

Solvent	Temp.	Yield
Neat	80°C	72%
n-Heptane	80°C	46%
Toluene	80°C	51%
EtOH	77°C	40%
DMF	80°C	78%
H_2O	80°C	98%

Figure 11.8 Synthesis of 4-cyano-1,2,3-triazoles in aqueous suspension.

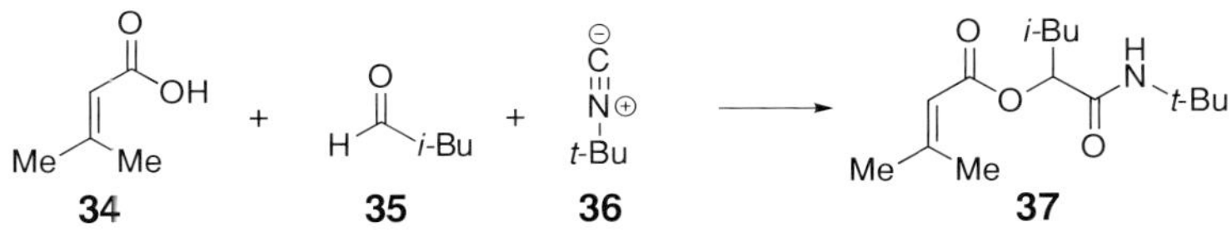

Solvent	Temp.	Relative rate
CH_2Cl_2	25°C	1
H_2O	50°C	10
H_2O	25°C	18
H_2O	4°C	20
0.5 M aq. glucose	25°C	129
1 M aq. LiCl	25°C	286

Figure 11.9 Aqueous solvent effect on the rate of Passerini reactions.

more reactants appear to be insoluble.[36,37] In particular, the rate of the Passerini reaction depicted in Fig. 11.9 is nearly 10-fold faster in water than in CH_2Cl_2. Moreover, the reaction rate in water increases with decreasing temperature, as well as in the presence of certain additives. On the basis of these and other observations, the authors suggest that the rate acceleration observed in water is primarily due to the unusually high cohesive energy density of water.[38]

11.2 The unique reactivity of azodicarboxylates on water

Methods that allow the direct introduction of heteroatom functionality, especially containing nitrogen functionality into hydrocarbons are especially valuable in organic synthesis.[39,40] In this regard, the cycloadditions of azodicarboxylates with dienes and other unsaturated hydrocarbons are extremely useful transformations. Azodicarboxylates also engage olefins in an ene reaction mode, which is a facile means of introducing allylic nitrogen functionality.[41,42] Due to the high energy content of the azo functionality, these reactions are highly enthalpically driven, and as a result, extremely reliable and versatile C—N bond-forming processes. Cycloadditions and ene reactions of azodicarboxylates allow direct access to hydrazines and, by extension, amine-containing compounds.[43] Consequently, we became interested in these transformations as part of our click chemistry program.

We have found that azodicarboxylates possess a unique level of reactivity in their reactions with various unsaturated hydrocarbons under 'on water' conditions. We first noticed this phenomenon in the context of our study on strained olefins. In particular, 1,2-diazetidines such as **40** were accessed via the $2\sigma + 2\sigma + 2\pi$ cycloaddition of quadricyclane (**38**) with dimethyl azodicarboxylate (**39**; Fig. 11.10).[44] This reaction typically requires prolonged reaction times and/or elevated temperatures when carried out in an organic solvent or in the absence of solvent.[45,46] However, it proceeds rapidly at room temperature or below when performed 'on water'.[8] In this case, the 'on water' reaction appears to be 2–3 orders

Solvent	Conc.	Temp.	Time	Yield
None	4.53 M	0°C	2 h	0%
None	4.53 M	23°C	48 h	85%
Toluene	1 M	80°C	24 h	74%
H_2O	4.53 M	0°C	1.5 h	93%
H_2O	4.53 M	23°C	10 min	82%

Figure 11.10 Dramatic 'on water' acceleration in the reaction of quadricyclane **(38)** and dimethyl azodicarboxylate **(39)**.

of magnitude faster than the corresponding neat reaction, while the reaction in nonpolar solvent is even slower.

We quickly came to a conclusion that this unique reactivity of azodicarboxylates 'on water' was a fairly general phenomenon. For example, the cycloaddition of diethyl azodicarboxylate (DEAD, **41**) with quadricyclane also shows a large magnitude of 'on water' acceleration (Fig. 11.11). The ene reaction of DEAD with β-pinene (**43**) is also considerably faster 'on water' – product formation is complete within 3 h, as compared to 1.5 days for the solventless reaction.

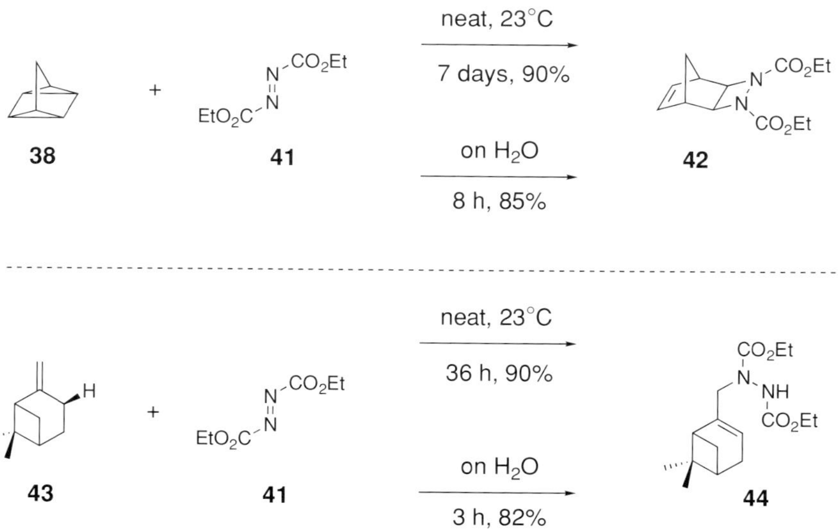

Figure 11.11 Cycloaddition and ene reactions of azodicarboxylates 'on water'.

Time	Conversion	
	Toluene	Toluene on H_2O
3 h	4%	42%
6 h	8%	56%
17 h	18%	69%

Figure 11.12 'On water' reaction monitored by NMR.

We also monitored the progress of many of these reactions by NMR in order to get a more accurate measure of rate acceleration. Accordingly, the cycloaddition of quadricyclane with DEAD (49% solution in toluene) was examined both 'on water' and in the absence of any other solvent (Fig. 11.12). A clear rate difference is observed between the two reactions, indicating that the 'on water' effect is operative even when a nonpolar solvent comprises part or most of the organic phase.[8]

In general, 'on water' reactions of nonpolar liquid substrates are the most efficient. However, reactions in which one of the components is solid can also be effectively carried out 'on water' if adequate mixing is provided. The ene reaction of cyclohexene with bis(trichloroethyl) azodicarboxylate (**45**) was shown to proceed at a much faster rate 'on water' than in the absence of solvent or when carried out in nonpolar solvent (Fig. 11.13).[8]

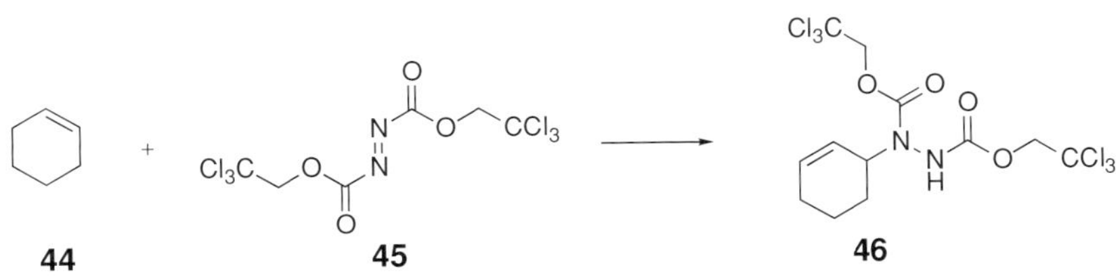

Solvent	Temp.	Time	Yield
Benzene	80°C	24 h	70%
Neat	50°C	36 h	62%
On H_2O	50°C	8 h	91%

Figure 11.13 Ene reaction of cyclohexene with bis(trichloroethyl) azodicarboxylate.

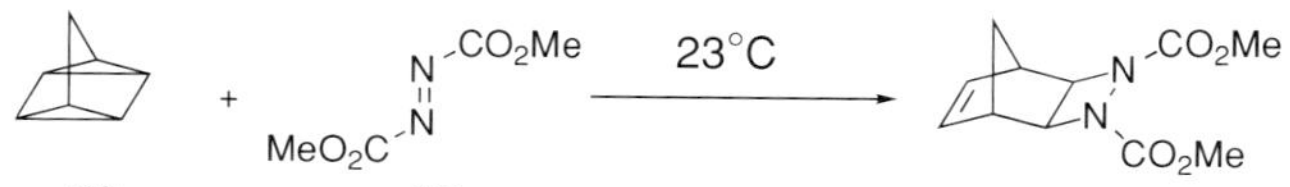

Solvent	Conc.	Time to completion	Solvent	Conc.	Time to completion
Toluene	2 M	>120 h	D_2O	4.53 M	45 min
EtOAc	2 M	>120 h	C_6F_{14}	4.53 M	36 h
CH_3CN	2 M	84 h	H_2O	4.53 M	10 min
CH_2Cl_2	2 M	72 h	MeOH•H_2O (3:1) (*homogeneous*)	2 M	4 h
DMSO	2 M	36 h			
MeOH	2 M	18 h	MeOH•H_2O (1:1) (*heterogeneous*)	4.53 M	10 min
None	4.53 M	48 h	MeOH•H_2O (1:3) (*heterogeneous*)	4.53 M	10 min

Figure 11.14 Solvent effect on the cycloaddition of quadricyclane **(38)** and dimethyl azodicarboxylate **(39)**.

The reaction of olefins with **45**, first reported by Leblanc, is a particularly simple and efficient method for allylic amination.[41] It is worth noting that the high reactivity of the azo reagent allows its use for the amination of relatively unactivated olefins such as cyclohexene; azodicarboxylates such as DEAD are not sufficiently reactive for this purpose.

We have carried out a number of control experiments in order to gain insight into the origin of these large rate accelerations. In case of the quadricyclane–DMAD reaction, we found that the reaction rate depends critically on solvent polarity – higher reaction rates are observed in polar aprotic solvents (Fig. 11.14). This trend suggests that rate acceleration is at least partly due to charge stabilization/dipolar effects as well as hydrogen bonding. However, this appears to be only part of the picture. Heterogeneity is also critically important for observing large rate acceleration. For instance, in homogeneous methanol–water mixtures, the reaction of **38** and **39** is considerably slower than in heterogeneous systems. We also observed that in comparison to water, the reaction is slowed down considerably in D_2O, and that perfluorohexane provides only a modest degree of acceleration over the neat reaction.[8]

In order to determine the precise degree of rate acceleration in the 'on water' reactions of azodicarboxylates, we have carried out preliminary rate measurements on the reaction of quadricyclane with DEAD (Fig. 11.12). We have found that the 'on water' reaction is nearly 86-fold faster than the reaction in toluene, and is about 20-fold faster than the neat reaction (S. Narayan and K.B. Sharpless, unpublished results). The precise mechanism of 'on water' acceleration in the reactions of azodicarboxylates is presently unknown; nevertheless, a number of observations point to the involvement of the water–hydrocarbon interface. Further studies along these lines are in progress.

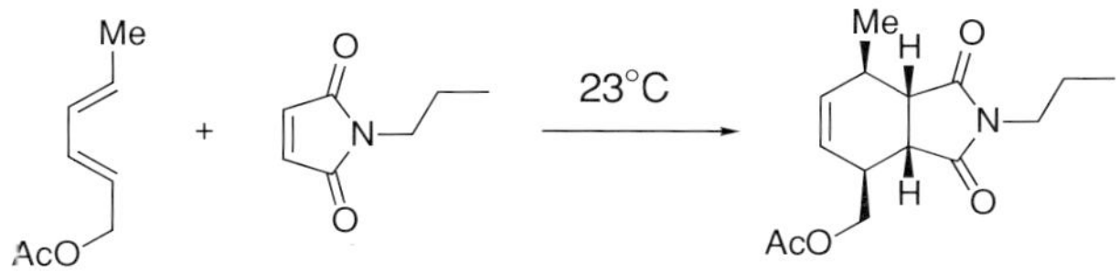

Solvent	Conc.	Time to completion	Yield
Toluene	1 M	144 h	79%
CH_3CN	1 M	>144 h	43%
MeOH	1 M	48 h	82%
None	3.69 M	10 h	82%
H_2O	3.69 M	8 h	81%

Figure 11.15 Comparative data for a Diels–Alder reaction in various solvents and 'on water'.

11.3 Other examples from our work

Although a number of researchers have taken advantage of the unique properties of water as a medium (*vide supra*), it is fair to say that a considerable barrier exists to its widespread use. Moreover, detailed comparisons of reaction rate and efficiency have not been carried out for preparative-scale reactions. The discovery of the dramatic 'on water' acceleration of the reactions of azodicarboxylates with unsaturated hydrocarbons prompted us to study other reaction classes in order to verify the generality of this phenomenon. Figures 11.15 and 11.16

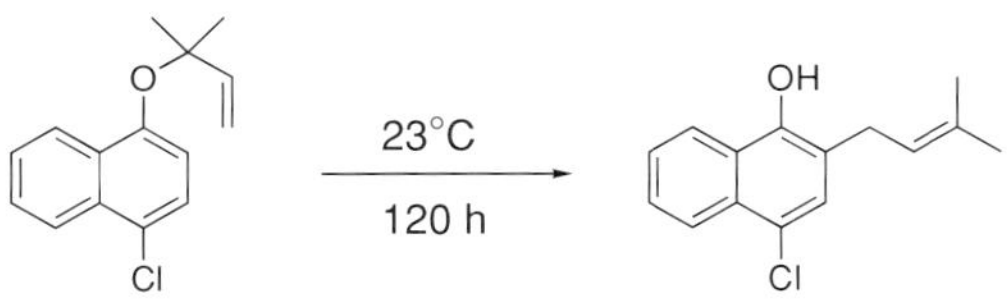

Solvent	Yield
Toluene	16%
DMF	21%
CH_3CN	27%
MeOH	56%
None	73%
H_2O	100%

Figure 11.16 'On water' acceleration of an aromatic Claisen rearrangement.

show comparative data for two synthetically important reaction classes – Diels–Alder reactions and Claisen rearrangements.[8] In both cases, the degree of 'on water' acceleration over the corresponding solventless reactions is small. Nevertheless, many reactions (especially those involving solid components) are much more reproducible in aqueous suspension than in the absence of any solvent. Water provides for efficient mixing in these reactions without the dilution cost of a true solvent.

11.4 Applications of the 'on water' method

Since the publication of our first report describing the 'on water' phenomenon, a number of other research groups have used this method to their advantage. The reported examples encompass a broader range of reaction classes than before, and provide evidence for the broad utility and versatility of the 'on water' method.

An elegant experimental and theoretical study on the influence of water on transition states of Huisgen 1,3-dipolar cycloadditions was reported by Butler and coworkers.[47] The authors concluded that the polarity of the transition state in cycloadditions of phthalazinium ylides such as **52** (Fig. 11.17) with various olefins remains nearly the same in acetonitrile and 9:1 water–acetonitrile. A few representative preparative reactions were also outlined such as the example shown in Fig. 11.17.

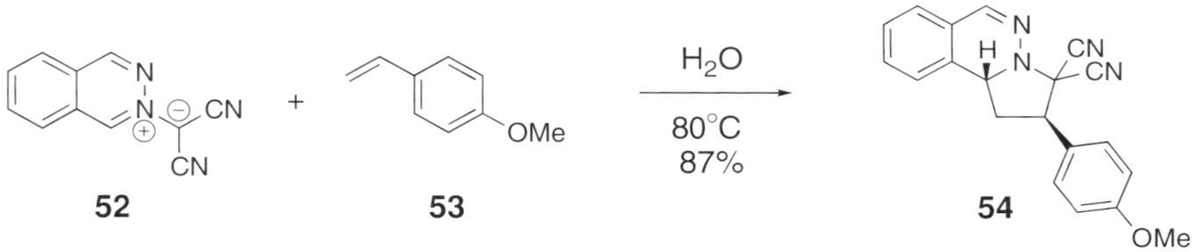

Figure 11.17 Example of 'on water' 1,3-dipolar cycloaddition reported by Butler.

Bergdahl and coworkers have reported the successful use of 'on water' conditions for Wittig reactions of stabilized ylides.[48] Although the protic stability of stabilized ylides is well known, these types of Wittig reactions are generally carried out in aprotic solvents. The authors describe a number of examples where rate acceleration is apparent in reactions carried out in water. For instance, the olefination of aldehyde **55** was reported to be unsuccessful in organic solvents (Fig. 11.18); on the other hand, the reaction in water afforded a 67% yield of α,β-unsaturated ester **56**.

Two interesting reports describing the advantages of water in radical reactions have recently appeared. In the first, Chatgilialoglu and coworkers demonstrate that radical

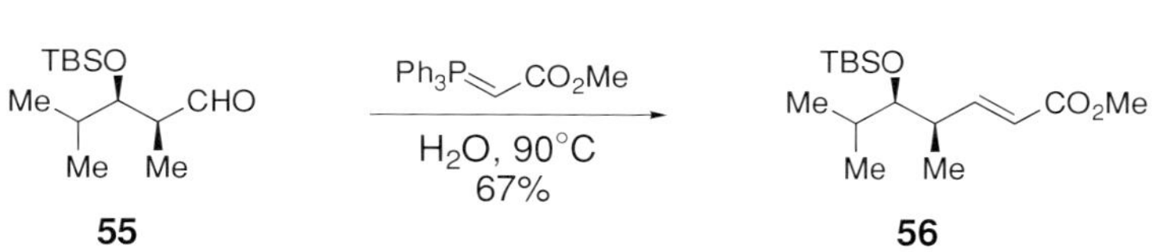

Figure 11.18 Wittig reactions of stabilized ylides performed using water as a medium.

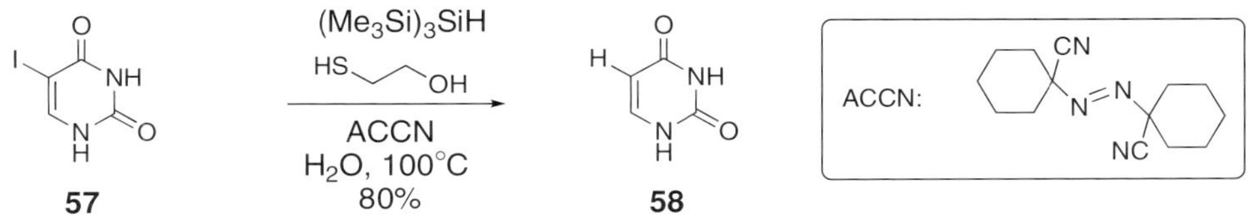

Figure 11.19 Radical reduction carried out 'on water'.

reductions of bromides and iodides occur efficiently under 'on water' conditions using the combination of tris(trimethylsilyl)silane and 2-mercaptoethanol.[49] A case in point is the reduction of 5-iodouracil (**55**; Fig. 11.19). Also notable was their use of the water-insoluble initiator ACCN, which was shown to be more efficient than AIBN in many of these reductions.

Radical brominations also take place readily in aqueous media, without the requirement of any cosolvents. In a recent paper, Iskra and coworkers report a number of cases of efficient benzylic bromination with NBS, using visible light (from sunlight or an incandescent lamp) as the initiator (Fig. 11.20).[50] However, the authors also report that in certain activated cases such as mesitylene (**61**), electrophilic bromination of the benzene ring predominates.

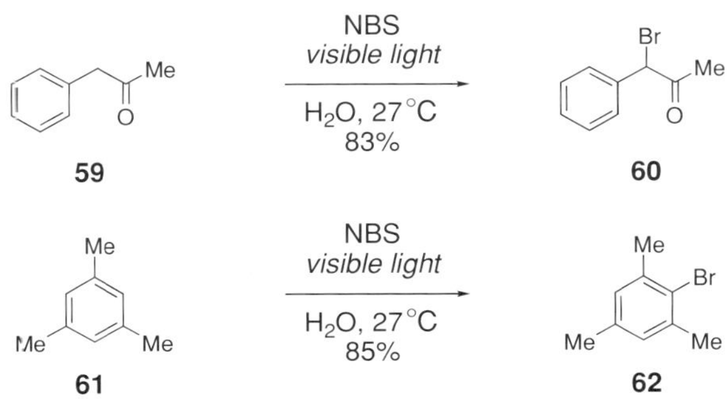

Figure 11.20 'On water' radical brominations.

11.5 Perspective and conclusion

For reasons of cost, convenience, and safety, it is desirable to carry out organic reactions using water as a medium. Moreover, many examples of the unique effect of water on chemical reactivity have been noted. Unfortunately, concerns regarding solubility have precluded the widespread use of water in synthesis. Nevertheless, a number of examples both from our laboratory and from others provide convincing evidence that water may provide the optimal medium for many *preparative* reactions, even when the reactant(s), reagent(s), and/or product(s) appear to be essentially insoluble. In this chapter, we have illustrated a number of these examples encompassing a broad range of reaction classes – pericyclic reactions such as cycloadditions and rearrangements, nucleophilic displacements, radical reactions, etc. In many instances, water provides the best results in terms of speed and efficiency of the reactions. In addition, several practical advantages accompany the use of water in this manner – ease of product isolation, convenience, and safety. We are optimistic that water will find widespread use in preparative organic chemistry, and that more instances of unique chemical reactivity will continue to be uncovered in this context.

References

1. Anastas, P.T. and Williamson, T.C., *Green Chemistry: Frontiers in Benign Chemical Syntheses and Processes,* Oxford University Press, Oxford, 1998.
2. Anastas, P.T., Heine, L.G. and Williamson, T.C. (Eds.), *Green Chemical Syntheses and Processes,* American Chemical Society, Washington, DC, 2000.
3. Matlack, A.S., *Introduction to Green Chemistry,* Marcel Dekker, New York, 2001.
4. Li, C.-J. and Chan, T.-H., *Organic Reactions in Aqueous Media,* Wiley, New York, 1997.
5. Grieco, P.A. (Ed.), *Organic Synthesis in Water,* Blackie, London, 1998.
6. Lindstrom, U.M., Stereoselective organic reactions in water, *Chem. Rev.*, 2002, **102**(8), 2751–2771.
7. Itami, K. and Yoshida, J.-I., The use of hydrophilic groups in aqueous organic reactions, *Chem. Rec.*, 2002, **2**(4), 213–224.
8. Narayan, S., Muldoon, J., Finn, M.G., Fokin, V.V., Kolb, H.C. and Sharpless, K.B., 'On water': unique reactivity of organic compounds in aqueous suspension, *Angew. Chem. Int. Ed.*, 2005, **44**(21), 3275–3279.
9. Kolb, H.C., Finn, M.G. and Sharpless, K.B., Click chemistry: diverse chemical function from a few good reactions, *Angew. Chem. Int. Ed.*, 2001, **40**(11), 2004–2021.
10. Otto, S. and Engberts, J.B.F.N., Hydrophobic interactions and chemical reactivity, *Org. Biomol. Chem.*, 2003, **1**(16), 2809–2820.
11. Fringuelli, F., Piermatti, O., Pizzo, F. and Vaccaro, L., Ring opening of epoxides with sodium azide in water. A regioselective pH-controlled reaction, *J. Org. Chem.*, 1999, **64**(16), 6094–6096.
12. Rispens, T. and Engberts, J.B.F.N., A kinetic study of 1,3-dipolar cycloadditions in micellar media, *J. Org. Chem.*, 2003, **68**(22), 8520–8528.
13. van Mersbergen, D., Wijnen, J.W. and Engberts, J.B.F.N., 1,3-Dipolar cycloadditions of benzonitrile oxide with various dipolarophiles in aqueous solutions. A kinetic study, *J. Org. Chem.*, 1998, **63**(24), 8801–8805.
14. Wijnen, J.W., Steiner, R.A. and Engberts, J.B.F.N., 1,3-Dipolar cycloaddition of phenyl azide to norbornene in aqueous solutions, *Tetrahedron Lett.*, 1995, **36**(30), 5389–5392.
15. Sonntag, N.O.V., The reactions of aliphatic acid chlorides, *Chem. Rev.*, 1953, **52**(2), 237–416.
16. Padwa, A. (Ed.), *1,3-Dipolar Cycloaddition Chemistry,* Wiley, New York, 1984.
17. Padwa, A. and Pearson, W.H. (Eds.), *Synthetic Applications of 1,3-Dipolar Cycloaddition Chemistry toward Heterocycles and Natural Products,* Wiley, New York, 2002.
18. Chuang, T.H. and Sharpless, K.B., Applications of aziridinium ions. Selective syntheses of β-aryl-α,β-diamino esters, *Org. Lett.*, 1999, **1**(9), 1435–1437.
19. Chuang, T.H. and Sharpless, K.B., Applications of aziridinium ions. Selective syntheses of α,β-diamino esters, α-sulfanyl-β-amino esters, β-lactams, and 1,5-benzodiazepin-2-one, *Org. Lett.*, 2000, **2**(23), 3555–3557.
20. Chuang, T.H. and Sharpless, K.B., Applications of aziridinium ions: selective syntheses of pyrazolidin-3-ones and pyrazolo 1,2-α-pyrazoles, *Helv. Chim. Acta*, 2000, **83**(8), 1734–1743.
21. Gajewski, J.J., Is the *tert*-butyl chloride solvolysis the most misunderstood reaction in organic chemistry? Evidence against nucleophilic solvent participation in the *tert*-butyl chloride transition state and for increased hydrogen bond donation to the 1-adamantyl chloride solvolysis transition state, *J. Am. Chem. Soc.*, 2001, **123**(44), 10877–10883.
22. Converso, A., Burow, K., Marzinzik, A., Sharpless, K.B. and Finn, M.G., 2,6-Dichloro-9-thiabicyclo[3.3.1]nonane: a privileged, bivalent scaffold for the display of nucleophilic components, *J. Org. Chem.*, 2001, **66**(12), 4386–4392.
23. Converso, A., Saaidi, P.L., Sharpless, K.B. and Finn, M.G., Nucleophilic substitution by Grignard reagents on sulfur mustards, *J. Org. Chem.*, 2004, **69**(21), 7336–7339.
24. Demko, Z.P. and Sharpless, K.B., Preparation of 5-substituted 1*H*-tetrazoles from nitriles in water, *J. Org. Chem.*, 2001, **66**(24), 7945–7950.
25. Schotten, C., The oxidation of piperidines, *Chem. Ber.*, 1884, **21**, 2544–2547.
26. Baumann, E., A method for the synthesis of benzoyl esters, *Chem. Ber.*, 1886, **19**, 3218–3222.
27. Georg, G.I., Boge, T.C., Cheruvallath, Z.S., Harriman, G.C.B., Hepperle, M. Park, H., et al., Schotten–Baumann acylation of *N*-debenzoyltaxol: an efficient route to *N*-acyl taxol analogues and their biological evaluation, *Bioorg. Med. Chem. Lett.*, 1994, **4**(2), 335–338.
28. Guss, C.O. and Rosenthal, R., Bromohydrins from olefins and *N*-bromosuccinimide in water, *J. Am. Chem. Soc.*, 1955, **77**(9), 2549.

29. Garner, P.P., Diels–Alder reactions in aqueous media, In: Grieco, P.A. (Ed.), *Organic Synthesis in Water*, Blackie, London, 1998, pp. 1–46.
30. Hopff, H. and Rautenstrauch, C.W., Production of addition products from dienes and compounds having an unsaturated carbon linkage activated by substituents, US Patent US 2262002, 1939.
31. Rideout, D.C. and Breslow, R., Hydrophobic acceleration of Diels–Alder reactions, *J. Am. Chem. Soc.*, 1980, **102**(26), 7816–7817.
32. Breslow, R., Maitra, U. and Rideout, D., Selective Diels–Alder reactions in aqueous solutions and suspensions, *Tetrahedron Lett.* 1983, **24**(18), 1901–1904.
33. Breslow, R., Hydrophobic effects on simple organic reactions in water. *Acc. Chem. Res.*, 1991, **24**(6), 159–164.
34. Grieco, P.A., Garner, P. and He, Z., Micellar catalysis in the aqueous intermolecular Diels–Alder reaction – rate acceleration and enhanced selectivity, *Tetrahedron Lett.*, 1983, **24**(18), 1897–1900.
35. Portmann, R., Process for preparing 1-substituted 4-cyano-1,2,3-triazoles, World Patent WO 9802423, 1998, Novartis AG.
36. Pirrung, M.C. and Das Sarma, K., Multicomponent reactions are accelerated in water, *J. Am. Chem. Soc.*, 2004, **126**(2), 444–445.
37. Pirrung, M.C. and Das Sarma, K., Aqueous medium effects on multi-component reactions, *Tetrahedron*, 2005, **61**(48), 11456–11472.
38. Pirrung, M.C., Acceleration of organic reactions through aqueous solvent effects, *Chem. Eur. J.*, 2006, **12**(5), 1312–1317.
39. Ricci, A., *Modern Amination Methods*, Wiley-VCH, Weinheim, 2000.
40. Togni, A. and Grützmacher, H., *Catalytic Heterofunctionalization: From Hydroamination to Hydrozirconation*, 1st edn. Wiley-VCH, Weinheim, 2001.
41. Leblanc, Y., Zamboni, R. and Bernstein, M.A., Amination of olefinic compounds with bis(2,2,2-trichloroethyl) azodicarboxylate, *J. Org. Chem.*, 1991, **56**(5), 1971–1972.
42. Brimble, M.A. and Heathcock, C.H., Allylic amination by the Lewis-acid-mediated ene reaction of diethyl azodicarboxylate with alkenes, *J. Org. Chem.*, 1993, **58**(19), 5261–5263.
43. Brown, B.R., *The Organic Chemistry of Aliphatic Nitrogen Compounds*, Oxford University Press, Oxford, 1994.
44. Rieber, N., Alberts, J., Lipsky, J.A. and Lemal, D.M. Δ^1-1,2-Diazetines, *J. Am. Chem. Soc.*, 1969, **91**(20), 5668–5669.
45. Platz, R., Fuchs, W., Rieber, N., Samuel, U.-R. and Jung, J., Polycyclic nitrogen-containing compounds, German Patent DE 2615878, 1977, BASF AG.
46. Platz, R., Fuchs, W., Rieber, N. and Jung, J., Polycyclic nitrogen-containing compounds, German Patent DE 2742034, 1979, BASF AG.
47. Butler, R.N., Coyne, A.G., Cunningham, W.J., Moloney, E.M. and Burke, L.A., Water and the Hüisgen cycloaddition reaction: a focus on polar contributions to the transition state in the reactions of dicyano(phthalazinium)methanide with substituted styrenes and benzylidene acetones, *Helv. Chim. Acta*, 2005, **88**(7), 1611–1629.
48. Dambacher, J., Zhao, W., El-Batta, A., Anness, R., Jiang, C.C., and Bergdahl, M., Water is an efficient medium for Wittig reactions employing stabilized ylides and aldehydes, *Tetrahedron Lett.*, 2005, **46**(26), 4473–4477.
49. Postigo, A., Ferreri, C., Navacchia, M.L. and Chatgilialoglu, C., The radical-based reduction with $(TMS)_3SiH$ 'on water', *Synlett*, 2005, (18), 2854–2856.
50. Podgorsek, A., Stavber, S., Zupan, M. and Iskra, J., Visible light induced 'on water' benzylic bromination with *N*-bromosuccinimide, *Tetrahedron Lett.*, 2006, **47**(7), 1097–1099.

Chapter 12

Water as a Reaction Solvent – An Industry Perspective

Ernst Wiebus and Boy Cornils

During the 1960s and 1970s, four developments of catalysis met each other: an increasing importance of homogeneously catalyzed reactions with a need for biphasic operation, the introduction of new central atoms and ligands, the technique of ligand modification, and the possibility of tailoring of new ligands, even water-soluble ones.

The importance of homogeneous catalysis increased with the tremendous application of their products – terephthalic acid for polyesters via oxidation of xylene or oxoalcohols for plasticizers via hydroformylation are the most significant examples in terms of tonnages. Because of the inherent disadvantages of homogeneous catalysis,[1] a biphasic operation was highly advisable.[2] Despite intense research, the obvious heterogenization of the homogeneous catalysts by immobilization onto solid supports did not succeed.[3] Specially the recommendations of Ruhrchemie, the inventors of the oxo synthesis, and Exxon for the use of rhodium as central atom for homogeneous catalysis[4] and the developments of Shell[5] for a successful phosphine-modified hydroformylation paved the way for alternative and highly effective new homogeneously catalyzed processes (e.g. low-pressure oxo units by the former Celanese and Union Carbide[6]). In the course of developments to alternatives for the Shell process, the tailoring of ligands became more familiar and ended up with the design of water-soluble compounds that enabled aqueous-phase processes. This was astonishing since the combination of, for example, metal carbonyl catalysts and an operation in water seems to be highly contradictory.[7] The developments of Joó[8] and Kuntz[9] were very important in this respect. All mentioned work culminated in the development of the aqueous-phase hydroformylation process of Ruhrchemie/Rhône-Poulenc,[10] which is the only high-tonnage process of this kind that has been applied in the chemical industry.

12.1 Hydroformylation as the master development

12.1.1 General

Biphasic techniques for recovery and recycle are among the recent improvements of homogeneous catalysis, and they are the only developments that have been recently and successfully applied in the chemical industry. Among the hydroformylation processes they form a 'fourth generation' of oxo processes.[10c,11] They are established as the 'Ruhrchemie/Rhône-Poulenc process' (RCH/RP process), with annual production rates of approximately 800,000 tons/year (tpy).

Acccording to the early scientists recommending and developing this process – Manassen/Whitehurst[2] and Joó,[8] *biphasic* stands for the use of two immiscible liquid phases, one containing the catalyst and the other containing the unreacted substrate and the reaction products. The tremendous progress in homogeneous catalysis is that both liquids, i.e. phases, can be separated off after the reaction is complete by simply separating the second phase from the catalyst solution, thus making it easy to recirculate the latter without any thermal or chemical stress (which is the case in classical separation methods). Therefore, this technique allows full utilization of the inherent advantages of homogeneous catalysis and avoids the costly recycling procedures of traditional homogeneous catalysts. In this respect, biphasic catalysis is the most successful variant of an *immobilization* technique, i.e. an 'anchoring' of the catalyst onto a *liquid support*.[3,10d] The application of biphasic catalysts simplifies the process of homogeneous catalysis considerably.

Early experimental work on biphasic techniques came up with large-scale processes for the above-mentioned hydroformylation, oligomerization of ethylene, telomerization of butadiene, and a couple of small-scale processes for the manufacture of fine chemicals. Whereas the oligomerization proceeds in the system organic/organic (SHOP process of Shell),[12] the other reactions use *aqueous* biphasic conditions which offer an especially advantageous mode of operation. Speaking of 'green chemistry' and 'green catalysis' and their definitions (Ken Seddon: 'Green chemistry is all about reducing the number and amount of harmful chemicals that are used and/or generated in research and industry.... This new field is all about minimizing the amount of waste...'), reacting substrates in aqueous biphasic operation is the most progressive example[13] – far beyond other biphasic techniques such as ionic liquids, operation in supercritical solvents or fluorous phases.[14]

This chapter will concentrate on the industrial-scale hydroformylation of propylene by means of rhodium catalysts, modified by water-soluble ligands such as TPPTS (triphenylphosphine *m*-trisulfonate). Some other developments will also be mentioned.

12.1.2 Immobilization with the help of liquid supports

Hundreds, maybe thousands, of publications and patent applications deal with *supported catalysts* and the numerous unsuccessful attempts to heterogenize oxo-active transition metal complexes onto solid supports: the difficulty of catalyst/product separation appeared to have been solved; however, it was discovered during long-term tests to verify the utility of the concept (tests that do simulate industrial conditions in terms of uninterrupted long runs and numerous catalyst cycles, turnover numbers and turnover frequencies, changes of temperatures, pressures, loads, etc.) that all supported oxo catalysts tended to leach out. This means that the active catalyst metal and also the modifying ligand slowly, but steadily, became detached from the heterogeneous support and were carried away after an uneconomically short time: the problem of homogeneous catalysis may have been delayed but has not been solved. Even hybrid techniques such as SLPC or SAPC (supported liquid [or aqueous] phase catalysis[3,15]) provide no improvement, probably because of the tremendous stress on the support/transition metal bond during the repeated change between tetrahedral and trigonal-bipyramidal metal carbonyls over the course of a single catalyst cycle. Only recent publications[15p,16] report on successful realization of supported homogeneous hydroformylation catalysts, but so far there is no confirmation by practice-oriented tests – not to mention by commercial applications. Only the biphasic method, specially of *aqueous* biphasic catalysis, has provided a fundamental remedy to the problem of stressfree and economical recovery and recycle of homogeneous oxo catalysts.[17] The fact that the catalyst, which still acts homogeneously, is dissolved in water and thus in a polar solvent and remains dissolved enables it to be separated from the nonpolar products without problems and with minimal effort after reaction.

The decisive step was the development of ligands with two concurrent properties: they are simultaneously water-soluble and they are modifying the catalytically active complex.[8,9,18] There was pioneering work done by Joó (mainly concerning hydrogenations), Kuntz (then with Rhône-Poulenc), and the former Ruhrchemie AG (summarizing reviews and details seen in Refs. 2, 11, 17f, 19, and 20) – and thus in the industry – which led to the first large-scale utilization of the aqueous homogeneous catalysis technique at the beginning of the 1980s.[12,17] The generally used embodiment of two-phase catalysis (e.g. as practiced in Shell's SHOP method[12]) was thus joined by *aqueous* two-phase catalysis.

In the Shell case, the classical homogeneous catalysis, an organic compound, i.e. a real liquid phase (a solvent) that dissolves all reactants, catalysts, and products, is used. The role of the solvent is underlined by the fact that it has to be separated from the reaction products by an additional and costly step, for example by distillation.

Olivier-Bourbigou and Hugues[21] define the importance of solvents as follows:

> The role of the solvent in organic reactions is of the utmost importance.[22] Its effect can just be limited to a 'physical effect' in making possible the solubilization of the reactants, with no direct interaction with the active center. More interesting are the cases in which the solvent interacts through specific forces, such as hydrogen bonds, thus altering the mechanism, the rate, and eventually the selectivity of the reaction in stabilizing certain reaction intermediates. In homogeneous catalysis, the solvent effect is often difficult to explain due to the large number of reaction intermediates involved. However, for catalytic applications involving coordinatively unsaturated cationic metal centers, the 'ideal' solvent should be able to solubilize the metal ion, while maintaining its ionic character, and should create weak and labile metal-solvent bonds. It should be highly polar and non-coordinating for the active metal center. Most of the classical organic solvents are covalent and do not satisfy the above requirements.

Especially the necessity of an additional and costly separation step is a heavy burden for bulk chemical processes operating with real solvents, diluents, etc.

In the case in question, the 'solvent' is water which shows pronounced solvent effects (cf. Section 12.1.3) but on the other hand does not have to be separated by special stressing means either from the reactants or from the products. Therefore, the water of the aqueous-phase processes has to be regarded as 'supporting fluid' rather than a real solvent, although there are many cases known in which water acts as accelerator for organic and even organometal-catalyzed reactions. The key for this behavior is the role of water in both, the influence on activity and selectivity of the desired reaction *and* the suitability to act as a phase-separating agent[23] – presumably enhanced by its tendency to form micelles, microemulsions, or other surface-based aggregates. Not recognized by all academian scientists,[24] the phase-separating power is the decisive advantage of water.

12.1.3 Principles

Aqueous biphasic catalysis is a special case of the two-phase processes of homogeneous catalysis. Despite the academic literature's provocative question 'Why water?'[25,26] the advantages of water as the second phase and the 'liquid support' are numerous. On the one hand, the search for the necessary solubility gap is much easier with water than with various organic-phase liquids (Fig. 12.1).

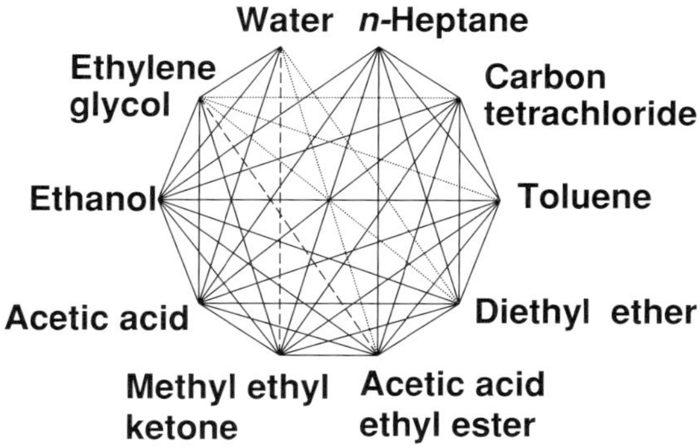

Figure 12.1 Miscibility diagram (and solubility gaps) of water and organic-phase liquids. Solvents not connected by a binding line are immiscible; solvents of unlimited miscibility are connected by a solid line, those of limited miscibility by a dotted line.[22,27]

Additionally, water has many properties which predestine it as a liquid support in homogeneous catalysis. These properties are compiled in Table 12.1.[17i,26]

Water has several anomalous features (e.g. density, being the only nontoxic and liquid 'hydride' of the nonmetals, melting point varying with pressure). Of direct importance for the aqueous biphasic process are the physiological (entries 2 and 4 of Table 12.1), economical (entries 1, 3, 6, 9), ecological/safety-related (entries 2, 3, 4, 9), process engineering (entries 1, 6, 7, 9, 10, 11, 12), and chemical and physical properties (entries 1, 5, 6, 8, 11, 13) of water. The different properties interact and complement each other. Thus water, whose high Hildebrand parameter and high polarity advantageously influence organic chemical reactions (such as hydroformylation), has sufficiently high polarity and density differences compared to

Table 12.1 Properties of water as a liquid support[17i]

1.	Water is polar and easy to separate from nonpolar solvents or products; its polarity may influence (i.e. improve) reactivity
2.	Water is inflammable and incombustible – a decisive advantage in terms of safety and occupational health
3.	Water is ubiquitous and available with suitable quality
4.	Water is odorless and colorless, which makes a contamination easily detectable
5.	The physical and physicochemical characteristics (e.g. hexagonal 2D surface structure, tetrahedral 3D molecular network) influence the mutual (in)solubility significantly; chaotropic compounds lower the order by H-bond breaking
6.	The Hildebrand parameter as the unit of solubility of nonelectrolytes in organic solvents/reaction products is high
7.	Water's density of 1 g cm^{-3} provides a sufficient difference to the density of most organic substances
8.	The dielectric constant ε is very high; the refraction index n_D is low
9.	The high thermal conductivity, the high specific heat capacity, and the high evaporation enthalpy of water make it suitable as solvent *and* heat removal fluid
10.	Water has a high solubility for gases, especially CO_2
11.	Water may form hydrates and solvates
12.	Water is highly dispersible and it has a high tendency of micelle and/or microemulsion formation, which both may be enhanced by additives such as surfactants
13.	Water has an amphoteric behavior in a Brønsted sense

organic (reaction) products to enable separation of the phases after the homogeneously catalyzed reaction is completed (see Section 12.2). Compared with this, the high solvent power for many compounds and gases, in some cases boosted by solvate or hydrate formation or by H-bonding, facilitates reactions in the two-phase system.

The chaotropic properties of many chemical compounds prevent the H_2O cage structures necessary for the formation of solvates and thus facilitate the transfer of nonpolar molecules from nonaqueous and aqueous phases. Water is noncombustible and inflammable, odorless and colorless, and is universally available in any quality – important prerequisites for the solvent of choice in catalytic processes. ε and n_D can be important in particular reactions and are advantageously used for analysis and control of substrates and products. The favorable thermal properties of water make it highly suitable for its simultaneous dual function as a mobile support *and* heat transfer fluid, a feature that is utilized in the RCH/RP process (see below).

Compared to the inexpensive and ubiquitous solvent and support water, with its unique combination of properties, other alternative solvents may well remain unimportant. Others make the same comments using different words (Table 12.2).

Operating homogeneously catalyzed conversions under aqueous biphasic conditions is certainly a brand-new technique but does not necessarily require newly designed apparatuses and exorbitant new sets of reaction conditions. To 'do the solubility split',[23] the catalyst should be soluble only in one of the liquids (i.e. water), while the products (and occasionally the reactants) should be mainly soluble in the other – preferentially the phase of the reaction products itself. In this case, the separation of the catalyst from the products can simply be done by decantation (or temporarily by other means such as extraction, etc.). In all cases, the reactor design ought to maintain biphasic operation by proper mixing of the reactants. After decantation, the phase containing the catalyst can simply be recycled to the reactor and the products can be separated and/or purified without the need to consider any stability or reactivity problems for the catalyst or the product. Basically, the flow sheet of an aqueous biphasically, homogeneously catalyzed oxo process is as simple as shown in Fig. 12.2(b) in comparison to the scheme of a classical hydroformylation unit shown in Fig. 12.2(a). The comparison gives an impression of the tremendous savings in investment the aqueous operation makes possible.

Table 12.2 Advantages and disadvantages of water as a reaction fluid[16]

Advantages	Disadvantages
Not inflammable	Large heat of evaporation[a]
Nontoxic	Detection in case of leakage[b]
No smell	Low solubility of many nonpolar substrates
Good separation with many organic compounds	Hard to collect in case of spills[c]
Inexpensive	
Unique fluid properties	No incineration of bleed streams[d]
Stabilization of certain organometal complexes	Decomposition of water-sensitive compounds[e]

[a] Which is an advantage within the economics of the heat compound.
[b] Which in contrast is an advantage since contaminated water smells intensively; by the way the danger occurring in case of spills is severely overestimated. Additionally, it must be underlined that the danger in case of spills is identical with biphasic or with homophasic (conventional) operation.
[c] Which is true for all 'solvents' and all organic liquids and is no specific disadvantage of water.
[d] Which in contrast is advantageous when containing the water-soluble catalyst – the system is self-extinguishing.
[e] This is true but a matter of evaluation: nobody will recommend water in case of decomposition of water-sensitive compounds.

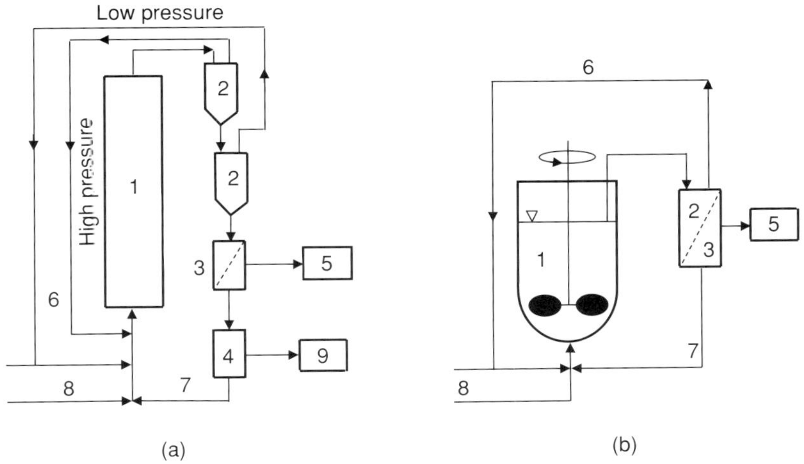

Figure 12.2 Basic flow sheets of (a) a conventional, homogeneously catalyzed process and (b) an aqueous biphasically, homogeneously catalyzed process. 1, Reactor; 2, separator(s); 3, catalyst separator; 4, make-up; 5, further purification and processing; 6, gas recycle(s); 7, catalyst recycle; 8, feed of reactants; 9, withdrawal of high boilers.

An important feature of biphasic hydroformylation is the separability due to density differences. Because of the differences in density of the polar compound water (1.00 g cm^{-3}) and the hydrophobic oxo products (with average 0.8) no problems occur. Additionally, the hydroformylation products are not sensitive against water. Another important question is to what extent water and the reactants are mixed. Therefore, the reactor in Fig. 12.2(b) – a continuously stirred tank reactor (CSTR) – normally contains usual installations to guarantee excellent mixing. For the lower alkenes with their significant water solubility (propene, butene) this is no problem. In these cases, the hydroformylation reaction takes place at the interfacial region.[28,29a]

So far only propene and butene are hydroformylated commercially using RCH/RP's process. A reason that has been postulated for this is the decreasing solubility in water with increasing number of C atoms in both the starting alkene and the reaction products (Fig. 12.3) and the associated mass-transfer problems in the relatively complicated gas/liquid/liquid three-phase reaction.

Up to now only limited kinetic data (and even mechanistic details) of aqueous-phase operation (and thus rate models, etc.) are available. So, in many cases only estimates and experimentally found data are at the disposal for reaction engineers' work.[31] The state of the art of the hydroformylation of higher alkenes ($>C_5$) comprises additions of supplemental solvents/diluents or extraction fluids, surface-active agents (detergents), intensity and mode of stirring[32] (and power of agitation; cf. Fig. 12.4), operation in micelles,[33] thermoregulated systems, etc.[33–35a] Up to now, it has to be noted that many measures may improve conversion and yield but are not sufficient to cover the additional expenses. The same is true for new and exotic ligands, although this measure would be the ideal solution of the problem when no additional solvents/extracting agents and their costly recycling are permitted.[34,36,37]

Quite new ideas for the reactor design of aqueous multiphase fluid/fluid reactions were realized by researchers from Oxeno. In packed tubular reactors and under unconventional

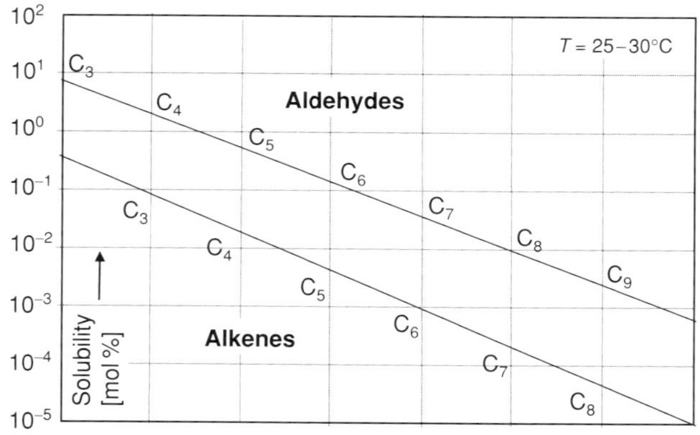

Figure 12.3 Solubility of alkenes and corresponding aldehydes in water.[11,30]

reaction conditions, they observed very high space–time yields that increased the conventional figures by a factor of 10 due to a combination of mass transfer area and kinetics.[38] Thus the old question of aqueous biphasic hydroformylation 'Where does the reaction take place?' – i.e. at the interphase or the bulk of the liquid phase[28,33f] – is again questionable, at least under the mentioned conditions (packed tubular reactors, other hydrodynamic

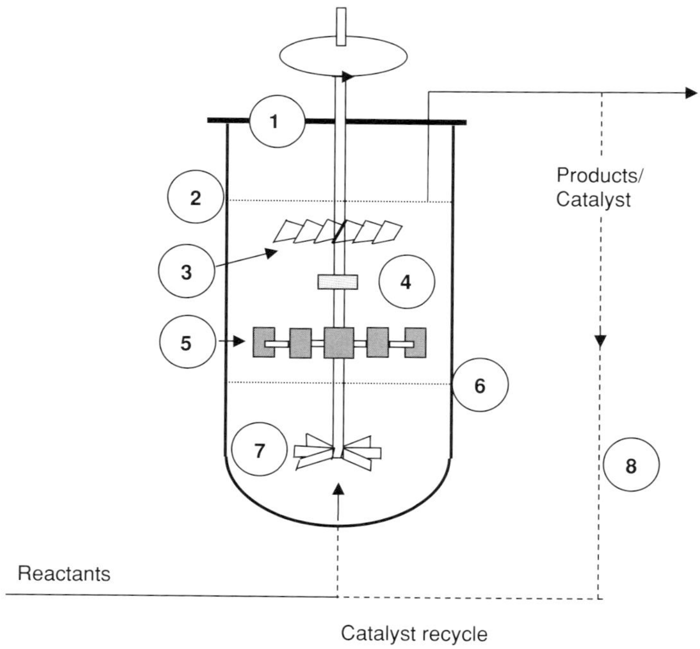

Figure 12.4 Example of a highly sophisticated CTSR [see Ref. 31a; for other examples, see Ref. 31c]. 1, CTSR; 2, interphase gas/liquid; 3, self-rotating floating baffle; 4, annulus for position limiting; 5, Rushton disk turbine; 6, interphase liquid/liquid; 7, pitched blade turbine upward (mixer/stirrer); 8, aqueous-phase catalyst recycle.

conditions, in mini plants, and in the unusual – and costly – presence of ethylene glycol) and not in harsh industrial operation. The considerable reduction of the laminar boundary layer in highly loaded packed tubular reactors increases the mass transfer coefficients; thus, the Oxeno researchers claim the successful hydroformylation of 1-octene.[31a,33j,33k,38,35,39a,39e] The search for a new reactor design may also include operation in microreactors.[40]

The hydroformylation reaction is highly exothermic, which makes temperature control and the use of the reaction heat potentially productive and profitable (e.g. steam generation). The standard installation of RCH/RP aqueous phase processes is heat recovery by heat exchangers done in a way that the reboiler of the distillation column for workup of the oxo products is a falling film evaporator incorporated in the oxo reactor itself.[17,41] The heat of the oxo reaction is thus recovered as the reboiler heat source. This is a great advantage over the classical (nonaqueous) hydroformylation, which simply discards parts of the oxo reaction heat. The RCH/RP process is a net steam exporter. In the case of the above-mentioned Oxeno developments the large catalyst flow is said to serve as a heat transfer medium itself, thus making the process nearly isothermal.[38g]

There are many proposals how to overcome the problem of proper phase separation (and reducing the leaching of the catalyst) of aqueous-phase processes when partial mixing of the two phases occurs (or is necessary because of reaction engineering reasons). In this case the simple, rapid, and perfect phase separation as in the RCH/RP process without any leaching (the losses are in the order of magnitude of ppb[17d]) needs considerable effort, for example extraction with an appropriate additional solvent which either is added after the catalytic conversion or is already present during reaction. In all cases, the reaction loop has to be completed by additional stages such as extraction devices and separation, recovery, and recycle of the extraction fluid. Two examples are given in Fig. 12.5; they should be compared with the simple flow sheet of RCH/RP oxo process (cf. Section 12.2) to recognize the expenses that are connected with those solutions.

Theoretically, it is possible to control chemical reactions by the different solubilities of the reactants or of special intermediates (e.g. according to their nucleophilicity) in different solvents or various catalyst fluids. The principle has been proven but is not yet used for special applications in industry.[42]

The search for new ligands (which in a restricted sense is no unit operation) that would combine various duties such as modification of the central atoms of the complex catalyst, creating and modifying surfactant properties, offering chiral properties, etc., is still going on. There are some results but no ligand better than TPPTS has been developed so far ('better' includes simultaneously water solubility, activity, selectivity, and price[43,44]). In contrast, all proposed new ligands are more expensive and shift the level of their costs to that of the precious catalyst metals such as rhodium, palladium, or platinum. The search for 'novel catalysts' for multiphase reactions must be done as an integrated task in close relationship with reactor design – colloidal suspensions included.

12.2 Examples of aqueous-phase catalyses

12.2.1 Hydroformylation (RCH/RP process)

The RCH/RP process converts propylene to *n*- and isobutyraldehydes (or butenes to valeraldehydes) in the presence of $HRh(CO)(TPPTS)_3$ (with TPPTS = tris[sodium-*m*-sulfonatophenyl]phosphine as a water-soluble ligand) according to Fig. 12.6.

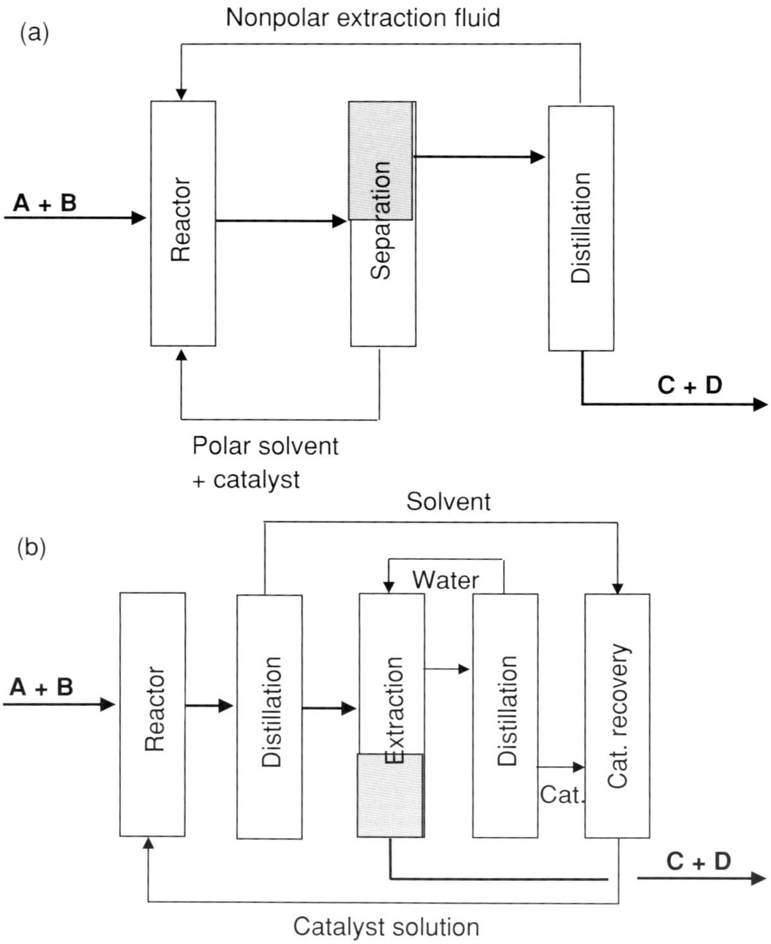

Figure 12.5 Process flow sheets of biphasic reaction A + B → C + D and extraction. 1, Simultaneous reaction and extraction within the reactor; 2, separate reaction and catalyst extraction.

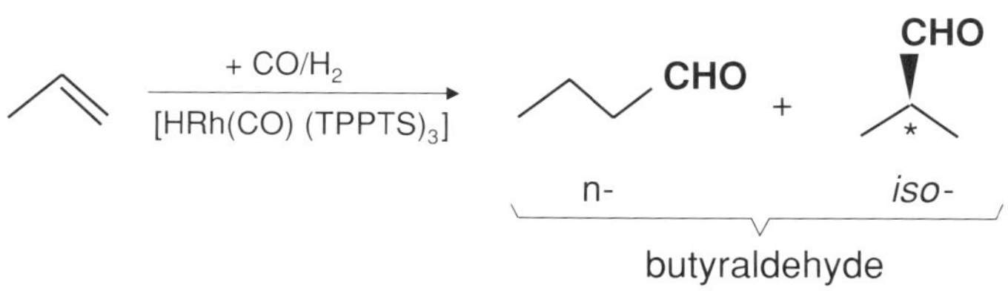

Figure 12.6 Hydroformylation of propylene.

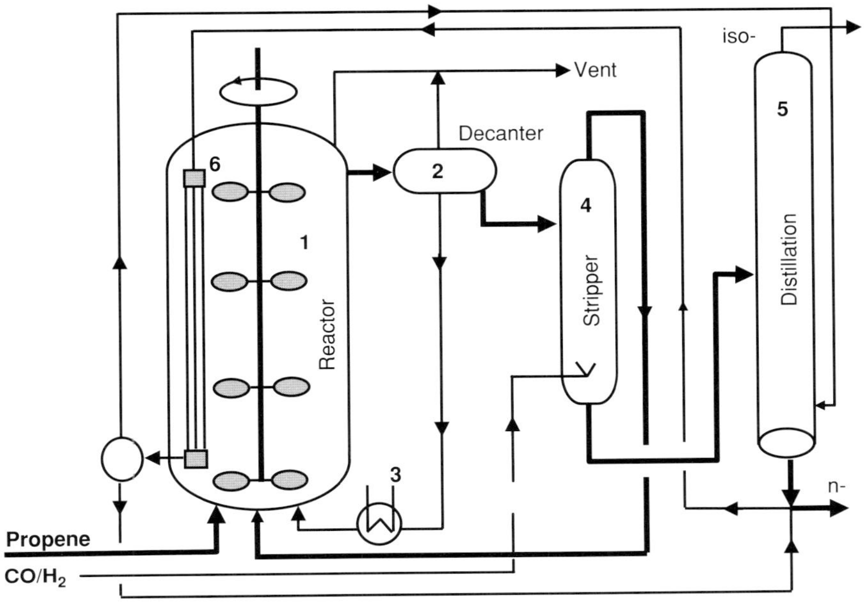

Figure 12.7 Ruhrchemie/Rhône-Poulenc oxo process.

The process scheme is shown in Fig. 12.7. The reactor 1 is stirred and supplied with reactants and the catalyst. The catalyst $HRh(CO)(TPPTS)_3$ is prepared by mixing an Rh salt and aqueous TPPTS solution (which may partly be recycled; see Section 12.2) in a simple 'preforming' step (much simpler than the preformation in other oxo processes[29]). Temperature is controlled via the heat recovery system, with the reaction heat from the exothermic hydroformylation being used in the reboiler of the distillation column 5. The reboiler is the internal cooler 6 of the reactor. The cooling medium is the reaction product, *n*-butyraldehyde. From the reactor, the reaction products pass through a downstream phase separator (decanter 2) and stripping column 4. The major part of the catalyst solution remains in the reactor and only a smaller part is separated off in the decanter and returns directly to the reactor. In the phase separator, the crude aldehyde formed is freed of gases and further separated into mutually insoluble phases. This decanter ensures the essentially spontaneous phase separation. The reaction heat that is retained in the catalyst solution is recovered by a heat exchanger 3, and the catalyst solution, supplemented by an amount of water equivalent to that carried off with the isobutyraldehyde (since the *n*-aldehyde is practically anhydrous), is recirculated to the reactor. The crude aldehyde is freed of any unreacted alkene in the stripper 4 by means of syngas flowing countercurrently. It is essential that this stripper is accomplished in the absence of the oxo catalyst and, therefore, no selectivity- and/or yield-reducing secondary reactions of the crude aldehyde occur. Typical for the RCH/RP process are high yields and selectivities (99% butanals with an *n/iso* ratio of up to 98:2; selectivity toward C_4 products: >99.5%) at mild conditions (120°C at 5 MPa).[17,41]

The process is highly effective and, as the process data illustrate, represents an economic advantage as well 12.4 and 12.6 as 12.5. In ecological terms, the RCH/RP process also provides a considerable improvement (Section 12.6). Recovery and recycle of the catalyst are described in Section 12.3.

12.2.2 Other industrially used aqueous biphasic processes

Hydroformylation comprises the state of the art of bulk chemical production via aqueous biphasic processes. At present five plants produce worldwide some 800,000 tpy of oxo products.[10c] Another bulk process – the hydrodimerization of butadiene and water, a variant of telomerization – is run by Kururay with a capacity of 5000 tpy (Fig. 12.8).[45b,46] The reaction product of telomerization is 2,7-octadiene-1-ol.

In subsequent steps this dienol may be converted to 1-octanol by hydrogenation or hydrogenated/dehydrogenated to 1-octenal. This unsaturated aldehyde again can be hydroformylated to yield nonanedialdehyde and then hydrogenated to nonanediol.

The remarkably versatile C_1 building block HCN may be used in the aqueous biphasic hydrocyanation, too (Fig. 12.9).[47] Also, some fine chemicals are manufactured at an industrial scale using this technology such as intermediates for vitamins and phenylacetic acid (Figs. 12.10 and 12.11).[17d,45b,48]

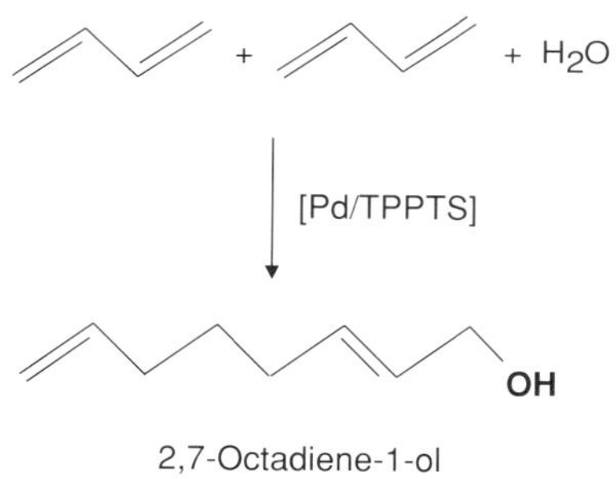

Figure 12.8 Hydrodimerization of butadiene and water.

Today, the Suzuki cross-coupling of aryl halides and arylboronic acids is also accomplished in aqueous biphasic operation starting from chlorinated derivatives instead of their more costly bromo or iodo equivalents (Fig. 12.12).[49]

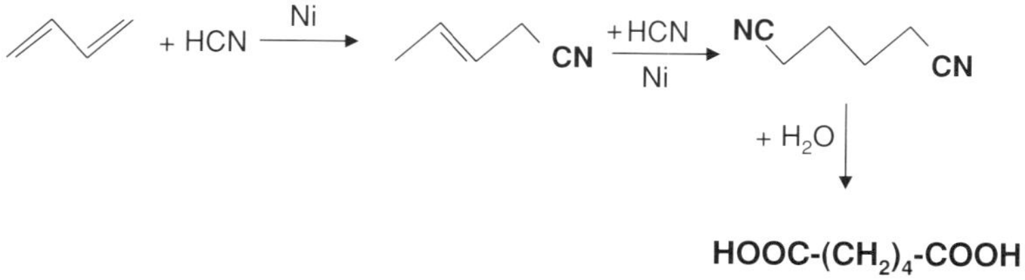

Figure 12.9 Hydrocyanation of butadiene.

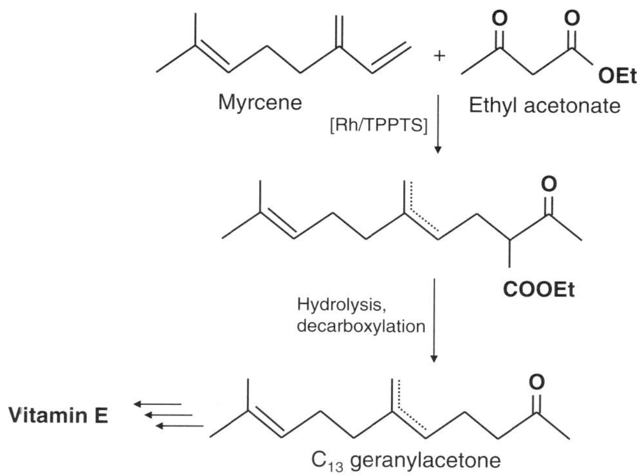

Figure 12.10 Manufacture of vitamin E precursors.

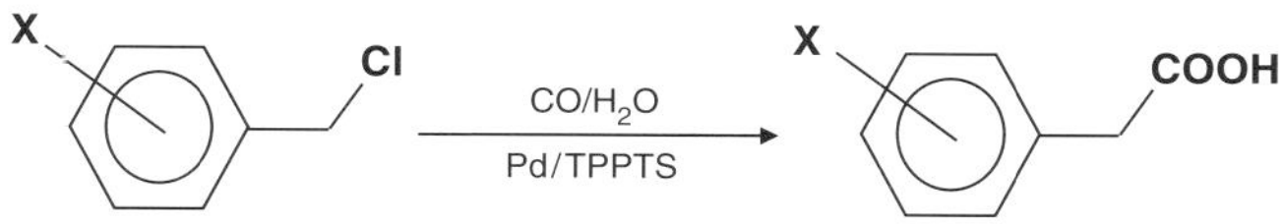

Figure 12.11 Manufacture of phenylacetic acid.

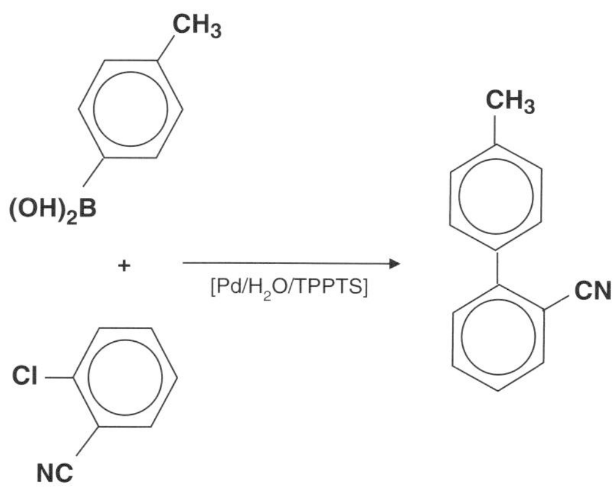

Figure 12.12 Suzuki coupling to yield aromatic biphenyls.

Table 12.3 Commercial biphasic processes

Process/catalyst	Phases	Reaction products	Capacity [tpy][a]
Shell SHOP/Ni-P,O ligand	o/o	α-Olefins and internal olefins	900,000
Ruhrchemie/Rhône-Poulenc (now Celanese)/ Rh-TPPTS	w/o	*n*-Butyraldehyde	800,000
Kururay Co. Ltd./Pd-TPPMS	w/o	*n*-Octanol, nonanediol	5,000
Clariant AG/Pd-TPPTS	w/o	Substituted biphenyls	<1,000
Rhodia (former Rhône-Poulenc)/Rh-TPPTS	w/o	Vitamin precursors[b]	

[a] o: organic; w: water.
[b] Exact production figures not known.

The commercially applied biphasic processes are compiled in Table 12.3. Tests to produce economically interesting profens or other analgesics by biphasic hydrocarboxylation[50] remain industrially unsuccessful.

12.2.3 Short overview of other (laboratory-scale) reactions

A multitude of other reactions are compiled in Table 12.4.

A proper choice of ligands and reaction conditions will make many other reactions available to the aqueous biphasic operation. Presently, a number of these reactions are in the stage of pilot-plant testing. This may also be true for various types of chemical engineering alternatives.

With the RCH/RP process, it is possible to hydroformylate propene up to pentenes with satisfying space–time yields. On the other hand, heavier aldehydes such as C_{10} (isodecanal) or higher from the hydroformylation of nonene(s), decenes, etc., cannot be separated from the oxo catalysts by conventional means such as distillation due to thermal instability at the required temperatures (and thus especially needs the careful aqueous biphasically separation technique). There are numerous attempts to overcome the problem of low reactivity of higher alkenes, which is due to low miscibility of the alkenes in water.[33j,34b,35,52a] These proposals can be briefly summarized as follows:

1. The use of amphiphilic water-soluble ligands which influences the alkene solubility or increase the catalyst concentration at the interface area of the phases[10b,39]
2. The use of cosolvents such as alcohols or detergents to improve interface exchange of the feedstock and the reaction products[15,16,33a–c,35b,52,53g]
3. Substitution of the water biphasic procedure by supported aqueous-phase catalysis (SAPC)
4. The separation of the reaction products from the catalyst solution by membrane techniques (Fig. 12.13)[54]
5. In continuation of Bayer/Schurig's work from 1975/1976,[36a,36b] the use of polymeric water-soluble supports as 'smart ligands' (see also the work of Bergbreiter and Mecking)[39,53m,55–57]
6. The addition of surfactants to improve the formation of micelles, microemulsions, etc.[33,35e]

Table 12.4 Recently described examples of aqueous biphasically operated reactions[11,51]

Type of reaction	Catalyst metal involved
Addition reactions	Rh
Aerobic oxidation	Mn
Aldehyde allylation	Sn
Aldolization	Sc, In, Cu, Ln
Alkylation	K, Pd, organocatalysts
Alkynylation	Pd
Allylation	Ru, Pd, Sn
Alternate copolymerization of CO–ethylene; arylation	Pd; Rh, Cu
Asymmetric allylic amination	Pd
Aziridination	Cu
Barbier-type allylation	Zn
Baylis–Hillman reaction	Organocatalysts
Bromination	Mo
Carbonylation/various Claisen rearrangement/Al conjugate addition	In, Zn
Cross-coupling	K
Crotylation of aldehydes	Organocatalysts
Diels–Alder reaction; disulfide exchange reaction	Si, Cu; Rh
Epoxidation; four-component reaction	W, Re, Mn, oxone; Ti
Friedel–Crafts reaction; haloaryl coupling	Sc; Pd
Heck reaction; hydrogenation; hydrosilylation	Pd, organocatalysts; Ru, Rh, Pd, Ir, Pt; Pt
Hydroxycarbonylation; isomerization	Pd/Ni; Ru
Ketone reduction	Enzyme
Mannich reaction; metathesis; Michael reaction	Zn; Ru; Ln, Rh
Mukaiyama aldol reaction	Zn
N-Oxidation	V
Oxidation; bimetallic Bi/Pt, Ru/Pt, Ru/Pd oxidative coupling	Ru, Pd, Os, W, Mn; Cu, Ce
Oxidative dimerization	Enzyme
Pauson–Khand reaction	Co, Triton
Polymerization; reductive amination	Cu, Ti, Ni, Pd, Rh, Co; Rh
Reformatsky reaction; reforming of ethylene glycol	Zn; Pt, Pd
Sharpless dihydroxylation	Os
Sonogashira cross-coupling	Pd
Sulfide oxidations	Fe, Cu
Suzuki reaction	Pd, organocatalysts
Transfer hydrogenation	Rh, Ru
Trifluoroacetylation	Na
Tsuji–Trost reaction	Pd

7. To introduce specially developed ligands
8. Measures for the controlled switch of the catalyst system from the two-phase system (suitable for the *separation* of products from the catalyst) to a monophasic system which supports the *reaction* itself (Fig. 12.14)[33d,34]

The method under point 8 may be applied by the use of thermoregulating or separating means by taking advantage of a temperature-dependent 'cloud point' associated with P-bound poly(alkylene glycol ether) as ligands, i.e. oxo catalysts which combine water solubility together with phase-separating properties depending upon temperature. Developments of Fell and Jin based on ethoxylated phosphines give the first pointers to such a procedure.

According to Fig. 12.14, at the cloud point, the ligand (and thus the catalyst complex) loses its hydration shell, just as in the case of other compounds of this type, causing the two-phase reaction mixture normally obtained when (higher) alkene is added to the catalyst solution to

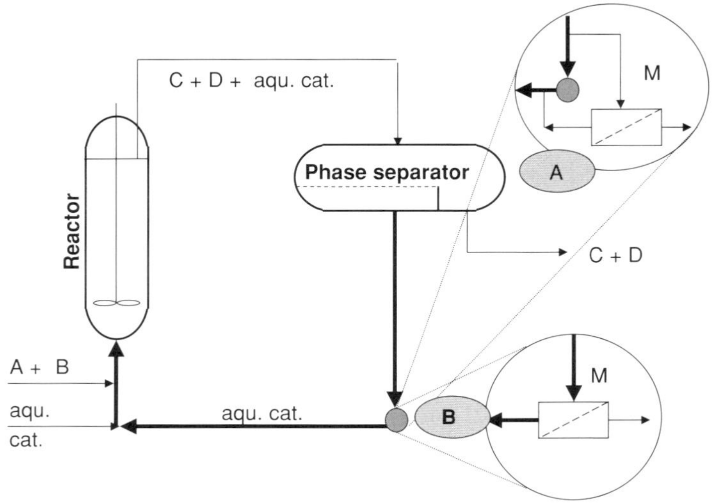

Figure 12.13 Membrane steps as a constituent part of aqueous biphasic hydroformylation A + B → C + D.

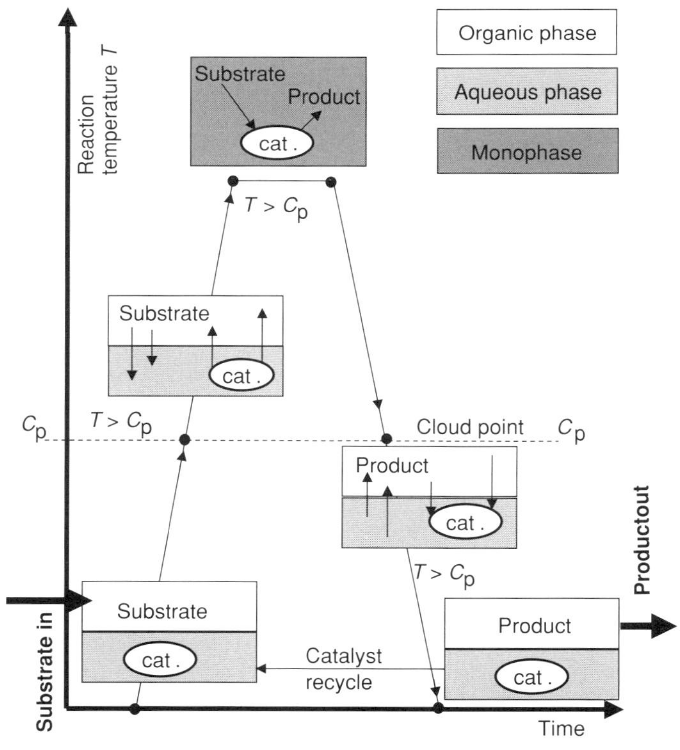

Figure 12.14 Phase change of thermoregulating ligands.

merge into a single phase, thereby initiating a rapid conversion that is no longer impeded by mass transport problems. Subsequent lowering of the temperature causes the hydration shell to be reversibly restored, inducing the catalyst complex solution once again to separate out as an independent phase, this time from reaction products, viz. the desired higher aldehydes. The disadvantage of this higher cost technique is the resulting lower linear/branched (*n/iso*) ratio. Jin refined this technique into a highly sophisticated method, including transitions to other variants (e.g. water-soluble polymers, pH-dependent operation).[58]

The membrane process, incorporated into the catalyst recycle either in the main stream or in the side stream, may also separate the reaction products from the remaining catalyst.

As an alternative to the heterogenization of homogeneous catalysis, there are some proposals to realize a solid catalyst with an immobilized species in aqueous/organic media. This concept, a continuation of the SLPC as mainly published and highlighted by Scholten et al.,[15k] consists of a thin film of catalytic material that resides on a high-surface-area support such as controlled-pore glass, silica, zeolites. Thus this concept of supported aqueous phase catalysis (SAPC) contains both a hydrophilic liquid and a hydrophilic organometallic catalytic complex on a solid support as shown in Fig. 12.15.[15n]

Reactions take place at the water/film organic phase interface and are catalyzed by the phase-immobilized complex catalyst which consists also of $HRh(CO)(TPPTS)_3$. Other hydrophilic solvents, e.g. glycols or suitable liquids, adjusted to the requirements of the respective reaction, can be used instead of water for the formation of the immobilized liquid layer. As in all SPC variants, the catalytic process proceeds homogeneously in the supported film (SLPC) or in the interface (SAPC), thus avoiding the problem of the separation of the reaction products from the catalyst. It is believed that the hydrophilicity of the ligands and the support creates interaction energies suffient to maintain the immobilization.

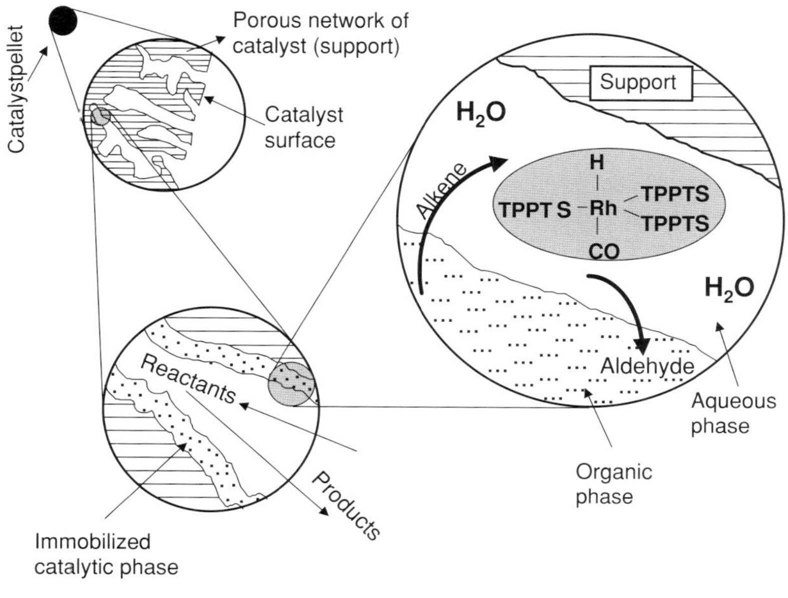

Figure 12.15 Schematic view of a supported aqueous-phase catalytic system.

So far, various reactions have successfully been accomplished with SAPC, such as hydrogenations (asymmetric hydrogenations as well), hydroformylations, alkylations, and Wacker oxidations. However, no continuous runs with long-term experience – not to mention commercial tests – have been carried out. As a process requiring relatively complex processing, SAPC technique is designated for the manufacturing of sophisticated and high-priced products (e.g. pharmaceuticals, chiral intermediates). It is likely that rare metals and unusual and especially designed ligands are needed to fully meet the demands of those processes and products. The lifetime of SAPS catalysts under continuous operation and industrial load is not as yet known.

Advantageously, SAPC as a technique with immobilized catalysts does not need devices for catalyst separation and recycling. On the other hand, the presumed processes for the workup of the constituents of the catalyst, the ligand (and – maybe – the support), will be demanding and expensive, too.

There are nearly no publications about organic/organic operations.

12.3 The 'aqueous' recycle and recovery of biphasic catalysts

Recovery and recycle of homogeneous catalysis are said to be the focal point of any new generation of, for example, hydroformylation technologies. Additionally, the costs of the new process are determined by the losses – an argument which is specially tabled at the occasion of discussions about the use of high-price rare metal catalysts such as rhodium or palladium. Obviously this is true because a new mode of recycling characterizes a new process. But on the other hand, metal losses (and supplementary ligand costs) are only a (minor) part of the overall cost as will be demonstrated in Section 12.4.

In recent years, the evaluation of environmentally benign, so-called 'green', processes has been a focal point of oxo developments, according to the definitions and targets of the OECD Workshop on Sustainable Chemistry (1998)[13c]:

> Within the broad framework of sustainable development, we should strive to maximize resource efficiency through activities such as energy and non-renewable resource conservation, risk minimization, pollution prevention, minimization of waste at all stages of a product life-cycle, and the development of products that are durable and can be re-used and recycled. Sustainable chemistry strives to accomplish these ends through the design, manufacture and use of efficient and effective, more environmentally benign chemical products and processes.

In this respect, risk minimization and life cycle assessments play a major role.

12.3.1 Recycle

As has been mentioned in Section 12.2.1, the part of the aqueous catalyst solution that leaves the oxo reactor accompanying (but not dissolved in) the reaction products passes a phase separator (decanter **2** in Fig. 12.7), which is a characterizing part of the plant shown in Fig. 12.16. In this decanter, which ensures the essentially spontaneous phase separation, the crude aldehyde formed by hydroformylation according to Fig. 12.6 is freed of gases and separated into mutually insoluble phases. The catalyst solution, supplemented by an amount of water equivalent to the water content of the crude aldehyde, is recirculated to the reactor. During its active life, the Rh catalyst is mainly situated in the oxo reactor, it is not moved severely, and no aliquot parts are withdrawn as in other processes. For this reason,

Figure 12.16 Large-scale oxo plant using the water-soluble RCH/RP catalyst $HRh(CO)(TPPTS)_3$. (Note the horizontal decanter in front of the reactor.)

rhodium losses are low – in the range of parts per billion (ppb) – and thus the background for the high economy (see Section 12.4).[17d,17k] Like every technically used and thus 'real' catalyst, the complex $HRh(CO)(TPPTS)_3$ and the excess ligand TPPTS undergo a degree of decomposition that determines the catalyst's lifetime as measured in years. The catalyst deactivation mechanism has been clarified in detail (Figs. 12.17–12.19).[59]

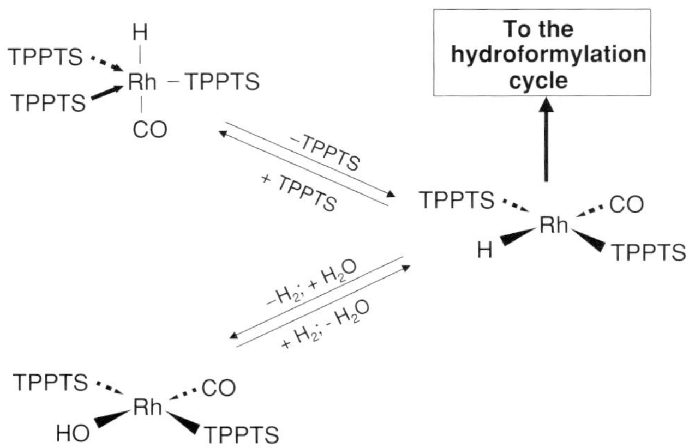

Figure 12.17 From the precatalyst to the oxo-active species.

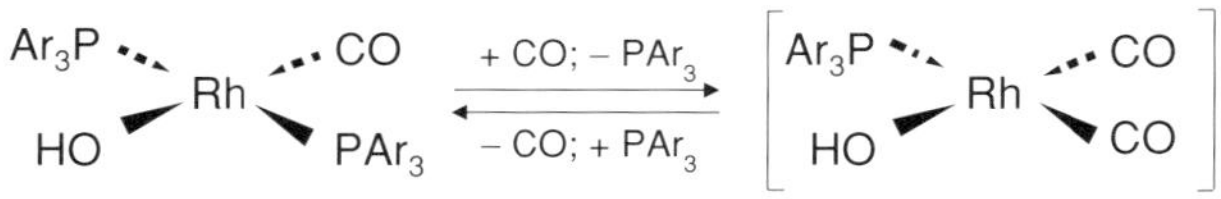

Figure 12.18 The influence of CO.

The compound HRh(CO)(TPPTS)$_3$ is something like a 'precatalyst' and dissociates into the 16e species HRh(CO)(TPPTS)$_2$. This oxo-active stage initiates the hydroformylation cycle. Under oxo conditions (presence of CO/H_2, H_2O, and a surplus of TPPTS) the hydroxo complex (HO)Rh(CO)(TPPTS)$_2$ may be formed and again reversibly converted to HRh(CO)(TPPTS)$_3$ (on which side the equilibrium lies almost completely). However, higher carbon monoxide partial pressures may cause the displacement of TPPTS by CO according to Fig. 12.18. The hydroxo complex (HO)Rh(CO)$_2$TPPTS starts the deactivation cycle as shown in Fig. 12.19[59] (for graphical simplification, Ar$_3$P is TPPTS and Ar$_2$P is bis[*m*-sulfophenyl]phosphine).[60a,60b] The single steps will not be discussed here. Other decomposition products such as the reductively eliminated bis(*m*-sulfophenyl)phosphinous acid Ar$_2$POH (cf. Fig. 12.19) and the phosphine oxides Ar$_2$P(OH)(=O) and Ar$_2$P(=O)(CH[OH]C$_3$H$_7$) have been identified.

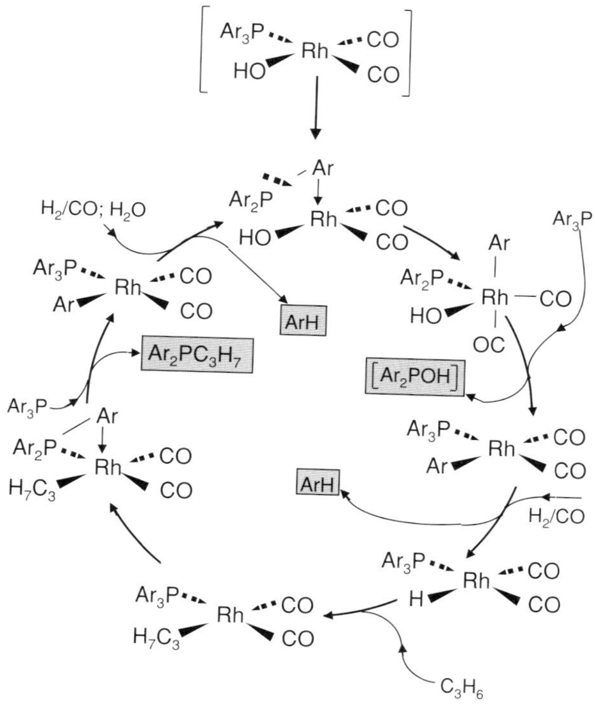

Figure 12.19 Deactivation mechanism of Rh^1-TPPTS catalyst.

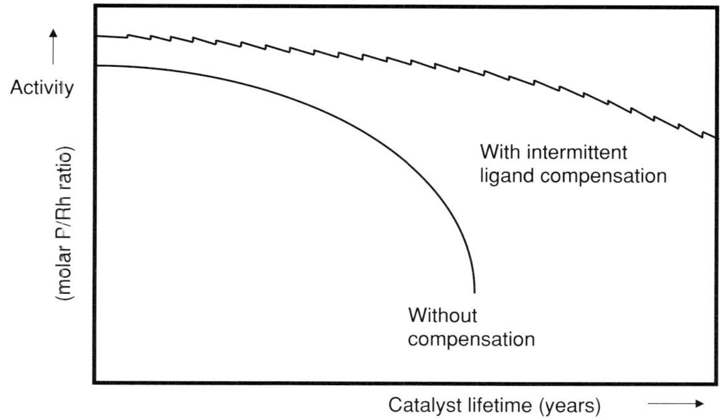

Figure 12.20 The effect of excess ligand on the catalyst's lifetime.

According to Fig. 12.20, the intermittent addition of excess ligands extends the catalyst's lifetime in a sawtooth curve. This addition of ligand compensates for the system-immanent formation of deactivating substances which are brought into the system by the feedstocks.

Most of the gaseous poisons are removed with the heavy ends of the stripper 4 in Fig. 12.7. Filters, guard beds, or special precautions to avoid larger sulfur or oxygen inputs and their concentrations (as in other processes)[60c,60d] are not necessary. Additionally, other activity-lowering oxo poisons may be separated with the organic product phase of the decanter and are thus continuously removed at the very point of their formation from the system: any accumulation of activity-decreasing poisons in the catalyst solution is prevented. It might be worth mentioning that the Ruhrchemie plant has been supplied over longer periods with syngas manufactured from coal by the TCGP (Texaco coal gasification process).[17d]

Catalyst deactivation includes (among other reactions) the formation of inactive Rh species, ligand decomposition, or P–C cleavage by direct oxidative insertion of the rhodium metal for formation of PDSPP (propyl di[*m*-sulfophenyl]phosphine) acting as strong electron donor reducing the amount of active Rh catalyst. It turned out to be beneficial to control the P^{III}/Rh ratio and the CO partial pressure very carefully.

Catalyst solutions after years of use typically contain 20 mg L^{-1} iron and 0.7 mg L^{-1} nickel, thus showing no corrosivity. The Rh content of crude aldehyde is in the ppb range; this corresponds to losses of less than 10^{-9} g kg^{-1} *n*-butanal, totaling some kilogram rhodium over a 20-year period and a production of considerably more than 5 million metric tons of *n*-butyraldehyde.

12.3.2 Recovery

Eventually, the spent catalyst solution has to leave the oxo loop for workup. The Ruhrchemie works of Celanese AG in Oberhausen (Germany) operate several rhodium-based oxo processes: besides the well-known RCH/RP process (the described low-pressure oxo process with TPPTS-modified Rh catalyst) the 'Ruhrchemie process' with an unmodified Rh catalyst at high pressure (comparable to the late ICI process[61]; this variant is for the benefit of a high *n/iso* ratio in cases where this strange ratio is required), and a high-pressure technique using ligand-modified rhodium catalysts. The various processes deliver aldehydes

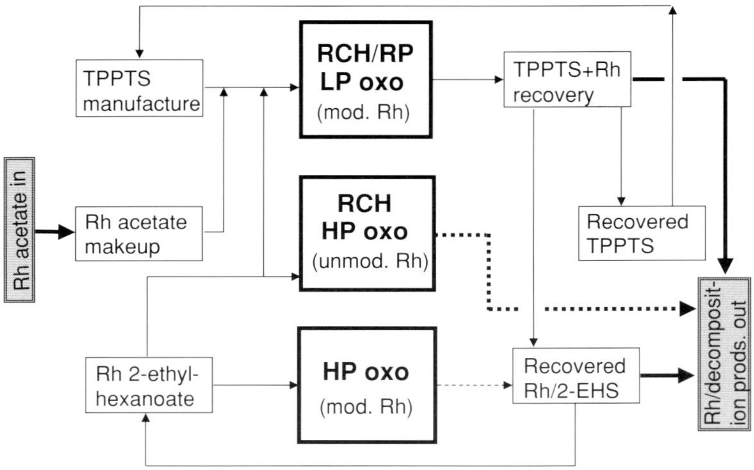

Figure 12.21 Compound of various oxo processes at the Ruhrchemie works.

from C_4 up to C_{10} and additionally fine chemical aldehydes, starting from bicyclic alkenes, functionalized olefins, etc.

The compound of the three distinct oxo processes, all rhodium-based, enables a highly efficient recovery system to be achieved (Fig. 12.21).

Figure 12.21 describes the TPPTS manufacture and its use for the preparation of the rhodium catalyst, using either freshly introduced Rh acetate or recycled Rh 2-ethylhexanoate. The recycle technique of the RCH/RP process and its performance is depicted earlier. Spent Rh-TPPTS solutions are worked up (see Figs. 12.22 and 12.23), and the resulting TPPTS returns to the RCH/RP process. The rhodium portion also passes a workup stage and is reformulated as Rh 2-ethylhexanoate. This Rh salt may serve all various oxo processes of the oxo loop and will compensate for possible Rh losses as mentioned earlier.

According to Fig. 12.21, the only external input is rhodium(III) acetate besides TPPTS from the own manufacture unit. Rhodium compounds in residuals too difficult to be worked up (catalyst poisons, metal aggregates, clusters, etc.) together with decomposition products, residuals, etc., leave the recovery steps and will be processed by external precious-metals refiners. They have their own expertise and trade (and production) secrets as far as the technical know-how is concerned. Depending upon the status and the 'quality' of these fractions (if necessary, following a referee's check, various samplings, etc.), the credit for the rhodium is >95% of its content.[62]

The workup scheme as described in Fig. 12.21 takes into account the decomposition rate of TPPTS (and thus the ligand losses), the different volume of the consumed Rh streams and their residual activity, and – last but not least – the Rh losses. Advantageously, these losses may be minimized by proper use of the different sources from the various oxo stages. Most steps are covered by patents,[19] and the same is true for similar processes developed by competitors.[63]

The economics of the TPPTS recovery (cf. Fig. 12.22) depend on the costs and its relation to the Rh price. Since the procedure of *workup* is identical with some steps within the *manufacture* of fresh TPPTS,[63] existing devices of the TPPTS manufacture unit can be used advantageously.

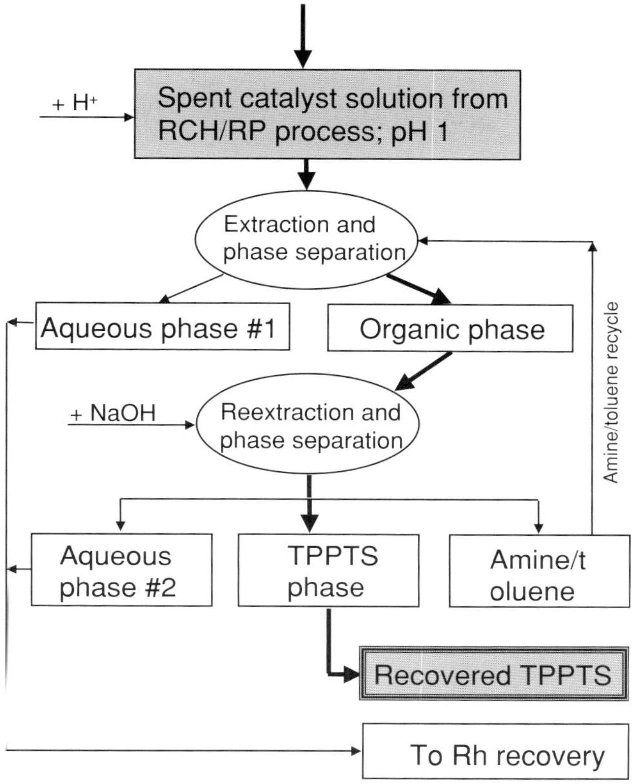

Figure 12.22 TPPTS recovery.

According to Fig. 12.22, the workup of rhodium follows an acidification of the spent catalyst and the extraction with a mixture of a tertiary alkylamine and toluene. The resulting two phases (aqueous phase #1, containing rhodium, and the organic phase with amine/toluene and TPPTS) are further processed. The organic phase is reextracted with aqueous NaOH (which purifies existent Rh(III) compounds and TPPTS), yielding three pH-dependent fractions: an aqueous solution #2 of rhodium compounds, the TPPTS fraction (TPPTS as sodium salt, ready for recycle and reuse; recovery rate: approximately 80%), and the amine/toluene phase which is recycled either to the manufacture of fresh TPPTS or to the TPPTS recovery. Both rhodium-containing fractions of the TPPTS recovery (#1 and #2) are subject to the rhodium recovery as depicted in Fig. 12.23. These Rh-containing phases are mixed with NaOH and 2-ethylhexanoic acid and subsequently oxidized with air, H_2O_2, or other oxidizing agents. The Rh compounds in phases #1 and #2 are thus converted to Rh 2-ethylhexanoate and are extracted in a cross-flow manner by means of 2-ethylhexanol. The recovery rate of rhodium is 97–98%. The residual water phase ends up in the fraction for external workup as described above.

It may be expected that higher valued ligands (which might be used for the conversion of long-chain alkenes) or even coligands, cosolvents, modifiers, surfactant promoters, etc., can

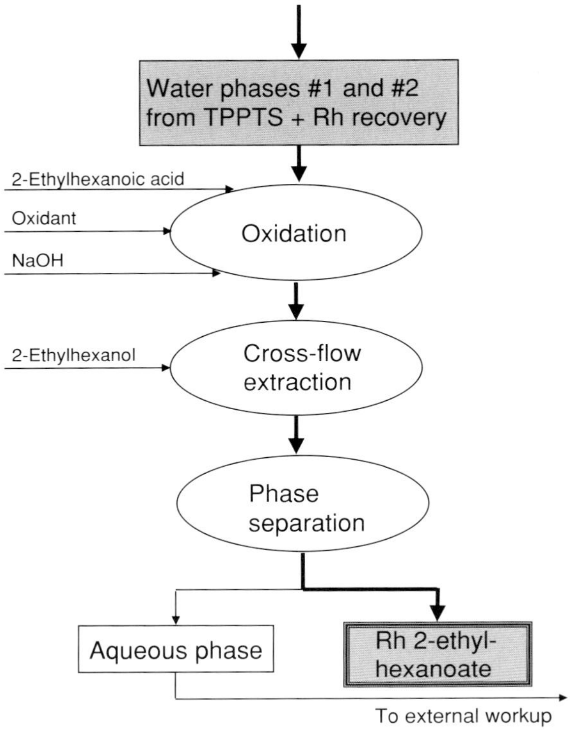

Figure 12.23 Rhodium recovery.

be processed the same way. There are already small-scale processes employing aqueous-phase catalysis which use 'exotic' ligands such as di-, tri-, or multidentate phosphines where the situation demands drastic measures including a high-sophisticated management for rhodium and ligand recycle and recovery.[64]

The 'real' oxo precatalyst [$HRh(CO)(TPPTS)_3$] is easily made in the oxo reactor by reacting suitable Rh salts (e.g. rhodium acetate or rhodium 2-ethylhexanoate) with TPPTS – both components freshly prepared or recovered and recycled – without any additional preformation step. After formation of the active species and adjustment of the whole system with water to the desired P/Rh ratio (ensuring the stability of the catalyst and the desired *n/iso* ratio) the reaction starts.

12.4 Economics of the process

In comparison to other oxo variants the RCH/RP process is highly economical, *inter alia*, reflecting the technological progress of the aqueous phase operation, the better energy compound, and the higher selectivities, which are a consequence of the TPPTS ligand. Under equal conditions (based on market prices without internal clearing prices) the RCH/RP

Table 12.5 Manufacturing costs of 100 kg *n*-butanal (RCH/RP = 100)

Costs	RCH/RP process	Other rhodium-catalyzed processes
1. Raw materials	88.9	89.0
2. Energies	1.5	9.7
3. Credits (isobutanal, *n*-butanol, others)	−7.7	−11.4
4. Costs for materials (sum 1 + 2 + 3)	82.7	87.3
5. Fix cost plus license fee[a]	17.3	22.1
6. Sum 4 + 5 = manufacturing costs[a]	100.0	109.4

[a] For updates, see the *Chemical Engineering Plant Cost Index* (CEPCI).[65]

process offers clear advantages of approximately 10% over other ligand-modified processes (Table 12.5).[17d]

12.5 Environmental aspects

Catalyzed organic reactions in aqueous media have received significant attention as a result of environmental and economic considerations.[10d,21]

The fundamental advance represented by the RCH/RP process in terms of the environment, conservation of resources, and minimization of risks such as environmental pollution – aiming at sustainable 'green' processes – can be demonstrated by various criteria and proved by means of the *atom economy* (according to Trost[66]) and – more convincingly and constructive – by the *environmental factor E* according to Sheldon.[67] The latter defined the *E* factor as the ratio of the amount of waste ('waste' is everything except the desired product) produced per kilogram of 'target' products and he specified the *E* factor for every segment of the chemical industry (Table 12.6).[67,68]

As expected from Table 12.7, this environmental factor for conventional oxo processes (basis: Co catalysts) and for the manufacturing of the bulk chemical *n*-butanal is actually about 0.6–0.9, depending on the definition of the term 'target' product.[68] The range 0.6–0.9 indicates that the by-product isobutanal occurring with conventional oxo processes is further processed by some producers (to isobutyric acid, neopentylglycol, etc.) so that the isoaldehyde thus becomes a target product and the *E* factor falls from 0.9 to 0.6. Strictly speaking, this observation is included in Sheldon's wider assessment, according to which the *E* factor is refined and becomes the environmental quotient EQ, depending on the nature of the waste. Since such quotients are 'debatable and will vary from one company to another and even from one production to another' (Sheldon), they will not be discussed here. The crucial point is that on the same basis (taking into account all by-products, including those produced in ligand manufacture, etc.) which exhibit for conventional oxo processes an

Table 12.6 The environmental factor *E*

Industry segment	Product tonnage (tpy)	*E* factor
Oil refining	10^6– 10^8	~0.1
Bulk chemicals	10^4–10^6	<1–5
Fine chemicals	10^2–10^4	5–50+
Pharmaceuticals	10^1–10^3	25–100+

Table 12.7 *E* factors of different oxo processes[68]

	E factor for different modes of operation based on the hydroformylation of propylene	
	Isobutanal as value product	Isobutanal as by-product
Catalyst on the basis of cobalt	>0.6	>0.9
Rh catalyst; RCH/RP process	<0.04	<0.1

E factor of 0.6–0.9, this factor falls to below 0.1 in the RCH/RP process – an important pointer to the environmental friendliness of the RCH/RP process (Table 12.7).

Whereas this important factor is calculated solely from the product spectrum, process simplifications are a consequence of combining the rhodium catalyst with the special two-phase process. Compared with the conventional oxo process and with other variants (which, for example, include disadvantageously thermal separation of the oxo reaction products from the catalyst), the procedure is considerably simplified (as shown in several papers; e.g. Refs. 17 and 41).

The conservation of energy resources with the RCH/RP process is dramatic. Note should be taken of the much milder reaction conditions and of the fact that the RCH/RP process is an energy exporter because of an intelligent, integrated heat network – an unusual occurrence for conventional oxo processes, competing processes with PPh_3-modified Rh catalysts included. Furthermore, the steam consumption figures for the older cobalt-based processes are very much higher than those for the Rh process and power consumption was twice as high as that of the RCH/RP process – both of these factors represent an environmental burden. The relative compression costs alone for the required syngas are 1.7:1 (Co versus Rh process). The volume of wastewater for the RCH/RP process is 70 times lower than that from the cobalt-based high-pressure process – convincing evidence of an environmental benign and 'green' process and one of the reasons of advantageous results from the life cycle assesssment. It may be added that the process thus meets the requirements of the 'production-integrated environmental protection', sometimes known as 'cleaner production', which has the aim of avoiding and reducing residues and using resources carefully.[13c,69]

The solvent water reliably averts the risk of fire inherent in the older cobalt-based process as a result of leaking highly flammable, metal carbonyls. The technique with its 'built-in extinguishing system' reliably prevents such fires, and the painstaking measuring and monitoring procedure necessitated by the valuable rhodium-based catalyst, accompanied by constant simultaneous balancing of the RCH/RP process, permits any leaks from the aqueous system to be detected much earlier than was ever possible with the expensive mass and liquid balance of the older cobalt-based process. This also applies for the cooling system, in which any leak from the falling film evaporator would be noticed immediately.

12.6 Concluding remarks

Taking all criteria into consideration, aqueous biphasic techniques are very sound methods for homogeneously catalyzed processes such as hydrogenations or hydroformylations. In the series of the various alternatives to the conventional (and solvent-free) processes, most

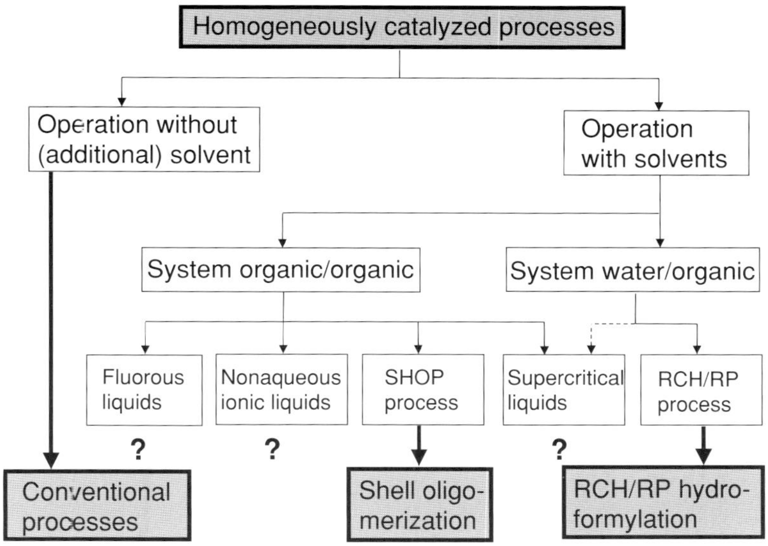

Figure 12.24 The various possibilities of biphasic operation of homogeneous catalysis.

progress in terms of ecologics and economics has been attained by the water/organic biphasic operation (Fig. 12.24).

So far, only the systems organic/organic and water/organic have emerged with cost-effective, commercial processes. It is believed that the latter system has the larger potential compared to organic/organic system because its solubility patterns are more promising. The other possibilities – fluorous liquids, supercritical fluids, and nonaqueous ionic liquids – are labeled by a question mark to underline the uncertainty of their development. The same is true for recent work using water-soluble polymers.[57] Additionally, all other proposed processes need costly media (fluorous solvents, nonaqueous ionic liquids), high-priced ligands (fluorous processes or processes with supercritical CO_2), and/or additional means such as phase-separating fluids and cosolvents.

The aqueous biphasic processes are 'good-natured' from the handling aspects, thus emphasizing the great environmental compatibility. They have great development potentials whether in terms of varying the ligand TPPTS (and thus the activity and selectivity of the conversions chosen, e.g. the hydroformylation[68]) as well as the possibility of achieving asymmetric conversions, or in terms of other feedstocks. The adoption of the aqueous biphasic reaction for hydroformylation in particular (but also for other homogeneously catalyzed conversions in general) shows the striking and unequaled advantages of this special neat type of 'immobile' catalyst and its 'heterogenization' by the catalyst carrier, water, as a result of which it is handled in similar fashion to a heterogeneous catalyst. This affects the costs as well as the environmental compatibility.

The successful waste management and avoiding by-products are further good points of the process, which ultimately lead to the higher cost-effectiveness as a result of the mentioned advantages together with less downtime – further proof of the likelihood that even in the medium term only environmentally sound processes will remain the most effective. The fact that Sheldon's *E* factor is basically inversely proportional to a successful life cycle assessment underlines the superior quality of the water/organic system.

References

1. Cornils, B. and Herrmann, W.A. (Eds.), *Applied Homogeneous Catalysis with Organometallic Compounds*, 2nd edn., Wiley-VCH, Weinheim, Germany, 2003.
2. (a) Manassen, J. and Whitehurst, D. D., In: Basolo, F. and Burwell, R.L. (Eds.), *Catalysis Progress in Research*, Plenum Press, London, 1973, p. 183, and US 4,415,500, 1983 (YEDA); (b) Joó, F. and Beck, M.T., *Magy. Kém. Folyóirat*, 1973, **79**, 189; (c) Joó, F., *Aqueous Organometallic Catalysis*, Kluwer, Dordrecht, 2001, p. 3; (d) Joó, F., Papp, E. and Kathó, A., *Top. Catal.*, 1998, **5**, 113.
3. Keyword 'immobilization' and other keywords. In: Cornils, B., Herrmann, W.A., Muhler, M. and Wong, C.H. (Eds.), *Catalysis from A to Z*, 3rd edn., Wiley-VCH, Weinheim, Germany, 2007.
4. (a) Schiller, G., DE 953,605, 1956 (Chemische Verwertungsgesellschaft mbH Oberhausen); *Chem. Abstr.*, 1959, **53**, 11226; (b) GB 801,734, 1956 (Esso); *Chem. Abstr.*, 1959, **53**, 7014.
5. (a) Mullineaux, R.D., BE 603,820, 1961 (Shell); (b) Cannell, L.G., Slaugh, L.H. and Mullineaux, R.D., DE 1,186,455 1965 (Shell); *Chem. Abstr.*, 1965, **62**, 16054; (c) Slaugh, L.H. and Mullineaux, R.D., *J. Organomet. Chem.*, 1968, **13**, 469.
6. (a) Hughes, O.R. and Douglas, M.E., DE 2,125,382, 1970 (Celanese Corp.); (b) Celanese Corporation, Annual Report, 1974, p. 9; (c) Fowler, R., Connor, H. and Baehl, R.A., *CHEMTECH*, 1976, 772; (d) Evans, D., Osborn, J.A. and Wilkinson, G., *J. Chem. Soc.*, 1968, **2660**, 3133.
7. Cintas, P., *Chem. Eng. News*, 20 March 1995, 4.
8. (a) Joó, F. and Beck, M.T., *Magy. Kém. Folyóirat*, 1973, **79**, 189; (b) Joó, F., *Aqueous Organometallic Catalysis*, Kluwer, Dordrecht, 2001, p. 3; (c) Joó, F., Papp, E. and Kathó, A., *Top. Catal.*, 1998, **5**, 113.
9. (a) Kuntz, E., FR 2,314,910, 1975 (Rhône-Poulenc Recherche); (b) FR 2,349,562, 1976; FR 2,338,253, 1976; FR 2,366,237, 1976, and other patent applications; (c) Cornils, B. and Kuntz, E.G., *J. Organomet. Chem.*, 1995, **505**, 177; (d) Kuntz, E.G., US 4,087,452, 1978 (Rhône-Poulenc).
10. (a) van Leeuwen, P.W.N.M. and Claver, C. (Eds.), *Rhodium Catalyzed Hydroformylation*, Kluwer, Dordrecht, 2000; (b) Cornils, B., et al. (Eds.), *Multiphase Homogeneous Catalysis*, Wiley-VCH, Weinheim, Germany, 2005; (c) Bohnen, H.-W. and Cornils, B., In: Gates, B.C. and Knözinger, H. (Eds.), *Advances in Catalysis*, Vol. 47, 2002, p. 1; (d) Cornils, B. and Herrmann, W.A. (Eds.), *Aqueous-Phase Organometallic Chemistry*, 2nd edn., Wiley-VCH, Weinheim, Germany, 2004.
11. Cornils, B. and Wiebus, E., In: Cole-Hamilton, D.J. and Tooze, R.P. (Eds.), *Catalyst Separation, Recovery and Recycle*, Springer, London, 2006, p. 105.
12. (a) Baur, S.R., Chung, H., Glockner, P.W., Keim, W., van Zwet, H., Camel, D. and Mason, R.F., US 3,635,937, 3,637,636, 3,644,563, 3,644,564, 3,647,914, 3,647,915, 3,661,803, and 3,686,159, 1972 and 1973 (Shell Dev. Corp.); (b) Keim, W., *New J. Chem.*, 1987, **11**, 531; (c) Keim, W., Kowaldt, F.-H., Goddard, R. and Krüger, C., *Angew. Chem.*, 1978, **90**, 493; (d) Vogt, D., In: Cornils, B. and Herrmann, W.A. (Eds.), *Applied Homogeneous Catalysis with Organometallic Compounds*, 2nd edn., Wiley-VCH, Weinheim, Germany, 2003, p. 245.
13. (a) Baker, R.T. and Tumas, W., *Science*, 1999, **284**, 1477; (b) Ritter, S.K., *Chem. Eng. News*, 20 May 2002, 38; (c) Curzons, A.D., Constable, D.J.C., Mortimer, D.N. and Cunningham, V.L., *Green Chem.*, 2001, **3**(1), 1; (d) Swatloski, R.P., Holbrey, J.D. and Rogers, R.D., *Green Chem.*, 2003, **5**(4), 361; (e) Centi, G. and Perathoner, S., *Catal. Today*, 2003, **77**, 287.
14. Cornils, B., et al. (Eds.), *Multiphase Homogeneous Catalysis*, Wiley-VCH, Weinheim, Germany, 2005, Chapters 3 (D. Vogt, Ed.), 4 (I.T. Horváth, Ed.), 5 (H. Olivier-Bourbigou, Ed.), 6 (W. Leitner, Ed.), and 7 (S. Mecking, Ed.).
15. (a) Li, Z., Peng, Q. and Yuan, Y., *Appl. Catal. A*, 2003, **239**, 79; (b) Benaissa, M., Jáuregui-Haza, U.J., Nikov, I., Wilhelm, A.M. and Delmas, H., *Catal. Today*, 2003, **79/80**, 419; (c) Jáuregui-Haza, U.J., Pardillo-Fontdevila, E., Kalck, Ph. and Delmas, H., *Catal. Today*, 2003, **79/80**, 409; (d) Riisager, A., Eriksen, K.M., Hjortkjaer, J. and Fehrmann, R., *J. Mol. Catal. A*, 2003, **193**, 259; (e) Dessoudeix, M., Jáuregui-Haza, U.J., Heughebaert, M., Wilhelm, A.M., Delmas, H., Lebugle, A. and Kalck, Ph., *Adv. Synth. Catal.*, 2002, **344**, 406; (f) Li, Z., Peng, Q. and Yuan, Y., *Appl. Catal. A*, 2003, **239**, 79; (g) Yuan, Y., Zhang, H., Yang, Y., Zhang, Y. and Tsai, K., *Catal. Today*, 2002, **74**, 5; (h) Li, Z., Peng, Q. and Yuan, Y., *Appl. Catal. A*, 2003, **239**, 79; (i) Fremy, G., Monflier, E., Carpentier, J.-F., Castanet, Y. and Mortreux, A., *Angew. Chem.*, 1995, **107**, 1608; (j) Botteghi, C., et al., *J. Mol. Catal. A*, 2003, **200**, 147; (k) Scholten, J.J.F., et al., *J. Mol. Catal.*, 1980, **9**(Special issue); (l) Frohning, C.D., Kohlpaintner, C.W. and Bohnen, H.-W., In: Cornils, B. and Herrmann, W.A. (Eds.), *Applied Homogeneous Catalysis with Organometallic Compounds*, 2nd edn., Wiley-VCH, Weinheim, Germany, 2003, p. 31; (m) Strohmeier, W., et al., *J. Organomet. Chem.*, 1981, **221**, 361; (n) Davis, M.E., In: Cornils, B. and Herrmann, W.A. (Eds.), *Aqueous-Phase Organometallic Chemistry*, 2nd edn., Wiley-VCH, Weinheim, Germany, 2004,

p. 313; *CHEMTECH*, 1992, **498**; (o) Li, Z., Peng, Q. and Yuan, Y., *Appl. Catal. A*, 2003, **239**, 79; (p) Reek, J.N.H., Sandee, A.J., Schreuder Goedheijt, M., Kamer, P.C.J. and van Leeuwen, P.W.N.M., *Erdöl, Erdöl, Kohle*, 2001, **117**, 134; (q) Anderson, J.R., Campi, E.M., Jackson, W.R. and Yang, Z.P., *J. Mol. Catal. A*, 1997, **116**, 109.

16. (a) van Leeuwen, P.W.N.M., Kamer, P.C.J. and Reek, J.N.H., *Cattech*, 1999, **3**(2), 64; (b) Schreuder Goedheijt, M., Kamer, P.C.J. and van Leeuwen, P.W.N.M., *J. Mol. Catal. A*, 1998, **134**, 243; (c) Botteghi, C., et al., *J. Mol. Catal. A*, 2003, **200**, 147.
17. (a) Bach, H., Gick, W., Wiebus, E. and Cornils, B., In: *Int. Symp. High-Pressure Chem. Eng.*, Erlangen, Germany, September 1984, Preprints, p. 129; (b) Bach, H., Gick, W., Wiebus, E. and Cornils, B., In: *8thICC*, Berlin, 1984, Preprints, Vol. V, p. 417 [*Chem. Abstr.*, 1987, **106**, 198,051; cited also in Behr, A. and Röper, M., *Erdgas Kohle*, 1984, **37**(11), 485]; (c) Bach, H., Gick, W., Wiebus, E. and Cornils, B., In: *1st IUPAC Symp. Org. Chem.*, Jerusalem, 1986, Abstracts, p. 295; (d) Wiebus, E. and Cornils, B., *Chem. Ing. Tech.*, 1994, **66**, 916; (e) Cornils, B. and Wiebus, E., *CHEMTECH*, 1995, **25**, 33; (f) Kuntz, E.G., *CHEMTECH*, 1987, **17**, 570; (g) Kohlpaintner, C.W., Fischer, R.W. and Cornils, B., *Appl. Catal. A*, 2001, **221**, 219; (h) Cornils, B., *J. Mol. Catal. A*, 1999, **143**, 1; (i) Cornils, B., *Org. Proc. Res. Dev.*, 1998, **2**, 121; (j) Frohning, C.-D., Kohlpaintner, C.W. In: Cornils, B. and Herrmann, W.A. (Eds.), *Aqueous-Phase Organometallic Chemistry*, 2nd edn., Wiley-VCH, Weinheim, Germany, 2004, p. 294; (k) Wiebus, E. and Cornils, B., In: *Hydrocarb. Proc.*, March 1996, p. 63; (l) Cornils, B. and Herrmann, W.A., In: Cornils, B. and Herrmann, W.A. (Eds.), *Aqueous-Phase Organometallic Chemistry*, 2nd edn., Wiley-VCH, Weinheim, Germany, 2004, Introduction.
18. (a) Kalck, P., Dessoudeix, M. and Schwarz, S., *J. Mol. Catal. A*, 1999, **143**, 41; (b) Deshpande, R.M., Divekar, S.S., Bhanage, B.M. and Chaudhari, R.V., *J. Mol. Catal. A*, 1992, **75**, L19.
19. (a) Dämbkes, G., Hahn, H.D., Hibbel, J. and Materne, W., DE 3,347,406, 1983 (Ruhrchemie AG); (b) Dämbkes, G., Hahn, H.D., Hibbel, J. and Materne, W., EP 0,147,824, 1984 (Ruhrchemie AG); (c) Cornils, B., Konkol, W., Bahrmann, H., Bach, H.W. and Wiebus, E., DE 3,411,034, 1984 and EP 0,156,253, 1985 (Ruhrchemie AG); (d) Bahrmann, H., Konkol, W., Weber, J., Bach, H.W. and Bexten, L., DE 3,534,314, 1985 (Ruhrchemie AG); (e) Bexten, L. and Kupies, D., DE 3,626,636, 1986 (Ruhrchemie AG); (f) Bexten, L. and Kupies, D., EP 0,255,673, 1987 (Hoechst AG); (g) Diekhaus, G. and Kappeser, H., EP 0,322,661, 1988 (Hoechst AG); (h) Lappe, P., Bexten, L. and Kupies, D., DE 3,822,037, 1988, EP 0,348,833, 1989, and US 5,294,415, 1994 (Hoechst AG); (i) Weber, J., Bexten, L., Kupies, D., Lappe, P. and Springer, H., EP 0,367,957, 1989 (Hoechst AG); (j) Lappe, P. and Springer, H., DE 3,934,824, 1989 and EP 424,736, 1990 (Hoechst AG); (k) Lappe, P. and Springer, H., DE 4,025,074, 1990 and EP 0,475,036, 1991 (Hoechst AG); (l) Lappe, P. and Springer, H., EP 0,510,358, 1992 (Hoechst AG); (m) Förster, I. and Mathieu, K., EP 0,538,723, 1992 (Hoechst AG); (n) Konkol, W., Bahrmann, H., Herrmann, W.A. and Kohlpaintner, C.W., EP 0,544,091, 1992 (Hoechst AG); (o) Diekhaus, G. and Kappeser, H., EP 0,584,720, 1993 (Hoechst AG).
20. (a) Gärtner, R., Cornils, B., Bexten, L. and Kupies, D., DE 3,235,029, 1982, EP 0,103,845, 1983, and EP 0,107,006, 1983 (Ruhrchemie AG); (b) Bexten, L., Cornils, B. and Kupies, D., EP 0,175,919, 1985 (Ruhrchemie AG); (c) Bahrmann, H., Konkol, W., Weber, J., Bach, H.W. and Bexten, L., EP 0,216,315, 1986 (Ruhrchemie AG); (d) Herrmann, W.A., et al., EP 0,352,478, 1989 (Hoechst AG).
21. (a) Olivier-Bourbigou, H. and Hugues, F., In: Rogers, R.D., Seddon, K.R. and Volkov, S. (Eds.), *Green Industrial Applications of Ionic Liquids*, NATO Science Series, Vol. 92, Kluwer, Dordrecht, 2002.
22. Reichardt, C., *Solvents and Solvent Effects in Organic Chemistry*, 3rd edn., Wiley-VCH, Weinheim, Germany, 2002.
23. Hope, E.G. and Stuart, A.M., *Chem. Ind.*, 4 October 1999, 755.
24. Tzschucke, C.C., Markert, C., Bannwarth, W., Roller, S., Hebel, A. and Haag, R., *Angew. Chem. Int. Ed.*, 2002, **41**, 3964.
25. (a) Lubineau, A., *Chem. Ind.*, February 1996, p. 123 [not p. 125, as printed]; (b) Roesky, H.W., Walawalkar, M.G. and Murugavel, R., *Acc. Chem. Res.*, 2001, **34**, 201; (c) Asmis, K.R., et al., *Science*, 2003, **299**, 1375; (d) Köddermann, T., Schulte, F., Huelsekopf, M. and Ludwig, R., *Angew. Chem. Int. Ed.*, 2003, **42**, 4904.
26. (a) Buurma, N.J., Blandamer, M.J. and Engberts, J.B.F.N., *Adv. Synth. Catal.*, 2002, **344**, 413; (b) Itami, K., Nokami, T. and Yoshida, J., *Adv. Synth. Catal.*, 2002, **344**, 441.
27. Duve, G., Fuchs, O. and Overbeck, H., *Lösemittel Hoechst*, 6th edn., Hoechst AG, Frankfurt, Germany, 1976, p. 49.
28. (a) Wachsen, O., Himmler, K. and Cornils, B., *Catal. Today*, 1998, **42**, 373; (b) Lindner, E., Schneller, T., Auer, F. and Meyer, H.A., *Angew. Chem. Int. Ed.*, 1999, **38**, 2155; (c) Kalck, Ph., Dessoudeix, M. and Schwarz, S., *J. Mol. Catal. A*, 1999, **143**, 41.
29. (a) Larpent, C., Bernard, E., Brisse-le Menn, F. and Patin, H., *J. Mol. Catal. A*, 1997, **116**, 277; (b) GB 2,075,857, 1981 and EP 096,987, 1983 (Davy McKee/Union Carbide); (c) EP 055,487, 1982 and EP 068,480, 1082 (Union Carbide); (d) JP 5-8186-443A, 1983 (Daicel); (e) EP 053,688, 1982 (BASF).
30. (a) Stephenson, R.M., *J. Chem. Eng. Data*, 1993, **38**, 630; (b) McAuliffe, C., *J. Phys. Chem.*, 1966, **70**, 1267; (c) Davis, P.L., *Gas Chromatogr.*, 1968, **6**, 518.
31. (a) Yang, C., Bi, X. and Mao, Z.-S., *J. Mol. Catal. A*, 2002, **187**, 35; (b) Önal, Y., Baerns, M. and Claus, P., In: Cornils, B. and Herrmann, W.A. (Eds.), *Applied Homogeneous Catalysis with Organometallic Compounds*,

2nd edn., Wiley-VCH, Weinheim, Germany, 2003, p. 151; (c) *Chem. Ing. Tech. Plus.*, 2003, September issue, p. 46; (d) Zhang, Y., Mao, Z.-S. and Chen, J., *Catal. Today*, 2002, **74**, 23; (e) Yang, C., Mao, Z.-S., Wang, Y. and Chen, J., *Catal. Today*, 2002, **74**, 111.

32. (a) Stokes, R.J. and Evans, D.F., *Fundamentals of Interfacial Engineering*, Wiley-VCH, New York; (b) DE 3,220,858, 1983 (BASF).
33. (a) Riisager, A. and Hanson, B.E., *J. Mol. Catal. A*, 2002, **189**, 195; (b) Chen, H., Li, Y., Chen, J., Chen, P., He, Y. and Li, X., *J. Mol. Catal. A*, 1999, **149**, 1; (c) Roucoux, A., Schulz, J. and Patin, H., *Adv. Synth. Catal.*, 2003, **345**, 222; (d) Borrmann, T., Roesky, H.W. and Ritter, U., *J. Mol. Catal. A*, 2000, **153**, 31; (e) Loch, J.A., Borgmann, C. and Crabtree, R.H., *J. Mol. Catal. A*, 2001, **170**, 75; (f) Larpent, C., Bernard, E., Brisse-le Menn, F. and Patin, H., *J. Mol. Catal. A*, 1997, **116**, 277; (g) Reger, D.L. and Habib, M.M., *J. Mol. Catal.*, 1978, **4**, 315; (h) Grassert, I., Kovács, J., Fuhrmann, H. and Oehme, G., *Adv. Synth. Catal.*, 2002, **344**, 312; (i) Nardello, V., Hervé, M., Alsters, P.L. and Aubry, J.-M., *Adv. Synth. Catal.*, 2002, **344**, 184; (j) Haumann, M., Yildiz, H., Koch, H. and Schomäcker, R., *Appl. Catal. A*, 2002, **236**, 173; (k) Haumann, M., Koch, H. and Schomäcker, R., *Catal. Today*, 2003, **79/80**, 43; (l) Haumann, M., Koch, H., Hugo, P. and Schomäcker, R., *Appl. Catal. A*, 2002, **225**, 239; (m) Giménez-Pedrós, M., Aghmiz, A., Claver, C., Masdeu-Âultó, A.M. and Sinou, D., *J. Mol. Catal. A*, 2003, **200**, 157; (n) Liu, X., Jiang, M., Yang, S., Chen, M., Chen, D., Yang, C. and Wu, K., *Angew. Chem. Int. Ed.*, 2002, **41**, 2950; (o) Riisager, A. and Hanson, B.F., *J. Mol. Catal. A*, 2002, **189**, 195; (p) Chen, H., Li, Y., Chen, J., Cheng, P., He, Y. and Li, X., *J. Mol. Catal. A*, 1999, **149**, 1; (q) Dwars, T., Haberland, J., Grassert, I., Oehme, G. and Kragl, U., *J. Mol. Catal. A*, 2001, **168**, 81.
34. Among a multitude of publications: (a) Fell, B., et al., *J. Prakt. Chem./Chem. Ztg.*, 1996, **338**, 124; *J. Mol. Catal.*, 1997, **116**, 55; *Fett/ Lipid*, 1996, **98**, 393; (b) Jin, Z. et al., *J. Mol. Catal. A*, 1999, **147**, 131 and **149**, 113; 2000, **157**, 111 and **159**, 37; 2001, **171**, 85; 2002, **188**, 79; 2003, **195**, 133; 2003, **198**, 23; *Catal. Today*, 2002, **74**, 85; *Catal. Commun.*, 2003, **4**(3), 129; (c) Paetzold, E., Oehme, G., Fischer, C. and Frank, M., *J. Mol. Catal. A*, 2003, **200**, 95; (d) Loch, J.A., Borgmann, C. and Crabtree, R.H., *J. Mol. Catal. A*, 2001, **170**, 75; (e) Bhanage, B.M., Shirai, M. and Arai, M., *J. Mol. Catal. A*, 1999, **145**, 69; (f) Ritter, U., Winkhofer, N., Schmidt, H.-G. and Roesky, H.W., *Angew. Chem., Int. Ed. Engl.*, 1996, **35**, 524.
35. (a) Karakhanov, E.A., Kardesheva, Y.S., Runova, E.A. and Semermina, V.A., *J. Mol. Catal. A*, 1999, **142**, 339; (b) Reek, J.N.H., Sandee, A.J. and Schreuder Goedheijt, M., Kamer, P.C.J. and van Leeuwen, P.W.N.M., *Erdöl, Erdgas, Kohle*, 2001, **117**, 134; (c) Chen, H., Li, Y., Li, R., Cheng, P. and Li, X., *J. Mol. Catal. A*, 2003, **198**, 1; (d) Ritter, U., Winkhofer, N., Schmidt, H.-G. and Roesky, H.W., *Angew. Chem. Int. Ed.*, 1996, **35**, 524; (e) Li, M., Li, Y., Chen, H., He, Y. and Li, X., *J. Mol. Catal. A*, 2003, **194**, 13; (f) Huang, Y., Min, L., Li, Y., Chen, H., Cheng, P. and Li, X., *J. Mol. Catal. A*, 2002, **185**, 41; (g) Zhang, Y., Mao, Z.-S. and Chen, J., *Catal. Today*, 2002, **74**, 23; (h) Chen, H., Li, Y., Chen, J., Cheng, P. and Li, X., *Catal. Today*, 2002, **74**, 131; (i) Li, Z., Peng, Q. and Yuan, Y., *Appl. Catal. A*, 2003, **239**, 79; (k) Monteil, F., Queau, R. and Kalck, Ph., *J. Organomet. Chem.*, 1994, **480**, 177; (l) Giménez-Pedrós, M., Aghmiz, A., Claver, C., Masdeu-Bultó, A.M. and Sinou, D., *J. Mol. Catal. A*, 2003, **200**, 157; Wang, L., Chen, H., He, Y., Li, Y., Li, M. and Li, X., *Appl. Catal. A*, 2003, **242**, 85.
36. (a) Bayer, E. and Schurig, V., *CHEMTECH*, March 1976, 212; (b) Bayer, E. and Schurig, V., *Angew. Chem., Int. Ed. Engl.*, 1975, **14**, 493; (c) Giménez-Pedrós, M., Aghmiz, A., Claver, C., Masdeu-Bultó, A.M. and Sinou, D., *J. Mol. Catal. A*, 2003, **200**, 157; (d) Paetzold, E., Michalik, M. and Oehme, G., *J. Prakt. Chem.*, 1997, **339**, 38; (e) Kumar, A., Oehme, G., Roque, J.P., Schwarze, M. and Selke, R., *Angew. Chem.*, 1994, **106**, 2272; (f) Ganguly, S., Mague, J.T. and Roundhill, D.M., *Inorg. Chem.*, 1992, **31**, 3500; (g) Shimizu, S., Shirakawa, S., Sasaki, Y. and Hirai, C., *Angew. Chem.*, 2000, **112**, 1313; (h) Schreuder Goedheijt, M., Kamer, P.C.J. and van Leeuwen, P.W.N.M., *J. Mol. Catal. A*, 1998, **134**, 243; (i) Katti, K.V., Gall, H., Smith, C.J. and Berning, D.E., *Acc. Chem. Res.*, 1999, **32**, 9; (k) Beller, M., Krauter, J.G.E., Zapf, A. and Bogdanovic, S., *Catal. Today*, 1999, **48**, 279; (l) Shimizu, S., Suzuki, T., Shirakawa, S., Sasaki, Y. and Hirai, C., *Adv. Synth. Catal.*, 2002, **344**, 370; (m) Bertucci, C., Botteghi, C., Giunta, D., Marchetti, M. and Paganelli, S., *Adv. Synth. Catal.*, 2002, **344**, 556; (n) Monflier, E., Fremy, G., Castanet, Y. and Mortreux, A., *Angew. Chem.*, 1995, **107**, 2450; (o) Baxley, G.T., Weakly, T.J.R., Miller, W.K., Lyon, D.K. and Tyler, D.R., *J. Mol. Catal. A*, 1997, **116**, 191; (p) Gong, A., Fan, Q., Chen, Y., Liu, H., Chen, C. and Xi, F., *J. Mol. Catal. A*, 2000, **159**, 225; (q) Peng, Q., Yang, Y., Wang, C., Liao, X. and Yuan, Y., *Catal. Lett.*, 2003, **88**, 219; (r) Aghmiz, A., Orejón, A., Diéguez, M., Miquel-Serrano, M.D., Claver, C., Masdeu-Bultó, A.M., Sinou, D. and Laurenczy, G., *J. Mol. Catal. A*, 2003, **195**, 113; (s) Liu, X.Z., Kong, F., Zheng, X. and Jin, Z., *Catal. Commun.*, 2003, **4**, 129; (t) Mudalige, D.C. and Rempel, G.L., *J. Mol. Catal. A*, 1997, **116**, 309; (u) Riisager, A., Erksen, K.M., Hjortkaer, J. and Fehrmann, R., *J. Mol. Catal. A*, 2003, **193**, 259; (v) Ganguly, S., Mague, J.T. and Roundhill, D.M., *Inorg. Chem.*, 1992, **31**, 3500; (w) Shimizu, S., Shirakawa, S., Sasaki, Y. and Hirai, C., *Angew. Chem.*, 2000, **112**, 1313; (x) Amengual, R., Genin, E., Michelet, V. and Savignac, M., *Adv. Synth. Catal.*, 2002, **344**, 393; (y) Karakhanov, E.A., Kardasheva, Y.S., Runova, E.A. and Terenina, M.V., *EuropaCat-VI*, Innsbruck, Austria, 31 August – 4 September 2003, Preprints; (z) Chécot, F., Lecommandoux, S., Gnanou, Y. and Klok, H.-A., *Angew. Chem. Int. Ed.*, 2002, **41**, 1340.
37. Ooi, T., Uematsu, Y. and Maruoka, K., *Adv. Synth. Catal.*, 2002, **344**, 288.
38. (a) Protzmann, G. and Wiese, K.-D., *Erdöl, Erdgas, Kohle*, 2001, **117**, 235; (b) Protzmann, G., Wiese, K.-D., Koch, J., Trocha, M. and Röttger, D., DE 199,25,384, 2000 (Oxeno); (c) Wiese, K.-D., Protzmann, G., Koch, J., Röttger, D. and Trocha, M., EP 1,057,524, 2000 (Oxeno); (d) Protzmann, G., Wiese, K.-D., Büschken, W. and

Röttger, D., DE 199,57,528 and 199,57,522, 1999 (Oxeno); (e) Protzmann, G., Wiese, K.-D. and Büschken, W., EP 1,057,538, 2001 (Oxeno); (f) Wiese, K.-D., Olbrich, P. and Gabriel, J., DE 199,08,320, 2000 (Oxeno); (g) Wiese, K.-D., Möller, O., Protzmann, G. and Trocha, M., *Catal. Today*, 2003, **79/80**, 97.

39. (a) Buhling, A., Kamer, P.C.J., van Leeuwen, P.W.N.M. and Elgersma, J.W., *J. Mol. Chem. A*, 1997, **116**, 297; (b) Buhling, A., Kamer, P.C.J. and van Leeuwen, P.W.N.M., *J. Mol. Catal. A*, 1995, **98**, 69; (c) Hanson, B.E., Ding, H. and Kohlpaintner, C.W., *Catal. Today*, 1998, **42**, 421; (d) Robert, F., Oehme, G., Grassert, I. and Sinou, D., *J. Mol. Catal. A*, 2000, **156**, 127; (e) Peng, Q., Yang, Y., Wang, C., Liao, X. and Yuan, Y., *Catal. Lett.*, 2003, **88**, 219; (f) Uozumi, Y. and Nakazono, M., *Adv. Synth. Catal.*, 2002, **344**, 274.
40. (a) de Bellefon, C., Tanchoux, N., Caravieilhes, S., Grenouillet, P. and Hessel, V., *Angew. Chem.*, 2000, **112**, 3584; (b) Hodiamont, R., Leitner, W. and Driessen-Hölscher, B., In: *36th Meeting of German Catalysis*, Weimar, 19–21 March 2003, Preprints, p. 77.
41. (a) Cornils, B., Kohlpaintner, C.W. and Wiebus, E., In: McKetta, J.J. and Weismantel, G.E. (Eds.), *Encyclopedia of Chemical Processing and Design*, Vol. 66, Marcel Dekker, New York, 1999, p. 273; (b) Cornils, B., In: Horváth, I.T., et al. (Eds.), *Encyclopedia of Catalysis*, Vol. 1, Wiley, New York, 2003, p. 805; (c) Cornils, B., In: Knochel, P. (Ed.), *Topics in Current Chemistry*, Vol. 206, Springer, Berlin, 1999, p. 133; (d) Cornils, B., *Ind. Catal. News*, 1998, No. 2; (e) Chaudhari, R.V., Bhattacharya, A. and Bhanage, B.M., *Catal. Today*, 1995, **24**, 123.
42. Prinz, T., Keim, W. and Driessen-Hölscher, B., *Angew. Chem.*, 1996, **108**, 1835.
43. (a) Grieco, P.A. (Ed.), *Organic Synthesis in Water*, Blackie Academic, London, 1998; (b) Fuhrmann, H., Dwars, T. and Oehme, G., *Chem. unserer Zeit*, 2003, **37**, 40; (c) Li, C.J. and Chan, T.H., *Organic Reactions in Organic Media*, Wiley, New York, 1997; (d) Li, C.J., *Chem. Rev.*, 1993, **93**, 2023; (e) Li, C.J. and Chan, T.H., *Tetrahedron*, 1996, **52**, 5643; (f) Groß, M., *Chem. unserer Zeit*, 2003, **37**, 170; (g) Nomura, K., *J. Mol. Catal. A*, 1998, **130**, 1; (h) Buurma, N.J., Blandamer, M.J. and Engberts, J.B.F.N., *Adv. Synth. Catal.*, 2002, **344**, 413; (i) Reynal, F., Barhdadi, R., Périchon, J., Savall, A. and Toupel, M., *Adv. Synth. Catal.*, 2002, **344**, 45; (j) Ribe, S. and Wipf, P., *Chem. Commun.*, 2001, 299.
44. (a) Bahrmann, H., Bach, H., Frohning, C.D., Kleiner, H.J., Lappe, P., Peters, D., Regnat, D. and Herrmann, W.A., *J. Mol. Catal. A*, 1997, **116**, 49; (b) Beller, M., Cornils, B., Frohning, C.D. and Kohlpaintner, C.W., *J. Mol. Catal. A*, 1995, **104**, 17; (c) Cornils, B., Herrmann, W.A. and Eckl, R.W., *J. Mol. Catal. A*, 1997, **116**, 27; (d) Yuan, M., Chen, H., Li, R., Li, Y. and Li, X., *Appl. Catal. A*, 2003, **251**, 181.
45. (a) Behr, A., *Chem. Ing. Technol.*, 1998, **70**, 685; (b) Auch-Schwelk, B. and Kohlpaintner, C., *Chem. Unserer Zeit*, 2001, **35**, 306.
46. (a) Monflier, E., US 5,345,007, 1994, (Elf Atochem); (b) Behr, A. and Urschey, M., *J. Mol. Catal. A*, 2003, **197**, 101; (c) Yoshimura, N., In [6], p. 408; (d) Prinz, T., Keim, W. and Driessen-Hölscher, B., *Angew. Chem.*, 1996, **108**, 1835; (e) JP 080642, 1981, JP 5-8131-994, 1982, and JP 58216138, 1984 (Kuraray); (f) Matsumoto, M. and Tamura, M., *J. Mol. Catal.*, 1982, **16**, 209; (g) EP 088,955, 1983 (Kuraray); Chauvin, H., US 6,525,228, 2001 (IFP).
47. (a) Drinkard, W.D., et al., US 3,536,748, 1970 (DuPont), 3,579,560, 1971 (DuPont), 3,676,481, 1972 (DuPont), and 3,739,011, 1973 (DuPont); (b) Kuntz, E.G., US 4,087,452, 1978 (Rhône-Poulenc); (c) Bryndza, H.E. and Harrelson, J.A., In: Cornils, B. and Herrmann, W.A. (Eds.), *Applied Homogeneous Catalysis with Organometallic Compounds*, 2nd edn., Wiley-VCH, Weinheim, Germany, 2003, p. 393.
48. (a) Chansarkar, R., Mukhopadhyay, K., Kelkar, A.A. and Chaudhari, R.V., *Catal. Today*, 2003, **79/80**, 51; (b) Kohlpaintner, C.W. and Beller, M., *J. Mol. Catal. A*, 1997, **116**, 259; (c) Lapidus, A.L., Petrovskii, K.B. and Beletskaya, I.P., *Russ. J. Org. Chem.*, 1997, **33**(8), 1083; (d) Jayasree, P., Gupta, S.P. and Chaudhari, R.V., *Stud. Surf. Sci. Catal.*, 1998, **113**, 875; Paetzold, E. and Oehme, G., *J. Mol. Catal. A*, 2000, **152**, 69.
49. (a) Dupuis, C., Adiey, K., Charruault, L., Michelet, V., Savignac, M. and Gênet, J.-P., *Tetrahedron Lett.*, 2001, **42**, 6523; (b) Leadbeater, N.E. and Marco, M., *Angew. Chem. Int. Ed.*, 2003, **42**, 1407; (c) Paetzold, E. and Oehme, G., *J. Mol. Catal. A*, 2000, **152**, 69; (d) Haber, S., In: Cornils, B. and Herrmann, W.A. (Eds.), *Aqueous-Phase Organometallic Chemistry*, 1st edn., Wiley-VCH, Weinheim, Germany, 1998, p. 440.
50. Bertoux, F., Tilloy, S., Monflier, E., Castanet, Y. and Mortreux, A., *J. Mol. Catal. A*, 1999, **138**, 53.
51. (a) Rout, L. and Punniyamurthy, T., *Adv. Synth. Catal.*, 2005, **347**, 1958; (b) Crochet, P., Díez, J., Fernández-Zúmel, M.A. and Gimeno, J., *Adv. Synth. Catal.*, 2006, **348**, 93; (c) Zhang, H.-B., Liu, L., Chen, Y.-J., Wang, D. and Li, C.-J., *Adv. Synth. Catal.*, 2006, **348**, 229; (d) Son, S.U., Lee, S.I., Chung, Y.K., Kim, S.-W. and Hyeon, T., *Org. Lett.*, 2002, **4**, 277; (e) Lautens, M., Yoshida, M., *Org. Lett.*, 2002, **4**, 123; (f) Deffernez, A., Hermans, S. and Devillers, M., *Appl. Catal. A*, 2005, **282**, 303; (g) Li, F-W., Xu, L.-W. and Xia, C.-G., *Appl. Catal. A*, 2003, **253**, 509; (h) Uozumi, Y., Tanaka, H. and Shibatomi, K., *Org. Lett.*, 2004, **6**, 281; (i) Bhattacharya, S. and Sengupta, S., *Tetrahedron Lett.*, 2004, **45**, 8733; (j) Mukhopadhyay, S., Rothenberg, G., Gitis, D. and Sasson, Y., *Org. Lett.*, 2000, **2**, 211; (k) Wu, H., Xu, L.-W., Xia, C.-G., Ge, J., Zhou, W. and Yang, L., *Catal. Commun.*, 2005, **6**, 221; (l) Choudary, B.M., Shomeshwar, T., Kantam, M.L. and Reddy, C.V., *Catal.Commun.*, 2004, **5**, 215; (m) Bhattacharjee, M. and Patra, B.N., *J. Organomet. Chem.*, 2004, **689**, 1091; (n) Wolf, C. and Lerebours, R., *Org. Biomol. Chem.*, 2004, **2**, 2161; (o) Nájera, C., Gil-Moltó, J. and Karlström, S., *Adv. Synth. Catal.*, 2004, **346**, 1798; (p) Zha, Z., Hui, A., Zhou, Y., Miao, Q., Wang, Z. and Zhang, H., *Org. Lett.*, 2005, **7**, 1903; (r) Legros, J. and Bolm, C., *Angew. Chem. Int. Ed.*, 2004, **43**, 4225; (s) Mévellec, Roucoux, A., Ramirez, E., Philippot, K. and Chaudret, B., *Adv. Synth. Catal.*, 2004, **346**, 72; (t) Gacia-Verdugo, E., Venardou, E.,

Thomas, W.B., Whiston, K., Partenheimer, W., Hamley, P.A. and Poliakoff, M., *Adv. Synth. Catal.*, 2004, **346**, 307; (u) Botella, L. and Nájera, C., *J. Org. Chem.*, 2005, **70**, 4360; (v) Genin, E., Amengual, R., Michelet, V., Savignac, M., Jutand, A., Neuville, L. and Genêt, J.-P., *Adv. Synth. Catal.*, 2004, **346**, 1733; (w) Jiang, Y., Xi, C. and Yang, X., *Synlett*, 2005, 1381; (x) Konovets, A., Penciu, A., Framery, E., Percina, N., Goux-Henry, C. and Sinou, D., *Tetrahedron Lett.*, 2005, **46**, 3205; (y) Xi, C., Jiang, Y. and Yang, X., *Tetrahedron Lett.*, 2005, **46**, 3909; (z) Velusamy, S., Kumar, A.V., Saini, R. and Punniyamurthy, T., *Tetrahedron Lett.*, 2005, **46**, 3819; (aa) Liao, M.-C., Duan, X.-H. and Liang, Y.-M., *Tetrahedron Lett.*, 2005, **46**, 3469; (bb) Nishiura, K., Urawa, Y. and Soda, S., *Adv. Synth. Catal.*, 2004, **346**, 1679; (cc) Gröger, H., Hummel, W., Buchholz, S., Drauz, K., Nguyen, T., Rollmann, C., Hüsken, H. and Abokitse, K., *Org. Lett.*, 2003, **5**, 173; (dd) Gallivan, J.P., Jordan, J.P. and Grubbs, R.H., *Tetrahedron Lett.*, 2005, **46**, 2577; (ee) Hickey, M.R., Nelson, T.D., Secord, E.A., Allwein, S.P. and Kress, M.H., *Synlett*, 2005, 255; (ff) Cannella, R., Clerici, A., Pastori, N., Regolini, E. and Porta, O., *Org. Lett.*, 2005, **7**, 645; (gg) Masuyama, A., Yamaguchi, T., Abe, M. and Nojima, M., *Tetrahedron Lett.*, 2005, **46**, 213; (hh) Luo, S., Mi, X., Wang, P.G. and Cheng, J.-P., *Tetrahedron Lett.*, 2004, **45**, 5171; (ii) Resende, G.O., Aguiar, L.C.S. and Antunes, O.A.C., *Synlett*, 2005, 119; (jj) Krafft, M.E., Wright, J.A. and Bonaga, L.V.R., *Synlett*, 2005, 71; (kk) Ma, Y., Liu, H., Chen, L., Cui, X., Zhu, J. and Deng, J., *Org. Lett.*, 2003, **5**, 2103; (ll) Aijou, A.N. and Pinet, J.-L., *J. Mol. Catal. A*, 2004, **214**, 203; (mm) Tan, X., Hou, Y., Shen, B., Liu, L. and Guo, Q., *Tetrahedron Lett.*, 2004, **45**, 5525; (nn) Uozumi, Y. Nakai, Y., *Org. Lett.*, 2002, **4**, 2997; (oo) Leadbeater, N.E. and Marco, M., *Org. Lett.*, 2002, **4**, 2973; (pp) Jankowska, J. and Mlynarski, J., *J. Org. Chem.*, 2006, **71**, 1317; (qq) Huber, G.W., Shabaker, J.W., Evans, S.T. and Dumesic, J.A., *Appl. Catal. B*, 2006, **62**, 226; (rr) Arisawa, M., Suwa, A. and Yamaguchi, M., *J. Organomet. Chem.*, 2006, **691**, 1159; (ss) Liu, S.-Y. and Nocera, D.G., *Tetrahedron Lett.*, 2006, **47**, 1923; (tt) Chanthavong, F. and Leadbeater, N.E., *Tetrahedron Lett.*, 2006, **47**, 1909; (uu) Sharma, A., Kumar, V. and Sinha, A.K., *Adv. Synth. Catal.*, 2006, **348**, 354; (vv) Itoh, T. and Mase, T., *J. Org. Chem.*, 2006, **71**, 2203; (ww) Gross, T., Seayad, A.M., Ahmad, M. and Beller, M., *Org. Lett.*, 2002, **4**, 2055; (xx) Lautens, M. and Mancuso, J., *Org. Lett.*, 2002, **4**, 2105; (yy) Thadani, A.N. and Batey, R.A., *Org. Lett.*, 2002, **4**, 3827.

52. (a) Monteil, F., Queau, R. and Kalck, P., *J. Organomet. Chem.*, 1994, **480**, 177; (b) Trinkhaus, S., Kadyrov, R., Selke, R., Holz, J., Götze, L. and Börner, A., *J. Mol. Catal. A*, 1999, **144**, 15; (c) Lavenot, L., Roucoux, A. and Patin, H., *J. Mol. Catal. A*, 1997, **118**, 153; (d) Tin, K.C., Wong, N.-B., Li, R.-X., Li, Y.-Z. and Li, X.-J., *J. Mol. Catal. A*, 1999, **137**, 113; (e) Hanson, B.E., In [6], 2nd edn., 2003, p. 181; (f) Deshpande, R.M, Purwanto, P., Delmas, H. and Chaudhari, R.V., *Ind. Eng. Chem. Res.*, 1996, **35**, 3927; (g) Bertoux, F., Monflier, E., Castanet, Y. and Mortreux, A., *J. Mol. Catal. A*, 1999, **143**, 23; (h) Paetzold, E., Michalik, M. and Oehme, G., *J. Prakt. Chem.*, 1997, **339**, 38; (i) Chen, H., Li, Y., Chen, J., Cheng, P. and Li, X., *Catal. Today*, 2002, **74**, 131; (j) Deshpande, M., Divekar, S.S., Bhanage, B.M. and Chaudhari, R.V., *J. Mol. Catal.*, 1992, **75**, L19; (k) Roucoux, A., Schulz, J. and Patin, H., *Adv. Synth. Catal.*, 2003, **345**, 222; (l) Kalck, Ph., Miquel, L. and Dessoudeix, M., *Catal. Today*, 1998, **42**, 431; (m) Paganelli, S., Zanchet, M., Marchetti, M, and Mangano, G., *J. Mol. Catal. A*, 2000, **157**, 1.
53. (a) Baricelli, P.J., Rodríguez, G., Rodríguez, M., Lujano, E. and López-Linares, F., *Appl. Catal. A*, 2003, **239**, 25; (b) Parmar, D.U., Bhatt, S.D., Bajaj, H.C. and Jasra, R.V., *J. Mol. Catal. A*, 2003, **202**, 9; (c) López-Linares, F., Gonzalez, M.G. and Páez, D.E., *J. Mol. Catal. A*, 1999, **145**, 61; (d) Mudalige, D.C. and Rempel, G.L., *J. Mol. Catal. A*, 1997, **116**, 309 and **123**, 15; (e) Kolaríc, S. and Sunjic, V., *J. Mol. Catal. A*, 1996, **111**, 239; (f) Garcia Fidalgo, E., Plasseraud, L. and Süss-Fink, G., *J. Mol. Catal. A*, 1998, **132**, 5; (g) Robert, F., Oehme, G., Grassert, I. and Sinou, D., *J. Mol. Catal. A*, 2000, **156**, 127; (h) Larpent, C., Bernard, E., Brisse-le Menn, F. and Patin, H., *J. Mol. Catal. A*, 1997, **116**, 277; (i) Dwars, T. and Oehme, G., *Adv. Synth. Catal.*, 2002, **344**, 239; (j) Joó, F., *Acc. Chem. Res.*, 2002, **35**, 738; (k) Jóo, F., *Aqueous Organometallic Catalys*, Kluwer, Dordrecht, 2001; (l) Sinou, D., *Adv. Synth. Catal.*, 2002, **344**, 221; (m) Fan, Q.-H., Deng, G.-J., Chen, X.-M., Xie, W.-C., Jiang, D.-Z., Liu, D.-S. and Chan, A.S.C., *J. Mol. Catal. A*, 2000, **159**, 37; (n) Malmström, T. and Andersson, C., *J. Mol. Catal. A*, 2000, **157**, 79; (o) Yang, Z., Ebihara, M. and Kawamura, T., *J. Mol. Catal. A*, 2000, **158**, 509; (p) Nomura, K., *J. Mol. Catal. A*, 1998, **130**, 1; (q) Papp, G., Elek, J., Nádashdi, L., Laurenczy, G. and Jóo, F., *Adv. Synth. Catal.*, 2003, **345**, 172.
54. (a) Livingston, J.R., Mozeleski, E.J. and Sartori, G., US 5,288,818, 1994 (Exxon) and WO 93/04,029, 1993 (PCT); (b) Müller, Th. and Bahrmann, H., *J. Mol. Catal. A*, 1997, **116**, 39; (c) Gosser, L.W., Knoth, W.H. and Parshall, G.W., *J. Mol. Catal.*, 1977, **2**, 253; (d) Dwars, T., Haberland, J., Grassert, I., Oehme, G. and Kragl, U., *J. Mol. Catal. A*, 2001, **168**, 81; (e) Fuhrmann, H., et al., *Adv. Synth. Catal.*, 2003, **345**, 202; (f) Liu, X., Jiang, M., Yang, S., Chen, M., Yang, C. and Wu, K., *Angew. Chem. Int. Ed.*, 2002, **41**, 2950; (g) Dixon, A.G., In: Spivey, J.J. (Ed.), *Catalysis*, Vol. 14, RSC, London, 1999, Chap. 2, p. 40; (h) Bahrmann, H., Haubs, M., Müller, Th., Schöpper, N. and Cornils, B., *J. Organomet. Chem.*, 1997, **545/546**, 139; (i) Fuhrmann, H., Dwars, T., Michalik, D., Holzhüter, G., Grüttner, C., Kragl, U. and Oehme, G., *Adv. Synth. Catal.*, 2003, **345**, 202.
55. (a) Kuroda, K. and Swager, T., *Chem. Commun.*, 2003, 26; (b) Buchmeiser, M.R., *Polymeric Materials in Organic Synthesis and Catalysis*, Wiley-VCH, Weinheim, Germany, 2003; (c) Huang, X., Witte, K.L., Bergbreiter, D.E. and Wong, C.-H., *Adv. Synth. Catal.*, 2001, **343**, 675; (d) Mecking, S., Held, A. and Bauers, F.M., *Angew. Chem. Int. Ed.*, 2002, **41**, 545; (e) Bergbreiter, D.E., *CHEMTECH*, November 1987, 686; (f) Bergbreiter, D.E.,

Franchina, J.G. and Case, B.L., *Org. Lett.*, 2000, **2**, 393; (g) Malmström, T. and Andersson, C., *J. Mol. Catal. A*, 2000, **157**, 79 and *Chem.Commun.*, 1996, 1135; (h) Bergbreiter, D.E., *Catal. Today*, 1998, **42**, 389; (i) Dickerson, T.J., Reed, N.N. and Janda, K.D., *Chem. Rev.*, 2002, **102**, 3325; (j) Lu, Z., Lindner, E. and Mayer, H.A., *Chem. Rev.*, 2002, **102**, 3543; (k) Fan, Q.-H., Deng, G.-J., Chen, X.-M., Xie, W.-C., Jiang, D.-Z., Liu, D.-S. and Chan, A.S.C., *J. Mol. Catal. A*, 2000, **159**, 37.

56. (a) Huang, X., Witte, K.L., Bergbreiter, D.E. and Wong, C.-H., *Adv. Synth. Catal.*, 2001, **343**, 675; (b) Bergbreiter, D.E., *CHEMTECH*, November 1987, 686; (c) Bergbreiter, D.E., Franchina, J.G. and Case, B.L., *Org. Lett.*, 2000, **2**, 393; (d) Bergbreiter, D.E., *Catal. Today*, 1998, **42**, 389.
57. (a) Zuideveld, M.A., Bauers, F.M., Thomann, R. and Mecking, S., In: *36th Meeting of German Catalysis*, Weimar, 19–21 March 2003, Preprints, p. 385; (b) Mecking, S., Held, A. and Bauers, F.M., *Angew. Chem. Int. Ed.*, 2002, **41**, 545; (c) Xing, B., Choi, M.-.F. and Xu, B., *Chem. Eur. J.*, 2002, **8**, 5028.
58. Paetzold, E., Oehme, G., Fischer, C. and Frank, M., *J. Mol. Catal. A*, 2003, **200**, 95.
59. Herrmann, W.A. and Kohlpaintner, C.W., *Angew. Chem., Int. Ed. Engl.*, 1993, **32**, 1524.
60. (a) Herrmann, W.A., Kulpe, J.A., Konkol, W. and Bahrmann, H., *J. Organomet. Chem.*, 1990, **389**, 85; (b) Herrmann, W.A., Kulpe, J.A., Kellner, J., Riepl, H., Bahrmann, H. and Konkol, W., *Angew. Chem., Int. Ed. Engl.*, 1990, **29**, 391; (c) EP 060,523, 1982 (Union Carbide); (d) Neftekhim, *Neftekhimija*, 1981, **21**, 879.
61. Kirk, F.A., In: *Proceedings of the DGMK-Conference on 'Synthesis Gas Chemistry'*, Dresden, Germany, 27–29 September 2000, Synetix, Billingham, UK, p. 131.
62. Beirne, K.M., *Chem. Eng.*, January 2002, 87.
63. Among a multitude of proposals: (a) Billig, E. and Stanton, D.B., EP 0,049,781, 1981 and EP 0,083,094, 1982 (Union Carbide Corp.); Dougherty, S.J., US 4,263,218, 1981 (UCC); Billig, E. and Stanton, D.B., US 4,605,780, 1986; (b) Kneuper, H.-J., Aron, A., Korgitzsch, F.-M., Nilles, M., Harder, W., Roeper, M. and Paciello, R., US 5,919,987, 1999 (BASF AG); Paciello, R., Kneuper, H.-J., Geissler, B. and Roeper, M., US 6,107,524, 2000 (BASF AG); Geissler, B., Roeper, M., Paciello, R., Decker, J., Voss, H. and Mahr, N., US 6,310,261, 2001 (BASF AG); (c) Gregorio, G. and Montrasi, G., US 3,968,134, 1976 (Montedison SpA); (d) Silverman, G.S. and Mercando, P., US 5,773,655, 1998 (Elf); (e) Carey, J.L., US 5,364,822, 1994 (BP); (f) Homeier, E.H., Imrai, T., Mackowiak, D.E. and Guzolek, C.E., US 4,329,521, 1981 (UOP Inc.); Imrai, T. and Homeier, E.-H., US 4,374,287, 1983 (UOP Inc.); Imrai, T., US 4,298,499, 1980 (UOP Inc.); Quick, H.M., US 4,388,279, 1981 (UOP Inc.); (g) Barnes, R.L., US 4,364,907, 1988 (Eastman-Kodak), Cooper, J.L., US 4,388,476, 1983 and WP 82-03,856, 1982 (Eastman-Kodak); Dawes, J.L. and Devon, T.J., US 4,400,547, 1981 (UOP Inc.); (h) Tsunoda, Y., Tomita, S., Miyazawa, C. and Omori, Y., GB 2,085,747, 1982 and JP 81-02,994, 1981 (Mitsubishi) [*Chem. Abstr.*, 1982, **96**, 68297q]; (i) Ali, Z. and Radder, J., FR 2,474,333, 1981 (Shell); (j) Wiese, K.-D., Trocha, M., Röttger, D., Toetsch, W., Kaizek, A. and Büschken, W., US 6,500,991, 2002 (Oxeno); Gubisch, D., Armbrust, K., Scholz, B. and Nehring, R., US 6,015,928, 2000 (Oxeno); (k) Demay, C., US 4,306,086, 1981 (Ets. Kuhlmann); (l) Chang, T., US 5,290,743, 1994 (ARCO).
64. (a) McCarthy, M., Stemmer, H. and Leitner, W., *Green Chem.*, 2002, **4**, 501; (b) Kainz, S., Koch, D., Baumann, W. and Leitner, W., *Angew. Chem.*, 1997, **109**, 1699.
65. *Chem. Eng.*, January 2002, 62.
66. Trost, B.M., *Science*, 1991, **254**, 1471 and *Angew. Chem., Int. Ed. Engl.*, 1995, **34**, 259.
67. Sheldon, R.A., *CHEMTECH*, 1994, **24**(3), 38; *Chem. Ind.*, 1992, (12), 903 and 1997, (1), 12; *Green Chem.*, 1999, **1**(1), G3.
68. Cornils, B. and Wiebus, E., *Recl. Trav. Chim. Pays-Bas*, 1996, **115**, 211.
69. Christ, C., *Chem. Ing. Tech.*, 2000, **72**, 42.

Index of Organic Reactions in Water

This index covers literature up to early 2006. Key references have been selected with preference given to recent publications, which often contain citations to previous work. In line with the focus of this book only reactions that do not employ substantial amounts of co-solvents are included.

Reaction	References
Acylation	Eur. J. Org. Chem. **2004**, 1254.
Addition to alkenes	See specific reaction types.
Addition to alkynes	See also *Enyne cyclization* and *Pauson-Khand reaction.*
- Nucleophilic addition	J. Org. Chem. **2003**, *68*, 762; Tetrahedron Lett. **2001**, *42*, 8467.
- Electrophilic addition	See *Kharasch reaction.*
- Hydrocarboxylation	J. Mol. Catal. A **1999**, *204–205*, 133.
- Hydrometalation	Lett. Org. Chem. **2004**, *1*, 122; Chem. Commun. **2003**, 1668.
Aldol reaction	Green Chem. **2004**, *6*, 583 (Review).
- Direct aldol reaction	Angew. Chem. Int. Ed. **2006**, *45*, 958; J. Am. Chem. Soc. **2006**, *128*, 734; Tetrahedron Lett. **2005**, *46*, 5617.
- Mukaiyama aldol reaction	Chem. Commun. **2002**, 2994; Tetrahedron **2002**, *58*, 8263.
Alkylation	
- Addition to C=O	J. Am. Chem. Soc. **2003**, *125*, 4062; see also *Conjugate addition.*
- Addition to C=N	Org. Lett. **2002**, *4*, 131; J. Org. Chem. **2000**, *65*, 5043.
- Substitution	Adv. Synth. Cat. **2003**, *345*, 357; Adv. Synth. Cat. **2002**, *344*, 370.
Alkynylation	
- Addition to C=O	Org. Lett. **2006**, *8*, 1953.
- Addition to C=N	Proc. Natl. Acad. Sci. U.S.A. **2004**, *101*, 5749; Org. Lett. **2004**, *6*, 1001.
- Conjugate addition	Tetrahedron Lett. **2004**, *45*, 2771.
- Direct coupling with C-H	Org. Lett. **2004**, *6*, 4997.
Allenylation	Chem. Commun. **1998**, 449.
Allylation (Barbier-type)	
- Addition to C=O	Chemistry Lett. **2006**, *35*, 498; J. Am. Chem. Soc. **2003**, *125*, 2958.
- Addition to C=N	Tetrahedron Lett. **2003**, *44*, 667.
Allylic amination	Org. Lett. **2004**, *6*, 281.
Amides from aldehydes	J. Org. Chem. **2003**, *68*, 1158.
Amides from nitriles	Angew. Chem. Int. Ed. **2004**, *43*, 1576.
Aminals from aldehydes	Tetrahedron **2004**, *60*, 3205.
Amination of aryl halides	Chem. Commun. **1998**, 1509.
α-Aminoallylation	Chem. Commun. **2005**, 104.
Aminocarbonylation	Org. Lett. **2005**, *7*, 3327.
Aminohydroxylation	Angew. Chem. Int. Ed. **2001**, *40*, 3455.
Arylation	
- Grignard-type	J. Am. Chem. Soc. **2001**, *123*, 7451.
- Heck-type	Tetrahedron **2004**, *60*, 4097; Green Chem **2004**, *6*, 280; Chem. Eur. J. **2004**, *10*, 1501; Synthesis **2004**, 2006.
- N-Arylation	Synthesis **2006**, 1868; J. Am. Chem. Soc. **2003**, *125*, 6653.
Aziridination	Angew. Chem. Int. Ed. **2004**, *43*, 79.
Baylis-Hillman reaction	Tetrahedron Lett. **2005**, *46*, 8125.
Benzoin condensation	J. Am. Chem. Soc. **1995**, *117*, 6601.
Bromination	Tetrahedron Lett. **2006**, 4707.
Buchwald-Hartwig coupling	Chem. Commun. **1998**, 1509.

Reaction	References
Carbonylation	J. Organomet. Chem. **2000**, *602*, 173; J. Org. Chem. **1993**, *58*, 4798.
Cyanation	J. Am. Chem. Soc. **1998**, *120*, 1688.
Cyclopropanation	Org. Lett. **2002**, *4*, 4531; Chem. Commun. **2001**, 59.
Claisen rearrangement	Angew. Chem. Int. Ed. **2005**, *44*, 756; J. Chem. Soc. Perkin 1 **1992**, 1631.
Conjugate addition	
- Addition of α-carbonyl compounds	See *Michael reaction.*
- Addition of alkyl groups	Org. Lett. **2004**, *6*, 3349; Chem. Eur. J. **2003**, *9*, 4179.
- Addition of vinyl and aryl groups	J. Am. Chem. Soc. **2001**, *123*, 7451.
Cope rearrangement	J. Org. Chem. **2002**, *67*, 6725; Synlett **1992**, 847.
Cross-dehydrogenative coupling	J. Am. Chem. Soc. **2005**, *127*, 6968.
Cycloaddition reaction	
- 1,3-dipolar	Chem. Commun. **2004**, 394; J. Chem. Soc. Perkin 2 **2002**, 1807.
- [3 + 2]	J. Am. Chem. Soc. **2003**, *125*, 3192.
- [4 + 2]	See *Diels-Alder reaction.*
- [4 + 3]	Tetrahedron Lett. **1997**, *38*, 8031.
- [5 + 2]	Synlett **2003**, 1295.
Dehydrogenation of amines	Eur. J. Org. Chem. **2005**, 3060.
Deprotection reactions	
- O-Acetyl	Tetrahedron **2003**, *59*, 1049; J. Org. Chem. **2003**, *68*, 8723.
- O-Allyloxycarbonyl (Alloc)	Tetrahedron Lett. **1993**, *34*, 4189.
- Methoxymethyl (MOM)	J. Org. Chem. **2003**, *68*, 8723.
- Oximes and imines	J. Org. Chem. **2005**, *70*, 1934.
- Silyl ethers	J. Org. Chem. **2003**, *68*, 8723.
- Thioesters	Org. Lett. **2003**, *5*, 101.
- Trityl	J. Org. Chem. **2003**, *68*, 8723.
Diels-Alder reaction	Tetrahedron **2005**, *61*, 7087; J. Am. Chem. Soc. **2002**, *124*, 2458; Eur. J. Org. Chem. **2001**, 439 (Review).
- Aza Diels-Alder	Adv. Synth. Cat. **2003**, *345*, 475; J. Org. Chem. **2001**, *66*, 4661.
- Inverse electron demand	J. Org. Chem. **1996**, *61*, 2001.
Dihydroxylation of alkenes	Org. Lett. **2001**, *3*, 2649; J. Chem. Soc. **1949**, 2988.
Electrophilic aromatic substitution	See *Friedel-Crafts reaction.*
Enyne cyclization	J. Am. Chem. Soc. **2001**, *123*, 10511.
Epoxidation	Org. Biomol. Chem. **2005**, *3*, 3883; Chem. Commun. **2005**, 4592; Acc. Chem. Res. **2004**, *37*, 488 (Review).
Esterification	Adv. Synth. Cat. **2002**, *344*, 270.
Etherification	Chem. Lett. **2004**, *33*, 940; J. Am. Chem. Soc. **2002**, *124*, 3622.
Friedel-Crafts reaction	Synlett **2004**, 555; Adv. Synth. Cat. **2001**, *343*, 174.
Grignard reaction	J. Am. Chem. Soc. **1998**, *120*, 9102.
Heck coupling	Tetrahedron **2004**, *60*, 5563; Tetrahedron **2004**, *60*, 4097; Green Chem **2004**, *6*, 280; Chem. Eur. J. **2004**, *10*, 1501.
Henry reaction (nitro-aldol)	J. Org. Chem. **1997**, *62*, 425.
Hydroformylation	Tetrahedron **2001**, *57*, 1631.
Hydrogenation (H_2)	Adv. Synth. Cat. **2002**, *344*, 239 (Review).
- Hydrogenation of aldehydes	Angew. Chem. **1998**, *37*, 969.
- Hydrogenation of ketones	Tetrahedron Lett. **2001**, *42*, 663; see also *Reduction.*
- Hydrogenation of olefins	Org. Biomol. Chem. **2006**, *4*, 613; Adv. Synth. Cat. **2006**, *348*, 471.
α-Hydroxyallylation	Tetrahedron **2004**, *60*, 11725.
Iodocyclization	Green Chem. **2006**, *8*, 522.
Kharasch reaction	Adv. Synth. Cat. **2002**, *344*, 261.
Knoevenagel condensation	Lett. Org. Chem. **2006**, *3*, 297; Tetrahedron Lett. **2005**, *46*, 6453.
Mannich reaction	J. Am. Chem. Soc. **2004**, *126*, 7768; Tetrahedron Lett. **2004**, *45*, 8949.
Metathesis reaction	J. Am. Chem. Soc. **2006**, *128*, 3508.
Michael reaction	
- C nucleophiles	J. Am. Chem. Soc. **2006**, *128*, 4966; J. Org. Chem. **2006**, *71*, 352; Chem. Eur. J. **2005**, *11*, 288.
- S nucleophiles	J. Org. Chem. **2006**, *71*, 3634; Adv. Synth. Cat. **2005**, *347*, 655; Can. J. Chem. **1999**, *77*, 624.
- N nucleophiles	Tetrahedron Lett. **2005**, *46*, 8329; Adv. Synth. Cat. **2005**, *347*, 655.

Reaction	References
Nucleophilic substitution	Eur. J. Org. Chem. **2004**, 1254; Tetrahedron Lett. **2003**, *44*, 5519; J. Am. Chem. Soc. **2002**, *124*, 3622; see also *Acylation, Alkylation, Etherification* and *Sulfonylation.*
Olefin migration	Eur. J. Org. Chem. **2003**, 998.
Oxidation of alcohols	J. Org. Chem. **2005**, *70*, 729; Angew. Chem. Int. Ed. **2003**, *42*, 194.
Passerini reaction	J. Am. Chem. Soc. **2004**, *126*, 444.
Pauson-Khand reaction	Angew. Chem. Int. Ed. **2003**, *42*, 2409; Tetrahedron Lett. **2003**, *44*, 3417.
Pinacol cross-coupling	Tetrahedron Lett. **2002**, *43*, 8967.
Pinner reaction	Green Chem. **2006**, *8*, 22.
Polymerization of olefins	Angew. Chem. Int. Ed. **2002**, *41*, 544.
Prins reaction	Org. Lett. **2003**, *5*, 4521; Green Chem. **2003**, *5*, 80.
Propargylation	J. Org. Chem. **1998**, *63*, 7472.
Radical addition	J. Org. Chem. **2003**, *68*, 5618; Synlett **2002**, 674.
Radical cyclization	Tetrahedron **2002**, *59*, 77; J. Am. Chem. Soc. **2000**, *122*, 11041.
Radical substitution	Tetrahedron **1999**, *55*, 6109.
Reduction	
- Hydride reduction of ketones	Org. Lett. **2004**, *6*, 4331; J. Am. Chem. Soc. **2001**, *113*, 5956.
- Hydride reduction of imines	See *Reductive amination.*
- Transfer hydrogenation of ketones	J. Org. Chem. **2005**, *70*, 9424; Tetrahedron Lett. **2005**, *46*, 7341.
- Transfer hydrogenation of imines	Chem. Commun. **2006**, 1766.
- SET reduction of ketones	Tetrahedron: Asymm. **2004**, *15*, 1735.
Reductive amination	J. Am. Chem. Soc. **2004**, *126*, 3020; Tetrahedron **2004**, *60*, 7899.
Reductive coupling of imines	Green Chem. **2000**, *2*, 117.
Reformatsky reaction	Can. J. Chem. **2003**, *81*, 1406.
Ring-opening of strained heterocycles	
- Alcoholysis of epoxides	J. Org. Chem. **2002**, *68*, 726.
- Aminolysis of epoxides	Org. Lett. **2005**, *7*, 3649; Org. Lett. **2005**, *7*, 4593; Green Chem. **2005**, *7*, 708.
- Aziridine ring-opening	Adv. Synth. Cat. **2006**, *348*, 696; Angew. Chem. Int. Ed. **2004**, *43*, 79; J. Org. Chem. **2002**, *68*, 726.
- Thiolysis of epoxides	Org. Lett. **2005**, *7*, 4411.
Sonogashira coupling	Tetrahedron Lett. **2006**, *62*, 31; J. Org. Chem. **2005**, *70*, 391; Org. Lett. **2004**, *6*, 3151.
Staudinger reaction	Science **2000**, *287*, 2007.
Stille coupling	Tetrahedron Lett. **2006**, *62*, 31; J. Org. Chem. **2003**, *68*, 7551.
Strecker reaction	Org. Biomol. Chem. **2004**, *2*, 2567; Angew. Chem. Int. Ed. **2000**, *39*, 567.
Sulfonylation	Green Chem. **2005**, *7*, 711.
Suzuki coupling	Tetrahedron Lett. **2006**, *47*, 197; J. Org. Chem. **2005**, *70*, 161; Angew. Chem. Int. Ed. **2003**, *42*, 4856 (Review).
Tsuji-Trost reaction	J. Org. Chem. **2005**, *70*, 6441; Org. Lett. **2004**, *6*, 4085.
Ugi reaction	J. Am. Chem. Soc. **2004**, *126*, 444.
Ullmann coupling	Org. Proc. Res. Dev. **2003**, *7*, 641; Adv. Synth. Cat. **2002**, *344*, 399.
Vinylation	
- Addition to C=O	Synthesis **2002**, 717; J. Am. Chem. Soc. **2001**, *123*, 7451.
- Addition to C=N	Tetrahedron Lett. **2004**, *45*, 2995.
Wittig reaction	Tetrahedron Lett. **2005**, *46*, 4473.
Wurtz coupling	Tetrahedron Lett. **1998**, *39*, 2499.

Index